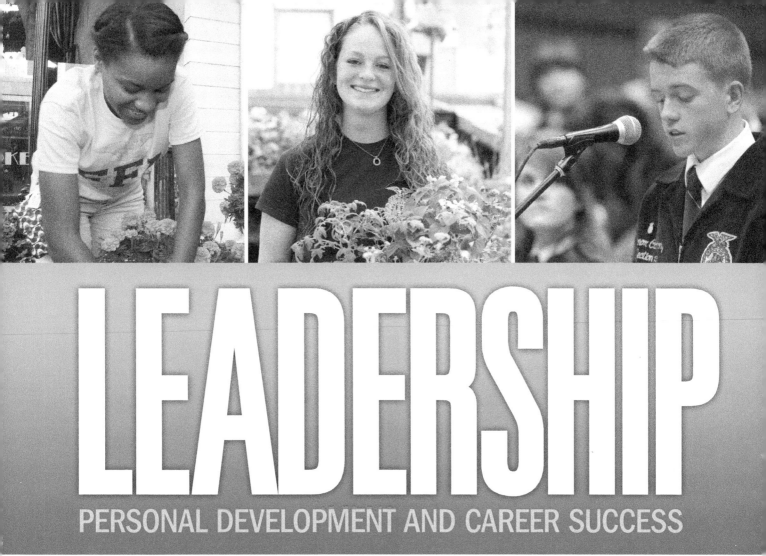

LEADERSHIP

PERSONAL DEVELOPMENT AND CAREER SUCCESS

Fourth Edition

Cliff Ricketts and John C. Ricketts

CENGAGE
Learning·

Australia • Brazil • Mexico • Singapore • United Kingdom • United States

CENGAGE
Learning®

Leadership: Personal Development and Career Success, **Fourth Edition**
Cliff Ricketts and John C. Ricketts

SVP, GM Skills & Global Product Management: Jonathan Lau

Product Director: Matthew Seeley

Product Manager: Nicole Robinson

Senior Director, Development: Marah Bellegarde

Senior Product Development Manager: Larry Main

Senior Content Developer: Anne Orgren

Product Assistant: Deborah Handy

Vice President, Strategic Marketing Services: Jennifer Ann Baker

Director, Product Marketing, Sales: Trish K. Bobst

Senior Production Director: Wendy Troeger

Production Director: Andrew Crouth

Senior Content Project Manager: Betsy Hough

Managing Art Director: Jack Pendleton

For product information and technology assistance, contact us at
Cengage Learning Customer & Sales Support, 1-800-354-9706
For permission to use material from this text or product, submit all requests online at **www.cengage.com/permissions**
Further permissions questions can be e-mailed to
permissionrequest@cengage.com

Library of Congress Control Number: 2016935771

ISBN: 978-1-3059-5381-9

Cengage Learning
20 Channel Center Street
Boston, MA 02210
USA

Cengage Learning is a leading provider of customized learning solutions with employees residing in nearly 40 different countries and sales in more than 125 countries around the world. Find your local representative at **www.cengage.com**

Cengage Learning products are represented in Canada by Nelson Education, Ltd.

To learn more about Cengage Learning, visit **www.cengage.com**

Purchase any of our products at your local college store or at our preferred online store **www.cengagebrain.com**

Notice to the Reader
Publisher does not warrant or guarantee any of the products described herein or perform any independent analysis in connection with any of the product information contained herein. Publisher does not assume, and expressly disclaims, any obligation to obtain and include information other than that provided to it by the manufacturer. The reader is expressly warned to consider and adopt all safety precautions that might be indicated by the activities described herein and to avoid all potential hazards. By following the instructions contained herein, the reader willingly assumes all risks in connection with such instructions. The publisher makes no representations or warranties of any kind, including but not limited to, the warranties of fitness for particular purpose or merchantability, nor are any such representations implied with respect to the material set forth herein, and the publisher takes no responsibility with respect to such material. The publisher shall not be liable for any special, consequential, or exemplary damages resulting, in whole or part, from the readers' use of, or reliance upon, this material.

Printed in the United States of America
Print Number: 06   Print Year: 2021

This book is dedicated to our former students. It is also dedicated to the patriarch and the matriarch of our family, Hall and Louise Ricketts, who educated us about entrepreneurial, servant, and community leadership. This dedication is also extended to our loving and supportive spouses, who are leaders and role models in their own right, and to our children, who inspire us to do what we do.

Cliff Ricketts & John C. Ricketts

Contents

Section 4 Managerial Leadership Skills

Chapter 13 Problem Solving and Decision Making 376

Chapter 14 Goal Setting 398

Preface

The mission statement for agricultural education and, specifically, the National FFA Organization, includes the enhancement of leadership, personal growth, and career success of its students and members. Educating the total person is one of the objectives of this book. In addition to the cognitive (academic) and technical (psychomotor) skills that secondary schools are presently teaching students, employers want employees with leadership and human relations skills. Schools, as a rule, are not teaching affective skills except through youth organizations. One of the purposes of this book, *Leadership: Personal Development and Career Success, Fourth Edition*, is to provide a formal text to help teach affective skills for a stand-alone course or a supplementary text to existing agricultural education courses. As a result, students will be better prepared to enter the workforce. Why? Employers tell us that most people lose their jobs not because of poor cognitive or technical skills, but because they lack affective (leadership and human relations) skills.

This book is intended to serve as an instrument to help students become more successful in life and the workplace. *Leadership: Personal Development and Career Success, Fourth Edition* achieves this by teaching students to learn and enhance personal development and communication skills; by helping students select a job, get a job, and attain career success; and by helping students attain any desired leadership positions,

in both their careers and their communities. In short, the intent of this book is to help the student attain professional and personal success.

Leadership: Personal Development and Career Success, Fourth Edition is divided into six sections. Section 1 introduces students to leadership and helps them begin to understand the concept of leadership. Chapters 1 through 5 cover leadership categories and styles, personality types and their relationship to leadership and human behavior, learning styles, the development of leaders, and leading teams and groups.

Section 2 deals with communicating with and speaking to groups. Personal communication, overcoming stage fright by reciting the FFA Creed, prepared public speaking, and extemporaneous speaking are covered in Chapters 6 through 9. The rules of FFA Career Development Events (CDEs) are included as appropriate in each chapter.

Section 3 is about leading individuals and groups. Topics include parliamentary procedure and conducting successful meetings. These topics are included in Chapters 10 through 12. The FFA parliamentary procedure CDE rules are included.

Section 4 investigates managerial leadership skills. The topics in this section include problem solving and decision making, goal setting, time management, motivation of others (emphasizing positive reinforcement), and conflict resolution. These topics are covered in Chapters 13 through 17.

Section 5 concerns personal development. Chapters 18 through 20 cover the topics of self-concept, attitude, and ethics in the workplace. Section 6, the final section, is about the transition to work skills. Topics in this section include selecting a career and finding a job, procedures for applying and interviewing for a job, and the employability skills necessary for the student who is making the transition to the world of work.

Appendix E

New to this Edition

The latest research findings concerning leadership (including updated theories) are reflected in an easy-to-understand format that readers of any level can comprehend. Updated tables and figures reflecting more recent events were updated. Updated rules of the National FFA CDEs have been added to the chapters on Parliamentary Procedure, Creed, Public Speaking, and Extemporaneous Speaking. A new Supervised Agricultural Experience appendix explains how SAE relates to leadership development. The heavily updated National FFA Job Interview Career Development Event appendix, now re-titled National FFA Employment Skills Career and Leadership Development Event, provides a description and the rules of the brand new Employability Skills Career and Leadership Development Event.

Pedagogical Features

Each chapter begins with a list of performance objectives, which reflect the main topics in each chapter. The objectives are followed by Terms to Know, which enable the students to better understand the chapter content. The definitions of all these terms appear both in the chapter and in the glossary. Almost every chapter also includes vignettes that make the discussion more relevant by illustrating a particular point or objective.

Each chapter ends with a summary as well as various types of exercises. The Take It to the Net section allows students to develop and practice research skills using the Internet to gain more recent information. Review questions encourage students to read the whole chapter. Completion and matching exercises are designed to reinforce students' understanding and retention of important ideas. Activities enable students to apply newly learned concepts and practices, as well as to draw on their own experiences and share those experiences with others. Critical thinking is also a component of the activities section.

About the Authors

Name: Dr. Cliff Ricketts

Affiliation: Middle Tennessee State University, Murfreesboro, Tennessee

Credentials: PhD

Author Biography: Dr. Cliff Ricketts came through the ranks of agricultural education and the Future Farmers of America; he is a product of the program. He has achieved every degree offered by the FFA, including the Greenhand, Chapter, State, and American FFA degrees; Honorary Chapter; Honorary State; and Honorary American FFA, which was received the same year by former President George W. Bush and Lee Iacocca. During his time as a high school agricultural education instructor, Dr. Ricketts' students won the State FFA Parliamentary Procedure Contest and the State FFA Creed Speaking Contest (four times), and were awarded runners-up in Public Speaking and National Extemporaneous Speaking. Dr. Ricketts was also a recipient of the National Vocational Agriculture Teachers Association's National Award during his tenure as a high school teacher.

Dr. Ricketts received his B.S. and M.S. degrees from the University of Tennessee and his Ph.D. from The Ohio State University in Agricultural Education, Career-Technical Education and Teacher Education. While teaching at Middle Tennessee State University (MTSU), he received both the MTSU Foundation's Outstanding Public Service Award and its Outstanding Teacher Award twice, He was also a finalist for its Outstanding Research Award and Career Achievement Award. He was also one of only two faculty members at MTSU to receive the Presidents' Silver Column Award.

Dr. Ricketts is still involved in production agriculture with a beef cattle operation on his family farm 20 miles east of Nashville, Tennessee. His children, John, Mitzi, and Paul, and seven grandchildren are the fifth and sixth generations on the family farm. Dr. Ricketts is also involved with alternative fuels research and has run engines off ethanol from corn; methane from cow manure; soybean oil; and hydrogen from water. He and his students from Middle Tennessee State University held the world's land speed record for a hydrogen-fueled vehicle for 15 years. He has driven coast to coast three times, first on 95 percent ethanol, then on biodiesel from chicken fat, and finally on hydrogen from water produced by solar energy in the MTSU alternative fuel lab. Dr. Ricketts has made approximately 1,000 presentations on leadership and alternative fuels in his 40 years at MTSU.

Name: Dr. John C. Ricketts

Affiliation: Tennessee State University

Credentials: PhD

Author Biography: Dr. John C. Ricketts is Professor of Agricultural and Extension Education at Tennessee

State University. He used his experiences as a secondary agriculture teacher, assistant football coach, and part-time farmer to contribute practical strategies for this book. He used his training in Agricultural Education from Middle Tennessee State University (MTSU) and the University of Florida to add to the philosophical framework of the text, and he used his research program in leadership development to strengthen the proven strategies being shared with every student who engages with the text.

Every day Dr. Ricketts works with students seeking to develop their leadership potential at different levels. He is the Coordinator of Agricultural Education in the Department of Agricultural Sciences, where he teaches, conducts research, and sponsors Extension outreach activities for agriculture teachers and extension educators.

He has been a recipient of the following honors: Outstanding Extension Educator at Tennessee State University, 2016; Outstanding Young Agricultural Educator, American Association of Agricultural Education (AAAE), Southern Region, 2008; three-time recipient of UGA's Outstanding Teaching Faculty Award; UGA College of Agricultural and Environmental Sciences (CAES) Outstanding Academic Advisor Award, 2007; and the Gamma Sigma Delta Jr. Achievement in Teaching Award, 2006. John has also received the American FFA Degree, the highest degree available in the FFA. He has taught leadership at the secondary, post-secondary, and graduate levels of education. He is a prolific author of scholarly research on leadership development and critical thinking/decision-making in Agricultural Education, with over 225 publications and $4.4 million of external funding over the years to support a research program looking at issues of leadership development in Agricultural Education.

John lives in Mount Juliet, TN on the family farm. He and his wife, Jennifer, and their two kids Jamey and Jackson raise, show, and sell market goats. Ricketts seeks to instill each and every concept outlined in this text in his own kids, his farm operation, and his job as a professor.

Acknowledgements

The authors wish to express their appreciation to the reviewers for the fourth edition of this book: Deby Cahan, John Bowne Agriculture Program, Queens, NY; Farrah Johnson, Deltona High School, Deltona, FL; and Stephanie Joliff, Ridgemont Agricultural Education Teacher and FFA Advisor.

Text Design Image Credits

Chapter 13–17: canoeing: © iStock.com/Terry J Alcorn; checking grapes: © goodluz/Shutterstock.com; farmer working on laptop: © Budimir Jevtic/Shutterstock.com; FFA members planting flowers: Courtesy of the National FFA.; working on team project: © Jacob Lund/Shutterstock.com; farmer checking field: © Ivonne Wierink/Shutterstock.com; agricultural worker using tablet: © SpeedKingz/Shutterstock.com; two FFA members working in a garden: Courtesy of the National FFA.; business people discussing a project: © g-stockstudio/Shutterstock.com

Chapters 18–20: mountain view: © iStock.com/sankai; farmer sitting by cows: © goodluz/Shutterstock.com; students at the computer in the library: © Golden Pixels LLC/Shutterstock.com; FFA student in a greenhouse: Courtesy of the National FFA.; Confident teen: © goodluz/Shutterstock.com; farmer: © erwinova/Shutterstock.com; standing in a vineyard: © Kinga/Shutterstock.com

Chapters 21–23: research: © pixelrain/Shutterstock.com; FFA student in the field: Courtesy of the National FFA.; Bee keeper: © Janis Smits/Shutterstock.com; cultivator: © oticki/Shutterstock.com; agricultural worker in a cowshed: © Ljupco Smokovski/Shutterstock.com; farmer and businessman shaking hands: © SpeedKingz/Shutterstock.com; farmer: © Phovoir/Shutterstock.com

Back Matter: Vermont farm: © Richard Cavalleri/Shutterstock.com;

Section 1 UNDERSTANDING LEADERSHIP

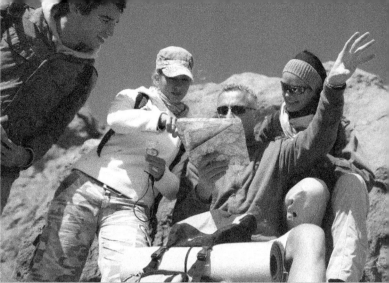

1 LEADERSHIP CATEGORIES AND STYLES

Leadership is the ability to move or influence others toward achieving individual or group goals. Leadership is not a magical trait. It can be developed, improved, and mastered if you have the motivation to study and apply its many known principles. In other words, you can become a leader if you have the determination to develop the abilities that make a leader.

Objectives

After completing this chapter, the student should be able to:

- Analyze various definitions of leadership
- Discuss the contributions of agricultural education and the FFA to leadership development
- Explain why leadership is challenging
- Explain why effective leaders need supportive followers
- Explain the difference between leadership and management
- Evaluate myths about leaders and leadership
- Describe the various leadership categories
- Explain the "big picture" relationship between leadership categories, behavior, and employment
- Describe authoritarian, democratic, and laissez-faire behavioral leadership styles
- Describe the situational (contingency) leadership behavioral style

Terms to Know

- leadership
- goals
- traits
- objectives
- trait leadership
- skills
- power leadership
- emulate
- influence leadership
- behavioral leadership
- authoritarian
- democratic
- situational leadership
- traditional leadership
- popularity (perceived) leadership
- combination leadership
- structural frame
- human resource frame
- political frame
- continuum
- laissez-faire leadership
- participative leadership
- self-esteem
- superleadership

eadership is the ability to move or influence others toward achieving individual or group goals. Leadership is not a mystical quality or trait that one individual has and another does not. It is learned behavior that anyone can acquire or improve by study and application. You can become a leader if you have the determination to develop the abilities of a leader.

Leaders are needed in organizations, communities, states, and nations. *Leaders* serve. This is the key to developing your own leadership abilities. Benefits to yourself are incidental when you as a leader are assisting others.

Persons trained to lead have characteristics that most people want: respect, poise, confidence, and the ability to think and take on responsibility. They are mature because development as a leader helps one to grow in areas such as communicating with, motivating, and understanding others. Learning leadership helps you prepare to take your place in society as a useful, productive, well-adjusted citizen (Figure 1–1).

Every member of a group participates in leadership when he or she contributes an idea. Leadership passes from person to person as each contributes to achievement of group goals. A democratic group reaches its height when leadership is spread throughout the membership. There is no limit to the number of leaders within a group. In fact, the more leaders the better, because the very act of leading develops the initiative, creativity, and responsibility that the group needs from each of the members.

A true leader helps the group achieve what the group as a whole believes is important. Leaders' thoughts and feelings are consistent with their speech and actions. In addition to putting forth their own ideas, they can make suggestions about, modify, or expand others' ideas; they can receive and implement others' ideas, too. Sometimes people confuse leadership with being bossy, but there is a difference.

FIGURE 1–1 FFA Career Development Events (CDEs) are an excellent venue to practice and exhibit leadership skills. These students and their adviser are about to compete in an FFA floriculture team CDE. *(Photo courtesy of Georgia Agricultural Education.)*

BOSS	LEADER
Assigns tasks	Sets the pace
Says, "I"	Says, "We"
Says, "Go"	Says, "Let's go"
Drives people	Guides people
Depends on authority	Depends on good will
Creates fear	Develops confidence[1]

DEFINITIONS OF LEADERSHIP

Research has yielded many definitions of leadership because it is complex, and because it is analyzed in a variety of ways that require different descriptions.[2] Even though leadership is hard to define and explain, it is easy to recognize. In a way, leadership is like love. We cannot define it or teach it, yet we can always identify it. We know when it is present and when we need it, but we cannot ensure its expression or continuation. The late leadership guru Warren Bennis once wrote, "To an extent, leadership is like beauty: it's hard to define, but you know it when you see it."[3]

The term *leadership* means different things to different people. People usually define leadership according to their individual perspective or the context that has meaning to them. Warren Bennis identifies leaders as people who are able to express themselves fully. They know who they are, what their strengths and weaknesses are, and how to fully display their strengths and compensate for their weaknesses. They also know what they want, why they want it, and how to communicate what they want to others to gain their cooperation and support. Also, Leaders are known for working hard to achieve their goals.[4]

In this book, we define **leadership** as the ability of a person—the leader—to move an organization or group toward the achievement or accomplishment of its goals and objectives, using whatever style is the most effective in each situation. Refer to Figure 1–2 for other definitions of leadership. Now, let us further investigate goal achievement.

Goal Achievement

Regardless of the exact definition, leadership involves the achievement of goals and objectives. **Goals** are the ends toward which effort is directed. **Objectives** are measurable goals. Leadership is a process through which group members are influenced to achieve group goals. The group may be a high school club such as FFA, a business, or a nation. Think carefully about your own behavior. To reach some of your personal goals, do you try

---

Leadership **Defined**

A process whereby an individual influences a group of individuals to achieve a common goal. —*Peter Northouse, university professor*

A function of knowing yourself, having a vision that is well communicated, building trust among colleagues, and taking effective action to realize your leadership potential. —*Warren Bennis, author*

A dynamic process of persuasion by which an individual induces a group to take action that is in accord with the leader's purpose, or the shared purpose of all. —*John W. Gardener, politician*

Leaders have to master . . . managing for the mission, managing for innovation, and managing for diversity. —*Frances Hesselbein, former CEO, Girl Scouts of America*

When one person attempts to influence the behavior of an individual or group, regardless of the reason. —*Paul Hersey and Kenneth Blanchard, authors*

Any person who influences individuals and groups within an organization, helps them in the establishment of goals, and guides them toward achievement of those goals, thereby allowing them to be effective. —*Afsaneh Nahavandi, author*

A relationship founded on trust and confidence. —*Jim Kouzes and Barry Posner, authors and leadership researchers*

Mental toughness, humility, and love. —*Vince Lombardi, former football coach for the Green Bay Packers*

A matter of how to be, not how to do it. —*Stephen Covey, author*

[Willingness] to take risks, be [a] good communicator, be decisive, and perhaps most importantly, [be] intellectually honest. —*Ann Winbald, entrepreneur and technology leader*

FIGURE 1–2 Leadership is defined in many ways. These descriptions of leadership come from successful individuals who have studied and practiced the science of leadership. What will be your definition of leadership?

to influence the behavior of other people so that something specific is accomplished? The ability to influence behavior is a key element of leadership.[5]

"Leadership is a process or a reasonably systematic and continuous series of actions directed toward group goals. Leadership is not usually a single act or even a few acts performed only in certain situations. It is a pattern of behaviors that is demonstrated consistently over time with specific objectives."[6] We discuss several leadership categories and styles in this chapter and throughout this book. Nevertheless, almost any behavior of leaders is directed toward the achievement of goals and objectives.

AGRICULTURAL EDUCATION, THE FFA, AND LEADERSHIP

It is in this context that agricultural education and the FFA (formerly the Future Farmers of America) enter the leadership picture. Bennis used the words *communication, cooperation,* and *support* in conjunction with those who have become leaders. The aims and purposes of the FFA are the development of leadership, citizenship, and cooperation. Agricultural education and the FFA can give students experiences and expose them to a variety of situations similar to those they may encounter in their chosen occupations.

Many people think of public speaking and parliamentary procedure when they hear the word *leadership.* Although leadership consists of much more than these skills, they are a first step toward leading groups and influencing others. Those involved in agricultural education and the FFA do many things that contribute to leadership development, such as communicating in many forms to various individuals and groups, guiding group discussions, setting goals, conducting meetings, solving problems and making decisions, developing social skills, and planning careers.

Importance of Youth Organizations

The National Association of Agricultural Educators (NAAE) believes that to ensure quality, agricultural education programs must be responsive to the needs of the individual for job-entry skills and the compatible skills of communication, citizenship, leadership, decision making, positive attitude toward learning, and personal occupational responsibility. It has been found that students who were more active or participated more frequently in educational student organizations (e.g., FFA) were rated higher by the students' instructors at the time of graduation and were rated higher by parents and employers six months after graduation, on leadership, citizenship, character, willingness to accept responsibility, confidence in self and work, and cooperative spirit and effort.[7]

Leadership and Personal Development Abilities

Agricultural education and the FFA can provide students with opportunities to relate positively to others in a variety of situations, as well as to participate in activities that develop leadership and increase occupational competency. Youth organizations, especially the FFA, can be one of the most effective ways of maximizing these important learning opportunities.

A well-balanced program of FFA activities supplementing formal classroom instruction and supervised agricultural experiences can constitute a desirable educational program for students. Agricultural education and the FFA can expose students to a variety of situations similar to those they may encounter in agribusiness and agriscience careers. In particular, the FFA encourages leadership and personal development through a variety of activities and roles. In the process of exercising initiative, accepting responsibility, and evaluating progress, members gain leadership and personal development skills.[8] Ten areas of the FFA that contribute to leadership and personal development are briefly discussed here.

Parliamentary Procedure Both members and leaders who want to take part in and guide group meetings need the ability to use parliamentary procedure effectively. The rules of parliamentary procedure make meetings more democratic. Learning parliamentary procedure has other practical applications as well. Practice in parliamentary procedure helps one learn to stay on the subject, respect the rights of others, and move to a decision. Essentially, while learning parliamentary procedure, one learns to be a democratic leader. Parliamentary procedure also aids in developing communication skills and public speaking ability.

Leading Individuals and Groups The FFA organization provides many opportunities for members to practice and improve their skills in democratic leadership. One function of a leader is to help groups work in a democratic way to ensure that individuals participate in making decisions. Participation and learning by doing are important to FFA leadership programs. Certainly, not all FFA members are born (natural) leaders. It takes study, practice, mistakes, and perseverance to develop chapter, state, and national officers. The FFA gives its members opportunities to become leaders at an early age, providing a testing ground where they can practice their skills and become experienced leaders.

Conducting Meetings Conducting and participating in FFA meetings offer excellent opportunities for developing leadership skills. These skills will be valuable when students graduate and participate in professional, business, and civic organizations. Through the FFA, members can practice speaking in public, facilitating groups, planning effective meetings, developing meeting agendas, establishing rapport to obtain group cooperation, following democratic procedures, supporting majority decisions of the group, and respecting the opinions of others.

Duties and Selection of Officers All members of an organization are responsible for electing capable officers. During an officer selection process, FFA members learn what is expected from leaders such as the officers, what the qualifications are for each position, and how to nominate and elect individuals to office properly. By learning the duties of officers and the proper ways to select them, FFA members can become better contributors to other civic and private organizations, as well as advisers to youth clubs and activities (Figure 1–3).

Managing Financial Resources The knowledge and abilities developed during work with FFA financing will help members perform their responsibilities as participants in other groups and organizations. Earning money and managing its distribution can be a truly meaningful educational experience. This is particularly so for members who keep official record books in the course of their supervised agricultural experience programs. Well-planned and well-executed fundraising activities give members a chance to acquire and use leadership abilities and learn how an organization manages money (Figure 1–4).

Developing Good Work Habits Supervised agricultural experience programs help members become more responsible people by developing

FIGURE 1–3 Being a chapter officer and speaking in front of groups may lessen a student's fear of public speaking. This student mastered his fears to become a state FFA officer and speak in front of thousands. *(Photo courtesy of Georgia Agricultural Education.)*

FIGURE 1–4 As part of being involved in agricultural education and the FFA, you may keep records for the supervised agricultural experience program. This will help you learn to manage financial resources in the future. (© iStock.com/JackF)

the good work habits necessary for success and advancement in the job market. Supervised agricultural experience is foundational for advancement in the FFA and for guidance in choosing the FFA member's agricultural career. Members can explore and practice agricultural and personal skills that prepare them for the future. Some of these skills include responsibility, personal integrity, initiative, willingness to accept supervision, a positive attitude toward work, dependability, and taking pride in one's work.

"Learning to do" and "doing to learn" (as the FFA motto puts it) are the basis of the supervised agricultural experiences that can help members develop good work habits. The FFA degree program is based on the member's supervised agricultural experience program and leadership development. Degrees of membership are contingent on definite accomplishment in the school's agricultural education program. However, one cannot attain specific levels of achievement with respect to agricultural work experience unless one develops good work habits in the process.

Individual Adjustment As used in this context, individual adjustment refers to personal

development abilities that are a part of the maturation process facilitated by FFA contests and other activities. These personal development skills include, but are not limited to, following established procedures and regulations, setting competitive goals, developing the ability to perform under pressure, and learning to accept success and failure. FFA Career Development Events (CDEs) provide practical experience in applying what is learned in agricultural education courses, as well as opportunities and challenges for personal growth and leadership development. All CDEs should be a natural outgrowth of the instructional program in agricultural education. Other measurable skills include using appropriate public relations approaches, managing use of time, and demonstrating poise (Figure 1–5).

Participating in a Committee No matter what a student's plans may be, he or she needs practice in committee work. Learning to work with others in accomplishing tasks is an important part of an individual's education. Committee participation provides a means of getting every member involved in some area of interest for maximum chapter effectiveness and member growth.

Participating in committees and groups requires cooperation. Committee activities should be designed to provide the best possible experience for members, whether they plan to become production agriculturists, enter nonfarm agricultural business, join industry, or go for advanced training at a technical school or university. Skills developed through committee work include identifying different kinds of committees, identifying purposes of committees, selecting members for committees, being an effective committee chairperson, presenting a committee report, promoting committee member participation, delegating responsibility, and setting time frames and timelines for accomplishment of the committee's purpose.

Developing Social Skills Members must become confident, responsible people before they can become confident, responsible leaders. Social and recreational activities, as well as educational activities, are necessary for the development of well-integrated and healthy people. Although it is not the primary responsibility or purpose of the FFA to teach social skills, the FFA can supplement and perhaps make more meaningful some of the learning acquired elsewhere and assist members

FIGURE 1–5 As these students compete in the Forestry Judging Career Development Event, they learn to follow established rules and procedures, accept success and failure, and develop many other skills. *(Photo courtesy of Georgia Agricultural Education.)*

in preparation for adult life. There is social value in almost every FFA activity that provides members with opportunities to experience and practice real-life situations for personal growth. This includes properly presenting an award, selecting the correct fork at a banquet, dressing appropriately for an occasion, meeting and greeting people, and making proper introductions (Figure 1–6).

Developing Citizenship Skills Community service activities and service learning help members develop and practice the good citizenship and leadership abilities they will need now and in the future. The focus is on making the community a better place to live and work. Personal development can be analyzed in terms of desirable personal characteristics, acceptable social behavior, and good citizenship. Measurable skills are the ability to plan and establish community service projects; develop a respect for national symbols and customs; understand that "if we belong, we pay dues"; respect the rights and views of others; and cooperate with others in group activities.

Citizenship development helps members become informed about their civic responsibilities, such as voting, paying taxes, and abiding by the laws of society. FFA activities strongly emphasize learning and personal development for responsible citizens. Volunteering to serve others is one form of civic engagement in which an FFA member can participate. FFA members can volunteer through a variety of service learning activities associated with programs such as the Million Hour Challenge, Earth Day, or the National Day of Service. These types of programs develop a sense of pride in one's community and in individual FFA members—because someone took the initiative to make a difference.

Do Agricultural Education and the FFA Make a Difference?

In a study intended to measure leadership and personal development abilities, researchers from the Ohio State and Middle Tennessee State Universities compared agricultural education students and FFA members to students who had never taken any agricultural education classes. The agricultural education students and FFA members had significantly more (57 percent) leadership ability than did nonagricultural education students (41 percent). The evaluation also revealed that the more active students were in FFA activities, the higher their level of leadership and personal development.[9]

FIGURE 1–6 By attending the FFA awards banquet, students can learn the importance of social skills such as public relations, dressing appropriately for an occasion, and proper etiquette for meals. *(© istock.com/kristian sekulic)*

LEADERSHIP IS CHALLENGING

Leading groups effectively is a tremendous challenge, a great opportunity, and a serious responsibility.[10] Today's organizations need effective leaders who understand the complexities of our changing global environment. Leaders need the intelligence, sensitivity, and ability to empathize with others, because leaders must motivate their followers to strive to achieve the group's goals and objectives. Leadership is a complex process that results from the interactions among a leader, followers, and the situation. All three of these elements are keys to the leadership process.[11]

Examples of Challenges

We are often fascinated by those who lead us. When we study historical figures or some of the leaders of our own times, we are amazed by their accomplishments. However, many extraordinary leaders have found themselves rejected by the people who once admired them. Consider the challenges of the following leaders:

- Julius Caesar experienced many conflicts with the members of the Roman Senate.
- President Charles de Gaulle's road to the leadership of France was fraught with failures. After coming to office as a hero following World War II, he was forced out of office twice.

- Former British Prime Minister Winston Churchill was removed from office twice and endured long periods during which his opinions were neither valued nor wanted.
- Nelson Mandela, a proponent of racial equality in South Africa, was imprisoned for 27 years before becoming president of that country in 1994.
- The popularity and reputation of former British Prime Minister Margaret Thatcher varied with the mood of the British public and the economic situation in Europe.
- Former President George W. Bush experienced both record highs and record lows in public approval during his time in office. More challenging, though, may have been the daily decisions he had to make about the war in Iraq and the stalled economy.
- Rudolph ("Rudy") Giuliani, former mayor of New York City, was both loved and hated throughout his tenure as mayor; after the September 11 terrorist attack on the World Trade Center, though, he rallied a city and a nation.
- Henry Cisneros, a former mayor of San Antonio, Texas, and a secretary of Housing and Urban Development under President Clinton, fell into disfavor with the electorate before reestablishing himself as a leader in the investment industry.

- George Watson Jr. was booted out of office after successfully leading IBM for many years.[12]
- Despite President Clinton's impeachment and Senate trial, he maintained his popularity in opinion polls until his term of office expired in 2001.
- Rick Pitino was once a great college basketball coach for the University of Kentucky. Today he still is a great coach for the University of Louisville. However, he resigned from the Boston Celtics, a professional team, without achieving a single winning season.

If the powers of each of these leaders were so outstanding, why did these leaders appear to lose those powers at times and experience so many setbacks? Why were these leaders not effective all the time? Could it be that for our groups and organizations to be effective and for our society to progress successfully, we must be able to select the right leaders and help them succeed?[13] What about the followers of these great leaders?

EFFECTIVE LEADERS NEED SUPPORTIVE FOLLOWERS

"Effective leadership can inspire and motivate followership, but there are characteristics and behaviors that followers should embrace and implement in order to make their leaders more effective."[14] These characteristics include collective responsibility, trust, communication, caring, and pride. Without supportive followers, even effective leaders can accomplish very little. The strength of a good coach is proven by a winning team. The same is true in an FFA Career Development Team. It is the hard work of all the members that allows a team to learn, win, and succeed.

In the workplace, leaders "demonstrate their competence by setting worthwhile and challenging goals with followers, by showing confidence in followers and supporting their efforts to perform well." Leaders also strive to get followers to improve by giving recognition when they do a job well.[15] Goal setting and recognition should be practices that permeate a successful FFA chapter as well.

F. E. Fiedler's Contingency Model of leadership defines leadership effectiveness in terms of follower performance. According to this view, a leader is effective when his or her group performs well. R. J. House's Path-Goal Theory considers followers' satisfaction a primary factor in determining leadership effectiveness: specifically, leaders are effective when followers are satisfied. Others studying the transformational and visionary leadership models define effective leaders as those who successfully implement large-scale change in organizations. In short, *an effective leader is one whose followers achieve their goals and objectives, can function well together, and can adapt successfully to changing external forces and demands.*[16]

Servant Leadership

Notice in the following statements how the terms *followers* and *group* are used in conjunction with leadership:

- Leadership is a group phenomenon. There are no leaders without followers. Therefore, leadership always involves interpersonal influence or persuasion.
- Leaders use influence to guide groups of people through a certain course of action or toward the achievement of certain goals.
- A leader helps followers establish goals and guides them toward achievement of those goals, enabling and allowing the followers to be effective.[17] In this respect, some people consider leaders to be servants of their followers—hence the term *public servant*.[18]

LEADERSHIP VERSUS MANAGEMENT

The differences between leadership and management appear to relate to effectiveness rather than the terminology or titles delineating the two concepts. In reality, the same person will be at times a leader and at other times a manager. For example, any manager who guides a group toward the accomplishment of goals and objectives is actually acting as a leader. Much of the distinction between management and leadership seems to spring from the assumption that the term *leader* implies competence; that is, that the person is good at what he or she does. Therefore, an effective and successful manager could be considered a leader, but a less competent manager would not be perceived as a leader.[19]

SOMETIMES A LEADER, SOMETIMES A MANAGER	
LEADERS	**MANAGERS**
Get companies and people to change	Know how to write business plans
Deal directly with people and their behavior	Make decisions based on the data and numbers
Initiate actions and expect change	Are concerned with issues such as keeping the organization running
Are charismatic; create a sense of excitement and purpose in their followers	Bring order and consistency through planning, budgeting, and controlling
Have (or are believed to have) attributes that allow them to energize their followers	Take care of the routine details
Do the right things	Do things right
Deal with changing and developing more effective organizations	Deal with mechanical and administrative activities
Find a niche in which an organization can function and grow	Are organizational engineers working to achieve well-known goals
May need to charge up followers, creating commitment, inspiration, growth, and adaptation	Tend to mundane matters such as modifying rules and regulations, allocating resources, and assigning tasks
Focus on the future	Focus on the present
Create a culture based on shared values	Implement policies and procedures
Use personal power	Use position power
Establish an emotional link with followers	Maintain objectivity

FIGURE 1–7 The terms leader and manager are often used interchangeably, but there are differences between the two roles. The effective leader usually acts as a leader, but may also occasionally take on the role of manager, as necessary.

Distinctions between Leaders and Managers

This book does not dwell on the fine line between leaders and managers; rather, it uses the terms interchangeably. However, the role of the leader/manager changes according to the situation. Figure 1–7 presents some distinctions between leadership and management. Whether the person is sometimes called a leader or sometimes called a manager, the purpose of the leader/manager is to influence followers to accomplish the group's or organization's goals.

MYTHS ABOUT LEADERS AND LEADERSHIP

There are several myths or misunderstandings regarding leadership and leaders:

Myth 1 Little is known about leadership.
Myth 2 All leaders are born with unique traits and characteristics.
Myth 3 Leaders make all the decisions for a group.

Myth 4 Every leader is a popular, charismatic individual.
Myth 5 To be a leader, one must have been elected or appointed.[20]
Myth 6 Leadership is a rare skill or ability.
Myth 7 Leaders control, direct, and produce.
Myth 8 Leadership exists only at "the top" (a high position in an organization).

In fact, information on leaders and leadership is readily available. Businesses, educators, not-for-profit groups, and others use this leadership information to improve their organizations. Current thinking on leadership rejects the proposition that leaders are born with special qualities or inherent abilities. Rather, researchers have found that almost anyone can learn and develop leadership skills. In particular, studies of leadership have revealed the importance of involving members in the decision-making process. New techniques help leaders determine how to facilitate this involvement. Leadership is not a place but rather a process; therefore, those at the top or with the titles may not be the only leaders in a group. When members participate in group decisions, they are leading.[21]

LEADERSHIP CATEGORIES

Major categories of leadership are (1) trait, (2) power and influence, (3) behavioral, (4) situational, (5) traditional, (6) popularity, and (7) combination.[22] These categories help explain why there are so many different definitions of leadership. Different cultures, groups, organizations, and nations have different perceptions of leadership. Just as people have different views of a religion or political party, they have different views of leadership. This section discusses each of these categories and supplies a schematic illustration of how various views of leadership fit into "the big picture" of the broad range of leadership categories.

Trait Leadership

One of the earliest approaches to studying leadership was the trait approach. **Traits** are distinctive qualities or characteristics. The **trait leadership** approach is based on a person's distinctive qualities or characteristics that predispose the person to be a leader. It assumes that some people are "natural" leaders who possess certain traits that others do not. Trait theories search for and try to pin down the exact qualities that leaders exhibit, in an attempt to prove that some people are born leaders. Individual physical attributes, personality, social background, abilities, and skills are among the traits that have been examined.[23]

Traits Scholars have researched the personalities of leaders, focusing on traits such as age, weight, height, appearance, physique, energy, health, speech fluency, intelligence, scholarship, knowledge, insight, judgment and decision, originality, adaptability, introversion/extroversion, dominance, initiative, persistence, ambition, responsibility, integrity, and conviction. Although certain traits increase the likelihood that a leader will be effective, they do not guarantee effectiveness. The relative importance of various traits depends on the nature of the leadership situation.[24]

Skills Remember, leaders are not born. Therefore, traits alone do not determine leadership. Skills, which are learned, are different from traits. **Skills** are practiced abilities (Figure 1–8). Intelligence, social skills, speaking ability, ability to work within group dynamics, and organizational ability are skills most frequently linked to leadership ability. Other skills and traits are discussed in detail in Chapter 2.

FIGURE 1–8 This student is demonstrating her skill in public speaking. Verbal and public speaking ability have a high correlation with career advancement. *(© iStock.com/Steve Debenport)*

Power and Influence

The power and influence theories explain leadership effectiveness in terms of the amount and type of power a leader possesses and how he or she exercises power.

Power Leadership Power is important not only for influencing subordinates but also for influencing peers, supervisors, and people outside the organization. **Power leadership** is leading by force, with the group submitting to the leader (perhaps against its will). A person's power often depends to a considerable extent on how the person is perceived by others. There are many sources and types of power, and more than one may be a factor in any given power leadership situation.

Formal authority Power stemming from formal authority is sometimes called **legitimate power**. Authority is based on perceptions about the goals, obligations, and responsibilities associated with particular positions in an organization.[25]

Control over resources and rewards Potential influence based on control over rewards is sometimes called **reward** or **position power**. This control stems in part from formal authority. The higher a person's position in the organization, the more control over scarce resources the person is likely to have. An example of reward power is influence over wages and promotions.

Control over punishment This form of power is sometimes called **coercive power**. It is the capacity to prevent someone from obtaining deserved rewards or to take away something the person already has.

Control over information This kind of power involves access to vital information and the distribution of information to others.[26] Can you imagine what would happen if the Internet totally quit working in a place of business?

Ecological control This form of power is sometimes called **situational engineering**. It means control over the physical environment, technology, and any organization of the workplace. Manipulation of these physical and social conditions allows one to influence, directly or indirectly, the situations of others.

Expertise This form of power is sometimes called **expert power**. Experts solve problems and perform important tasks. However, expertise is a source of power for a person only if others depend on the person for advice, knowledge, or assistance.

Friendship and loyalty Sometimes called **referent power**, this source of power is the desire of others to please a person for whom they feel strong affection or to whom they are loyal. Success in developing and maintaining referent power depends on interpersonal skills such as charm, tact, diplomacy, empathy, and humor. People with friendship and loyalty power often do not hold elected or appointed positions, but, because of the respect they have earned (**reference**), they can exert great power over those who do have appointive or elected posts.[27]

Charisma Charismatic leaders are excellent communicators who are charming, persuasive, and able to engage followers on a deep and emotional level. They are able to rally followers around a common goal and vision, and to that motivate commitment to their method of doing business.[28] Leaders with charisma fight for what is right for their followers. Their charisma generates excitement and greater contributions to the common goals of the group.[29]

Many years ago, Sigmund Freud theorized that charismatic leaders have such strong effects on followers because followers resolve inner conflicts between their self-image and what they think they should be by making the charismatic leader a representative of their desired state. To followers, the leader becomes an ideal whose behavior they can **emulate**, or copy or simulate. By completely accepting the leader and his/her ideas, followers fulfill a natural human desire to become more noble and worthy. Through imitation of the leader, followers thus become their ideal selves.[30] "The process by which followers connect with charismatic leaders is called **personal identification**, and it helps explain why they will actively defend the leader against critics or other attackers. They are really defending their own ideal self, which they are striving to become."[31]

Charismatic leaders have an enormous impact on their followers because they are able to do three main things. First, they are able to exert idealized influence, which means that they have a vision of great meaning and purpose, and they are able to get followers to buy in to their idealization.

Second, they are able to be inspirational leaders: they have the ability to influence others by getting followers excited about the ideals. Third, charismatic leaders are able to make their followers think; they stimulate followers to form their own opinions and make independent judgments.

Charismatic leadership can be good or bad, depending on the leader's goals. We call "good" charismatic leaders **authentic** leaders, and we call "bad" charismatic leaders **inauthentic** leaders. The terms *ethical* and *unethical* are also used in place of *authentic* and *inauthentic*, respectively. All charismatic leaders are concerned with bringing about big changes, and all charismatic leaders have a group of followers who believe in their leader above all else. The difference lies in the moral nature of the charismatic leader's proposed goal.[32]

For example, charismatic leaders often get results, but they may also put their followers at risk. Charismatic leaders such as Adolf Hitler, Jim Jones, David Koresh, and Charles Manson have carried out evil or immoral missions and brought ruin and death to their followers. Refer to Figure 1–9 for the qualities of ethical and unethical charismatic leaders and their effects on followers.

Political power Political power is actually a process whereby groups and group members work to ensure power or protect existing power sources. This requires remaining in control of one's group, whether it is a political party or an organization. People do many things to solidify their power base (Figure 1–10).

Influence Leadership The exercise of influence is different from the exercise of raw power. When one exercises raw power, one forces a group to submit, perhaps against its will. When one influences others, one shows them why an idea, a decision, or a means of achieving a goal is superior in such a way that they freely follow that person's lead. Members will continue to be influenced as long as they are convinced that what they have agreed to is right or is in their best interest as individuals or as a group.

The essence of leadership is influence over others, but influence flows in both directions. Leaders influence followers, but followers also have influence over leaders. For example, in an FFA chapter, the effectiveness of the president depends on his or her influence over the adviser and fellow officers (peers). However, the

KEY CHARACTERISTICS AND BEHAVIORS OF CHARISMATIC LEADERS	
ETHICAL (AUTHENTIC) CHARISMATIC LEADER	**UNETHICAL (INAUTHENTIC) CHARISMATIC LEADER**
• Uses power to serve others	• Uses power for personal gain or impact
• Aligns vision with followers' needs and aspirations	• Promotes own personal vision
• Considers and learns from criticism	• Censures critical or opposing views
• Stimulates followers to think independently and question the leader's views	• Demands that own decisions be accepted without question
• Uses open, two-way communication	• Uses one-way communication
• Coaches and develops followers; shares recognition with followers	• Is insensitive to followers' needs
• Relies on internal moral standards to satisfy organizational and societal interests	• Relies on convenient external moral standards to satisfy self-interests
• Develops followers' ability to lead themselves	• Selects and produces obedient, dependent, and compliant followers
• Uses crises as learning experiences, to develop a sense of purpose in the organization's/group's mission and vision, and to emphasize the leader's intention to do right	• Uses crises to solidify own power base, minimize dissent, and increase dependence of followers
• Avoids the trappings of success; shares credit with followers and stays humble	• Succumbs to delusions of invincibility, greatness; places extreme emphasis on image management

FIGURE 1–9 Charismatic leaders come in different forms. This table outlines the differences between ethical and unethical charismatic leaders.
Source: Understanding Behaviors for Effective Leadership (2d ed.), by Jon P. Howell and Dan L. Costley, p. 214 (Upper Saddle River, NJ: Pearson Prentice Hall, 2006).

FIGURE 1–10 Politicians often speak to different groups like the FFA or other youth or agricultural organizations. Besides being a noble gesture, this type of activity helps to enhance a politician's political power base.
(© istock.com/Steve Debenport)

VIGNETTE 1-1	THE TWO SIDES OF CHARISMATIC LEADERSHIP

OSAMA BIN LADEN ALLEGEDLY MASTERMINDED the attacks on the World Trade Center and the Pentagon on September 11, 2001; at the least, we know that he supported and encouraged the act. He had an army of terrorists willing to die for the cause (idealized influence), because he related to them. He slept and ate in caves with his followers, and formed many relationships with them. He stands as an example of an inauthentic or unethical charismatic leader.

Then-President George W. Bush proved himself a charismatic leader in the immediate aftermath of the September 11 terrorist attacks, as he inspired and influenced many in the United States to believe that things were going to be all right despite the tragedy. He too had strong followers and a big change in mind (destroying terrorism), but the moral difference between Bush's goals and bin Laden's goals indicate that George W. Bush was an authentic or ethical charismatic leader.

effectiveness of the president also depends on the members' acceptance, implementation, and loyalty—in short, their influence.

Leadership consists of exerting influence. **Influence leadership** is leading by convincing others of an idea, so that they will follow of their own free will. It is the ability to bring about changes in the attitudes and actions of others. Influence can be indirect (unconscious) or direct (purposeful). We often influence others without being aware of it. If you get a new hairstyle, wear a new suit, or purchase a flashy new truck, you may influence someone who sees you to copy your hairstyle or buy a similar suit or truck. In a group, a leader can influence members indirectly by serving as a role model.

Behavioral Leadership

Behavioral leadership is leadership according to one's personality style. Behavioral leadership theories assume that there are distinctive styles that effective leaders use consistently. The two basic styles are **authoritarian** (theory X), in which the leader leads and makes decisions regardless of the wishes of the group, and **democratic** (theory Y), in which the leader leads and makes decisions based on the input of the group. All other behavioral styles are variants or combinations of the authoritarian and democratic styles of leadership. Because of the importance of these behavioral leadership styles, they are discussed in depth later in this chapter.

Situational Leadership

Some people refer to **situational leadership** as **contingency leadership**. The term *contingency* used in this leadership theory relates to the fact that the emergence or effectiveness of any style depends (is contingent) on the situation or immediate circumstances in which the leader is operating. One's assumptions about, behavior toward, and attitudes regarding people are difficult to change, but leaders can change their styles to meet the needs of a particular group or task. For example, you may prefer the democratic style, and hold democratic ideals and philosophies; but if you are leading a low-motivated group that needs close supervision, you may use an authoritarian supervisory style for greater effectiveness and efficiency.

As a leader, one not only must be aware of one's attitudes but also must use the leadership style appropriate to the situation. The leader who acquires a large number of leadership skills can exercise different abilities to meet the requirements of different situations. Situational leadership is discussed in greater detail later in this chapter.

Traditional Leadership

Some people see **traditional leadership** as culture based and highly dependent on symbols.[33] One way to sum up this leadership style is "This is the way we do things here." We do many things out of tradition: it may not be the only way, but it is the group, organization, or company way. Another way to describe this style is "playing the game." If one wants to become a leader in some groups, certain rules or customs have to be followed—that is, games have to be played. For example, you may want to be a leader in a contact sport like football, but what if you don't like running into other people? If you are ever going to be a successful leader at football, you must "play the game" and run into people as hard as you can.

An understanding of organizational culture is vital to leadership effectiveness. Depending on the culture, the leader may be the most intelligent person, the best speaker, the hardest worker, or the best dresser. Some examples follow to illustrate traditional, cultural, and symbolic leadership.

Public Officials Public officials are captives to the demands of cultural and symbolic leadership. For example, if the mayor of a city did not attend ribbon-cutting ceremonies, award keys to the city, or speak often to clubs, convention groups, and fund-raisers, she might not be perceived as a

---

VIGNETTE 1-2 | **THE POWER OF INFLUENCE**

WILLIAM H. COLEY WAS A RESPECTED TEACHER

The school administration respected Mr. Coley. The students respected Mr. Coley, and the people in the town revered him as well. He was an exemplary agriculture teacher.

After several years of teaching, Mr. Coley took a job with the State Department of Education. The respect he had earned as a teacher carried forward into his new position: Just as he had been respected as a high school agriculture teacher, he became respected as an educational administrator with the Department of Education. He was honest, fair, and intelligent, and his human relations skills were unmatched. Soon his influence spread across the state.

Mr. Coley eventually rose to the top position in state government for education. However, due to a family illness and other developments beyond his control, he had to move to a lesser post. Nevertheless, Mr. Coley's legacy as an educator continued even though he was no longer the top administrator in his field.

An interesting phenomenon occurred. Quarterly, the State Department of Education staff in Mr. Coley's educational area would meet. The state staff consisted of supervisors from the State Department of Education, regional supervisors, college professors, and high school teachers who were elected officers in teachers' organizations. As topics were discussed, almost everyone would give their input—everyone but Mr. Coley. Discussion might continue for a long time. Finally, after everyone had stated their opinions, Mr. Coley would speak—and members of the assembled teams would listen.

Although Mr. Coley was no longer a top administrator, whatever he said was usually the direction the team chose to follow. Why? Not because he had formal power, but because his many years of doing the right thing and working hard as an educator made him a person of influence. When William H. Coley spoke, people listened.

leader even though she had been an expert manager of finances and a terrific promoter of city growth. Certain things are done just because it is traditional to do so.

FFA Tradition and Culture The FFA is rich in traditional symbols and organizational culture. Many FFA advisers and members believe that leadership consists of being in official dress (standard uniform for FFA members), knowing the opening and closing ceremonies, reciting the FFA Creed, and participating in contests. Although these traditions may foster leadership development, much of what is done in the FFA is cultural and symbolic. FFA members who appear at official contests, especially speaking contests, without official dress are seen as not taking participation seriously and not giving the organization the proper respect.

Does this mean that people cannot become leaders if they do not wear the official FFA uniform? No, but they may not be perceived as leaders, because of the FFA symbolic culture. Remember, perception is reality: if we are not perceived as leaders by our group, we will not become leaders. This explanation is not intended to be negative or critical. Its purpose is simply to illustrate the cultural and symbolic nature of the FFA.

FFA symbols Many symbols exist within the FFA. The FFA emblem, banner, and paraphernalia are all symbolic (Figure 1–11). These things can instill pride, create motivation, and stimulate goal setting. If cultural symbols can enhance these characteristics and activities, we can again thank the FFA. What about the name *FFA*? The National FFA Organization and its members are no longer officially called Future Farmers of America. However, due to organizational culture and symbolic tradition, it has not changed "FFA" to another name. Should it? It is for you to decide. You are a part of the culture, and its symbols belong to you, too.

Corporate Symbols Coca-Cola® destroyed a national as well as a corporate symbol when it introduced a new Coke® formula some years ago. The power of the symbol was demonstrated by the public's refusal to change. The symbol seemed to be owned by the public in addition to the Coca-Cola corporate shareholders.[34]

American Flag Another example of a powerful symbol is our "Stars and Stripes," the American flag. No matter what may be perceived to be wrong with our country, when we raise this symbol and sing the national anthem, people stand in allegiance and seem to unite in recognition that the symbol represents our ideals and highest aspirations.

FIGURE 1–11 How many FFA symbols can you find in this picture?

Educational Symbols High schools and universities exhibit symbols that can elicit strong feelings of loyalty. During graduation ceremonies, retirement banquets, inaugurations, homecomings, and athletic events, for example, symbols are everywhere. Graduation caps and gowns, crowns, and flowers at homecomings are just a few of many possible examples.

Motivational Symbols A person in a newspaper office invented a "taking off" symbol to represent the paper's efforts to overtake the competition. Posters of spaceships were used to create the image, and the president of the newspaper installed and used an airplane seat belt on his office chair.[35] The symbol and the behaviors it inspired resulted in a more cohesive group at the newspaper office. Symbols are powerful tools for inspiring and motivating organizational members. The FFA symbol has a similar motivational effect on many present and past members.

Popularity (Perceived) Leadership

The **popularity (perceived) leadership** category relates to symbolic leadership in that this theory concerns perceived leadership. In other words, leadership is bestowed on someone because of celebrity or group perceptions rather than the leader's actual ability. The decision as to who should lead within an organization may be irrational.[36]

Falsely Positive Self-Concept The popularity or perceived leadership theory contends that leader effectiveness is bestowed by others. Effectiveness in this case relates to follower perceptions and not necessarily to the leader's actual ability. Followers often acknowledge or confer leadership where no proof of ability exists. Baseless praise of the leader creates or reinforces a false self-concept within the leader. Self-esteem is fine, but a falsely inflated self-concept can destroy a person's judgment. A person could develop a baselessly positive self-concept by being elected to an office because he is good-looking, even though he is not actually qualified to hold the office or carry out its responsibilities.

FFA Members FFA members should be aware—and beware—of the popularity or perceived leadership category. For example, just because a boy is captain of the football team does not mean he will be a good FFA president. However, if football is "king" at a particular school, the football captain might be perceived as a good leader, and

---

VIGNETTE 1-3   **TRADITIONAL LEADERSHIP OR LACK OF IT**

DR. RICK FUQUA WAS SELECTED AS PRESIDENT OF NORTHWEST STATE UNIVERSITY

Dr. Fuqua was a man of high integrity. His academic credentials had few equals. Thirty-seven other candidates had applied for the job, but Dr. Fuqua was the unanimous choice.

Dr. Fuqua was energetic and excited to be in his new position. He was a forward thinker and was studying changes that should be made at the university. Like any university president, he was looking particularly for ways to save money.

The first change was to cut the graduation ceremonies. Of course, students would still receive their diplomas, but he thought the graduation caps, tassels, gowns, and all the administrative costs were unnecessary. He believed that this tradition should be abandoned. This seemed like an ideal place to cut costs without affecting the university's high academic standing.

Dr. Fuqua's second plan was to cut expenses in the athletic program. He thought about cutting the football program, but it was breaking even. He did decide to cut the cheerleading squad, perceiving it as an unnecessary entity and another drain on scarce funds.

After Dr. Fuqua made a few other changes to university traditions, the fact that he was honest and a man of great integrity did not seem to matter anymore. His academic credentials were less appreciated. The students began to complain. The faculty and the donors who gave money to the university did not like the decisions he'd made. The alumni were especially outraged. The Student Government Association and the Faculty Senate organization both formed committees to voice their displeasure with Dr. Fuqua. Both groups met with the Higher Education Board of Trustees to ask for his dismissal.

Dr. Fuqua was very confused as to why so many people were upset with him. He had saved the university money as expected. He had also raised the academic standing of the university. What more could be expected?

What was the problem? This was not "the way we do things around here."

students may come to believe that good leaders are FFA presidents who are football captains. A tradition can even develop based on false assumptions. For instance, a charismatic club leader who delivers great speeches does not automatically qualify as a good club leader or administrator—but tradition may force her into those jobs anyway.

Organizations Organizations find ways to legitimize the choice of leaders. Organizational members might select leaders based on salary or experience. The following vignette describes a situation in which the organization legitimized the selection of a popular person as leader. What was the basis for the leader's election?

Combination Leadership

The **combination leadership** category unites the previous leadership models and styles. In this section, we view the combination category through four **frames**, or images, of leadership: authoritarian, democratic, political, and traditional (Figure 1–12).[37]

The dynamics of leadership range across all four frames or categories; problems and solutions are not restricted to any single one. This is the foundation of some basic assumptions about leadership effectiveness. The leader who can make positive use of the frames or categories will succeed where others may fail. Leaders who do not respond appropriately or who fail to use the correct frame or category can jeopardize their leadership ability. The problems that can arise often make solutions seem impossible.

Authoritarian or Structural Frame The authoritarian or **structural frame** relates to relationships and formal roles in the organization. Organizational charts, policies, procedures, authority, and responsibility guide the leader's decisions and behavior. The emphasis is on goals, roles, and formal relationships. Leaders do the following:

- Develop a strategy
- Focus on implementation
- Continually experiment, evaluate, and adapt
- Do their homework[38]

How does the FFA compare to the authoritarian structural frame? Structure does not imply an utterly rigid or strict hierarchy. The FFA is a structured organization, yet it is flexible (within certain boundaries) as members demand and situations require. It is structured from the local to the state and national levels, but the organization is flexible because it uses democratic principles.

COMBINATION LEADERSHIP CATEGORY	
Authoritarian Frame Authority Goals Task oriented Rules Roles Formal relationship Focus on implementation of procedures Policy and organization charts	**Democratic Frame** Meet needs Provide support Feeling oriented Motivation Believe in people Visible and accessible procedure
Political (Power and Influence) Frame Power Conflict Coalitions Backroom politics Bargaining Negotiation Compromise Cooperation with rivals	**Tradition (Symbolic) Frame** "This is the way we do it here" Culture (the best fighter is the leader) Ceremonies (opening ceremony) Rituals (ribbon-cutting ceremonies) Symbols (FFA symbols, American flag)

FIGURE 1–12 Leadership is a combination of various categories and styles. *Source: Adapted from work by Tom Burks, Director of the Leadership Institute at Middle Tennessee State University; and* Reframing Organizations: Artistry, Choice, and Leadership, *by Lee G. Bolman and Terrence E. Deal (New York: John Wiley & Sons, 1991).*

IT WAS MID-NOVEMBER AND THE STUDENT COUNCIL was getting ready to elect officers for the next semester. Although this was a good group with many outstanding students, its officer selection process took the form of spontaneous nominations from the floor. Many youth organizations, like the FFA, sometimes have students complete an officer application. It is also fairly common for a nominating committee to interview potential candidates and then present a slate of officers for organizational approval. A floor nomination can still be accepted as long as the nominee completes the application and interview process.

On the Friday night before the big election, Josh Courtney rushed for 268 yards, helping his team win the state football championship in its division. The team and fans were elated. The students carried Josh off the field and celebrated afterward.

The following Monday, the Student Council met to hold officer elections. As expected, Josh was nominated for president and easily won the election.

The following semester, Josh took over as president of the Student Council. By then, the excitement of the football championship had faded. Although Josh was a popular student, his academics, administrative, and human relations skills did not equal his ability to play football. The Student Council just was not getting the things done that it usually did. The senior prom was approaching, spring activities were in full swing, and deadlines were not being met. The group began to wonder whether it had selected the right leader for the organization.

Democratic or Human Resource Frame The democratic or **human resource frame** relates to the needs of members, without the strong emphasis on procedure and policy found in the structural frame. Within this frame or leadership style,

- Democratic leaders believe in people and communicate that belief.
- Democratic leaders are visible and accessible.
- Effective democratic leaders delegate; they increase participation, provide support, share information, and spread decision making among as many members in the organization as possible.[39]

Fitting the organization to the people and meeting the needs of followers become the keys to effectiveness for the democratic frame. You may recognize this as the democratic style, which states that organizations should design conditions that allow people to accomplish their own goals along with organizational objectives.

How does the FFA adhere to the democratic or human resource frame? Programs, projects, goals, objectives, and tasks are important, but not at the expense of people. As a saying attributed to Zig Ziglar goes, "People don't care how much we know until they know how much we care." The FFA is a democratic human resource (people) organization.

Political Frame The **political frame**, not to be confused with the political leaders or structure in a democracy, is the persuasive process within groups that involves efforts by group members to gain or consolidate power or to protect existing power sources. It focuses on the struggle for scarce resources in an organization. Power and influence are part of it. Sometimes things seem so difficult that it is "every person for himself or herself." Special-interest groups and coalitions form and dissolve as the need arises. Negotiation, conflict, and compromise are part of everyday life; the ability to persuade is essential. Whether we like it or not, when resources are scarce (as they usually are), the situation and organization may demand political leaders who can

- clarify what they want and what they can get;
- assess the distribution of power and interests;
- cooperate with rivals; and
- persuade first, negotiate second, and use coercion only if necessary.

However, a politically effective leader often persuades others to follow the leader's self-interest rather than focusing on organizational interests. Also, backroom political deals that serve the leader's own interests may be made, as a result of bargaining with interest groups, which can impose goals that prove to be constraints on the organization. Therefore, because of the leader's bargaining, organizational goals may be replaced by the special-interest group's goals, thus degrading the organization. If organizations are to diminish political influence, they must have shared vision and organizational values throughout.

How can the FFA or you as part of another organization overrule a person who has too much political influence? To promote shared vision and organizational values, the FFA can fully utilize its program of activities, which is prepared by the executive committee and voted on by the members. There are times when leaders have to do what is necessary to get the job done. However, if autocratic rule becomes more than situational, the members of the group can act to ensure that democracy remains the form of leadership to which they adhere.

Traditional Frame Refer to the previously discussed traditional leadership category for further explanation. This frame emphasizes, among other things, shared vision and values. Organizational ceremonies, traditions, and symbols are part of this frame. Traditional and symbolic leaders interpret experience, use symbols to capture attention, and discover and communicate a vision.

Figure 1–13 summarizes the categories of leadership.

CATEGORIES OF LEADERSHIP (TRADITIONAL UNDERSTANDING AND AGRICULTURAL EDUCATION–FFA RELATIONSHIP)		
LEADERSHIP CATEGORIES	**TRADITIONAL UNDERSTANDING**	**AGRICULTURAL EDUCATION–FFA RELATIONSHIP**
A. Trait	Some people assume that certain traits make you a leader. However, leaders are not born; they are made. Leadership skills are learned through FFA activities.	Public speaking Parliamentary procedure Leading group discussions Conducting meetings

B. Power and Influence	Forms of power:
	    Formal of legitimate
	    Position
	    Coercive
	    Ecological
	    Expert
	    Referent
	    Charisma
	    Influence
		FFA officers
		Chairperson, Program of Activities
		    Committee
		Influence through contest and awards
		    recognition

C. Behavioral	Authoritarian (theory X)
	Democratic (theory Y)
	Laissez-faire
		Democratic leadership learned through
		    parliamentary procedure and conduct of
		    meetings

D. Situational	Using the leadership style appropriate for
	    the situation.
		Parliamentary procedure: suspend the rules
		    in certain situations; rules have to be
		    suspended to get immediate action

E. Traditional	"This is the way things are done here."	Official FFA dress, FFA paraphernalia, FFA
		    emblem, FFA banner, opening and closing
		    ceremonies, FFA awards banquet, contest
		    participation

F. Popularity	Leadership is bestowed on those who
	    may not deserve it or be able to handle it
	    capably.
		Spur-of-the-moment FFA officer elections,
		    without applications and interviews,
		    result in most popular student being
		    elected regardless of leadership ability

G. Combination
  • Structural
  • Human Resource
  • Political
  • Symbolic
	Unites or combines the previous leadership categories. No leadership category stands alone.	Structural: FFA constitution and bylaws
		Human Resource: Democracy through par-
		    liamentary procedure and officer elections
		Political: Voting
		Symbolic: Refer to segment E in this table

FIGURE 1–13 Leadership can be divided into seven categories. This table provides the traditional understanding of each category, along with its relationship to agricultural education and the FFA. *Source: Adapted from work by Tom Burks, Director of the Leadership Institute at Middle Tennessee State University.*

THE BIG PICTURE

Figure 1–14 provides an overview of the leadership categories and their connections with each other. All these areas are discussed in this book. "The big picture" illustrates the connections between, relationship of, and importance of leadership and personal development to employment and leadership in your home and community. It also suggests why leadership has so many different definitions and why it can become complicated. Refer to Figure 1–14 as you read this book; it may help you fit the pieces into the leadership puzzle.

Democratic Leadership Is in the Center of the Big Picture

There are many leadership styles, but the basic style is democratic, which is closely connected to situational leadership. Although democratic leadership is the preferred style for most, there are

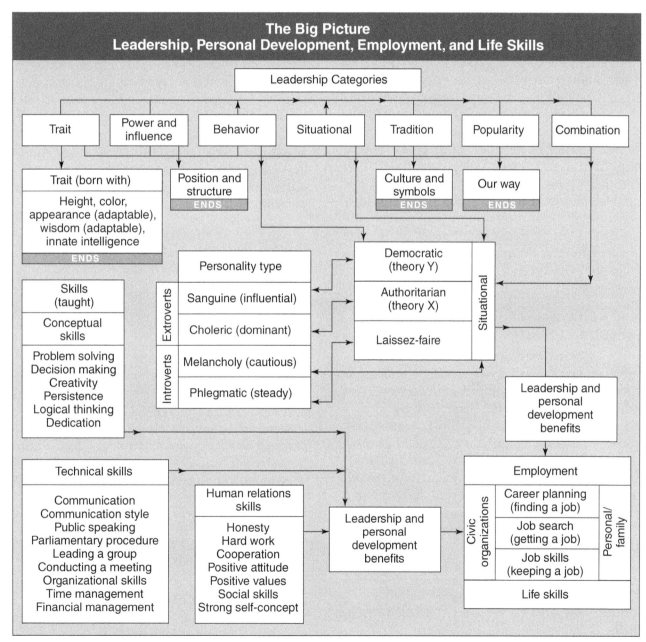

FIGURE 1–14 Defining leadership is not easy. Some leadership categories are related to personality type, personal development, employment, and life in general; other leadership categories do not relate to any other traits or categories.

situations in which authoritarian leadership must be used.

Leaders Are Not Born; They Are Made

People are born with certain traits, such as height, eye color, innate intelligence, and personality type. However, our traits and abilities have to be channeled toward goals. Most of the skills we possess are taught either informally through our environment or formally through our parents and schools.

Personality Types and Leadership Styles

There are many different types of leaders, all of whom may contribute to an organization or cause. Personality is one way in which people differ. Basically, Sanguine (influencing) personality types tend to be democratic leaders, whereas Choleric (dominating) personality types tend to be authoritarian leaders. Phlegmatic (laid back) personality types tend to be laissez-faire leaders, and Melancholy leaders are all business. However, through leadership training, any personality type can be taught to use the appropriate leadership style for the situation. Personality types are discussed in Chapter 2.

Leadership Skills Are Learned in Agricultural Education and the FFA

Public speaking, parliamentary procedure, leading of groups, conduct of meetings, and other skills learned in agricultural education classes and the FFA promote the growth of leadership and personal development skills. Parliamentary procedure is a key element. When one has mastered parliamentary procedure, one has mastered democratic leadership. In this system, all views are presented, and everyone has an equal chance for input, but the wishes and decisions of the majority group take precedence over the desires of any one individual.

Employers Want People with Good Leadership and Personal Development Skills

Employers want persistent, dedicated people who can think, solve problems, and make decisions. They want people who can communicate and manage their time wisely; who are honest, hardworking, and cooperative; and who have positive attitudes and values, as well as good social skills. They also want people with a positive self-concept. If you do not like yourself, you will not like other people. You may find yourself putting others down, working under the misconception that this will lift you up or make you feel better about yourself.

Communication Style Relates to Personality Type and Leadership Style

Certain communication styles are associated with each of the personality types and leadership styles. Briefly, aggressive communicators tend to be authoritarian leaders with choleric personalities. Passive communicators tend to be laissez-faire leaders with phlegmatic personalities. Assertive communicators tend to be democratic leaders with sanguine personalities. (See Chapter 2 for discussion of these personality types.)

Power, Traditional, and Popularity Leadership Styles Stand Alone

The power and influence, traditional, and popularity leadership styles basically stand alone, without direct connections to other leadership categories. In some cases, leadership is attained by knowing somebody, by following tradition, or by "playing the game." Obtaining the position of leader is not necessarily connected to skills and abilities formally developed for democratic leadership. Of course, there are always informal connections between categories; for example, human relations or power/influence skills can be used to get appointed to a leadership position for which one does not qualify.

Leadership and Personal Development Skills Help the Total Person

Besides leading to employment, leadership and personal development skills are important for family and personal relationships. Leadership and personal development skills also help one become a better member of one's community and civic organizations.

AUTHORITARIAN, DEMOCRATIC, LAISSEZ-FAIRE, AND SITUATIONAL LEADERSHIP: BEHAVIORAL LEADERSHIP THEORIES

Many leadership theories can best be explained by visualizing a leadership **continuum** (Figure 1–15), a series connecting two extremes with an infinite number of variations in between. On the extreme left is a leader who wants complete control (authoritarian). On the extreme right is the leader who lets the followers completely lead themselves (laissez-faire). Every leadership theory will fall somewhere within this leadership continuum.

Many leadership theory experts believe that the study of human nature is the most effective way of explaining how to influence, change, and lead people. For the most part, if one can (1) predict what motivates people to do what they do, as well as (2) understand why one does what one does, then leadership becomes much easier. A wise man once said that "knowing is half the battle." This section discusses the major leadership styles or theories: authoritarian, democratic, laissez-faire, and situational.

All of these are behavioral leadership theories. If one believes either that a great leader is someone who leads by authority or that a great leader leads by listening to followers, then one can best explain leadership by using behavioral theories. There are many different theories of leadership, all of which attempt to explain the leader, the follower, and even leadership development. This section investigates and summarizes behavioral theories of leadership in the hope that you will be able to choose the one(s) that will aid in the development of the best possible leadership skills and abilities for you personally.

Democratic and authoritarian leaders fall into the behavioral leadership category mentioned earlier. Most descriptions of leadership styles, categories, and types center around whether a person has an authoritarian or democratic philosophy. The late Douglas McGregor, while a professor at MIT, published a book in which he set forth what were to become his famous theory X (authoritarian) and theory Y (democratic) concepts.[40] Although some people have generalized and applied these ideas inappropriately, a brief review at this point will be helpful in achieving a better understanding of leaders.

Theory X (Authoritarian)

Each theory, as McGregor defined it, is a set of assumptions made about people in general—a way some of us look at humanity. Theory X is traditional and may seem quite familiar if one has ever worked for anyone with this style. The main characteristics of a theory X person include the following:

- The average human being has an inherent dislike of work and avoids it if possible.
- Because of this human characteristic of dislike for work, most people must be coerced, controlled, directed, or threatened with punishment to get them to work toward the achievement of organizational objectives.
- The average human being prefers to be directed, wishes to avoid responsibility, has relatively little ambition, and wants security above all.[41]

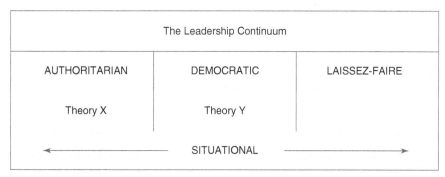

FIGURE 1–15 The different styles of leadership are generally associated with a particular set of behaviors the leader displays, either naturally or (sometimes) by choice.

FIGURE 1–16 What characteristics or actions of this man might lead one to believe that he follows a theory X philosophy? *(© iStock.com/imtmphoto)*

Those who hold such views expect their coworkers to be lazy, to require close control, and to "goof off" at every opportunity. The gentleman in Figure 1–16 appears by his facial expression to be primarily a theory X individual.

The authoritarian leader makes decisions and closely supervises or instructs people. This can be related to theory X assumptions. Authoritarian leaders use power and position to force their decisions on their coworkers. Coworkers are expected simply to do as they are told. They are not asked for any opinions and could even be reprimanded for making suggestions.

Saddam Hussein of Iraq was an authoritarian leader. Hitler was an authoritarian leader. Joseph Stalin ruled the Soviet Union in an authoritarian manner for 30 years, using force, power, and fear to get his way. However, there are times when even democratic leaders have to exert authoritarian leadership, as in time of war. The U.S. military forces use an authoritarian style of leadership even though they are part of a country that is a democracy. It would not be appropriate, for example, to have soldiers vote on whether or when to charge or retreat in the midst of a battle.

Theory Y (Democratic)

Theory Y reflects a totally different set of values and expectations of people. Here is McGregor's description of theory Y:

- The expenditure of physical and mental effort in work is as natural as in play or rest.
- External control and threat of punishment are not the only means for bringing about effort toward organizational goals.
- People will exercise self-direction and self-control to achieve objectives to which they are committed.
- Commitment to objectives depends on the rewards associated with achievement of those objectives.
- Under proper conditions, the average human being learns not merely to accept but actually to seek responsibility.
- Imagination, ingenuity, and creativity are widely distributed in the human population.
- Under the conditions of modern industrial life, the intellectual potential of the average human being is only partially used.[42]

Theory Y states that people work because work is natural for them and, under the proper conditions, they want to achieve the goals of the group. In general, the theory Y approach offers a leader more options in working with people. By their non-verbal expressions, the two turfgrass students in Figure 1–17 appear to be theory Y individuals.

Contrasting the X and Y Theories

Theory X and theory Y may seem somewhat simplistic. However, they are important in that they dramatize how we may feel about coworkers. If you view your coworkers as lazy people who must be coerced and controlled in order for you to properly perform your leadership functions, this attitude will affect your behavior toward them.

You may unconsciously be causing mistrust, suspicion, rebellion, and many other forms of nonproductive behaviors. In contrast, acceptance of, confidence in, and trust in coworkers show that you expect them to want to do a good job. This attitude adds to their self-esteem, self-respect, and motivation to do a good job, thus contributing to group effectiveness. Refer to Figure 1–18 for other assumptions of theory X and theory Y leaders.

Except when the situation demands other styles, most of us prefer theory Y or democratic leadership. American democracy was founded on the concepts of shared leadership and wide participation by many members of a group in making and carrying out group decisions. These concepts

FIGURE 1–17 What characteristics, especially in the two women shown here, might lead one to believe that some people have a theory Y philosophy? *(Photo courtesy of University of Georgia—College of Agricultural & Environmental Sciences.)*

MOTIVATING SUBORDINATES (TWO SETS OF ASSUMPTIONS ABOUT PEOPLE)	
TRADITIONAL X (AUTHORITARIAN)	**PROGRESSIVE Y (DEMOCRATIC)**
People are naturally lazy; they prefer to do nothing.	People are naturally active; they set goals and enjoy striving.
People work mostly for money and status rewards.	People seek many satisfactions from work: pride in achievement, enjoyment of process, sense of contribution, pleasure in association, and the stimulation of new challenges.
The main force keeping people productive in their work is fear of being demoted or fired.	The main force keeping people productive is the desire to achieve their personal and social goals.
People remain children; they are naturally dependent on leaders.	People normally mature beyond childhood; they aspire to independence, self-fulfillment, and responsibility.
People expect and depend on direction from above; they do not want to think for themselves.	People close to the situation see and feel what is needed and are capable of self-direction.
People need to be told, shown, and trained in proper methods of work.	People who understand and care about what they are doing can devise and improve their own methods of doing work.
People need supervisors who will watch them closely enough to be able to praise good work and reprimand errors.	People need a sense that they are respected and are mature enough to do good work without constant supervision.
People have little concern beyond their immediate, material interests.	People seek to give meaning to their lives by identifying with nations, communities, religions/churches, unions, companies, or causes.
People need specific instruction on what to do and how to do it; they ignore larger policy issues.	People need ever-increasing understanding; they need to grasp the meaning of the activities in which they are engaged; they have unlimited cognitive hunger.
People appreciate being treated with courtesy.	People crave genuine respect from their peers.
People are naturally compartmentalized; work demands are entirely different from leisure activities.	People are naturally integrated; when work and play are too sharply separated, both deteriorate.
People naturally resist change; they prefer to stay in the old ruts.	People naturally tire of monotonous routine and enjoy new experiences; everyone is creative to some degree.
Jobs are primary and must be done; people are selected, trained, and fitted to predefined jobs.	People are primary and seek self-realization; jobs must be designed, modified, and fitted to people.
People are formed by heredity, childhood, and youth; as adults, they remain static; old dogs don't learn new tricks.	People constantly grow; it is never too late to learn; they enjoy learning and increasing their understanding and capability.
People need to be "inspired" (pep talk), pushed, or driven.	People need to be encouraged and assisted.

FIGURE 1–18 Traditional X represents older, authoritarian beliefs about leading others. Progressive Y represents a newer, more democratic way of leading. Some situations, however, may require use of the traditional X approach. *Source: Adapted from* Horace Small Manufacturing Training Manual *(Nashville, TN: Horace Small Manufacturing Co., 1992).*

are essential to the successful operation of the FFA or any other club, group, or organization.

Democratic leadership is based on the participation of members of a group in making decisions. Each person is recognized as important by the leader, whose job is to help the group get things done. The democratic leader also protects the rights of the minority while the wishes of the majority are being developed and pursued. If procedures have been established, it is the democratic leader's duty to see that they are followed.

Most Americans hold certain beliefs about leadership, just as they do about other areas in life. The democratic style of leadership is a part of our heritage and has been important to the success of American democracy. Most of us believe that democratic leadership is superior to autocratic leadership because we think that each individual has something to contribute to the group or the nation. We believe the collective judgment of a group is better and wiser than the opinion of any one individual.

Sociologists who study how groups function report that participation in making group decisions increases the morale of individuals in the group. Group members are more likely to accept and carry out decisions that they helped make than those that are imposed on the group.

Some leaders act autocratically in one situation and democratically in another. It cannot be said that democratic leaders do not use authority. To accept leadership means to accept the responsibility to use authority for the best interest of the group, but that use of authority is tempered by democratic beliefs and practices.

Group Involvement Because democratic leaders cannot simply order that things be done, they must involve group members. They are responsible for directing the activities of the organization. An effective democratic leader works with the group to accomplish its objectives.

Democratic leaders do not closely supervise followers.[43] They share leadership with others. They need to encourage the growth of informal leadership within their groups. In fostering this process, the democratic leader ensures that all opinions are heard, all sides of an issue are discussed, and all members are encouraged to express their ideas. If the leader fulfills these functions, the action taken by the group will reflect the desires of the majority.

Characteristics A democratic leader encourages

- Sharing of everyone's ideas in the group
- Full engagement by everyone on the group
- Creativity
- Honesty
- Intelligence
- Courage
- Competence
- Fairness
- Diverse thinking[44]
- Critical thinking

Understanding authoritarian and democratic philosophies is the key to understanding leadership. Several sophisticated models of leadership exist, but they are all based on or center on authoritarian and democratic attitudes. In Chapter 2, you will see how these philosophies relate to personality types.

Laissez-Faire Leadership

A leader who uses the **laissez-faire leadership** style believes that the group can make its own decisions without the leader or, at least, with very little input from the leader. Although it may appear that these leaders are avoiding responsibility, they recognize that members are often in a better position to make a decision than the leader.

For example, elementary school teachers usually use an authoritarian style of leadership. High school teachers tend to use a situational style of leadership. College professors teaching undergraduate students can take a more democratic approach, and graduate students (those working on a master's or doctoral degree) can be led using the laissez-faire leadership style, because of their intelligence, experience, and competence. The laissez-faire leadership style is appropriate only when the group's level of maturity and intelligence is close to that of the leader. Whatever the situation, the leader makes the final choice of which leadership style to use.

Although there are situations in which the leader does not interfere at all, and lets the group make all the decisions, it is very rare for a leader to act in such a way. In reality, there are three modified versions of laissez-faire leadership: participative leadership, delegation, and superleadership.

In recent years, employee or team involvement has become a hot concept within management circles. These styles (participative, delegation, and superleadership) assume that individual members of a group who take part personally in the decision-making process will have a far greater commitment to the goals and objectives of the group or organization. Although the three styles are similar, a brief discussion of each follows so that you can discover the distinctions among them.

Participative Leadership **Participative leadership** does not necessarily assume that leaders do not make any decisions. On the contrary, it is leading by gathering and considering input from group members. Leaders should understand and communicate what the goals and objectives of the group or organization are so that they can draw upon the knowledge of the members of the group, organization, or workplace.

Effective leaders who use participative leadership in planning, effecting change, or solving problems will meet with the affected group or

DAWNYA, A SENIOR AT HILLSMAN, decided to use democratic leadership after analyzing the situation with which she was dealing. Shortly after being elected president of the senior class, she realized that one of her major responsibilities as president would be making sure the senior class had a homecoming float. Now, this was not the reason Dawnya ran for president, but she willingly took on the task.

Dawnya knew that she had a core group of seniors who had helped out with the float in previous years. She also had an enthusiastic group of senior officers who wanted to make sure that the seniors won the float competition. Because Dawnya believed in and trusted the people she led, she chose a democratic style of leadership. She knew that her team's experience and enthusiasm would be the edge the senior class would need to win the float competition.

Dawnya immediately "put the ball in their court." She informed the group that she wanted all of the decisions, such as the title of the float, what it should look like, and even who would pull it, to be made by the group. The members debated various possibilities for a while, but eventually they all came to an agreement and the float plans got underway. Dawnya was very pleased that she had such a great group to work with, and she knew that attaining group consensus and using the democratic style of leadership would work best. Because the entire membership of the group voted on and agreed with all of the plans made, it was natural for them to follow through wholeheartedly with implementing those plans. It was truly a team effort that year, as Dawnya and her senior class took home the "float gold."

Did Dawnya choose the right leadership style? Do you think that Dawnya and the other seniors would have been as successful if Dawnya had decided to make all of the decisions herself? What do you think would have happened if Dawnya had not let the majority rule?

organizational members and inform them of the problems, needs, goals, and objectives. Next, the participative leader will ask for the group's ideas about implementing changes.

Advantages of participative leadership In certain situations, the participative approach can be extremely effective. Consider the following:

- Group and organization members like to feel that their ideas are important and tend to feel considerably more committed to changes and decision making in which they have participated.
- Group and organization members develop greater feelings of **self-esteem**, belief in one's abilities and respect for oneself, when they perceive that they have been trusted to make competent decisions.
- Often the combined knowledge and experience of the members of a group or organization exceed those of the leader.
- Problems that members work on collectively often generate new ideas, created as a result of the interpersonal exchanges and discussions of various options.[45]
- Participation allows members of a group or organization to learn more about implementing new programs or procedures after decisions are made.[46]

Disadvantages of participative leadership The participative leadership approach makes certain assumptions that, when false, can result in complications for the group or organization. Consider the following disadvantages of the participative approach:

- The approach assumes a considerable commonality of interest between the leaders and their group or organization members; this might not be the actual situation.
- Some individual members may be uninterested, resulting in apathy.
- Group or organization members might perceive the participative approach as an attempt to manipulate them.
- The participative approach assumes that group or organization members have the necessary knowledge and skill to implement and participate in the decision-making process.
- Some leaders feel uncomfortable using a participative style because of their personality type.
- Some leaders hesitate to use participative leadership for fear that they will lose control over their group or organization members.[47]

As can be seen from this list, participative leadership is not the answer in every instance; it is

FIGURE 1–19 The summer camp leader in this photo is about to take a group of young boys on an environmental education field trip. Do you think this leader should employ participative leadership or some other form? *(Photo courtesy of University of Georgia—College of Agricultural & Environmental Sciences.)*

more appropriate in some situations than others. Do you think the situation in Figure 1–19 might allow participative leadership or require a different approach?

Role of culture in use of participative leadership In many countries, values other than participation are emphasized, such as obedience, submission, and respect for authority. In these countries, participative leadership may not be effective.[48] These cross-cultural differences in team behavior create considerable challenges for the leaders of culturally diverse teams. Success depends on accurate perceptions and careful reading of cross-cultural cues. Consider the various cultural values of the following countries:

- The Mexican culture has a well-established tradition of autocratic leadership without a history of participative leadership.[49]
- In France, the focus has traditionally been on performance through obedience and respect for legitimate authority.[50]
- An effective team member in Japan is, above all, courteous and cooperative, and Japanese team members avoid conflict and confrontation.[51]

- German employees are taught early in their careers to seek technical excellence.
- In Afghanistan, a team member is obligated to share his or her resources with others, making generosity an essential team behavior.
- In Israel, values of hard work and contribution to the community drive team members.
- The Swedes pay more attention to verbal and written communication than to the context, such as situational factors or body language; they value direct, specific messages.[52]
- Groups in the United States and Australia place a high value on participative leadership, but they also highly value individual contributions and appreciate getting credit for their input.

Delegative Leadership Delegation is similar to participative leadership. The difference can be shown by a continuum (see Figure 1–20). Authoritarian leaders (on the left-hand side) seldom delegate, but as one moves to the right of the continuum, one finds that laissez-faire leaders use a high degree of delegation.

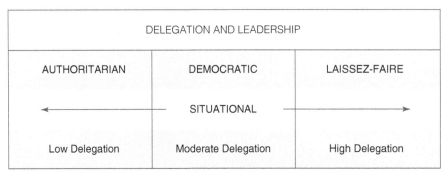

FIGURE 1–20 The amount of delegation a leader is able to employ depends on his or her tendency toward authoritarian or participative leadership. *(Delmar/Cengage Learning)*

Appropriate levels of delegation In reality, somewhere in the middle is probably ideal for delegation. The better leaders will vary their delegation, using different amounts according to the situation. Some leaders use delegation only after specifying guidelines and setting limits on the decision options; they may require that the final decision be submitted for their approval before it is implemented.[53] Other leaders give followers complete freedom to find and carry out a solution. However, most leaders use different combinations of participation and delegation at different times, adapting them to each situation and group.[54]

Advantages of delegation Delegation is a basic tool for a leader's success. The potential benefits of delegation include

- Giving the leader time to take on new tasks and strategic activities
- Providing employees with opportunities to learn and develop
- Allowing employees to be involved in tasks
- Allowing observation and evaluation of employees as they do new tasks
- Increasing employee motivation and satisfaction[55]

Delegation allows members of a group or organization to attempt new tasks and learn new skills, thereby possibly enriching their jobs and increasing their satisfaction and motivation. This in turn increases follower involvement and commitment. Furthermore, when members of a group or organization perform new tasks, the leader can also observe them and gather performance-related information for use in further development and evaluation. Leaders may thus use delegation as a tool in succession planning for their organizations: members who consistently handle new tasks well and are eager to accept more responsibility may be the best future leaders of the organization. If members have no opportunities to grow or perform outside of their current jobs, the leader will not have data with which to accurately predict their performance in other positions.[56]

Effective delegation There are several guidelines for effective delegation. One primary issue for leaders is to make sure that delegation does not become "dumping." The leader needs to delegate varied levels of tasks to group or organization members. If the leader consistently gives only unpleasant, difficult, and unmanageable tasks to group or organization members, while he or she takes on high-profile, challenging, and interesting projects, delegation has mutated into dumping. This is the root of many complaints from group or organization members regarding delegation. For effective delegation, the leader must pass on a variety of tasks and should make sure their delegation is balanced.[57] Refer to Figure 1–21.

Excuses for not delegating Of course, there are situations in which delegation is not possible, but the underlying factors that stop many leaders from delegating are their personality style, their need for control, and their fear of losing control. Later in this book we will fully explore personality types and learn that some types have an extreme need for control and do not like to give up power. Delegation is difficult for some

GUIDELINES FOR GOOD DELEGATION	
GUIDELINE	**DESCRIPTION/EXPLANATION**
Delegate rather than dumping	Delegate both pleasant and unpleasant jobs; give followers a variety of experiences.
Clarify goals and expectations	Set clear goals and guidelines regarding expectations and limitations.
Provide support and authority	As a task is delegated, provide the necessary authority and resources, such as time, training, and advice needed to accomplish the task successfully.
Monitor and provide feedback	Keep track of progress and provide regular feedback during task performance and after completion of the task.
Delegate to different followers	Delegate tasks to those who are most motivated to complete them, as well as to those who have potential but no track record of performance.

FIGURE 1–21 This listing provides a set of guidelines for effective delegation. *Source: Adapted from* The Art and Science of Leadership *(4th ed.), by Afsaneh Nahavandi, p. 207 (Upper Saddle River, NJ: Pearson Prentice Hall, 2006).*

people. Refer to Figure 1–22 for other excuses for not delegating.

Superleadership The ultimate leadership is superleadership. **Superleadership** prioritizes leading people to lead themselves.[58] Members of a group or organization are encouraged to make their own decisions and accept responsibility for the consequences of those decisions—to the point where they do not need much leadership.

With superleadership, everyone on the team is involved in setting, reaching, and evaluating goals of the organization. Group or organization members themselves decide what they want to accomplish, what they need to do it, and how to achieve it. Superleadership consists of the following:

Developing positive, motivating thought patterns Members of a group or organization seek and develop environments that provide

EXCUSES FOR NOT DELEGATING	
EXCUSE	**COUNTERARGUMENT**
• Group members are not ready.	• The leader's job is to get followers prepared to take on new tasks.
• Group members do not have the necessary skills and knowledge.	• The leader's responsibility is to train followers and prepare them for new challenges.
• I feel uncomfortable asking other people to do many of my tasks.	• Only a few tasks cannot be delegated; this is not equivalent to shirking responsibility. Balancing delegation of pleasant and unpleasant tasks is appropriate.
• I can do it faster myself.	• Taking time to train followers frees up time in the long run.
• Group members are too busy.	• Both leaders and followers must learn to manage their workload by setting priorities.
• If group members make a mistake, I am responsible.	• Encouraging experimentation and tolerating mistakes are key to learning and development.
• My own manager may think that I am slacking off and not working hard enough.	• Doing "busy work" is not an appropriate use of a leader's time. Delegation allows time to focus on strategic and higher-level activities.

FIGURE 1–22 People make all kinds of excuses for not delegating. Here are some counterarguments to such excuses. *Source: Adapted from* The Art and Science of Leadership *(4th ed.), by Afsaneh Nahavandi, p. 209 (Upper Saddle River, NJ: Pearson Prentice Hall, 2006).*

positive cues and a supportive and motivating environment.[59]

Personal goal setting Members of a group or organization set their own performance goals and expectations.

Observation and self-evaluation Members of a group or organization observe their own and other team members' behaviors to provide feedback, critique, and evaluate each other's performance.

Self-reinforcement Members of a group or organization reward and support one another.[60]

With superleadership, the essential role of leaders is to help others to lead themselves. In the traditional view of leadership, the leader is expected to choose the questions, provide the answers, and guide or direct subordinates. In contrast, superleadership suggests that the leader's job is to develop his or her group or organization members to the point where they do not need the leader very much.[61]

SITUATIONAL (CONTINGENCY) LEADERSHIP

A common myth about leadership style is that one style is always superior to the others. In reality, all styles may be appropriate, depending on the situation and the group. Leaders need to answer three basic questions before selecting a leadership style:

1. Is there an obvious solution to the task?

2. Is it important for the leader to make the decision?

3. Does the decision have to be made immediately?

If the answer to any of these questions is yes, the leader should make the decision. Some input from members might be helpful, but the leader should be directive. If the answer to any of these questions is no, the leader assesses the situation and selects a leadership style appropriate to the situation at hand.[62]

How Leaders Assess a Situation Leaders' first move in assessing a situation is to determine the members' ability and willingness to complete the activity. If ability and willingness are low, the leader uses a telling style; if they are moderate to low, the leader uses a selling style; if they are moderate to high, the leader uses the participating style; and if ability and willingness are high, the leader uses a delegating style.[63] Refer to Figure 1–23 for a visual explanation.

Leaders need to evaluate the individuals who will be involved in each activity or task. An individual might be able and willing to perform one task but unable to perform another task. In this situation, the leader uses two different leadership styles with the same individual. The optimum situational leadership occurs when a leader uses an appropriate leadership style that corresponds to the level of follower readiness (the most desirable, of course, being willing and able group members).

There are four steps in assessing a situation.[64]

HOW LEADERS ASSESS A SITUATION		
ASSESSMENT OF FOLLOWER	**WHICH LEADERSHIP STYLE DO I USE?**	**WHO MAKES THE DECISION?**
Not able and not willing	Telling	Leader (authoritarian)
Willing, not able	Selling	Leader with group consultation (democratic)
Able, not willing	Telling	Leader (authoritarian)
Able and willing	Delegating	Group (laissez-faire)

FIGURE 1–23 For the most effective leadership, evaluate each situation as it arises. The leader assesses the followers, decides which leadership style to use, and chooses whether a leader or group decision is most appropriate. *Source: Adapted from "Leadership Styles," Effective Leadership series, No. 8783-B, p. 4 (College Station, TX: Instructional Materials Service, Texas A&M University, 1988).*

VIGNETTE 1-7 THE LAISSEZ-FAIRE LEADER

THE OLYMPIC BASKETBALL TEAM'S HEAD COACH COULD BE THE BEST EXAMPLE OF A LAISSEZ-FAIRE LEADER

When the Olympic committee decided to let professional athletes compete, the United States immediately had the edge with what became known as the "dream team." The first-ever dream team consisted of some of the best basketball players who have ever played the game. Their talent and ability were second only to their experience and knowledge of the game.

But even these best-of-the-best had a leader. They had a coach. There is a need for a leader even in the most motivated and talented groups or organizations. The dream team was highly skilled, was motivated, wanted to participate, wanted to work well together, and knew how to play; but the laissez-faire coach, who did not really seem to be in charge, was there only if he was needed and offered opinions only when they were requested. This is an excellent example of a laissez-faire leader at work.

Step 1 *Determine your preferred leadership style.*

There are several evaluations to help you determine which leadership style you prefer. Check on the internet for sites using the search term Leadership Styles. The style can then be applied as follows.

Telling—This provides specific instructions and supervises performance. Synonyms are **guiding**, **directing**, and **establishing**. This style tends to be authoritarian.

Selling—This explains decisions and provides opportunities for clarification. Synonyms are **explaining**, **clarifying**, and **persuading**. This style tends to be democratic.

Participating—This shares ideas and facilitates the making of decisions. Other descriptors are **encouraging**, **collaborating**, and **committing**. This style tends to be democratic.

Delegating—This turns over responsibilities for decision and implementation. Other descriptors are **observing**, **monitoring**, and **fulfilling**. This style tends to be laissez-faire.[65]

Step 2 *Once you have determined your preferred leadership style, decide when to use each style according to the task.*

As mentioned earlier, there is no "best" supervisory style for all situations. Instead, the effective leader adapts his or her style to meet the capabilities of the individual or group.

Step 3 *Determine the capabilities of your coworkers or followers.*

The capability of those you lead may be measured by indicators of willingness and ability, which you, as a leader, must evaluate. You select the capability level that best describes a coworker's ability and motivation for the specific task.

There are several ways to determine followers' willingness and ability. Leaders who know their followers well are best able to assess their followers' willingness and ability accurately.[66]

Indicators of willingness. The individual

- Has an interest in the activity
- Volunteers for the activity
- Discusses the activity with others
- Displays a positive attitude toward the group
- Follows through on commitments

Indicators of ability. The individual

- Has experience in the activity
- Has skills in related activities
- Is intelligent and can think through problems
- Can find and use resources effectively
- Is self-directed[67]

Step 4 *Use the appropriate leadership style.*

Once again, the "correct" leadership style depends on the situation and the individual; the situation is a function of capability. The better you are at matching your leadership style to followers' or coworkers' capabilities, the greater your chances of being a successful leader.

HAVE YOU EVER HAD A TEACHER DIVIDE ONE OF YOUR CLASSES INTO GROUPS?

Derek Jones was a young man who had a challenging group experience in his American History class. The instructor divided the class into groups of five and asked the groups to role-play a significant event in American history. Derek was not very sure about his group. Each of the members seemed to have a different amount of enthusiasm for the assignment.

Both he and Alexis loved history and were very excited about the chance to act it out. There was also Jeff, who was smart but did not care for this type of activity; Tina, who could be smart but did not appear to care about anything; and Al, who was excited about the activity but just was not very good at history. Derek, having studied group dynamics in a leadership workshop in the FFA, decided that such differences in the group would require him to use a situational approach.

With Tina, Derek had to do a lot of telling and directing just to get her to acknowledge and understand her role in the group. Because Tina was not really motivated about either the project or the people in the group, she wanted just enough to do to complete the assignment and be done. Derek was able to convince Tina to contribute by repeatedly giving instructions to her. Because Al was very willing but had trouble with history, Derek also gave him a lot of instructions, as well as doing a great deal of coaching. Derek used a high amount of relationship behavior and task behavior with Al. Because Jeff was willing, but not very excited about the task, Derek also used a lot of relationship behavior with him. Derek was successful in convincing Jeff to participate because he led by encouraging and participating, not by telling Jeff what to do or how to do it. Alexis was just like Derek. She and Derek were very willing and able, so Derek did not feel that he needed to direct her at all. He got out of her way and let her work while he attended to the needs of the rest of the team.

Because of Derek's ability to get everyone to contribute (by his use of the situational approach to leadership), the group received an A+ for its reenactment of the signing of the Declaration of Independence.

Continuum of Leadership Behavior

Another approach to describing leadership styles is to use a continuum, rather than assigning specific titles or names to the styles. On the continuum, styles vary from leader-centered decisions to group-centered decisions (Figure 1–24). All styles on the continuum may be appropriate and effective, depending on the situation. Leader-centered or autocratic leadership does not mean the leader is a dictator, or that she or he does it all. Similarly, group-centered or laissez-faire leadership does not mean that the leader does not care and is avoiding responsibility by leaving the decision to the group.

CONCLUSION

We can change or alter our leadership style. People are born with natural tendencies, but through education we learn what works best in particular situations. Just as we can alter our eating habits to our advantage, we can alter our leadership style to match the situation even though it is not our natural tendency. Great leaders care more than others think is wise, risk more than others think is safe, dream more than others think is practical, and expect more than others think is possible.[68]

SUMMARY

Leadership is the ability to move or influence others toward achieving individual or group goals. Leaders are needed in the workplace, organizations, communities, states, and nations. There is no limit to the number of leaders within a group. A true leader helps the group achieve what the group as a whole believes is important.

Leadership is many things to many people; however, leaders know who they are, what their strengths and weaknesses are, and how to fully display their strengths and compensate for their weaknesses. They also know what they want, why they want it, and how to communicate what they want to others to gain their cooperation and support. Leaders know how to achieve their goals.

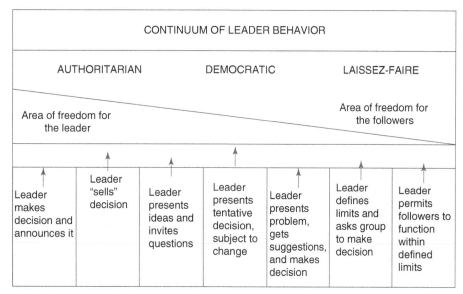

FIGURE 1–24 Continuum of leader behavior. *Source: Adapted from "How to Choose a Leadership Pattern," by Robert Tannenbaum and Warren H. Schmidt, p. 164,* Harvard Business Review (May/June 1973); Human Relations in Organizations: Applications and Skill Building *(4th ed.), by Robert N. Lussier, p. 225 (Boston: Irwin/McGraw-Hill, 1999); and "Leadership Styles," Effective Leadership series, No. 8737-B (College Station, TX: Instructional Materials Service, Texas A&M University, 1988).*

We do many things in agricultural education and the FFA that contribute to leadership and personal development. These include following parliamentary procedure, leading group discussions, setting goals, managing financial resources, developing social skills, and serving our community.

Leading groups effectively is a tremendous challenge, a great opportunity, and a serious responsibility. Leadership is a complex process that results from the interactions among a leader, followers, and the situation.

Effective leaders without supportive followers could accomplish very little. In groups or organizations, leaders demonstrate their competence by setting worthwhile and challenging goals with followers, by showing confidence in followers, and by supporting their effort to perform well. Many people consider leaders to be servants of their followers.

The difference between leadership and management relates more to effectiveness than to actual differences between the two concepts. The title of "leader" implies competence. Therefore, an effective and successful manager could be considered a leader, but a less competent manager would not.

Leaders are not born, they are made. There are seven leadership categories: trait, power and influence, behavioral, situational, traditional, popularity, and combination.

When one looks at "the big picture," it is easier to understand the leadership categories and their relationships to and connection with each other; one can also recognize why *leadership* has so many different definitions. The big picture also illustrates the connection, relationship, and importance of leadership and personal development leading to employment and leadership in one's home and community.

Of the many leadership styles, categories, and types, most center on whether a person has an authoritarian (theory X) or democratic (theory Y) philosophy. Authoritarian leaders believe people have to be pushed, whereas democratic leaders believe people can be pulled or led.

Except when the situation demands other means, most of us prefer democratic leadership. American democracy was founded on the concepts of shared leadership and wide participation by many members of a group in making and carrying out group decisions. These concepts are essential to the successful operation of the FFA or any other club, group, or organization.

Take It to the Net

1. Use an Internet search engine (such as Google or Yahoo!) to find a web page or article on leadership or leadership styles. Write a brief summary of the piece and present it to the class.

2. Using a search engine, identify two leaders in history. Write a two-page report in which you compare and contrast their leadership traits and styles, and then present your report to the class.

Chapter Exercises

REVIEW QUESTIONS

1. Define the Terms to Know.

2. What is the difference between a leader and a boss?

3. Do students really learn leadership and personal development skills by taking agricultural education courses and being active FFA members? Support your answer.

4. Regardless of the definition of leadership you use, what are two things leaders must consider?

5. What is the key word that distinguishes leaders from managers?

6. List eight myths about leaders and leadership.

7. List the seven categories of leadership.

8. Distinguish between traits and skills.

9. Contrast power and influence.

10. Distinguish between authentic and inauthentic charismatic leaders.

11. Name three FFA symbols that are part of the traditional organizational culture.

12. Give examples of three other symbols in our society (corporate, public, or educational) that either directly or indirectly signify leadership.

13. Why should your FFA chapter be aware of the popularity leadership category?

14. List 15 assumptions of theory X leaders and 15 assumptions of theory Y leaders.

15. Compare the democratic and authoritarian leadership styles.

16. What are five advantages of participative leadership?

17. What are six disadvantages of participative leadership?

18. List eight examples of various cultural values considered in using participative leadership.

19. List five advantages of delegation.

20. What are four elements of superleadership?

21. List four steps leaders use in assessing a situation so they can choose an appropriate leadership style.

22. List five indicators of coworker willingness and five indicators of coworker ability.

23. Briefly explain the continuum of leadership behavior.

COMPLETION

1. Effective _____ without supportive _____ could accomplish very little.

2. Power stemming from formal authority is sometimes called _____ power.

3. Potential influence based on control over rewards is sometimes called _____ power.

4. Control over punishment is sometimes called _____ power.

5. Control over the physical environment, technology, and organization of the work within a group is called _____ control or _____ engineering.

6. _____ power is the type of power a person has when she or he is the only one with the ability to solve a special problem or perform important tasks.

7. Friendship and loyalty are sometimes called _____ power.

8. _____ power structures involve people who are not elected or appointed to positions, but, because of their earned respect or deference, exercise power over appointed or elected leaders.

9. _____ power is a pervasive process that involves efforts by members of a group to ensure their power and protect existing power sources.

10. The combination leadership category includes four frames or images of leadership: _____, _____, _____, and _____.

11. A leader who uses the _____ leadership style believes the group can make its own decisions with little or no leader input.

MATCHING

_____ 1. Tall and distinguished looking.

_____ 2. Comes in many forms, such as legitimate, reward, conceived, expert, and referent.

_____ 3. Opening ceremony, FFA official dress.

_____ 4. Leadership bestowed on those who may not deserve it.

_____ 5. Use whatever or as many resources as needed to lead.

_____ 6. Authoritarian or democratic leaders.

_____ 7. Different leadership style for different occasions.

A. cultural and symbolic

B. behavior

C. power and influence

D. trait

E. popularity

F. situational

G. combination

ACTIVITIES

1. After reading the different definitions of leadership, write a definition that has the most meaning to you.

2. List 10 areas of the FFA that contribute to leadership and personal development. Select an activity for each area that you could perform or experience while in agricultural education and the FFA to enhance your leadership and personal development skills.

3. Write a one-page essay on the importance of learning and following parliamentary procedure for developing democratic leaders.

4. This chapter gave several examples of leaders who experienced "ups and downs" of leadership. Give an example of a contemporary leader (not in the book's list) who has experienced such positive and negative situations. Share with the class.

5. Write a paragraph either supporting or not supporting the following statement: *Effective leaders need supportive followers.* Share your answer with the class and be prepared to defend your position.

6. Your teacher is going to select students to be part of a debate. The statement you will debate is: *Charismatic people make great leaders.* Be prepared to debate this statement by arguing for or against it.

7. Give five examples of servant leaders. Write a short statement about why you believe each would be considered a servant leader.

8. Study "the big picture" (Figure 1–14) of leadership, personal development, employment, and life skills. Write a one-page essay explaining at least five connections, observations, or relationships among leadership, personal development, employment, and life skills.

9. Describe an example of when a leader would need to be each of the following: authoritarian, democratic, laissez-faire, and situational. Defend your choices.

10. Prepare a one-page case study or scenario describing a leader who uses theory Y, the democratic style of leadership.

11. Prepare a one-page case study or scenario describing a leader who uses theory X, the autocratic style of leadership.

12. Prepare a one-page case study or scenario describing a leader who uses the laissez-faire style of leadership.

13. Prepare a one-page case study or scenario describing a leader who uses the situational (contingency) style of leadership.

14. Select one of the following leadership categories, and write a one-page case study or scenario describing a person you know who fits the category. Do not mention the person's name.
    - Trait leadership
    - Power or influence leadership
    - Traditional leadership
    - Popularity leadership

15. Write a one-page essay determining and defending the leadership style of Donald Trump.

16. Write a one-page essay determining and defending the leadership style of Barack Obama.

NOTES

1. *Leadership of Youth Groups* (Kansas City, MO: Farmland Industries [no date]), p. 12.
2. R. N. Lussier and C. F. Achua, *Leadership: Theory, Application, & Skill Development, 5th ed.* (Mason, OH: South-Western, Cengage Learning), p. 5.
3. J. McGregor, On Leadership: Remembering Leadership Sage Warren Bennis, *The Washington Post.* (August 4, 2014).
4. R. N. Lussier and C. F. Achua, *Leadership: Theory, Application, & Skill Development, 5th ed.*, p. 53.
5. M. W. Drafke and S. Kossen, *The Human Side of Organizations,* 7th ed. (Reading, MA: Addison-Wesley, 1997), p. 368. (10th, ed., 2008).
6. J. P. Howell and D. L. Costley, *Understanding Behaviors for Effective Leadership,* 2d ed. (Upper Saddle River, NJ: Pearson Prentice Hall, 2006), p. 4.
7. L. R. Rathbun, *The Relationship between Participation in Vocational Student Organizations and Student Success* (unpublished dissertation, Ohio State University, Columbus, OH, 1974).
8. S. C. Ricketts, *Leadership and Personal Development Abilities Possessed by High School Seniors Who Are FFA Members in Superior FFA Chapters, Non-Superior Chapters, and by Seniors Who Were Never Enrolled in Vocational Agriculture* (doctoral dissertation, Ohio State University, Columbus, OH, 1982).
9. Ibid.
10. A. Nahavandi, *The Art and Science of Leadership,* 4th ed. (Upper Saddle River, NJ: Pearson Prentice Hall, 2006), p. xi. (7th ed., 2014).
11. Howell and Costley, *Understanding Behaviors for Effective Leadership.*
12. Nahavandi, *The Art and Science of Leadership*, p. xi.
13. Ibid., p. xii.
14. R. H. Jerry, II, "Leadership and Followership," *University of Toledo Law Review* Vol. 44 (2013), p. 345; University of Missouri School of Law Legal Studies Research Paper No. 2015-30. Available at SSRN: http://ssrn.com/abstract=2689243.
15. Howell and Costley, *Understanding Behaviors for Effective Leadership*, p. xiii.
16. Nahavandi, *The Art and Science of Leadership*, pp. 4–6.
17. Ibid., p. 4.
18. R. K. Greenleaf, *The Power of Servant Leadership* (San Francisco: Berrett-Koehler, 1998).
19. Nahavandi, *The Art and Science of Leadership*, p. 18.
20. "Leaders and Leadership," *Effective Leadership* series, No. 8737-A (College Station, TX: Instructional Materials Service, Texas A&M University, 1989), p. 1.
21. Ibid.
22. These seven categories are adapted from work by Dr. Tom Burks, director of the Leadership Institute at Middle Tennessee State University.
23. T. Burks, *Leadership in Higher Education* (unpublished dissertation, Vanderbilt University, Nashville, TN, 1992).
24. E. M. Bensimon, "The Meaning of 'Good Presidential Leadership': A Frame Analysis," *Review of Higher Education* 12(2):107–123 (December 1989).
25. J. R. P. French Jr. and B. Raven, "The Bases of Social Power," in *Studies in Social Power*, edited by D. Cartwright (Ann Arbor, MI: Research Center for Group Dynamics, Institute for Social Research, The University of Michigan, 1959), pp. 150–167.
26. A. M. Pettigrew, *The Politics of Organizational Decision Making* (London: Tavistock, 2008).
27. French and Raven, "The Bases of Social Power."
28. P. Spahr, "What is Charismatic Leadership? Leading Through Personal Conviction," *Leadership is Learned,* (St. Thomas University Online, 2016). Available at online.stu.edu/charismatic-leadership/.
29. Ibid, 2016.
30. Howell and Costley, *Understanding Behaviors for Effective Leadership*, p. 211.
31. Ibid., pp. 213–214.
32. B. M. Bass, *Transformational Leadership: Industrial, Military, and Educational Impact* (Mahwah, NJ: Lawrence Erlbaum Associates, 1998), p. 23. (Latest ed., 2006).

33. T. A. Deal and A. A. Kennedy, *Corporate Cultures: The Rites and Rituals of Corporate Life* (Reading, MA: Addison-Wesley, 1982).

34. Burks, *Leadership in Higher Education.*

35. Bennis and Nanus, *Leaders,* pp. 38–39.

36. Burks, *Leadership in Higher Education.*

37. Ibid. (The terms used by Burks were *structural, human resource, political,* and *symbolic*).

38. Burks, *Leadership in Higher Education.*

39. Ibid.

40. D. McGregor, *The Human Side of Enterprise* (New York: McGraw-Hill, 2006).

41. Ibid., pp. 33–34.

42. Ibid., pp. 47–48.

43. R. N. Lussier and C. F. Achua, *Leadership: Theory, Application, & Skill Development, 5th ed.,* p. 71.

44. K. Cherry, *What is Democratic Leadership?* (verywell.com, 2016).

45. Drafke and Kossen, *The Human Side of Organizations,* p. 378.

46. K. I. Miller and P. R. Monge, "Participation, Satisfaction, and Productivity: A Meta-analytic Review," *Academy of Management Journal* 29: 727–753, at 730 (1986).

47. Drafke and Kossen, *The Human Side of Organizations,* pp. 378–379.

48. Howell and Costley, *Understanding Behaviors for Effective Leadership,* p. 134; E. E. Lawler III, *High-Involvement Management* (San Francisco: Jossey-Bass, 1991).

49. P. W. Dorfman, J. P. Howell, S. Hibino, J. K. Lee, U. Tate, and A. Bautista, "Leadership in Western and Asian Countries: Commonalities and Differences in Effective Leadership Processes across Cultures," *Leadership Quarterly* 8(3):233–274 (1997).

50. F. Trompenaars, *Riding the Waves of Culture: Understanding Diversity in Global Business* (Chicago: Irwin Professional Publishing, 1994), pp. 154–163, 178. (3rd ed., 2012).

51. A. Zander, "The Value of Belonging to a Group in Japan," *Small Group Behavior* 14(1):3–14, at 7–8 (1983).

52. Nahavandi, *The Art and Science of Leadership,* p. 11.

53. Yukl, *Leadership in Organizations,* pp. 204, 206–207; Howell and Costley, *Understanding Behaviors for Effective Leadership,* p. 136.

54. Howell and Costley, *Understanding Behaviors for Effective Leadership,* p. 136.

55. Nahavandi, *The Art and Science of Leadership,* p. 205.

56. Ibid. (citing J. R. Hackman and G. R. Oldham, *Work Redesign* [Reading, MA: Addison-Wesley, 1980]).

57. Ibid., pp. 205–206.

58. C. L. Pearce and J. A. Conger. *Shared Leadership: Reframing the Hows and Whys of Leadership.* (Sage Publications, 2012).

59. C. C. Manz and C. P. Neck, *Mastering Self-Leadership: Empowering Yourself for Personal Excellence,* 3d ed. (Upper Saddle River, NJ: Pearson Prentice Hall, 2004), pp. 62–63. (6th ed., 2013).

60. Nahavandi, *The Art and Science of Leadership,* p. 213 (citing Manz and Neck, *Mastering Self-Leadership* [1999]).

61. Manz and Neck, *Mastering Self-Leadership* (2004), pp. 138–139.

62. "Leadership Styles," *Effective Leadership* series, No. 8737-B (College Station, TX: Instructional Materials Service, Texas A&M University, 1988), p. 4.

63. P. Hersey, K. H. Blanchard, and D. E. Johnson, *Management of Organizational Behavior: Utilizing Human Resources,* 7th ed. (Upper Saddle River, NJ: Prentice Hall, 1996), pp. 201–207. (10th ed., 2013).

64. Ibid., pp. 188–217.

65. Ibid.

66. "Leadership Styles," *Effective Leadership* series, No. 8737-B, p. 4.

67. Ibid.

68. U.S. Military Academy (West Point) cadet maxim, quoted in J. Kremer et al., *How to Live Your Dreams* (Taos, NM: Open Horizons, 2008); available at http://www.quotablebooks.com.

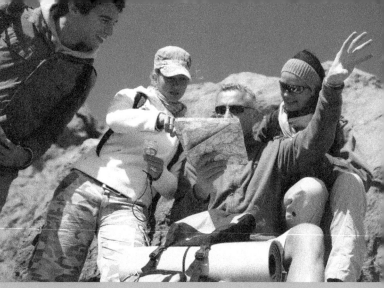

2 PERSONALITY TYPES AND THEIR RELATIONSHIP TO LEADERSHIP AND HUMAN BEHAVIOR

Personality consists of a relatively stable set of behavioral traits or characteristics. Personality theories and types aid in explaining and predicting individual behavior. Can we predict our own or someone else's leadership style based on personality type? Even though exact certainty is not possible, if we know ourselves and the personalities of others, we can determine how a leader will most likely react to a given situation.

Objectives

After completing this chapter, the student should be able to:

- Describe the four personality types
- Explain the importance of knowing and understanding personality types
- Correlate personality types with extroversion and introversion
- Understand the relationship between personality type and leadership style
- Describe the communication styles of the four personality types
- Identify ways to adapt your communication style to that of others
- Explain the communication continuum
- Define the relationship between personality type and group decision making, learning style, and career selection

Terms to Know

- sanguine
- choleric
- melancholy
- phlegmatic
- impetuousness
- perpetuating
- self-deprecating
- conceptualize
- subtlety
- authenticity
- vivacity
- socializers
- directors
- thinkers
- relaters
- reprimanding

Have you ever wondered why you loved agriculture while your friend hated it? Have you ever felt inspired by one particular teacher while your classmates thought she was boring? Have you ever been uncomfortable at a party where others seem to be having a wonderful time? Do you ever wonder why some people miss deadlines and never seem to have things organized? All these situations are connected to personality type, which encompasses one's preferred way of doing and viewing things.

Try this experiment. Hold out your arms as wide apart as you can. Bring them together as though you were clapping, and clasp your hands together. Look to see which thumb is on top: is it your right thumb or your left thumb? Now do the same thing again, but this time change thumbs so that the other one is on top. It usually feels awkward, even uncomfortable. Do it again with your favorite thumb on top. If you clasped your hands together thousands of times, you would probably place your favored thumb on top every time. This is what we call a preferred way of acting.

In addition to personality type, we also have personal preferences in leadership style, communication style, learning style, and careers. When you are in a classroom environment that matches your learning style, everything feels right. The teacher is stimulating, the material is exciting, and the work is enjoyable. If the environment does not match your preferred learning or communication style, you feel out of place, uncomfortable, and unable to do your best.

There are no right and wrong or good and bad styles, just preferences. Understanding learning styles and personality preferences has helped people succeed in class, in the FFA, and in the workplace. This understanding is an important dimension of self-discovery and personal growth throughout life.

The leader who recognizes his or her own personality type and other leadership characteristics—as well as the personality type, leadership style, communication style, and learning style of coworkers, followers, and friends—can channel each person's strengths for productivity and effectiveness in the group while minimizing or eliminating weaknesses that may harm the group or organization. In this chapter, we explore each personality type and communication style and discuss how each type correlates with leadership styles, learning styles, and career selection.

PERSONALITY TYPES

What personality type are you? Answering this question can be quite enlightening in terms of self-discovery, but it will also help you in relationships, in getting hired and staying employed, and in your overall leadership development. We will introduce one model of personality types in this text, but there are many other models and measures of personality. We encourage you to look up some of the popular personality assessments that are available free online. The *Four Temperaments* assessments, also found online, will utilize the model we discuss in this text. Others available online include the following:

- Myers–Briggs Type Indicator®
- Dominance, Influence, Steadiness, Conscientiousness (DISC) Profile
- California Psychological Inventory

We encourage you to study these assessments and look for ways to apply what you learn in different leadership situations when working with people.

People behave differently. Some people are usually self-motivated, whereas others are almost impossible to motivate. Some people are outgoing; others are not. Some people take pride in how they dress or in their grooming, and others do not seem to care how they look. We tend to use trait adjectives, such as *warm*, *aggressive*, and *easygoing*, to describe people's behavior. Fundamentally, personality consists of a relatively stable set of behavioral traits or characteristics. Theories and categorizations of personality—personality types—aid in explaining and predicting individual behavior. Our personalities are the product of both genetics and environmental factors. Knowing someone's personality traits can help you understand and predict that person's behavior in a given situation. Each individual's personality affects his or her human relations.

Can we predict our own or someone else's leadership style based on personality type? The authors of this book believe this is very possible. If we know ourselves and the personalities of others in a group, we can also determine with reasonable accuracy how we as leaders and each person in the group will react to a given situation.

Origin of the Identification of Four Personality Types

Hippocrates, the Greek physician and philosopher, identified four types of temperament. He mistakenly believed that these four temperaments were the result of body liquids that predominate in each individual. He identified these as

- Sanguine (blood)—lively temperaments
- Choleric (yellow bile)—active temperaments
- Melancholy (black bile)—dark temperaments
- Phlegmatic (phlegm)—slow temperaments

Obviously, the idea that temperament depends on body liquids has been discounted, but these four classifications of temperament are still in widespread use. **Sanguine** is a personality type with a warm, lively, cheerful temperament. **Choleric** is a personality type with a bossy, quick, active, and strong-willed temperament. **Melancholy** is a personality type that is analytical, self-sacrificing, gifted, and perfectionist. **Phlegmatic** is a personality type with a calm, cool, slow, easygoing, well-balanced temperament. Although modern psychology has developed many new classifications of temperaments or personality, few have become more accepted than those of Hippocrates.[1] Other names for each of the four basic personalities are shown in Figure 2–1.

Personality Profiles Personality profiles like the one available at http://www.oneishy.com measure or inventory a person's personality. After completing a survey and analyzing the results, you can better understand yourself; you can extrapolate this knowledge to better understand others and predict how they will react in certain situations. All of us have developed behavioral patterns or distinct ways of thinking, feeling, and acting. Our behavioral patterns tend to remain stable because they reflect our individual identities.

Understand Personality Type The demands of everyday living constantly require responses, and one's usual or habitual responses evolve into a behavior style or personality type. When you have heightened understanding of your own personality type, you can identify situations and approaches that will be most conducive to your success. At the same time, you should learn about the different styles others may possess and the situations they require for maximum productivity and harmony in the workplace and community. Research supports the conclusion that the most effective people are those who know themselves, know the demands of the situation, and adapt their strategies to meet those needs. Therefore, a personality profile enables you to

- Identify your behavior style or personality type
- Create situations that will give you the best chances of success
- Increase your understanding and appreciation of other personality types
- Predict and minimize potential conflicts with others[2]

Many organizations and/or businesses in which people must work together use personality learning instruments as tools; a popular one is the *Team Dimensions Profile*® originated by the Carlson Learning Company (now, Inscape Publishing). You can obtain a personality profile from one of the many websites that offer such instruments; see the Take It to the Net section at the end of this chapter.

HIPPOCRATES	MYERS-BRIGGS TYPE INDICATOR	CARLSON LEARNING COMPANY	TRUE COLORS	AUTHOR
Sanguine	Intuitive feelers	Influencing	Orange	Charismatic/influencer
Choleric	Intuitive thinkers	Dominant	Gold	My way
Melancholy	Sensing thinkers	Cautious	Green	Methodical
Phlegmatic	Sensing feelers	Steady	Blue	Laid-back/carefree

FIGURE 2–1 Although most people agree that there are four basic personality types, different researchers have coined different names for the types. Five ways of characterizing each personality type are presented here.

Ratio and Blend of Personality Types

The four temperaments or personality types are basic temperaments. No person has only a single temperament type. We all display a mixture of temperaments, though usually one predominates. There are varying degrees of temperament: for example, some may be 60 percent sanguine and 40 percent choleric. Some are a mixture of more than two. Many people have a mixture of all four personality types, such as 30 percent sanguine, 40 percent choleric, 20 percent melancholy, and 10 percent phlegmatic. In reality, determining ratios and blends is not important.[3] The important thing is determining your basic personality type, as well as being able to identify the basic personality type of others. If you know your own personality, as well as those of the people you lead, you can become a better leader, leading and directing according to the situation. Refer to Figure 2–2.

The Sanguine Personality Type Generally

Characteristics Sanguines need the freedom of immediate action. A zest for life and a desire to test the limits characterize these people. They take pride in being highly skilled in a variety of fields. They are master negotiators. They like adventure. A hands-on approach to problem solving and a direct line of reasoning create the excitement and immediate results that sanguines admire.

Sanguines value freedom and excitement. They think that skill is more important than structure or logic. They like being spontaneous and want to enjoy what they are doing; for them, planning sometimes takes the fun out of things. Sanguines like games and competition. They also like to learn things that they can put to use immediately.

Strengths Sanguines have a warm, lively, "enjoying" temperament; they are extroverted. They are receptive by nature, and external impressions easily influence them. Often they rely on feelings or gut instinct rather than rational thinking and logic. They tend to be cheerful people. Sanguines often become salespeople, teachers, politicians, actors, and public speakers.

Weaknesses When you look beneath the surface, the constant activity of the sanguine temperament often turns out to be little more than restless energy. Many are not good students because

---

VIGNETTE 2-1 | **GOT PERSONALITY?**

HAVE YOU EVER HEARD SOMEONE SAY, "That guy's got personality"? What do people mean when they use this phrase? Does it mean that someone is outgoing, likable, or fun to be around? We all have personality. Each individual's personality is useful in many ways. At least, that is a lesson that Nancy learned while volunteering in the local hospital.

Nancy loved working with the patients and the patients definitely appreciated having her around. The problem was that Nancy allowed her feelings to be hurt at least once a day by one or more of the patients. Nancy was a cheerful, warm, lively individual who enjoyed making people happy, and she sometimes wondered why she was not greeted with the same attitude and type of treatment.

Ron, a psychologist in the hospital, saw Nancy crying one day and stopped to ask her what the problem was. When she explained that she felt some patients didn't like her, he informed her that many of those patients actually adored her. This did not make sense to Nancy until Ron explained that everyone has a different way of viewing the world and responding to life. Some people are warm and lively (like Nancy); some are very analytical and self-sacrificing; some are happy, but passive; and some are very practical and strong-willed. Nancy began to understand that she should not take it personally when some people did not automatically return her cheerfulness in kind. Ron helped her realize that everyone has personality—and very different personalities at that!

Nancy has continued to work at the hospital, but she has also started to investigate personality types, using the Internet and reading books on the subject. By learning about and recognizing the basic differences among people, she has been able to become a better hospital volunteer. People who are not like her do not offend her anymore, and she better understands where they are coming from. The patients are still crazy about her, and Nancy now enjoys her volunteer experience more than ever.

Sanguine Emotions

- Appealing personality
- Talkative, storyteller
- Life of the party
- Good sense of humor
- Emotional and deomstrative
- Lives in the present
- Enthusiastic and expressive
- Cheerful and sincere at heart
- Curious
- Changeable disposition
- Compulsive talker
- Exaggerates and elaborates
- Scares others off
- Has restless energy
- Egotistical
- Naive
- Has loud voice and laugh
- Controlled by circumstances
- Gets angry easily

Sanguine at Work

- Great presenter
- Volunteers for jobs
- Thinks outside the box
- Has energy and enthusiasm
- Good at starting projects
- Inspires others to join
- Charms others to work
- Creative and colorful
- Would rather talk
- Forgets obligations
- Doesn't follow through
- Confidence fades quickly
- Can be undisciplined
- Priorities can be out of order
- Easily distracted
- Wastes time talking

Sanguine among Friends

- Makes friends easily
- Loves being around people
- Thrives on compliments
- Seems exciting
- Doesn't hold grudges
- Apoligizes quickly
- Prevents dull moments
- Likes spontaneous activities
- Hates to be alone
- Needs to be center stage
- Wants to be popular
- Looks for credit
- Dominates conversations
- Interrupts and doesn't listen
- Answers for others
- Forgetful
- Makes excuses
- Repeats stories

FIGURE 2–2 Each personality type has strengths and weaknesses. Remember, no person is only a single personality type, although one type may predominate; each of us displays characteristics from a combination of types. *Source: Adapted from Tim LaHaye,* Spirit-Controlled Temperament, *eBook (Wheaton, IL: Tyndale House, 2014). (© iStock/michaeljung; © iStock/RossHelen; © iStock/VladTeodor; © iStock/Ismail Çiydem; © iStock/Sergey Nivens; © iStock/Steve Debenport; © iStock/Steve Debenport; © iStock/LattaPictures; © iStock/MachineHeadz; © iStock/Mikolette; © iStock/sharpshutter; © iStock/visualspace)*

of this restlessness. Because of their emotional nature, sanguines can get instantly excited, but they tend to be impractical and disorganized. They seldom live up to their potential. Their habitual pattern of restless activity usually proves unproductive in the long run, as they frequently spend their lives running off on tangents.[4]

Sanguine Students Sanguines perform well in competition, especially when there is a lot of action. They love games and hands-on activities. Because sanguines crave fun and excitement, they dislike routines and structured presentations. Sanguines get a kick out of putting what they learn to immediate use. Sanguine students like classes that have contests, changes of pace, and variety. They learn by doing, seek immediate results, and like tools; they are impulsive, physical, and competitive.

Sanguines as Friends Planning ahead bores sanguines because they never know what they want to do until the moment arrives. They like to excite their dates with new and different things, interesting places to go, and romantic moments.

Sanguines with Family Sanguines need plenty of space and freedom. They want everyone to have fun. It is hard for them to follow rules, as they feel that family members should all just enjoy one another.

Sources of Personal Success for Sanguine Personalities

- The impulse to live fully
- Testing the limits

Choleric Emotions
- Born leader
- Dynamic and active
- Compulsive need for change
- Must correct wrongs
- Strong-willed and decisive
- Unemotional
- Not easily discouraged
- Independent and self-sufficient
- Exudes confidence
- Can run anything
- Bossy
- Impatient
- Quick-tempered
- Enjoys controversy and arguments
- Won't give up when losing
- Inflexible
- Is not complimentary
- Dislikes tears and emotions
- Is unsympathetic

Choleric at Work
- Goal oriented
- Sees the whole picture
- Organizes well
- Seeks practical solutions
- Moves quickly to action
- Delegates work
- Inisists on production
- Makes the goal
- Stimulates activity
- Thrives on opposition
- Little tolerance for mistakes
- Doesn't analyze details
- Bored by trivia
- May make rash decisions
- May be rude or tactless
- Manipulates people
- Demanding of others
- Work is the most important thing in their life
- Demands loyalty in the ranks

Choleric among Friends
- Has little need for friends
- Will work for group activities
- Will lead and organize
- Will lead in emergencies
- Is usually right
- Tends to use people
- Dominates others
- Decides for others
- Knows everything
- Can do everything better
- Is too independent
- Possessive of friends
- Can't apologize
- May be right, but unpopular

FIGURE 2–2 (continued)

- Variability
- Excitement, adventure
- Lightheartedness
- Being a natural entertainer
- Spontaneous relationships
- Taking off for "somewhere else"
- Being able to act in a crisis
- A love of tools
- Charm, wit, and fun
- Viewing defeats as only temporary[5]

Sanguine Uniqueness

Dream of: freedom
Like: opportunities, options, competition
Dislike: rigidity, authority, forcefulness
Express: optimism, impatie nce, eagerness, confidence, spontaneity, **impetuousness** (rushing into action with little forethought)
Foster: recreation, fun, enjoyment

Respect: skill, grace, finesse, charisma, artistic expression
Promote: stimulation, risk[6]

The Choleric Personality Type Generally

Characteristics Cholerics value order and cherish the traditions of home and family. They provide and support the structure of our society. Steadfastness and loyalty are their trademarks. Generous and parental by nature, cholerics show they care by making sure everyone does the right thing. It never occurs to them to shirk or disregard responsibility of any kind.

To cholerics, being responsible and following the rules are more important than excitement and emotions. They like family life and saving money, and they plan to really make something of themselves. They enjoy belonging to groups and want to

Melancholy Emotions
- Deep and thoughtful
- Analytical
- Serious and purposeful
- Genius-prone
- Talented and creative
- Artistic or musical
- Philosophical and poetic
- Appreciates beauty
- Sensitive to others
- Self-sacrificing
- Conscientious
- Idealistic
- Remembers the negatives
- Moody and depressed
- Has false humility
- Off in another world
- Low self-image
- Has selective hearing
- Self-centered
- Too introspective
- Guilt feelings
- Persecution complex

Melancholy at Work
- Schedule oriented
- Perfectionist, high standards
- Detail conscious
- Persistant and thorough
- Orderly and organized
- Neat and tidy
- Economical
- Sees the problems
- Finds creative solutions
- Needs to finish what he or she starts
- Likes charts, graphs, figures, lists
- Not people oriented
- Depressed over imperfections
- Chooses difficult work
- Hesitant to start projects
- Spends too much time planning
- Prefers analysis to work
- Self-deprecating
- Hard to please
- Standards often too high
- Deep need for approval

Melancholy among Friends
- Makes friends cautiously
- Content to stay in the background
- Avoids attracting attention
- Faithful and devoted
- Will listen to complaints
- Can solve others' problems
- Deep concern for other people
- Moved to tears with compassion
- Seeks ideal companion
- Lives through others
- Insecure socially
- Withdrawn and remote
- Critical of others
- Holds back affection
- Dislikes those in opposition
- Suspicious of people
- Antagonistic and vengeful
- Unforgiving
- Full of contradictions
- Skeptical of compliments

FIGURE 2–2 (continued)

help make those groups run smoothly. They enjoy learning about things that are useful to them.

Strengths The choleric has a quick, active, practical, and strong-willed temperament. These people are often self-sufficient and very independent. They tend to be decisive and opinionated and find it easy to make decisions for others as well as themselves.

Weaknesses The admirable qualities of the choleric are accompanied by some serious weaknesses, notably cholerics' hard, angry, impetuous, self-sufficient traits. Their bossy, authoritarian ways often grate on coworkers. They do not give up when they are losing. They cannot admit to mistakes and find it very difficult to apologize. A most unfortunate tendency is to become bitter and end the relationship when friends cross them, even when the choleric is wrong.[7]

Choleric Students As students, cholerics do their best when course content is structured and clearly defined. Abstract ideas and concepts should not be introduced until the foundations of a subject are plainly presented. Cholerics always want to know when they are on the right track; rules and direction are a great help to them. These people believe students should share in the responsibilities and duties of the classroom. Choleric students prefer useful subjects, thrive on routine and orderliness, are punctual and dependable, think problems through before making a decision, have a strong sense of right and wrong, and respect the school rules.

Cholerics as Friends Cholerics prefer people who are careful with their money and make plans ahead of time. They like their dates to be loyal, dependable, and punctual. Cholerics are serious about love and show it in many practical ways.

Phlegmatic Emotions
- Low-key personality
- Easygoing and relaxed
- Calm, cool, and collected
- Patient, well-balanced
- Consistent life
- Quiet but witty
- Sympathetic and kind
- Keeps emotions hidden
- Happily reconciled to life
- All-purpose person
- Unenthusiastic
- Fearful and worried
- Indecisive
- Avoids responsibility
- Quiet will of iron
- Selfish
- Too shy and reticent
- Too compromising
- Self-righteous

Phlegmatic at Work
- Competent and steady
- Peaceful and agreeable
- Has administrative ability
- Mediates problems
- Avoids conflicts
- Good under pressure
- Not goal oriented
- Lacks self-motivation
- Hard to get moving
- Resents being pushed
- Lazy and careless
- Discourages others
- Would rather watch

Phlegmatic among Friends
- Easy to get along with
- Pleasant and enjoyable
- Inoffesive
- Good listener
- Dry sense of humor
- Enjoys watching people
- Has many friends
- Has compassion and concern
- Dampens enthusiasm
- Stays uninvolved
- Is not exciting
- Indifferent to plans
- Judges others
- Sarcastic and teasing
- Resists change

FIGURE 2–2 (continued)

Cholerics with Family Stability and security are important to the choleric personality type, and cholerics enjoy traditions and frequent celebrations. They like to spend holidays with family members and may plan such gatherings for months in advance.

Sources of Personal Success for Choleric Personalities

- Generosity
- Strong work ethic
- Parental nature
- Ceremony
- Sense of history
- Dignity, culture
- **Perpetuating** (continuing in existence; keeping alive or active) heritage
- Steadfastness

- Orderliness
- Predictability
- Home and family
- Establishing and organizing institutions[8]

Choleric Uniqueness

Dream of: assets, wealth, influence, status, security

Like: service, dedication

Dislike: disobedience, nonconformity, insubordination

Express: concern, stability, purpose

Foster: institutions, traditions

Respect: loyalty, obligation, dependability, accountability, responsibility

Promote: groups, ties, bonds, associations, organizations[9]

SAMMY SANGUINE IS A WARM, outgoing, energetic, lively, and happy soul. He is receptive by nature and is easily swayed by external impressions, especially when they trigger an outburst of emotion. Feelings rather than reflective thought predominate in his decision making.

Sammy has an unusual capacity to enjoy himself and usually passes on his emotions and mood to others. When he comes into a room full of people, he tends to brighten up the place by his exuberant flow of conversation. He is a thrilling story-teller because his emotional nature allows him almost to relive the experience, which makes the story exciting for listeners. Sammy, like most sanguines, does not believe that faults or overattentiveness to facts should get in the way of a good story.

Sammy never lacks for friends. His naïve, spontaneous, friendly nature opens doors. Sammy genuinely feels the joys and sorrows of the people he meets, and he has the capacity to make them feel important, as though he were a very special friend, even when he is in fact just a passing acquaintance. Sammy enjoys people, dislikes solitude, and is at his best surrounded by friends where he is the life of the party. His stories and good nature make him a favorite with both children and adults.

Sammy is never at a loss for words. He often speaks before thinking, but his open sincerity has a disarming effect on most of his listeners, causing them to respond to his mood rather than his sometimes awkward words. His freewheeling, exciting way of life often makes him the envy of more introverted people.

Sammy's noisy, blustering, friendly ways make him appear more confident than he really is, but his energy and lovable disposition help him squeeze by the challenges of life. People frequently excuse his weaknesses by saying, "That's just the way Sammy is."

The world is enriched by cheerful, sanguine people like Sammy. They make good salespeople, hospital workers, teachers, conversationalists, actors, and public speakers.

The Melancholy Personality Type Generally

Characteristics Those with melancholy personalities feel best about themselves when they are solving problems and their ideas are recognized, especially when they feel that those ideas are innovative or ingenious. The melancholic personality seeks to express himself or herself by becoming an expert in everything. A melancholic's idea of a great day is to use knowledge to create solutions, because these people are complex individualists of great analytical ability. Although melancholy personality types do not express their emotions openly, they experience deep feelings.

Melancholic personalities value knowledge—knowledge is their strength. Discovering solutions and using their brains are more important than feelings, rules, or excitement. Melancholics like to know how and why things work the way they do. Melancholics prefer to work alone and also need room to think.

Strengths The melancholic type is often referred to as the dark temperament. A stereotype of this personality is a researcher, scientist, lab technician, or mathematician. Melancholics are analytical, self-sacrificing, gifted perfectionists with very sensitive emotions. No one gets more enjoyment from the fine arts than the melancholic.

Weaknesses Because of their perfectionism and (overly) analytical nature, melancholics tend to be pessimistic and quite critical. Because it is more self-centered than any other personality type, the melancholy personality type is also more **self-deprecating** (devaluing or downplaying one's own abilities or accomplishments) and self-critical than the others. Melancholics tend toward strict and constant self-examination through a perfectionist lens that magnifies the bad to the point that they become chronically dissatisfied with reality. Daydreaming and nostalgia can consume their time and energy.[10]

Melancholic Students Melancholics perform best when exposed to the driving force or overall theory behind a subject. They prefer to work independently. New ideas and concepts arouse their curiosity, and they enjoy interpreting ideas before adding them to a personal knowledge bank. Melancholics are gratified by probing abstract concepts. Melancholics respond positively to recognition and appreciation of their competence.

VIGNETTE
2-3 COLIN CHOLERIC

COLIN CHOLERIC IS THE DOMINEERING, aggressive, authoritarian, strong-willed type. Colin is self-sufficient and highly independent. He tends to be decisive and opinionated, easily making decisions for himself and others.

Colin thrives on activity; to him, life *is* activity. Colin does not need to be stimulated by his peers; instead, he stimulates them with his endless ideas, plans, and ambitions. This is not aimless activity, because Colin has a practical, keen mind and is capable of making sound decisions for both the long and the short term. Colin does not waver under pressure, nor is he swayed by what others think. He takes a definite stand on issues and puts heart and soul into his position.

Colin is not frightened by adversity or adversaries. In fact, opposition tends to encourage him. Colin has dogged determination and often succeeds where others fail—not because his plans are better, but because he continues working after others have become discouraged and quit.

Colin's emotional nature is the least developed part of his temperament. He does not sympathize easily with others, nor does he naturally show or express compassion. Colin is often embarrassed or disgusted by the tears of others. He has little appreciation for the fine arts. Colin Choleric's primary interest is in the more utilitarian, practical aspects of life.

Colin is quick to recognize opportunities and equally quick to figure out the best way to make use of them. He has a well-organized mind, though details usually bore him. Colin is not given to analysis, but rather makes quick, almost intuitive appraisals. Therefore, he tends to look at the goal toward which he is working without seeing the potential pitfalls and obstacles—including people—in the path. Once Colin has started toward his goal, he may run over individuals who get in his way. Colin tends to be domineering and bossy and does not hesitate to use people to accomplish his ends.

Many of the world's great generals and leaders have been cholerics like Colin. Cholerics make good executives, idea women, producers, or dictators, depending on their moral standards.

Students with a melancholic personality are logical, theoretical, and curious. Melancholic students **conceptualize** (visualize an abstract concept) well, are driven to understand, learn best independently, and need immediate challenges.

Melancholics as Friends Melancholic personalities may seem cool and unexpressive. They are uneasy about being controlled by emotions and do not want relationships to be complex. Once they have expressed their feelings, talking about their emotions causes doubt and discomfort.

Melancholics with Family In the family, the melancholic personality is often seen as a loner because this type likes a lot of private time to think. Sometimes melancholics find family activities boring and have difficulty following family rules that do not make sense to them.

Sources of Personal Success for Melancholy Personalities

- Developing models
- Abstract thinking
- Analytical processes
- Exploring ideas
- Variety of interests
- Striving for competency
- Admiring intelligence
- Storing wisdom and knowledge
- Perfectionism
- Eliminating redundancy
- Using precise language
- Handling complexity[11]

Melancholy Uniqueness

Dream of: truth, perfection, accuracy
Like: efficiency, increased output, reduced waste
Dislike: injustice, unfairness
Express: coolness, calm, reserve
Foster: inventions, technology
Respect: knowledge, capability, intelligence, explanations, answers, resolutions
Promote: effectiveness, competence, know-how[12]

The Phlegmatic Personality Type Generally

Characteristics The phlegmatic type seeks to express his or her inner self with authenticity and honesty above all else. Sensitive to **subtlety** (having or showing a keenness about small differences, particularly in meaning), phlegmatic

MICHELLE MELANCHOLY COULD BE CALLED the "methodical one." Michelle is an analytical, self-sacrificing, gifted, perfectionist type, with a very sensitive emotional nature and great appreciation for the fine arts.

By nature Michelle is prone to introversion, but because her feelings predominate, Michelle experiences a variety of moods. Sometimes Michelle's good moods cause her to act more extroverted. However, at other times Michelle is gloomy and depressed, and during these periods she is definitely withdrawn and can be quite difficult.

Michelle is a very faithful friend, but unlike Sammy Sanguine, she does not make friends easily. Michelle will not put herself out to meet people; instead, she allows people to come to her. Michelle's natural resistance to putting herself forward is not an indication that she does not like people. Like the rest of us, Michelle not only likes others but also has a strong desire to be loved by them. However, disappointing experiences make her reluctant to take people at face value. Michelle thus tends to become suspicious when others seek her out or give her too much attention. Michelle Melancholy is perhaps the most dependable of all the types, as her perfectionist tendencies do not permit her to be a slacker or let others down when they are depending on her.

Michelle's exceptional analytical ability allows her to accurately identify the obstacles and dangers of any project in which she takes part. This is in sharp contrast to Colin Choleric, who rarely anticipates problems or difficulties and who is confident that he can cope with whatever problems arise. This analytical characteristic often makes Michelle reluctant to initiate new projects. Occasionally, when Michelle is on an emotional high, she may produce some great work of art or even genius. These accomplishments are often followed by periods of deep depression.

Michelle is hard on herself and often finds her greatest meaning in life in a difficult vocation involving significant personal sacrifice. Once she chooses it, Michelle tends to be very thorough and persistent in her pursuit of a goal or vocation and is more than likely to accomplish great good. Many of the world's great geniuses—artists, musicians, inventors, philosophers, educators, and theoreticians—are of the melancholy temperament.

individuals create roles in life's drama with a special flair. They enjoy close relationships with those they love, and there is a strong spirituality in their nature. Making a difference in the world comes easily to the phlegmatic type because these people cultivate the potential in themselves and others.

Phlegmatics value people. Being liked and having everyone around them get along is more important than facts, rules, adventure, or logic. Phlegmatics are sensitive to others and become uncomfortable when there is conflict or competition. They like socializing and working with other people.

Strengths Phlegmatic personality types get their name from the body fluid that Hippocrates thought produced a calm, cool, slow, easygoing, well-balanced temperament. Life for phlegmatics is an unexcited, pleasant experience in which they avoid involvement as much as possible. They are passive and extreme introverts.

Weaknesses The primary weakness of phlegmatics is their apparent slowness and laziness. They often seem reluctant because they resent having

been urged into action against their will; their lack of self-motivation results in their doing as little as possible and proceeding as slowly as they can.[13]

Phlegmatic Students Phlegmatic students feel best in an open, interactive atmosphere. They like to feel that the teacher really cares about them and gives the classroom a personal touch. It is important that their teachers value them as people and respect their feelings. Phlegmatics appreciate a lot of genuine human feedback. They thrive in a humanistic, people-oriented environment. Phlegmatics turn off when conflicts arise, but flourish in an atmosphere of cooperation. Phlegmatics are verbal and social, work best in a group setting, are sensitive to rejection and conflicts with teachers and fellow students, and need to feel valued and reassured.

Phlegmatics as Friends Phlegmatic types look for perfect love. They are very romantic and enjoy touching, holding hands, exchanging love poems and notes, flowers, and quiet talks.

Phlegmatics with Family In family life, phlegmatic people like to be happy and loving. They are very sensitive to rejection by family members and to family conflicts. They really like to be well thought of and need frequent reassurance. Phlegmatics love intimate talks and warm feelings.

Sources of Personal Success for Phlegmatic Personalities

- **Authenticity** (genuineness; realness) as a standard
- Seeking reality
- Devotion to relationships
- Cultivating potential in others
- Assuming creative roles in life's drama
- Writing and speaking with poetic flair
- Self-searching
- Having a life of significance
- Sensitivity to subtlety
- Spirituality
- Making a difference in the world
- Seeking harmony[14]

Phlegmatic Uniqueness

Dream of: love, affection, authenticity
Like: meaning, significance, identity
Dislike: hypocrisy, deception, insincerity
Express: **vivacity** (being spirited, full of life and energy; liveliness), enthusiasm, inspiration
Foster: potential growth in people, harmony
Respect: nurturing, empathy, shared feelings, compassion, sympathy, rapport
Promote: growth and development of others[15]

IMPORTANCE OF PERSONALITY TYPES

Understanding the four personality types will help you be more accepting and tolerant of others. When you realize that people are victims of their personality type, things they do will then amuse you rather than hurt you, offend you, or make you angry. As we mature, we learn not to take things personally. Also, once we learn why we do things the way we do, we can take steps to curb our undesirable behaviors.

Our temperaments or attitudes can never be entirely eliminated or changed, but we quickly learn that it is not good to exhibit certain behaviors in particular situations. For example, a person with an autocratic (choleric) attitude may realize that it is more appropriate to follow the democratic process when conducting meetings—and in many other situations. A president of a major university once faced this dilemma. Being very educated, he knew that the acceptable way to run a university was with a democratic style. The faculty and students appreciated him because they had a voice. However, when one on one with mid-level administrators, his authoritarian attitude resurfaced; he became hostile when challenged. Nevertheless, because he was aware of his undesirable personality traits and behaviors and generally was able to control or moderate them, he was a successful administrator.

Although we are born with certain temperaments and personalities, we can be taught how to become leaders by adapting our personality types to varying situations.

Learning Personality Types in the Workplace

Companies have become aware that mere warm bodies filling positions is not enough. Rather, they must hire the right person for each job, and employees must perform well to maintain productivity and enhance the company's competitive edge. Thus, personality tests are increasingly being used for hiring staff and for building teams among current staff.[16] Literally millions of people in the United States have taken the Myers-Briggs Type Indicator (MBTI) test, for example. Because of this demand, personality tests are being generated in greater and greater numbers. Though their proponents and validity vary tremendously, most of them attempt to measure or identify similar characteristics, abilities, and/or behaviors.

Many companies find that such testing promotes mutual understanding among coworkers and between supervisors and their staff.[17] When used properly, tests make staff members aware of others' styles so they can work better together. Personality tests can also improve communication and productivity. The majority of tests are administered by businesses seeking to better understand their employees, in an effort to improve leadership within the company. Corporations such as Apple, AT&T, Citicorp, Exxon, General Electric, Honeywell, and 3M have all used personality profile tests for leadership development.[18]

PHIL PHLEGMATIC HAS A COOL, slow, well-balanced temperament. Life for Phil is a happy, unexcited, pleasant experience in which he is not overly involved.

Phil is so calm and easygoing that he never seems to get ruffled, no matter what the circumstances. Phil has a very high boiling point and seldom explodes in either anger or laughter; usually he maintains control of his emotions. People with Phil's temperament type are consistent every time you see them. Beneath Phil's cool, apparently shy personality is a very capable combination of abilities. Phil feels much more emotion than he allows to appear on the surface, and he has a good appreciation of the fine arts and the better things in life.

Phil attracts friends because he enjoys people and has a natural dry sense of humor. Phil is the guy who can have a crowd of people laughing hysterically and never crack a smile himself. Phil has a unique ability to find humor in others and the things they do. He has a keen mind and is often quite a good imitator. One of his great sources of delight is making fun of the other personality types. Phil is annoyed by the aimless, restless enthusiasm of the sanguine and often confronts that type. He is disgusted by the gloomy moods of the melancholic and is prone to ridicule her. He also delights in throwing ice water on the bubbling plans and ambitions of the choleric.

Phil tends to be a spectator of life and tries not to get too involved in the activities of others. In fact, Phil is seldom motivated to undertake any activity beyond his daily routine, and he does so with great reluctance. This does not mean that he cannot appreciate the need for action or empathize with the difficulties of others. He and Colin Choleric may see the same social injustices, but their responses will be entirely different. The crusading spirit of Colin will cause him to say, "Let's get a committee organized and do something about this!" Phil is more likely to say, "These conditions are terrible! Why doesn't someone do something about this?" Once Phil is roused to action, though, he can be very capable and efficient. Phil will not usually ask for a leadership role, but if fate thrusts a leadership role upon him, he always rises to the challenge as a good leader. Phil has a soothing effect on others and is a natural peacemaker.

The world has greatly benefited from the gracious nature of the efficient phlegmatic. Phil is usually kindhearted and sympathetic but seldom conveys his true feelings. The Phils of the world make great diplomats, accountants, teachers, leaders, and scientists.

Many companies, such as Farm Credit Mid-America, use personality tests in considering whom they should hire. Each job requires people with certain personalities for optimum success. This will be discussed later in the text.

Results of Using Personality Tests in the Workplace

It is often argued that people of different personality types have difficulty working together. However, through personality testing and the sharing of test results, people can become aware of which types they and their coworkers are; thus, both communication and productivity improve.[19] Knight-Ridder's *Charlotte Observer* successfully used the MBTI as a basis for team building in a newsroom that was undergoing near-constant conflict.[20] Understanding personality differences helps to build more productive teams who communicate better, and helps reduce stress from poor relationships at work.[21]

Many companies (e.g., Transamerica Corporation, Farm Credit) use personality assessments to encourage openness, trust, the use of affirming language, diversity, and paying attention to each other's strengths.[22] Companies also use assessments like MBTI to understand what motivates their employees and to gauge their attitudes. This type of information is vital for a company because personality has a direct influence on how employees treat each other on a team and how they interact with customers.[23] Personality can also impact decision making, change management, and conflict resolution in an organization.

Explanation of the Myers-Briggs Type Indicator (MBTI)

People with different personality styles approach problem solving and decision making differently. Some studies have linked MBTI to behaviors such as time urgency, academic problems, and even a person's ability to function in life. For example, most scientists are ST (melancholy) or NT (choleric), whereas research and development managers tend to be either SF (phlegmatic) or NF (sanguine). People with different dominant styles prefer certain types of information, have different time orientations, and generally approach life in different ways. The assumption is that individuals with different MBTI styles are likely to approach their tasks and their groups differently (Figure 2–3).

As another example, a sensing-thinker (ST) or melancholic leader is likely to require his or her group to seek out facts and figures and communicate alternatives and solutions. Because the feeling (F) dimension is secondary for such a leader,

focus on interpersonal issues is likely to be either absent or an unimportant factor. This may hinder the resolution of interpersonal conflict in some situations. In contrast, an intuitive-feeler (NF; sanguine) leader will have little interest in his or her group's factual analysis and is likely to prefer hearing about creative possibilities and alternatives to problems.

Infusion of the Personality Types

All four of the basic personality types are needed to provide variety and purposefulness in this world. No personality type is "better" than another. Each type has strengths and advantages, and each has weaknesses and disadvantages.

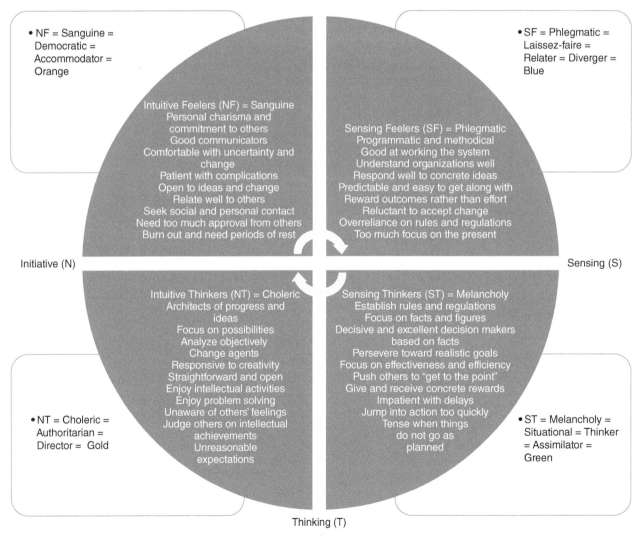

FIGURE 2–3 Using temperaments (i.e., sanguine, choleric, melancholy, phlegmatic), MBTI, and other classifications, this figure summarizes the discussion of personality types. It also provides a profile of different generalized tendencies and behaviors for groups of individuals.

Consider the following sequence of positive events involving the four personality types: the hard-driving choleric produces the inventions of the genius-prone melancholic, which are sold by the personable sanguine and enjoyed by the easygoing phlegmatic. In contrast, consider the shortcomings of the four personality types in their relationships to other people: the sanguine type enjoys people and then forgets them. The melancholic is annoyed with people but lets them go their own crooked ways. The choleric makes use of people for his or her own benefit and then ignores them. The phlegmatic regards people with benign indifference.

HOW PERSONALITY TYPE CORRELATES WITH EXTROVERSION AND INTROVERSION

Generally, people are divided into two broad categories: extroverts and introverts. The previous discussion noted that the sanguine and choleric personality types tend to be extroverted, whereas the melancholy and phlegmatic personality types tend to be introverted. There are gray areas, however. For example, a person may exhibit an extroverted personality when in comfortable or familiar surroundings, but in a less comfortable or unfamiliar environment, that same person may act introverted.

Activities through the FFA, such as leadership camps, Made for Excellence programs, the Washington Leadership Conference, and classes or lessons on leadership and personal development, can help students develop the skills to become more outgoing (extroverted) when their natural tendency is to remain reserved (introverted). Of course, extroverts have to learn that there are times when it is better to be introverted; part of wisdom is speaking when you need to speak and remaining silent when you should remain silent.

Interesting cross-cultural studies point to national differences regarding extroversion and introversion according to the MBTI. For example, Chinese managers are higher in introversion than their European counterparts, who tend to be more extroverted. For both European and Chinese groups, though, managers' extroversion

is linked to better performance and higher satisfaction. Perhaps differences exist among many different kinds of groups. Is it possible that FFA members who participate in parliamentary procedure are more extroverted than FFA members who judge soil? Regardless, it is important for leaders to understand people's differences and similarities and to build upon their group members' strengths.

Although we all have inadequacies, we still control our destiny. *Our attitude controls our altitude*: how far we go in life or what we accomplish depends on our attitude, along with how hard we are willing to work to overcome any natural weaknesses.

RELATIONSHIP BETWEEN PERSONALITY TYPE AND LEADERSHIP STYLE

Some personality types correlate with either a democratic or an authoritarian leadership style. For example, sanguine personalities have a more democratic attitude, whereas choleric personalities have a more authoritarian attitude. Melancholy personality types tend to be situational leaders, and phlegmatic types tend to be laissez-faire leaders.

Can we change? Absolutely! Thus arises the debate about whether leaders are born or made. People are born with natural tendencies, but through education and experience we learn what works best in particular situations. Even though this is a democratic nation, there are times when authoritarian decisions must be made. For example, once it is democratically decided that an interstate highway is needed, authoritarian actions are needed to obtain property from landowners. Remember, no one personality type is best. All types are needed for a complete, well-rounded society, group, or team. Figure 2–4 provides a summary of the four types, along with affirmation statements for each to build self-esteem. (An **affirmation statement** is a positive declaration about who we are and what we can become.)

Changing or altering personality type can be compared to eating and weight control. Some people, due to metabolism, can eat as much of

PERSONALITY TYPE	STRENGTHS	WEAKNESSES (LIMITATIONS)	AFFIRMATION STATEMENTS
SUMMARY OF THE FOUR PERSONALITY TYPES (STRATEGIES FOR STRENGTHENING SELF-ESTEEM)			
Sanguine (influencing)	Enthusiastic Good communicator Optimistic Involved Spontaneous Persuasive People person Imaginative Goal oriented Confident Gets results	Excitable, emotional Talks too much Unrealistic Disorganized Impulsive, undisciplined Manipulative Goes along with peers Daydreamer Impatient Self-reliant Never slows down	I am enthusiastic about life. I am good at expressing my thoughts, opinions, and ideas. I am a positive person. I am eager to participate in everything that is going on. I join in on things quickly. I have a unique ability to motivate people. I like people and want them to like me. I have a lot of creative ideas. I am able to set my mind on something and then go after it. I am capable of handling many things on my own. I can accomplish a lot when I make up my mind.
Choleric (dominating)	Competitive Decisive, determined Courageous Direct, straightforward Responds quickly Analytical Cautious, intense Conscientious	Attacks first Stubborn Reckless Blunt, harsh Lacks empathy Critical Unsociable Worries too much	I am a winner. I know what I want and go after it. I am courageous. I honestly express how I feel about things. I am quick to respond to a situation and seek a solution. I carefully think things through. I focus a lot of energy on getting things right. I am a hard worker. I always do my best.
Melancholy (conscientious)	Sensitive, intuitive Strives for excellence Does things correctly High personal standards Curious, questioning Steadfast Stable Systematic	Easily hurt Perfectionist Fears criticism of work Judgmental Nosy Resists change Boring Slower paced	I am attentive to what others say and feel. I am inspired by excellence. I take the time to do things correctly and without mistakes. I am a person with high standards. I like to understand all I can about what I am planning to do. I like things to stay the same. I stick with things that I know work well.
Phlegmatic (steady)	Agreeable Easygoing Good listener Softhearted Reliable	Indecisive Lacks initiative Too accommodating Uncommunicative Easily manipulated Overly dependent	I make it a point to get along with everyone. I am relaxed and pleasant to be around. I am a good listener. I am a compassionate person. I follow through on things I begin.

FIGURE 2–4 No person has only a single temperament type, but most people's behavior patterns indicate that they lean toward one or that a particular type predominates in that person. *Source: Adapted from Carlson Learning Company.*

whatever they want without gaining weight. Others must constantly exercise and be very mindful of what they eat or they will gain weight. Similarly, we can override our personality-type tendencies to identify and employ the best leadership style for the situation, even though that style may not be our natural choice. If you know your own personality type, as well as those of your coworkers, you can become a better leader.

COMMUNICATION STYLES OF THE FOUR PERSONALITY TYPES

We have discussed personality types and their relationship to leadership styles. Remember that sanguine personality types tend to be democratic leaders, and choleric personality types tend to be autocratic leaders. Taking this one step further, leadership style and personality type are both closely related to communication style. We can become better communicators if we adapt our communication style to the leadership style we need and the personality type of the person with whom we are speaking.

Tony Alessandra and Phil Hunsaker, in their book *Communicating at Work*, identified four styles of communicators: socializers, directors, thinkers, and relaters. Each type of communicator has particular personality characteristics that affect communication with other people. The reason for examining communication styles is, of course, to improve communication. You do this by changing your own communication style so you can relate to people who have any of the other styles; the situation may also require adapting your style to mesh with the style of a single other person with whom you are communicating.

When you learn the four communication styles and how to adapt to them, you are learning to improve your communication skills. You are not merely imitating another style or being insincere, but simply relating to the other person in the style that person prefers. By doing so, you are better able to convey your messages, ideas, and beliefs while at the same time reducing the possibility of being misunderstood or experiencing conflict. Refer to Appendixes B and C for further insight into the four personality and communication styles.

In this section, we describe each communication style so that you can identify your own category. We then suggest how to adapt your style to the other three styles in general. Finally, we offer ways to adapt yourself to each of the four styles individually.

Socializers

Socializers are relationship oriented, seeking the approval of those around them. They move, act, and speak quickly and avoid or ignore details whenever possible. They are risk takers who thrive on excitement and change. They enjoy being in the spotlight and have good persuasive skills (Figure 2–5).[24] Socializers may be public relations specialists, talk-show hosts, trial attorneys, social directors, or hotel personnel. You may recognize this as the sanguine personality.

FIGURE 2–5 Former state FFA officer, United States Army technician, and agriculture teacher Chaney Mosely poses with one of his students. Mr. Mosely has a socializer communication style. *(Photo courtesy of Georgia Agricultural Education.)*

If you are a socializer, you can adapt your style by recognizing and reducing your need for approval from other people or groups and concentrating on developing more directive skills and actions, such as conflict resolution, self-assertion, and negotiation.[25]

If you must deal with a person who is a socializer, you may adapt to that person by

- Demonstrating sincere interest in the person
- Validating the person's opinions, ideas, and dreams
- Complimenting the person's appearance, creativity, persuasiveness, and charisma
- Avoiding conflict and arguments
- Agreeing on and making notes of the specifics of any agreement
- Allowing the person to speak freely and fully
- Allowing the discussion to develop naturally and occasionally go off the topic
- Being entertaining and moving quickly[26]

Increasing the Self-Esteem of Socializers (Sanguines) You can increase the self-esteem of socializers by praising and rewarding them in ways most meaningful to them.

- Give immediate, favorable responses
- Emphasize their behavior and performance more than the finished product
- Appreciate their cleverness and spontaneity
- Comment on skills they demonstrate
- Note the speed of their actions
- Recognize the impact and effects of their performance
- Give a variety of tangible rewards[27]

Directors

Directors tend to be task oriented. Like socializers, they move, act, and speak quickly. They want to be in charge and oversee the actions of others as a way to achieve results. They make decisions quickly and display good administrative skills. Directors may be investigative newspaper reporters, stockbrokers, freelance consultants, corporate CEOs, or drill sergeants.[28] You may recognize this as the choleric personality. Would you agree that the gentlemen in Figure 2–6 could very well have a director communication style?

If you have a director style, you may wish to adapt to the other styles by relinquishing or reducing your emphasis on control of other

FIGURE 2–6 What nonverbal characteristics could lead someone to believe that these men use the director communication style? *(Photo courtesy of Georgia Agricultural Education.)*

people and conditions and by focusing on relationship-type skills and actions, such as listening, questioning, and positive reinforcement.[29]

If you must deal with a director-style person, you may adapt to that person by

- Supporting the person's goals and objectives
- Recognizing the person's ideas rather than him or her as an individual
- Emphasizing the desired results
- Keeping your communication businesslike
- Being precise, efficient, and well organized
- Providing detailed options and supporting analyses
- Making arguments based on facts, rather than feelings, when disagreements arise[30]

Increasing the Self-Esteem of Directors You can increase the self-esteem of directors by praising and rewarding them in ways most meaningful to them.

- Be honest and sincere
- Specifically praise their actual accomplishments
- Note accuracy, efficiency, and thoroughness in their performance
- Acknowledge their sense of responsibility
- Describe how the task they completed affects the well-being of others
- Recognize significant contributions from their efforts[31]

Thinkers

Thinkers, like directors, tend to be task oriented, but they move, act, and speak slowly. They enjoy solitary, intellectual, and philosophical challenges and prize accuracy. They are cautious decision makers who demonstrate good problem-solving skills. Thinkers may enter professions such as accounting, engineering, computer programming, the hard sciences (e.g., poultry science, soil science, biochemistry), systems analysis, and architecture.[32] You may recognize this as the melancholy personality type. The student researcher shown in Figure 2–7 looks as though he might have a thinker communication style.

If you have a thinker style, try soft-pedaling your unnecessary perfectionism and tendency to focus on weakness. Instead, exercise supportive skills and actions, such as active listening, positive reinforcement, and involvement with others whose strengths complement your own.[33]

You can adapt to a person with a thinker style by

- Being systematic, exact, organized, and thorough, and well prepared
- Validating the person's organized, thoughtful approach
- Complimenting the person's "efficiency, thought processes, and organization"
- Demonstrating through actions rather than explaining with words
- Asking questions and letting the person display his or her depth and breadth of knowledge
- Describing a process in detail and explaining how it will produce results
- Answering questions and providing details and analysis

FIGURE 2–7 What nonverbal characteristics could lead someone to believe that this student has a thinker communication style? *(© istock/AlexRaths)*

- Listing advantages and disadvantages of any plan
- Providing solid, tangible, factual evidence[34]

Increasing the Self-Esteem of Thinkers You can increase the self-esteem of thinkers by praising and rewarding them in ways most meaningful to them.

- Sincerely and appreciatively acknowledge their ideas and competent performance
- Verbally state your recognition of the value and usefulness of their work
- Recognize their specific knowledge and skills, especially in dealing with concepts and abstractions
- Acknowledge their intellectual capacities for analysis and precise explanations
- Compliment their creativity and ingenuity
- Acknowledge their ability to complete tasks independently
- Give only deserved positive feedback
- Provide the person with further opportunities to demonstrate competence; this may be a thinker's best reward for a job well done[35]

Relaters

Relaters are relationship oriented. They move, act, and speak slowly. They avoid risk and seek tranquility, calmness, and peace. They enjoy teamwork and show good counseling ability and skills. Relaters may enter the helping professions, such as "counseling, teaching, social work, the ministry, psychology, nursing, and human resource development."[36] You may recognize this as the phlegmatic personality type. The individuals in Figure 2–8 appear to have a relater communication style.

If you have a relater style, you may adapt to the other styles by decreasing your resistance to new things and by trying to develop directive skills and actions, such as negotiation and divergent thinking.[37]

In communicating with relaters, the following strategies are helpful:

- Being warm and sincere
- Showing interest in them personally
- Assuming that they will take everything personally

FIGURE 2–8 What characteristics of the individuals in this photo could lead someone to believe that they have a relater communication style? *(Photo courtesy of University of Georgia—College of Agricultural & Environmental Sciences.)*

- Allowing them time to learn to trust you
- Proceeding slowly and informally
- Listening actively
- Discussing personal feelings if you disagree with them
- Discussing, supporting, and maintaining your relationship with the person
- Complimenting the person's teamwork and ability to get along with and form relationships with others[38]

Increasing the Self-Esteem of Relaters You can increase the self-esteem of relaters by praising and rewarding them in ways most meaningful to them.

- Frequently tell them that their achievements and contributions make you feel good
- Frequently point out and acknowledge their unique personal characteristics
- Demonstrate that you care how they feel

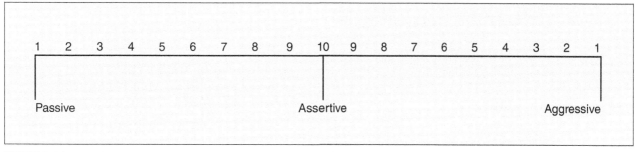

FIGURE 2–9 In most situations, leaders should be assertive communicators. Passive and aggressive styles are typically undesirable.

- Emphasize their importance
- Respond in kind to their honesty and sincerity
- Publicly recognize their participation in successful group sessions[39]

THE COMMUNICATION CONTINUUM

Your behavior and demeanor often communicate many things to those around you. Think of passive, assertive, and aggressive as points on a continuum; try to spend less time using the passive and aggressive styles at the ends of this continuum (Figure 2–9). Aggressive behavior may signal disrespect for others. Passive behavior may imply that you are holding back your true feelings and ideas. With assertive behavior, in contrast, you honestly express your feelings and thoughts without threatening others or causing anyone to experience anxiety. Assertive behavior is desirable. Nonverbal assertiveness includes confident body position and movement, positive eye contact, and use of comfortable, natural hand gestures.[40]

An example of aggressive behavior that might display disrespect for others is glaring at someone or using a rude or dismissive hand gesture. An example of passive behavior is laughing at (but not answering) a question that seems to pry too deeply into your personal beliefs or feelings.

Passive

When a leader uses a style at the passive end of the continuum, the characteristics and consequences shown in Figure 2–10 occur. The goals of a passive leader are to be liked, to be nice and friendly, and to appease others. Underlying these

goals is the desire to avoid conflict at all costs. A passive leader wants to be liked rather than admired. The unspoken message of passive leaders is, "What I think doesn't matter; what I feel is unimportant. I don't respect myself and I don't expect you to, either." People accept our own subconscious evaluation of ourselves.

Passive communicators do not take charge of the moment, which results in things not getting done. Passive communicators exhibit little self-esteem, deny or do not express feelings, keep thoughts and ideas inside, and allow others to make choices for them. The thought patterns of passive leaders include

- I'm not okay, but I'm not sure about you.
- Everyone has rights but me.
- I can survive if everyone likes me and approves of what I do, say, and feel.
- Nice people don't disagree.
- Don't make waves, don't rock the boat; people won't like you.
- Peace at any price.
- What I think is unimportant.
- I won't offer my opinion. People might laugh at me.
- It's not my place to speak up.

CHARACTERISTICS	CONSEQUENCES
Indecisive	Frustration
Not clear about expectations	Lack of vision
Unable to delegate	Things do not get done
Will not take a stand	Lack of leadership
Never gives correction	Does not know what to change

FIGURE 2–10 Characteristics and consequences of a passive communication style.

PAYOFFS	COSTS
Avoid confrontation	Fear
Occasionally is liked	Does not meet people's needs
Few open conflicts	Perceived as weak

FIGURE 2–11 Results of passive leadership.

Because passive communicators do not take charge or deal with situations directly, they do not meet people's needs and are perceived as weak. The results of passive leadership are listed in Figure 2–11.

Aggressive

When leaders move to the aggressive end of the continuum, the characteristics and consequences listed in Figure 2–12 occur. The goals of aggressive leaders are to win, dominate, intimidate, overpower, and get what they want when they want it. They often move toward their goals by insulting, **reprimanding** (giving a severe, usually formal, reproof, rebuke, or correction), and humiliating others. The basic message of aggressive leaders is, "You will never have to wonder what I think because I am going to tell you! You will never have to wonder what I feel because I am going to tell you! I guarantee that you are going to do what I want you to do, even if I have to use fear and intimidation to get you to do it! You are even more stupid than I thought if you disagree!"

Aggressive communicators like to exert authority. They go overboard when they communicate, intent on having things their way. Aggressive communicators diminish self-esteem in others, express their own negative feelings, say what they think regardless of whom it hurts, like to be in control, and make choices for others. They also

CHARACTERISTICS	CONSEQUENCES
My way—PERIOD!	Blunts creativity
Humiliates others	Avoidance, sabotage
Thinks people need constant surveillance	Distrust; people perceive that leader does not trust them
Emphasizes what people do wrong	Frustration, rejection

FIGURE 2–12 Characteristics and consequences of an aggressive communication style.

reach goals at the expense of others. The thought patterns of aggressive leaders include

- I'm okay. You're not.
- I have rights. You don't.
- People should do what I want without questioning me.
- Personnel does not send me good people anymore!
- If more people were like me, we wouldn't have the problems we have.
- I am never wrong!
- My feelings are more important than yours.
- I don't need to listen to other people. They have nothing to offer me.

Because aggressive communicators are authoritarians, they tend to get what they want—at the cost of fear, anger, and lack of teamwork. Figure 2–13 summarizes the results of aggressive leadership.

Aggressiveness might be your most effective response in the following situations:

- When there is confusion
- When it is time to move ... and no one is moving
- When questionable ethics are involved
- When you need to get people's attention and emphasize the importance of a matter
- When there is great tension and you need to clear the air

Assertiveness

When leaders communicate assertively, the characteristics and consequences in Figure 2–14 occur. The goals of assertive leaders are to get the work done at a high level of excellence while enhancing the growth and development of those doing the work, to communicate in a style that is accurate and respectful of the dignity of all persons involved, and to encourage those they work with to do the same. The basic message

PAYOFFS	COSTS
Win a lot in the short run	Fear
Get what they want	Distance
Protect self and space	Anger
Domination	Lack of teamwork

FIGURE 2–13 Results of aggressive leadership.

CHARACTERISTICS	CONSEQUENCES
Competent	Trusted
Communicates goals	Knows what is expected
Listens	Cooperation
Does not denigrate or put others down	Good feelings about relationship
Helps others feel good about themselves	Teamwork—working together rather than merely alongside one another
Ability to praise and correct	You know where you stand

FIGURE 2–14 Characteristics and consequences of an assertive communication style.

PAYOFFS	COSTS
Little fear	More conflict because of freedom of expression
Live your own life	Less fear of ridicule and rejection
Improved confidence	Some may be threatened or intimidated by leader's confidence
Do what needs to be done	Unpopular with some

FIGURE 2–15 Results of assertive leadership.

of assertive leaders is, "You will never have to wonder what I think because I will tell you. You will never have to wonder what I feel because I will share that with you. I guarantee that I have no interest in being critical of you for what you think or feel. Indeed, I invite you to share these things with me. We are here to get the job done and to create and contribute to a positive work environment."

Assertive communicators have high self-esteem, which builds others' esteem for them. They make their own choices and accept responsibility for their decisions. Thought patterns of assertive leaders include

- I'm okay, and you're okay.
- I have rights, and so do others.
- It is all right to make mistakes if you learn from them.
- I am a valuable and worthy person, and so are you.
- I have choices in nearly all situations, and I am responsible for the consequences of my choices.
- I trust you, and you can trust me.
- I am not a helpless victim.
- I will not allow others to decide for me how I will behave.
- Conflicts provide opportunities to grow and are not something to be avoided.
- I want to find a way we can all win.

Because assertive communicators take charge by doing what must be done, while being free and open, they permit those around them to be creative and expressive, without fear of ridicule or rejection. The results of assertive leadership are listed in Figure 2–15. Remember, though, that even assertive people cannot please everyone. There are, of course, situations in which assertiveness might not be your most effective response; in some instances, it is more appropriate to be passive or aggressive.

RELATIONSHIP BETWEEN PERSONALITY TYPE, GROUP DECISION MAKING, LEARNING STYLE, AND CAREER SELECTION

Once you have identified your personality type, you can determine many other things about yourself. We have discussed how leadership styles relate to personality type and also how communication style and preference relate to personality type. In other chapters, we discuss how group decision making, learning style, and career selection relate to personality type.

Group Decision Making

Group leaders, whether they are conscious of it or not, have leadership styles they use in directing their groups. How leaders involve group members in the problem-solving and decision-making processes reveals their leadership style. In Chapter 13, we show the connection between leadership styles and group problem solving and decision making. Figure 2–16 indicates that autocratic decision makers tend to be choleric, consultative decision makers tend to be sanguine, participating decision makers tend to be melancholic, and laissez-faire decision makers tend to be phlegmatic.

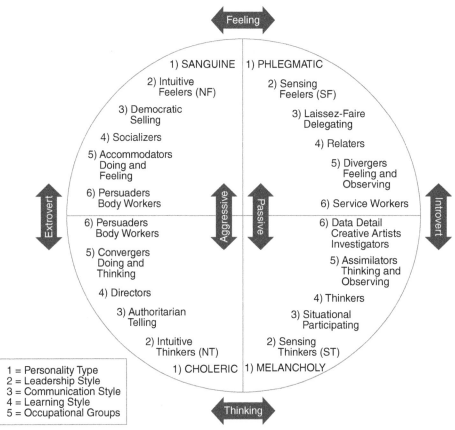

FIGURE 2–16 Personality types and their correlations with leadership styles, communication styles, learning styles, and occupational groups.

Learning Style

In Chapter 3, people are again categorized into one of four groups, this time relating to learning style. Your learning style correlates to your personality type, and knowledge of learning style helps you understand why we behave and react to things as we do. Figure 2–16 shows the following correlations: those with accommodator learning styles are sanguine, those with the converger learning style are choleric, those with an assimilator learning style are melancholic, and those with a diverger learning style are phlegmatic. Knowing colleagues' and/or followers' learning styles and placing them in situations compatible with those styles should lead to higher productivity for a company or organization.

Career Selection

In Chapter 21, we also show the correlation between occupational groups and personality type. From Figure 2–16, you can see that body workers tend to be sanguine or choleric; data detailers, artists, and investigators tend to be melancholic; persuaders tend to be sanguine or choleric; and service workers tend to be phlegmatic.

Your career choice should correlate with your personality type and learning style. Otherwise, you may end up in the wrong occupation and wonder why you are not happy with your job.

CONCLUSION

Which came first: the chicken or the egg? Does personality come from learning style, or does learning style come from personality? Whatever the order, there are close correlations among personality type, communication style, learning style, leadership type, group decision making, and even the careers we choose. Once we discover these correlations and relationships, we can communicate better, enhance our personal development, and become better leaders.

SUMMARY

There is a relationship between personality type and leadership style. The four personality types are sanguine, choleric, melancholy, and phlegmatic. Sanguine personalities have a more democratic attitude, whereas choleric personalities have a more authoritarian attitude. Melancholic personality types tend to be more situational leaders, and phlegmatics tend to be laissez-faire leaders. Knowing personality types will help you understand others better and help you choose appropriate leadership styles.

People are born with natural tendencies, but through education we learn what works best in particular situations. You can modify the natural tendencies arising from your personality type to adopt the appropriate leadership style.

Understanding the four types will help you be more accepting and tolerant of others, once you realize that they are merely expressing their personality types. Organizations make extensive use of personality tests and knowledge of personality types in hiring decisions and team building.

Generally, people are considered either extroverted or introverted. Sanguine and choleric personality types are more extroverted; melancholic and phlegmatic types are more introverted.

Leadership style and personality type are closely related to communication style. The four styles of communicators—socializers, directors, thinkers, and relaters—have personal characteristics and work habits that determine how they communicate with others. Learning how to adapt to each style is essential to being an effective communicator.

Your behavior communicates many things to those around you. Think of passive, assertive, and aggressive as points on a continuum, and try to emphasize use of the assertive style.

Just as personality relates to leadership and communication styles, it also correlates with psychological and personal development. Group decision making, learning style, and career choice also relate to and are affected by personality type.

 ## Take It to the Net

1. Explore personality styles on the Internet. The websites listed here provide personality tests. Browse as many of the sites as you want and take as many of the tests as you want. Print the results of two tests. Write whether or not you agree with the results and explain why.

   http://www.9types.com

   http://www.keirsey.com

   http://www.personalitytype.com

   http://www.personalityquiz.net

   http://www.queendom.com

2. There are many other fun and informative websites that can help you to understand yourself and others. Web addresses can change, so if the sites listed in the preceding activity do not work for you, you can find several free personality tests by using your favorite search engine. The following search terms may prove useful:

   personality types

   personality test

   personality test online

   free personality test

Chapter Exercises

REVIEW QUESTIONS

1. Define the Terms to Know.
2. List five personal success traits of each of the four personality types.
3. Give three key characteristics each of sanguine, choleric, melancholic, and phlegmatic students.
4. Why is it important to learn personality types?
5. Which personality types tend to be extroverts?
6. Which personality types tend to be introverts?
7. Match the four personality types with the four leadership styles.
8. Briefly discuss the four communication styles.
9. Match the four communication styles with the four personality types.
10. Briefly describe each of the three communication behaviors on the communication continuum.
11. Which of the three communication behaviors is the most desirable? Explain.
12. What are three other areas of life that correlate with one's personality type and communication style?

COMPLETION

1. _____ have a warm, lively, "enjoying" temperament.
2. _____ have a bossy, quick, active, practical, and strong-willed temperament.
3. _____ are analytical, self-sacrificing, gifted perfectionists.
4. Life for _____ is an unexcited, pleasant experience in which they avoid as much involvement as possible.
5. _____ and _____ personality types tend to be extroverts.
6. _____ and _____ personality types tend to be introverts.
7. _____ personalities have a more democratic attitude.
8. _____ personalities have a more authoritarian attitude.
9. _____ personalities tend to be laissez-faire leaders.
10. On the communication continuum, _____ behavior consists of honestly expressing your feelings and thoughts without threatening others or causing anxiety.

MATCHING

Answer choice may be used more than once.

_____ 1. Communication style that enjoys being in charge.

_____ 2. Communication style that favors solitary intellectual work.

_____ 3. Communication style that tends to enter the helping professions.

_____ 4. Communication style that describes risk takers who enjoy the spotlight.

_____ 5. Avoid conflict at all costs.

_____ 6. Goals are to win, dominate, intimidate, overpower, and get what they want when they want it.

_____ 7. Goals are to get the work done at a high level of excellence while enhancing the growth and development of those doing the work.

_____ 8. I'm okay, and you're okay.

_____ 9. I'm okay, and you're not.

_____ 10. I'm not okay, but I'm not sure about you.

A. relaters

B. aggressive

C. directors

D. passive

E. assertive

F. thinkers

G. socializers

ACTIVITIES

1. Determine your primary personality type. You can do this by one of three methods.

    a. Complete a survey, supplied by your teacher, from *Personality Plus* by Florence Littauer of Fleming H. Revell Publishing or from http://72244.netministry.com/images /PersonalityScoreSheet%2Epdf on the Internet.

    b. Complete a survey from the *Team Dimensions Profile* (Carlson Learning Company, now Inscape Publishing, supplied by your teacher) or a similar one found on the Internet.

    c. Study the four personality types summarized in Figure 2–3. Evaluate your own personality type and/or blend, and then evaluate the type or blend of each member of your class. Compare your self-evaluation with the evaluations made by others.

2. Give each personality type a fictitious name that correlates to the name and personality. Write a one-page case study or scenario describing each.

3. Select four people (friends, acquaintances, celebrities, public figures, or historical personages) to match with each personality type. Write a brief description of each person, describing why you believe each of your choices matches the personality type you selected.

4. Study the four personality types described in Figure 2–4. Identify your areas of strength and weakness. Write five strengths that you can build on and five weaknesses of which you need to be aware.

5. Explain how knowing the strengths and weaknesses of each personality type will help you become a better leader.

6. On a piece of paper, make four columns. Label one column "socializer," one "director," one "thinker," and one "relater." Under each column, list the characteristics of each as described in the text. Try to identify which style of communication you feel you practice most. Compare your results with those of a classmate. See whether your classmate agrees or disagrees with you.

7. Study Figure 2–9. Describe where you believe you are on the continuum. Where would you like to be? Name the five biggest disadvantages that you will have to overcome to achieve the most appropriate assertive communication level.

8. Draw a chart, picture, or matrix showing the relationship between personality types, extroversion/introversion, leadership styles, communication styles, learning styles, and career (occupational) groups.

NOTES

1. T. LaHaye, *Spirit-Controlled Temperament* [e-book] (New York: Tyndale House, 2014), pp. 10–11.
2. *Personality Profile* (Minneapolis, MN: Carlson Learning Company, 1992).
3. LaHaye, *Spirit-Controlled Temperament*, pp. 11–12.
4. Ibid., pp. 45–46.
5. D. Lowry, *True Colors: Keys to Successful Teaching Profile* [booklet] (Laguna Beach, CA: Communication Companies International, 1988).
6. Ibid.
7. LaHaye, *Spirit-Controlled Temperament*, pp. 51–53.
8. Lowry, *True Colors*.
9. Ibid.
10. LaHaye, *Spirit-Controlled Temperament*, pp. 56–60.
11. Lowry, *True Colors*.
12. Ibid.
13. LaHaye, *Spirit-Controlled Temperament*, p. 63.
14. Lowry, *True Colors*.
15. Ibid.
16. K. M. Evans and R. Brown, "Reducing Recruitment Risk through Preemployment Testing," *Personnel* 65(9): 55–64 (September 1988).
17. P. Thome, "Psychological Testing in Manager Assessment," *International Management* 44(7):50 (July/August 1989).
18. T. Moore, "Personality Tests Are Back," *Fortune* 115(7):74–82, at 80 (March 30, 1987).
19. Ibid., p. 75.
20. Ibid., p. 74.
21. S. Tremper, "Encouraging Collaboration Between Different Personality Types," TratifyBlog, retrieved from https://medium.com/traitify/ (June 29, 2016).
22. J. Terkelsen, "Team Building, Team Development and Team Effectiveness Using the MBTI," retrieved from http://www.janterkelsen.com/ (July 2, 2016).
23. "Introduction to Personality Tests," *Psychometric Success*, retrieved from http://www.psychometric-success.com/personality-tests/personality-tests-introduction.htm (July 5, 2016).

24. T. Alessandra and P. Hunsaker, *Communicating at Work* (New York: Fireside, 1993), pp. 36, 44.
25. Ibid., p. 46.
26. Ibid., p. 47.
27. Lowry, *True Colors.*
28. Alessandra and Hunsaker, *Communicating at Work*, pp. 38, 44.
29. Ibid., p. 46.
30. Ibid., p. 47.
31. Lowry, *True Colors.*
32. Alessandra and Hunsaker, *Communicating at Work*, pp. 39–41, 44.
33. Ibid., p. 46.
34. Ibid., p. 48.
35. Lowry, *True Colors.*
36. Alessandra and Hunsaker, *Communicating at Work*, pp. 41–43, 44.
37. Ibid., p. 46.
38. Ibid., p. 48.
39. Lowry, *True Colors.*
40. Adapted from Dr. Joe Townsend's leadership class at Texas A&M University, College Station, Texas. Appreciation is also extended to my friend and associate John Grogan of John M. Grogan Company, 49 Chandler Road, Mt. Juliet, TN 37122, for information on passive, assertive, and aggressive communication.

3 LEARNING STYLES AND LEADERSHIP

Once leaders know the learning style of each of their group members, they can help the members reach their full potential. Members who are working at their full potential benefit themselves and the organization to which they belong. Unless leaders recognize different styles of learning, though, they may confuse intelligence with a particular learning style. By understanding learning styles and discovering all group members' learning styles, leaders can increase the knowledge level, effectiveness, and productivity of a team.

Objectives

After completing this chapter, the student should be able to:

- Explain common misconceptions about intelligence
- Discuss the various intelligences of and factors in the cognitive learning domain
- Discuss the various intelligences of and factors in the affective learning domain
- Discuss the various intelligences of and factors in the psychomotor learning domain
- Explain how human beings learn
- List characteristics of the four learning styles

Terms to Know

- intelligence
- self-concept
- cognitive learning
- linguistic intelligence
- logical-mathematical intelligence
- spatial intelligence
- affective learning
- interpersonal intelligence
- intrapersonal intelligence
- mentalist
- existential intelligence
- psychomotor learning
- bodily-kinesthetic intelligence
- musical intelligence
- naturalistic intelligence
- synergy
- mnemonic
- accommodator
- converger
- assimilator
- diverger

As leaders, we need to know how to get the best out of the people we lead. Some people can learn or do a task easily, whereas others of equal intelligence have difficulty with the same task. Why is this? In this chapter, we attempt to answer this question so that you can assign roles and tasks appropriately to people when you lead.

For example, certain people, because of their personality and learning style, are naturally good at starting conversations and getting to know others. However, some interpret such behavior as aggressive, which can lead to miscommunication, hurt feelings, and distrust. Similarly, individuals who naturally display passive behavior may be seen as disinterested or even discriminatory. By understanding personality, behavior, and learning styles, and by recognizing others' natural way of doing things, we can often avoid misreading people.[1] When you know learning styles, you can appreciate the rich differences between people and harness their talents and strengths once you are in a leadership position. Here are a few things that you will be better able to do, in a leadership role, when you understand the various learning styles:

- Match people with job requirements by understanding individual strengths and weaknesses
- Resolve conflicts by realizing that a learning-style problem, rather than a personal one, could be the root of the discord
- Make work flow more smoothly by allowing each person to work according to his or her own style
- Reduce stress levels by remembering that what excites and energizes one person can stress, exhaust, or annoy another
- Meet deadlines better by realizing that different types of people deal with time in different ways[2]

All people are smart, but they have different learning styles and abilities. As a future leader, you should first identify your own learning style. For starters, are you a visual, auditory, or tactile learner? We analyze what these mean later in the chapter, but you can find out now by taking the short quiz at http://www.educationplanner .org/students/self-assessments/learning-styles .shtml.

Knowing your learning style, regardless of which assessment you use, allows you to understand your personal preferences and discover how you are similar to and different from those in your group. Although we may think we cherish difference and diversity, in reality few of us make allowances for them or handle them well. Once you understand learning styles, though, you will tend to appreciate the advantages of the differences and employ them for the benefit of individuals as well as the entire group.[3]

Knowing how people learn and their psychological and mental strengths are keys to unlocking the great potential in every individual. However, unless leaders know the different areas of learning and various styles of learning, they may confuse learning style with intelligence, equating intelligence with only one or two sorts of ability. This chapter discusses three major areas (**domains**) of learning: cognitive, affective, and psychomotor. (Cognitive learning is typically the area emphasized in school.) First, though, let us take a closer look at intelligence.

MISCONCEPTIONS ABOUT INTELLIGENCE

Intelligence Defined

During the 20th century, we have come to associate high intelligence with studious behavior or academic proficiency. However, by definition, **intelligence** is "the ability to respond successfully to new situations and the capacity to learn from one's past experiences." Intelligence varies according to the context, the tasks, and the demands that we confront in life; it cannot be determined only by IQ scores or by rankings on a particular test.[4] Some people are "book smart" and some people are mechanically smart.

Intelligence Tests and Real-World Success

Psychologist Robert Sternberg of Yale University said that the people who make the greatest difference to society are often not the people with the highest intelligence quotients (IQs). Test scores and criteria tend to mislead teachers, he claimed.[5] Although intelligence tests are reasonably good indicators of school success, they fail miserably at predicting how students will fare in the real world. One study of highly successful professionals found that more than a third of them had low IQ scores.[6] Just as tests are not good at measuring workers in the workforce, they also do poorly at measuring practical intelligence, because what is actually being measured could very well be perception and/or cultural conditioning. "IQ tests have been measuring something that might be more properly called **schoolhouse giftedness**," whereas real intelligence encompasses a much broader range of skills: cognitive, affective, and psychomotor.[7]

This does not mean that it is not good to score high on tests and be a good student as measured by test scores. However, poor test scores do not necessarily indicate that a person is unintelligent. In fact, there is not just one type of intelligence, so leaders need to understand, foster, and appropriately use all aspects and kinds of intelligence.[8]

Smart people find a way to succeed. We need to recognize both people who are smart (intelligent) in the conventional sense and people who are smart because they figure out what they are good at and make the most of their abilities. Leaders need to realize that the most practically intelligent people are not always the ones who make high grades on tests, but rather those who have determined their strengths and capitalized on them. Being smart in the real world means making the most of what you have, not conforming to a stereotypical pattern of what others may consider intelligent.[9] For example; we have a friend who has three children. The oldest son has a doctorate degree, and is a college professor. One has her Master's degree and teaches at a high school. The third child did not go to college at all, but he is a mechanical genius. Is he not smart?

Intelligence and Self-Concept

Why do we need to know this? Why is this important? The answer lies in the link between self-concept and the perception of intelligence. Many people are judged to be slow or stupid when, in reality, they are very smart. They may be weak in the cognitive areas that schools use to measure intelligence but exceptionally gifted in the affective and psychomotor areas.

Self-concept is the way you perceive and feel about yourself. A lack of understanding of intelligence can cause serious harm to this concept.

Many people who are, in fact, brilliant have been told, either directly or indirectly, that they were not very smart. Young people in particular tend to accept and internalize these judgments and perceptions, and as a result of their damaged self-concept, they do not believe in themselves and do not achieve their full potential. It is not uncommon for university professors to have to write special letters in support of quality potential students, just so they can get into graduate school, because these students' scores on cognitive admissions tests were not high enough. It is also common for these students eventually to bring more positive attention to, and reflect better on, the university than many of the professors! Students are able to shine when they understand their strengths and abilities, even if their strength is not test taking.

People Possess Many Kinds of Intelligence

Howard Gardner, a Harvard professor, published compelling evidence that each human being possesses many different types of intelligence. Each type appears to be housed in a different part of the brain. Although other intelligence areas may exist, Gardner and his coauthor T. Hatch originally identified seven.[10] Later, they added two more areas, for a grand total of nine different intelligence types, all of which are discussed later in this chapter in relation to the cognitive, affective, and psychomotor domains of learning.

COGNITIVE LEARNING

Cognitive learning is learning based on theoretical symbols (numbers, words) as well as logical reasoning. Those who are successful in the cognitive learning domain are usually thought of as "book smart" or "test smart," and most people make judgments as to whether an individual is intelligent according to performance in this domain (Figure 3–1). Students who are cognitive learners are analytical. Though they are considered bright, they are not always good at coming up with their own ideas. This learning domain relies on symbols, such as words, numbers, and graphics, and sensory stimuli. Cognitive learning also includes the thought patterns and reasoning

FIGURE 3–1 Cognitive learners are book smart and are good at taking tests. These students' focus on studying should lead to good test scores. *(© iStock.com/PeopleImages)*

processes an individual prefers or uses most in problem solving and reaching conclusions. The cognitive learning domain encompasses three of the multiple intelligences: linguistic, logical-mathematical, and spatial.

Most of the standardized tests used in schools measure only these three areas of cognitive learning. Very seldom is learning style analyzed further. In measuring intelligence, most intelligence tests (cognitive tests) use only 4 of the 30 intelligence areas identified by various researchers.[11] Admittedly, these tests do measure something, but it is by no means certain that these scores gauge or even relate to success (intelligence) in life and the workplace. Remember the third child mentioned earlier in the text.

Linguistic Intelligence

Linguistic intelligence concerns verbal abilities and sensitivity to language, meanings, and the relationships among words. Those who possess high amounts of this kind of intelligence tend to be very good at writing, reading, speaking, and debating. These learners have highly developed auditory skills and are generally eloquent and elegant speakers. They think in words rather than pictures. Because most IQ tests draw heavily on linguistic abilities, linguistically intelligent people tend to be considered very smart.[12] On the other hand, some folks do very well in some subjects in school while struggling in linguistic courses like English.

Linguistic job skills include talking, informing, teaching and instructing, writing, verbalizing, speaking foreign languages, interpreting, translating, lecturing, discussing, debating, researching, listening to words, proofreading, editing, word processing, and reporting. Occupations that use linguistic intelligence include librarian, archivist, curator, editor, translator, speech pathologist, writer, radio/television announcer, journalist, legal assistant and lawyer, secretary, typist, proofreader, and English teacher.

Logical-Mathematical Intelligence

Logical-mathematical intelligence refers to an individual's abilities with numbers, patterns, abstract thought, precision, counting, organization, and logical structure. Learners who are strong in this area think conceptually, in logical and numerical patterns, making connections among pieces of information. Always curious about the world around them, these learners ask lots of questions and like to do experiments. Those who are naturally gifted in this area may become scientists, mathematicians, and philosophers.[13] Most people must have—or must develop—at least a small measure of this type of intelligence, just to balance a checkbook successfully or understand the basics of economics. Logical-mathematical job skills include financing, budgeting, doing economic research, estimating, accounting, calculating, using statistics, reasoning, analyzing, classifying, and sequencing. Logical-mathematically intelligent learners would thrive as auditors, accountants, purchasing agents, underwriters, mathematicians, scientists, statisticians, computer analysts, economists, bookkeepers, or computer programmers.

Spatial Intelligence

Those with high **spatial intelligence** have the ability to think in vivid mental pictures, make keen observations, and mentally revise or re-create a given image or situation. They enjoy looking at and using maps, charts, pictures, videos, and movies. People with good spatial intelligence can look at something and instantly identify how a change to one part would alter its appearance. Job skills include visualizing, designing, inventing, imagining, creating visual presentations, illustrating, coloring, drawing, painting, mapping, photographing, decorating, and filming. Highly spatially oriented professions include architecture and drafting;[14] other occupations that use spatial intelligence include engineer, surveyor, urban planner, graphic artist, interior decorator, photographer, art teacher, inventor, pilot, navigator, and sculptor.

AFFECTIVE LEARNING

Affective learning has to do with intelligence in the personality or human relations arena. People who are affective learners have characteristics such as persistence, curiosity, risk taking, and social motivation.[15] Basically, they are "people persons." Affective learners would probably score high on personality tests, but they may not score particularly high on IQ tests. Affective learning involves communication, often through body language; appreciation of beauty; persuasiveness; knowledge of one's own strengths; commitment; and individuality (Figure 3–2). Three types of affective intelligence are interpersonal, intrapersonal, and existential. As you will see, many of the job skills in the affective learning area are not measured or developed in a traditional school setting. However, these skills can become the basis of very successful lives.

Interpersonal Intelligence

Interpersonal intelligence is the ability to relate to, understand, appreciate, and get along with other people. It is natural for people with interpersonal intelligence to see things from others' points of view and to understand how others think and feel.[16] Interpersonal intelligence is not usually measured in the traditional academic setting, even though teachers may recognize and appreciate students who are adept in this area.

People with strong interpersonal intelligence are great organizers, although they sometimes resort to manipulation. Generally, they encourage cooperation and try to maintain peace in group settings. They use both verbal (speaking) and nonverbal (eye contact, body language) means to open and maintain communication channels with others. Interpersonal intelligence also includes sensitivity to others, ability to read the intentions and desires of others, and being considerate. People with highly developed interpersonal intelligence have a natural gift for "reading" another person. For example, they can tell when something is wrong even without linguistic communication.[17] A person extremely gifted in this area could be called a **mentalist**.

Workplace skills include hosting, communicating, trading, tutoring, coaching, counseling, assessing others, motivating, selling, recruiting, publicizing, encouraging, supervising, coordinating, delegating, negotiating, and interviewing. Other strengths include seeing things from other perspectives, listening, using empathy, understanding other people's moods and feelings, cooperating with groups, building trust, resolving conflicts peacefully, and establishing positive relations with other people. Appropriate occupations

FIGURE 3–2 Communicating is an affective skill. This student is practicing her affective skill of public speaking.
(© iStock.com/Steve Debenport)

include administrator, manager, school principal, personnel worker, arbitrator, social worker, counselor, psychologist, nurse, public relations person, politician, salesperson, travel agent, and social director.

Intrapersonal Intelligence

Intrapersonal intelligence involves understanding ourselves, being sensitive to our own values, and knowing who we are and how we fit into the greater scheme of the universe. A person weak in this area might have difficulty making important personal decisions, such as selecting a career. People with strong intrapersonal intelligence enjoy reflection, meditation, and time alone. These learners try to understand their inner feelings, hopes and dreams, relationships with others, and strengths and weaknesses. They

often possess more positive self-concepts than most people, and they do not set their life goals or change their aspirations according to others' ideas or demands.[18]

Job skills include implementing decisions, setting goals, attaining objectives, initiating, evaluating, appraising, planning, organizing, and discerning opportunities. Other skills include recognizing personal strengths, self-reflection and analysis, being aware of inner feelings, envisioning what is possible, evaluating personal thinking patterns, self-reasoning, and understanding personal roles in relationships with others. Occupations to which intrapersonal learners are attracted include psychologist, therapist, counselor, psychology teacher, clergy, theologian, program planner, theorist, philosopher, and entrepreneur.

Existential Intelligence

Existential intelligence relates to the capacity to raise and reflect on philosophical questions about life, death, and ultimate realities. Those with this type of intelligence have the sensitivity and capacity to discuss deep questions concerning human existence, such as the meaning and purpose of life, why we die, and how we got here. An example of this is trying to determine the relationship between religion and science. With existential intelligence, one learns by seeing the "big picture." Learners with high existential intelligence base their leadership on a firm understanding of their purpose and role in a group (family, school, FFA, community, team, etc.). People with this intelligence seek real-world connections to and applications of what they are learning.[19]

PSYCHOMOTOR LEARNING

Psychomotor learning refers to learning that has to do with intelligence in the area of manual dexterity; that is, using one's hands and/or body in performing a mechanical skill. It includes a wide variety of talents or abilities, such as overhauling an engine; doing carpentry, plumbing, or electrical work; and exercising other skills usually associated with a trade. Playing a piano, typing, and using a computer also fall into this category. Gardner identified three types of intelligence in the psychomotor area: bodily-kinesthetic, musical, and naturalistic.

Leaders need to be aware of the people who are talented in this domain. Whereas the cognitive learner may research technical aspects of a project, and the affective learner may sell or promote the product, the psychomotor learner is the one who actually constructs the product (Figure 3–3). Again, remember the third child mentioned earlier who is a mechanical genius.

Bodily-kinesthetic Intelligence

Bodily-kinesthetic intelligence involves the skillful control of one's body, especially when handling objects (e.g., soccer ball, welding rod), and includes the timing and reflexes related to manipulating those objects. These learners express themselves through movement. They have a good sense of balance and eye-hand coordination

FIGURE 3–3 Creating floral arrangements is a necessary psychomotor skill for work in the floriculture industry. This is just one of the many floriculture skills students can perform. *(© iStock.com/skynesher)*

(e.g., ball play, balance beams). They remember and process information by interacting with the space around them.

Although schools are highly enthusiastic about physical education and sports activities, bodily-kinesthetic ability is not often considered a way of being smart. For example, we can consider a baseball pitcher who can throw a 100-mph fastball accurately to be smart, especially if he has figured out how to make millions of dollars playing a fun game. We should recognize and value kinesthetic intelligence, as well as academic (cognitive) intelligence, as an indicator of accomplishment.[20] The student and other volunteers in Figure 3–4 are learning by bodily kinesthetics as they do carpentry work while building a house.

Workplace skills of the bodily-kinesthetic intelligent person include crafting, restoring, shipping,

FIGURE 3–4 "Learning by doing" has always been a basic philosophy of agricultural education. This student's learning by doing is also an act of service: he is assisting Habitat for Humanity in building a house. *(© iStock.com/FangXizNuo)*

Musical intelligence consists of sensitivity to pitch and rhythm, and to the emotional power and complex organization of music. We find musical intelligence in "any individual who has a good ear for music, can sing in tune, keep time to music, and listen to different musical selections with some degree of discernment."[21] Job skills include singing, playing an instrument, recording, conducting, composing, orchestrating, transcribing, arranging, listening, and tuning. Examples of good occupations for musically intelligent persons are performer, composer, conductor, recording engineer, musical-instrument maker, disc jockey, musician, piano tuner, music therapist, instrument salesperson, songwriter, and music teacher.

Naturalistic Intelligence

After he identified the original seven intelligences, Gardner later added **naturalistic intelligence** to his list. This intelligence deals with observing, understanding, and using patterns in the natural environment.[22] A **naturalist** has expertise in the recognition, investigation, and classification of plants and animals. This could be anyone from a molecular biologist to a veterinarian. People with naturalistic intelligence have a great sensitivity to nature and their place within it. They have the ability to nurture and grow things, and they are skilled at caring for, training or taming, and interacting with animals. Examples of famous people with naturalistic intelligence include author and syndicated television personality Jack Hannah and the late "Crocodile Hunter" himself, Steve Irwin.

Naturalistic intelligence is directly related to recognition, knowledge, appreciation, and understanding of the natural world around us. It includes the capacity to interact with the natural world, using skills such as species or breed identification, recognition and classification of various plants, and even sensitivity to weather patterns. You may have naturalistic intelligence if you find yourself drawn to and fascinated by animals and their behavior. You may notice the effect on your mood and sense of well-being when someone brings plants or cut flowers into an otherwise sterile, human-made environment. Those with naturalistic intelligence learn best if they are outside, as this is the environment in which they are most comfortable.[23]

delivering, manufacturing, repairing, assembling, installing, operating, performing, signing, dramatizing, and modeling. Some occupations that use these abilities are dancer, athlete, actor, surgeon, physical therapist, recreational worker, fashion model, farmer, mechanic, carpenter, craftsperson, physical education teacher, factory worker, choreographer, forest ranger, firefighter, and jeweler.

Musical Intelligence

Musical intelligence does not fit exactly into any single one of the three learning domains. However, playing a musical instrument would fall in the psychomotor domain. Appreciation of music could be considered an affective skill. Writing or analyzing music could fit in the cognitive domain.

Workplace skills include collecting data and objects from the natural world, labeling and mounting specimens, organizing collections, observing nature, doing experiments in the field, noticing changes in the environment, categorizing objects, learning about natural phenomena and characteristics of the natural world, drawing or photographing natural objects, implementing wildlife protection projects, caring for animals (both wild and domestic), and studying others' reports and descriptions of nature. Good occupations for people with strengths in this area include park ranger, production agriculturist, field researcher, and other positions with organizations such as the Audubon Society, Sierra Club, Boy and Girl Scouts, National Forest Service, National Park Service, natural history museums, observatories, parks, planetariums, and zoos.

THE WHOLE PERSON

The nine areas of intelligence must be considered together. Only by looking at all areas can you discern someone's learning style or make a judgment about intelligence. A person's intelligence should never be assessed on the basis of a single characteristic; more specifically, no one should be labeled unintelligent because he or she is weak in one particular area. When you broaden your learning style to include cognitive, affective, and psychomotor learning, you become a completely intelligent person. Understanding the different learning styles as a leader allows you to become good at "educating the whole person." Once people have been made aware of all three major domains of intelligence and the subcategories within each, they can identify and tap into their own natural strengths to achieve personal and team success.

Effect of Genetic and Environmental Factors on Intelligence

Over the years, some psychologists have argued that intelligence is genetically determined and exists at birth. Research has shown, for example, that intelligence develops through a series of stages. Other psychologists believe that intelligence is primarily determined by the environment in which an individual is raised. However, it is generally agreed that intelligence is a product of both genetics and environment. Children raised in a learning-rich environment (e.g., a home in which learning and education are encouraged; good schools) are in an ideal setting for the development of intelligence.[24] Both genetic and environmental factors also affect the multiple intelligences. Leaders should consider these fundamentals as they work with their groups.

Building a Team

As a leader building a team, you must be aware of the three major learning domains and the different ways of being smart. Study the group you lead if you want to unleash its full potential. You will build the best team possible if you place people in positions and roles, and in charge of specific duties, wherein they can perform best.

Synergy is the power of a group of people working together. Even more specifically, synergy makes the total group output of a team greater than the output that would be possible if the team members were working individually. Part of the reason synergy works is that when individual abilities or learning areas (cognitive, affective, and psychomotor) are used appropriately, each individual is able to concentrate on the area in which he or she performs best. A person working alone must do many tasks that are not within his or her strong ability or learning area, so productivity is not as great. The key is teamwork. You, as a leader, must assemble that team and assign tasks that play to members' strengths.

WAYS OF LEARNING

In addition to types or areas of intelligence, there are various ways in which people prefer to learn. Do you learn best when you see, hear, or touch the information? This section describes visual, auditory, and kinesthetic learners.

Visual Learners

Visual learners learn by seeing. These learners need to see the speaker's body language, gestures, and facial expressions to understand fully the content of a lecture or lesson. They tend to sit at the front of the room to avoid obstructions in their line of sight (such as heads of people in front of them). They may think in pictures and learn best from visual displays such as diagrams, illustrated textbooks, colorful PowerPoint slides, videos, flip charts, and handouts. During a lecture or classroom discussion, visual learners often take detailed notes as a way of absorbing the information; some may draw pictures or use graphic symbols and depictions to capture information.

If you learn best by seeing, you should read the material to be discussed. When you are reading, remember not only to read the words but also to look at any pictures, charts, maps, or graphs. Take notes about the information by clustering, outlining, or even drawing pictures. Writing information into maps, graphs, or the margin of your book will also help you to remember what you have read.[25]

Visual learners characteristically

- Ask for both verbal and written instructions
- Follow written instructions better than oral ones
- Get lost when they get only verbal directions
- Can understand and follow directions on maps
- Watch speakers' facial expressions and body language
- Take notes for later review
- Remember best by writing things down several times or drawing pictures and diagrams
- Are good spellers
- Prefer information to be presented visually (e.g., flip charts or slides)
- Are skillful at making graphs, charts, and other diagrammatic displays
- Remember best by picturing something in their heads
- Are good at solving jigsaw puzzles
- Are good at the visual arts

Auditory Learners

Auditory learners learn by listening. They learn best through verbal lectures, discussion, talking things out, and listening to what others have to say. Auditory learners comprehend the underlying meanings of speech by listening to and interpreting tone of voice, pitch, speed, and other nuances. Written information may convey little meaning until they hear it; these learners often benefit from reading text aloud and using a tape recorder.

If you learn best by hearing, you should listen carefully. Ask permission to record presentations. Listen to the recording several times if the material is new or difficult, and keep the recording for later review. Another study technique is to discuss the material with another person. If no one is available, turn on the recording device and verbalize what you are thinking about the recorded material.[26]

Auditory learners usually

- Follow oral directions better than written ones
- Would rather listen to a lecture than read the material in a textbook
- Prefer to listen to a daily podcast rather than read a newspaper
- Understand better when they read aloud
- Struggle to keep notebooks neat
- Frequently sing, hum, or whistle to themselves
- Talk to themselves
- Dislike reading from a computer screen
- Require explanation of diagrams, graphs, or maps
- Enjoy talking to others
- Use musical jingles to learn or remember things
- Would rather listen to music than view a piece of art
- Like to tell jokes and stories or make verbal analogies to illustrate a point

Tactile/Kinesthetic Learners

Tactile/kinesthetic learners learn by moving, doing, and touching. Tactile/kinesthetic persons learn best through hands-on methods, actively exploring the physical world in which they live. They may find it hard to sit still for long periods and may become distracted by their need for activity and exploration.[27] Agricultural education is a great place for this type of learner because of the many opportunities for learning on the school farm or in the agriscience lab. However, there are some learning tips that might help such learners in the classroom as well.

If you learn best by actually doing something, you should write notes about what you are to learn. As you read, underline the important ideas after you have read a section, or write notes in the margin. It may also help you to make flash cards or maps, or to create charts, crossword puzzles, word bingo, picture puzzles, or mnemonic devices. Another effective strategy is to make a practice test about the material to be learned. Making a sample test will help you pinpoint important areas that you need to study in more depth. You will learn best by using as many activities as possible.

Tactile/kinesthetic learners characteristically

- Reach out to touch things
- Collect things
- Talk fast, using their hands and body language to communicate what they want to say
- Fidget constantly (e.g., tapping pen, playing with keys or change in pocket, swinging a foot)
- Are good at sports
- Take things apart and put them back together
- Prefer to stand while working
- Like to have music in the background while working
- Enjoy working with their hands and making things
- Like to chew gum or eat in class
- Learn through movement and hands-on exploration of the world around them
- May be considered hyperactive
- Are good at finding their way around
- Prefer to do things rather than watch demonstrations or read about it in a book

FORMATION OF LEARNING STYLES

Our capacity to learn new things is important, and, as we have just read, we do not all learn things in exactly the same way. All of this information may make you wonder how these differences in learning arise. There are two major differences in how we learn: how we **perceive** and how we **process** information and experiences.

Perceiving

The way you perceive reality—whether you favor sensing and feeling (concrete) or thinking things through (abstract)—is one of the two major determinants of your learning style (Figure 3–5).

Those who perceive in a **sensing** or **feeling** way make meaning from the experience and

LEARNERS PERCEIVE INFORMATION

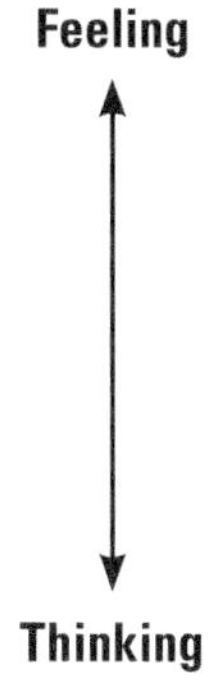

Feeling

↕

Thinking

FIGURE 3–5 All people hover somewhere on a continuum from feeling to thinking. Both feelers and thinkers have their own strengths and weaknesses.

the information. They learn through empathy. People who sense and feel pay more attention to the actual experience itself. They do not look for hidden meanings or try to discern relationships between ideas and concepts. People who perceive in this manner place value on relating to people, being involved in real situations, and taking an open-minded approach to life.

Those who **think** through an experience prefer to deal with the concept of reality rather than reality itself. They initially analyze what is happening, using rational intellect to make the first appraisal. They evaluate their experience through the lens of reason and logic. Thinkers look beyond what they cannot actually see. People with this orientation value precision, the rigor and discipline of analyzing ideas, and the quality of a neat and organized system.

Processing Experience and Information

In processing experience and information, some of us are **observers** and others are **doers** (Figure 3–6).

Observers reflect on new things. They create meaningful connections by filtering novelties through their own experience. People reflect on and observe experiences from many perspectives. Observers use a sequential method of thought that organizes information in a linear, programmatic manner. Their rule is "follow the steps."[28] People with this orientation value patience, impartiality, and thoughtful judgment.

LEARNERS PROCESS INFORMATION

Doing ←——————→ **Observing**
Some people   Other people
jump right in   observe what
and try it.   is happening and
                    reflect on it.

FIGURE 3–6 With regard to processing information, some are observers and some are doers. Both ways of processing information and experiences are equally valuable.

Doers act out new information immediately and think about it or reflect on it only after they have tried it. To integrate it, they need to physically and actively *do* it.[29] Doers work with unorganized information in no particular sequence, sometimes even skipping steps in a procedure, but still produce the desired result. Doers may seem impulsive or spontaneous. They are concerned with what works as opposed to what is absolute truth. Their key phrase is "just get it done."[30]

Merging Perceiving and Processing

Both kinds of perceiving (feeling and thinking) are equally valuable, as are both kinds of processing (observing and doing).[31] We must learn to feel and to think, appreciating both modes of perceiving. We also must learn to observe and to do. Once you merge perceiving and processing, the way you learn begins to take shape, as shown in Figure 3–7.

MERGING PERCEIVED AND PROCESSED INFORMATION

Concrete Experience
You sense or *feel* a real experience.

Active Experimentation
You try it.
"Just *do* it."

Reflective Observation
You reflect on what you *observe*.

Abstract Conceptualization
You *think* about a concept.

FIGURE 3–7 Once feeling, thinking, observing, and doing are merged, the basis for the four learning styles is established.

LEARNING STYLES

When perceiving and processing are combined, four distinct learning styles result. Remember, no individual learns using only one style. Nevertheless, each of us has a dominant style or styles that combine for a unique blend of strengths and abilities.[32] The four learning styles that emerge from these patterns are

- **Accommodator** (person who learns by doing and feeling)
- **Converger** (person who learns by doing and thinking)
- **Assimilator** (person who learns by observing and thinking)
- **Diverger** (person who learns by observing and feeling)[33]

By learning some of the common characteristics of each of the four learning styles (Figure 3–8), we can recognize and value what we most like to do and what comes naturally. We can also identify our weaknesses. As leaders, we must first recognize our own natural learning style(s). As we begin to understand the various ways in which people perceive and process

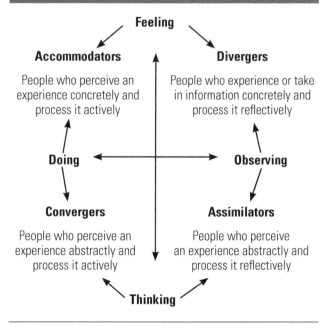

MERGING PERCEIVED AND PROCESSED INFORMATION

Feeling

Accommodators
People who perceive an experience concretely and process it actively

Divergers
People who experience or take in information concretely and process it reflectively

Doing ←——————→ **Observing**

Convergers
People who perceive an experience abstractly and process it actively

Assimilators
People who perceive an experience abstractly and process it reflectively

Thinking

FIGURE 3–8 The combination of how we perceive (feeling and thinking) and how we process (observing and doing) creates the four unique learning styles: accommodator, converger, assimilator, and diverger.

information, we can better understand what comes naturally to us and to members of our group. We can then identify the differences between leaders and group members that may cause frustration and misunderstandings.

The following discussion of these four learning styles characterizes each as learners and leaders and highlights the advantages and disadvantages of each style.

Accommodator

Accommodators prefer learning by doing and feeling. They tend to learn primarily from personal, hands-on experience and may act on gut feelings rather than logical analysis. They also learn by trial and error. They are believers in self-discovery and are enthusiastic about new things. They prefer to learn only what is necessary to get the job done. Accommodators believe that the strength of their competence lies in getting things done, carrying out plans and tasks, and seeking new experiences. Opportunity seeking, risk taking, and action characterize this style. This style is called **accommodation** because it is best suited to situations in which one must adapt to quickly changing circumstances.

When in situations in which plans do not fit the facts, those with an accommodative style will most likely discard the plan or theory. Accommodators often reach accurate conclusions in the absence of logical justification. They act on the spur of the moment. When making decisions, they tend to rely more heavily on other people for information than on their own technical analysis. They enhance reality by taking what exists and adding something of themselves to it. Accommodators are generally at ease with people, but may be viewed as impatient and aggressive. People with this learning style are found in action-oriented jobs, such as marketing, sales, public relations, management, politics, social professions, and entertainment. It is not unusual for accommodators to have several careers in a lifetime or even two careers at once.

Accommodators tend to be democratic leaders. They are not deterred by the word *impossible* if they have decided that the goal is a worthy one. However, they may not attempt even an obviously achievable goal if they have decided it

is just not worth the trouble. As leaders, accommodators thrive on crisis and challenge; they lead by energizing and presenting a vision of what could be. They work hard to establish their organizations as pacesetters, but they need staff that can follow up with details and implement plans.

Accommodators are usually good leaders. They are willing to take necessary risks, and they get things done. They inspire others to action, contribute unusual and creative ideas, visualize the future, accept many types of people, and think fast on their feet.

Accommodators do not always set clear goals or develop practical plans. They often waste time on unimportant activities. It is hard for accommodators to work under restrictions and limits, do or redo routine work and reports, keep detailed records, show how they got an answer, or accept that they have no options. They become stressed by constraints and limitations (especially if the restrictions appear excessive), and are often distressed if they are not appreciated as unique or are not given credit for knowing the right thing to do.[34]

Accommodators' natural abilities tend to be

- Experimenting to find answers
- Searching out a variety of options
- Creating or discovering new approaches and processes
- Finding possibilities
- Creating change
- Being independent
- Considering solutions
- Taking calculated risks
- Exercising a high degree of curiosity

Accommodators work best when they

- Can try new approaches and solve problems
- Are self-directed
- Are competitive
- Create their own answers
- Use trial-and-error approaches
- Do brainstorming
- Produce imaginative products
- Have hands-on experiences

What do accommodators do best?

- Inspire others to take action
- See many options, solutions, and possibilities

- Contribute unusual and creative ideas
- Visualize the future
- Accept many different types of people
- Think fast on their feet
- Take risks

What makes sense to accommodators?

- Using insight and instinct to solve problems
- Working within general time frames
- Using real-world experiences to learn
- Trying things for themselves

What is hard for accommodators?

- Abiding by restrictions and limitations
- Formal reports
- Routines
- Redoing anything
- Keeping detailed records
- Showing how they got an answer
- Choosing only one answer
- Having no options

While learning, accommodators tend to ask "How much of this is really necessary?"[35]

Converger

Convergers learn by doing and thinking. They seek practical uses for information and always attempt to integrate theory and practice. They learn by testing theories and applying common sense. The greatest strength of this learning style lies in problem solving, decision making, and the practical application of ideas: if it works, use it. Convergers do not stand on ceremony, but instead get right to the point. They prefer always to do things the same way. This learning style is called **converging** because a person with this style seems to do best in situations such as conventional intelligence tests, wherein there is a single correct answer or solution to a question or problem.

Convergent persons control their expressions of emotion, sometimes seeming abrupt, bossy, or impersonal. They prefer dealing with technical tasks and problems to handling social and interpersonal issues. Convergers often specialize in the physical sciences. This learning style is characteristic of many engineers, technical specialists, surgeons, production supervisors, computer workers, and managers.

| VIGNETTE 3-1 | AUGUSTUS ACCOMMODATOR |

AUGUSTUS, better known as "Gus," was a very conscientious ninth grader who seemed to care a lot about his community and the things that were going on in it. Early in the year, he participated in some service learning activities with his agriscience class. One of the most memorable things they did was to serve a holiday supper in a local soup kitchen for the homeless. When Gus went back to school after the break, he really surprised his teachers when he told them about his additional soup kitchen experiences.

Gus was usually rather quiet, but he became very excited as he explained that he had been helping out there every evening since the class did the group project. He told story after story of the people he had met and the things he had learned about them. Gus begged his classmates to help out in the evenings, too. Gus's teacher knew that Gus was an accommodator, so she assisted Gus with recruitment of interested individuals. Mrs. Moore knew that Gus had learned a great deal from his experiences, and she helped him organize a nightly schedule of students to serve at the soup kitchen.

As Gus grew older, he learned to think things through a little more, and he discovered the importance of planning; but he still loves just to jump in and help when an opportunity presents itself.

Why do you think Gus was so excited about helping out in the soup kitchen? Why did Mrs. Moore feel that she needed to help Gus with organizing additional volunteers?

As leaders, convergers thrive on plans and timelines, and they are production oriented. They lead by force of personality, inspiring quality even though they tend to be authoritarian. Convergers work hard to make the organization productive and stable. They need employees or group members who are task oriented and will not hesitate to take immediate action.[36] They lead best with organization, routines, predictability, tangible rewards, and schedules.

Convergers are usually very good at thinking critically, solving problems, and making decisions. They fine-tune ideas to make them more efficient and economical. They produce concrete products from abstract ideas and work well within time limits.

On the downside, convergers tend to make hasty decisions without reviewing all the possible alternatives. They often implement ideas without testing them first. It is hard for convergers to work in groups, take part in discussions that have no obvious point, work in a disorganized environment, work with abstract ideas, or attempt to answer questions that have no clear right or wrong answers. They become stressed by not knowing expectations, by vague or general directions, and by not seeing examples.[37]

Convergers' natural abilities tend to be

- Carrying out tasks in a step-by-step manner
- Planning or organizing their time
- Following directions
- Getting correct answers
- Looking for usable results
- Creating practical products
- Working in a structured environment
- Using performance standards
- Applying facts and information
- Focusing on details and specific results

Convergers work best when

- There is an orderly, quiet environment
- They are given guidelines
- They have exact directions and examples
- They can be consistent and efficient
- They can predict situations
- They have approval for specific work
- They can apply practical, active learning
- They can trust others to follow through

What do convergers do best?

- Apply ideas in a practical way
- Organize
- Fine-tune ideas to make them more efficient
- Produce concrete products from abstract ideas
- Work within time limits[38]

What makes sense to convergers?

- Working systematically, step by step
- Paying attention to details
- Following schedules
- Making literal interpretations
- Knowing what is expected
- Adhering to routines, habits, and established ways of doing things[39]

What is hard for convergers?

- Working in groups
- Discussions that do not seem to have a specific purpose
- Working in a disorganized environment
- Incomplete or unclear directions
- Working with unpredictable people
- Dealing with abstract ideas
- Using imagination
- Questions with no right or wrong answers[40]

While learning, convergers tend to ask these sorts of questions: "What are the facts I need?" "How do I do it?" "What should the result look like?" "When is it due?"[41]

Assimilator

Assimilators prefer to learn by observing and thinking. The assimilator learning style is less focused on people and more concerned with ideas and abstract concepts. Assimilators are effective at consolidating or synthesizing information and putting it into concise, logical form. They seek continuity and want to know what the experts think. They need details, are thorough, and prize intellectual competence and personal effectiveness. It is more important to them that an idea or theory have logical soundness than practical value.[42] This learning style is more characteristic of individuals in the basic sciences and mathematics than of those in the applied sciences. In organizations, assimilators generally are found in the research and planning departments. They often seek careers as mathematicians, scientists, researchers, and planners.

As leaders, assimilators thrive on gathering facts and turning them into theories. They attack problems with critical thinking, reason, and logic. Assimilators lead by principles, procedures, and facts. They need well-organized followers who will accurately and conscientiously write things down and follow through on agreed decisions.[43] Assimilators lead best with organization, logical outcomes, plenty of time to work, credible sources of information, opportunities for analysis, and appreciation of their input.

They are skilled at creating models and developing theories and plans. Assimilators work through an issue thoroughly, use facts to prove

CORRIE WAS JUST A JUNIOR AT HIGHLAND HIGH SCHOOL when she was hired by a local businessman, Mr. Foutch, to keep all his computers in working order. Corrie had been working with computers most of her life, and she knew that things did not always go as planned when working with technology. She found out about this job from a friend, and she has been solving Mr. Foutch's computer problems for more than two years now.

Corrie is good at what she does because she learns by doing. Because she has a converger learning style, she is always figuring out how a program or piece of hardware should be fixed by diving in and "doing" it. Corrie shows up to work at the same time every day, does the same routine checks, and fixes any problems that she finds. Mr. Foutch has been so impressed with her work that he wants to hire some students from another local high school to work for Corrie, which would free up time for Corrie to assume more responsibility with his business partners. Corrie is not as good with people as she is with computers, and she is dubious about working with other students. She finally decides to take Mr. Foutch up on his offer, but she makes it clear that the people who would work for her will need to work hard, stick to her schedule, and do things right the first time.

Do you think Corrie is setting the ground rules too early? She is obviously a converger, but what about the people who work for her? What if their learning styles are not the same as hers? What could she do, and what adjustments might she have to make, to guide her group to perform effective, efficient work?

or disprove theories, and analyze the means to achieve goals.

They tend to be too idealistic, however, and not practical enough. It is hard for them to rush a decision, deal with rules and regulations, express emotions, be diplomatic when convincing someone else of their point of view, and not monopolize a conversation about a subject that interests them.[44] They become stressed by being rushed through anything, not having hard questions answered, abiding by sentimental decisions, or being asked to express emotions or feelings.

Assimilators' natural abilities tend to be

- Debating specific points
- Organizing ideas in a logical way
- Gathering information and analyzing ideas
- Thinking in a structured way

- Being patient learners
- Judging value or importance
- Examining key points and forming theories
- Conducting research

Assimilators work best when they

- Have references and expert sources
- Rely on notes and written material
- Are sure of themselves
- Follow traditional procedures
- Have time to learn thoroughly
- Can work alone
- Are respected for their intellectual ability
- Write analytically

What do assimilators do best?

- Gather information (preferably complete) before making a decision
- Analyze ideas
- Research
- Establish logical sequence
- Use facts to prove or disprove theories
- Figure out what ought to be done[45]

What makes sense to assimilators?

- Using exact, well-researched information
- Learning more by watching than by doing
- Using logical reasoning
- A teacher who is an expert in the subject
- Abstract ideas
- Taking time to work through an issue thoroughly[46]

What is hard for assimilators?

- Being forced to work with those of differing views
- Having too little time to deal with a subject thoroughly
- Frequently repeating the same tasks
- Dealing with many specific rules and regulations
- Engaging in emotion-based, "sentimental" thinking
- Expressing emotion
- Being diplomatic when convincing others
- Not monopolizing conversations[47]

While learning, assimilators tend to ask "How do I know this is true?" "Are there any possibilities we have not considered?" "What do we need to accomplish this?"[48]

Diverger

As learners, divergers prefer to receive information by observing and feeling. This style is called **diverging** because people of this type perform better in situations that call for generating alternative ideas and implications, such as brainstorming sessions. People with a strong diverger style are typically interested in other people, tend to be creative and oriented to feelings, and often specialize in the liberal arts and humanities. They may seek careers in counseling, teaching, the service sector, design, social work, nursing, consulting, and personnel management.

As leaders, divergers thrive on developing good ideas. They handle problems by first reflecting alone and then seeking the advice of others. They exercise authority with trust and participation. They work to keep the organization unified, and they need staff that supports their goals. They lead best with frequent, honest praise; reassurance of worth; opportunities to work collaboratively; opportunities to use creativity and imagination; and acceptance of personal feelings and emotions.

Divergers tend to be imaginative and able to recognize problems. They listen sincerely to others, understand feelings and emotions, focus on themes and ideas, bring harmony to group situations, have good rapport with almost anybody, and recognize the emotional needs of others.

They tend to overanalyze problems, however, and be slow to act. Divergers often miss opportunities. It is hard for them to explain or justify emotions, be competitive, work with unfriendly people, give exact details, accept criticism (even if positive), and focus on only one thing at a time.[49] They become stressed when they do not feel liked or appreciated.

Divergers' natural abilities tend to be

- Reflecting upon feelings
- Being flexible and adaptable
- Relating to others
- Having sensitivity
- Appreciating the arts, beauty, and nature
- Personalizing information
- Using imagination to create
- Interpreting feelings and ideas
- Seeing a holistic view

VIGNETTE 3-3 ANDREW ASSIMILATOR

ANDREW, a recent graduate of a local university, has been a huge professional basketball fan for most of his life. One of his major goals was to be associated with professional basketball in some way, but he did not really have the physical attributes or the kinesthetic intelligence to be a player or the interpersonal intelligence to be a coach. Thus, he decided that he would design basketball shoes for a living. Andrew secured the resources of some local investors and began to research ways of making the best basketball shoes on the market.

Andrew learned as much as he could about the physics, biology, and chemistry of the athlete's movements, the shoes, and the environment in which a shoe and a particular athlete operated. This fact-gathering mission yielded some interesting findings that led him to develop new ideas for basketball shoes.

The problem was that Andrew only had ideas. He needed some help in making the concepts become a reality. He was aware of his assimilator learning style, so he enlisted the help of an individual who was a little more application oriented (meaning she could make the ideas come to life). He hired Jennifer and some associates, who created a couple of prototypes based on the models Andrew had developed, and began to market them to different companies.

Andrew Assimilator became a very successful shoemaker for basketball players, because he could observe and think well enough to generate the ideas and found the right people to work with to fulfill his dreams.

Divergers work best when they

- Can work and share with others
- Have assignments that involve interpretation
- Get personal attention and emotional support
- Are engaged in social activities to balance work
- Have freedom from control by others
- Have a noncompetitive atmosphere
- Have open communication with others
- Use personal, individual, or artistic expression

What do divergers do best?

- Listen to others
- Understand feelings and emotions
- Focus on themes and ideas

- Bring harmony to group situations
- Establish positive relationships with others
- Recognize and meet other people's emotional needs[50]

What makes sense to divergers?

- Personalized learning
- Broad, general guidelines
- Maintaining friendly relationships
- Participating in projects they believe in
- Emphasizing high morale
- Making decisions with the heart instead of the head[51]

What is hard for divergers?

- Having to explain or justify feelings
- Competing
- Working with dictatorial/authoritarian personalities
- Working in a restrictive environment
- Working with people who do not seem friendly
- Concentrating on one thing at a time
- Giving exact details
- Accepting criticism[52]

While learning, divergers tend to ask "What does this have to do with me?" "How can I make a difference?"[53]

A Summary of Learning Styles

The ability to learn is the most important skill we can acquire.[54] When faced with new experiences or learning situations, we may need to change our learning style from getting involved (feelings), to listening (observing), to creating an idea (thinking), or to making a decision (doing). However, because we each have preferred learning styles, we tend to be stronger in some modes and weaker in others.[55] Figure 3–9 summarizes the important characteristics of each learning style.

Learning Styles and Personality Types

In Chapter 2, we compared personality types and leadership styles. We can also find some similarities between learning styles and personality types. Accommodators, being people-oriented risk takers and operating more on feelings than theory, tend to be sanguine personality types.

VIGNETTE 3-4 — DAWN DIVERGER

DAWN, a high school senior, was considered a leader by many of her peers and teachers at Linwood High. She had many friends because of her sensitive, creative, and all-around people-oriented disposition. Because Dawn was a senior, and well known in the school, she decided to run for class president, a job that would entail being in charge of her class's high school reunions in the coming years. Dawn was elected senior class president, and with that post she also assumed the responsibility of organizing the upcoming class reunions.

For the next 10 years, Dawn learned by observing how the classes that graduated before her handled their reunions. She watched as other reunion coordinators succeeded and failed with different ways of doing things. She talked with several of the reunion coordinators who preceded her to get their feelings on how to do a good job. She also asked members of her own class what they expected from a 10-year reunion.

Although Dawn occasionally encountered unfriendly people, and had to deal with her own issue of trying to do too many things at once, the first reunion that she planned turned out to be a huge success. More than half her class showed up for the reunion, and everyone she talked to really seemed to appreciate all the work she had done. Dawn and her staff were very proud of how things turned out, and Dawn knew that it was a success because of the teamwork and excellent attitude of her helpers.

Dawn is a classic example of a diverger. She learned by observing and feeling. She was slow to act, but her creativity, imagination, and planning paid off!

Like choleric personality types, convergers seek a single correct answer or solution to a question or problem. Assimilators are similar to melancholy personality types in that they create models, analyze things, and develop plans. Divergers are similar to phlegmatic personality types, as both are often slow to act and therefore miss opportunities.

As with personality types, there is no "best" learning style; each has its own advantages and disadvantages. You have probably realized that you have one preferred learning style, even if you have not specifically identified it, but you also have characteristics of other styles and use them depending on the situation. Figure 3–9 also correlates learning style with personality type.[56]

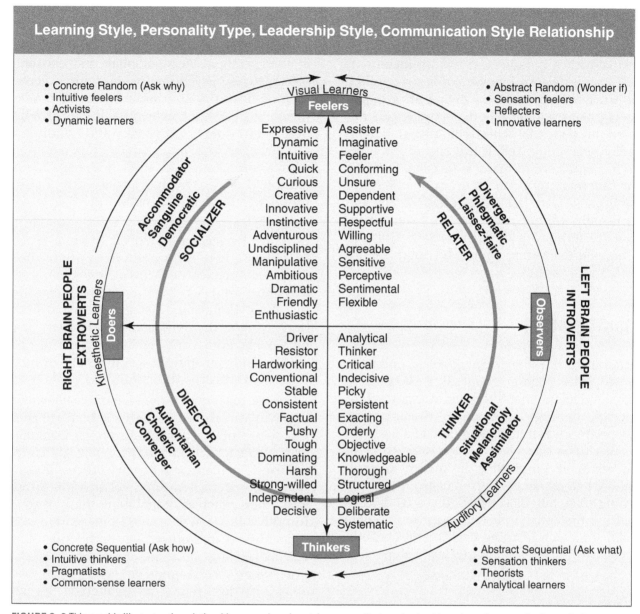

FIGURE 3–9 This graphic illustrates the relationship among learning style, personality type, leadership style, and communication style. In most people, one or two of the styles or types tend to predominate.

Leadership and Learning Styles

Leadership styles relate to learning styles as well as to personality types. Accommodators tend to have a democratic style of leadership. Democratic leaders are people oriented, as are accommodators; generally, they view the interests of the group as more important than the task. Convergers are task oriented and tend to be authoritarian; to them, the task is much more important than people's feelings. Assimilators and divergers tend to be situational leaders with no overwhelmingly democratic or autocratic

tendencies. However, as shown by their placement on the perception and processing learning chart (refer to Figure 3–9), divergers slightly favor laissez-faire leadership and assimilators slightly favor situational leadership.

Leadership, Learning Styles, and Decision Making

People with similar learning styles tend to behave in the same way, and people with different learning styles tend to act differently. Team productivity and harmony depend on how people are grouped as they work together. When working with a person who has a compatible learning style, you tend to get along well. Likewise, when working with a person who has an incompatible learning style, you may encounter a conflict. Conflicts can be prevented if people are grouped correctly. For example, if both parties are assimilators, they generally tend to be theoretical and therefore get along well. If two convergers work together and have to make a decision, but they have different opinions, typically conflict will arise because both believe they are correct, and they usually will not negotiate. However, a converger (bossy) and a diverger (happy with others making decisions) should be able to work together successfully.

CONCLUSION

In examining the four learning styles, one finds that people have characteristics that lead them to behave and learn differently. This is important for leadership because followers will respond more appropriately when leaders understand and provide situations that allow for followers to be at their best in accomplishing the goals of an organization. Remember, no person displays only one learning style, but most people have tendencies toward or preferences for one. Once you become familiar with your learning style and those of others, you can assume a successful leadership role.

SUMMARY

Once leaders know the learning styles of their group members, they can help the members reach their full potential, thereby benefiting both the members and their organization. Some people can learn and do certain things quickly and easily, whereas others of equal intelligence have trouble with the same task.

Understanding learning styles and psychological strengths is critical to unlocking the great potential in every individual. Unless leaders recognize different styles of learning, they will confuse learning style with intelligence. The three major domains of learning are cognitive, affective, and psychomotor. Cognitive learning, usually thought of as "book smarts," includes linguistic, logical-mathematical, and spatial intelligences. Affective learners have abilities in the personality or human relations arena, displaying strength in interpersonal, intrapersonal, and existential intelligences. Psychomotor learning refers to manual dexterity and performance of physical, mechanical tasks. People gifted in this area have bodily-kinesthetic, musical, and naturalistic intelligences. The totally educated person is one who is competent in all three domains. A leader can build a team by appropriate use of all the group members' intelligences.

People learn in three major ways. Visual learners learn by seeing; they need to see a speaker to understand fully the content of what is being said. Auditory learners learn by listening; they learn best from verbal lectures, discussions, talking about things, and listening to what others have to say. Tactile/kinesthetic learners learn by moving, doing, and touching; they learn best with a hands-on approach, actively exploring the physical world around them.

Genetic and environmental factors affect learning style. We cannot do anything about our genetics, so we must concentrate on environmental learning factors. There are two major differences in how we perceive and how we process information. Some people sense and feel their way through a situation, whereas others think things through. In processing experience and information, some of us observe and others act. Both kinds of perceiving (feeling and thinking) and both kinds of processing (observing and doing) are equally valuable. Once these four areas are merged, our intelligence area or learning style begins to take shape.

There are four major learning styles: accommodator, converger, assimilator, and diverger. Accommodators prefer to learn by doing and feeling; they are creative, innovative, instinctive, and adventurous. Convergers prefer to learn by doing and thinking; they are hardworking, dependable, dominant, and somewhat bossy. Assimilators prefer to learn by observing and thinking; they are analytical, thorough, structured, and systematic. Divergers learn by observing and feeling; they are sensitive, compassionate, sentimental, and flexible.

There are similarities between learning styles and personality types. Leadership styles and learning styles are also similar. By knowing all group members' learning styles, leaders can increase the productivity of the group.

Take It to the Net

1. Explore learning styles on the Internet. Go to your favorite search engine and enter "learning styles test" into the search field. Browse some of the websites that offer learning style tests. Take as many of the tests as you want. Print out the results of two tests and explain whether you agree with the results, and why or why not.

2. Go to the following website: literacynet.org/mi/. Click on *Assessment* and then *Find your strengths!* Here you will find a multiple intelligences test. After you complete the test, you will see a wheel describing your rating in eight of the nine intelligences discussed in this chapter. Write a one-page paper or develop a skit describing the areas in which you are most intelligent.

Chapter Exercises

REVIEW QUESTIONS

1. Define the Terms to Know.

2. List five ways in which understanding the various learning styles can help leaders.

3. How do smart people find ways to succeed?

4. What three types of intelligence are grouped in the cognitive learning domain?

5. List 15 jobs or occupations that would suit people with strong linguistic intelligence.

6. List 11 jobs or occupations for people with strong logical-mathematical intelligence.

7. List 12 jobs or occupations for people with good spatial intelligence.

8. What three types of intelligence are grouped in the affective learning domain?

9. List 14 jobs or occupations for people with good interpersonal intelligence.

10. List eight jobs or occupations for people with strong intrapersonal intelligence.

11. What three types of intelligence are grouped in the psychomotor learning domain?

12. Why is it hard to place musical ability within a single one of the three domains of learning?

13. List and briefly discuss the three major ways in which people learn.

14. What is meant by reaching the "whole person"?

15. What are the two kinds of perceiving and the two kinds of processing?

16. List seven advantages and nine disadvantages of accommodators.

17. List four advantages and eight disadvantages of convergers.

18. List five advantages and seven disadvantages of assimilators.

19. List eight advantages and ten disadvantages of divergers.

20. Why do leaders need to be aware of learning styles?

21. What are nine leadership traits of accommodators?

22. What are nine leadership traits of convergers?

23. What are 10 leadership traits of assimilators?

24. What are six leadership traits of divergers?

COMPLETION

1. Only the _____ learning domain is typically recognized in schools and universities.

2. _____ is the ability to respond successfully to new situations and the capacity to learn from past experiences.

3. Your _____ can be damaged by a misconception or lack of understanding of intelligence.

4. Although intelligence tests consistently predict school success, they _____ to indicate how students will do after they graduate.

5. The _____ learning domain deals with attitudes and human relations skills; the _____ learning domain deals with manual dexterity skills.

6. _____ is what happens when a group of people's total output as a team, working together, is greater than their output would be if they were working individually.

7. It is generally agreed that intelligence is a consequence of both _____ and the _____.

8. There are two major differences in how we learn: how we _____ and how we _____.

9. Some people sense and feel their way through a situation, whereas others _____.

10. In processing experience and information, some of us are observers and some of us are _____.

MATCHING

Terms may be used more than once.

_____ 1. Phlegmatic personality tendencies.

_____ 2. Sanguine personality tendencies.

_____ 3. Melancholy personality tendencies.

_____ 4. Choleric personality tendencies.

_____ 5. Democratic leadership tendencies.

_____ 6. Authoritarian leadership tendencies.

_____ 7. Likely to have a job in social work.

_____ 8. Likely to have a job in mathematics.

_____ 9. Likely to have a job in production supervision or management.

_____ 10. Likely to work in marketing or entertainment.

A. diverger

B. assimilator

C. converger

D. accommodator

ACTIVITIES

1. Name one person in your class who exhibits a cognitive learning style, one who has an affective learning style, and one with a psychomotor learning style. Justify your choices in writing. (You may complete this assignment in pairs and share with each other.)

2. Select four people in your class who show the four learning styles: accommodator, converger, assimilator, and diverger. Justify your choices in writing. (You may complete this assignment in pairs and share with each other.)

3. Write a one-page essay on how people can be smart yet unintelligent at the same time.

4. Study the sections on cognitive, affective, and psychomotor learning. Determine which of the three learning styles most closely describes you. Write a short essay and read it to the class to see whether class members agree with your self-assessment.

5. Within the cognitive, affective, and psychomotor learning areas are nine areas of intelligence: linguistic, logical-mathematical, spatial, interpersonal, intrapersonal, existential, bodily-kinesthetic, musical, and naturalistic. Rank each of these according to your perception of your own intelligence areas or strengths. Share your self-ranking with class members and see whether they agree.

6. Write four one-page case studies of four people you know, other than classmates, who each exhibit one of the learning styles: accommodator, converger, assimilator, and diverger.

7. Go to the Ways of Learning section in this chapter. Put a check mark by each bullet point in the lists of items that describe visual, auditory, and tactile/kinesthetic learners that you feel applies to or describes you. Of the three ways you learn, the one with the most check marks is probably the way you learn best.

NOTES

1. O. Kroeger and J. M. Thueson, *Type Talk at Work* (New York: Delacorte Press, 1992), p. 9.
2. Ibid., p. 11.
3. Ibid., pp. 11–12.
4. T. Armstrong, *Seven Kinds of Smart* (New York: Plume, 1999), p. 8.
5. R. J. Sternberg, quoted in *Learning Styles: Putting Research and Common Sense into Practice* (Arlington, VA: American Association of School Administrators, 1991), p. 22.
6. Armstrong, *Seven Kinds of Smart*, p. 8.
7. Ibid.
8. *Learning Styles*, p. 22.
9. Ibid.
10. H. Gardner and T. Hatch, quoted in *Learning Styles*, p. 23; H. Gardner, *Frames of Mind: The Theory of Multiple Intelligence* (New York: Basic Books, 1993).
11. J. Thorpe, *Learning Style Inventory Manual* (Vinemont, AL: Route 1, Box 2685, 1978).
12. C. U. Tobias, *The Way They Learn* (Wheaton, IL: Tyndale House, 1994), p. 132.
13. Ibid.
14. Tobias, *The Way They Learn*, p. 133.
15. *Learning Styles*, p. 16.
16. Tobias, *The Way They Learn*, p. 135.
17. Ibid., pp. 135–136.
18. Ibid., p. 136.
19. W. McKenzie, *It's Not How Smart You Are, It's How You Are Smart! Howard Gardner's Theory of Multiple Intelligences* (1999), retrieved October 2, 2009, from http://surfaquarium.com/MI/overview.htm

20. Tobias, *The Way They Learn*, p. 135.
21. Armstrong, *Seven Kinds of Smart*, p. 10.
22. Wikipedia, *Theory of Multiple Intelligences*, retrieved October 2, 2009, from http://en.wikipedia.org/wiki/Theory_of_multiple_intelligences.
23. Ibid.
24. J. B. Miner, *Organizational Behavior: Performance and Productivity* (New York: Random House, 1988), p. 130.
25. Beginner's Guide, *What Are the Different Learning Styles?* retrieved October 2, 2009, from http://beginnersguide.com/college/learning-styles/what-are-the-different-learning-styles.php.
26. Ibid.
27. Ibid.
28. B. McCarthy, *The 4MAT System: Teaching to Learning Styles with Right/Left Mode Techniques* (Barrington, IL: EXCEL, 1987), p. 10.
29. Tobias, *The Way They Learn*, p. 16.
30. McCarthy, *The 4MAT System*, p. 10.
31. Tobias, *The Way They Learn*, p. 16.
32. McCarthy, *The 4MAT System*, p. 11.
33. Tobias, *The Way They Learn*, p. 18.
34. The terms *accommodator, converger, assimilator*, and *diverger* were coined in D. A. Kolb, I. M. Rubin, and J. M. McIntyre (Eds.), *Organizational Psychology: An Experiential Approach*, 3rd ed. (Englewood Cliffs, NJ: Prentice-Hall, 1979).
35. Tobias, *The Way They Learn*, p. 68.
36. D. W. Mills, *Applying What We Know: Student Learning Styles* (July 2, 1999), http://www.heraldpress.com (January 31, 2010).
37. McCarthy, *The 4MAT System*, p. 41.
38. Tobias, *The Way They Learn*, p. 35.
39. Ibid., p. 22.
40. Ibid., p. 23.
41. Ibid., p. 24.
42. Tobias, *The Way They Learn*, p. 25; Mills, *Applying What We Know: Student Learning Styles* (July 2, 1999), http://www.heraldpress.com (January 31, 2010).
43. R. N. Lussier, *Human Relations in Organizations: Applications and Skill Building* (Boston: Irwin/McGraw-Hill, 1999), p. 51.
44. Ibid.
45. Tobias, *The Way They Learn*, p. 24.
46. Ibid., p. 22.
47. Ibid., p. 23.
48. Ibid., p. 24.
49. Tobias, *The Way They Learn*, p. 25; Mills, *Applying What We Know: Student Learning Styles* (July 2, 1999), http://www.heraldpress.com (January 31, 2010).
50. Tobias, *The Way They Learn*, p. 24.
51. Ibid., p. 22.
52. Ibid., p. 23.
53. Ibid., p. 24.
54. Ibid., p. 25; Mills, *Applying What We Know: Student Learning Styles* (July 2, 1999), http://www.heraldpress.com (January 31, 2010).
55. Gardner, *Frames of Mind*, esp. pp. 60–61.
56. Lussier, *Human Relations in Organizations*, p. 52.
57. Ibid., p. 51.

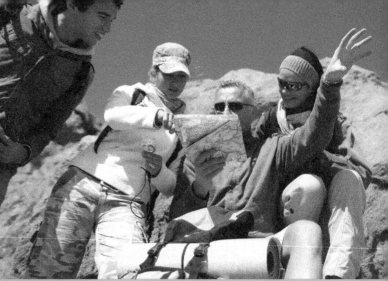

4 DEVELOPING LEADERS

There is no one correct way to lead, and successful leaders come in many personality types. It only makes sense, then, that there are many ways to develop leaders. Some people may have natural abilities called for by certain leadership roles, but leadership development is mostly a result of one's motivation, experiences, and education in and about leadership. Leadership is not something you learn once and for all! It is an ever-evolving pattern of skills, abilities, and qualities that grow and change as you change. To develop into a leader, you must have an understanding of leadership qualities, a purposeful vision and mission, and an honest desire for personal improvement.

Objectives

After completing this chapter, the student should be able to:

- Explain the components of the leadership development model
- Discuss the importance of personal leadership development
- Explain how to attain group acceptance
- Describe the types of individuals who emerge as a group's leader
- Discuss the relationship among ability, experience, and opportunity
- Discuss the improvement of workplace leadership
- List practical tips to help a person develop as a leader
- Explain the types of leadership traits, abilities, and skills
- Name qualities of successful leaders
- Describe human relations leadership qualities and skills
- Describe technical-human relations leadership qualities and skills
- Describe technical leadership qualities and skills
- Describe conceptual-technical leadership qualities and skills
- Describe conceptual leadership qualities and skills
- List 10 ways to identify a potential leader

Terms to Know

- proclivities
- initiative
- qualities
- multidimensional
- ambitious
- abilities
- attributes
- conceptual leadership skills
- technical leadership skills
- human relations skills
- innate
- value system
- portfolio
- opportunists
- brainstorming
- discontent
- status quo

Leadership involves an observable, learnable set of practices. It is not something mystical that ordinary people cannot comprehend. Given the opportunity for feedback and practice, those with a desire to lead and the determination and persistence to learn how can substantially improve their leadership abilities.

Anyone can learn to lead: an FFA officer, a team facilitator (someone who works with a group to make things run effectively and efficiently), a middle-level manager, an account executive, an athletic team captain, a mail clerk, or just about anyone else who has a good reason to learn how to lead. To an enormous degree, leadership skills determine how much success a person will achieve and how happy he or she will be. Families, charity groups, sports teams, civic associations, and social clubs all need dynamic leaders.

There is no one correct way to lead, and many personality types become successful leaders, as discussed in Chapter 2. Leaders may be loud or quiet, funny or sincere, tough or gentle, boisterous or shy. They can be any age, any race, and either sex. The leadership techniques that work best for you are the ones you should nurture.

Leadership helps people arrive at a better understanding of themselves, others, and the issues at hand. Leaders use this understanding to accomplish the goals that bring members of a group together. Whether leadership is planned or unplanned, it always has a purpose and a goal. It is a process of human interaction.

There can be no leadership without someone to lead. The relationship is successful only as long as the coworkers wish to follow the leader. A leaderless society does not exist; whenever two or more people come together, there is no such thing as uncontrolled, unrestricted, or uninfluenced behavior.

Leadership is a group effort. The existence of any group is evidence of members' willingness to work together rather than individually. Working together is a give-and-take business, and the leader is the catalyst of the process. He or she is successful when the members' accomplishments as a group are greater than those that could have been achieved by each member working individually.

Some people have greater natural tendencies or proclivities toward leadership than others, but there is sufficient research evidence to prove that leadership can also be created, trained, and developed in people of normal intellectual ability and emotional stability who are willing to learn. A leader is a person who, on the whole, best lives up to the standards or values of the group.

Remember, leadership is an action, something you do, not merely a collection of personality traits. The leader helps others develop their skills and share their knowledge. Psychologists say that coworkers determine (set the limits for) what they want their leaders to be and where they want to be led.[1] Figure 4–1 provides some excellent quotations about leader/coworker relationships.

THOUGHTS ON COWORKER RELATIONSHIPS

If your actions inspire others to dream more, learn more, do more, and become more, you are a leader.
— *John Quincy Adams,*
U.S. President

When a genuine leader has done his work, his co-workers will say, "We have done it ourselves," and feel that they can do great things without leaders.
— *Eric Hoffer,*
longshoreman-philosopher

Outstanding leaders go out of the way to boost the self-esteem of their personnel. If people believe in themselves, it's amazing what they can accomplish.
— *Sam Walton,*
founder of Wal-Mart
and Sam's Club

You take people as far as they will go, not as far as you would like them to go.
— *Jeannette Rankin,*
Congresswoman

In the world people must deal with situations according to what they are, and not what they ought to be; and the great art of life is to find out what they are, and act with them accordingly.
— *Charles F. Greville,*
English political writer

I make progress by having people around me who are smarter than I am—and listening to them. And I assume that everyone is smarter about something than I am.
— *Henry Kaiser,*
industrialist

No man will make a great leader who wants to do it all himself or to get all the credit for doing it.
— *Andrew Carnegie,*
industrialist

FIGURE 4–1 The leader and coworker constitute a team. These sayings show how leaders depend on coworkers.

THE MODEL FOR LEADERSHIP DEVELOPMENT

Have you ever tried to drive somewhere unfamiliar and found that you could get there only by using a map? In this chapter, we present a map for your leadership development. Think of Figure 4–2 as a road map to leadership success. It identifies the stages, phases, and areas of leadership that you might go through on your way to becoming the best leader you can be: one who influences others in a positive way for school, community, or job.[2] The three rectangles at the bottom of the figure each represent a different stage of your leadership development process.[3] The next section discusses these stages, using communication leadership skills as an example.

Stages of Leadership Development

Awareness At this point, leadership is not really part of your life, but you know it exists and feel that it could be important. In this stage, you may be aware of the value of good communication skills, but you do not put much thought into how they might affect you.

Interaction In the interaction stage, you think about leadership and want to explore and understand it. In this stage of leadership development, you may begin wondering how to strengthen your communication skills, so you read textbooks, search the Web, or talk to others about ways to improve those skills. You may start practicing some of the methods you find to enhance your communication abilities.

Integration In the integration stage, you work on improving leadership skills and abilities. For example, you may focus on becoming the best communicator in your school. You then search out opportunities and put yourself in situations in which you can use your communication skills to help others.

Areas of Leadership Development

You will move through the stages of the leadership development process for each particular area of leadership.[4] In Figure 4–2, the circles represent these different areas, which are briefly discussed in this section.

Leadership Knowledge and Information This area concerns what you know about leaders and leadership. The goal of this stage is to find and process accurate leadership information. What do you think about when you think of leadership? Who do you think of as a leader? Your answers to these questions more than likely are influenced by what you have learned in school

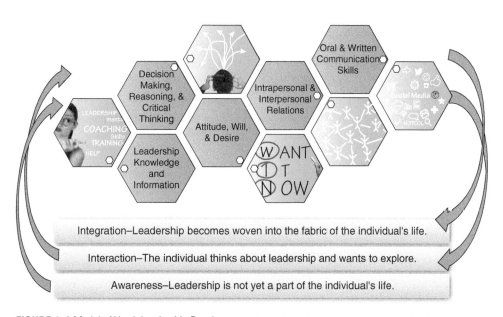

FIGURE 4–2 Model of Youth Leadership Development. *(Source: J. C. Ricketts and R. D. Rudd, "A Comprehensive Leadership Education Model to Train, Teach, and Develop Leadership in Youth," Journal of Career and Technical Education 191:7–17 [2002].)*

and the experiences you have had in the FFA, a performing arts group, or a sports team. Among the many sources of leadership information are your personal experiences and materials you have read.[5]

When you begin developing as a leader, pay attention to people you consider leaders. Focus on specific skills and abilities that these people demonstrate and that you feel are leadership skills. Decide which things are leadership skills and abilities that you will model and which are skills about which you would like more information. Take the information and knowledge that you gain and learn more; use your knowledge, perfect it, and teach it.

Leadership Attitude, Will, and Desire This area of leadership development is critical. For the most part, your knowledge, your eventual skill level, and everything else you read about in this book depend on your attitude. John Maxwell, leadership author, speaker, and teacher, defines **attitude** as "an inward feeling expressed by behavior."[6] What this means is that even if you know everything there is to know about leadership and have excellent ability, your leadership knowledge and abilities will not function if your internal compass—your will and desire—is not correct. Your poor attitude and lack of determination will simply bury them.

This leadership development area highlights the importance of motivation, self-realization, and health in fulfilling your leadership capacity. **Motivation** is what the coach provides to a team before the game, but it is also something that you as a leader will need to develop in your followers. Different levels of achievement, power, and sociability will motivate your followers. It is up to the leader to discover what motivates people and then supply that needed motivation. Furthermore, self-realization and self-knowledge are required because you need to find out what motivates you, as well.

Importantly, this area also includes health. Yes, your health can greatly affect your leadership development! Research indicates that the best development occurs when a person is healthy, keeps fit both physically and mentally, has a positive self-image, and possesses appropriate and adaptive coping skills.[7]

Decision Making, Reasoning, and Critical Thinking During your development as a leader, you will have to make many decisions. Reasoning and critical thinking serve as very useful tools for making good leadership decisions. **Reasoning** involves using logic to come up with the best possible solution to a problem. You use **critical thinking** to examine ideas, opinions, alternatives, and situations; to reflect on your leadership experiences; and to make judgments and decisions that often must be based on incomplete evidence.

The best leaders are those who can make the tough decisions. After you have progressed past the awareness and interaction stages of leadership development, you will inevitably be faced with some major decisions to make as a leader. Consider the following story:

> A railroad bridge operator took his son to work. He and his son were having a great time together, but while the bridge operator took a phone call, the boy temporarily disappeared. The operator eventually heard his son cry out: He had fallen into the gear system of the movable bridge and was trapped there. The problem was that the train was on its way. The bridge operator had to decide whether to put the bridge down and lose his son; or not put the bridge down, thereby saving his son but allowing the hundreds of people on the train to plunge to their deaths in the river below. What would you do?

We hope the decisions you will have to make are not quite as difficult, but the point is that no one can really be considered a leader until he or she can be trusted to make the tough decisions. Leaders at the integration stage think critically and logically, putting aside their personal biases, to choose the best solution possible. (Specific aspects of this area are discussed later in this text.)

Oral and Written Communication Skills John Gardner, former U.S. Secretary of Health, Education, and Welfare, claimed that if he had to name one "all-purpose instrument of leadership, it would be communication." Oral and written communications are our only method for sharing the knowledge, attitudes, opinions, feelings, interests, and ideas with which we influence and ultimately lead others.[8] Just think about it:

without communication, how would we get people to do anything?

Communication takes many forms. To be effective leaders, we need to be well versed in as many of those different forms as possible. Good communicators express themselves in such a way that people accept their ideas and become willing to follow. You may express yourself through speeches, written reports, active listening, and technological methods, in one-on-one contacts or in group or team settings. Start thinking about how good communication skills—everything from sending appropriate e-mails to delivering quality speeches—can help you develop into a quality leader.

Intrapersonal and Interpersonal Relations
Intrapersonal relations include the skills you will need to evaluate yourself. As leaders, we need to be self-reflective, which means that we must know what our personality type, communication style, and leadership style are so that we can feel confident about our decisions and work well with others. **Interpersonal relations** refers to the skills that you need to get along with others. Without these skills, you will have difficulty leading people in a democratic manner. A large portion of this text concentrates on your ability to assess yourself and then use that knowledge to work effectively and efficiently with other people (both leaders and followers) to get things done.

This particular facet of leadership development will help you with many of the other responsibilities that come with being a leader. Leaders who are competent in both the intrapersonal and interpersonal areas can master issues such as conflict resolution, problem solving, time management, group dynamics, and team building.

IMPORTANCE OF PERSONAL LEADERSHIP DEVELOPMENT

Leadership is needed in all occupational areas and at all levels. John, a state FFA Creed contest winner and two-year state finalist in public speaking, worked as a freight loader with United Parcel Service (UPS) to help pay his college expenses. At a group meeting, his supervisor mentioned that UPS was looking for someone to speak to the corporate executives at a convention on improvements and efficiency in the freight-moving system. John accepted the assignment because of his confidence in public speaking. His speech caught the attention of the UPS executives, and John was marked as having future leadership potential in the company.

It is a challenge to learn leadership skills, but these skills will earn you respect from others. Leadership helps you understand others as well as yourself. Leadership helps individuals mature and develop their self-concepts. When we like ourselves, we like others. We are more secure; therefore, we are happier. The result is that life is improved.

Personal Leadership, Direction, and Success

Personal leadership concerns your ability to establish a specific direction for your life; commit yourself to moving in that direction; and take considered, determined action to acquire, accomplish, or become whatever that goal demands. It boils down to this attitude: "If it's going to be, it's up to me."

Success depends on our own actions. Your attitude determines your altitude. Obviously, it is harder for some to succeed because of financial constraints, innate intelligence levels, environmental factors, or unmotivated peers. Although physical limitations may limit choices in certain areas (for example, not many short people play professional basketball), people can accomplish just about anything they want to in life if they are willing to make sacrifices. We must find our strengths, deal with our weaknesses, and have the persistence and determination to accomplish our goals. Winners find a way. Losers find an excuse.

If you try to please everyone except yourself, you are destined for failure. Success consists of attaining our own personal goals, not other people's goals. In this day of peer pressure, trends, and fads, we need to realize and accept that each individual is custom-made, with unique personal traits. We should strive to be ourselves and not copy other people—although we certainly can learn positive traits from others and use exceptional persons as role models. People are born originals, but most die copies.

Becoming a Successful Leader

Successful leaders come in different packages, but if they had certain competencies in common, what would they be? One is character. It's been said that "leadership is always about character," and "the best way to define a man's character is to seek out the particular mental or moral attitude in which, when it came upon him, he felt himself most deeply . . ." What mental or moral attitude defines your character?[9]

To become a truly good leader, you must also remember that "you are your own best teacher." No one but you can teach you how to become your best self. An extremely helpful way to learn leadership is to place yourself in situations requiring leadership action. This does not mean that you dominate every situation, but rather that you act on a desire to serve, achieve goals, and leave things better than they were when you found them. Take on new projects, investigate new ideas, and undertake challenging assignments, so you can practice your knowledge and skills every day.[10]

A successful leader also has to accept responsibility and never cast blame. A good leader learns much by reflecting on negative experiences, such as losing a job or failing to reach a goal. Such setbacks can actually be more valuable than success if you are willing to learn from them.[11]

Here are some other practices that will help you become a good leader:

1. Study the qualities of recognized leaders and learn from their mistakes. Watch, listen, and learn, but do not imitate.
2. Identify and analyze your weak and strong points. Set goals for improvement of weaknesses and enhancement of strengths.
3. Learn how to take direction (how to follow). If you cannot take directions, you may not be able to give directions.
4. Learn as much as you can about groups in general and how they function. Identify the types of people in your group.
5. Make and follow a plan to develop your personal leadership skills.

Steps of Your Personal Leadership Plan

Step 1 *Develop a vision and focus your thinking.*

Successful leaders, whether they are leaders of others or simply effective leaders of their own lives, all have one thing in common: **vision**, which is a clear picture of what one wants to attain. Examples of individuals who had a strong vision include

- Martin Luther King Jr., the great civil rights leader, who had a vision of racial equality for all people in America. He is still remembered for his contributions to the equality we now enjoy.
- Mother Theresa, who had a vision for compassion and service; she became an international symbol of these values.
- Lee Iacocca, who had a vision of turning an almost bankrupt Chrysler Corporation into a profitable company; in doing so, he saved hundreds of thousands of jobs.
- Thomas Edison, who had a vision of providing light by harnessing the power of electricity.
- Samuel C. Ricketts, who for decades had a vision of producing hydrogen and other clean fuels from natural resources; he drove a vehicle from the east coast to the west coast of the United States on hydrogen fuel produced from water using the sun/solar energy to power the process of decomposing water (H_2O).

Without commitment to a vision, we cannot lead ourselves in a definite direction. Vision separates the average from the great. Vision provides

- direction
- a worthwhile destination
- motivation
- enthusiasm
- a sense of achievement
- fulfillment of one's purpose in life

Step 2 *Set goals.*

Once you have your vision, set goals to make it reality. Develop a plan of action and pursue it. Goal setting is the key to accomplishment. (Goal setting is discussed fully in Chapter 14.) What do you want to be like in 5, 10, and 20 years from now? Remember, few people plan to fail; they just fail to plan. Set goals in the following areas: physical, family, social, financial, educational, ethical, and career.

Step 3 *Develop initiative.*

Be energetic and persistent in accomplishing your goals. Take the **initiative** (the ability to take action and be a self-starter). There are three types of people: those who make things happen, those who ask what's happening, and those who ask what happened. Do not let life pass you by; take action, participate, and lead in life. Look at the way things can be done rather than the ways they cannot be done. George Bernard Shaw said, "Some men see things as they are and say why; I dream things that never were and say why not."[12]

Step 4 *Develop self-confidence.*

The only way we gain self-confidence is to expose ourselves to risk. By taking risks, we gain experience, which only builds confidence. Thomas Edison said that he learned 999 ways the lightbulb did not work before he found one way it did. A modern-day inventor echoed that idea, saying that there are always a thousand ways and reasons that something will not work; it takes effort to find the one way that will. When you are low on confidence, do not give up. As Elbert Hubbard said, "There is no failure except in no longer trying."[13] Although failure can be valuable for what it teaches us, most failure comes from quitting. Therefore, to eliminate most failure in life, do not quit.

Step 5 *Develop personal responsibility.*

Take personal responsibility for your own thoughts, actions, and feelings. A basic problem with society today is that most people refuse to accept responsibility or to be accountable for their actions. True leaders take control of their own destinies. If we blame others for our mistakes or claim that the way we feel is not our fault, we are not in charge of our lives. If we are not in charge, we will never exhibit real personal leadership. If you make mistakes, admit them. Be willing to take

correction and criticism without becoming defensive, vengeful, or sulky. Defensiveness is a sign of insecurity.

Step 6 *Develop a healthy self-image.*

"Self-image is how you perceive yourself. It represents a number of self-impressions that have built up over time . . ."[14] There are a number of laws that govern our results in life. The law of gravity works every time, whether we like it or not. The self-image law works just as surely: we never rise above the picture we create for ourselves. People cannot exceed their self-imposed limits.

In May 1954, Englishman Roger Bannister, a part-time athlete, pictured himself breaking the four-minute barrier for the mile race. Athletes before him could not run such a time because they believed it was impossible. Bannister did it because he saw himself doing it. It was only a matter of time before the Australian athlete John Landy also broke the barrier. Why could he do it? Because someone else had already done it, Landy accepted that it was possible. Within a year, more than 10 other runners had also run sub-four-minute miles.

Let your self-image run free. Let it be creative and innovative. Let it inspire you to success. If you can dream it, you can achieve it.

Step 7 *Develop self-organization.*

One of the best traits of personal leadership is that of self-organization. You must know what to do next if you are to move toward achievement of your goals. Consider the following points in self-organization:

Write down everything—An old Chinese proverb says, "The shortest pencil is better than the longest memory."

Use a good diary system—Use the style that works best for you. Many organized professionals like to keep notes on a smartphone; others still prefer to keep a small paper journal that fits in a pocket.

Make a daily to-do list—Each night, list the tasks to be done the following day; then arrange them in order of priority. There are some great to-do list apps for your smartphone or computer if you prefer to keep your lists electronically.

Use good time management skills—"If you want something done, ask a busy person." Why? Busy people know how to work on priorities and overcome the habit of procrastination. Establish a weekly planner in which you allocate specific tasks to specific time blocks during that week. Do exactly what you planned to do. If you have set short- and long-range goals, you will find it easier to decide what to do on a daily basis.

Step 8 *Eliminate procrastination.*

Procrastination is the tendency to put something off rather than taking action on it now. To overcome procrastination, develop constructive habits to replace the destructive ones.

Following are some suggestions for overcoming procrastination:

1. Develop Awareness – Recognize that you are a procrastinator, and think about the reasons you procrastinate.
2. Assess – Identify and analyze the feelings and emotions that lead to procrastinating.
3. Develop an Outlook – Tackle big projects in smaller chunks. This makes the project seem less scary.
4. Commit and Set Goals – Make your mind up that you will not be beaten. Stay focused. Write down your goals. Make a "to-do" list, and reward yourself when you achieve your goals.
5. Choose Your Surroundings Wisely – When you have a big project, choose a work environment that fits your personality type. Choose surroundings that you know help you focus.
6. Be Realistic – Don't mess yourself up by setting goals that are not achievable.
7. Use Positive Self-talk – Speak positive messages to yourself about finishing your task at hand.[15]

Step 9 *Study.*

People who wish to be good at a task need to study. If you want to be a good student, you must study. If you want to be a good doctor, you must study medicine. If you want to be a good parent, you must study the art of parenting and child rearing. If you want to be a good leader of yourself and others, you must study

leadership. This means talking to successful people and asking them how they achieve results, and reading books on leadership and success. Throughout history, the remarkable achievers in life have been students who were eager to learn and improve.

Step 10 *Magnify your strengths.*

The first step toward success is identifying your own leadership strengths. Ask yourself what personal **qualities** (characteristics, innate or acquired, that determine the nature and behavior or a person or thing) or strengths you possess that could be turned into the qualities of leadership. Whatever talents you have (and we all have them, whether we have discovered them or not), use them to your advantage. Work in ways that best suit your personality type (discussed in Chapter 2) and take advantage of your learning style and strengths.

Consider the following:

1. For coach Pat Summitt, the all-time winningest coach in NCAA basketball history, competitive drive was a strength.
2. For Oprah Winfrey, celebrity media proprietor and former talk-show host, communication is a strength.
3. For Peyton Manning, Super Bowl–winning quarterback, thinking and studying are some of his strengths.

Dale Carnegie said that whatever strengths you have—"a dogged persistence, a steel-trap mind, a great imagination, a positive attitude, or a strong sense of values—let them blossom into leadership. And remember, actions are far more powerful than words."[16]

ATTAINING GROUP ACCEPTANCE

Regardless of personal success, most of us like to be accepted by others. In fact, acceptance is a basic human need. Leadership effectiveness depends to a great extent on whether the group accepts you. You are likely to attain group acceptance if you do the following:

Become Knowledgeable about the Purpose of the Group Group members are more willing to follow when the leader appears to be well informed. The more knowledgeable you are, the more credibility you will gain.

Become Committed to the Group and Its Purpose If group members detect lack of commitment in you, they will lose confidence in both you and the group, and withdraw their support of the group's purpose.

Work Harder than Anyone Else in the Group Leadership often consists of setting an example. When others in the group see a person doing more than his or her fair share for the good of the group, they are likely to support that person. The person who wishes to lead must be willing to make personal sacrifices.

Be Willing to Be Decisive When leaders are unsure of themselves or unwilling to make decisions, their groups may veer off course, become frustrated, or become contentious and irritable. Sometimes, of course, leaders' decisions will be resented or cause conflict. Nevertheless, people who are indecisive will not be able to maintain leadership for long. An old Arab proverb says, "An army of sheep led by a lion would defeat an army of lions led by a sheep."

Interact Freely with Others in the Group Participate fully in group discussions, but do not dominate the discussions. Share your ideas, feelings, and insights and allow others to do so as well. Before you try to gain leadership status, participate fully in the early stages of group work to find out whether you are able to influence others. This is why political candidates announce that they are considering running for an office: they want to test their influence before making a commitment. The group in Figure 4–3 is engaged in and enjoying a freely interacting group discussion.

Exhibit Human Relations Skills Effective leaders make others in their group feel good, contribute to group cohesiveness, and give credit where it is due. Although a group may have both a task leader and a human relations leader, the primary leader is often the one who demonstrates good human relations skills.[17]

FIGURE 4–3 This group of freely interacting young leaders looks poised and ready to achieve their goals for success.
(© iStock/Petar Chernaev)

TYPES OF INDIVIDUALS WHO EMERGE AS GROUP LEADERS

Once you become accepted in a group, you will be considered for group leadership. If a group is to be effective and achieve its goals, it must have successful, effective leadership.

Within any group, three types of individuals may emerge as leaders: member-oriented, goal-oriented, and self-oriented people. The member-oriented individual becomes a leader when a conflict occurs within the group and the individual can solve the difficulty, mediate, or referee the conflict. When the group's goals and rewards are attractive, a goal-oriented member who thrives on accomplishment of tasks usually emerges as the leader. If, however, the leadership attempt fails, this person will probably not try to lead again. Self-oriented leaders emerge when group leadership is prestigious or provides some direct personal reward. This type of leader enjoys the recognition that comes with a leadership position.

Each leader usually combines all three types to some extent. Good group leaders are willing, motivated, energetic, and goal oriented. They exhibit originality and initiative and often are more able to respond successfully to new situations and learn from past experiences than other group members.[18]

ABILITY, EXPERIENCE, AND THE OPPORTUNITY TO LEAD

To realize maximum leadership potential, three elements—ability, experience, and opportunity—must occur together. Given the proper experience and an opportunity, anyone who truly desires to can become a leader.

Ability Whether a person's ability consists of skills learned through formal or informal leadership training, or arises from a hereditary trait, such as personality type, competence is needed for leadership. In class-structured societies, the close association between leadership and high social status has led many people to assume that leadership is an inherited trait. However, geneticists have never discovered a gene they could label the "leader gene," and the genetically based

theory of leadership has been refuted by the many people of all classes, at all levels of social status, who have become successful leaders.

Experience Although we cannot control our heredity, we can enhance our abilities through experience. You develop your leadership abilities and skills by participating in various activities, such as training and practice at school, camps, and seminars, as well as by reading and learning on your own.

Opportunity Even if you have abilities and experience, you cannot lead until you get the opportunity to lead. Sometimes this occurs because you are in just the right place at the right time. More often, though, it occurs because you are prepared to lead and are ready to act when an opportunity arises.[19]

Relationship of the Three Elements

An experienced or skilled person may never get the opportunity to lead. This may be due to never being selected as the leader or, as noted earlier, never being in the right place at the right time. For example, many students would make excellent officers if only someone would nominate or select them. (Note that in situations such as this, skilled and experienced persons can do much to create their own opportunities.) Conversely, circumstances or necessity may force an inexperienced or unskilled person into a leadership position. Unfortunately, a person with no prior experience or skill in leading a group will probably not accomplish at a high level, and the group is likely to suffer for this leader's shortcomings.

IMPROVEMENT OF LEADERSHIP IN THE WORKPLACE

Business and industry use three basic approaches to improving leadership within the organization: selection, situational hiring or engineering, and training.

Selection The company simply chooses or assigns the person with the skills most needed by the group; this is the quickest way to improve workplace leadership. In an effective selection process, the business analyzes its needs and then assesses each candidate's record of past performance. Finally, it selects a leader. For example, when a company wants to improve its meetings, it should look for someone with experience in parliamentary procedure and conduct of effective, efficient meetings. Your ability and experience in this area can combine with this opportunity and put you at the top of the list for this leadership position.

Situational Hiring or Engineering With this approach, the situation is matched with a leader. Manipulation (or engineering) may be done before the leader actually faces a problematic situation, to enhance the leader's probability of success. Wise companies avoid throwing individuals into situations in which failure is likely. However, the amount of authority the leader is assigned may depend on the situation he or she is expected to handle. If the leader proves to be capable, the company can place more responsibility on, and leave more of the decision making to, the chosen leader.

Training When they decide to train for leadership, companies usually work with the individuals they have rather than hiring new people. Organizations send employees and members to seminars, workshops, retreats, and camps to improve their leadership skills.[20]

PRACTICAL TIPS FOR IMPROVING YOUR LEADERSHIP

Developing leadership is important in all areas of life, not just the workplace. The best way to develop into a better leader in all areas of life is to master a wide array of skills, ranging from implementing and administering processes to inspiring others to achieve excellence. Consider the following a cookbook of practical strategies for developing into a better leader.[21]

Gain Experience

Excellent leaders are developed, not born. To be the best, you must learn the essential skills of leadership through formal leadership courses, workshops and conferences, and leadership experiences in the FFA or on the job. You learn by doing!

Learn from Others

Almost every successful singer has a singing coach, and top singers often teach classes. The principle is just the same for leaders. You learn better leadership skills by being coached, and you develop those skills further by coaching others. Are you learning from other FFA members? Are you mentoring any other FFA members?

Master the Various Leadership Roles

Leadership is a **multidimensional** (having several sides or parts) function, requiring knowledge and understanding of many different organizational processes and needs. As a leader, you must master the various roles required to handle, with skill and efficiency (producing the desired effect without waste of time or materials), the people and circumstances you will encounter.

Develop Strengths

You can develop all the attributes required of a leader if you have the necessary drive and energy. Self-confidence and self-determination, combined with your unique set of strengths, will make you into an effective leader who is able to achieve desired goals.

Form the Best Team

The leader is responsible for establishing a team or appointing new team members. Find the best candidates to form a balanced and dynamic team (with diverse personality types, leadership styles, communication styles, and so on), through either internal promotion or external recruitment. Help each member to feel that he or she is an integral part of the team.

Exercise Authority

The leader must ensure that everyone understands instructions and carries them out effectively. Rarely does everything go according to plan, so establish reporting systems that alert you to problems and enable you to deal with any issues swiftly.

Delegate Tasks

As leader, concentrate on activities that nobody else can do. Delegation is a form of time management. It is a way of exercising control and

VIGNETTE 4-1	BEING A GOOD COACH

MICHAEL ASKED DORIS, one of his second-level managers, to produce a report that involved a degree of financial knowledge. He took it for granted that she understood the basics of management accounting and was unpleasantly surprised to find that Doris had made many errors through ignorance. Because a deadline was approaching, and this was work that came easily to him, Michael rewrote the report and passed it on.

Doris asked for a meeting with Michael. She was angry, and Michael assumed that this was because he had taken over the report writing. But Doris was cross for a different reason. As she put it, "How do you expect me to learn if you don't tell me what I've done wrong?" Michael realized that he had failed Doris. He then set aside time to coach her in management accounting and also sent her to take a course in finance.

Taking over the task that he had assigned to Doris seemed like the easy option to Michael. He learned from this experience, though, that he had actually missed the important issue. He was looking at the problem in the short term rather than focusing on helping Doris to perform better in the future. He realized that training people was far more productive in the long term than doing everything himself.

meeting your own responsibilities more effectively, while developing the skills of your staff or members. Remember, you cannot do it all!

Communicate Clearly

The ability to communicate well with other members of your team is essential. Use your knowledge of the communication process and communication styles to ensure that messages are received and understood by all. Treat each team member as an equal, and let everyone know that his or her ideas and contributions are valuable.

Set Ambitious Goals

Goals are the essence of planning, whether for the long, intermediate, or short term. They should be **ambitious** (showing a desire to achieve or obtain power, superiority, or distinction) but achievable. Set "stretch" goals that challenge followers, but also set feasible subgoals to help your team hit the targets.

Foster Teamwork

For a team to work well, several different roles must be filled. The leader's role is to develop a team that thinks and acts together, with individual and team interests and efforts aligned in pursuit of the common goal.

Lead Discussions

Whether they are formal or casual, involve groups of people, or are conducted on a one-to-one basis, discussions are how a great deal of work gets done. By leading and facilitating them skillfully, you can keep discussions productive and meaningful, while allowing people to share ideas or concerns freely.

Use Meetings

Meetings are a staple of almost all organizations, but often they lack any clear purpose. If you want a meeting to be effective, ensure that you have defined a valid goal for that meeting. Never bring people together just to rubber-stamp decisions that have already been made. Do not hold a meeting just to have a meeting.

Analyze Problems

Leaders will always face problems that demand attention and must be addressed. The trick is to be able to analyze the issues underlying any particular problem so that you can make an informed decision. By being positive and doing the necessary analysis, you can overcome even large obstacles and find a solution to the problem.

Give Support

Trust is difficult to build and easy to lose. People often start with a distrustful or skeptical mindset. As a leader, you need to work hard at earning trust and then foster that trust by showing loyalty and supporting your team fully.

Inspire Excellence

The difference between leadership and management lies in the leader's ability to inspire followers to excellence. To mobilize a team's inner drive,

VIGNETTE
4-2   DONAVAN THE DELEGATOR

DALLEN WAS WORKING FOR A NEW BOSS IN A NEW COMPANY

As one of his first assignments, he was given the task of organizing a new team project. But problems quickly arose. With existing resources, there was no hope of meeting the production targets from internal supply. He went back to his boss, Donavan, with the problem and was disconcerted by the response.

"I don't want people bringing me problems without solutions," said Donavan. "I want you to present me with at least two possible solutions every time we meet, and I want to hear your recommendation on which one to take and why. Don't come back to me until you have identified the solutions and possible recommendations, or I will not be very happy." Dallen went away and returned with two solutions: subcontract some of the work or ask for more finances and people. He preferred the first option and so did Donavan.

Being a good delegator, the leader in this case did not want his people to become dependent on him and his decisions. Therefore, he forced them to make up their own minds. The boss was still prepared to discuss the issue, but his insistence helped his subordinate to function as a real leader.

enthusiasm, and vigor effectively, you need to be a credible leader who sets an inspiring example. Encourage people to achieve their best by setting that example and using appropriate motivational strategies. Believe that your people are capable of remarkable achievement, and show that you trust them to perform.

Establish a Vision

Human beings find it easier to look backward than forward, but effective leaders look into the future. If you establish a vision of where you want to be in the long term and communicate that vision in a compelling manner, your vision will inspire others to follow.

Generate Ideas

The leader need not be the most inventive person on the team. However, the leader does need to tap each individual's potential for generating ideas. This will help both in achieving the group vision and in resolving day-to-day issues.

Manage Openly

Information sharing positively affects performance. Information withholding has the opposite effect. By trusting your staff with information and by being open and honest with them, you will inspire them and assist them to perform better.

Be Competitive

Leaders understand that achievement of anything worthwhile necessarily carries a risk of failure. If you desire success for your team, you have to take risks and compete.

TYPES OF LEADERSHIP TRAITS, ABILITIES, AND SKILLS

Chapter 1 discussed seven categories of leadership: trait, power and influence, behavioral, situational, traditional, popularity, and combination. Actually, when most people think about leadership, they think of leadership traits, abilities, and skills. **Abilities** are competencies in an activity or occupation. These are also referred to as leadership **attributes** (characteristics) or qualities.

Using these abilities and skills effectively, a leader gets extraordinary things done in an organization, whether it is the FFA or the workplace. There are three major types of leadership abilities and skills: conceptual, technical, and human relations skills (refer to Figure 4–4 for examples of each).

Conceptual (Thinking) Skills Conceptual **leadership skills** are thinking skills that can be taught, such as problem solving, decision making, and delegation. They are needed in analyzing situations and generating ideas. These skills are more difficult to develop than technical skills. One good way to work on these skills is to give the potential leaders a dry run, placing them in situations they might face as leaders. Games that use thinking abilities (such as logic puzzles,

TYPES OF LEADERSHIP TRAITS, ABILITIES, AND SKILLS		
TYPE OF SKILL	**EXAMPLES**	**SKILLS THAT YOU MIGHT POSSESS**
Conceptual (Thinking) Conceptual-Technical	Analyzing a situation Thinking logically Combining concepts and ideas into a workable relationship Generating ideas Helping solve a group or individual problem Anticipating change Recognizing opportunities and potential problems	Good imagination Dedication Combining concepts and ideas into a workable solution Good problem-solving skills Creativity Logical thinking Good decision-making skills Anticipating problems Ability to think independently Foreseeing change Open-mindedness Welcoming new opportunities Persistence
Technical (How-to) Technical-Human Relations	Knowing a variety of ways to do things Understanding how to get things done Ability to establish and follow procedures Techniques for conducting a special activity Knowledge gained from study and experience	Communication skills Prepared speaking skills Extemporaneous speaking skills Parliamentary procedure Group organization, group dynamics, and leading discussions Goal setting and program of activities Time management Financial management Conducting successful meetings Organizational skills
Human Relations (People)	Human behavior Interpersonal communication—how to get along with others Feelings about yourself and others The variety of attitudes and values people have Motives that others may have Good self-concept and self-esteem	Honesty Capacity for hard work Social skills Listening Cooperativeness Strong self-concept Enjoy working with people Sensitivity Positive attitude

FIGURE 4–4 Leadership skills and abilities are traditionally discussed by category (conceptual, technical, and human relations), but these categories overlap and are not always clearly separated. Thus, the conceptual-technical and technical-human relations categories were added. For example, we can be taught how to think logically and how to improve self-esteem.

Sudoku, chess, and brainteasers) and afford practice in solving problems, forecasting events, and making decisions can improve conceptual leadership skills.

Technical (How-to-Do-It) Skills Technical leadership skills involve doing. Improvement in technical skills may occur with practice in public speaking, time management, or use of parliamentary procedure. Of the three types of leadership skills, technical leadership skills are the most easily learned and retained.

Human Relations (People) Skills Human relations skills are those needed to understand and work with others. "People are the foundation of all groups, so leaders must be able to deal effectively with members." Therefore, leaders should seek training in working effectively with groups and group members.[22]

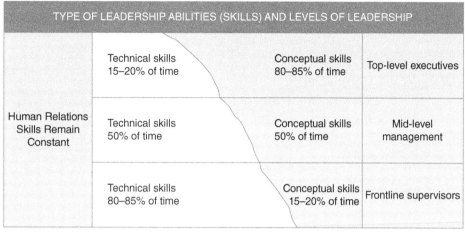

FIGURE 4–5 Percentage of use of leadership skills.

As noted, technical leadership skills are usually the easiest to obtain and remember. Conceptual skills are more difficult to learn than technical skills. Human relations (people) skills are a requirement for all leaders. They are not difficult to learn, but leaders tend to forget or set them aside when under pressure to complete an activity.

Each type of skill is important for all tasks. Some leaders are stronger with one type than the others. To maximize the group's productivity, leaders must be able to recognize and draw on both their own and other members' talents and abilities. Accurate assessment of members allows leaders to match members' skills with the jobs that must be done and the work that must be performed. Figure 4–5 shows the relationship between types of leadership abilities (skills) and levels of leadership. Top-level executives use conceptual (thinking) leadership skills 80 to 85 percent of the time. Frontline supervisors use technical (how-to-do-it) leadership skills 80 to 85 percent of the time. Notice that human relations (people) skills are used equally and are equally important for upper-, mid-, and lower-level (frontline) leaders.

QUALITIES OF SUCCESSFUL LEADERS

Many qualities, abilities, and skills are important in leading others. Together, they constitute a profile of a leader who has compassion and consideration for his or her colleagues or coworkers while at the same time holding them strictly accountable for results. Some psychologists call this "tough love." The makeup of a leader includes desired personal traits (such as honesty, integrity, and sensitivity) and the necessary abilities and skills, in addition to intangible factors such as determination and courage.

Leadership Is Not an Exact Science

The study of leadership is not an exact science, like biology, chemistry, or physics. The social science world is nowhere near as orderly as the physical world and does not function according to fixed, universally applicable laws. People are anything but uniform, and often they are hard to predict. Thus, there is no simple formula for leadership success.

In Pursuit of Leadership Perfection

It is not possible to reach perfection in each leadership skill, but the capacity to lead gradually builds over time. Leadership development can be accelerated, however, by constantly visualizing the ideal leadership qualities and modeling your behavior after them. Figure 4–6 lists more than 50 leadership attributes. Several sources were reviewed to arrive at this list, including books written by famous leadership researchers and authors such as Shaw, Bennis, Nanus, Kouzes and Posner, Bethel, Manske, Covey, and Maxwell. Each author has described various qualities that he or she feels strongly about.

QUALITIES OF SUCCESSFUL LEADERS				
HUMAN RELATIONS SKILLS	**TECHNICAL-HUMAN RELATIONS SKILLS**	**TECHNICAL SKILLS**	**CONCEPTUAL-TECHNICAL SKILLS**	**CONCEPTUAL SKILLS**
Honesty and integrity	Listening	Staff selection	Intelligence	Vision
Sensitivity	Team (group) building	Competence	Problem-solving ability	Ability to motivate
Cooperation	Earning loyalty	Communication	Decision making	Decisiveness
Flexibility	Rewarding accomplishments	Time management	Using cooperative genius approach (synergy)	Straightforwardness
Supportiveness	Coaching and strengthening others	Goal-setting and planning	Delegation	Creativity
Self-confidence		Achievement	Opportunism	Courage
Responsibility and dependability	Planning short-term goals and wins	Response to failure	Insistence on excellence	Risk taking
Emotional maturity	Administrative ability	Leading individuals and groups	Holding coworkers accountable	Commitment, determination, and persistence
Taking pleasure in work			Broad interests and abilities	Sense of urgency and initiative
Servanthood				Follow-through
Setting an example				Mastery of change
Availability and visibility				Wisdom
Showing confidence in people				Desire to lead
Using power wisely				Critical thinking

FIGURE 4–6 The qualities of successful leaders fall into five categories.

Many leadership abilities and skills are referred to by similar names. For example, **inspiring** is also known as **motivator, people-mover, enthusiastic, charismatic,** and **enlists others.** For Figure 4–6, we selected and combined representative terms to arrive at the listed leadership qualities. If you study these qualities, you will encounter numerous ideas that will help you improve your strengths and eliminate your weaknesses. You can also do the following:

- Equip yourself with skills, tools, and techniques to maximize your natural leadership talents.
- Identify abilities and skills that will build the confidence you need to make leadership decisions.
- Sharpen your curiosity and reinforce your commitment to develop and demonstrate leadership.
- Challenge yourself with questions that will help you discover how to be the best leader you can be.

Leadership is not something you learn once and for all. It is an ever-evolving pattern of skills, abilities, and qualities that grow and change as you do. To become a leader, you must first have a mission, an honest desire to improve yourself, and a clear picture of desirable leadership qualities. Remember, we lead by example. Everything we say or do sends a message, sets a tone, or teaches people what to do or what not to do.

Five Leadership Quality Categories

The authors attempted, in Figure 4–6 and the remainder of this chapter, to place the leadership abilities into five categories: human relations skills, technical-human relations skills, technical skills, conceptual-technical skills, and conceptual skills. The proper category for any given skill is, however, open to discussion, and some skills fall into more than one. Certainly, the skills are interdependent to a large degree. For example, critical thinking may not be possible unless the mind is educated and cultivated (technical); likewise, learning is enhanced by a good attitude (human relations).

HUMAN RELATIONS SKILLS

The following human relations skills tend to reflect a person's **innate** (inborn or inherent) personality type or **value system** (set of beliefs). Human relations skills are needed to understand and work with people.

Honesty and Integrity

Successful leaders live by the highest standards of honesty and integrity. In response to surveys that asked business and government leaders what quality they thought was most important to their success as leaders, their unanimous answer was **integrity**. It is one thing to talk about integrity and quite another to live that way. The best way you can develop honesty and integrity is not only by talking about its importance, but also by modeling honesty and integrity in your everyday life for your followers.[23]

A good name is seldom regained. When character is gone, one of the richest jewels of life is lost forever.

—J. Hanes

Without integrity you do not have trust, and without trust you have nothing.

—Fred A. Manske Jr.[24]

It is more important to do the right thing than to do things right.

—Peter Drucker

Sensitivity

A sensitive leader inspires loyalty. An effective leader is a flexible leader—someone who can adapt his or her style of leadership to fit the needs of others and the situation. Sensitive leadership is not leadership that lacks strength or courage; it does not imply being soft or having less power. A sensitive leader is, fundamentally, acutely aware of the issues, values, and people in an ever-changing society. Sensitive leaders are keenly aware of the personal needs of their followers, but they are also sensitive to the organization they lead and the systems of the organization. A good leader uses this skill to find problems and fix them in support of accomplishing team goals.[25]

Cooperation

Americans have always believed strongly in self-reliance, individuality, and competitiveness. However, in today's complex, interdependent world, cooperation is much more important than competition.[26] Smart people realize that much of their own success depends on others succeeding as well; therefore, they orient themselves toward helping each other to perform effectively. They encourage each other because they understand others' priorities and want to help everyone be successful. Compatible goals promote trust. Nothing is more energizing than a leader who is able to meld talented individuals into a team, wherein each is willing to do his or her very best for the good of the group and achievement of the group's goals.

Flexibility

Chapter 1 discussed situational leadership. Situational leaders are flexible. They adapt to the occasion, using an authoritative style if followers are unwilling or a democratic style if the group is competent. A flexible leader can adapt the organization to meet changing needs with minimal opposition. Flexibility includes a willingness to hear and try new ideas, even if they are untested or unfamiliar to the group.

Supportiveness

Leaders support their followers or coworkers. When a coworker knows the leader will be supportive, there is a healthy relationship. If workers or employees are afraid to make a mistake, they will not risk being creative or innovative. Supportive leaders encourage followers who make mistakes in an effort to find a better way. People respond in very positive ways to encouragement, and when you support your team, you will find it easier to move them toward accomplishing the team's goals. Celebrating accomplishments and giving proper recognition are thus crucial in leading successful groups and organizations.

This does not mean that leaders support or accept everything, especially violation of policies, laziness, or insubordination. It would be foolish to tolerate bad attitudes or wrong behaviors. However, leaders at any level will, sooner or later, encounter a situation in which they either need support themselves or find it necessary to support others. If followers have received support from the leader, they will in turn support the leader when the need arises.

Self-Confidence

Self-confidence is belief in oneself. If we do not believe in ourselves, we cannot lead. If we have no self- confidence, we will not set goals or vision. We will not plan. Why should we? We do not have enough confidence in ourselves to make it happen.

Self-confident leaders display confidence in their capabilities and in the correctness of their positions and decisions. They project this confidence to followers and coworkers. They know they will find a way to reach their goals and complete their tasks.

Responsibility and Dependability

You must be able to follow directions before you can become a good leader. Workers must be responsible and dependable; therefore, leaders must be, too. A group likes to know that the leader is getting the job done. Dependability involves getting to work on time; completing assignments; following through on promises; giving an honest day's work; not missing school, work, or meetings without notification; and completing tasks on your own. The willingness to accept responsibility is equally important. Responsibility means being reliable and accountable for your actions and fulfilling your obligations. Think about how effective a leader or boss would be if they never made it to work on time or if they couldn't be trusted with certain tasks. They wouldn't be in charge for long, would they?

Emotional Maturity

Some people never grow up. They are emotionally immature. Their feelings are hurt easily and they get angry if they do not get their way. They may pout, get into verbal battles, slam doors, refuse to talk, and indulge in self-pity. They do not ask for anything because they fear rejection.

They are afraid to confront someone when an issue arises. We once had a brilliant colleague who was very good at his job, but when he was confronted or challenged by others he exhibited anger and the emotional maturity of a child. Ultimately, this cost him his job.

Maturity balances personal courage with consideration for others. A person is mature if he or she can express feelings and convictions with courage and honesty, tempered by consideration for the feelings and convictions of others, particularly if the issue is important to both parties.[27] It is not enough to be nice. You must be not only brave, assertive, and confident but also empathetic, considerate, and sensitive.

Takes Pleasure in Work

If a person does not like the job he has, he should get one he does like. If a person does not like children, he should not teach. If an airplane pilot is afraid of flying, she should not fly. Similarly, if you are not comfortable leading, do not lead.

Successful leaders derive pleasure from what they do. They enjoy leading, directing, and organizing. They set goals and get pleasure from achieving those goals—and from setting new goals once the old ones have been reached. If they do not see any further goals, successful leaders may become bored or dissatisfied. Bill Walsh, Bill Parcells, and Joe Gibbs, all former professional football coaches, retired after multiple Super Bowl wins only to take new coaching positions with new teams later in their careers. Why? They had accomplished their goals and felt that they were growing stale in jobs that they did not like any more. Changing jobs offered an opportunity to set new goals or renew former goals, thus providing another chance to receive pleasure from work.

Servanthood

Leaders must serve their followers or coworkers, either directly or indirectly. For example, elected officials are referred to as public servants; many leaders have to speak at public meetings and maintain a schedule dictated by others.

Servant leaders are sensitive to the needs and feelings of their people. They are supportive of, helpful to, and concerned for the well-being of their coworkers. Such leaders put the needs of others above their own interests. Because getting a fair deal is of great concern to employees at all levels, morale and productivity are highest in work groups where the leader shows a high degree of consideration, servanthood, and concern. Servant leadership does not mean sacrificing your goals; in fact, the more you give of yourself, the more you receive and the more able you will be to influence others positively toward achievement of common goals.

> *The time is always right to do what is right.*
>
> —Martin Luther King Jr.[28]

Sets an Example for Others to Follow

What you do as a leader exerts far more influence on employees or colleagues than what you say. Around your coworkers, you are constantly on stage. Everything you do is interpreted as acceptable behavior for them to follow. People are not changed by force or intimidation, but by example. Great leaders never set themselves above their coworkers except in carrying responsibilities. Figure 4–7 suggests some ways for leaders to lead by example.

Availability and Visibility

If others are going to follow your example, they have to be able to see you. Face time with your followers is very important—even if it's just eating lunch together or walking as a group to the next meeting. When the opportunity arises, share examples with your employees about how you and they have collectively solved problems together. You can also increase your visibility by helping your followers connect with others in your network and by training them on important tasks yourself. Last but not least, don't forget to use the power of social media to increase your availability and visibility. Let followers see your leadership vision and human side via social media, and their loyalty will grow.[29]

Empowerment

One of the biggest mistakes leaders make is not having faith in people. You must trust people even if it involves some risk. Research has proven

---

LEADERSHIP: SET THE EXAMPLE FOR OTHERS TO FOLLOW

1. Be totally committed to doing the best job possible.

2. Support yourself with the highest-quality people.

3. Maintain a cheerful, pleasant attitude.

4. Take the heat for your own errors—don't blame others.

5. Be willing to say, "I made a mistake" or "I don't know."

   There are no mistakes so great as that of being always right.
   —Samuel Butler

6. Avoid indiscreet, negative criticism of co-workers, press, and superiors.

7. Work hard and smart.

8. Stand up for the principles you believe in.

   Keep true, never be ashamed of doing right; decide on what you think is right, and stick to it.
   —George Eliot

9. Keep an open mind.

10. Be diplomatic.

11. Maintain a positive mental attitude.

    Nothing can stop the man with the right mental attitude from achieving his goal; nothing on earth can help the man with the wrong mental attitude.
    —W. W. Ziege

12. Develop a professional, energetic image.

13. Be a team player.

14. Be enthusiastic.

15. Treat your co-workers as you would like to be treated—with respect and dignity.

    Example is not the main thing in influencing others. It is the only thing.
    —Albert Schweitzer

FIGURE 4–7 As leaders, our walk is more important than our talk: we lead by example. Each of these attributes helps make a leader more successful in setting examples. *Source: Adapted from: Brent Gleeson, "7 Simple Ways to Lead by Example," April 23, 2013. Retrieved July 7, 2016 from www.inc.com/brent-gleeson/.*

---

that a sense of trust from a leader improves task performance and behavior on a team. In other words, followers are more comfortable and they do a better job.[30] Many leadership experts refer to trusting others as **empowerment**. It is called this because you are giving power away. Many times, empowered individuals will perform far beyond expectations. Empowered followers also take initiative in their work, feel that making

THE COACH WHO SET THE EXAMPLE

ROB TODD got his most powerful lesson in leadership through an example from a young basketball coach, Todd Hess, in rural Cannon County. Hess made a lasting impression on Todd and the lucky few athletes who were given the opportunity to play under his guidance.

Fresh out of college at the age of 26, Coach Hess began his coaching career at Cannon County High School. On the first day of tryouts, 45 eager boys awaited the introduction of the school's new coach. Coach Hess promptly emerged from his office and began sizing up the rough-looking bunch like a military drill sergeant. He began his introductory speech by going over his policies for the team and his new rules for the basketball program. Hess admitted to being a strict authoritarian who enforced every rule with no exceptions. He also proudly announced that he would not ask any member of his team or anyone trying out to do anything that he himself would not do. Like many introductory speeches from coaches in the past, Coach Hess's speech drifted in one ear and out the other. When the whistle was blown to signal the start of a scrimmage, the coaching staff began assessing the talent of those trying out, and the speech was forgotten.

After two and a half hours of brutal scrimmaging, Hess blew his whistle again, signaling what the potential team members thought was the end of the first day of tryouts. As the team gathered around for a few final words, they were ordered to line up for wind sprints. Shocked, they made their way to the baseline. Disgruntled and tired, they mumbled among themselves about doing conditioning running before the team had been picked or final cuts made. Suddenly, the whistle blew and they began a grueling battle to see who could last the longest at the wind sprints. After 10 repetitions, some weakhearted participants began filtering out the door of the gymnasium. Just as Rob was about to follow the crowd, he looked over to see Coach Hess panting on the sideline with the rest of the team, gasping for air. As they watched, the coach blew the whistle again and took off, leading the small group that had had enough desire to stay. Rob stayed—and was glad he did.

That first day set the stage for Coach Hess's coaching career at Cannon County. They ran 25 wind sprints that day, with the coach running right along with them. Out of 45 boys who began tryouts, only nine showed up the second day. Coach Hess remained true to his promise: from that day on, he ran every wind sprint with the team and participated in every drill. That year, the nine players became a team. Not a coach and his players, but a team.

Coach Hess's actions spoke much louder than his words. He is the only athletic coach that Rob still refers to as "coach." For Rob, Coach Hess also was the first person who truly taught a leadership lesson by example—not only as a sports coach but also for life in general.

---

positive change is their responsibility, and give extra effort.[31] If you make an effort to show confidence in your followers by creating a culture or environment of empowerment, then follower performance will improve.[32] In Figure 4–8, the blindfolded FFA member in the center of the circle is showing confidence in her team members.

Uses Power Wisely

Leaders need to be strong enough to tackle tough problems and gentle enough to keep the solutions humane. They must be demanding enough to challenge others not to settle for easy answers and patient enough to know that progress takes time.

Wise use of power brings all the other leadership qualities together. "It energizes people and resources." When you use power wisely, you show that you are trustworthy. People know you are sensitive and democratic and will not take advantage of them. When your coworkers trust you, you gain their loyalty and respect.[33]

TECHNICAL-HUMAN RELATIONS SKILLS

Because the preceding group of human relations skills tends to tap into our innate personality type or value system, those skills may not come naturally to everyone; they may require quite a bit of effort to develop and maintain. In contrast, the following technical-human relations skills can be learned rather easily.

Listening

Did you ever wonder why we were created with two ears and one mouth? Perhaps to help us listen twice as much as we talk! Listening

FIGURE 4–8 The FFA member in the middle of the circle is showing confidence in fellow team members.

attentively and with concern for others is one of the best ways to show respect. It demonstrates that you believe the individual who is communicating with you has worthwhile thoughts and therefore is a valued member of the team. When people feel needed, they tend to take more interest in what they do. When coworkers no longer believe that their leader listens to them, they start searching for someone who will. (Listening is discussed in detail in Chapter 6.)

Leaders who make a habit of drawing out the thoughts, ideas, and suggestions of their coworkers, and who are receptive even to disappointing news, will be properly informed. Coworkers who are not afraid of the leader's reaction will not lie or hide unpleasant truths. Having all the appropriate information is very important to a leader's ability to make good decisions.

Team (Group) Building

A leader builds group cohesiveness and pride. The thing that motivates "all successful organizations is the pride and sense of accomplishment that come from striving for and achieving high levels of performance. Employers want to be a part of a winning team. Successful teamwork is exciting . . . and offers personal rewards." "Deep within all of us is the desire to be a winner—to be somebody."[34] Well-led coworkers feel needed, important, and part of a worthwhile enterprise.

Successful leaders seeking a cohesive team will also focus on the following:

1. Selecting the right members of a team
2. Making sure all team members understand team goals and objectives
3. Keeping communication free and open
4. Promoting trust
5. Assisting in conflict resolution
6. Encouraging feedback
7. Making time for fun[35]

Earning Loyalty

Earning the loyalty of followers is important because "all teams hang together or fall apart because of loyalty."[36] Effective leaders make followers and/or coworkers feel good about themselves and their work. People thrive on the appreciation you show them—smiles, thanks, gestures of kindness. Ann Mabry, an elementary school teacher, says that students have to learn to like you before they will like to learn.

Coworkers will never be totally loyal until they feel you are sincerely devoted to the good of all and are doing everything in your power to help

WAYS TO EARN LOYALTY

1. Develop warm, person-to-person relationships with your co-workers.

2. Be aggressive in serving your co-workers.

3. Confide in the people you trust.

4. Provide special rewards for your outstanding performers.

5. Recognize and show appreciation for other co-workers, too.

6. Give your co-workers the benefit of the doubt.

7. Never say you will do something for an employee, then change your mind.

FIGURE 4–9 Earning loyalty is a key characteristic of successful leaders. If a leader has earned loyalty, coworkers will be supportive when the going gets tough. *Source: T. Krause, "Leadership: Loyalty from Leadership: Holding the Team Together," Leader Values, 2006. Retrieved July 7, 2016 from www.leader-values.com.*

them succeed both as individuals and as a group. Figure 4–9 lists important guidelines for earning the loyalty of your followers and coworkers.

Rewarding Accomplishments

Many believe that money is the greatest motivator, but studies have shown that this is not true. The greatest morale booster for any group or individual is simply being appreciated.

Appreciation provides encouragement. The root of the word *encouragement* is the Latin *cor*, which means "heart." You can provide encouragement as a reward to individuals or groups. It is also important for a leader to reward by giving due recognition and celebrating accomplishments.[37] Remember that leadership is heart and encouragement; some would even say love. In fact, James Kouzes and Barry Posner, authors of *The Student Leadership Challenge*, believe that love is what sustains a leader.[38] When accomplishments are rewarded, followers feel the love—and the productivity that ensues is contagious.

Coaching and Strengthening Others

Leaders coach their followers to improve performance. "The most important responsibility of a leader is to develop personnel." Your success as a leader depends on how well your people perform.[39] There is no greater proof of a leader's effectiveness than the performance of individuals who have responded to that leader's coaching

and achieved their full potential.[40] Leaders create the kind of organizational climate in which personal growth is expected, recognized, and rewarded. There are five common reasons why coaching to improve performance is necessary:

1. Workers do not know what they are supposed to do.
2. Workers do not know how to do a job.
3. Workers do not know why they should do the job.
4. There are obstacles beyond the workers' control.
5. Workers have an attitude problem.[41]

Encourage your followers or coworkers to explore new options instead of settling for the obvious. If they do not make an occasional mistake, you have a sure sign that they are playing it too safe. The role of the coach is to teach players to stand on their own two feet.

Good coaches know that real learning does not occur unless people are challenged to do their own thinking. Coaching for the playoffs is great, but the payoff is even greater for leaders who coach winners by asking the proper questions, encouraging and reinforcing positive behavioral changes, and setting an example of excellence. The coach in Figure 4–10 is motivating and challenging his players to do their best "and then some."

> *Good leadership consists in showing average people how to do the work of superior people.*
>
> —John D. Rockefeller[42]

Planning Short-Term Goals and Wins

Leaders create "visible, unambiguous success as soon as possible" for their followers. Good leaders convince followers that the impossible is possible for documenting the "wins." Small wins lead to success and motivate coworkers. Small victories create momentum and move teams to greater heights.[43] Successful leaders pursue a strategy of getting first downs or hitting singles rather than always trying for the home run.

Successful leaders find ways to make it easy to succeed. They notice and celebrate movement in the right direction.[44] When the senior author

FIGURE 4–10 Mark Richt is a winning football coach who strengthens his players by instilling character, persistence, commitment, and moral responsibility. *(Photo by Steve Colquitt.)*

worked with students in researching how to run engines with hydrogen from water, our first goal was to run a small lawn mower engine for 10 seconds. The students accomplished this goal, and the win felt tremendous. They had several other short-term goals and wins, including running a large tractor with hydrogen, running a car with hydrogen, and making hydrogen from water on board a vehicle. The biggest win came four years later when they set the world's land speed record for a hydrogen-fueled vehicle at the Great Salt Flats in Wendover, Utah, at the World Finals. Figure 4–11 shows the newest hydrogen hybrid car, which runs off electricity and hydrogen from water. It is a product of attaining many small goals and wins.

Administrative Ability

Leaders must have their priorities in order. They need to give attention to detail, especially regarding finances and personnel. Leaders must have the ability to move their followers and the organization forward. They have to be aware of the 80–20 rule: twenty percent of the people in any group do 80 percent of the work. Four critical administrative abilities follow:

Identify the Real Problem Think about the symptoms and causes of the problem, and gather information to help you identify the real problem.

Discuss the problem situation; then write down your key findings and make a conclusion.

Manage Time Effectively and Shift Priorities as Necessary Apply the 80-20 rule to your time, assigning priorities to the activities that will benefit you the most. Make a to-do list and assign a priority to each item on it.

Explain Work Explain in detail what you expect of your followers. Keep in mind expected results, delegated authority, and progress reports.

Listen Actively Develop a genuine interest in your employees and their activities. When they come to you with suggestions, really listen.

TECHNICAL SKILLS

Technical skills are the teachable "how-to" activities that leaders need to be able to execute. Improvement in the technical skills, such as time management and setting goals, may occur with practice. Technical skills are the most easily obtained and retained of the major types of leadership skills.

Selecting Staff

Leaders need to surround themselves with excellent people. Every person hired should be good enough to replace the leader. It takes a secure

FIGURE 4–11 This vehicle is a result of many small wins. By setting and accomplishing many short-term goals over a period of five years, the researchers ultimately attained the long-term goal of an alternative to oil.

leader to do this, however. The opposite approach is to hire weak people who will make the leader look superior and feel indispensable. Obviously, in this situation, everyone loses. No matter what leadership training you have had or what kind of a leadership coach you are, staff members who are lazy or have bad attitudes will make your life as a leader miserable.

Mr. Hollie Sharpe, a business executive, told the authors about a job interview with Ohio Casualty Insurance. The vice president who was interviewing Sharpe asked him where he wanted to be in five years. After thinking for a moment, Sharpe said, "Sitting in your chair." The vice president was briefly taken aback by the answer, then replied, "I like that." The company knew it was getting a good, aggressive person; Sharpe was hired.

Competence

At the very least, leaders should know what they are doing. Education, either on the job by coming up through the ranks or at a trade or technical school, college, or university, is necessary for successful leaders. You cannot get there if you do not know where you are going. Become competent before you take on a leadership position.

Communication

Verbal fluency is the number-one predictor of promotion within an organization. It is also a major predictor of success in life. If you speak poorly, you will be perceived as unintelligent. If your communication ability is excellent, you will be perceived as very intelligent. "Perception is reality": you are what people think you are.

Chapter 6 discusses communication in detail. Chapters 7, 8, and 9 discuss giving speeches: the FFA Creed, public speaking, and extemporaneous speaking. Chapters 10 and 11 discuss parliamentary procedure. After studying those chapters and performing the activities suggested there, your communication ability should be greatly enhanced.

Time Management

The technical skill of time management is also important. Make every minute count. Visualize the future and prepare for it. Time is your most valuable personal resource. Use it wisely, because it cannot be replaced. Time management skills can be developed, leading to substantial improvement in your productivity. One aspect of wisdom is knowing when to say no. Chapter 15 discusses time management in detail.

Sets Goals and Plans

We have already mentioned the importance of goals, and Chapter 14 delves further into this leadership quality. Lack of goal attainment can cause sadness and depression. Truly successful leaders develop and pursue challenging goals; their goals are their driving force.

Achievement

Leaders are winners. They are achievers. This is why employers want to see a résumé when hiring: résumés provide a list of achievements. Many employers also want to see a **portfolio**. Portfolios are a collection of tangible (and usually visual) evidence of a person's accomplishments and performance, such as samples of a journalist's articles or a photographer's pictures.

Achievement is habit-forming. Leaders who achieve once will achieve again. Achievers find a way to win, even if it means multitasking, getting up earlier, or working harder (Figure 4–12).

Responds to Failure

For many, the word *failure* carries with it a sense of discouragement and finality. For the successful leader, though, failure is just a beginning. Perhaps the most impressive and memorable quality of leaders is how they respond to failure. Successful leaders put all their energies into their task. They simply do not think about failure. They use words such as *mistake, false start, mess, setback,* and *error;* they never use the word *failure.* They believe a mistake is another opportunity to do something better—to learn.

Leads Individuals and Groups

Actual leading of individuals and groups is one of the basic steps in becoming a leader. As a matter of fact, it is so basic that it is taken for granted. One way to learn to lead groups is by learning and practicing parliamentary procedure. Once you have mastered parliamentary procedure, you can conduct successful meetings. Learning group dynamics is also important for future leaders. Chapters 5 and 12 cover the topic of leading individuals and groups.

FIGURE 4–12 Achievement does not just happen. Nonachievers ask what's happening; achievers make things happen. This student is studying hard for his plant lab test in order to achieve. *(Courtesy of University of Georgia—College of Agricultural & Environmental Studies.)*

CONCEPTUAL-TECHNICAL SKILLS

The following leadership qualities fall somewhere between conceptual and technical skills. They are thinking skills, but they can be taught, so the authors of this book call them conceptual-technical skills.

Intelligence

Less has been written about intelligence and leadership than about any other leadership quality. We do know there are so many different types of intelligence that no one type guarantees the ability to lead.

Consider the following dilemma: Keith has much wisdom. He always seems to know just what to do, how to solve problems, and how to make excellent decisions. However, Keith struggled to complete high school because of family problems. Kim was an honor student and scored very well on the ACT, but she does not seem to know how to get things done. Sandra is an excellent speaker with a winning personality, but she never even made the dean's list in school. Randy is a near-genius mechanic, but his college entrance test scores were too low to win him admission to college. Who should fill the leadership role?

Any of these people could get the leadership role, but they are all intelligent in different ways. Part of being intelligent is knowing your strengths and weaknesses. The former president of the University of Tennessee, Dr. Andy Holt, once said, "Know a little about everything, but know much [be an authority] about something." Capitalize on your strengths and use the intelligences in which you are strong; then surround yourself with people who excel in areas in which you are weak. With this combination, you can build a winning team.

Problem-Solving Ability

Problem solving is a methodical process of finding solutions and making decisions. Xerox Learning Systems identified problem solving and decision making as two of the top six critical skills for the success of supervisory leaders. Problem solving is the second most time-consuming task of a supervisor.

With all problems, certain steps are required to reach solutions. To solve a problem, follow the steps of the scientific method: define the problem, determine and evaluate alternatives, gather data, select a workable solution, implement the solution, evaluate the solution, and implement the decision. Chapter 13 discusses these steps in detail.

Decision Making

Decision making is the process of choosing between alternative solutions to a problem. It is a basic function of leadership. Decisions must be made when you are faced with a problem. The first decision is whether or not to take action to solve the problem.

Making poor decisions can get leaders removed. The decisions that leaders make can affect the health, safety, and well-being of employees and the general public. Ethical decision making is becoming more important as new technologies become available. As with most leadership qualities, decision-making skills can be developed. Chapter 13 discusses the decision-making process.

Cooperative Genius Approach (Synergy)

A technique that has proved very useful to many leaders is the **cooperative genius approach**. If 10 people focus on the same goal, and each person is one-tenth as smart as Thomas Edison or Albert Einstein, the group can accomplish great things. Leaders can assign each person on a team different parts of a puzzle to solve. This process makes for a group with unlimited intelligence that would make even Einstein jealous.

This process can also be called **synergy**. Simply put, it means the whole is greater than the sum of its parts. The relationship the parts have to each other is a part in and of itself. In *The 7 Habits of Highly Effective People*, Stephen Covey points out that

> [s]ynergy is everywhere in nature. If you plant two plants close together, the roots co-mingle and improve the quality of the soil so that both plants

will grow better than if they were separated. If you put two pieces of wood together, they will hold much more than the total of the weight held by each separately. The whole is greater than the sum of its parts.[45]

Synergy is teamwork at its best.

Master of Delegation

The leader who attempts to "do it all" is heading for disaster. Successful leaders know that they cannot know everything; as humans, they can know and do only so much. No matter how dedicated a leader may be, each individual's time and energies are limited. The successful leader delegates authority along with responsibility. Nothing is more frustrating than being assigned responsibility without the power needed to carry out the task.

A leader must feel secure before committing to delegation. Resistance to delegation comes from fear of losing control and the habit of holding on to work. Consider the following as you delegate:

- Planning is the first step in delegation; it largely determines the success and efficiency of the action.
- A key to effective delegation is delegating the right task to the right person.
- The delegator should specify the *what*, *when*, and *why* in delegating—but not the *how*.
- To delegate effectively, the leader must give authority and responsibility to followers; however, the ultimate responsibility remains with the delegator.
- Delegated tasks should be evaluated by results rather than the means used to accomplish them.[46]

Opportunist

Opportunists take advantage of unplanned situations, coincidences, and serendipity and turn them into positives. Opportunists do not see problems; they see opportunities. Risk and possibility go hand in hand when every problem is an opportunity. "Reasons first, answers second" is the first rule of an opportunist. If you have

---

VIGNETTE 4-4 | **SYNERGY: 1 + 1 = 3**

SYNERGY IS NOT A NEW CONCEPT; it is just a new word that has been applied to a way of doing things. It has been practiced for a long time, but now people are making an effort to learn it so that they can get more done. Some call it new math because it is like saying 1 + 1 = 3, but it is essentially a way of learning how to get more out of what you have now, whether it is time or knowledge.

Synergy can be as simple or as complicated as the person trying to achieve it. It is a process of combining two or more activities, tasks, brains, or lessons and getting more from doing them together than you would by doing them individually. For example, you can apply this to something as simple as doing daily chores. Make a list of everything that has to be done:

- Wash clothes
- Dust rooms
- Clean bathrooms
- Buy groceries
- Go to the bank
- Fill up gas tank

To get finished faster, group together items that can be done simultaneously. In this instance, you can do laundry while dusting and cleaning the bathrooms, stopping only long enough to put clothes in the washer or dryer. Instead of making multiple separate trips to the store, bank, and gas station, plan your route before you leave and make all those stops along the way in one outing.

The purpose of synergy in this example is to get through the daily chores faster so as to free up more time to do other things. The main focus of synergy is to find a better way, to achieve more toward an objective by multitasking.

---

enough reasons to seize the opportunity, you will find answers.

Effective leaders are opportunity oriented rather than problem oriented. They feed opportunities and starve problems. They think preventively. Also, when leaders see their coworkers' problems as opportunities to build a relationship instead of as burdensome irritations, the nature of the interaction changes. Leaders become more willing, even excited, about understanding and helping their coworkers.

Insists on Excellence

Great leaders are never satisfied with current levels of performance. They constantly strive for higher levels of excellence. If you insist on excellence as a leader, there is a high likelihood you and your followers will achieve it.

An effective leader brings out the best in people by stimulating them to achieve what they thought was impossible. In a classroom experiment with grade-school children, one teacher was told that she had a group of high academic achievers; the truth was that they were average students. Another leader was told that she had a group of learners who traditionally underachieved; actually, they too were average. The result of the experiment was that the so-called high achievers progressed much more rapidly than the group perceived to be underachieving.

Both student groups had equal intelligence levels. The difference in performance came from how the teachers perceived and therefore related to their students. The lesson here is that if you lead with high expectations, your people will attain them. Lead with low expectations, and your people will also attain them—and no more.[47]

Holds Coworkers Accountable

An effective leader is tough-minded about getting results from followers or coworkers. When goals are met, leaders should provide appropriate recognition. But when goals are not met, a good leader engages in meaningful conversations with followers to let them know the consequences. We all live and work in a data-driven society, and accountability is inevitable. It may be uncomfortable for a leader, but if a follower doesn't perform as expected, leaders have to have courage to hold followers accountable. Effective leaders seldom have to use this power, but the employees must know they will do so if necessary. Effective leaders can handle occasionally being unpopular or disliked by some people.

Broad Interests and Abilities

Leaders must have broad interests and abilities to obtain respect from their groups and to

---

VIGNETTE 4-5 INSIST ON EXCELLENCE

INSISTING ON EXCELLENCE IS OFTEN A DIFFICULT TASK FOR LEADERS

Effective leaders strive to bring out the best in everyone around them by stimulating others to achieve higher goals—even ones that seem impossible. Another way "insisting on excellence" is often presented to many young people is as the "I Can" principle: when you insist on excellence, you must always believe in yourself and never say "I can't." The correct definition of the word *insist* is to be firm in a demand or course of action. The definition of *excellence* is a state, quality, or condition of excelling or being superior.

Excellence can also be explained in relation to perfection. Perfection is what you always strive for, while knowing that true perfection is an impossibility. However, *striving* for perfection is not an impossibility; neither is insisting on excellence. If you do the very best you can under the conditions that exist day to day, you are striving for perfection. When you do so, others will follow your lead. As a leader, you should insist that your group make the effort to contribute in whatever ways possible.

People want to believe that their leaders are sincerely interested in them. If you establish high expectations and show interest in people for who they are and what they can do with a little motivation, they will know you are someone they can trust and follow.

Former UCLA basketball coach John Wooden was a philosopher, coach, and insister on excellence. His beliefs reached far beyond the court, ultimately addressing how to bring out the very best in yourself and others in all areas of life. Bill Walton stated of Wooden, "He is a master teacher who understands motivation, organization, and psychology. The skills he taught us on the court—teamwork, personal excellence, discipline, dedication, focus, organization, and leadership—are the same tools you need in the real world. John Wooden taught us how to focus on one primary objective: Be the best you can be in whatever endeavor you undertake." John Wooden insisted on excellence and got it.

---

be the best possible communicators. Can you imagine a business luncheon led by someone who has no interest in sports, politics, or international affairs? Good leaders are well-read and stay up to date on the news and current events.

CONCEPTUAL SKILLS

Conceptual skills are thinking skills. They are needed to analyze situations and generate ideas. Like technical skills, these skills can be developed and sharpened with practice. It is helpful to use scenarios, role-playing, and "thinking games" (problem-solving games) in developing conceptual skills.

Vision

Vision is a critical dimension of effective leadership. Without vision, there is little or no sense of purpose in an organization; efforts are aimless and uncoordinated. However, a vision and intellectual strategies are insufficient in and of themselves to motivate and energize followers. Substantial results come only if followers or coworkers take on the leader's vision as their own; only then will they accept responsibility for achieving that vision.

Outstanding leaders are visionaries. They have to dream about what could be and involve others in their dreams. Like small creeks that grow into mighty rivers, the dreams of these leader visionaries eventually shape the course of history. Do you know a visionary leader? Can you think of some visionary leaders throughout history?

Sam Walton, Jack Welch, Herb Kelleher, Frederick Smith, Walt Disney, and Bill Gates are just some examples of excellent leaders who set a vision for their organization and saw the vision actually happen. Being a visionary works in conjunction with many of the skills we are discussing, and it requires long-range, enterprise-level thinking rather than getting bogged down in the day-to-day issues that trouble many leaders.[48]

Ability to Motivate

Leaders must be able to motivate or inspire others. Some leaders motivate with enthusiasm or charismatic behavior; others inspire and move people through persuasion. Whatever the method, leaders must understand motivation and be aware of both psychological and physiological needs.

Motivation includes recognizing people, including them, encouraging them, and teaching them the required behavior. A leader must make followers feel loved, trusted, respected, and cared for. Above all, leaders must motivate people to excel for themselves and for the organization.

Decisiveness

Decisiveness means making appropriate decisions quickly and confidently. Some decisions must be made immediately. Successful leaders have the courage to take action when others hesitate. Some contend that a wrong decision is better than no decision. Whether or not this is always true, a leader who does not give an answer may delay someone else from pursuing alternatives or finding solutions. In other words, "Tell me yes, or tell me no. But tell me *something.*"

Another dimension of decisiveness is making hard decisions. Leaders also have to be willing to make the unpopular but appropriate decision. The leader must do what he or she thinks is right, not what someone else thinks should be done.

In today's work environment, it is also imperative that coworkers be involved in the decision-making process. People carry out decisions that they have participated in making much more enthusiastically than they carry out orders. It has been proven that involvement leads to commitment. Nevertheless, the decision-making burden ultimately belongs to the leader.

Straightforwardness

Somewhat related to being decisive is being straightforward. To be straightforward is to "tell it like it is." In *Principle-Centered Leadership*, Stephen Covey wrote, "When you are living in harmony with your core values and principles, you can be straightforward, honest, and up-front. And nothing is more disturbing to people who are full of trickery and duplicity [deceit] than straightforward honesty—that's the one thing they can't deal with."[49] The person in Figure 4–13 is being straightforward, as evidenced by her body language and facial expression.

Creativity

Leaders who make a difference are creative. They know that seeing things others cannot see is not just a quality of leadership; it is a responsibility.

FIGURE 4–13 We have to lead according to the situation and our personality type. This leader is being straightforward.
(Courtesy of University of Georgia—College of Agricultural & Environmental Studies.)

Creativity is as much an attitude as it is an aptitude. Walking on the moon and heart transplants were somebody's creative ideas that became realities. Without creativity, we would make very little progress.

Creativity, or **innovativeness** as some experts prefer to call it, is an attitude that sees people, places, and things bigger and better than they see themselves. As a creative leader, you will find positives where others do not, you will seek opportunity where others find only problems, and you will see answers where others have not yet asked the questions!

Creativity and imagination are abilities distinct from the capacity to acquire knowledge. Albert Einstein said that imagination is "more important than knowledge." To enhance creativity and productive innovation, leaders must let their imaginations flow.

Courage

Courage is demonstrated when a person exhibits perseverance and determination in the face of adversity, conflict, or an extraordinary challenge. Courageous leaders "do not suffer from the crippling need to be loved by everyone. For example, they are not afraid to say 'no' to unreasonable requests and demands placed on them and to take positions on issues of importance." Working through disapproval and even ridicule, the courageous leader eventually earns respect. In contrast, those who try to be the "good guy" by pleasing everyone end up losing the respect of the entire organization.

Success is never final and failure never fatal. General George S. Patton said that "courage is fear holding on a minute longer."[51] It is courage that counts. Leaders with courage face conflict openly and directly, follow a difficult path in the face of danger, and boldly support their beliefs and values.[52]

Risk Taking

Risk taking is another skill without which a leader cannot operate. Leaders who are making a difference take many risks during the process of visioning, managing, and improving a goal, project, product, or organization. Taking risks comes more naturally to some, but it is never easy. The opposite of taking a risk is being frozen in time, never trying anything, and thus, never accomplishing anything.

VIGNETTE 4-6   HOW DO WE DEVELOP CREATIVITY?

CREATIVITY CAN BE DEVELOPED IN SEVERAL DIFFERENT WAYS, and everyone who develops creativity does so in his or her own way. Consider the following options for developing creativity in yourself and others:

- **Exercise your brain.** Brains, like bodies, need exercise to stay fit. If you do not work your brain, it will get flabby. Talk to smart people and civilly disagree with people; arguing can be a terrific way to give your brain cells a workout. Note, though, that arguing contrasting positions about politics or the Farm Bill is good for you, whereas bickering over whose turn it is to wash the dishes is not.

- **Brainstorm.** If properly carried out, brainstorming, the unrestrained offering of ideas or suggestions, can not only help you come up with many new ideas but also help you decide which idea is best.

- **Record your ideas.** Always have a small notebook and a pen with you (or use an electronic device as a notebook). That way, if an idea strikes you, you can quickly jot it down. When you reread your notes, you may realize that about 90 percent of your ideas are dull, incomplete, or just plain nuts. That is quite normal. What is important are the 10 percent that are brilliant and intriguing.

- **Use a dictionary.** If you cannot seem to generate any ideas, randomly select a word from a dictionary and then try to formulate ideas incorporating that word. You might be pleasantly surprised at how well this works.

- **Define your problem.** Grab a sheet of paper, smartphone, computer, or whatever you use to make notes, and define your problem in detail.

- **Do physical exercise.** If you cannot think, take a walk. "A change of atmosphere is good for you, and gentle exercise helps shake up the brain cells."

- **Read.** Read as much as you can about anything and everything. "Books exercise your brain, provide inspiration, and fill you with information that allows you to make creative connections easily."[50]

These methods of stimulating creativity are not the only ways to enhance your innate creative abilities, but they are ones that can challenge most of us in productive ways.

A good leader not only takes risks, but she helps followers become more comfortable with taking them as well. Teams and organizations achieve more when everyone in the organization takes risks such as being creative, trying new things, and seeking innovative solutions to problems. And, somewhat paradoxically, a risk-taking leader who develops risk-taking followers must also be comfortable with failure: they have to be willing to learn from mistakes and take more risks.

Commitment, Determination, and Persistence

It is hard to discuss these three qualities of leadership separately. Commitment, determination, and persistence are all traits of a leader who finishes the race, completes battles, and attains goals. These leaders show the toughness of a bulldog, which has a nose that slants inward so it can breathe while it holds on.

Commitment is "the binding force that holds the other leadership qualities together" and makes them work. Commitment is "the inner strength that keeps you going when everyone else gives up," helping you through the tough times.[53] Faithfulness and persistence are two important ingredients of commitment.

Good leaders show exceptional determination in pursuing their objectives. They never give up until they succeed. Timing is also important. Sensitivity and knowledge of when to push for change and improvements are often keys to obtaining the desired change. People are not always psychologically ready to accept innovative ideas. With determination and timing, though, you cannot lose.

Sense of Urgency and Initiative

Without urgency, progress is slow. The attitude has to be "Let's do it, and let's do it now!" Leaders seize the moment. However, a particularly difficult challenge of leadership is to maintain that sense of urgency while accomplishing difficult goals year after year. Many believe it is easier to build a winning team from scratch than to keep it continuously on top. Here are three ways for leaders to keep everyone energized and inspired:

1. Periodically change or rotate coworkers' responsibilities
2. Take unusual or unexpected action
3. Set the example of urgency for others to follow

Initiative is recognizing one's responsibility to make things happen and then fulfilling that responsibility. Taking initiative does not mean being pushy, obnoxious, or aggressive. It does mean acting rather than allowing oneself to be acted upon; many people just wait for something to happen or someone to take care of them. The people who get the good jobs and eventually become successful leaders are the ones who actively seek solutions to problems. They seize the initiative to do whatever is necessary, consistent with high ethical standards, to get the job done.

Follow-Through

Some leaders are great at starting projects, but not at seeing them through. Once the project is launched, their enthusiasm fades quickly. Thereafter, if the new idea or project does not run as smoothly or work as well as first planned, they let it die a slow death. If ideas or projects are worth doing, they should be given a chance to work, even if the natural or realistic timetable for bringing a new idea or project to fruition is slower than

the leader expects. If a project really does not mature, the leader should follow through with a formal closure, not just let it disappear without proper evaluation from the group.

Master of Change

Change is going to happen. The trick for leaders is knowing how to manage change and how to lead followers through changes. Leaders have to realize that, basically, people do not like change. They like things as they are, if only because they are used to the status quo. Fear of change also comes from fear of the unknown, fear of being different, or lack of trust in the change. Everyone handles change differently, and some are more open to it than others.

In fact, people's adaptability to change follows a bell-shaped curve (Figure 4–14). For example, early adopters of change comprise about 13 percent of the population, whereas those resistant to change comprise another 16 percent. Creative thinkers create the changes. Early users are the first group of people to try the new idea, invention, or concept. The early coworkers watch an early adopter, who tends to be a leader, before they make any change. The late coworkers are openminded but require much convincing before they will change. Diehards change only because they have no other choice; either no other options are available or "everyone else is doing it."

Consider the introduction of the round hay baler. Very few farmers approved of the round

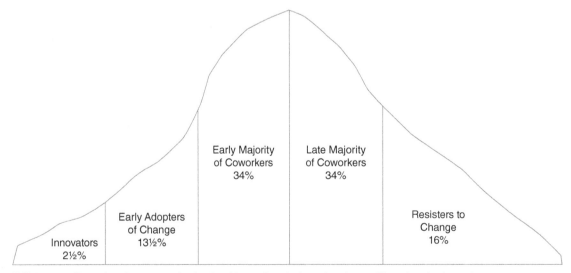

FIGURE 4–14 Change is a slow process. It takes 7 to 14 years from the inception of a new idea to its adoption and use.

hay baler at first, but after a few years went by, they adapted; acceptance of the round hay balers followed the chart pattern exactly. Nowadays, hardly anyone uses the traditional small square balers except for special purposes.

Here are three steps you can take to foster receptivity and openness to change:

1. Provide new information to expand your followers' thinking
2. Provide new ideas to spark their creativity and broaden their horizons
3. Provide new experiences to build a desire for and a belief in the value of change

Wisdom

Wisdom is knowing how and why human beings function and progress. To have wisdom is to have the ability to size up a situation quickly. Wisdom also includes having insight, awareness, and timing. A wise leader can differentiate between the authentic and the fake, between things worthwhile and things unworthy, between things lasting and fads that will fade quickly.

Knowledge for the sake of knowledge is not important; it *is* important that leaders have the wisdom to apply what they learn. The pursuit of wisdom builds character and gives you the inner tools to lead others.

Desire to Lead

Your potential as a leader is of no value if you have no desire to lead. With leadership comes responsibility. Many people desire a simpler, worry-free lifestyle. Someone may be offered a leadership position but, for family, economic, physical, or other reasons, decide to stay where he or she is and be quite happy with the decision, and this is fine.

However, each of us has leadership responsibilities beyond the workplace. Many of these leadership responsibilities are unavoidable, such as leadership within the family, and some are required as a part of public or community service. When you are faced with taking on leadership responsibilities, be assertive, study, review the leadership qualities presented here, and seek help from peers and role models. Refer to Figure 4–15 for categories and examples of organizations in which leadership skills may be used.

Critical Thinking

Critical thinking involves carefully considering all options before making a decision about what to believe or do. It can become second nature, but we have to practice critical thinking if we hope to be able to use it to lead or influence others.

| CATEGORIES OF ORGANIZATIONS IN WHICH LEADERSHIP SKILLS MAY BE USED ||
ORGANIZATION	EXAMPLES
Work	Supervisor, firefighter, union officer
School	FFA, FCCLA, SKILLSUSA-VTCA, DECA, FBLA, BPA, student council
Civic/service clubs	Kiwanis, Lions, Rotary, Exchange, Jaycees, Masons, Shriners, Elks, Eastern Star
Economic organizations	Cooperatives, Cattleman's Association, chamber of commerce
Professional organizations	National Association of Agricultural Educators, National Education Association, Business and Professional Women
Charitable organizations	Red Cross, United Way
Advisory committees	School advisory committees, community advisory committees
Special-interest groups	County fair boards, American Legion, ecology groups
Political organizations	Republicans, Democrats
Support groups	FFA alumni, Parent-Teacher Association, Band Boosters
Age-oriented groups	Boy Scouts, Girl Scouts, senior citizen groups
Athletic groups	Little League, riding clubs, ski clubs

FIGURE 4–15 Leadership skills are used in the workplace as well as in a variety of academic and civic organizations. *(Adapted from Curriculum Instructional Materials Center, Stillwater, OK: Oklahoma Department of Education.)*

Critical thinking has been defined by Richard Paul and Linda Elder as "the disciplined art of ensuring that you use the best thinking you are capable of in any set of circumstances." Better thinking could be clearer thinking, more accurate thinking, or more justified thinking. Leaders cannot afford to make decisions based on poor thinking. Learning critical thinking can help you make the best choices for you and your followers.[54]

TEN WAYS TO IDENTIFY A POTENTIAL LEADER

"The most gifted athletes rarely make good coaches. The best violinist will not necessarily make the best conductor. Nor will the best teacher necessarily make the best head of the department." It is "critical to distinguish between the skill of performance and the skill of leading the performance"—they are in fact two completely different skills. It is also important to identify those who are capable of learning to lead. Fred Smith, the founder of FedEx, offered 10 ways to identify a potential leader.[55] Consider the following traits to help you determine whether someone is capable of learning to lead.

Leadership in the Past

"The best predictor of the future is the past." If you find yourself the leader of a business some day, take note of anyone who was an active FFA member or a former Scout leader. People who were leaders in other organizations probably have leadership potential for the workplace.

The Capacity to Create Vision

Potential leaders need to be excited about the future of the organization and interested in building something great. "A person who doesn't feel the thrill of challenge is not a potential leader."

A Constructive Spirit of Discontent

Though some call this criticism, there is a world of difference between being constructively **discontent** (dissatisfied) and being critical.

When people ask about other ways of doing something, test their leadership potential by asking whether they have thought about what a better way might look like. If they have not come up with alternatives, they are merely being critical; if they have, they are exercising "a constructive spirit of discontent." "People locked in the **status quo** [the way it already is or has been so far] are not leaders." Ask a potential leader, "Do you believe there is always a better way to do something?"

Practical Ideas

Highly original people often cannot tell whether their ideas can become reality and be put into practice. "Not everybody with practical ideas is a leader, of course, but leaders seem to be able to identify which ideas are practical and which are not."

A Willingness to Take Responsibility

Many people equate responsibility with worry. Potential leaders, in contrast, do not feel the need to leave their responsibilities at work, because they enjoy accomplishment—"the vicarious feeling of contributing to other people"—that is an essential part of leadership. Leadership does, however, create somewhat of a gap between the leader and his or her peers, as a result of carrying responsibility that only the leader can bear.

Strong Commitment and Persistence

Someone who will not give up on a problem until it is solved has leadership potential. This ability to get things done (Smith calls it the "completion factor") is crucial for leaders, who must work through the times when nothing but willpower keeps them going.

Mental Toughness

No leader escapes criticism and discouragement. A tough-minded leader sees things as they are and is ready to do whatever it takes to fix problems and accomplish goals. A strong leader is able to withstand some loneliness while making decisions and formulating plans.

Peer Respect

Having the respect of one's peers is an indicator of character and personality. Look for people whose peers and associates are supportive and want them to succeed. "It isn't important that people like you. It's important that they respect you. They may like you but not follow you. If they respect you, they'll follow you, even if perhaps they don't like you."

Family Respect

How does a person's family feel about him or her? The presence or absence of family respect is an important indicator of potential leadership ability.

Commanding Attention: A Quality That Makes People Listen

When potential leaders speak, people listen. Other people may talk a great deal, but if nobody pays attention, they are not providing leadership; they are merely giving speeches. "[T]ake notice of people to whom others listen."[56]

CONCLUSION

No matter what your level of leadership, and no matter what your leadership accomplishes, it is exciting to know that you are in the process of becoming. You are an ever-changing composite of the things you say, the books you read, the thoughts you think, the company you keep, and the dreams you dream.

What you think is what you become. Your potential is endless and your ability to make a difference has no boundaries. It is said that one person with belief is equal to a hundred with only interest. If you think you can, you can. If you think you cannot, you cannot.

Many different qualities are needed for effective leadership. Even if you possess 95 percent of the required leadership qualities, the 5 percent you lack can prevent you from being a great leader. Some of the leadership qualities will come to you easily and naturally, but learning and developing some other qualities will require study, training, and experience. Learn all you can about leadership and the leadership qualities needed to be an effective, responsible, and empowering leader.

SUMMARY

Leadership is an observable, learnable set of skills and qualities. Leaders come in many personality types. Leadership has a purpose and goal. Leadership is a group effort.

To become a leader who can influence others in a positive way for your school, community, or job, you will probably go through certain stages and dimensions of leadership. One model for leadership development is divided into three stages: awareness, interaction, and integration. The five dimensions of leadership development include leadership knowledge and information; leadership attitude, will, and desire; decision making, reasoning, and critical thinking; oral and written communication skills; and intrapersonal and interpersonal skills.

Personal leadership development is important. It is a challenge to learn the leadership skills that will earn respect from others. Leadership helps you mature, understand yourself, and develop your self-concept; it also helps you understand others.

The best way to learn leadership is to place yourself in situations requiring leadership action. Five steps to becoming a good leader are studying, analyzing oneself, developing, learning, and following a definite plan.

Steps of your personal leadership plan include developing a vision; setting goals; developing initiative, self-confidence, personal responsibility, a healthy self-image, and self-organization; eliminating procrastination; studying; and magnifying your strengths.

Most of us like to be accepted. You are likely to attain group acceptance if you become knowledgeable about the purpose of the group, become committed to the group and its purpose, work hard, are willing to be decisive, interact freely with others, and exercise human relations skills. Three types of individuals emerge as leaders: member-oriented, goal-oriented, and self-oriented.

To realize maximum leadership potential, three elements must be present: ability, experience, and opportunity. Business and industry use three basic approaches to improving leadership within the organization: selecting, situational hiring or engineering, and training.

One of the many keys to leadership development is mastering a wide range of skills. Heller gave several practical tips for becoming an inspirational and confident leader: gaining experience, learning from others, mastering roles, developing strengths, forming the best team, exercising authority, delegating tasks, communicating clearly, setting ambitious goals, developing teamwork, leading discussions, using meetings, analyzing problems, giving support, inspiring excellence, establishing a vision, generating ideas, managing openly, and being competitive.

There are three major types of leadership traits, abilities, and skills: conceptual (thinking) skills, technical (how-to) skills, and human relations (people) skills. Two more categories are technical-human relations skills and conceptual-technical skills. Human relations skills include honesty and integrity; sensitivity; cooperation; flexibility; supportiveness; self-confidence, responsibility, dependability, and emotional maturity; deriving pleasure from work; serving others; setting an example; being available and visible; showing confidence in people; and using power wisely.

Technical-human relations skills include listening, team (group) building, earning loyalty, rewarding accomplishments, coaching and strengthening others, planning short-term goals and successes, and administrative ability.

Technical skills include selecting staff, being competent, being a communicator, managing time, setting goals and plans, achieving, responding appropriately to failure, and leading individuals and groups.

Conceptual-technical skills include intelligence, problem-solving ability, decision making, use of a cooperative genius approach (synergy), delegation, being an opportunist, insisting on excellence, holding subordinates accountable, and exhibiting broad interests and abilities.

Conceptual skills include vision, being a motivator, decisiveness, straightforwardness, creativity, courage, risk taking, commitment, determination, persistence, sense of urgency, initiative, follow-through, mastery of change, wisdom, the desire to lead, and critical thinking.

Fred Smith's 10 ways of identifying a potential leader include the following: proof of past leadership, capacity to create or catch a vision, a constructive spirit of discontent, generation of practical ideas, willingness to take responsibility, completion of work, mental toughness, peer respect, family respect, and compelling communication ability.

 Take It to the Net

Go to http://scholar.google.com/ and use the search terms listed here to find research articles on leadership development. Read some of the articles. Properly cite at least one article and write a short summary of it.

Search Terms

leadership

leadership skills

effective leadership skills

leadership development

developing leadership

leadership qualities

Note: You can use any of the leadership qualities discussed in this chapter as search terms.

Chapter Exercises

REVIEW QUESTIONS

1. Define the Terms to Know.
2. What are the stages of development in the model of leadership development?
3. What are the five areas within the model of leadership development?
4. What are five steps to becoming a good leader?
5. List 10 steps of a personal leadership plan.
6. Discuss the importance of a healthy self-image.
7. List four techniques to help develop self-organization.
8. List seven steps to help overcome procrastination.
9. What are six ways to attain group acceptance?
10. What three types of individuals may emerge as a group's leader?
11. Explain the relationship between ability, experience, and opportunity to lead.
12. What are three ways to improve leadership in the workplace?
13. List 12 practical tips for improving your leadership.
14. What are three major types of leadership qualities?
15. List 14 human relations skills.
16. List eight technical skills.
17. List 14 conceptual skills.
18. List five reasons why coaching is necessary to improve performance.
19. Describe a current problem you may be able to solve by utilizing the qualities of successful leaders.
20. This chapter lists 10 ways to identify a potential leader. Rank them in order of most important to least important, as you perceive them.
21. List seven technical-human relations skills.
22. List nine conceptual-technical skills.

COMPLETION

1. Leadership is an observable, _____ set of practices.
2. The leader is a person who, on the whole, best embodies the standards or values of the _____.
3. "If it's going to be, it's up to _____."
4. Success depends on our _____ actions.
5. Winners find a way. Losers find an _____.
6. People can accomplish just about anything they want to if they are willing to _____.

7. If you try to please everyone but yourself, you are destined for _____.

8. Some people see things as they are and ask why. Some dream of things that never were and ask _____.

9. Winners never quit, and quitters never _____.

10. If you can dream it, you can _____ it.

11. _____ leadership skills are usually the easiest leadership skills to develop and retain.

12. _____ leadership skills are needed to understand and work with people.

13. _____ leadership skills are needed to analyze situations and generate ideas.

14. Once lost, a good _____ is seldom regained.

15. Without _____ you do not have trust, and without trust you have nothing.

16. One of the biggest leadership mistakes is not having _____ in your people.

17. Less has been written about _____ and leadership than about any other leadership quality.

18. Great leaders believe that a _____ is another opportunity to do something better.

19. Your attitude determines your _____ in life.

20. Few people plan to fail. They just fail to _____.

21. Most failures come from _____.

22. People cannot exceed their _____. limits.

23. If you think you can, you can. If you think you cannot, you _____.

24. If you want something done, ask a _____ person.

25. If you want to be a good leader of yourself and others, you must study _____.

26. The difference between leadership and management is the ability to inspire _____ to excellence.

MATCHING

Terms may be used more than once.

_____ 1. People skills.

_____ 2. Thinking skills.

_____ 3. How-to-do-it skills.

_____ 4. People skills easily taught.

_____ 5. Thinking skills easily taught.

_____ 6. Honesty and integrity.

_____ 7. Listening.

_____ 8. Communication.

_____ 9. Problem solving.

_____ 10. Vision.

A. conceptual skills

B. technical skills

C. human relations skills

D. conceptual-technical skills

E. technical-human relations skills

ACTIVITIES

1. Set three goals in each of the following categories: physical, family, social, financial, educational, ethical, and career.

2. Write three goals each for where you want to be in 5, 10, and 20 years from now. List one sacrifice or obstacle that you anticipate for each goal. (For example, five-year goal: graduate from college; obstacle: studying and self-discipline.)

3. List five facts about modern-day leadership that you can use in the future. Share these with your classmates and get their reactions.

4. Select three conceptual skills that you believe to be your strongest. Write a paragraph on each, defending your reasoning.

5. Select three conceptual-technical leadership skills, and repeat the process in activity 4.

6. Select three technical leadership skills, and repeat the process in activity 4.

7. Select three technical-human relations skills, and repeat the process in activity 4.

8. Select three human relations leadership skills, and repeat the process in activity 4.

9. Draw the name of another student in your class from a container. Identify five leadership qualities that person exhibits. Write a paragraph on each quality, describing how that student demonstrates it, and present your essay to the class.

10. Identify your five weakest leadership qualities. Write a short plan describing how you plan to improve each of them.

11. Select one of the 50 leadership qualities and write a paper or lesson on how you would teach this to others.

12. Consider each of the following leadership qualities and the impact they have on people's lives. Select one of the following and present a report to the rest of the class on its potential impact.

    a. Honesty—A professor included questions in the final exam concerning material that he expressly said would not be on the exam. You fail the exam and the class. How do you react? How does misinformation shape our actions? How does dishonesty separate us from reality?

    b. Sensitivity—A teacher assigns students to seats using alphabetical order. A student then states that he cannot see the board. The teacher tells him to wear his glasses or get his eyes checked. The teacher then asks the student if he has glasses. When the student puts them on, the teacher laughs and tells him to take them off. The class laughs at the student. How should this have been handled differently? What effect could a different approach have had on each of the students in the class?

    c. Cooperation—Three students are on a budget committee. You want to purchase a new digital camera for the organization. This expenditure would make it impossible to balance the budget and still do all the traditional activities next year. No one will give in. The committee is eventually dissolved and a new budget committee takes over. Do you think you should have cooperated more? Why? What are the consequences of being uncooperative?

    d. Flexibility—The agribusiness manager shows his employees (including you) how to solve a certain type of statistics problem. The employees do not understand. You remember a way your statistics teacher showed you to work the same kind of problem. You show the manager and the other employees your way of solving the problem. The manager becomes very argumentative. What should the agribusiness manager's response have been? Why is it so important to be flexible?

e. Supportiveness—This week your agriculture teacher told you that at next Friday's awards banquet you will be giving the opening speech of welcome and introductions. You have never done anything like this before. What kind of support do you need? Who could support you?

f. Self-confidence—Someone you have always wanted to date has recently broken up with a boyfriend or girlfriend. You would like to ask this person out, but you delay because of a lack of self-confidence. While you hesitate, the person starts dating someone else. You find out later that the person liked you too, but thought you were not interested. How was your lack of self-confidence destructive? What role does fear play? How can you keep this from happening again?

g. Responsibility and dependability—You ask a friend to take notes for a class you must miss because of an FFA event. When you see your friend again, she tells you about an upcoming test on Friday. She cannot find the notes she took while you were absent. How will this lack of dependability affect your relationship?

h. Emotional maturity—A friend has yelled at you about something you thought was a minor matter. How will this lack of emotional maturity affect your relationship? How do children handle conflicts differently than young adults? What situations bring out a lack of maturity in young adults?

i. Pleasure from work—You are the boss of a crew in a landscaping company. One of the four workers on your team loves his job and is always positive. The others complain and make life miserable. Why is pleasure from work important to everyone on the crew? How will this affect the rest of their lives?

13. Listening activity. Find a partner and take turns listening to your partner as he or she responds to the following prompts:

- The best thing about me is …
- I think a lot about …
- I often worry about …
- I believe I am …
- Something I believe in strongly is …
- Something I really want is …
- I am brave when …
- When I am afraid, I …
- The worst thing about me is …

After both of you have responded to these prompts, tell your partner what you understood was said, using different words. Take turns.

14. The poor listener. Role-play the following poor listeners to illustrate the problems created by not listening.

- A husband who is reading the news on the Internet and grunts, "Yes, dear," in response to his wife's comments
- A girl who will not listen to her sister's explanation of why she borrowed the girl's clothes without asking

- A father who will not listen to his son's explanation of why he broke his curfew
- A student who will not listen to a teacher's instructions about the lesson
- A girl who will not listen as her boyfriend tries to explain why he was unable to call the night before
- An employer who will not listen to a worker's explanation of an unfriendly encounter with a customer
- A store manager who does not listen to a customer's complaint
- A wife who does not listen to her husband tell her about a new coworker
- A teenager who does not listen when his father cautions him about the brakes on the family car
- An FFA president who does not listen to her FFA adviser explain the details of the chapter fund-raiser[67]

15. Refer to the "10 ways to identify a potential leader." Select your three greatest strengths and explain why these are your strengths.

NOTES

1. B. Patton et al., *VICA: Learn, Grow, Become* (ED250575) (Stillwater, OK: Curriculum and Instructional Materials Center, State Board of Vocational and Technical Education, 1984).
2. J. Ricketts, *The Efficacy of Leadership Development, Critical Thinking Dispositions, and Student Academic Performance on the Critical Thinking Skills of Selected Youth Leaders* (doctoral dissertation, University of Florida, 2003), p. 15.
3. J. A. van Linden and C. I. Fertman, *Youth Leadership: A Guide to Understanding Leadership Development in Adolescents* (San Francisco: Jossey-Bass, 1998), p. 40.
4. Ibid.; Ricketts, *The Efficacy of Leadership Development.*
5. van Linden and Fertman, *Youth Leadership.*
6. J. C. Maxwell, *The Complete 101 Collection: What Every Leader Needs to Know* (Nashville, TN: Thomas Nelson Publishers, 2010), p. 15.
7. R. M. Lerner, *America's Youth in Crisis* (London: Sage Publications, 1994), pp. 5–17.
8. J. W. Gardner, *Leadership Development* (Washington, DC: Leadership Studies Program, Independent Sector, 1987).
9. W. Bennis, *On Becoming a Leader* (Philadelphia, PA: Basic Books, 2009), pp. xxvi.
10. Ibid, pp. 52.
11. Ibid.
12. Retrieved October 2, 2009, from http://www.quotedb.com.
13. Retrieved October 2, 2009, from http://www.cybernation.com.
14. *Positive Self Image and Self Esteem* (Mountain State Centers for Independent Living), para. 2. Retrieved July 7, 2016 from http://www.mtstcil.org/skills/image-1.html
15. *Understanding and Overcoming Procrastination* (The McGraw Center for Teaching and Learning, 2012). Retrieved July 7, 2016 from http://www.princeton.edu/mcgraw/library/for-students/avoiding-procrastination/procrastination.pdf.
16. S. R. Levine and M. A. Crom (Dale Carnegie & Associates, Inc.), *The Leader in You: How to Win Friends, Influence People, and Succeed in a Changing World* (New York: Simon & Schuster, 1993), p. 27.
17. *Plans for Training Professionals in Community Development* (Alexandria, VA: The National FFA Organization, 1980).

18. "Personal Leadership Potential," *Effective Leadership* series, No. 8738-A (College Station, TX: Instructional Materials Service, Texas A&M University, 1989), p. 2.

19. Ibid., pp. 1–2.

20. Ibid., p. 2.

21. R. Heller, *Effective Leadership* (Essential Managers series) (New York: DK Publishing, 2005).

22. "Personal Leadership Potential," p. 2.

23. J. Stern. "Teaching Your Kids to be Honest: What's the Best Strategy for Teaching Your Kids Honesty?" *Psychology Today*, March 23, 2011. Retrieved July 7, 2016 from https://www.psychologytoday.com/.

24. F. A. Manske Jr., *Secrets of Effective Leadership* (Columbia, TN: Leadership Education and Development, Inc., 3rd ed., 1999), pp. 33, 34, 36.

25. R. Herman, "Lessons in Leadership," *The Herman Group*, 2001. Retrieved July 7, 2016 from http://www .hermangroup.com/.

26. B. Nanus, *The Leader's Edge: The Seven Keys to Leadership in a Turbulent World* (Chicago: Contemporary Books, 1991).

27. E. W. VanderBerg, *Leading in Conflict Resolution*. Retrieved October 3, 2009, from http://knol.google .com/k/evert-w-vanderberg/leading-in-conflict rcsolution/2r13bsw2hlyje/3#.

28. Martin Luther King, Jr. Quotes. Retrieved July 2016, from www.brainyquote.com.

29. P. Perkins, "5 Ways to Improve your Visibility as a Leader," *Black Enterprise: Wealth for Life*, 2010. Retrieved July 7, 2016 from www.blackenterprise.com.

30. S. B. Dust, C. J. Resick, and M. B. Mawritz, 2014, "Transformational leadership, psychological empowerment, and the moderating role of mechanistic-organic contexts," *Journal of Organizational Behavior*, 35, pp. 413–433.

31. Ibid.

32. Ibid.

33. Bethel, *Making a Difference*, p. 169.

34. Manske, *Secrets of Effective Leadership*, p. 25.

35. E. Schreiner, *Seven Strategies for Developing Cohesive Teams*. Retrieved July 7, 2016 from http://smallbusiness.chron.com/.

36. T. Krause, "Leadership: Loyalty from Leadership: Holding the Team Together," *LeaderValues*, 2006. Retrieved July 7, 2016 from www.leader-values.com.

37. J. M. Kouzes and B. Z. Posner, *The Student Leadership Challenge: Five Practices for Becoming an Exemplary Leader*, 2nd ed. (Jossey-Bass, 2014).

38. Ibid.

39. Manske, *Secrets of Effective Leadership*, p. 41.

40. Ibid., p. 47.

41. Ibid., p. 44.

42. Quotes: The Web's Largest Resource for Famous Quotes & Sayings, para 1. Retrieved July 7, 2016 from www.quotes.net.

43. J. Kotter, "The 8-Step Process for Leading Change." Retrieved July 5, 2016 from http://www.esc20.net /users/0077/docs/TTESS/Lead%20Meetings/05.11.16/The%208%20Step%20Process%20for%20Leading%20 Change.pdf.

44. Kouzes and Posner, *The Student Leadership Challenge*.

45. S. R. Covey, *The 7 Habits of Highly Effective People: Restoring the Character Ethic* (New York: Free Press, 2004), p. 263.

46. J. Townsend, notes from "Leadership" class, Texas A&M University (College Station, TX).

47. F. A. Manske Jr., *Secrets of Effective Leadership*, p. 49.

48. R. Selladurai (2006), "Leadership effectiveness: A "new" integrative leadership model," *Journal of Business & Leadership*, 2 (1), p 7.

49. S. Covey, *Principle-Centered Leadership* (New York: Simon & Schuster, 1991), p. 51. (Unabridged ed., 2012).

50. J. Baumgartner, *10 Steps for Boosting Creativity*. Retrieved October 9, 2009, from http://www.jpb.com.

51. "Quotations about Courage," *The Quote Garden,* n.d. Retrieved October 4, 2009, from http://www.quotegarden.com.

52. Manske, *Secrets of Effective Leadership,* pp. 67–68.

53. S. M. Bethel, *Making a Difference: Twelve Qualities that Make You a Leader* (New York: G. P. Putnam's Sons, 1990), pp. 259–260.

54. R. W. Paul and L. Elder, *Critical Thinking: Tools for Taking Charge of Your Professional and Personal Life,* 2nd ed. (Upper Saddle River, NJ: Pearson Education, 2014), p. 9.

55. "Leadership Qualities: Ten Ways to Identify a Promising Person," *Leadership Training Skills,* p. 1. Retrieved July 7, 2016, from http://www.leadershipskills.us/leadership_articles/ten_ways_identify_a_promising _person.html.

56. Adapted from *The Basics Guide 2* (n.p., n.d.: North Carolina Department of Public Instruction).

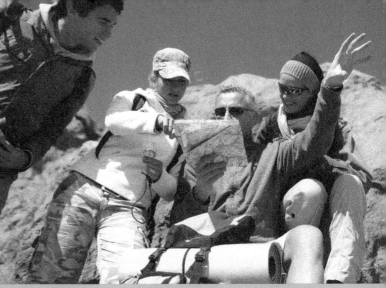

5 LEADING TEAMS AND GROUPS

Democratic leadership involves groups and the processes of working together. A leader has to mobilize and motivate group members to strive toward a common goal with strong commitment and a sense of excitement and anticipation.

Objectives

After completing this chapter, the student should be able to:

- Discuss the importance of democratic group leadership
- Explain the importance of groups
- Explain the importance of leading teams
- Describe the three types of groups and their subgroups
- Explain how groups are organized
- Explain group dynamics
- Analyze the five stages of group development
- Describe the various types and forms of group discussion
- Demonstrate how to lead a group discussion
- Discuss common mistakes made by leaders
- Explain useful strategies to employ upon taking over as a new leader in an organization

Terms to Know

- participative management
- confront
- tasks
- functional groups
- task groups
- affinity
- group dynamics
- norms
- cohesiveness
- intragroup
- intergroup
- status
- task roles
- maintenance roles

Most of you are leaders, would like to become leaders, or have been leaders. At the least, you would like to understand how you are being led. Therefore, you need a good idea of how groups function. As you develop personal and leadership skills, it is natural to apply them in group or team situations. Examples of groups are your FFA chapter, your group at work, or your group of friends (Figure 5–1).

"Anthropological evidence suggests that people have always lived in groups in which leader–follower relationships quickly and naturally emerge."[1] We use groups in nearly every part of our lives to solve problems. We do this because we understand that as a group we will always make better decisions and perform at higher levels than one person alone.

DEMOCRATIC GROUP LEADERSHIP

In Chapter 1, we concluded that in most situations, democratic leadership is preferred. It is not merely the American way; it is the most productive way. Therefore, the formal democratic group is neither laissez-faire nor autocratic. The laissez-faire group is characterized by its lack of organization. The autocratic group is under the domination of an individual or a power clique. Congress is a democratic group, and the Capitol building (Figure 5–2) where Congress meets is a symbol of democracy.

Why Democratic Group and Team Leadership?

When compared to autocratic groups, democratic groups have greater motivation to work, more member satisfaction, and higher productivity. There is less discontent among democratic group members and less evidence of frustration and aggression. There is more friendliness, cooperation, and group-centered spirit when democracy prevails. Also, more individual initiative is displayed in democratic groups. There is evidence that a leadership style emphasizing democratic team building is positively associated with high productivity and profitability. Teamwork ensures not only that the job gets done, but also that it gets done efficiently, thoroughly, and harmoniously.[2]

FIGURE 5–1 As human beings, we are group oriented. Membership in a group meets a basic psychological need of all humans. Groups give rise to synergy, which allows us to accomplish things we could not do alone.
(Courtesy of University of Georgia—College of Agricultural & Environmental Sciences.)

FIGURE 5–2 The Capitol is a symbol of democracy. Democratic groups have greater motivation to work, higher member satisfaction, and greater productivity than other types of groups. *(Courtesy of University of Georgia—College of Agricultural & Environmental Sciences.)*

Why Not Autocratic Group Leadership?

The autocratic group cannot compete with the democratic group in all-around productivity. Autocratic groups often engender excessive irritability, hostility, and aggression, directed at fellow members as well as the autocratic leader. The members of such a group are apt to display an apathetic general attitude even when secretly discontented. Group members are much more dependent and show little creativity. When the leader is absent, little or nothing gets done.

The stress of trying to do it all can wear autocratic leaders down quickly. There is also evidence that a lack of team function negatively affects the physical and psychological well-being of leadership personnel. More than just a good idea, team building and teamwork appear to be a necessity of biological life.[3]

Democratic Group Leadership and Attitudes

In industry, in the classroom, and even in the military, it has been demonstrated that involving people in groups and reaching decisions in a democratic atmosphere lead to more favorable attitudes regarding the decision. Studies reveal that participation in decision making by all members of the group results in greater productivity, less resistance to change, and a lower turnover rate than either committee-made decisions or decisions made by individual leaders (no matter how carefully they are explained).[4]

Democratic Group Leadership and Peer Pressure

When group members commit themselves to act in a certain way, their decision is strengthened by the knowledge that others are similarly committed. They do not wish to lose status or let others down by failing to follow through on a decision or promise made to and witnessed by their peers. One of the strongest motivating forces for any individual is to be respected and to have status in the eyes of the members of groups to which the individual considers it important to belong. This principle is one of the most important factors establishing the superiority of group (democratic) action over individual (autocratic) action.

Importance of Group Leadership Skills

The research that has been conducted on group effectiveness makes clear that "leaders have a strong influence over group effectiveness—for better or for worse."[5] The leader who understands how to manage a diverse group of individuals and get them to operate as one unit accomplishing a collective goal represents the "'gold standard' for evaluating leadership" in modern group and team dynamics.[6] Regardless of title or formal rank, an individual is a leader when his or her ideas and actions influence others in the group.

IMPORTANCE OF GROUPS

People like to belong. A sense of belonging is a basic human need that everyone shares to some degree.

Why People Belong to Groups

In our quest for satisfaction, we often find that many of our needs and wants are best met through group affiliation and action. We hope you want to be a part of the FFA. Some of you want to be in marching band or on a sports team. You like to choose the people with whom you socialize. Joining groups helps you meet your psychological needs. Other reasons people belong to groups are affiliation, proximity, attraction, activity, assistance, and tradition.[7]

Affiliation We all have a need to belong to a group. When there is a choice, we spend time around people we like. People who like their coworkers tend to have high levels of job satisfaction. Studies have shown that today's young workers value affiliation with strong brands (e.g., Google, Apple, the United States Department of Agriculture) and a teamwork approach.[8] As a matter of fact, they go to work with the expectation of meeting new friends and joining a group or team. The individuals in Figure 5–3 work together as state officers, but they are also friends, which increases the productivity of their group.

Proximity We tend to form groups with the people we see often. People who have classes together, participate in extracurricular activities together, work at the same place, and live in the same dormitory tend to associate with each other.

FIGURE 5–3 These state FFA officers are also friends. They are from various high schools, but they are unified as a group because of their shared goal of improving agricultural education and the FFA. *(Courtesy of Georgia Agricultural Education.)*

Attraction We tend to be attracted to people who have attitudes, personalities, social class, income, and interests similar to our own. We also like to associate with people for whom we have physical, cognitive, or emotional attraction.

Activity We join a group because of the activity it performs—what it does. For instance, FFA is an integral part of agricultural education. Agricultural education is a part of the agriculture community, including production agriculture, agribusiness, and agriscience. Similar to Activity is Interest. Social media groups form when individuals have the same interests and wish to engage in discussions about those interests. A common interest in greenhouse management may be one reason the individuals in Figure 5–4 have come together.

Assistance We often join groups for the assistance they provide. Assistance with leadership development, travel, and awards may be a reason for joining the FFA. Once you graduate, you may wish to join a civic club such as the Kiwanis, Lions, Sertoma, or Rotary to develop business contacts and to participate in the services these clubs provide to a community.

Tradition You may join a group purely because of tradition (for instance, your mother or father was

an active member and it was expected that you would also become a member). Various communities develop rich traditions surrounding certain groups: athletic teams, band, chorus, FFA, and so on. If you have the opportunity, you join because everybody in the community expects you to become a part of that group or team.

Why People Do Not Join Groups

There are many reasons why people do not belong to groups.

Unaware of the Group People may not know that a group focused on their interests exists. If they do know about a potential group, they may not join because they do not understand what the group does or represents.

Insecurity and Feelings of Inferiority Students may feel insecure about joining a group. They may worry about whether they will be accepted by the existing group members. Some people fear that they lack the human relations skills to get along with other group members. They may feel inferior to them for reasons such as status or educational attainment.

Cannot Meet Expectations People may hesitate to join a group because they are not sure what

FIGURE 5–4 People belong to groups because of affiliation, proximity, attraction, activity, assistance, and tradition. Interest in greenhouse management has brought the people in this picture together. (© iStock/Steve Debenport)

FIGURE 5–5 Listening is part of being a good group leader. The supervisor in this photograph appears to be using participative management. *(© iStock/Geber86)*

the group expects from its members. They may believe that the other group members know much more than they do.[9]

Why Group Leadership Skills Are Important

Besides being necessary for school, civic, and most other types of groups, leadership skills are needed in the workplace. In fact, they are highly prized and sought after, and those who have such abilities and exercise them skillfully may be well compensated for their work.

Leader Evaluation Traditionally, managers or leaders concentrated on supervising individual employees rather than a work team. Today, however, supervisors' performance is often rated or evaluated on the basis of the department's or group's performance as a whole, rather than or in addition to each individual employee's results.

Big Part of Job Job supervisors report spending 50 to 90 percent of their time in some form of group activity. The current trend is to have employees or coworkers more involved in decision making/problem solving. Industry calls this **participative management.** Workers at all levels, from the beginning to the end of the work process, provide input to improve the company and its products or services. Gathering and integrating this input often consumes much of the leader's time while he or she functions as a group leader or facilitator. However, this time is well spent if productivity is increased.[10] The supervisor in the middle of Figure 5–5 is employing participative management.

Better Group Performance Technologies or new ideas for a work group can fall short if people do not feel that they are part of the team. Input from everyone in the group is needed. Group participation in decision making results in better decisions and more commitment to implementation of those decisions.

LEADING TEAMS

Some say that **TEAM** is an acronym for "<u>T</u>ogether <u>E</u>ach <u>A</u>ccomplishes <u>M</u>ore." In the 21st century, teamwork is increasingly the normal operating procedure for business, industry, and organizations. Tomorrow's leaders will most often have to take on the role of facilitator or coach. If you have ever played in or watched an athletic contest, you can understand the importance of teamwork:

If the team loses, every individual on the team loses. No chain is stronger than its weakest link.

Another positive outcome of teamwork is an increase in synergy. Synergy is when two separate individuals or pieces working together can accomplish exponentially more than they could apart.

Effective teams **confront** (approach a problem head-on or face to face), question, set policy, and, above all, participate. A team, including its leaders and followers, also continuously analyzes itself to make sure it is functioning properly. Everyone on the team must also work together to set and achieve group goals and solve current and forecasted problems. The bottom line is this: people, organizations, and departments that work well as a team outperform those that depend on individual efforts. Effective leaders, however, are necessary for motivating and managing teams for teams to reach their full potential. Consider the following as you build your team:[11]

Create a Shared Sense of Purpose A shared vision unites team members' efforts and spurs them to accomplish tremendous things in furtherance of that vision.

Establish Team Goals When the team wins, each team member wins. Likewise, each individual on the team loses if the team does not achieve a goal. Note that this truth does not prevent you from achieving individual or personal goals in the course of contributing to your team. If you do your very best, and your interests are aligned with the team's focus, both personal and group goals are more likely to be reached.

Recognize that People Are Unique Individuals and Treat Them Accordingly Individuals have different personalities, abilities, dreams, and fears. "A talented leader will recognize those differences, appreciate them, and use them to the advantage of the team."[12]

"Make Each Member Responsible for the Team Product"[13] If people do not think that their contributions are valued or important, they will not remain committed and will not give their best efforts to their **tasks** (things to be done).

Spread the Glory but Collect the Blame When a team does well, the leader is responsible for making sure that each team member gets his or her share of any benefits and rewards. When the team is criticized, the good leader accepts whatever blame or complaints are aired publicly; he or she then works privately with team members on improving and learning from mistakes.

Get Involved and Stay Involved Know what is happening and what is going on around you. Use your intuition and people skills to develop a feel for the team's mood and attitudes.

Mentor Your Team Members Today's team members are tomorrow's leaders. Part of the leadership role is recognizing, supporting, and enhancing the talents and strengths of the people on the team. Set the standards, help team members to achieve goals, and build their confidence in their own abilities by showing that you trust and appreciate them.

Research on Teams

The movement toward work teams appears to be a major and continuing trend in organizations. Many organizations are establishing teams of rank-and-file workers who take on tasks previously done only by leaders or managers. Research shows that the use of teams increases productivity, improves the quality of work, produces more positive attitudes in members, reduces absenteeism, and creates a desire to stay on the team.[14]

The use of teams in this country was stimulated by the American adaptation of a particular Japanese management style called *Kaizen*, which we will examine in more depth in a moment. The Japanese credit much of their success in the global economy to their participative, team-based decision making and management style.[15]

Purpose of Teams

A **work team** is a group (two or more) of individuals with a common purpose. The very existence of the team implies and assumes that members will work together cooperatively to achieve a common objective or goal.[16] It takes teamwork to make the team work.

Characteristics of Effective Teams

Team members must be carefully selected, both for compatibility and to ensure that each person brings something unique to the team. Teams are difficult to start and must be managed or led well if they are to get up and running quickly. Teams have to be given a charge, mission, or goal. The team also needs enough power and authority to accomplish its tasks and implement its ideas.[17] Consider the following four characteristics of teams:

- Team members must be fully committed to a common goal and approach. Members must agree that the goal is worthwhile and decide on a general approach for reaching it.
- If the team is to succeed, members must be accountable to one another and to the organization for the outcome of their work.
- Teams must develop a culture based on trust and collaboration. Team members must be "willing to compromise, cooperate, and collaborate."
- Teams must share leadership among all members.[18]

Figure 5–6 lists the characteristics of an enthusiastic team.

Establishing an effective team is a challenging process that requires both interpersonal skills and extensive technical support. The development of trust, a common vision, and good collaboration

ability depend on appropriate interpersonal skills. Once trust and goals are established, team members must be trained to take on complex tasks. Equipping followers with interpersonal skills and training them for their roles is the leader's responsibility.[19]

The Leader's Role in a Team

In a team environment, the role of a leader changes to that of facilitator and/or coach. Leaders act as caretakers of their teams, helping them to achieve their goals by giving instructions, encouragement, and resources. "Leader/facilitators still fulfill many of the functions of traditional leaders, but they do so to a lesser extent and only when asked." They assist the team by obtaining the resources needed to solve problems and implement solutions, intervening only when they must. Primary roles of a leader include assessing team members' abilities and skills and helping them develop the necessary skills; often this means making sure that team members get the right kind and amount of training.[20]

Another leadership function is to make the team aware of its boundaries and the limits of its power. "Many teams fail because they take on too much or ignore organizational realities and constraints." The team leader is the one who must keep the team focused on its specific task or align the team's efforts with those of others who could help implement its broader recommendations.[21]

Potential Problems with Teams

The team concept does not always work well in the United States. Japan, where much of the team management concept originated, has a more collectivist culture. United States culture tends to be more individualistic, and this emphasis often conflicts with team-type approaches.

Australians have devised a concept called **collaborative individualism**, in which individuals are not bound by the group's limits. Rather, they cooperate with other team members while maintaining their personal, internal motivation and conflict management skills. Some believe that such an approach might be more suitable to U.S. culture than the Japanese approach of aiming for consensus and conformity in a team.[22]

ELEVEN COMMANDMENTS FOR AN ENTHUSIASTIC TEAM

1. Help one another be right—not wrong.
2. Look for ways to make new ideas work, not for reasons they will not.
3. If in doubt, check it out! Do not make negative assumptions about one another.
4. Help one another win, and take pride in one another's victories.
5. Speak positively about one another and about your organization at every opportunity.
6. Maintain a positive mental attitude no matter what the circumstances.
7. Act with initiative and courage, as if it all depends on you.
8. Do everything with enthusiasm; it is contagious.
9. Whatever you want, give away.
10. Do not lose faith; never give up.
11. Have fun!

FIGURE 5–6 Eleven commandments for an enthusiastic team.

VIGNETTE
5-1   LEADERS MAKE MISTAKES, TOO

LEADERS AND MANAGERS MAKE MANY MISTAKES when they are setting up and leading work teams. Some of the common mistakes are listed here. Try not to replicate them when you find yourself in charge of a team in a work situation!

Mistake 1: Use a team for work that is better done by individuals.

Not all job responsibilities are best performed by a team, even though *team* is currently the hot buzzword in business. One of those tasks is creative writing, whether it is a song, a poem, a book, or a play. Even committee reports, dry and boring as they often are, frequently turn out better when a single person writes them. Can you think of other tasks or activities that would be best done by a "single intelligence"?

Mistake 2: Call the group a team, but really manage members as individuals.

Teams should be built, trained, and then trusted to act and perform as a team. That means giving them power to make decisions and the confidence to do their job. It also means allowing them to take risks and sometimes fail. When leaders claim to have a team but then micromanage the members individually, the whole team concept is destroyed. Members feel powerless, devoid of confidence, distrusted, and unappreciated. This sort of environment turns individuals in an effective team into unproductive and negative members.

Mistake 3: Maintain a balance of authority.

Democratic leadership is the goal. We want our teams to have power to make decisions and confidence that their members will do a great job. However, you must also be careful not to abandon all your authority as a leader of that team. Research has shown that giving a team clear direction and instruction empowers a team and focuses its efforts on goals set by the larger organization.

Mistake 4: Take apart or ignore existing organizational structures so that the team has all the authority for accomplishing its work.

Many experts in participative leadership say that true teams have no structure. This type of thinking may be empowering at first, but eventually teams operated in this way run into process problems, both internally and externally. Some structure and direction are necessary, especially for linking the team and its work to the rest of the larger organization.

Mistake 5: Set challenging team objectives, but then do not support the team in achieving them.

This is one of the biggest mistakes you can make. Your team may need physical resources, information, and training to operate competently. Also, you must give your team praise and rewards, no matter how good and mature the team members appear to be. They will not continue to work hard if they do not feel appreciated.

Mistake 6: Assume that members already have all the skills they need to work as a team.

Train, train, and retrain your team members. Often leaders assume that the team knows what they want and then get angry when the team does something incorrectly or differently. Take the time to train your team initially and to coach team members appropriately as they do their work.[23]

TYPES OF GROUPS

There are three types of groups. The first two, functional groups and task groups, are both formal groups. The third type is informal groups.[24]

Functional Groups

Functional groups are formal groups with an indeterminate life span. The group's objectives and members may change over time, but its basic functions remain the same. Many of the social groups, teams, and clubs with which you are familiar—FFA, 4-H, Jr. MANRRS, various civic clubs, and many others—are functional groups.

Functional groups also exist within the workforce and within each workplace. Most companies have marketing, finance, production, and personnel or human resources departments. Each department constitutes a functional work group. The supervisors of each department provide a link to managers, who link to a vice president, who links to the president. Most colleges and universities use the same type of organizational system, with subject departments under deans or chairpersons linking to an administrative hierarchy. Organizational structures like this form the traditional pyramid, with a president or CEO at the top of the pyramid.

Nowadays the pyramids are crumbling. Many companies are moving away from the pyramid structure because of the communication barriers it creates. The traditional rigid structure is being reworked to allow people to exercise their creativity and to develop fully the talent in an organization or team.[25] Innovative, well-led organizations are using the team approach. Here are some examples.

Problem-Solving Teams The most widely known type of problem-solving group is the **quality circle**, first used by the Japanese, in which employees focus on ways to improve quality in a production process. These problem-solving groups consist of rank-and-file employees who have hands-on jobs in a production process, who meet to discuss ways of improving product quality, operational efficiency, and the work environment. These recommendations are then proposed to management for approval. The Japanese call this *Kaizen*. If you walk into any Toyota factory in the United States, you can witness Kaizen at work.

Self-Managing Teams These multiskilled employees rotate jobs to produce an entire product or service; the team also handles all the management duties for its project, such as scheduling, hiring, and ordering.[26] The result is a well-rounded, highly experienced, competent team.

Special-Purpose Teams These teams are often involved in "developing new work procedures or products, devising work reforms, or introducing new technology." They provide a link among separated functions, such as production, distribution, finance, and customer service.[27]

Task Groups

Task groups, more commonly called **committees**, are formal groups established by an organization to perform specific functions. Task groups have proved to be an effective problem-solving tool. Unlike work groups, they may include members from different departments who participate in the task group in addition to their regular jobs. Commonly used task groups include standing committees, ad hoc committees, and boards.[28]

Standing committees are permanent groups that exist to deal with year-to-year issues. Members usually serve a one-year term on such a committee. When their terms are up, others take their places. To keep things running smoothly, committees usually replace members on a rotating basis.[29] Examples of standing committees in the FFA are the Finance Committee, Scholarship Committee, and Recreation Committee.

Ad hoc committees are temporary task groups formed for a specific purpose. They are temporary because they exist only until the task has been accomplished.[30] For example, in an FFA meeting, someone makes a motion to buy a vending machine for the agricultural mechanics lab. Often such a motion is referred to an ad hoc committee. Why? Decision making on this matter requires investigation of many questions and issues, such as school principal approval, projected income and expenses, and management. Once appointed, an ad hoc committee does the necessary investigation and then reports back at the next FFA meeting; after giving its report, the ad hoc committee disbands.

Corporations have **boards** (a type of standing committee) of directors or trustees whose task is to oversee management of the organization's assets, set or approve policies and objectives, and review progress toward the objectives.[31] The National FFA Organization has a board of directors, as do most state FFA associations. FFA alumni groups, which serve the active FFA members, also each have a board of directors.

Informal Groups

Functional and work groups are intentionally created by an organization; informal groups are not. Informal groups arise when people come together voluntarily because of similar interests. Such groups exist primarily to satisfy the psychological need of belonging. Informal groups are sometimes referred to as **cliques**. These groups may help or hurt the organization as they pursue their small-group objectives.

In the workplace, people from different departments may get together for breaks; in schools, various students may meet for lunch and after school. Membership in these informal groups tends to be open and to change faster than it does in formal groups.[32] An informal group starts without a purpose or agenda, but if its members share common concerns, they may establish goals and objectives. As complexity and formalities grow, the once-informal group may evolve into a formal or functional group.

ORGANIZING GROUPS

Levels of Group Membership

Figure 5–7 illustrates group membership. The outer line of the circle is the boundary that indicates membership. Those who meet membership requirements are inside the circle, and those not meeting the requirements remain outside the circle.

The group itself defines this membership boundary. Some groups might require only an interest in joining; others might require payment of dues or sponsorship by an existing member. Sometimes membership is passive and automatic. For example, persons who graduate from your high school become alumni. Those who did not graduate from your particular school cannot become alumni and therefore cannot be a part of that specific group.[33]

Within the membership circle, members fall along a spectrum of activity, commitment, and involvement. In this representation, those who are most active are toward the center; members whose involvement and activity are weaker, sporadic, or minimal remain along the periphery.

Nucleus As Figure 5–7 shows, members' degree of activity within a group varies. Some members, such as officers or committee chairs, are extremely active and thus are at the center of activity. They form the nucleus of the group. A group's achievement depends on the motivation and leadership supplied by these members.[34] Groups are good illustrations of the 80-20 rule, in which 20 percent of the group membership does 80 percent of the work. This is true of many civic clubs and high school groups.

Faithful Members The second level of membership includes individuals who are loyal to the group and faithfully attend all the meetings. However, they still are not part of the nucleus (the inner circle), either because they do not want to be or because they do not have the necessary abilities. It may also be that their abilities have not yet been discovered or tapped.

Fair-Weather Members The third or outer circle represents fair-weather status: fringe or marginal individuals whose membership is mostly nominal. They may jump on the bandwagon when things are going well, but they are often the first to criticize or leave when the group has tough times. Most athletic fans are like this. If the team is going for a championship, they say, "Look at what *our* team is doing." If the team is having a losing season, they say, "Look at what *that* team is doing." A primary challenge for the group leader is to structure the group and guide its activities so that all members are interested and involved enough to fall within the inner two categories of membership.[35]

Group Structure

The structure of a group strongly affects its productivity. People use interaction with other group members "to satisfy their needs for love, security,

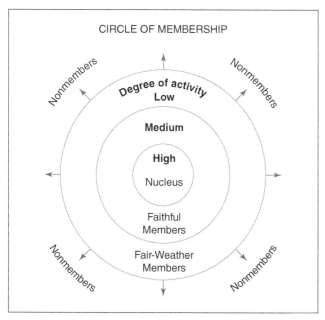

FIGURE 5–7 The three levels of group membership—nucleus, faithful members, and fair-weather members—each display a different amount of commitment to and interest in the group. Because of your priorities, time commitments, and interests, you may be in the nucleus of one group and a fair-weather member of another.

recognition, sense of accomplishment, and power." Therefore, people seek out and choose to join groups that allow them to meet these needs. Strong "groups develop by identifying and gratifying" people's needs. Group leaders must remember that most members join a group for social or emotional reasons rather than because of its activities, even though the activities are the reason the group exists.[36]

Maintaining Strong Groups

Several factors help maintain strong groups. First, people join, attend meetings, and assist the group in reaching its objectives if the group satisfies their personal needs. Second, nothing succeeds like success: a group is strengthened when it is successful and accomplishes its objectives. People go where the action is. Time is too valuable to be wasted: if groups and meetings are not productive, members will drop out. Third, groups "are strengthened when the members have clearly defined goals" and each member recognizes and accepts his or her role in achieving these goals.[37]

The FFA Program of Activities is an excellent example of setting goals and objectives and establishing the ways and means to achieve them (refer to the *Program of Activities Handbook*). A fully functioning FFA chapter assigns a task to every member. Pride in accomplishment and involvement is a great motivator.

Teamwork is another important factor in maintaining a strong group. Lack of direction and competition among members impairs teamwork. The leader can avoid these problems by assigning clearly defined roles and tasks to each member; this promotes teamwork as members each fulfill their functions in working together toward a common goal.[38] Strong groups in turn promote group interaction; just as importantly, they also provide an environment in which everyone has opportunities to make individual contributions and earn higher status.[39]

A group flourishes only when its members commit time and energy to it. By demanding some investment, groups both increase their attractiveness to members and weed out those with insufficient interest or motivation. For example, strict or arduous initiations screen out people who are unwilling to make a commitment, but also heighten a sense of belonging in those who are. Members who invest time, energy,

and money in the group develop a vested interest in the group's well-being and success, and most likely will feel strong **affinity** (attraction among or similarity of group members) for the group.

A group must recognize and celebrate the contributions of all participants in group activities. Making members feel needed is an important part of maintaining a group's attractiveness. A successful group allows and encourages members to make significant contributions,[40] and then rewards them when they do.

Group Goals

If it exists, a group has objectives, whether or not they are explicitly stated. However, some groups do not seem to know why they exist, what they are trying to accomplish, or why they do what they do. They exist mainly because they always have and the members continue to hold meetings.[41]

Organized groups have well-defined purposes, goals, and objectives for both the long and the short term. To enhance commitment, members should participate as much as possible in choosing and setting the goals. If goals are to be accomplished, members must understand the goals and their roles in achieving the goals. There are several ways to clarify a group's goals:

- Clearly define goals and ensure that members understand the intent of each one.
- Commit the goals to writing, using "concise, simple, and measurable terms."
- Review past experiences that relate to the goals. Examine both successes and failures. Adjust the goals as needed to reflect what you learned and what you foresee for the future.
- Do regular progress checks while the group works toward achieving the goals.
- Make sure the group stays on track by using the goals as guides for conduct and completion of the group's activities.[42]

Establishing clear, concise goals is critical in building a strong group. All members should participate in developing and establishing goals, because this enhances the likelihood that each member will understand, support, and work toward achievement of these goals. In turn, their participation enhances teamwork and group success. The FFA Program of Activities format is an excellent way to clarify goals.

The late **WARREN BENNIS,** leadership expert and chairman of the Leadership Institute at the University of Southern California, believed that leaders of "great groups" possess certain traits. If you as a team leader could develop each of these traits, you would be very effective as a team leader at work or in any other position.

According to Bennis, the leaders of great teams always do the following:

- *Provide direction and meaning.* They remind people of what's important and why their work makes a difference.
- *Generate and sustain trust.* The group's trust in itself and its leader allows it to deal with the problems that arise and end up a better and stronger team.
- *Encourage action, risk taking, and courage.* Leaders of great groups have a sense of urgency and are willing to risk failure in the course of achieving the group's goals.
- *Act as sources of hope.* Leaders of great groups find many different ways of getting a team to understand that it can win.[43]

GROUP DYNAMICS

Group dynamics involves the study of groups, especially the patterns of interactions that emerge as groups develop. As a leader, you need to understand how group dynamics affect performance. Groups consist of several processes or dynamics. Twelve of the most important are discussed here (Figure 5–8).

Objectives

To be effective, groups must agree on clear objectives and commit to achieving those objectives. Although agreement and commitment are different things, they are highly interrelated: the more that group members participate in defining agreed-upon objectives, the more committed they become to attainment of those goals. This is a primary reason why the leader should allow group input in setting objectives:[44] acceptance of specific objectives improves performance.

All objectives should be

- Cooperatively determined and accepted by the group

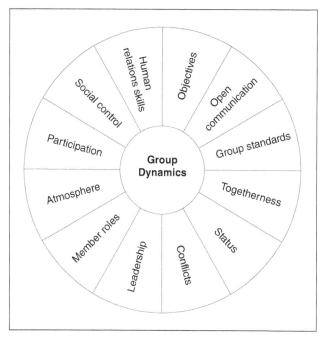

FIGURE 5–8 Group dynamics refers to the pattern of interactions or processes within a group. Twelve ingredients of group dynamics are shown here.

- Dynamic and likely to promote action
- Clearly stated, with specific identification of people, groups, and the types of behavior or behavior changes expected
- Compatible with the general aims and purpose of the group
- Achievable with the maturity of and resources available to the group
- Able to guide the group to increasingly higher levels of achievement
- Varied enough to meet the needs of individuals in the group
- Limited enough to match the resources available and avoid diluting achievement efforts
- Evaluated for evidence of actual progress[45]

Open Communication

Open communication—the opportunity and ability to freely express one's thoughts—is an essential element of good group dynamics. **Self-disclosure** refers to a group member's willingness to engage in open communication with one or more other group members. Feedback is also important for the growth and success of a group. For feedback to be effective, someone must give it and someone must receive it.

Successful leaders often achieve their results by paying attention not only to the members as individuals but also to members' relationships, interactions, and communication. When formal communication is suppressed or ignored, informal means of communication usually spring up. A leader should be able to answer the following questions about communications within his or her group:

- Does the group actually work at ensuring open communication internally?
- Are there definite avenues and means for sharing knowledge, plans, and decisions?
- Is there really two-way communication, or is most communication one-way?
- Have weaknesses in the communication system encouraged the development of cliques?[46]

Group Standards

Groups tend to form their own written or unwritten rules about how things are done. Group standards, which are sometimes referred to as **norms**, are the group's shared expectations regarding member behavior. Standards determine what should or must be done for the group to maintain consistent and desirable behavior. Group standards can be formal or informal, depending on the group.

Formal Groups Standards "provide structure and order for the group. They help coordinate and focus the members' efforts on the goals and tasks of the group." The leader may establish group norms, or the nature of the group may imply them. Group members take on roles and behave in ways that fulfill the expectations established by the group.[47]

Standards must be realistic and attainable. They should be understood by all group members. The group should be involved in setting standards, and the standards should be evaluated periodically and revised as necessary.

Informal Groups Group norms within informal groups form spontaneously as group members interact. Each group member has cultural values and past experiences. Their beliefs, attitudes, and knowledge influence the types of norms developed. If a group member violates the norm, the other members usually let him or her know that the behavior is out of line and thus enforce compliance. This communication may be subtle or overt, verbal or nonverbal, direct or indirect, but it usually boils down to "That's not acceptable" or "That's not how we do things here."

An interesting thing about group norms is that one member's higher standards can be as poorly received as someone else's lower standards. Following is an example of what we mean. While in college, Jackson took a job with a freight trucking company, loading and unloading trucks. Because Jackson was raised on a dairy farm, his norm was to work until the immediate job or task was done. On his first day on the job, when break time came, he kept working because the truck had not yet been fully loaded. He found out quickly that the labor union norm was that everybody stops working at break time.

Togetherness

When a group works together, it demonstrates group cohesiveness. **Cohesiveness** refers to the bonds among members as well as to members' bonds with and closeness to the group as a unit. At the signing of the Declaration of Independence, Ben Franklin told fellow patriots, "We must all hang together, or assuredly we shall all hang separately." This was a very practical call for group cohesiveness—and a warning of the danger of trying to work without it!

Groups promote the cohesiveness that helps them function well when they provide services their members want and need. Togetherness and cohesiveness develop when groups address issues that members think are important, recognize members' contributions, facilitate friendly interaction among members, and encourage cooperation. It is important for members to feel that they belong, that they are making meaningful contributions, and that their contributions further the group's purpose and achievement.[48]

Togetherness can be defined as group solidarity, morale, or *esprit de corps*. Group members feel a common concern for, and have a stake in, what happens to other members and the group as a whole. Individual members feel that they belong to, are a part of, and have a common concern with the group; they naturally use the word *we*. Other factors influencing togetherness and cohesiveness include the following:

Objectives The stronger the agreement and commitment made to achieving the group's objectives, the greater the group's cohesiveness.

Size Generally, the smaller the group, the greater the togetherness. "The larger the group, the more difficulty there is in gaining consensus on objectives and norms. Three to nine members seems to be a good group size for cohesiveness."

Similarity (Homogeneity) Generally, the more similar the group members are, the greater their cohesiveness. Individuals tend to be attracted to people who are similar to themselves.

Participation Generally, a group is more cohesive when the members' participation levels are approximately equal. "Groups dominated by one or a few members tend to be less cohesive."

Competition The focus of members' competitive instincts affects togetherness. If the group focuses on **intragroup** competition, in which each member tries to outdo the others, cohesiveness will be low. If the group's competitive focus is **intergroup**, the members tend to come together as a team to outdo their rivals. "It is surprising how much a group can accomplish when no one cares who gets the credit."

Success "The more successful a group is at achieving its objects, the more cohesive it tends to become. Success tends to breed cohesiveness, which in turn breeds more success. People want to be on a winning team." You may notice that losing teams have more internal arguments and complaints than do high-achieving teams.[49]

Leaders should strive to develop a cohesive, winning team that propels the group to higher levels of success and productivity. Stress competition with outside groups, not within the group.

Status

Status "is the perceived ranking of one member relative to other members of the group."[50] Some group members may attain more power, authority, or status than other members, and may express their status through their behaviors toward other members. In turn, that behavior may influence other members' expression of status. Members

with higher status generally receive greater respect than those at lower status levels.[51]

Status Development Status is based on several factors, including job title, pay rate, seniority, expertise or knowledge, interpersonal skills, appearance, education, and sometimes race, age, and sex. Members who conform to the group's standards tend to have higher status than members who do not.[52]

Status Leader High-status members have a great deal of influence on the development of a group's norms. A group is often willing to overlook a high-status member's violation of a standard, even if a member with lower status would suffer adverse consequences for the same behavior. Lower-level members tend to copy high-status members' behavior and standards; therefore, high-status members have a major impact on the group's performance. To be effective, the leader needs to have high status in the group and to set a good example.[53]

Conflicts

No group can avoid occasional conflict. Nor should it try to: conflict can benefit a group if resolution of the problem stimulates growth, change, and cohesiveness. When members express their hostility and anger through open confrontation, often communication barriers can be overcome, and the resulting dialogue leads to understanding and accommodation.[54]

Leadership

The group's leader exerts tremendous influence on the effectiveness of the group. Leadership behavior, along with the group's standards, assists in directing the group toward accomplishment of its goals and objectives.

Both leaders and members are needed for successful groups. Even a dynamic, creative leader can do little if the group members are apathetic, unmotivated, or uncommitted. Conversely, ineffective leadership can dampen enthusiasm and stifle initiative, whereas effective leadership motivates the group to achieve its goals and objectives.

A democratic atmosphere and approach promote dynamic interaction between leaders and members. Democratic leadership is the most effective way of arriving at decisions in groups with voluntary membership. Democratic leadership is also superior at building group cohesiveness and cooperation. Democratic leaders make everyone a part of the team, give the team direction and encouragement, and support the team efforts.[55]

Atmosphere

Group **atmosphere** is the mood, tone, or feeling that permeates the group. A good atmosphere is essential for successful operation of a group. Effective leadership is needed to create a positive atmosphere.

Collective-Whole Atmosphere When individuals meet and work together, they respond as a group to the prevailing atmosphere. A warm, permissive, democratic atmosphere fosters good motivation, high member satisfaction, and high productivity. Less discontent, frustration, and aggression are evident in these groups, compared to other types. In Collective-Whole Atmosphere groups there is more friendliness, cordiality, and cooperation.[56] These types of groups also tend to foster individual thinking and creativity for the good of the group/team/organization.

A group member's behavior is determined to a considerable extent by the group's reaction to him or her. Individuals who feel secure and believe that their group skills are adequate will take the lead in group activities more often and will participate more fully in those activities. The total resources of the group are enhanced and can be accessed more fully when all individuals feel free to contribute and ask questions. Motivation and morale reach high levels in a democratic, permissive atmosphere in which both the leaders and the members participate actively.[57]

Authoritarian Atmosphere In an authoritarian atmosphere, the responsibility lies with the leader, and no one may participate or initiate action except with the leader's permission or order. It is presumed that the leader knows best what the group should believe and do. Group member behavior is directed toward the leader's predetermined goals.[58]

Democratic Atmosphere In a democratic atmosphere, all group members share in leadership, and individuals do what is necessary for group productivity. Both the nominal leader and other group members are responsible for "creating conditions—including group atmosphere—under which group members are best able to work together to accomplish chosen ends."[59] The group members in Figure 5–9 are working in a democratic, productive atmosphere.

FIGURE 5–9 The responsibility of a formal group leader is to create a positive group atmosphere. It appears that this objective has been accomplished in this group! *(Courtesy of University of Georgia—College of Agricultural & Environmental Sciences.)*

Participation

Participation includes involvement through speaking and entering into discussions. **Breadth** of participation indicates how many members take part. **Intensity** of participation is a measure of how often individual members take part and how emotionally involved they become. Along with attending meetings, being on committees, and serving as an officer, participation may also include helping with finances and fund-raising, being part of work groups, hosting and cleaning up after events, and writing publicity materials and grant applications.[60]

Patterns and Examples Participation patterns are simply how people respond to each other. When a particular person enters the discussion, do certain others usually follow him or her? Do a few people monopolize the discussion, or are there opportunities for all to participate? Does the group help and encourage everyone to participate? Is the participation pattern leader centered or distributed throughout the group?[61]

Why Emphasize Participation? Research indicates that individual and group productivity are directly related to opportunities for member participation. Such opportunities may include setting goals, deciding on means of attaining goals, and other decision making. Even when their initial ideas did not match the group's final decision, people are happier and more productive when they have had an opportunity to participate in and express their opinions during the decision-making process.[62]

Social Control

The process of ensuring conformity to the expectations of group members is called **social control**. This type of control often rewards group members for meeting group standards. Such rewards may take the form of tangible recognition, such as a certificate or pin. Other, less tangible rewards may include group acceptance or recognition, a verbal expression of encouragement or gratitude, or just a smile or pat on the back.

Every group has standards and uses varying degrees of social control to enforce them. Some groups use incentives or rewards for control, whereas other groups use fear or punishment. "If groups are to be productive, members need to know what the standards of the group are and the means used to enforce those standards—the methods of control."[63]

Human Relations Skills

Human relations skills are the skills needed to work with and get along with people. We frequently assume that we are good at human relations simply because we have lived all our lives with other people. Most of us, for example, at least have the ability to disagree without creating open hostility. However, true skill in human relations differs greatly from the minimum needed to function acceptably in a society.[64] Leaders who understand and facilitate good human relations within their groups are the most successful. "Some studies suggest that it is more important for leaders to understand and be skillful in human relations, individual motivation, and group process than to be *highly* proficient in the subject matter."[65]

Member Roles

Member role is a term for the general expectation of the group member within the group. (Do not confuse this with the role of group members in a discussion.) Organizations that specify the roles of group members are likely to have greater goal achievement than organizations that do not. If roles are not clearly defined, if there are overlapping roles, or if the defined roles leave responsibility for important functions unspecified, fewer goals are achieved. A member's understanding of his or her role, how that role relates to other roles in the group, and the importance of the role all determine the member's sense of responsibility to the group and motivate the member to contribute.[66]

GROUP DEVELOPMENT

Each group is unique, and its dynamics change over time. All groups go through the same stages as they become smoothly operating and effective teams, chapters, clubs, organizations, or departments. However, the time a group spends in each of the developmental stages is different. The stages of group development are as follows (the identifications in parentheses refer to names in the original Tuckman Model of Group Development):

- Orientation (forming)
- Dissatisfaction (storming)
- Resolution (norming)
- Production (performing)
- Termination (adjourning)

An understanding of group dynamics and the stages of development provides managers, consultants, and other leaders with realistic expectations of group behavior. Recognizing group behavior traits and being able to predict the course of development enable one to guide groups effectively and promote group growth. Figure 5–10 gives an overview of the five stages of group development.

Orientation (Forming)

When people form a group, they tend to bring a moderate to high commitment to the group. Because they have not yet worked together, however, they do not have the competence to achieve the goal; they are not yet a team. When first interacting, members tend to be anxious about whether they will fit in, what will be required of them, what the group will be like, and the purpose of the group, among other things.

The leader must create, foster, and maintain an open atmosphere within the group, as he or she promotes interaction among group members and orients their efforts toward the group goals and objectives. "An open atmosphere is essential during the forming stage to promote the sharing of ideas." Especially in initial brainstorming sessions, group members must be receptive to all ideas and not pass judgment on anything submitted for the group's consideration.[67]

Dissatisfaction (Storming)

After members have worked together for a while, they tend to become dissatisfied with the group

STAGES OF GROUP DEVELOPMENT			
STAGE	**TASK**	**RELATIONSHIP**	**LEADER**
Orientation (forming)	Set boundaries Clarify task Define task High commitment and low competence	High anxiety Size up one another Establish turf Protect turf	Helps set boundaries Be prepared Establishes open atmosphere
Dissatisfaction (storming)	Identify alternatives Look at "big" picture Task more complex than anticipated	Frustration Differences emerge Conflicts develop Confrontation	Mediates conflicts Keeps communication open
Resolution (norming)	Make decisions Select alternatives	Become more satisfied Listen to understand Compromise Share information	Decision making Strengthens relations Increases morale Builds team
Production (performing)	Productive Conduct task Follow through	Commitment Competence Work as team Positive group structure	Delegates information gathering Keeps things moving
Termination (adjourning)	Make decisions Closure of task Evaluate results	Evaluate process and roles Experience feeling of leaving	Interprets results for future action

FIGURE 5–10 Most groups go through these stages as they grow to become smoothly operating, effective teams.

and its progress. They start to question the reasons for their membership, wonder whether the group will really accomplish anything, and criticize the participation of other group members. Often the task is more complex and difficult than anticipated, and members get frustrated and feel incompetent.[68] They may even leave the group.

Conflict is a normal part of group development. A leader should maintain open communication channels during this stage and keep conflict at a controllable level, using whatever conflict resolution skills and techniques are available. After a while, the members will realize that such internal conflict is self-defeating and unproductive. They accept the need to make decisions and move on to the next stage of development.[69]

Resolution (Norming)

Given time, members often bring their initial expectations and the realities of the group's goals into alignment. As members gain competence, they frequently become more satisfied with the group. Relationships develop that satisfy group members' togetherness needs. They learn to work together as they develop a group structure with acceptable standards and cohesiveness, share information, and reach compromises. Commitment will vary as the group interacts and various tasks are undertaken, but team-building efforts and success at reaching some objectives can increase group morale.[70]

Production (Performing)

At the production stage, commitment and competence do not fluctuate much. The group works well as a team and yields high satisfaction of members' affiliation needs. The group maintains a positive structure and relationships. Members' productivity leads to positive feelings. Group dynamics may change over time, but when the group reaches this stage, any issues are resolved quickly and easily; members are open with each other.[71]

Termination (Adjourning)

During termination, members may feel sad that the group is disbanding (if it was successful) or may be relieved (if the group did not develop appropriately).[72] In this stage, the group makes decisions and evaluations concerning the roles of members and the procedures used in accomplishing the group's goals and objectives.

GROUP DISCUSSION

Many people have tried to describe what occurs during a group meeting. Sometimes we try to review a meeting to determine why it was a success or failure. The personality types of individual members and the roles of members are prominent factors in any group discussion.

Regardless of your personality type, you can learn to perform any of the following roles. Some will come naturally. For others, you will have to work harder to develop and perform the roles. Familiarity with the roles will enable group leaders and members to recognize and analyze, more or less automatically, the roles being played by group members. Also, the leader can better lead discussions when he or she recognizes the members' roles. Each group member is in a position to play the roles needed for group productivity, to encourage others, and to suppress roles that are not contributing to the group's work.

"For a group to be productive, each member must contribute. Group members contribute to a group through the roles or jobs they perform. No one performs the same role" throughout a group's entire life span. Also, the same person may play a completely different role in another group. Three categories of common group member roles are task roles, maintenance roles, and individual roles. An understanding of these roles will increase a group leader's effectiveness.[73]

Task Roles

Task roles are parts that members in a group play in an effort to get things done.

Task roles of discussion group members are all aimed at defining and solving problems. These roles include seeking information, presenting ideas and information, and summarizing information. The following roles are necessary for accomplishing group tasks:

- **Initiator-contributors** propose new ideas or different ways of approaching group problems or goals.

- **Information-seekers** ask for clarification of suggestions, authoritative information, and facts relevant to the problem being discussed.
  - **Opinion-seekers** ask what group values have a bearing on the project the group is undertaking or the problem it is addressing.
  - **Information-givers** provide authoritative facts or generalized rules. As part of sharing information, they may relate personal experiences to the group problem.
- **Opinion-givers** express their personal beliefs or opinions rather than facts. They may submit these opinions as suggestions or alternative solutions to the group problem.
  - **Elaborators** give examples of suggestions or develop meanings. Elaborators may offer a rationale for suggestions and try to project how an idea or suggestion would work out if adopted by the group.
  - **Summarizers** pull together the ideas and suggestions under discussion to help determine where the group is in its thinking or action process.
- **Coordinator-integrators** clarify the relationships between ideas and suggestions, extract key ideas from member contributions, and integrate or synthesize those ideas into a meaningful whole.
- **Orienters** define where a group is with respect to its goals, point to departures from previously agreed-on directions or goals, or raise questions about the direction being taken.
- **Disagreers** state a different point of view, argue against propositions, or point out errors in reasoning. They may also disagree with opinions, values, sentiments, discussions, or procedure.
- **Evaluator-critics** judge accomplishments according to some set of standards regarding group functioning.
- **Energizers** urge the group to action or decision, attempting to move the group to greater or higher-quality activities or discussions.
- **Procedural technicians** perform tasks such as distributing materials, arranging the seating, or operating equipment for the group.

- **Recorders** write down suggestions, ideas, and group decisions. The recorder, in essence, is the secretary, whether formally or informally.[74]

Maintenance Roles

Whereas task roles are concerned with the content of discussion, **maintenance roles** help the group members work together smoothly as a unit. People who play maintenance roles support other members, smooth over tensions, manage conflict, and give everyone a chance to talk.[75]

- **Encouragers** accept other members' contributions, praising the contributions and finding points of agreement in them. Their attitude toward other group members is warm, as they show solidarity, offer encouragement, and in various ways indicate understanding and acceptance of other points of view, ideas, and suggestions.
- **Harmonizers** bring the group together. They help reduce tensions and straighten out misunderstandings, disagreements, and conflicts. They recognize when the group is not moving forward or is tired. They have a good feel for when to tell a joke and get the group to loosen up a little before returning to a task.
- **Compromisers** reevaluate their positions or opinions, "admit any errors or unfairness, and agree to settle tensions by meeting halfway."[76]
- **Gatekeepers** help maintain open communication. Gatekeepers ensure balanced participation, keeping those who tend to dominate in check and encouraging those who tend to be shy. Examples: "We haven't heard from Ms. Davis yet," or "Let's keep our comments brief so everyone will have a chance to speak."
- **Standard-setters** express standards or ideals that guide group functioning or apply standards in evaluating the group discussion.
- **Group observers** keep records of various aspects of the group process. These records are then used in the group's evaluation of its own procedures.
- **Tag-alongs** just go along with the group. "They basically accept the ideas of others

and serve as an audience for the group discussion and decision making."[77]

Individual Roles

When group members attempt to satisfy personal psychological needs that are irrelevant to the group task or team building, they reduce the effectiveness and productivity of the group. If he or she is aware of these roles, the group discussion leader can often head off conflict among members and keep the group focused on its main task.

- **Aggressors** are almost always on the offensive, demeaning others or expressing disapproval of another member's values, acts, or feelings. Aggressors attack the group or the problem it is working on and may express jealousy or envy by trying to take credit for another's contribution.
- **Dominators** try to assert and enhance their personal authority and superiority within the group. Frequently they do so by manipulating, flattering, projecting the assumption that they are superior, and acting authoritative. They may downplay or denigrate the contributions of others in an attempt to attract or keep attention on themselves.
- **Blockers** tend to have a negative, stubbornly resistant, disagreeable attitude. They do not like change. Blockers may oppose others' ideas for no apparent reason and may try to reopen or keep an idea alive after the group has already considered and rejected it.
- **Anecdoters** relate situations and tasks to their own experiences. The experiences that an anecdoter shares may merely distract group attention; at other times, the stories may affect the attitudes of other group members (either positively or negatively).
- **Recognition-seekers** like to hear themselves talk. They seek any type of acknowledgment or attention from the group. They boast, report personal achievements, and generally focus attention on themselves to avoid being ignored or placed in what they perceive as an inferior position.
- **Help-seekers** try to garner sympathy from other group members, usually through

expression of insecurities, confusion, or self-pity. Sometimes this is simply a ploy to attract attention; at other times, it may be a disguised way of asking to be relieved of responsibility.
- **Players** lack involvement in the team effort. Their distracting behavior may take the form of inattention, frequent wisecracks or inappropriate humor, or horseplay.
- **Special-interest pleaders** focus on whatever meets their individual needs, whether or not it is relevant to the group's task or topic.[78]

Leading a Group Discussion

Group discussion is rapidly becoming one of the most widely used forms of communication. With team building and functions now being an integral part of the workplace, leaders may spend 50 percent of their time (or more) in meetings. A leader who understands the various roles that members may assume can devise strategies for dealing appropriately with individual members. The objective, of course, is to proceed with leading a productive discussion, like the food marketing team in Figure 5–11 that is having a group discussion. Consider the following factors when leading a group discussion.

Planning Planning is crucial for a successful discussion. Planning includes choosing a topic and questions pertinent to that topic; preparing an outline; and studying the background knowledge, experience, and ideas of the participating team members.

Beginning the Discussion A group discussion should start and end on time. If they are not already familiar with one another, introduce group members to each other; this is a logical starting point to build group cohesiveness.

Introducing the Topic Stating the topic at the start of the discussion prevents uncertainty about the subject to be discussed.[79]

Questioning Do not call on a group member to answer a question before you have actually asked the question. Other group members will not focus on a question directed to someone else. Pose the question; then allow about 10 seconds for thought and reflection before you select an

FIGURE 5–11 Group discussion is becoming a common practice among workplace teams. This photo shows a food marketing team having a group discussion. *(Courtesy of University of Georgia—College of Agricultural & Environmental Sciences.)*

individual group member to respond. This procedure keeps the group's attention on the question.[80]

Regulating Communication During the discussion, the leader has several tasks. The leader must not only invite but also encourage members to contribute to the discussion. All members should feel free to speak their minds fully and frankly. Group leaders may assist members in expressing their ideas clearly, but should do so without introducing their own ideas and opinions.

The idea that each individual's ideas are valuable should be stressed, but participation has to be balanced. Do not let one person dominate the discussion. A leader must also be prepared to step in if members begin to argue. Furthermore, the leader is expected to keep the discussion on track, constantly guiding the discussion toward relevant and important matters and not letting the group get sidetracked.[81]

Concluding the Discussion When the leader believes that the group has thoroughly examined the discussion question, or when a preset time limit is approaching, the leader should "summarize the major ideas and outcomes of the discussion." Do not use the summary as a way of merely reintroducing your own ideas; make sure it really

does briefly recap all of the important points in the discussion. Always reserve enough time for members to add to or disagree with the summary or to express minority positions if they so desire.[82]

After the meeting, it is very helpful if the group leader (or someone appointed by the leader) sends out a memo or e-mail to remind the group of key action items that were developed during the meeting or steps that should be taken as a result of the discussion.

Conflict Resolution

Because people are unique individuals, personality conflicts will inevitably arise within groups. Those with Choleric personalities believe they are always right, so when two Cholerics have a difference of opinion, sparks will fly. If we are truly convinced the other person is wrong, we enter into conflict with one of two options: fight or flight. Epictetus said, "What concerns me is not the way things are, but rather the way people think things are."[83] When we are aware of our personality tendencies and weaknesses, though, we can better cope with conflict.

Communication "Communication is the key element affecting conflict: both its cause and remedy. Open communication is the means by which disagreement can be prevented, managed, or resolved. The lack of open communication can drive conflict underground and create a downward spiral of misunderstanding and hostility."[84]

Problem (Conflict)-Solving Resolution Plan Once open communication is attained, resolve conflicts by using the problem-solving process discussed in Chapter 13. Most groups find that having a plan to follow for resolving conflicts is valuable. A step-by-step plan helps everyone stay focused on solving the problem and preserves the self-esteem of all involved. The problem-solving plan for conflict resolution has six steps:

1. Define the problem
2. Collect facts and opinions
3. Consider all proposed solutions
4. Define the expected results
5. Select the solution(s)
6. Implement the solution(s)[85]

Conflict-Resolution Behaviors Good conflict-resolution behaviors help you resolve conflict in almost any situation. They allow you to "benefit from positive disagreement without having those disagreements escalate into out-of-control personality conflicts that damage the morale and productivity of the organization." The five basic behaviors are as follows:

Openness "State your feelings and thoughts openly, directly, and honestly without trying to hide or disguise the real object of your disagreement."

Empathy "Listen with empathy. Try to understand and feel what the other person is feeling and to see the situation" from the other person's point of view.

Supportiveness "Describe the behaviors you have difficulty with rather than evaluating them.... Be willing to support the other person's position if it makes sense to do so."

Positiveness "Try to identify areas of agreement and emphasize those.... Be positive about the other person and your relationship" with that person.

Equality "Treat the other person and his ideas and opinions as equal. Give the person the time

and space to completely express his ideas. Evaluate all ideas and positions logically and without regard to ownership."[86]

Discussion Techniques

Several techniques can be used to get a group to act. Techniques have been likened to the vehicle that helps move a group toward its goals. These meeting or session techniques are designed either to bring information and understanding to a group or to move a group to action.

- **Small-group discussion** is face-to-face mutual interchange of ideas and opinions among members of a relatively small group (usually 5 to 20 people).[87] Refer to Figure 5–12 for a room arrangement that is conducive to small-group discussion.
- The name of the **huddle/discussion 66 method** comes from the concept of six people discussing a subject for six minutes.[88] Figure 5–13 shows a room arrangement that works well for this discussion technique.
- **Buzz groups** are an alternate method of breaking a large group into smaller, more intense groups to facilitate discussion. Often, only two to three people are in a buzz group (Figure 5–14).[89]
- A **symposium** is a group of talks, lectures, or speeches presented by several individuals on the various aspects of a single subject.[90]
- **Panel discussions**, which take place in front of an audience, are held by a selected

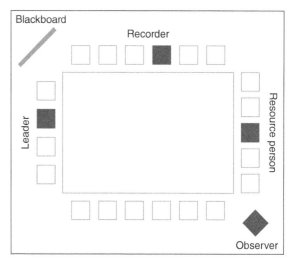

FIGURE 5–12 Suggested room arrangement for small-group discussion.

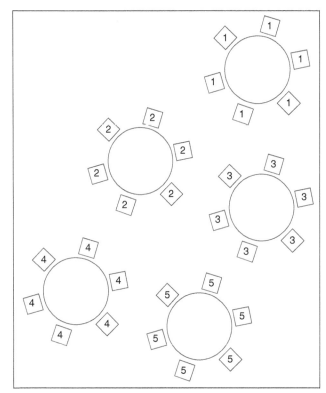

FIGURE 5–13 Seating arrangement for huddle/discussion 66 method.

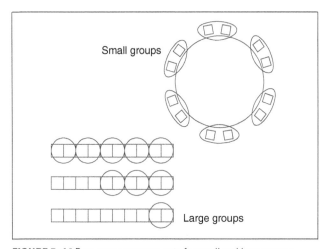

FIGURE 5–14 Buzz-group arrangements for small and large groups.

group of people (usually three to six) under the control and direction of a moderator.[91]

- **Interrogator panels** are an interrogation-discussion interchange between a small group of knowledgeable individuals (the panel) and one or more questioners (interrogators); such panels are often conducted under the direction or facilitation of a moderator.[92]

- **Committee hearings** involve several people questioning an individual.[93]
- **Dialogues** are discussions between two knowledgeable people, carried on in front of a group.[94]
- **Interviews** involve the questioning of an expert on a given subject by an interviewer who represents the group.[95]
- **Brainstorming** is a group activity that calls on the free flow of ideas with no limitations or penalties. Usually a group leader will pose a question, and then the group will offer up any and every solution that comes to mind. As long as all the ideas are captured, brainstorming is working. A common follow-up activity is discussion of the ideas and making meaning of them through grouping and the identification of themes.
- **Role-playing** is the dramatization of a problem or situation in the general area of human relations.[96]
- **Nominal technique** is an alternative form of group communication designed to reduce two problems: (1) the amount of time group discussions take and (2) the domination of a group discussion by one or a few members. A group using the nominal technique begins by asking each member to write down a list of possible solutions to the group's problem. Each member is then asked to state one idea from his or her list. Each idea stated is written down so that all participants can see the entire list. The third step consists of a brief discussion of the ideas on the board. This discussion is intended primarily to clarify but also may involve some evaluation of ideas. Next, a secret vote is conducted, with each member ranking the ideas in order of personal preference. Finally, the rankings are tabulated and the solution with the highest ranking is adopted as the group's solution.[97]

During your first 100 days in a new leadership role, you will be constantly scrutinized as people search for information about you: how you think, how you build your team, how you lead, the focus of your efforts, and your character. A six-step program for your new position includes maintain focus, build momentum, build your team, execute your plans, understand the existing politics and culture of the group to build support coalitions, and use effective fact-finding methods and data analysis.

TWENTY LEADERSHIP MISTAKES

Why do some talented, well-meaning people fail so consistently, and sometimes spectacularly, when they try to act as leaders? Although these folks may hold high-level positions, with job titles that make people expect great things, somehow they "never seem to deliver the goods when it comes to actually inspiring and leading others.... What is it that stands in the way of otherwise intelligent, motivated people realizing their full leadership potential?" Drawing on responses to a nationwide research question, Phillip Van Hooser compiled the following "20 Leadership Sins," any of which could "silently sabotag[e] your leadership growth and development."[98]

Obvious Lack of Self-Discipline

Actions speak louder than words to group members. No matter what the would-be leader says is important, followers will judge the acceptability of behaviors by what the leader actually does. For example, a leader stresses the importance of appearance in customer relations, but comes to meetings dressed casually or with indifferent grooming. What inference do you think group members will make? A "do as I say, not as I do" attitude undercuts the leader's authority and integrity.

Poor Judgment

When in leadership positions, some people act or speak rashly; make decisions hastily, without due consideration of the available information; or simply react emotionally. In short, they display bad judgment. One of the best techniques to use when called upon to take an action or make a decision is to consider the following three questions: Is this good for the organization? Is this good for my coworkers? Is this good for my future as a leader? If the answer to any of the three is no, the leader would be well advised to do a little more thinking before proceeding.

Insensitivity to the Needs of Others

The best leaders care deeply about their followers as individuals, even while they ask for and expect superior work and commitment from each team member. One of the biggest mistakes would-be leaders make is failing to acknowledge and address followers' needs.

Being Too Strict or Too Lenient

If a leader goes too far in either direction, their leadership credibility will be undermined. Rigidity concerning conduct and insistence on strict observation of rules and procedures, can make you into a heavy-handed micromanager. Conversely, a completely laissez-faire approach gives followers no guidance or structure within which to work. Establish reasonable goals and expectations, and hold group members accountable for their actions.

Being Cold, Aloof, or Arrogant

As noted, good leaders care about their people, and do everything possible to make them feel that they and their contributions are important and valued. An impersonal attitude or arrogance from a leader pushes people away rather than inviting and encouraging them. If you, the leader, are not there for your followers, why should they be there for you?

Doing Too Much and Leading Too Little

The leader's job is to get it done, not to do it—even though sometimes it would be faster and easier to do it personally. A true leader coaches, guides, supports, and encourages, but assists directly only when absolutely necessary. Delegation is a critical skill that some in leadership positions develop only slowly, but they must learn it if the group is to flourish. Learn to accomplish goals through others' efforts. You cannot afford the mind-set of "If you want it done right, do it yourself." The leader's job is to empower followers to do it successfully.

Micromanaging

Leaders should lead rather than trying to do it all themselves. This means that they must empower and trust group members to get the work done in whatever way suits the members best—even if it is not the way the leader would have done it. Leaders who consistently question, critique, and second-guess followers' every move will eventually dishearten and destroy the confidence of even the best team members.

Appearing to Play Favorites

Leaders often rely heavily on team members who consistently perform well, but leaders need to be careful. If followers think you are unfairly showing favoritism, they may try to sabotage your efforts. A leader must be even-handed with everyone, treating all group members fairly and equitably.

Betraying Trust

Trust is hard to earn and easy to lose. A leader must behave with scrupulous integrity, faithfully keeping promises and confidentiality. In this area too, actions speak louder than words. If group members do not trust the leader, that person will lose any influence he or she had with them.

Holding Grudges

Conflict that is not resolved can leave hostility, resentment, and fear to fester below the surface interactions in a group. If followers think that the leader is holding a grudge, they will not trust the leader to act fairly and impartially.

Inability to Think Strategically

Strategic thinking allows you to solve problems, make decisions, and accomplish many goals in a timely manner. If a leader does not make a master plan, or does not adequately communicate that plan, group members may not perceive their contributions as important and thus will lose individual motivation.

Failure to Staff Effectively

Good leaders take much care to select good people, even when the pool of candidates seems to offer little choice. Leaders need to get it right the first time. If a staff member has the right personal and leadership skills and attitude, they can be trained to do certain skills for a certain job. Invest the time and effort it takes to hire the right person, and then invest in that person so they know what is expected of them. It will pay off. Ineffective leaders do not follow this procedure.

Inflexibility and Reluctance to Adapt

Rigidity and inflexibility are usually fatal to a group effort. A leader who cannot adapt his or her style and methods to the myriad of attitudes and abilities that members bring to the group is doomed to fail. The leader who understands and appropriately leverages the differences in followers—in personality type, learning style, communication style, and other ways—will gain the loyalty and trust of group members and be able to elicit their best efforts in the manner most comfortable for each individual.

Disregard of Organizational Policies and Procedures

Once again, actions speak louder than words. Followers tend to reflect the attitudes and behavior of their leaders. Leaders should never publicly voice disapproval of or contempt for the organization of which they are a part. Disagreements, concerns, and questions should be directed to a person in the organization who can actually do something to help. Otherwise, it's just gossip.

Establishing Unclear or Vague Parameters

Unclear explanations of expectations can cause real trouble for a leader. Followers need the freedom to create, confront, and innovate, but leaders need to provide distinct expectations, guidelines, and boundaries within which to work. Vagueness can cause the group to go off course very quickly.

Failing to Act When Necessary

Procrastination and inaction are common leadership failures. Many would-be leaders go to great lengths to avoid conflict, confrontation, and disagreements, hoping that they will not have to deal with problems if they put it off long enough. This almost never works; more often the issue just becomes more complex when it is not handled in a timely manner.

Offering Personal Advice

The best advice regarding personal advice is to not give any. Even though you care about your followers as the individual people they are, you should not insert yourself into their lives in this way. You can empathize with them, ask what you can do to assist, or suggest task or work reassignments that might be helpful—but do not weigh in on their personal dilemmas. (What if someone follows a piece of advice that turns out to be bad or unwise?) Stay within the limits of the leadership role: give encouragement, praise, correction, suggestions, and directions pertaining to the group vision, goals, and tasks.

Being Overly Ambitious

Some ambition is fine; it is a good motivator and a driving force for accomplishment. However, get-there-at-any-cost ambition that rides roughshod over anything in its way is a distinct negative, and your followers will correctly perceive it as such. "Remember that there are two ways to get to the top.... [Y]ou can get there by climbing over people," or you can "get there by being lifted up by people."[99] For a good leader, especially one who values the democratic style and process, there is no question about which approach is better, both ethically and practically.

Letting Performance Problems Slide

If leaders notice a recurring problem, they must act to solve it. Allowing performance problems to continue hinders the group's goal attainment, erodes morale, and also undermines the leader's credibility and authority. A leader cannot let merely average performance become the norm; the leader's job is to inspire group members to excellence, remember? The leader in Figure 5–15 is taking the time to encourage some of his employees to improve their performance. In addition, only leaders who keep working on improvement of their own weaker performance areas can legitimately expect their followers to do so.

Allowing Power to Corrupt

Power and a high-status position can be intoxicating. Some leaders use their authority wisely and grow personally as they mature into extended leadership responsibility. For others, increased status and power just cause heads and egos to swell. Always remember the ideal of servant leadership; never believe that you are somehow superior to the people you lead just because of your title.

FIGURE 5–15 This leader has pulled some of his employees aside to encourage them. If you notice someone who needs some coaching and you look the other way, efficiency could suffer. *(Courtesy of University of Georgia—College of Agricultural & Environmental Sciences.)*

TAKING OVER A NEW GROUP, ORGANIZATION, OR BUSINESS

The first hundred days in your leadership of a new group, organization, or business often make or break you. Both followers and the people to whom you report will be watching your every move, "looking for clues to your thinking, your team building, your leadership, your focus, and your character."[100] The material in this section, adapted from IdeaBridge's *100-Day Success Plan for Crisis Recovery and New CEOs*, can help you establish yourself quickly and successfully in a new leadership role while you lay a strong foundation for future endeavors with your new group. These points are just as pertinent and applicable to school, team, or club leadership as they are to business executives!

Maintain a Sharp Focus

As a new leader, you must focus your initial efforts solely on a very limited number of strategic priorities that you have identified as crucial for the team or group (no more than four). You can be flexible about how they are implemented, but you must insist on the primacy of these few items and the objectives supporting them. You always have to make your goals and objectives known, explain why you chose these goals, and repeat them constantly to keep everyone's thinking and efforts aligned and on target.

Build Momentum

Initial successes invigorate your followers, grab their attention, and build their trust in and acceptance of your leadership. Plan goals so that you can achieve some wins early on and build momentum for your longer-range initiatives.

Your first 100 days are the time to establish a base for your program of growth and change. Tangible improvements and successes that occur early in your tenure encourage coworkers, heighten motivation, and build support for your other goals.

Build Your Team

One of your first tasks is to choose your team members and decide how roles will be assigned. This may require structural and personnel changes—the sooner you implement them, the easier they tend to be. Usually, a few key people can make all the difference; your job is to find those people and get them on board. Remember that although you inherit what you start with, "after 100 days, you have bought into everything" that you allow to continue.[101]

Execute Your Plans

Implementation of changes requires a lot of work and a lot of oversight. To support your goals, you should draw up an action plan that includes subgoals, objectives, timelines, and personnel assignments. Let your team know, clearly and in detail, what has to be done, who is responsible for doing it, and when it has to be completed. Keep tabs on progress through meetings and regularly scheduled communication; you may want to delegate this supervision to someone with a knack for keeping people on task. If team members fall behind, get an explanation of why and details about how they intend to catch up. Remember that part of your leadership role is to get your team members the resources and training they need to succeed.

Understand the Politics and Build Coalitions of Support

As a new leader, you must quickly learn about the organization's existing culture and politics. Determine where you fit in, what changes you want to see, and who will support or oppose you. As noted earlier, be visible and available: people will want to meet you, talk to you, get to know you, and find out whether they can influence you.

For each of your priority plans, identify likely allies, opponents, and potential obstacles. Develop ways to enlist support and reduce resistance.

Use Effective Fact-Finding Methods and Analyze Data

Facts and objective data are powerful. Insist that decision making be data driven and that your people be able to support their decisions, suggestions, and critiques with facts and fact-based analyses. You can move your team forward in a powerful way if you show them how to analyze facts and data.

CONCLUSION

Leaders must be able to lead groups and teams. A leader has to mobilize and motivate groups to work toward a common goal with a strong sense of excitement and anticipation. The way a leader conveys the purpose of a task to his or her group can help instill this positive attitude. A leader should also emphasize the fact that the group has a specific purpose and that the particular skills and participation of each individual member are fundamental to the success of the group. This helps people to identify with the group goals and empowers them to use their creativity.

SUMMARY

Understanding the group process makes a group leader better. Leaders should know why people join groups, why people refrain from joining groups, why group leadership skills are important, and why teamwork is so important.

There are three types of groups: functional groups, task groups (both of which are formal groups), and informal groups. Regardless of the type of group, the democratic approach to leading groups provides greater productivity and increased member satisfaction and commitment. Status within any type of group has to be earned.

Not all group members are equally committed to the group. There are nucleus members, faithful members, and fair-weather members. Group membership will remain strong if the group is cohesive, accomplishes its objectives, works as a team, and provides recognition.

Group dynamics refers to the patterns of interactions or processes in a group. The dynamics include objectives, open communication, group standards or norms, togetherness, status, conflicts, leadership, atmosphere, participation, social control, human relations skills, and member roles. Groups go through five stages: forming, storming, norming, performing, and adjourning.

The leader must be able to lead group discussions. Members assume three roles during a discussion: task roles, maintenance roles, and individual roles. Task roles have the mission of selecting, defining, and achieving goals. These roles include the initiator-contributor, information-seeker, opinion-seeker, information-giver, opinion-giver, elaborator, summarizer, coordinator-integrator, orienter, disagreer, evaluator-critic, energizer, procedural technician, and recorder.

The maintenance role focuses on the cohesiveness of the group. These roles include the encourager, harmonizer, compromiser, gatekeeper, standard-setter, group observer, and tag-along.

Individual roles meet the member's personal psychological needs. These roles include the aggressor, dominator, blocker, anecdoter, recognition-seeker, help-seeker, player, and special-interest pleader.

Various techniques can be used to secure group action. These include small-group discussion, huddle/discussion 66 method, buzz groups, symposiums, panel discussions, interrogation panels, committee hearings, dialogues, interviews, brainstorming, role-playing, and nominal technique.

Take It to the Net

Go to http://scholar.google.com/ and use the search terms listed here to find research articles related to teams and groups. Read some of the articles. Write a short summary of one article, making sure to cite your source completely and properly.

Search Terms

group development

group work

leading groups

group dynamics

team building

Chapter Exercises

REVIEW QUESTIONS

1. Define the Terms to Know.
2. List six reasons people are attracted to groups.
3. Why not use autocratic group leadership?
4. List three reasons why people do not belong to groups.
5. List five considerations for team building.
6. List five things that research has revealed about teams.
7. What is the purpose of teams?
8. What is the role of the leader in using teams?
9. What are some potential problems with using teams?
10. What are the three types of groups?
11. What are three types of task groups?
12. Name 12 dynamics or processes typical of most groups.
13. In what ways can conflict benefit a group?
14. Name the five stages of group development.
15. What are the three levels of membership?
16. What five things must exist to maintain strong groups?
17. Name three categories of group member roles.
18. List 11 discussion techniques.
19. What are five ways to clarify group goals?
20. List 11 factors that influence togetherness and cohesiveness.
21. What are six steps a leader should follow when leading a group discussion?
22. List 20 mistakes made by many group leaders.
23. What six steps should a leader take during the first 100 days with a new group or organization?

COMPLETION

1. _____ involves the study of groups, including the interactions patterns that emerge as groups develop.
2. Written or unwritten rules about how things are done are group _____.
3. When a group works well together, it is displaying group _____.
4. _____ is the perceived ranking of one member relative to another member.
5. The behavior of the group's leader _____ the effectiveness of a group.

6. _____ is working with people and getting along with people.

7. _____ effectively removes communication barriers and leads to understanding and acceptance.

8. When a group secures conformity to the expectations of its group members, it is termed _____.

9. Group _____ is the prevailing mood, tone, or feeling that permeates the group.

10. _____ can be brought about by communication and problem solving, along with openness, empathy, supportiveness, positiveness, and equality.

MATCHING

_____ 1. Maintains a negative attitude.

_____ 2. Keeps group on task.

_____ 3. Develops suggestions further.

_____ 4. Wants sympathy from others.

_____ 5. Draws group ideas together.

_____ 6. Provides authoritative facts.

_____ 7. Defines the group's position.

_____ 8. Group mediator.

_____ 9. Lacks group involvement.

_____ 10. Wants others' opinions clarified.

A. information-giver

B. elaborator

C. opinion-seeker

D. harmonizer

E. blocker

F. summarizer

G. gatekeeper

H. help-seeker

I. player

J. orienter

ACTIVITIES

1. Remember a time when you participated in a group. Recall any group processes and developmental stages that became apparent in that group. List the group processes or developmental stages that you experienced and explain why you categorize your experiences this way.

2. Observe a group and make a checklist of the various roles performed by members.

3. Before attending a meeting, list what you expect to gain from the meeting. Afterward, think about whether your expectations were met. If not, explain why.

4. Observe someone leading a group discussion. Make a list of the leader's strengths and weaknesses.

5. Consider a group meeting you have attended. What was the atmosphere during the meeting? Were you comfortable or uncomfortable? Explain your feelings and what could be done to make you feel better about the group.

6. Your teacher will divide you into groups of four to seven members. Give your group a name, or use your FFA chapter as the group. Each member of the group will answer the following questions.

   a. What is the purpose of our group?

   b. Why do individuals join our group?

   c. What problems have we experienced in our group?

Next, select a recorder and a person to report to the total group by sharing your ideas on each of those three questions.

7. Write a one-page case study or scenario describing a situation in which the leader should use a team or group to accomplish a task or achieve a goal.

8. Refer to the "20 leadership mistakes" of group leaders. Select three mistakes that could be a potential problem for you. State three things you could do to improve in each area. Share these with the class.

9. If you were to take over a new group, organization, or business, what would your biggest fear be? Share this fear with the class.

10. Your teacher will put you in groups of four to five and assign you a topic or problem for discussion. Nominate and vote on a facilitator for your group. The facilitator will lead the group through discussion of the topic using information from this chapter. Take 10 to 20 minutes to record your observations, and then share them with the class.

NOTES

1. G. Thomas, R. Martin, and R. E. Riggio (2013). Leading groups: Leadership as a group process, *Group Processes and Intergroup Relations, 16* (1) p. 4.

2. B. L. Reece and R. Brandt, *Effective Human Relations in Organizations,* 7th ed. (Boston: Houghton Mifflin, 1999), pp. 304–305. (12th ed., 2014).

3. G. Lippitt, R. Lippitt, and C. Lafferty, "Cutting Edge Trends in Organization Development," *Training and Development Journal* 38(7):59–62, at p. 62 (July 1984).

4. R. N. Lussier, *Human Relations in Organizations: Applications and Skill Building* (Boston: Irwin/McGraw-Hill, 1999), p. 482. (9th ed., 2012).

5. G. Thomas, R. Martin, and R. E. Riggio, *Group Processes and Intergroup Relations*, p. 4.

6. Ibid., p. 10.

7. G. M. Beal, J. M. Bohlen, and J. N. Raudabaugh, *Leadership and Dynamic Group Action* (Ames, IA: Iowa State University Press, 1980), p. 64.

8. Understand your Student: What College Graduates Look for in Jobs. (n.d.). Retrieved January 4, 2016, from https://www.universityparent.com/topics/career-planning/understand-your-student-what-college-graduates-look-for-in-jobs-2/

9. Beal et al., *Leadership and Dynamic Group Action*, p. 67.

10. Lussier, *Human Relations in Organizations*, pp. 482–488.

11. S. R. Levine and M. A. Crom (Dale Carnegie & Associates, Inc.), *The Leader in You: How to Win Friends, Influence People, and Succeed in a Changing World* (e-book) (New York: Simon & Schuster, 2010).

12. Ibid., p. 113–115.

13. Ibid., p. 113–115.

14. J. P. Howell and D. L. Costley, *Understanding Behaviors for Effective Leadership,* 2nd ed. (Upper Saddle River, NJ: Pearson Prentice Hall, 2006), p. 267.

15. A. Nahavandi, *The Art and Science of Leadership,* 4th ed. (Upper Saddle River, NJ: Pearson Prentice Hall, 2006), p. 216. (7th ed., 2014).

16. T. Holmes, "Ten Characteristics of a High Performance Work Team," *The 2005 ASTD Team & Organization Development Sourcebook.* Ed. M. Silberman (Alexandria, VA: ASTD), p. 179.

17. Nahavandi, *The Art and Science of Leadership*, p. 211.

18. Ibid., pp. 209–210.

19. Ibid., pp. 211–212.

20. Ibid., p. 215.
21. Ibid., p 216.
22. Ibid., p. 217.
23. J. R. Hackman, "Why Teams Don't Work," in *Leader to Leader: Enduring Insights on Leadership from the Drucker Foundation's Award-Winning Journal,* edited by F. Hesselbein and P. Cohen (San Francisco: Jossey-Bass, 1999), pp. 338–344. (1st edition, 2009, edited by F. Hesselbein and A. Shrader)
24. L. H. Lamberton and L. Minor, *Human Relations: Strategies for Success* (Chicago: Irwin/Mirror Press, 1995), p. 131. (5th ed., 2013)
25. Levine and Crom, *The Leader in You.*
26. Reece and Brandt, *Effective Human Relations in Organizations*, p. 306; D. F. Jennings, *Effective Supervision: Frontline Management for the '90s* (Minneapolis/St. Paul: West, 1993), p. 231.
27. Reece and Brandt, *Effective Human Relations in Organizations*, p. 306.
28. Lussier, *Human Relations in Organizations*, p. 323.
29. Ibid., p. 324.
30. Ibid., p. 323.
31. Ibid., p. 324.
32. Ibid.
33. "Organizing Groups," *Efficiency* series, No. 8742-A (College Station, TX: Instructional Materials Service, Texas A&M University, 1989), p. 1.
34. Ibid.
35. Ibid.
36. "Organizing Groups," p. 2.
37. Ibid., p. 2.
38. Ibid., p. 3.
39. R. F. Verderber, *Communicate!* 14th ed. (Belmont, CA: Wadsworth, 2013).
40. "Organizing Groups," p. 3.
41. Beal et al., *Leadership and Dynamic Group Action*, p. 43.
42. "Organizing Groups," p. 2.
43. W. Bennis, "The Secrets of Great Groups," in *Leader to Leader: Enduring Insights on Leadership from the Drucker Foundation's Award-Winning Journal*, edited by F. Hesselbein and P. Cohen (San Francisco: Jossey-Bass, 1999), pp. 320–321. (1 edition, 2009, edited by F. Hesselbein and A. Shrader).
44. Lussier, *Human Relations in Organizations,* pp. 194, 391.
45. Beal et al., *Leadership and Dynamic Group Action*, p. 143.
46. Ibid., p. 87.
47. "Working with Diverse Groups," *Communication* series, No. 8741-G (College Station, TX: Instructional Materials Service, Texas A&M University, 1988), p. 1.
48. B. R. Stewart and M. L. Amberson, *Leadership for Agricultural Industry* (New York: McGraw-Hill, 1978), pp. 65–66.
49. Lussier, *Human Relations in Organizations,* pp. 358–359.
50. Ibid., p. 359.
51. "Working with Diverse Groups," p. 2.
52. Lussier, *Human Relations in Organizations,* pp. 359–360.
53. Ibid., p. 360.
54. "Working with Diverse Groups," p. 2.
55. Stewart and Amberson, *Leadership for Agricultural Industry*, pp. 71–72.
56. Beal et al., *Leadership and Group Action*, pp. 82–83.
57. Ibid., p. 82.
58. Ibid., p. 83.
59. Ibid.
60. Beal et al., *Leadership and Group Action*, pp. 88–89.
61. Ibid., pp. 88–89.

62. Ibid., p. 89.

63. Ibid., pp. 93–94.

64. Ibid., p. 110.

65. Ibid., p. 112.

66. Ibid., pp. 99–100.

67. "Working with Diverse Groups," p. 3.

68. Lussier, *Human Relations in Organizations*, p. 364.

69. "Working with Diverse Groups," p. 3.

70. Ibid.; Lussier, *Human Relations in Organizations*, p. 364.

71. Lussier, *Human Relations in Organizations*, p. 365.

72. Ibid., p. 365.

73. "Group Discussions," *Communication* series, No. 8741-H (College Station, TX: Instructional Materials Service, Texas A&M University, 1988), p. 1.

74. Beal et al., *Leadership and Group Action*, pp. 103–105; "Group Discussions," pp. 1–2.

75. Verderber, *Communicate!* p. 261.

76. "Group Discussions," p. 2.

77. "Group Discussions," p. 2; Beal et al., *Leadership and Group Action*, pp. 106–107.

78. "Group Discussions," p. 3; Beal et al., *Leadership and Group Action*, pp. 107–109.

79. J. R. O'Connor, *Speech: Exploring Communication*, 4th ed. (Lincolnwood, IL: National Textbook Company, 2001), p. 140.

80. "Group Discussions," p. 4.

81. Ibid.

82. O'Connor, *Speech*, p. 142.

83. Quote retrieved July 8, 2016 from www.thinkexist.com.

84. T. Alessandra and P. Hunsaker, *Communicating at Work* (New York: Fireside, 1993), p. 92.

85. Reece and Brandt, *Effective Human Relations in Organizations*, pp. 344–347.

86. Alessandra and Hunsaker, *Communicating at Work*, pp. 106–107.

87. Beal et al., *Leadership and Dynamic Group Action*, p. 181.

88. Ibid., p. 191.

89. Ibid., p. 197.

90. Ibid., p. 200.

91. Ibid., p. 206.

92. Ibid., p. 214.

93. Ibid., p. 222.

94. Ibid., p. 230.

95. Ibid., p. 235.

96. Ibid., p. 251.

97. Lussier, *Human Relations in Organizations*, p. 333.

98. P. Van Hooser, *What You Can Learn about Being a Leader*. Retrieved January 14, 2016, from http://www.leader-values.com/.

99. Ibid.

100. IdeaBridge, *100-Day Success Plan for Crisis Recovery and New CEOs* (The IdeaBridge White Paper Series, 1999). Retrieved January 14, 2016 from http://www.lesaffaires.com/.

101. Ibid., p. 3.

Section 2 COMMUNICATION AND SPEAKING TO GROUPS

6 COMMUNICATION SKILLS

A person who has mastered the art of communication can enhance personal satisfaction and fulfillment. Communication of all types can be improved by understanding the problems associated with each type and by making a conscious effort to practice techniques designed to foster and improve communication skills.

Objectives

After completing this chapter, the student should be able to:

- Define communication
- Explain the relationship between communication and leadership
- Explain the purposes of communication
- List the forms communication takes
- Explain the communication process
- Recognize and overcome communication barriers
- Describe and use techniques to improve listening, reading, writing, and speaking
- Define nonverbal communication and discuss its impact
- Develop skill in using feedback (both giving and receiving)
- Discuss the importance of self-communication and interpersonal communication

Terms to Know

- communication
- output-based communication
- input-based communication
- nonverbal communication
- message
- sender
- encoded
- channel
- receiver
- decoding
- perception
- feedback
- interference
- barriers
- semantics
- situational-timing barriers
- hearing
- listening
- body language
- kinesics
- proxemics
- doodling
- self-communication
- interpersonal communication barriers
- social media

Words are only one small part of communication. When people speak, only 7 percent of the message communicated is attributable to the words they use. Voice expression accounts for 38 percent and body gestures account for 55 percent. Therefore, we can communicate just about anything we want to just about anyone if we know how to "say" it. The 4-H members in Figure 6–1 must communicate to keep the boat going in the right direction.

In this chapter, we discuss why communication is important, what purposes it serves, and what forms it takes. You will discover that communication is a complex process with distinct elements that work together to convey accurate messages. We will examine barriers to communication and learn ways to overcome them. We will develop an understanding of the communication styles people use and learn how to adapt to each style. The chapter concludes with suggestions on how to improve listening, reading, writing, and nonverbal communication skills, and investigations of feedback techniques and interpersonal communications.

COMMUNICATION AND LEADERSHIP

Communication may be defined as the process of sending and receiving messages through which two or more people reach understanding.[1] The message itself may come from an oral source, a written source, or a gesture. It is important to realize that if the person sending the message and the person receiving it do not agree on the meaning of the message, true communication has not taken place.[2] Miscommunication causes misinterpretation, which can create misunderstanding, conflict, confrontation, hurt feelings, and other problems.

You will quickly find that all forms of communication are important when you are working with others in a group. The American democratic system depends on skill in group communication. Parliamentary procedure is used in local, state, and national governments. Public speaking is a necessity for people in leadership positions. All leaders need group communication skills and skills in conducting meetings. Agricultural education and the FFA help develop communication skills.

FIGURE 6–1 There are many ways to practice and enhance your communication skills through agricultural education and leadership organizations such as the FFA or 4-H. *(Courtesy of University of Georgia—College of Agricultural & Environmental Sciences.)*

FIGURE 6–2 Effective communication in the workplace is an essential life skill. How could miscommunication cause problems for the people in this photograph? *(Courtesy of University of Georgia—College of Agricultural & Environmental Sciences.)*

Members recite the Creed (Chapter 7) and are involved in public speaking (Chapter 8), extemporaneous speaking (Chapter 9), parliamentary procedure (Chapters 10 and 11), leading group discussions (Chapter 5), and conducting meetings (Chapter 12).

1. Seventy-five percent of each workday is used talking and listening. Seventy-five percent of what we hear, we hear imprecisely; and 75 percent of what we hear accurately, we forget within three weeks. Communication, the skill we need the most at work, is the skill we most lack.[3]
2. Managers spend 70 percent of their time communicating in some way. That time breaks down as follows: 9 percent reading, 16 percent writing, 30 percent talking, and 45 percent listening.
3. The majority of problems are caused by poor communication—by people who are unable or unwilling to communicate.

The ability to communicate can be learned. It *must* be learned by people who want interesting and rewarding careers or who want to take part in community affairs. Positions of responsibility, pay increases, and promotions go to those who

can communicate well, particularly those who can speak effectively. These skills are essential for success in business, politics, or community service.

The development of improved communication skills is crucial to your success as a student, as an employee, and as a social person. The coworkers in Figure 6–2 are communicating as they perform their daily tasks.

THE PURPOSES OF COMMUNICATION

Communication is an important and ever-present aspect of life. We are expected to communicate with our family, friends, teachers, and coworkers daily. How well we communicate helps determine how successful we are in school, in the workplace, and in relationships with others. There are five primary purposes or goals of communication:

To Inform Communication may simply seek to give information to another person. This information may be necessary to solve problems, make decisions, or increase understanding. A

YouTube video documenting the high-quality care that farmers and ranchers give to their livestock is an example of communication used to inform.

To Influence Communication may also seek to influence another person's behavior. Examples of communication used to influence others are persuasive speeches, commercials, and instructions or directions. Remember, influence is part of leadership.

To Express Feelings Communication may be used to express how a person feels. For example, a message may indicate that a person is happy, angry, or discouraged.[4]

To Motivate Communication is paramount for motivating followers to improve performance and complete tasks. Setting a vision and clarifying expectations, rewards, or even consequences of poor performance or unfinished tasks are all examples of communication that motivate.

To Interact Some believe that humans cannot survive without social interaction and communication with others. Communication is used to build relationships and teams that are more powerful than any one individual.[5]

Note that although we have separated the communication purposes into categories, any message could have one, two, or all of these purposes. For example, a frustrated supervisor reprimands an employee who is late for work and explains that company policy allows her to fire employees who consistently arrive late. In this communication, four of the five purposes are present: information is given concerning company policy; influence is exerted by the supervisor who wants the employee to arrive on time; the correction from the supervisor should motivate the employ to be on time; and a feeling (frustration) is expressed through the supervisor's reprimand.

FORMS OF COMMUNICATION

Communication is output based for the person producing the communication and input based for the person receiving the communication. Nonverbal elements may be part of either output- or input-based communication, when behavior or the physical environment conveys part (or all) of the message. Forms of verbal communication or **output-based communication** are speaking (formally and casually) and writing (all forms). **Input-based communication** involves listening and reading.[6]

Nonverbal communication consists of messages conveyed by physical behaviors, including gestures and body language. The physical environment, such as arrangement of furniture, use of space, and building design, also can be a form of nonverbal communication.[7] Interestingly, social media platforms can include all forms of communication. Output-based communication may include posts about what a particular group is doing in celebration of a team achievement, and the same post can have input-based communication as friends or fans of the group read or listen to the post. Nonverbal communication may also be used with social media images and/or video.

THE COMMUNICATION PROCESS

The process by which a message is sent from one person to another may seem relatively simple, but a number of elements may intervene in that apparently straightforward process and either enhance or distort the intended message or the understanding of the message. The elements that exist in any form of communication are the (1) situation, (2) message, (3) sender, (4) channel, (5) receiver, (6) feedback, and (7) interference.[8] A scenario demonstrating the communication process follows; after the stage has been set, each element of the communication process is defined and described.

Setting

Stephen is doing his homework one evening. He begins thinking about the big ball game coming up next weekend. Stephen desperately wants to ask Chelsea to go to the game, but he is not at all sure that she would want to go with him. After thinking about it, Stephen decides to call Chelsea with his smartphone and ask. When Chelsea answers his call, Stephen begins by talking about how important the game is and how many of their mutual friends will be there. Stephen finds, however, that he has to repeat himself frequently because there is a bad connection. Also, Stephen is so nervous that he cannot seem to come right out and ask Chelsea to go to the game with him. Finally, he comments that he would like to go to the game if someone would accompany him. Chelsea understands Stephen's hint and says she would be glad to go to the game with him. Figure 6–3 shows the nervousness on Stephen's face as he talks to Chelsea.

Situation

The **situation** is when and where a communication takes place.[9] In this example, the communication takes place one evening in Stephen's and Chelsea's homes. In any communication, it is important to ensure that the occasion and the physical setting are appropriate for the communication. If Stephen had tried to ask Chelsea to the game during a math test, the situation obviously would not have been appropriate.

Message

The **message** is anything one person communicates to another.[10] In the example, the intended message is Stephen's desire to have Chelsea attend the game with him. It is important to make communication clear and precise. In our example, Stephen's message could have been misunderstood because he was not precise and definite about asking Chelsea to go to the game.

Messages are more effective if they are of reasonable length; if they are accurate, concise, and interesting; and if they are delivered in a timely and orderly manner.[11] Remember that words are not the only carriers of a message: voice, gestures, and facial expressions are equally important (if not more so) in conveying meaning. Awareness of all the ways in which your message is being transmitted helps you be sure the message you want to communicate is the one the other person actually receives.

Sender

The **sender** is the person who wishes to communicate with (send a message to) someone else. In the example, the sender is Stephen. Stephen must first formulate the message he wishes to communicate. The message is then **encoded**, or put into a form that the sender believes the receiver will understand. Messages may be encoded in words, gestures, sounds, pictures, or numbers.[12] Stephen's message was to ask Chelsea to go to the game. He encoded his message in words when he made the phone call.

A sender's effectiveness can be influenced by several factors:

Purpose and Motive The sender's purpose and motive (reason) for sending the message are

FIGURE 6–3 Asking for a date causes communication difficulties for many young people. This student is obviously a bit nervous about his call.

important. Generally, the better the motive, the better the message will be received.

Status or Position The sender's status or position may affect how well the message is received. For example, we are more likely to listen closely to what an employer tells us to do than to what a coworker tells us to do.

Personality Personality factors, such as attitude, credibility, reputation, and abilities, affect how others view the sender and accept that sender's message. Obviously, if the receiver appreciates the personality characteristics of the sender, the chances of the sender and the message being accepted are improved.[13]

Channel

The **channel** is the means through which the sender communicates a message.[14] The channel Stephen used to send his message was his smartphone. The appropriateness, efficiency, and dependability of the channel determine its effectiveness. The sender's choice of channel may change based on the time available for communication (a phone call is faster than a letter), the importance of the communication (an emergency requires quick action and therefore the fastest form of communication available), and the need for response (if you need an answer immediately, a phone call is much quicker than getting in the car and going to see the person).[15] Texting might be even quicker, but this type of channel can distort the message because of perceptions that the channel is impersonal or challenges with decoding abbreviated texts. Be careful when using channels that were created for annotated messages.

Receiver

The **receiver** is the one for whom the message is intended. The receiver takes the message and converts it into a form that he or she understands; this is called **decoding** the message. Chelsea decoded all of the talk about the game and realized that Stephen wanted her to go to the game with him. It is essential to realize that each receiver has her own **perception**, or way of understanding the message, based on her own beliefs, knowledge, and ways of organizing information[16]—thus the saying "Perception is reality." Whether we are right or wrong, the way we perceive something is the way it is, at least in our own minds. The receiver may not understand or interpret a word, phrase, or thought in the same way the sender does. Make every effort to ensure that both the sender (you) and your receivers agree on the message.

Feedback

Because each person has his or her own way of understanding a message, it is important that the sender get feedback. **Feedback** is the receiver's verbal or nonverbal response to the message. It allows the sender to know whether the message was clearly understood. If the receiver does not provide feedback, the sender may check for understanding by asking questions or interpreting body language, such as puzzled looks.[17] Chelsea's telling Stephen she would be glad to go to the game with him was a form of feedback. From the feedback, Stephen was able to determine that Chelsea understood the intent of his call, even though he had had difficulty in expressing his true message.

Because feedback helps the sender determine the effectiveness of the message, time and opportunities for feedback from both the receiver and the sender are essential. Some organizations even provide training in feedback techniques.

Interference

Interference is anything that hinders the sender's ability to make a message understood. Interference may come from sources external to the receiver (such as bad WiFi) or from the receiver (not paying attention, or doing another activity at the same time).[18] In our scenario, interference came from a poor phone connection and from Stephen's nervousness. Obviously, to convey a message that the receiver will clearly understand, you should minimize interference.

The Communication Process at a Glance

When all the elements are present, the communication process is complete. As you have discovered, what initially seemed to be a simple process actually consists of several complex choices, circumstances, and assumptions. A diagram of the communication process is provided in Figure 6–4.

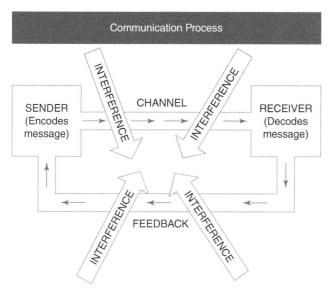

FIGURE 6–4 The communication process by which a message is sent from one person to another seems relatively simple at first glance; however, a number of elements that may intrude into the process can enhance or distort understanding of the message.

BARRIERS TO EFFECTIVE COMMUNICATION

Much of what is communicated is misunderstood or understood only in part. Sometimes this may simply be a result of a poor smartphone connection or bad handwriting. More often, though, the reasons for ineffective communication are more complex. An understanding of the **barriers** (obstructions or hindrances) to effective communication can help us avoid problems with our own communications. It may also help us understand why others are unable, or seemingly unwilling, to communicate.

Usually, one of four major types of barriers hinders communication: (1) language barriers, (2) interpersonal (sender and receiver) barriers, (3) situational-timing barriers, and (4) organizational structure and procedural barriers.[19] Of course, some messages may have to overcome more than one type of barrier in the course of the communication process.

Language Barriers

Language barriers affect the way our words are understood. Examples of language barriers are unfamiliarity with a language, use of jargon, ungrammatical constructions, and variable meanings of words and phrases.

The most obvious language barrier arises when two people do not share a common language. If you speak only English and your receiver speaks only Chinese, you face a rather large obstacle to clear communication.

Even within a single language, a particular word or phrase may not mean the same thing to one person as it does to someone else. This study of the shifting meanings of language is called **semantics**. For instance, to a learner the word *slowly* may indicate a pace twice as long as that intended by an instructor.[20] In this case, the instructor might do better to specify a time range ("take at least 10 seconds to do this") rather than saying "Do this slowly." As a leader, you must make sure receivers understand what you say.

When words are incorrectly used, are used out of context, are too specialized, are too vague, have too many different meanings, or have emotional overtones, it stands to reason that the intended message can be misunderstood. Understanding and interpretation can also be hindered if the entire message is too complex or is presented in an illogical or grammatically incorrect way.

Interpersonal Barriers

Interpersonal barriers are the differences and personal characteristics of the sender and the receiver that hinder communication.

Age, status, role, and cultural differences often make communication difficult. Consider, for example, how hard it can be to communicate with a small child, and how closely we pay attention to a person we respect.

People also have varying levels of communication skills. Some people are naturally good communicators, and others have taught themselves to communicate well. Confident people, and those who speak well or have a persuasive manner, tend to be good communicators.[21] Judgmental, dishonest, and insincere people, as well as those who have withdrawn, insecure, defensive, or apathetic (uncaring) personalities, may not be good communicators because of these traits.[22]

Remember, people tend to view the message you send in light of their own experiences, background, and personality (perception). People often hear only what they want to hear. Knowing this is essential for improving communication.

Understanding the personalities of others can make our communication easier or at least explain why communication so frequently seems difficult or goes awry. The effects of leadership style, personality type, and communication style are discussed later in this chapter.

Selective Perception "Selective perception occurs when people block out information they do not want to hear. This is usually information that conflicts with their values, beliefs," or desires. Selective perception and stereotyping often go hand in hand: for a person who holds preconceived ideas about someone or something, selective perception may allow only messages that confirm those ideas to get through. For example, environmentalists might have trouble getting a community to support additional environmental controls on a local factory. Community members who believe that any environment-related change will negatively affect jobs and the local economy may never really hear or understand the idea that is actually being proposed. In short, people hear what they want to hear.[23]

Bad Listening Poor listening skills can also be a major barrier to communication. Poor listening may be caused by a number of things. For example, a receiver might be preoccupied with something else or already be thinking about how she will respond. Perhaps the receiver does not care about the subject of the message or has little knowledge of or interest in the subject. Whatever the case, poor listening skills can reduce the effectiveness of the communication process.[24]

Sender's Credibility The credibility of the sender affects how people receive a message. If the receiver perceives the sender as dishonest, deluded, irresponsible, or ignorant, or simply has negative feelings about the sender, communication barriers will arise. This type of barrier can also be raised by status differences between communicators. As a leader, you must build your credibility, so that your receivers become people who trust and respect you.[25]

Filtering Filtering means manipulating a message (information) to reflect the receiver's desire or objective. In some instances, this is called **spin**. This occurs often with communication in organizations that have several levels of management. Each level filters what it does not want the next

level to know or changes the message to express particular ideas or feelings.[26] Sometimes the distortion is so severe that the original sender cannot even recognize the communication as his or her own! A leader can avoid filtering or spin by self-evaluating messages or seeking assistance to audit information for bias.

Inability to Process Details Many human minds are somewhat detail-averse. A lot of us cannot remember many details if they are communicated in a short time. Read these numbers: 10, 17, 9, 43, 0, 87, 4, 12. Now, without looking, how many can you repeat in proper order?[27] A good leader will work hard to remember details, especially as they pertain to their followers, vision, or bottom line.

Past Experience with Communicator Past experiences with others also may affect our communication. If a person has been sarcastic, untrustworthy with confidential information, unwilling to cooperate, or resistant to new ideas and change, future attempts at communication will likely be difficult.[28]

Situational-Timing Barriers

Situational-timing barriers deal with hindrances to understanding raised by time and place: when and where a communication takes place. People in today's society are so hurried and so bombarded with information that they do not have time to pay close attention to everything they hear or read. An individual may receive more than 1,500 messages during a single day.[29] There simply is not enough time to process so many messages.

The timing of communication, however, may be unrelated to hours and minutes. If a person is under a great deal of stress or is in a personal crisis, any effort to communicate with that person may be ineffective. Be sensitive to such people's situations. You have probably learned that it is best not to ask for something if your parents or teachers are in a bad mood. If you want a positive response, wait until the time is right (when the mood changes). Likewise, a person in a hurry may not have time or be able to spare brain cells to focus on your communication. A forced or rushed answer is usually negative.

The place where communication occurs may make successful communication difficult. An environment that is hostile, lacks privacy, or is uncomfortable will negatively affect communication. The amount of noise (e.g., loud announcements being made while you speak), volume of messages (e.g., receiving too many e-mails per day), and interference (e.g., overly sensitive spam filters for e-mail) in the environment also affect how well we understand and can be understood.

Do your best to match your communication and leadership style with the situation. Earlier in the book, we studied situational leadership; the same principles apply here. For example, the sender who initiates a strong presentation with a closed communication style is using autocratic leadership, whereas mild presentations by open communicators form signals of the laissez-faire leadership style. The mild-to-strong presentation spectrum exists for any of the styles, but strong messages tend to be used more often in the autocratic and democratic styles.[30]

Organizational Structure and Procedural Barriers

Organizational structure and procedural barriers deal with how and through what structure a message passes on its way from the sender to the receiver. Barriers certainly exist in the workplace, but also can be found in schools, churches, or any place where people have varying roles or levels of authority in an organization.

Communication may be hindered by something as simple as the fact that offices are spaced too far apart for people to communicate easily or that an organization cannot afford wireless communication devices for its employees.

Organizations may also lack policies that promote effective communication. Especially in large organizations, a message may have to be forwarded through many organizational layers and seemingly endless numbers of people before it gets to the intended recipient—the person who acts on the message. The more levels there are in an organization and the farther the receiver is from the sender, the harder it is to communicate a message accurately and effectively. The potential for distortion exists when a message is transmitted through a series of individuals at different management levels in the organization. Parts of the message may be ignored, reinterpreted, omitted, or misunderstood. You may have played the

game "Gossip" or "Telephone" when you were a child. Remember the completely different message that emerged by the time it reached the tenth person in the circle? Communication in organizations is not much better if messages are not transmitted in writing as well as verbally.

Today, people tend to rely far too heavily on impersonal forms of communication (such as texting), forgetting that any one message can easily be lost or overlooked in the sheer volume of communications that most of us receive each day. Furthermore, the impersonality of these messages can make people feel isolated and detached rather than like an important, contributing part of the organization.[31]

OVERCOMING COMMUNICATION BARRIERS

We have seen that there are numerous barriers to effective communication. This section presents some general guidelines for improving overall communication. Later, we look at specific ways in which the most common forms of communication—listening, reading, speaking, and writing—can be improved.

Barriers can be overcome in at least three primary ways: (1) improve perception, (2) improve the physical processes of communicating, and (3) improve relationships (communicate ethically).[32]

Improving perception involves putting yourself in another person's place rather than focusing only on your own experiences, emotions, and background. Essentially, this means you try to understand the point of view of those with whom you communicate. Perceptions can be influenced by age, status, personality, culture, and feelings. If you understand how your intended receiver perceives and experiences the world, you may be able to put your communications in a form he or she can readily understand. In other words, talk or write at the level of your intended receiver. Remember also that you have your own perceptions; take time to examine them to be sure they are accurate and are not based on false information or prejudices.[33]

To improve the physical processes of communication, pay attention to the elements of the communication cycle described earlier. Be sure the situation (time and place) is appropriate. As a sender, make an effort to encode messages clearly by selecting words and gestures carefully. Use an appropriate channel to send your message. Pay close attention to the feedback (both verbal and nonverbal) the receiver gives so that you can tell whether your message was understood. As a receiver, make an effort to decode messages accurately and provide feedback to the sender so that misinterpretations can be corrected.

Improving relationships involves building trust between yourself and those with whom you communicate. Trust, honesty, and confidence are essential to good communication; therefore, follow through on promises you make, do not lie or shade the truth, and maintain confidentiality whenever you promise it. Once again, step into the other person's shoes to understand how that person feels and thinks. Craft your message so that your receivers can easily and accurately decode your communication.

Speaking Ethically

Communication is a form of power and therefore entails ethical responsibilities. What would happen if an influential leader who was an effective communicator used that communication to influence followers for negative purposes, rather than for positive purposes? Communication can be so powerful that it could be disastrous. Hitler is an example. He used his strengths in speaking to convince thousands of followers to commit horrible acts.[34]

Make sure your goals are ethically sound. Use your speaking skills for good, not evil. Following are some steps you can take to make sure you speak ethically when you are communicating:

- Be well informed and knowledgeable about your subject.
- Be honest in whatever you say.
- Use sound evidence.
- Employ valid reasoning.[35]

IMPROVING COMMUNICATION SKILLS

Listening

Listening is an important skill for following directions, avoiding mistakes, getting along with others, and learning.

Hearing is simply receiving sound. Most of what we hear is ignored.[36] **Listening** is a conscious mental effort to understand what is heard. For example, when you truly listen to something, you receive the message, and then you interpret it by putting it in a form you can understand. Next, you evaluate the message to see whether you agree or disagree; finally, you respond.

Although statistics show that the average person can speak from 125 to 200 words per minute, the brain can actually process at least four times that many words.[37] This might make it seem that listening is easy—but studies indicate that although listening consumes up to 95 percent of all the time we spend communicating, listeners receive only 50 percent of the information sent and remember 25 percent or less.[38] In other words, although listening takes up a major portion of our communication time, it is a poorly developed skill.

Barriers to Listening What barriers make it difficult to develop good listening skills and habits? Often, the environment is simply too noisy for good communication; in other instances, the words used by the speaker may be too difficult or unfamiliar for listeners to comprehend. However, substantial barriers are also created by our own poor listening habits. In *The Art of Public Speaking*, Stephen Lucas discusses the following poor habits.[39]

- Lack of concentration is a common barrier to good listening. Something the speaker says may trigger an unrelated thought in the listener, who focuses on that thought while shutting out the speaker. The listener may have pressing problems or worries that are distracting, or the listener may be physically uncomfortable or not very awake. In either case, it is not that the listener is purposely not paying attention; it is just that the mind has wandered.
- Listeners may concentrate too hard. When listeners try to remember every single thing that is said, the main message may become lost in all the details.
- Listeners may jump to conclusions. They may anticipate what the speaker is going to say, or they may react strongly to what the speaker initially says. When these things happen, listening may stop before the entire message is completed and/or understood.

- Focusing on the speaker's appearance or way of speaking may distract listeners. Grooming, clothes, speech defects, accents, or mannerisms may so irritate listeners that they do not pay attention to what is being said.
- Other poor habits that hinder good listening are completing the speaker's sentences (mentally or aloud), not making eye contact, preparing responses rather than listening to what is being said, interrupting, and doing other things while the speaker is talking.

Effective Listening Skills Overcoming listening barriers requires a conscious effort to understand, remember, and evaluate what is said. Here are some guidelines for developing effective listening skills:[40]

- Eliminate noise and other distractions that may draw attention away from the speaker.
- Be quiet. You cannot listen effectively if you are talking.
- Put the speaker at ease by being friendly and attentive.
- Let the speaker know you are interested in what is being said. Make eye contact and ask questions. Repeat what is said in your own words. In other words, provide feedback.

- Make notes or record presenters if there is a great deal of information to remember.
- Listen for main ideas.
- Listen to the entire message, even if you think you object to what is being said. Be patient and attempt to see the speaker's point of view. Do not worry about how you will respond, or become angry to the point of arguing or criticizing.
- Try to put aside your opinions of the speaker's appearance, mannerisms, or accent. Focus on the message, not the person.

Remember, being a good listener requires effort and practice. Mastering these skills will pay off: they are essential to becoming a good communicator. The students in Figure 6–5 are exhibiting good listening skills. Refer to Figure 6–6 for a summary of ways to improve your communication skills.

Reading

Reading is important for success in school, as much of the information you are taught comes from textbooks, articles, and Internet pages. A basic ability to read is necessary in most

FIGURE 6–5 Notice the listening skills of the group members. What is each person doing that makes you think he or she is a good listener? *(© iStock/annebaek)*

IMPROVING COMMUNICATION TECHNIQUES					
LISTENING	**READING**	**WRITING**	**SPEAKING**	**NONVERBAL**	**FEEDBACK**
Eliminate distractions Be quiet Show interest Make notes Be objective Observe nonverbal cues	Concentrate Find overall meaning Summarize periodically Think critically Avoid prejudice Use a dictionary Make relevant to yourself	Know audience Know purpose Know subject Be precise, clear, brief Stay on topic Use good grammar Use correct style Proofread	Speak clearly Make eye contact Use pleasant tone of voice Use good grammar Observe nonverbal cues Stay on topic Be brief, but thorough Be assertive	Understand meaning of nonverbal behaviors Know your own nonverbal behaviors Interpret body language Be open with your own nonverbal actions Use with care Be sensitive to physical environment	Encourage feedback Observe nonverbal cues Ask specific questions Use paraphrasing techniques Take responsibility for being understood

The left margin of the table reads "SKILLS" at top and "Techniques" along the side.

FIGURE 6–6 Good communication skills and techniques are an asset to any leader. Six areas of communication, with key things to remember for each, are summarized here.

occupations. In the workplace, professionals keep current on new innovations and important changes in their fields by reading journals or training materials. Many people regularly read some kind of website, blog, newspaper, or magazine, such as *FFA New Horizons*, *Progressive Farmer*, or *Successful Farming*. To be a well-informed consumer, you must be able to read information in advertisements and on labels, to compare claims and prices and determine important features of products. Can you imagine trying to apply pesticides without being able to read the manufacturer's directions? Simply put, how well you read can determine your success in school, in the workplace, and in life in general.[41]

VIGNETTE 6-1   THE GREATEST LISTENING EXERCISE ON EARTH

THE FOLLOWING STEPS CONSTITUTE THE GREATEST LISTENING EXERCISE ON EARTH:

- With classmates, or with friends and family, form groups of five.
- Using small strips of paper, write down as many good things as you can think of about each member of your group.
- Make sure everyone has a large piece of paper labeled "Me" at the top. Have lots of tape on hand.

After everyone has finished writing, the listening and compliments begin.

- One person must sit quietly—not talking or responding—and just listen to all of the good things each member of the group has to say about him or her.
- As each good trait is mentioned, the person listening to the comments said about them, takes the comments which were written on the small stripes of paper and tapes them on a sheet of paper labeled "Me".
- When each member of the group has given his or her compliments and pieces of paper to the person who was listening quietly, it becomes another person's turn to listen.

At the end of the exercise, everyone should have not only a great description of what others think of him or her but also a realization of how hard it really is to listen.[42]

Some of the same problems that hinder our ability to listen can also hinder our ability to understand what we are reading. Distractions in our surroundings and from other thoughts can keep us from comprehending what we read. Almost everyone has, at one time or another, been reading only to realize that while one's eyes may have been moving across the page, other thoughts have kept the information from sticking. Some personality types concentrate well by reading with white noise (e.g., a fan or heater) in the background, whereas others concentrate best in complete silence. Everyone is different. Knowing your personality type and learning style will help you identify the setting that affords you optimal reading comprehension.

If information read is written in professional or technical terms or uses unfamiliar words, our ability to understand may be reduced. When the reading is not interesting to us personally, it is difficult to concentrate on the meaning and relevance of the information. As a leader, you need to do everything possible to focus—even on the boring material, if it is important. Also, if you are developing materials to be read by followers, make it as interesting as possible—entertaining, even!

Another common problem for readers is believing or assuming that everything in print is true. This is most problematic when statistics or figures are presented or when the author is well known or considered an authority. People should read skeptically, and make every effort to find supporting evidence for the claims made in what they are reading, before accepting a piece of information as fact. This is absolutely critical for information read on the Internet; before you trust it, believe it, or cite it, check out the source!

Finally, readers may also infer more from a selection than the author intended. Extrapolating a line of reasoning may be appropriate, but you must keep in mind what the author actually wrote. Do not attribute viewpoints or positions to an author unless you can back up your claims with evidence from the piece you are reading.

Improving Reading Skills Reading requires receiving the printed words and processing them in your mind until you understand them. The following suggestions can help you become a more effective reader:

- Concentrate on what you are reading. Eliminate noise and other distractions, and try to focus your thoughts on the material being read.
- Begin by reading introductions, section headings, and summaries first to get an overall idea of the author's purpose and the scope of the material.
- At the end of a section, stop and think about whether you have understood what the author wrote. If not, reread the section.
- Look up unfamiliar words in a dictionary, or try to derive their meaning based on the context (the words surrounding the new word).
- Become familiar with the *jargon*, also sometimes known as *terms of art*, used in the subject area; these are specialized words or word meanings that are common to a specific subject, occupation, or field of study.
- Read critically and without prejudice. However, do not assume that everything you read is true. Look for false logic, factual errors, and faulty reasoning.[43]
- Make the reading meaningful by relating the piece to yourself, your job, or your interests.
- Use any graphs, charts, and other visual aids to clarify, summarize, or simplify what is presented by the text.
- Have a conversation with the written word. "Listen" as you read, by highlighting; or "talk back" to the text by taking notes and paraphrasing what you read.

A conscious effort to improve your reading skills will benefit you in school, in the workplace, and as a well-informed person. The researcher in Figure 6–7 reads constantly as part of her job.

Writing

Writing is a form of communication in which messages are put into words. The ability to write effectively is almost as important to success as the ability to read well.

Poor writing usually occurs for one or more of the following reasons:

- The intent or purpose of the writing is unclear. Sometimes we begin writing before we really know what we want to write about. This lack of clarity and focus keeps us from communicating accurately and effectively with our intended audience.

FIGURE 6–7 Reading is an essential life skill. For animal science researcher Dr. Gabriela S. Brambila, extensive reading is part of her job. *(Courtesy of University of Georgia—College of Agricultural & Environmental Sciences.)*

- The writer's thinking seems confused. When the writer has not thought sufficiently about the subject, the ideas will not be presented in an organized manner. Sometimes contradictory ideas or statements will appear in a piece of writing if the author has not thought out his or her position on the subject.
- Bad sentence or paragraph structure, poor grammar, and bad spelling may be a further sign of disorganized thinking. These problems may also be the result of carelessness, a lack of review or proofreading, or a lack of knowledge about the basics of good writing. Bad writing mechanics can obscure even the most brilliant ideas and cause communication to fail entirely.
- The author does not know who the audience is, the subject matter or material is inappropriate for the audience, or the style of writing is inappropriate for the occasion.[44]

Guidelines for Effective Writing Regardless of whether you are writing an essay, a term paper, a business letter, or a memorandum, the following guidelines will help you convey your message effectively:

- Know your audience. Know your intended reader's probable level of understanding about the subject. Use words and phrases the reader will understand.
- Know why you are writing. What is the purpose of your communication? Are you giving information? Making a request? Confirming information? Lodging a complaint?
- Be knowledgeable about your subject. Although you need not be the world's foremost expert on the subject, you cannot write effectively unless you understand what you are writing.
- Present your ideas clearly, in a logical order. Be sure your ideas flow naturally and smoothly from point to point. To assist with clarity and sequence, make an outline of what you want to write before you write it.
- Be precise and concise. Do not try to impress your readers with long sentences and big words.
- Stay on the topic. Everything you include should relate directly to your purpose for

writing. Here again, an outline will be useful; outlining before you begin writing helps you avoid digressions and stick to your topic.

- Use correct grammar and spelling. Check a writing handbook, textbook, or dictionary if you are unsure.
- Use the correct style. For example, the format for a business letter is quite different from the format for a friendly letter. Ask teachers or employers what formats they require or prefer for written communications.
- Proofread your writing. Reading out loud or having someone else read your work can be helpful. It is surprisingly difficult to proofread your own writing. Because you already know what you intended to write, your eye is fooled into seeing what you know *should* be there. Be sure you have made your point in a way your audience can understand. Check for errors in spelling and grammar.[45]

Always support what you write with facts, experiences, and/or credible findings or research concerning the subject about which you are writing. Even if you are an expert who is completely knowledgeable about the topic, you can make your argument and claims stronger by providing support from sources other than yourself.

In the workplace, written communications such as memos, business letters, and e-mails provide a record of the message sent to a specific person on a particular date. People act more readily in response to written business communication than to spoken conversations or orders. In school, essay examinations, term papers, and theses all require good writing skills to demonstrate how well you understand a body of material.[46]

Remember, writing is one form of communication that can be kept and reviewed. Written communication is a permanent record that reveals how well you develop, organize, and present your thoughts.

Speaking

Oral communication is a and quick way to send a message. Oral communication may take place face to face or through a variety of types of digital conferencing; it may involve conducting a meeting, leading a group discussion, or making presentations. Nearly every day you talk to teachers, friends, family, employers, and coworkers. You may be called on to speak formally in a business situation or informally in conversations with friends. In any of these situations, your success depends on how well you express yourself orally.

Problems with speech may arise from not speaking so listeners can understand, not keeping your attention on the person to whom you are speaking, using a tone of voice that angers or makes the listener defensive, using poor grammar, pronouncing words incorrectly, not giving opportunities to respond (not inviting feedback), and confusing listeners by not sticking to the subject.[47]

Practicing Effective Speaking Skills Later chapters discuss, in depth, the special instances of preparing and giving speeches, conducting meetings, and leading discussions. However, you can apply the following suggestions in any speaking situation to increase the effectiveness of your oral communication:

- Speak clearly. Concentrate on correct pronunciation, appropriate volume, and reasonable pace (speed).
- Make eye contact with the audience. You can hold a person's attention by looking directly at that person. If you are speaking to a large group, try to speak to all parts of the room and include all listeners.
- Use a pleasant tone of voice. Even if your message is critical, it will be better accepted if you sound positive and friendly. In many cases, the way you say something is as important as what you say.
- Avoid using slang in formal situations. Use good grammar and appropriate terminology. Not only do these make your communication clearer and more precise, they enhance the credibility that you have earned based on your education and knowledge.
- Be sure that listeners understand the words you use. Allow listeners to respond with questions or comments. Watch for nonverbal communication, such as puzzled looks or nods of affirmation.
- Keep to your subject. Avoid interjections, tangents, and rambling.
- Be brief but thorough.[48]

Good speaking skills are essential for everyday communication with people at school or at work. The impression you make may well affect future relationships with friends, teachers, employers, and customers.

NONVERBAL COMMUNICATION

Nonverbal communication is sending a message with body movements, behaviors, and the physical environment rather than with words. Some have estimated that 55 percent of communication is nonverbal, with voice accounting for 38 percent and the actual words only 7 percent. The *way* you send a message can be as important as—or even more important than—the message itself. Improving your nonverbal communication can help you relay messages to others more effectively and may help you understand how others feel and think about your message.

> **WARNING!**
>
> Nonverbal communication is difficult to understand accurately. *It may be true that "actions speak louder than words," but interpreting actions correctly is often difficult. Never use nonverbal cues alone to determine how a person feels or thinks.*

Nonverbal communication signs may support, refute, or add information to a verbal message.[49] For example, a person tells you that he will do something for you, but shakes his head negatively at the same time. This could indicate that the person will do what you have asked but is not happy about doing it.

There are many types of nonverbal communication. Although certain behaviors are thought to mean certain things, be aware that some behaviors may be habits rather than signs with deeper meaning. For example, sitting with your arms crossed in front of you is interpreted as shutting out incoming messages, even though it may simply be your preferred way of sitting, as illustrated in Figure 6–8. Also, you may occasionally encounter people who have sleep disorders: "When they stop, they flop." For these people, sleeping in class or meetings could be misinterpreted as disinterest.

FIGURE 6–8 Even though this may be the student's preferred way of sitting, it nonverbally communicates that he is not interested in the program. *(© iStock/IS_ImageSource)*

Despite the dangers of misunderstanding and misinterpretation, some easily communicated nonverbal signals are fairly reliable. Most people nonverbally display emotions and attitudes, such as excitement, joy, pain, boredom, anger, disappointment, sadness, enthusiasm, and interest, in similar ways. You can learn to "read" these signals as part of a person's overall communication.

Body Language

If you have ever seen a good pantomime performer in action, you know how much can be communicated through **body language**, the nonverbal way one communicates. **Kinesics** is the study of communication through body motion. There are entertainers who can create a mood or tell a story simply through body movement, gesture, and facial expression. Most people use such body language to some extent in their everyday conversation. Body movements can convey many messages to others. Do you form opinions about people based on the way they walk? A slow step may communicate a lack of confidence and lack of decision-making ability; a quick, lively walk generally communicates certainty, security, and happiness. Posture—the ways people stand or sit—can be signals to others.

Facial expressions are the most common nonverbal way of communicating emotions and feelings. A smile or a frown sends a clear message about your attitude. Tightly closed lips indicate disapproval. The very expressive facial area around the eyes can convey love, hate, interest, indifference, or anger. A raised eyebrow may indicate uncertainty, questioning, or disagreement; biting the lip may signal disagreement or concern. The eyes themselves can also provide clues: if a person's pupils enlarge, he or she is responding positively; smaller pupils indicate negative feelings. A speaker's apparent sincerity increases with eye contact: the more eye contact a speaker uses, the more trustworthy he or she seems to an audience.

Leaders or speakers can use the nonverbal signal of touch effectively in certain situations. However, touch is easily misunderstood or misinterpreted, so use it carefully. Some people will readily accept a pat on the back or a hug as a show of your approval or excitement. Others, however, might misconstrue touch as an inappropriate gesture. Be alert for indications that the other person misunderstands your intent, and immediately correct a miscue if you believe that your message was not received accurately.[50]

Proxemics is the science of how spacing and placement send messages. Leaders or people who command a great deal of respect are often given quite a bit of empty space around them. For example, political leaders often have more space around them than do the journalists crowding together to interview them; coworkers maintain less "personal space" than the company CEO in the company cafeteria. Leaders often choose how close to approach others.

Though some personal-space distances are culturally based, we each have our own physical comfort zone, and preferences vary with every individual. How close is too close when you are talking to others? At what distance from a speaker or leader are you comfortable? How do you feel when someone stands very close while talking to you? Be aware of regional and cultural diversity that contributes to these differences. A person whom you perceive as being pushy, aggressive, aloof, shy, or rude may simply be following his or her culture's norms of appropriate spacing.

Posture Standing or sitting erect may communicate interest and alertness. Slumping while sitting or standing is thought to indicate tiredness or a lack of interest. Nervousness or boredom can be conveyed by swinging a leg or tapping fingers on a desk. Boredom may also be communicated by fidgeting or continual shifting in a seat. Shrugging the shoulders is often a sign of indifference.[51]

Gestures and Emotions Putting a hand to the mouth or putting the head in the hands can be a way of showing objection to or surprise at a message. Hands placed on the hips with elbows pointing outward may be a sign of anger. Arms folded across the chest often indicates that a person is defensive or closed to the message. Conversely, arms at the sides may convey the idea of openness and relaxation. Pointing a finger while speaking may be used to communicate unhappiness or to show authority.

A crossed-legs gesture says many things. When someone crosses the legs and moves a foot in a slight kicking motion, the person is probably bored. Crossed legs often indicate a narrow line of thinking, or a person who is not open to negotiation. Individuals who cross their legs seem to be the ones who give you the most competition and need the greatest amount of attention. If crossed legs are coupled with crossed arms, you really have an adversary.

Boredom is demonstrated through several different forms of body language. One of the most recognized is drumming fingers or tapping the foot. Some psychiatrists believe that when we are impatient, we try to duplicate a prior life experience when we felt safe and secure, such as when we were in the womb. The drumming is supposedly an unconscious attempt to re-create the mother's heartbeat. Another gesture associated with boredom is putting one's head in one's hand (supporting the head with the hand). However, when coupled with drooping eyelids, this is a gesture of regret. (This subtle difference is a reminder of the misinterpretation danger inherent in nonverbal communication.) Another sign of boredom is **doodling**, drawing or scribbling idly. When a person doodles, interest is waning. For the most part, people who doodle let that activity interfere with or impair their listening, thus jeopardizing open communication.

Stroking the chin is a gesture of thinking and evaluating. It seems to say, "Let me consider." Chin stroking, as shown in Figure 6–9, occurs when people go through a decision-making process. A facial expression congruent with this gesture is a slight squinting, as if trying to see an answer to the problem in the distance.

We have learned that some gestures are stall tactics. One such gesture is that of removing eyeglasses and putting the earpiece of the frame in the mouth. Taking the glasses off and placing them on the table is also a stall. Both are usually used to give the person more time to think about a message or a response.

Like boredom, nervousness is also communicated most clearly through body language. Clearing the throat is often associated with nervousness and anxiety. When a person gets nervous, mucus forms in the throat. The natural thing to do is to clear it, making that familiar sound often heard from the speaker's podium. Sometimes throat clearing is simply a habit, but if continued, throat clearing may signal that the speaker is uncertain and apprehensive.

In addition to somewhat defensive body motions and positions, anxious people often make sounds that indicate their feelings. A hissing release of the breath, a sigh, and a sharp intake of breath can all indicate various emotional states related to anxiety or the relief of anxiety.

Frustration is revealed by many features of body language. Highly emotional people may take deep breaths and expel the air slowly, making long, sighing sounds. Other features denoting frustration (often mixed with anger or fear) are clenched fists, tightly clasped hands, and wringing hands. These are usually observed when someone is on the hot seat. People displaying these gestures tend to be tense and very difficult to relate to. Try to put them at ease before you attempt to communicate with them. Realize, though, that some people just do not want to be bothered when they are in certain moods.

FIGURE 6–9 This student's chin stroking is a nonverbal indication that she is thinking and evaluating. *(© iStock/pkujiahe)*

Physical Environment

Large offices with expensive decor may convey that the person who occupies the office is important in the organization.[52] Generally, the person seated behind a desk is perceived to be in charge. Certain desk arrangements in conference rooms or classrooms, such as horseshoe designs or desks and chairs grouped in small circles, may indicate a desire for open dialogue and collaboration, whereas rooms arranged with desks and chairs in straight rows facing the front may indicate a teacher's preference for lecturing (one-way communication).

Improving Nonverbal Communication

To enhance or improve your own nonverbal skills (for both recognizing and sending messages), consider the following:

- Be aware of and receptive to nonverbal communications and study the meanings of body language, gestures, and expressions.
- Remember that these meanings will vary among different cultural groups. When you speak, use appropriate gestures to emphasize and drive home your meaning.
- Become aware of your own nonverbal communication by having others watch you or by studying videotapes of yourself speaking.
- Work to convey open, friendly messages with your body language. Smiles, erect posture, and positive nods of the head are good ways to send such messages.
- Interpret nonverbal communication *only* as a way of confirming or refuting a verbal message.
- Realize that nonverbal communication is often imprecise. Interpret nonverbal messages with care, keeping in mind that any particular gesture or action may have more than one meaning. If you are unsure of what a person is communicating to you with body language, ask tactful questions that might help you understand.
- Be sensitive to the physical environment. Use seating arrangements that are appropriate to the type of communication environment you want. Be aware that if you sit behind a desk, the other person may feel like a subordinate (that you are his or her boss). This may make communication more difficult.

Being aware of the messages others send with body language can help you determine the effectiveness of a communication. Also, being aware of your own body language and using it to your advantage may help you better communicate your ideas to others.

FEEDBACK

Feedback is the receiver's response to a message; it helps the sender know whether the message was understood. If the receiver gives no feedback, the sender may have to determine whether the receiver understood the message by drawing out or directly asking for feedback. Getting feedback is a necessary step in the communication process; without it, the cycle is incomplete.

Some leaders have a special talent for reading their followers' faces and other body language, and are able to adjust their presentation accordingly. When leaders see looks of confusion, boredom, or lack of understanding, great communicators change tactics. The feedback they are receiving tells them that they have to do something else—quickly—because the communication is being received poorly or ineffectively.

False Feedback

There are problems with feedback that may reduce its effectiveness and usefulness. Not allowing time (or enough time) for feedback is an obvious problem; the communication process is short-circuited if the sender does not let receivers respond. Unfortunately, even when senders actively seek feedback, they frequently do so in this form: "Are there any questions?" When you ask this question, but get no responses, you may believe that the receivers have completely understood your message. In fact, depending on the situation, receivers may feel awkward or ignorant if they ask questions, or they may not have understood enough of the message to even form a question.[56] Furthermore, the generality of this feedback-seeking approach may also make receivers think that the sender is not sincerely interested in their responses. This perception is validated when the speaker pauses for only a few seconds after calling for questions and then concludes the session. It is better to say, "What questions do you have?" If you wait more than a few seconds for a response, it should elicit at least more than a yes or no response.

Receivers may give feedback that is vague or even untruthful. Responses such as "Interesting," "Yes, dear," "Thanks for sharing," "I'll get back to you on that," or even "Loved your speech!" do qualify as feedback; but they do not give the sender much useful information about how accurately or completely the communication was received.

Improving Feedback

To improve feedback, consider the following suggestions:

- As a sender, encourage feedback. Do this by allowing plenty of time. Be patient and encouraging while making others feel comfortable enough to respond to you.
- Be aware of nonverbal messages. Puzzled looks may tell you that your message was not understood. Also, be aware of your own nonverbal messages that may discourage communication, such as rolling your eyes or sighing when you are asked to repeat something you said earlier.
- Make it your responsibility to be understood. As a sender, you can actively seek (and as a listener you can provide) feedback instead of simply waiting for it to happen.

- Ask specific questions about the details of your message to be sure you were understood. Avoid the "Do you have any questions?" trap. Instead, say, "What questions do you have?"
- Use paraphrasing techniques. Use a statement such as, "Please explain to me in your own words what we need to do so I can be sure I have explained myself well." This type of question takes pressure off the listener because it is phrased so that you are checking your own performance, not doubting whether the listener has paid attention.[53]

SELF-COMMUNICATION AND INTERPERSONAL COMMUNICATION

Before we master interpersonal communication, a person must be able to self-communicate if he or she is going to effectively communicate with other people. Remember the advice of Socrates, "Know thyself." When you know who you are—when you understand your personality type, communication style, learning style, and their relationship—you are in touch with your feelings and have peace of mind about your life goals (Figure 6–10). We

FIGURE 6–10 The ultimate goal of interpersonal communication and self-communication is happiness. This dairy producer loves her career, and she knows what makes her happy. She will not have any problems communicating what she does for a living to the public. *(© iStock/JackF)*

call this **self-communication**. The following drill may help you communicate better with yourself.

Self-Communication

Visualize a day in your life five years from now. What do you want to be? Ask yourself the following questions.

1. What am I doing now to accomplish this objective?
2. What is keeping me from doing it?
3. What resources am I not using?
4. What do my actions now have to do with this goal?

Next, evaluate the past five years. Ask yourself the following questions.

1. What have I accomplished?
2. How did I do it?
3. Why did I succeed?
4. Why did I not succeed?
5. What was I doing right?
6. What was I doing wrong?
7. How can I alter my behavior?
8. What is happening to me in relationship to other people?

These may seem to be simple questions, but it is difficult to answer all of them. It is hard to be honest with ourselves and admit we have done some things we may regret. Furthermore, it is hard to admit that we could have accomplished much more if we had communicated more effectively with ourselves and set better goals. Also, if we had taken time to communicate with ourselves and define some timetable objectives for our lives, we would now be happier with ourselves. We would also have a feeling of self-fulfillment. Once a person has achieved self-communication, a major barrier to communicating with other people has been broken.

Interpersonal Communication

Good interpersonal communication is the next step in communicating effectively with others. How can you get close to people? How can you make people open up so that you better understand their feelings and beliefs? Do not let the following develop into **interpersonal communication barriers**, which are differences and personal characteristics of senders and receivers that hinder communication.

Honesty Honesty is crucial for interpersonal communication. If people you are supposed to be leading determine that you are dishonest, you can forget about being able to lead. Followers prefer credible leaders.

Awareness We have to be aware of our own experiences and be able and willing to make others aware of them. We have to send clear messages about our awareness, our experiences, our feelings, and our needs. However, we also must be aware of the messages that others send.[54] Not being aware of ourselves or the situation can interfere with interpersonal communications.

Clear Communication In addition to honesty and awareness, leaders seeking to master interpersonal communication need to communicate clearly. If a leader is unclear or vague, the relationship between the leader and the follower(s) can be strained. To communicate clearly with another person, particularly with verbal messages:

1. Send concise messages that are as clear as possible.
2. Pause periodically to allow the listener time to understand what you are saying. (It might be useful to think of yourself as talking in short paragraphs; do not spout out whole chapters at once.)
3. Be aware of your own feelings.
4. Know your intentions in communicating with your listener.
5. Express yourself as honestly as possible.
6. Maintain contact with the people to whom you are talking. Pay attention to their verbal and nonverbal responses, maintain frequent eye contact, and watch for cues to their understanding of your message (or lack of it).
7. Be aware of your own nonverbal messages (playing with your phone, leaning forward, laughing, hesitating, biting your lip, your voice quality), and notice whether these messages are consistent with the verbal communication you are sending. For example, are you saying that you are calm and comfortable with a decision, but your voice is shaking and you have crossed and uncrossed your legs a dozen times?

SELF-COMMUNICATION IS ONE OF THE MOST IMPORTANT THINGS YOU CAN DO AS A LEADER

Dawn learned that lesson the hard way. Dawn always wanted to win, which is not a horrible quality, but it did tend to get her into trouble. One Saturday she wanted to win a particular softball game so badly that she began to tell all her teammates what they were doing wrong. The more frustrated she got, the more she berated her friends, degrading them and making them feel terrible. By the end of the game, no one was happy and Dawn's friends were not speaking to her. They also lost the game.

Dawn's coach thought Dawn needed to have a nice conversation with herself. If she would just take the time to calm down and really reflect on her behavior, how it affected others, and how it helped her, she would be able to correct her mistakes and focus her energy in a positive manner. At first Dawn thought her coach was crazy: "I'm not going to talk to myself!" Eventually, even though it sounded silly, Dawn took her coach's advice and began to ask herself questions like these:

- "How have I been helping my team so far?"
- "Has my help been useful?"
- "How did my behavior look to others?"
- "What could I do differently?"
- "What can I do to make sure I do not mess up again?"
- "How can I get my team's trust back?"

As she reflected on it, Dawn began to be embarrassed by her behavior, but she used the self-communication process to improve herself and her team. Dawn is now a believer in talking to herself or self-communicating.

Not What, But How It has often been said that it is not what you say but how you say it. The "how" of the message is complex. In addition to the "face value" content (the literal meaning), every message contains information about how it should be interpreted. Cues at this level tell receivers whether the message from the sender is a question, a joke, a statement, or an opinion. Sarcasm, irony, and doubt are almost always expressed at this level. The subtext, which in verbal communication consists of body language, also conveys huge amounts of information about the sender's sincerity, emotions, and attitude toward the content of the message. All these peripheral bits of information can support, detract from, or distort the intended message.

Relationship to Receiver Every message contains information concerning how the sender perceives his relationship to receivers, whether he considers them friends, enemies, or equals. Information about the relationship is transmitted in many different ways, including (1) words, or verbal behaviors; (2) nonverbal behavior such as gestures, tone of voice, posture, and facial expression; and (3) the situation itself, or the context in which the communication occurs. A positive (or perceived positive) relationship between the sender and the receiver can only enhance interpersonal communication.

Directness One aspect of being direct involves the sender's "ownership" of the thoughts, feelings, ideas, evaluation, or expectations she conveys; the content of the message is personal rather than being attributed to some other source. This kind of personal responsibility for the message makes the difference between saying, "You are obnoxious," and saying, "I am really angry at what you're doing." In this instance, I can be direct about my own feelings, because I know them firsthand; therefore, I can send the second message legitimately. Sending the first message almost inevitably leads to interpersonal difficulty for a number of reasons. Of course, it is accusatory, denigrating, and perception based; typically it only makes the receiver angry or defensive. Another problem with the "You are obnoxious" message form is that it is usually the first move in the "blame game," a pattern of interaction that generally results in stalemate. It should go without saying that directness is not synonymous with rudeness or tactlessness.

Being Nonjudgmental Another primary aspect of interpersonal communication is knowing how to present one's thoughts and feelings in a way that is nonjudgmental and undemanding. If you are nonjudgmental, you will never say things like this:

- "You shouldn't interrupt." (you are breaking rules or not observing standards)
- "It isn't right to interrupt." (you are wrong)
- "Shut up and let me talk." (you are not as important as I am)

All such statements inherently pass judgment on another person's behavior or character.

Respect for Other Person's Position Still another aspect of interpersonal communication is being able to respect and understand the other person's position. As mentioned earlier, this involves knowing how to listen nonjudgmentally to what another person is thinking and feeling. Interestingly, this is one way we show our love for others.

Love is a state of complete attention, without intruding thoughts and motivations. Contrary to general belief, love is not just a feeling or emotion. The opposite of love is not hate; the opposite of love is . . . judgmental thinking.

—Marshall Rosenberg

COMMUNICATION TECHNOLOGY

Reading, writing, speaking, and listening are foundational concepts and aspects of communication, but technological advances have also changed how we communicate with our friends, family, and followers. For many years, leaders used letters sent through the mail or telephone calls. Now we also communicate with texts and various social media platforms, and tomorrow's leaders will be communicating with one another in ways that would have been hard to imagine just a few years ago. A study of four major American universities determined that a majority of college students had experience with the following technologies: Internet, e-mail, Facebook and other social networking websites, and handheld devices (for digitized music and videos). Several of the students surveyed also used blogs, podcasts, wikis, and satellite radio communication technologies.[55]

Social Media

Social media are defined as the "various electronic tools, technologies, and applications that facilitate interactive communication and content exchange."[56] Social media is something that leaders cannot ignore. Studies have determined that leaders who frequently use social networking sites are seen as more powerful and influential by their followers than leaders who are less frequent users. There are numerous studies documenting the impact of specific platforms, but as an example one study determined that using Twitter has helped many leaders to develop their transformational leadership skills among followers.[57]

Social media can "empower" leaders and help develop leadership abilities by helping with the following:

- Managing followers
- Creating vision in an organization
- Motivating others to achieve goals
- Communicating internally (followers in an organization) and externally (the public at-large)
- Providing knowledge for decision-making and problem solving
- Collaborating with other leaders and followers
- Evaluating the organizational culture[58]

Major Functions of Social Media

Leaders believe that the major functions of social media are broadcasting or promoting messages, building community, monitoring, managing and managing crisis.

Broadcasting Great leaders use the mostly free, but essential, platform of social media to push the vision, goals, and objectives of their organizations almost daily. Examples of messages that leaders can broadcast on social media sites such as Facebook, Twitter, YouTube, Instagram, etc. include "news releases, new product announcements, educational materials, organizational plans, success stories, policy updates, internal policies, personnel changes, or community involvement."[59]

Building Community Social media communications are also used by leaders and organizations to build digital communities that are "interactive, participative, user-centered, informative, creative, and geared toward problem solving."[60] As an example, your Chapter FFA Alumni could use a photo-sharing site such as Instagram or a video-sharing resource like Snapchat to creatively share memories and stories from their time as an FFA member as a way to build community. Exchanges on Twitter or Facebook could also be used to solve problems in a given community, build alumni membership, or increase participation in your club or organization.[61]

Monitoring Issues Leaders can also use social media to gather information about the attitudes, concerns, and needs of followers.[62] Information travels quickly in this type of communication, and leaders should learn to monitor these sites closely. This allows them to (1) quickly identify negative comments and inaccuracies that could seriously harm the goals and aspirations of their organization, and (2) address them via the same communication channel.

Managing Crisis Leaders can also use social media to manage crisis in an organization. For instance, a snowstorm might make roads impassable for employees, and a leadership team might think it best to have employees work from home for their safety. Social media could be used to inform them of the decision and even provide organizational goals and activities they can work on from home. In the same situation, social media could be used to organize food or shelter relief for employees in need as a result of the storm.

You are the next generation of leaders, and you will need to use, or at least understand, these different forms of communication to exchange messages both with individual followers and with the teams you will direct. One of the best things you can do as a leader is to embrace each new form of communication. Just remember that the

foundational skills of listening, reading, writing, and speaking are still the basis for effective use of these and other forms of communication technology.

CONCLUSION

A person who has mastered the art of communication is likely to live a satisfied, fulfilled life. It appears that happiness consists in large part of closeness to and friendship with other people. Such relationships can be achieved only when we communicate effectively, both with others and with ourselves.

SUMMARY

Communication is the process of sending and receiving messages through which two or more people reach understanding. Communication is important because so much of each day is spent in some form of communication and because many problems in our relationships occur because of miscommunication.

The purposes of communication are to inform, to influence, and to express feelings. We communicate when we send messages (speaking, writing, blogging, body language) or receive messages (listening, reading). The communication process involves a situation in which a sender provides a message through a channel to a receiver. The receiver responds with (or the sender seeks) feedback. The process may be derailed or broken down by interference from numerous sources.

Factors that may hinder communication are language barriers, interpersonal barriers, situational-timing barriers, and organizational structure and procedural barriers. These barriers may be overcome by improving perception, improving the physical processes of communication, and improving relationships. To communicate with other people, we must first be able to communicate with ourselves. The ultimate communication for most of us is interpersonal communication.

Communication of all types can be improved by understanding the problems associated with each type and by making a conscious effort to practice techniques designed to foster and improve communication skills.

Take It to the Net

Explore communication skills via the Internet using your favorite search engine. Search for articles using the following search terms. Choose an article on communications skill, and write a one-page summary of the article or design a concept map that summarizes the major points of the article.

Search Terms

communication skills

communication

effective communication

importance of communication skills

social media

Chapter Exercises

REVIEW QUESTIONS

1. Define the Terms to Know.
2. Define communication.
3. List three situations in which communication is important.
4. List and give examples of three primary purposes of communication.
5. List and give examples of three forms communication takes.
6. List and briefly describe seven elements of the communication process.
7. List four barriers to communication and give an example of each.
8. Explain three ways to overcome communication barriers. Give an example of each.
9. Give at least three suggestions for improving each of the following skills: listening, reading, writing, and speaking.
10. Explain how and when nonverbal communication should be used.
11. Is asking "Are there any questions?" a good method for checking understanding and soliciting feedback? Why or why not?
12. Name five characteristics of good interpersonal communication.
13. List and describe three forms of social media.
14. Explain how social media as a form of communication can help you as a leader.
15. List four major functions of social media.

COMPLETION

1. _____ barriers deal with the way words are understood.
2. _____ barriers deal with personal characteristics of the sender and receiver that hinder communication.

3. _____ barriers deal with how and through what structure a message travels from the sender to the receiver. Barriers of this type are common in the workplace.

4. _____ barriers are those that deal with when and where a communication takes place.

MATCHING

_____ 1. The process of sending and receiving messages to achieve understanding.

_____ 2. When and where communication takes place.

_____ 3. The person who wishes to start communication with someone else.

_____ 4. The responses a receiver gives the sender so the sender knows the message was understood.

_____ 5. The means a sender uses to convey encoded messages.

_____ 6. A person's understanding of a message based on personal beliefs, knowledge, and ways of organizing information.

_____ 7. Person for whom a message is intended.

_____ 8. That which is intended to be communicated from one person to another.

_____ 9. Anything that hinders the sender from making the message understood.

_____ 10. Process by which the sender puts the message in a form the receiver will understand, such as words, numbers, gestures.

A. perception

B. communication

C. receiver

D. encoding

E. message

F. sender

G. channel

H. feedback

I. situation

J. interference

ACTIVITIES

1. Write a one-page essay expressing your feelings about an issue in your life or something in the news. Have the class evaluate your ability to communicate your feelings.

2. Write an imaginary communication scene similar to the one in this chapter about Stephen and Chelsea. Have a partner identify each element of the communication process.

3. Read an article from a favorite magazine. Write a brief summary of the article, outlining the main points.

4. Read the summary you wrote for question 3 to a classmate. Have the classmate listen carefully and try to identify the main points of your summary. Check the classmate's listening skills by comparing

the answers to your summary. If there are misunderstandings or mismatches, reread the summary or rewrite it if your writing is what caused the misunderstanding.

5. Watch a classmate, teacher, or favorite actor. Write down the nonverbal forms of communication you see. Also note how the nonverbal communication matched (or did not match) the verbal communication used.

6. During a class discussion, jot down ways students provide feedback (verbally and nonverbally). Share your findings in small groups.

7. In 5 or 10 minutes, develop a brief role-play situation in which you and a partner demonstrate one of the barriers to communication (language barrier, interpersonal barrier, situational-timing barrier, or organizational structure and procedural barrier). Perform the role-play for the class. Have class members guess which barrier you are demonstrating. Have a class discussion about how the communication could be improved.

8. Visualize a day in your life five years from now. What do you want to be? Write an answer to each of the following questions:

    a. What type of job, educational accomplishment, and personal goals do I want to have achieved five years from now?

    b. What am I doing now to accomplish these objectives?

    c. What is keeping me from doing this?

    d. What resources am I not using?

    e. What do my actions now have to do with this?

9. Form a plan for using different forms of communication to promote an activity in your FFA chapter.

10. Evaluate three different social media platforms (e.g., Twitter, Instagram, etc.) by listing three different ways each one could be used to lead a group.

NOTES

1. J. R. O'Connor, *Speech: Exploring Communication* (Lincolnwood, IL: National Textbook Company, 1996), p. 5. (4th ed., 2001).
2. R. N. Lussier, *Human Relations in Organizations: Applications and Skill Building* (Boston: Irwin/McGraw-Hill, 1999), p. 105. (9th ed., 2012).
3. Ibid., p. 104.
4. Ibid., pp. 105–106.
5. MSG Experts. "Importance of Communication in an Organization," *MSG Management Study Guide.* Accessed January 19, 2016.
6. W. D. St. John, *A Guide to Effective Communication* (ERIC #ED057464) (Nashville, TN: Author, 1970).
7. D. F. Jennings, *Effective Supervision: Frontline Management for the '90s* (Minneapolis/St. Paul: West, 1993), p. 174.
8. Ibid., p. 172.
9. S. E. Lucas, *The Art of Public Speaking,* 12th ed. (New York: Random House, 2014). (12th ed., 2014).
10. Ibid.
11. St. John, *A Guide to Effective Communication.*
12. Jennings, *Effective Supervision,* p. 173.
13. St. John, *A Guide to Effective Communication.*
14. Lucas, *The Art of Public Speaking.*

15. St. John, *A Guide to Effective Communication.*
16. Lucas, *The Art of Public Speaking.*
17. Jennings, *Effective Supervision,* p. 173.
18. Lucas, *The Art of Public Speaking.*
19. St. John, *A Guide to Effective Communication.*
20. "Barriers to Communication," *Communication* series, No. 8741-B (College Station, TX: Instructional Materials Service, Texas A&M University, 1988), p. 2.
21. Jennings, *Effective Supervision,* p. 186.
22. St. John, *A Guide to Effective Communication.*
23. "Barriers to Communication," p. 1.
24. Ibid.
25. Ibid.
26. "Barriers to Communication," p. 2.
27. Ibid.
28. St. John, *A Guide to Effective Communication.*
29. J. R. Noe, *People Power* (Nashville, TN: Oliver Nelson, 1986), p. 124.
30. Lussier, *Human Relations in Organizations,* p. 157.
31. St. John, *A Guide to Effective Communication.*
32. "Barriers to Communication," p. 2.
33. Ibid.
34. Lucas, *The Art of Public Speaking.*
35. Ibid.
36. O'Connor, *Speech: Exploring Communication,* p. 59.
37. Lucas, *The Art of Public Speaking,* p. 29.
38. Ibid., pp. 28, 29.
39. Lucas, *The Art of Public Speaking,* pp. 29–32.
40. Ibid., pp. 33–40.
41. Jennings, *Effective Supervision,* p. 189.
42. P. Clark, "Leadership: The Individual" (exercise used in an educational leadership course) (Gainesville, FL: University of Florida, 2000).
43. Quoted in "Concise Writing Guide," *Garbl's Writing Center.* Retrieved October 24, 2009 from http://garbl .home.comcast.net/~garbl/stylemanual/betwrit.htm
44. St. John, *A Guide to Effective Communication.*
45. Ibid.
46. Ibid.
47. Lussier, *Human Relations in Organizations,* p. 108.
48. "Improving Oral Communication Skills," *Skills for Success* series, no. 50 (Nashville, TN: Tennessee Department of Education, 1993).
49. "Improving Nonverbal Communication Skills," *Skills for Success* series, no. 52 (Nashville, TN: Tennessee Department of Education, 1993).
50. "Non-Verbal Communication," *Communication* series, No. 8741-E (College Station, TX: Instructional Materials Service, Texas A&M University, 1988), pp. 1–2.
51. Ibid, p. 1.
52. Lussier, *Human Relations in Organizations,* p. 110.
53. Jennings, *Effective Supervision,* p. 174.
54. J. O. Stevens, *Awareness: Exploring, Experimenting, Experiencing* (Moab, UT: Real People Press, 1971), pp. 88–89. (Eden Grove ed., 1989).
55. E. Rhoades, C. Friedel, and T. Irani, "Classroom 2.0: Students' Feelings on New Technology in the Classroom" (paper presented at the American Association of Agricultural Education Research Conference, Reno, Nevada, May 21–23, 2008), *Proceedings of the 2008 AAAE Research Conference* 35:548–561. Retrieved October 9, 2009 from http://www.aaaeonline.org/conference_files/829108.proceedings.pdf.

56. B. A. Hamilton, *Social Media and Risk Communications During Times of Crisis.* (July 23, 2009). Retrieved from http://www.boozallen.com/insights/2009/07/42420696, p. 1.

57. Y. Luo, H. Jiang, and O. Kulemeka, "Strategic Social Media Management and Public Relations Leadership: Insights from Industry Leaders," *International Journal of Strategic Communication*, vol. 9, issue 23, 2015, p. 170.

58. Ibid., p. 170.

59. Ibid., p. 177.

60. Ibid., p. 178.

61. Ibid., p. 178.

62. Ibid., p. 179.

7 RECITING (FFA CREED)

A primary predictor of success for many leaders has proven to be verbal ability. Reciting the FFA Creed is one of the first exercises that FFA members can perform to help develop their speaking and verbal ability, in pursuit of the longer-term goal of becoming great leaders.

Objectives

After completing this chapter, the student should be able to:

- Recite the FFA Creed
- Explain why one should participate in reciting the FFA Creed
- Demonstrate basic principles of speaking
- Demonstrate practice methods
- Reduce stage fright and nervousness
- Establish rapport with an audience
- Present the Creed effectively
- Use practical, enjoyable ways to learn and say the FFA Creed

Terms to Know

- FFA Creed
- agricultural education
- curriculum
- attitude
- delivery
- convey
- winner
- affective skills
- self-confidence
- insecure
- monotone
- chapter
- rapport
- lectern
- spontaneity
- tone
- inflections
- articulation
- pronunciation
- force
- fluency
- intangible
- clone

Reciting the **FFA Creed**, a philosophy statement on agriculture written by E. M. Tiffany, is the first of many organized speaking activities in **agricultural education**, a program of instruction in agriculture in high schools. Although the FFA Creed is an excellent document to study and memorize, our purpose in this chapter is to use it as a tool to begin public speaking.

The FFA Creed outlines the FFA Organization's beliefs regarding the industry of agriculture, FFA membership and the value of citizenship and patriotism. As part of the requirements to earn the Greenhand FFA Degree, FFA members recite the Creed. The purpose of the Creed Speaking Career Development Event (CDE) is to develop the public speaking abilities of first-time members, improve their self-confidence, and help them to advance in the FFA degree program. The competition is usually among seventh-, eighth-, and ninth-grade FFA members, but many states offer Creed competitions for older first-time students too. Members must recite the FFA Creed from memory and answer questions about its meaning and purpose. The FFA provides updated information about the Creed Speaking CDE and other CDEs at https://www.ffa.org/participate/cdes.

Here is an excellent plan for agricultural education students to follow to master the skills required for good public speaking. Students may participate in these categories:

1. FFA Creed—9th or 10th grade
2. Parliamentary procedure—10th grade
3. Prepared public speaking—11th grade
4. Extemporaneous speaking—12th grade

Even if this plan of progression is not appropriate for you or your school, it is helpful to start with reciting the FFA Creed, as it can build confidence for later public speaking opportunities as a leader. In fact, speaking can begin at any grade level. All the speaking examples listed could be learned in the same class or in other parts of the **curriculum**, or course of study.

Public speaking is fun. With an open mind and a proper **attitude** (disposition toward others and ourselves), public speaking can be as natural and enjoyable as playing sports. The person in Figure 7–1 is enjoying herself as she practices reciting a speech, such as the FFA Creed.

FIGURE 7–1 Speaking can be fun if you develop and maintain the proper attitude. Loosen up, enjoy yourself, and remember that if you mess up, your classmates will laugh with you, not at you. *(© iStock/Steve Debenport)*

LEARNING THE FFA CREED

Again, a creed is a philosophy statement of about the beliefs of an individual or an organization. If an FFA member has the dedication to memorize the FFA creed and the courage to present it in a speech, this says volumes about their commitment to the organization. Memorizing and reciting the Creed is also an example of the symbolic leadership mentioned in Chapter 1, and is a requirement of the Greenhand degree, the second degree attainable in the FFA. Memorizing and public speaking are not easy, but as the old saying goes, "When the going gets tough, the tough get going." A great tool for memorizing the Creed is a game called "speed creed." In speed creed, FFA members compete to determine who can say each paragraph or the entire Creed the fastest. Make sure you have a stopwatch for this fun game.

Because the Creed script is already written, after you memorize it you have more time to work on **delivery**, the method of making a speech. Part of effective delivery is emphasizing important words. The Creed has many expressive words that should be said with feeling.

Key Words in the FFA Creed

There are no dull subjects, but there are dull people—do not become one of them when you speak! Almost every sentence in the FFA Creed revolves around one or more key words. In the first paragraph, these are *believe, future, agriculture, faith, words, deeds, achievements, present, past, generations, promise, better, enjoy,* and *struggles.* That is a lot of key words, but the Creed is powerful because of them.

It may be helpful to print out a paper copy of the Creed and work with a pen or pencil: start by selecting and underlining the key words in each paragraph of the Creed—the ones that seem the most important or meaningful.

Another helpful activity is to find a picture in an agricultural magazine or some other source that illustrates some of the key words, sentences, or paragraphs that you find. Choosing pictures that **convey** (communicate or make known) meaning to you may help you memorize the Creed and say certain important words with sincere feeling and expression.

After you've selected the key words and concepts in the Creed, practice saying them with feeling that reflects their meaning. For example, say *better* with positive facial and vocal expressions. Say *struggles* with negative facial and vocal expressions. This will help you memorize the words, and will help your presentation be more expressive and meaningful to your audience.

As you memorize the Creed, you may wish to develop a paragraph similar to the one in Figure 7–2. This accomplishes two things: first, it helps you memorize the Creed; second, it helps you emphasize the key words when you recite the Creed.

Consider the nonitalicized words in the FFA Creed, reprinted here. Write each paragraph with blanks for the key words (refer to Figure 7–2). You may use the key words indicated here, or make your own selections.

I *believe* in the *future* of *agriculture*, with a *faith* born not of words, but of *deeds—achievements* won by the present and past *generations* of *agriculturists*; in the promise of *better* days through better ways, even as the better things we now enjoy have come to us from the *struggles* of former years.

I *believe* that to live and *work* on a *good* farm, or to be *engaged* in other agricultural pursuits, is *pleasant* as well as *challenging;* for I know

"I (1) _____ in the
(2) _____ of
(3) _____, with a
(4) _____ born not of words, but of
(5) _____.
(6) _____ won by the present and past
(7) _____ of
(8) _____ in the promise of
(9) _____ days through better
ways, even as the better things we now enjoy have
come to us from the (10) _____
_____ of former years."

FIGURE 7–2 As you memorize the Creed, you may wish to develop a paragraph similar to the one shown here. This accomplishes two things: (1) it helps you memorize the Creed, and (2) it helps you emphasize key words when you recite the Creed.

the *joys* and *discomforts* of agricultural life and hold an inborn fondness for those *associations* which, even in hours of *discouragement*, I cannot *deny*.

I *believe* in leadership from ourselves and *respect* from others. I believe in my own ability to *work efficiently* and *think clearly*, with such *knowledge* and skill as I can *secure*, and in the ability of progressive agriculturalists to serve our own and the public interest in *producing* and *marketing* the product of our *toil*.

I *believe* in less dependence on *begging* and more *power* in *bargaining*; in the life *abundant* and enough *honest wealth* to help make it so—*for others* as well as *myself*; in less need for *charity* and more of it when needed; in being *happy* myself and playing *square* with those whose *happiness* depends upon me.

I *believe* that American agriculture *can* and *will* hold *true* to the best *traditions* of our national life and that I can *exert* an *influence* in my *home* and *community* which will stand *solid* for my part in that *inspiring task*.[1]

REASONS TO PARTICIPATE

Like many high school students, you may be dragging your feet about learning and reciting the FFA Creed, feeling that this is not for you. Maybe you feel it is not worth the effort. You may be thinking, "I can't win with speaking, so why try?"

You Win by Participating

It is common for students to feel that the winner will always be someone else, but this is not true. You can be a **winner**—someone who achieves victory over others in a competition or success in an endeavor. In fact, the only way you cannot be a winner is by refusing to participate. David Jameson, in his *Leadership Handbook*, tells us why. Most people mistakenly equate being a winner with being the single, number-one, first-place contestant. For example, we focus our attention on the athletes who are declared the champions and consider them the winners and all their competition losers. (This in itself is a mental mistake; athletes or teams do not rise to being second, third, or even fifteenth in the world unless they are amazingly good!) However, if you follow the so-called losing teams for two or three years, you will probably notice several star players who came from that team. In fact, many stars come from last-place teams—but as individuals, they were winners. They became winners because they participated and learned not only athletic skills but also the values of persistence, humility, and teamwork. We need to look beyond the immediate present when we think about who the winners really are.[2]

You Can Win in the Game of Life

Many occupations require speaking skills. Speaking is indicative of **affective skills**. Affective skills are those that help you behave appropriately in the social situations of life. Lawyers, salespeople, preachers, politicians, auctioneers, and those in many other occupations need speaking skills.

The winner of a public speaking contest may end up in an occupation that requires very little public speaking, whereas students who did not place first may have what it takes to enter the field of politics or assume a position of responsibility in the community. The skills they learn by participating in public speaking help them become winners.[3] Will Lewis, a former Tennessee State Department of Education leader, always used to address the FFA members participating in a CDE by saying, "You are all winners." He was correct.

When you say, "I can't win. Why should I try?" you are being very shortsighted. How do you know what you can do before you even attempt it? The truth is that, by not participating, you may be establishing a self-fulfilling expectation of failure that sets a pattern for lack of success in life.

By Participating, You Cannot Lose

Opportunities for learning many essential leadership skills arise only occasionally. Therefore, everyone who wants to be a long-term winner should be eager to participate when an opportunity does present itself, recognizing that the real wins may be several years in the future. Because of your involvement now, however, you will be ready for the day when the bigger contests come.[4] By participating, you cannot lose. You lose only when you quit or do not even try.

BASIC PRINCIPLES OF SPEAKING[5]

Reasoning and Feeling

Communication involves both reasoning and feeling. The skillful speaker both thinks and cares. The correct attitude is one of friendly goodwill, joined with a desire to help, inform, or inspire. If a speaker does not believe in the message, lacks **self-confidence** (belief in oneself or one's abilities), is apologetic, is immature, or is self-deprecating, listeners will place no confidence in what he or she says.

Rational and Creative Thinking

The able speaker thinks rationally and creatively. When both reason and imagination are used in crafting and delivering a speech, listeners view the speaker as having good judgment, common sense, competence in the subject, and a healthy self-image. Thus, the first task in developing into a good speaker is to become worthy of being heard. This means practicing, listening to good speakers, and working on your delivery.

Effective Speech Style

The skillful speaker develops an effective speech style by integrating appropriate language, vocal expression, and body action. Every speaker has a distinctive style that reveals many things about him or her as a person. An **insecure** speaker who lacks self-confidence uses impersonal language and speaks in a **monotone**, without fluctuations or inflections in the voice. Fear of the audience, or of embarrassment, leads the speaker to use this style as a shell into which to withdraw and hide from the audience. Most of the time, when an audience dislikes a speaker, it is not because of the statements made but because of an irritating voice or manner. Speakers must adapt their style and delivery to each audience.

Observe the Audience

The able speaker watches the audience while giving the speech, observing facial expressions and body movements that indicate interest and attention (or lack thereof). A skillful and practiced speaker can distinguish between the frown that indicates "You're making me think hard" and the frown that indicates "You are so wrong."

Persuade the Audience

If they are to persuade audiences, speakers must understand how the listeners feel, what they want, and what they need. A speech must begin with the listeners; otherwise, it is a waste of time. The greatest artists of persuasion are those who understand and genuinely like people. No technique can substitute for sincere concern and respect for others.

David Jameson, a past National FFA officer, suggests that you can easily master what he calls the "Big Four" speaking skills. These are the big four common-sense skills:

Volume Always speak loudly enough to be heard easily in the back row of the room.

Tempo Speak slowly enough that the audience can grasp what you are saying and think about it.

Diction Speak clearly enough to make every word distinct; the audience should hear each and every word.

Enthusiasm Speak with enthusiasm, so that people will know you mean what you are saying, find it easier to pay attention to you, and remember more of what you said.

What do we accomplish, in the minds of the audience, by mastering these four basic skills? First, we please the audience because they can easily hear what we are saying; they do not have to exert physical and mental efforts just to hear us. People are also impressed when we speak slowly and clearly enough for them to understand and digest what we are saying. Finally, and most importantly, enthusiasm makes a speech seem shorter and more memorable.[6]

PRACTICING

The old saying is "practice makes perfect." Is this really true? Consider most people's handwriting and you will agree that practice does not make perfect. In reality, practice makes perfect only if you practice perfectly; that is, "*perfect* practice makes perfect." Practice as well as you can each time. You will pursue perfection through self-evaluation and evaluation by others. Practice with the intention of doing a better job each time.

FIGURE 7–3 Recording your Creed presentation on video and evaluating it is an excellent way to improve delivery. *(© iStock/philipimage)*

There are a number of ways to practice the Creed. Use an audio recording device to make sure you are speaking loudly and clearly enough and that you have expression in your voice. Use a video recorder to rehearse your gestures, as shown in Figure 7-3. Evaluate how and what you emphasize with your hands and facial expressions. Once several members have been taped, the teacher and the students will evaluate each speaker.

You may be surprised to find that one of the most effective ways to practice is in front of a mirror (Figure 7–4). You may feel silly at first. You

FIGURE 7–4 Saying the FFA Creed in front of a mirror is a good way to help overcome stage fright and improve your facial expressions and hand gestures. *(© iStock/Antonio_Diaz)*

may be embarrassed even though no one else is around. You may even have trouble concentrating at first. Eventually, though, you will gain self-confidence, and your ongoing self-evaluations will allow you to be more comfortable speaking before others.

Practice the Creed in front of your family, before members of your FFA **chapter** (a local branch of an organization or club), your chapter adviser, or other adults. There is no substitute for frequent and regular practice with the intention of getting better each time. More is said about practicing later in this chapter and in Chapter 8.

STAGE FRIGHT AND NERVOUSNESS

Frequently, when students are asked to speak in front of a group, they say, "I *cannot* get up in front of a group and talk. My knees would be knocking and I would be so nervous that I would make a fool of myself. I'd like to do it, but I just can't." If you feel this way, you are not alone. Most adults, like most students, fear getting up in front of an audience. Your classmates are probably just as scared as you are. Their knees are weak and their voices quiver. Even professional speakers admit that they often become quite nervous before speaking, whether they are making a speech, making a sales presentation to a group, or participating in some other important occasion.

The ability to communicate to groups of people is a skill that can make a critical difference in your career and in your ability to share information, ideas, experience, and enthusiasm with others. A study conducted by AT&T and Stanford University revealed that the top predictor of [professional success and upward mobility] is how much you enjoy public speaking and how effective you are at it.[7]

The young woman in Figure 7–5 has mastered the skill of public speaking. She is focused. She is using hand gestures. She is dressed professionally, and she even has some notes handy. Notes are okay if you need to make sure your facts are correct, but they won't be necessary for the FFA Creed.

One survey asked more than 2,500 Americans to list their greatest fears. The number-one fear

FIGURE 7–5 This woman is making good use of hand gestures and facial expressions, probably to drive home an important point. She also has some notes handy. This is okay if you need to support a speech with evidence, but notes won't be necessary when presenting the FFA Creed. *(© iStock/GlobalStock)*

(41 percent) was speaking in public before a group. Ironically, only 18 percent included death as one of their greatest fears. Other studies also rank speech making near the top as one of people's greatest fears.[8]

The Primary Fear Is the Primary Predictor of Success: Nervousness Is Normal

The nervousness that we call stage fright is an entirely normal—and very common—reaction. Sir Winston Churchill compared his prespeech fear to a nine-inch block of ice sitting in the pit of his stomach.[9] Even powerful orators such as Abraham Lincoln and Franklin D. Roosevelt became nervous before speaking. Actually, most people tend to be anxious about doing anything important in public. Actors are nervous before performing in a play; politicians are nervous before giving a campaign speech; athletes are nervous before playing a big game. The ones who succeed have learned to use their nervousness to their advantage.

Although stage fright may never go away entirely, professionals know you can make the butterflies in your stomach "fly in formation"; that is, you can learn to manage your fear. It is perfectly normal—even desirable—to be nervous at the start of a speech.[10] What happens is this: your brain and muscles throughout your body become supercharged, as your body carries out its natural function of preparing you to meet a special situation. This is basically the fight-or-flight reaction, and good speakers can channel this energy appropriately. The trick is to control your nervousness and make it work *for* you rather than *against* you.

"*The only thing we have to fear is fear itself.*"

—Franklin D. Roosevelt

Controlling Your Nervousness

We have now learned that some stage fright will always be with us. However, this is good news. Without some stage fright, we probably would not be at our best. The key is to control your nervousness rather than trying to get rid of it completely. The following steps show how.[11]

Step 1 *Prepare well.*

Most stage fright comes from a fear of embarrassing yourself in public (performing poorly, memory lapses, stumbling over words, not pleasing the audience, and so on). Thorough and complete preparation, including practice in front of different people, can ensure that about 90 percent of your Creed delivery will go smoothly.

Thorough preparation gives you confidence both before-hand and after you have actually started speaking. If you miss a line or make a mistake, do not call attention to it. Correcting yourself does not help, and in any speeches other than Creed recitation, the mistake may go undetected.

Step 2 *Release physical tension before you speak.*

Some speakers feel their neck, stomach, or facial muscles tightening and tensing up. Remember, "these are symptoms rather than causes of stage fright."[12] O'Connor suggests the following relaxation techniques to help reduce physical symptoms of stage fright:

1. Force yourself to yawn widely several times. Fill your lungs with air each time by breathing deeply.
2. Let your head hang down as far as possible on your chest for several moments. Then slowly rotate it in a full circle, at the same time allowing your eyelids to droop lazily. Let your mouth and lower jaw hang open loosely. Repeat this rolling motion five or six times, very slowly.
3. Sit in a slumped position in a chair as if you were a rag doll. Allow your arms to dangle beside the chair, your head to slump on your chest, and your mouth to hang open. Then tighten all muscles one at a time, starting with your toes and working up your body to your neck. Next, gradually relax each set of muscles, starting at the top and working back down to your toes. Repeat this process several times.[13]

Step 3 *Remember that audiences are usually aware of and sympathetic to the anxious speaker's plight.*

Most listeners understand what you are going through and are sincerely empathetic. They would expect to feel the same way if they were speaking. Listeners usually try to give a speaker friendly support and encouragement when they see signs of performance anxiety or stage fright.

Step 4 *Develop the right attitude.*

Try to view stage fright as positive nervous energy that you can learn to harness and direct. Controlled stage fright helps you be a better, more successful speaker. Just before delivery of the Creed, use a little self-communication. Talk to yourself as follows:

> What I am feeling are symptoms of stage fright. It is great that my body prepares me for these events. This tense feeling is just what I need to sharpen my thinking, enliven my presentation, and help me reach my full potential once I start speaking.

Step 5 *Concentrate on the Creed.*

Stop thinking about yourself and concentrate on the Creed. You have a chance to persuade, entertain, and develop leadership. Don't think, "I *have to* give a speech today"; instead, think, "I *have a* speech to give today!"

Step 6 *Concentrate on the audience.*

When you are reciting the Creed, carefully scan the faces in the audience, whether it is your classmates or some other group. Look at the interactive listeners while speaking. "Their obvious appreciation for your speech can be a great confidence-builder for you."[14]

Avoid fixating on a couple of negative listeners. If you concentrate on trying to please them, you will likely lose all the rest of your audience—the obvious majority. When you concentrate intently on your audience, there is not much time to think about yourself. Thus, your stage fright will recede to the background and your confidence will increase.

Step 7 *Smile!*

A self-confident, pleasant smile relaxes you and the audience more than anything else. For presentations other than the Creed, using a bit of humor also relieves tension in both speaker and audience. Getting a laugh is a quick and certain confidence booster that shows the speaker he or she need not fear these listeners.

Step 8 *Move around.*

Although any type of movement helps reduce nervous tension and the physical symptoms of stage fright, "movements that help communicate your message nonverbally are the only kind" you should allow yourself during a speech.[15] Movements that you have actively decided to incorporate into your speech, and have practiced until they seem natural, do just as well for releasing nervous tension as distracting, unconscious ones, such as pacing, fiddling with a pencil, or frequently running a hand through your hair.

VIGNETTE 7-1   **NERVOUS NED**

NERVOUS NED WAS A FRESHMAN FFA MEMBER at OLA high school who was preparing for the FFA Creed contest. Ned was excited about participating because he really believed he could excel in the Creed. He was one of the first of his classmates to memorize all five paragraphs. He was crowned the "speed creed" champion by his agriculture teacher. Not only could he say the Creed faster than anyone in his school, but he thought he might also be able to become the champ at delivering the FFA Creed in a serious and professional manner as well.

The problem, as Ned saw it, was the whole speaking-in-front-of-others thing. He knew the Creed and he could speak like a seasoned politician, but he always was so nervous that he got sick. Ned's adviser gave him a few tips and reassured him that everything would be just fine. Ned wrote down the tips his adviser gave him. He even put the tips for calming down in his jacket pocket and practiced every one of them at his first speaking (Creed) contest.

The first thing Ned's agriculture teacher (and his notes) said to do was be prepared before you get there. Well, Ned had done that, so he kept reading each time he looked at his notes. The second thing he was supposed to do was relax before speaking. He decided to chat with some friends who were also at the contest. This really did help.

The next thing he was advised to do was to realize that others understand the fear and nervousness that most speakers suffer. That was somewhat reassuring, but he was really helped by the advice to keep a positive attitude! Ned just kept telling himself, "I know every part of the Creed and I'm going to live." This mantra seemed to help immensely.

Nervous Ned placed second in the district that year. Everyone he knew was proud of him, and he was proud of himself. Ned especially liked the celebration dinner after the event: nothing about a burger and milkshake made him nervous.

Step 9 *Deal with your most difficult stage fright problems.*

You may find that some manifestations of stage fright are more problematic than others. "Successfully handling symptoms that you find particularly annoying will help build your confidence."[16]

Step 10 *Speak in public frequently.*

More than anything else, experience is what builds your confidence in your speaking abilities. Practice, practice, practice! The more often you speak publicly, the better your delivery will become.

ESTABLISHING RAPPORT WITH THE AUDIENCE

Rapport is a feeling of warmth, harmony, and alignment between speaker and listener. Before an audience can be persuaded or stimulated, it must have positive feelings for the speaker. Whatever else happens during the speech, you must make that rapport occur. Although rapport is important throughout the speech, it is most important at the beginning.

The following are ways to establish rapport:[17]

- Walk to the platform or **lectern** (speaker's stand or podium) with quiet assurance. This display of confidence puts the audience at ease by demonstrating that you know what you are doing and are in control of the situation.
- Stand quietly before beginning. A nervous speaker tends to begin immediately and talk too fast. Pause before you start, to give the audience a moment to quiet down and yourself a chance to size up the situation. A brief silence pulls attention to you and indicates to the audience that something important is about to take place.
- Look directly at the audience. Lack of eye contact gives the impression that you are insecure, emotionally detached, or unprepared. You may glance at your notes every now and then, but look right at the audience, especially at the beginning of your presentation.

- The audience reflects the speaker. If you seem pleasant and friendly, the audience generally responds warmly. If you genuinely care about people, you have a natural advantage. The right touch of friendliness, casual approach, and **spontaneity** (action occurring or arising without apparent premeditation) invites audience members to think of themselves as welcome and valued guests.
- Dress appropriately. We live in a casual age, but casualness should not deteriorate into sloppiness. As a company president said, people who feel themselves to be above good grooming are also above the patience and attention that the work requires. Appearance matters.

For an official FFA Creed contest, wear the FFA official dress. Female members are to wear a black skirt, white blouse with official FFA blue scarf, black shoes and hose, and official jacket zipped to the top. Official dress for male members is black trousers, white shirt, official FFA tie, black shoes, black socks, and official jacket zipped to the top.[18] The FFA members in Figure 7–6 are in official dress.

Do not forget the proper uses of the FFA jacket that relate to official dress at speaking contests:

- The jacket should be worn on official occasions with the zipper fastened to the top. The collar should be turned down and the cuffs buttoned.
- School letters and insignia of other organizations should not be attached to or worn on the jacket.
- No more than three medals should be worn on the jacket. These should represent the highest degree earned, the highest office held, and the highest award earned by the member.[19]

DELIVERY OF THE CREED

The delivery of the Creed in a contest is based on a 100-point scale. Of course, most of you are not learning and reciting the Creed for a contest, but presenting the material in this section from a contest perspective helps us focus on what top-quality speakers strive to master.

FIGURE 7–6 These FFA members are in official dress. *(Courtesy of Tennessee FFA.)*

Figure 7–7 shows a sample National FFA Creed Speaking Career Development Event Score Card.

Voice

The voice is the sound effect of personality. A good personality without an effective, pleasant voice can be compared to a good movie without sound. Just as we are judged by appearance, so are we judged by voice.

Quality Quality refers to tone of voice. Voice is often considered the mirror of the mind. A friendly tone of voice makes conversation a pleasure. Tone refers to the particular or relative pitch of a word, phrase, or sentence and the modulations of the voice. No two people have the same tone, and there can be as much as four octaves' difference between male and female voices.

Tone of voice greatly affects the importance a listener attaches to what is said. In fact, tone of voice can cause the listener to pay very little attention to an important statement. The speaker projects mental and emotional attitudes to listeners by the tone of voice. Good posture is important to the tone of voice because voice tone and control depend on a constant flow of air from the lungs through the throat and out the mouth. Obviously, the tone of voice should be varied during a speech or conversation.

Pitch Pitch is the height or depth of the tone. Most people speak with a medium or low pitch. Pitch is a determining factor in voice quality. Listeners become uncomfortable when the pitch is extremely low or high. Variations or changes in pitch are known as **inflections**. Inflections give your voice a quality that people are drawn to. The inflection of your voice can indicate your emotions or whether you are making a comment or asking a question.[20]

Some people fall into a monotone without knowing they are doing it. You can become aware of this problem by recording your speeches as you practice them. If you notice a pattern in your voice inflection, simply practice different pitches for different words. Try attaching a pitch that matches a certain word.[21] For example, in the FFA Creed you could use a deep, somber tone as you speak about "struggles" associated with being in agriculture.

NATIONAL FFA
CAREER AND LEADERSHIP
DEVELOPMENT EVENTS

Creed Speaking LDE Presentation Rubric

100 points

Participant # _____

INDICATORS	Very strong evidence of skill is present 5–4 points	Moderate evidence of skill is present 3–2 points	Strong evidence of skill is not present 1–0 points	Points Earned	Weight	Total Points
Oral Communication – 30 points						
Pace	Speaks very articulately at rate that engages audience.	Speaks articulately but occasionally speaks too fast or has long unnecessary hesitations.	Speaks too slow or too fast to engage audience.		X 2	
Tone	Voice is upbeat, impassioned and under control.	Voice is somewhat upbeat, impassioned and under control.	Voice is not upbeat; lacks passion and control.		X 2	
Volume	Emitted a clear, audible voice for the audience present.	Emitted a somewhat clear, audible voice for the audience present.	Emitted a barely audible voice for the audience present.		X 2	
Non-verbal Communication – 30 points						
Eye contact	Eye contact constantly used as an effective connection. Constantly looks at the entire audience (90-100 percent of the time).	Eye contact is mostly effective and consistent. Mostly looks around the audience (60-80 percent of the time).	Eye contact does not always allow connection with the speaker. Occasionally looks at someone or some groups (less than 50 percent of the time).		X 2	
Mannerisms and gestures	Hand motions are expressive and used to emphasize talking points. No nervous habits.	Sometimes exhibits nervous habits. Hands are sometimes used to express or emphasize.	Displays some nervous Habits. Hands are not used to emphasize talking points; hand motions are sometimes distracting.		X 2	
Poise	Portrays confidence and composure through appropriate body language (stance, posture, facial expressions).	Maintains control most of the time; rarely loses composure.	Lacks confidence and composure.		X 2	

FIGURE 7–7A AND 7-7B The National FFA Creed Speaking Career Development Event Score Card. The score card includes five areas; except for accuracy, the other parts are basically the same as for the public speaking contest. *(Courtesy of the National FFA Organization.)*

NATIONAL FFA CAREER AND LEADERSHIP DEVELOPMENT EVENTS HANDBOOK 2017–2021 9

Creed Speaking

Creed Speaking LDE Presentation Rubric continued

INDICATORS	Very strong evidence of skill is present 5–4 points	Moderate evidence of skill is present 3–2 points	Strong evidence of skill is not present 1–0 points	Points Earned	Weight	Total Points
Question and Answer—40 points						
Response to questions	Is able to respond with organized thoughts and concise answers.	Is able to speak effectively and sometimes gets off topic. Answer lacks organization.	Response fails to answer question.		X 2	
Support	Always provides details which support answers/ basis of the question.	Usually provides details which are supportive of the answers/basis of the question.	Sometimes overlooks details that could be very beneficial to the answers/basis of the question.		X 3	
Knowledge of agriculture	Answer shows knowledge of agriculture.	Answer shows limited knowledge of agriculture.	Answer shows no knowledge of agriculture.		X 3	

Gross Total Points	
Time Deduction*	
Accuracy Deduction**	
NET TOTAL POINTS	
RANK	

* -1 point per second over, determined by the timekeepers

** - 20 points per word, determined from by the accuracy judges.

FIGURE 7–7A AND 7-7B (Continued)

Articulation Articulation refers to the way the tongue, teeth, palate, and lips are moved and used to produce the crisp, clear sounds of good speech. Although most of us are physically capable of producing vowel and consonant sounds clearly, we get lazy and do not put forth the necessary effort, especially in casual conversation. When a speaker allows these bad habits to continue during public speaking, the audience will probably react badly, becoming impatient with and disrespectful of the sloppy speaker.[22]

Many speakers have minor articulation problems, such as adding a sound to a word (ath-*a*-lete for athlete), leaving out a sound (li-*b*ary for li*br*ary), transposing sounds (re*va*-lent for rel*e*vant), and distorting sounds (tru*f* for tru*th*). Although some people have consistent articulation problems that require speech therapy (such as consistently substituting *th* for *s*), most of us are simply guilty of carelessness that is easily corrected. When practicing the Creed, therefore, "concentrate on moving your tongue, lips, and lower jaw vigorously enough to produce crisp, clear sounds."[23] Be especially careful of consonants that can be slurred or dragged. Consider the following examples.

Word(s)	Misarticulation
going to	*Gonna*
did you	*Didja*
specific	*Pacific*
ought to	*Otta*
will you	*Wilya*

Practice the activities at the end of the chapter. They are the ultimate articulation testers.

Pronunciation Pronunciation is the form and accent a speaker gives to the syllables of a word. What makes a certain pronunciation correct and another incorrect is usage. Once enough people agree to pronounce a word in a certain way, that pronunciation becomes the norm. If you are unsure of how to pronounce a word you are going to use, look it up in a dictionary. Along with word choice and articulation, pronunciation plays a large part in determining the degree of respect an audience gives to a speaker.[24]

Complicating matters, but making life more interesting, is the fact that different people speak in different ways based on region and ethnicity. For example, almost 10 percent of Americans say the word *aunt* with an "ah" sound, rhyming with *haunt*, while 75 percent say it the same as *ant*, the insect.[25].

Because constant mispronunciation suggests that a person is ignorant or careless (or both), you will want to try to correct any mistakes you make. Each language/dialect culture has a unique set of words that are mispronounced. Consider the following mispronounced words familiar to the authors of this text.

Word	Common Error	Correct Pronunciation
genuine	*gen'-you-wine*	*gen'-u-win*
theater	*thee-ate'-er*	*thee'-uh-ter*
athlete	*ath'-a-lete*	*ath'-lete*
family	*fam'-ly*	*fam'-ah-ly*
particular	*per-tik'-ler*	*par-tik'-you-ler*

Force Force refers, in speaking, to volume and variations in volume for emphasis. A speaker who gives everything equal force at first makes everything sound important, but soon the uniformity of volume high-lights nothing—even the critical points are disguised because they are given the same emphasis as everything else. The other side of that coin is the speaker who always talks so softly that he or she is difficult to hear at all. Besides being particularly thoughtless of those with hearing impairments, this habit soon annoys almost all listeners, who must strain to make out the speaker's words. Most audience members will quickly decide that it is not worth the effort.

Important ideas should be delivered forcefully enough so that listeners in the farthest reaches of the room can hear them easily. "Transitions to new sections of a speech, the start of the conclusion, and the parts where you wish to be dramatic with a kind of 'stage whisper' may be spoken more softly, but should still be audible to listeners in the rear of the room."[26]

Stage Presence Stage presence includes personal appearance, poise and body posture, attitude, confidence, personality, and ease before an audience.

Personal Appearance Personal appearance is a combination of grooming and dress. For the Creed recitation, wear official FFA dress, as discussed earlier. Good grooming is essential to a professional image. Styles vary from year to year, of course, but extremes of style and the appearance of rebellion against the norm will not make a good first impression on your audiences. Make sure your grooming is acceptable to your adviser and the majority of the chapter members.

Personal appearance plays an important role in speech making, simply because listeners always see you before they hear you. Just as you adapt your language to the audience and the occasion, so should you dress and groom appropriately. When you are well dressed and well groomed, you show respect for your audience and indicate that you take your subject seriously. No matter what the speaking situation, try to win a favorable first impression that will make listeners more receptive to what you say. The man in Figure 7–8 would make a

FIGURE 7–8 This man makes a favorable impression because of his clean, professional appearance. *(© iStock/Rich Legg)*

FIGURE 7–9 This leader's appearance and body language do not make a positive impression. *(© iStock/shironosov)*

favorable impression as he delivers a speech; the speaker in Figure 7–9 would not.

Poise and Body Posture Poise and body posture show self-confidence. To picture poise, think of an NFL quarterback standing in the pocket, calmly getting ready to deliver his pass on target as six 300-pound linemen are charging toward him. A speaker should exhibit the same self-confident poise as he or she stands to deliver a speech.

What should you do with your arms and hands? There is no simple answer to this. However, hand and arm positions, as well as any gestures you choose to make, must be both natural to you and suited to the audience and the total speaking situation. Consider the following possible positions:

One or b oth arms hanging naturally at your side
One or both hands resting on (not grasping) the speaker's stand
One or both hands held in front of your abdomen
Fingertips touching with hands held in front of your stomach

These positions look quite natural for some speakers but not for others. Use the reactions of your teacher and classmates to decide which work best for you.

Attitude The speaker's attitude greatly affects the audience members' attitude. If you exhibit a positive attitude, your audience will react positively to you. Act as though there is nowhere you would rather be and nothing else you would rather be doing right at this moment. Being positive is contagious.

Confidence Confidence is attained through knowledge, preparation, and practice. You nonverbally demonstrate confidence by good eye contact, a relaxed appearance, good body posture, and an alert, engaged facial expression that draws attention and commands respect.

Personality Personality involves being enthusiastic, lively, expressive, cheerful, inspiring, charming, and energetic. When you present your speech, really care about what you have to say: enthusiasm is by far the most important element of effective speaking. People will listen to a speaker who looks and sounds enthusiastic, and they will remember that speaker's ideas.

Ease before Audience Ease before the audience assures your listeners that you are calm and everything is under control. Being at ease means staying calm and confident. It means standing up straight and making sure you don't say your speech too fast. It means making sure you have everyone's attention before you start speaking—and then keeping it. Always remember, most people are amazed by people who can speak in front of others. You're a hero. Act like it.

Power of Expression

Power of expression includes fluency, emphasis, directness, sincerity, communicative ability, and the conveyance of thought and meaning.

Fluency Fluency is the smooth and easy flow of the speech. Do not proceed too fast or too slow. End a sentence, pause, and then start the next sentence. Although variety in loudness and speech is needed for emphasis, a smooth, easy-flowing speech still conveys meaning well.

Emphasis Earlier, we discussed key words in the FFA Creed. These key words should receive emphasis. Say words in a way that expresses their meaning and the feeling you want them to convey.

Directness Directness means using language that adapts to the needs, interests, knowledge, and attitudes of the audience. In most situations, the more direct you can make your language, the more appropriate it will be. Although you cannot do this when delivering the Creed, to increase directness in answering the questions, you can respond in a way that stimulates mental activity, shares common experiences, and creates stories to illustrate a point (hypothetically).

Sincerity Sincerity is being genuine, honest, and straightforward as you present your speech. Your speech or delivery of the Creed must be genuine to make a lasting positive impression. You must believe what you say. Politicians are often doubted because the audience cannot figure out whether they are sincere or only trying to get elected. The slightest hint of insincerity turns many listeners against a speaker instantly.

Communicative Ability Communicative ability refers to your skill in conveying your thoughts and feelings to the audience so that they can be understood. Eye contact lets us communicate nonverbally, conveying truthfulness, intelligence, attitude, and feelings.

You should look at the audience 80 to 90 percent of the time you are talking.

Conveyance of Thought and Meaning Conveyance of thought and meaning is a transfer of your feelings and perceptions to the audience. Perception is reality. When you speak, it is not what you think, it is what you convey that the audience judges.

Thoughts and meaning are transmitted largely by facial expression. Your facial muscles can be arranged in an almost infinite variety of positions to express a wide range of emotions. They can show the degree of interest you have in your speech and give certain information about your personality.

Audiences respond positively to honest and sincere expressions that reflect your thoughts and feelings. Think actively about what you are saying, and your face probably will respond accordingly. If not, practice in front of a mirror to improve your facial expressions.

General Effect

General effect includes the extent to which the speech is interesting, understandable, convincing, pleasing, and holds audience attention. If you combine the preceding delivery characteristics to suit your preferences and your material, your speech will be interesting, understandable, convincing, pleasing, and hold attention. It will also be specific, logical, organized, motivating, credible, and persuasive.

There is also an **intangible** trait (one that cannot be appraised for value) that occasionally contributes to general effect. It may be a quality that only you have that can be used to your advantage. If it is good or it works, use it.

Response to Questions

Studying and preparing to answer questions are an important part of your preparations for the FFA Creed Career Development Event (CDE),. There are many resources on the Internet hosted by different chapters, associations, and the National FFA Organization itself that have sample Creed questions you can use to prepare. Refer to Appendix J in this text for an example of FFA Creed questions.

In answering questions, be sure to stay fully detail-oriented. Always provide support for your answers, and speak with comfort and ease. You should speak as promptly as possible after the question is asked, and reply with organized thoughts and concise answers. Your examples should be vivid, precise, and clearly explained. Also, be sure your examples are logical and relevant. Perhaps most importantly, remember to speak with self-confidence and a positive facial expression.

PRACTICAL STEPS AND FUN TECHNIQUES FOR PERFECTING THE FFA CREED

A proper attitude toward speaking can be developed if you make it fun. Following are some fun techniques you might want to try. They may help you develop your confidence in reciting.

Rolled-Up Pants Legs

When the senior author starts teaching the Creed, he rolls up his pants legs nearly to his knees. This amuses students and makes them relax. It lets them know that they are going to have fun while learning. It helps to lessen stage fright because students know they cannot look more foolish than their teacher when they get in front of the class!

Why rolled-up pants legs? To fully understand and appreciate this trick, you have to have been around a barnyard during a spring thaw. For those who have not, the mud (and other material) gets very deep. It also gets very deep when we start speaking if we pour our emotions into the speech. We have to roll up our pants, wade in, and have fun.

Eggshell Analogy

Students tend to be bound up in an invisible shell when they start speaking: They are tight, nervous, and restricted. Visualize breaking out of a shell as if you were a newborn chick. Move your arms freely and stretch. When your teacher permits, take turns saying "I believe" in a fun and dramatic fashion.

Toothpaste Tube

Have you ever thought that you were out of toothpaste and set the empty tube aside? You forget to get a new tube, so you have to try to squeeze some more out of the old tube. To your surprise, you get more toothpaste to come out. The next day, you again forget to get a new tube. You go back again to the old tube and you squeeze a little harder. Again, you get more toothpaste. This can go on for a week. Finally, you tear open the tube and scrape the toothbrush on the inside of the tube to get that last bit . . . and don't forget the top.

Your intelligence and speaking ability are like the toothpaste. There is more within you. Just when you have done the best you think you can do, you can do better. Every one of you can learn to speak better than you'd ever thought possible if you keep squeezing. Just when you think it is all gone, there is still more "toothpaste"—speaking ability—within you.

Football Field/Gymnasium Practice

Some of us speak softly and shyly when we first start speaking. To cure this, go to the football field or gym. Stand beneath one football goalpost or basketball net while the teacher and the rest of the class stand beneath the other. Say one or more paragraphs of the Creed so they can clearly hear you.

Public speaking is not a shouting match. However, once you return to the classroom, you will be amazed at how much easier it is to speak louder. After this exercise, it is easy to tone down the volume a little and clean up the rough edges.

Model Speaker

Sometimes learning is best done by following a model—in this case, a model speaker. Maybe your teacher could show you videotapes of some excellent Creed speakers. The National FFA Supply Service sells videos of the National Public Speaking Contest. Learn from the best.

Face-to-Face Imitation

Another good technique for learning how to deliver the Creed is to stand opposite your teacher or someone else, with a desk or table between you. Put your fingers on the table, relax, and look the other person straight in the eye. Your teacher or a more advanced speaker says a line (sentence) using good voice, stage presence, power of expression, and general effect. You immediately follow, imitating him or her. Do not move to the next line until you have sufficiently practiced the first line.

The idea is not to be a clone (an exact copy) of your teacher, but to develop a speaking style and see what it feels like. Disc jockeys, ringmasters, play-by-play sports announcers, and newscasters each have their own styles. Creed speakers and FFA public speakers also have a certain style.

Note: The next steps are for those who are trying to win the "big one." However, you may wish to practice just for fun.

Hayloft

Most of you do not have a hayloft (Figure 7–10), but you can create your own "hayloft" out of a park, field, warehouse, or whatever large unpopulated space is available. Get in the hayloft and have fun speaking. Use exaggerated gestures, body positions, voice variations, and facial expressions. After being silly, get serious and pretend that you are speaking to 22,000 people at the National FFA Convention.

Balcony

Voice projection practice is best done from a high place, such as a balcony or a second-story window. It gives you a sense of power and command of the situation. Practice the Creed several times.

FIGURE 7–10 The hayloft is not only good for storing hay. You can also use it to practice the Creed. *(Courtesy of Country Side Studios, Gallatin, Tennessee.)*

It is both fun and challenging, but remember safety—do not get so carried away that you pitch yourself out the window! The only drawback is the funny looks you may get when cars pass by.

Turkey Test

The authors used to have turkeys that offered an evaluation of voice pitch, tone, and variation. (Really!) The male turkeys (gobblers) would gobble when they heard a loud or high-pitched noise. To pass the turkey test, the Creed candidate would have to emphasize certain words loudly enough to make the turkeys gobble. The louder and more expressive the candidate, the more the turkeys would strut and gobble. There are modern electronic devices that produce a sound or move a gauge to measure voice force, but they are not as much fun as turkeys.

It is the same with entertainers. The better they perform, the better the response or applause. If you are really good, the audience will give you a standing ovation.

NATIONAL FFA CREED SPEAKING CAREER DEVELOPMENT EVENT

The following are the rules for the National FFA Creed Speaking Career Development Event (CDE). Most states follow the same rules as the national contest.

Event Rules

- The National FFA Creed Speaking CDE is limited to one participant per state, who must qualify in grades 7, 8, or 9 and must compete at the next national convention following the state qualifying round.
- It is highly recommended that participants be in official FFA dress at each event.
- The National FFA Creed Speaking CDE follows the general rules and policies for all National FFA CDEs.
- The National FFA Officers and National Board of Directors are in charge of the event.
- Three to six competent and impartial persons are selected to judge the event. At least one judge should have an agricultural/FFA background. Each state with a speaker provides a judge for the national event.

Event Format

- The event includes oral presentation of the Creed, as well as answering questions directly related to the FFA Creed. Each contestant is asked three questions per round, with a five-minute limit. The questions posed change as the contestant progresses to semifinal and final rounds of competition. The questions are formulated annually by the Creed Speaking CDE committee. Sample questions are not available before the event, but the committee never asks two-part questions.
- Members present the FFA Creed as published in the current year's official FFA manual.
- The event is a timed activity, with four minutes for presentation. After four minutes, one point is deducted for every second over the set time.
- The national event is conducted in three rounds: preliminary (consisting of five to eight speakers per section), semifinals (two sections of eight speakers each), and finals (four participants). No ranking is given except for the final four.
- Event officials randomly determine the speaking order. The program chairperson introduces each participant by contestant number and in order of the drawing. No props are to be used. Applause is withheld until all participants have spoken.
- Each contestant must recite the FFA Creed from memory. Each contestant begins the presentation by stating, "The FFA Creed by E. M. Tiffany." Each contestant ends the presentation with the statement ". . . that inspiring task. Thank you."
- Contestants are held in isolation until their presentation. Contestants are not allowed to have contact with any outside persons.
- At the time of the event, the judges sit in different sections of the room in which the event is held. They score each participant on his or her delivery of the Creed, using the official score sheet.
- Timekeepers record the time each participant takes to deliver his or her speech. Timekeepers are seated together.
- The content accuracy judges record the number of recitation errors during delivery. The accuracy judges are seated together.
- When all participants have finished speaking, each judge totals the score for each speaker. The timekeepers' and accuracy judges' records are used in computing the final score for each participant. The judges' score sheets are then submitted to event officials to determine participants' final ratings.
- Participants are ranked in numerical order on the basis of the final score, to be determined by each judge without consultation. The judges' rankings of each participant are then added, and the winner is the participant whose total score is the lowest.

Tiebreakers

Ties are broken based on the greatest number of low ranks. The participant with the greatest number of low ranks is declared the winner. If a tie still exists, the event superintendent ranks the participants' response to questions. The participant with the greatest number of low ranks from the response to questions is declared the winner. If a tie still exists, the participants' raw scores are totaled. The participant with the greatest total of raw points is declared the winner (refer to Figure 7–8).[27]

CONCLUSION

Learning and reciting the Creed can be fun. Reciting the Creed is one of the first things done in an agricultural education program in the FFA in pursuit of leadership skills. It is hard to lead if you cannot communicate well, and communication includes speaking before groups. Principles learned for presentation of the Creed also apply to public speaking, extemporaneous speaking, and parliamentary procedure.

SUMMARY

Basic principles of speaking include using reason and feeling, using rational and creative thinking, using an effective speaking style, observing the audience, and persuading the audience. Every speaker should master delivery by using the correct volume, tempo, diction, and enthusiasm.

Polish your skills by using an audio recording device, video recorder, practicing in front of a mirror, and practicing in front of friends. Practice helps in overcoming stage fright and nervousness. You can also practice establishing rapport with your audience. As you rehearse your delivery, work on your voice, stage presence, power of expression, and general effect.

To control your nervousness, prepare thoroughly, relax before you speak, remember that audiences tend to be sympathetic, develop the right attitude, concentrate on the Creed and the audience, wear a self-confident smile, and move naturally. Speak as often as you can; experience also quells stage fright and builds confidence.

Establish rapport with the audience by walking to the platform or lectern with quiet assurance, standing quietly before beginning, looking directly at the audience, appearing friendly and pleasant, and dressing appropriately.

Remember that voice qualities include tone, pitch, articulation, pronunciation, and force. Stage presence includes personal appearance, poise and body posture, attitude, confidence, personality, and ease before an audience. Power of expression includes fluency, emphasis, directness, sincerity, communicative ability, and the conveyance of thought and meaning. General effect includes the extent to which the speech is interesting, understandable, convincing, pleasing, and holds attention.

Some practical and fun ways for saying the Creed include rolling up your pants legs; practicing on a football field, in a gym, or in a hayloft; watching a model speaker; face-to-face imitation; and passing the "turkey test."

 Take It to the Net

Go to the National FFA website: http://www.ffa.org. Browse the website and all the things going on in the FFA. Try to find the Spanish version of the FFA Creed. Practicing the Creed in another language might also be a fun way to practice volume and tone.

Chapter Exercises

REVIEW QUESTIONS

1. Define the Terms to Know.
2. What are the advantages of participating in recitation of the FFA Creed?
3. What are five basic principles of speaking?
4. Do you agree with the old saying, "Practice makes perfect"? Why or why not?
5. What did studies show is the top predictor of professional success and upward mobility (job promotions)?
6. Explain the statement, "Train the butterflies to fly in formation."
7. List 10 ways to help control nervousness.
8. What are five ways to establish rapport with the audience?
9. What is the official FFA dress for males? Females?
10. What is meant by *general effect?*
11. What are 10 practical and fun things you can do to make saying the Creed more effective and enjoyable?

COMPLETION

1. By participating, you cannot _____. You can only _____ when you quit or do not even try.

2. The number-one fear of most adults (even above death) is _____.

3. No more than _____ medals should be worn on the FFA official jacket.

4. _____ is a combination of grooming and dress.

5. _____ is attained by knowledge and preparation.

6. _____ is the smooth and easy flow of speech.

7. _____ means using language that adapts to the needs, interest, knowledge, and attitudes of the audience.

8. _____ is being genuine, real, honest, and straightforward.

MATCHING

Not all answer choices will be used

_____	1. Refers to the tone of voice.	A. force
_____	2. Height or depth of voice.	B. pitch
_____	3. Shaping of speech sounds into recognizable oral symbols.	C. pronunciation
_____	4. Refers to volume for the purpose of emphasis.	D. quality
_____	5. Self-confident and calm.	E. poise
_____	6. Enthusiastic, lively, expressive, cheerful, enthusiastic.	F. personality
_____	7. Form and accent a speaker gives to the syllables of a word.	G. articulation
_____	8. Conveying your thoughts and feelings to the audience in an understandable fashion.	H. personal appearance
_____	9. Combination of grooming and dress.	I. poise and body posture
_____	10. Transfer of your feelings and perceptions to the audience.	J. attitude of the speaker
_____	11. Can affect audience members' attitude.	K. confidence
_____	12. Attained by knowledge and preparation.	L. ease before audience
_____	13. Being genuine, honest, and straightforward as you present your speech.	M. fluency
_____	14. Assures the audience that you are calm and everything is under control.	N. emphasis
_____	15. Say words to express the feelings they convey.	O. directness
_____	16. Smooth and easy flow of speech.	P. sincerity
_____	17. Using language that adapts to the needs, interests, knowledge, and attitudes of the audience.	Q. communicative ability
		R. conveyance of thought and meaning

ACTIVITIES

1. Write/type the first paragraph of the FFA Creed. Cut out two or three pictures that illustrate or reflect the meaning of this paragraph, and tape or paste them beneath the paragraph. Write/type and illustrate the other four paragraphs following the same procedure.

2. Rewrite Figure 7–2 in your notebook. Fill in the key words.

3. Write out paragraphs two through five of the Creed and underline the key words.

4. Practice saying the Creed in front of a mirror. Write a short essay about what you felt and learned as you practiced.

5. Demonstrate three relaxation techniques to help reduce physical symptoms of stage fright.

6. Stand behind a lectern. Just before delivery of the Creed, talk to the audience in a loud and dramatic fashion. Say:

   What I am feeling are symptoms of stage fright. It is great that my body prepares me for events this way. This tense feeling is just what I need to sharpen my thinking and help me reach my full potential once I start speaking.

7. Practice articulating the following phrases:

   The big black bug bit the big black bear, and the big black bear bled.

   Peter Piper picked a peck of pickled peppers. A peck of peppers, Peter Piper picked.

   I would if I could, couldn't, how could I; you couldn't without your would, could you? I couldn't, could you?

   Sid said to tell him that Benny hid the penny many years ago.

   Fetch me the finest French-fried freshest fish that Finney fries.

   Three gray geese in the green grazing; gray were the geese and green was the grazing.

   Shy Sarah saw six Swiss wristwatches.

   One year we had a Christmas brunch with Merry Christmas mush to munch. But I don't think you'd care for such. We didn't like to munch mush much.

8. Compile a list of positive delivery characteristics exhibited by other members of the class.

9. Practice the Creed on the football field or in the gym. Half the class can stand beneath one goalpost (basketball net) and the other half of the class can stand beneath the other goalpost (net). There are several ways to practice. To add variety, you may wish to alternate paragraphs between the groups under each goalpost (net).

10. Practice saying the Creed using any of the other practical and fun techniques described in this chapter. Write a report on your experiences.

11. Recite at least one paragraph of the FFA Creed in front of the class while wearing official dress. Use as many of the techniques learned in this chapter as you can.

12. List 10 questions that you might be asked in an FFA Creed CDE.

NOTES

1. The Creed was written by E. M. Tiffany, and adopted at the 3rd National Convention of the FFA, 1990. It was revised at the 38th Convention and the 63rd Convention.
2. D. B. Jameson, *Leadership Handbook* (New Castle, PA: LEAD, 1978).
3. Ibid., p. 210.
4. Ibid., p. 211.
5. Adapted from Dr. Joe Townsend's "Leadership" class (Texas A&M University, College Station, TX).
6. Jameson, *Leadership Handbook*, p. 53.
7. T. Alessandra and P. Hunsaker, *Communicating at Work* (New York: Simon & Schuster, 1993), p. 169.
8. P. Bruskin, "What Are Americans Afraid Of?" *The Bruskin Report: A Market Research Newsletter*, July 1973, p. 73.
9. Alessandra and Hunsaker, *Communicating at Work*, p. 170.
10. Ibid., pp. 170–171.
11. J. R. O' Connor, *Speech: Exploring Communication,* 4th ed. (Lincolnwood, IL: National Textbook Co., 1996), pp. 164–172.
12. Ibid., p. 165.
13. Ibid., pp. 165–166.
14. Ibid., p. 168.
15. Ibid., p. 170.
16. Ibid., p. 171.
17. "Applying Prepared Speaking Skills," *Agricultural Communications* series, No. 8373-F (College Station, TX: Instructional Materials Service, Texas A&M University, 1989).
18. *Official FFA Manual 2009–10* (Alexandria, VA: National FFA Supply Service, 2009), p. 10.
19. Ibid.
20. S. E. Lucas, *The Art of Public Speaking,* 12th ed. (New York: McGraw Hill Education, 2015).
21. Ibid.
22. O'Connor, *Speech: Exploring Communication*, p. 270.
23. Ibid., p. 270.
24. Ibid., pp. 271–272.
25. B. Vaux and S. Golder, *The Dialect Survey,* Retrieved July 24, 2016 from http://www4.uwm.edu/FLL/linguistics/dialect/maps.html
26. O'Connor, *Speech: Exploring Communication*, p. 268.
27. *Career Development Events Handbook, 2006–2010* (Indianapolis, IN: National FFA Organization, 2006), pp. 99–102.

8

PREPARED SPEAKING (FFA PUBLIC SPEAKING)

Speaking skills increase a person's effectiveness in influencing the decisions of others. Certainly, an effective communicator provides people with information, but such a leader should also be able to influence others, changing their attitudes, opinions, or behaviors. Speaking in public is an art form that can be quite beautiful if performed effectively.

Objectives

After completing this chapter, the student should be able to:

- Plan a speech
- Analyze the audience
- Select a topic for a speech
- Gather information for a speech
- Record the ideas to be included in a speech
- Prepare an outline for a speech
- Write a speech
- Practice the speech
- Present the speech
- Answer questions about the subject matter of the speech
- Evaluate speeches
- Speak to special groups

Terms to Know

- empathize
- quotations
- statistics
- introduction
- body
- simile
- metaphor
- personification
- hyperbole
- irony
- ethical
- defamation of character
- pangs
- ardently
- salutation
- deliberate

Speeches are given to inform, to persuade, or to integrate the audience members into a group (as in pep talks, welcome speeches, and introductions). People also listen for the same reasons. They want to be informed, persuaded, and included.

Through persistence, you can achieve your speaking goal. The student in Figure 8–1 is practicing speaking before her friends. They will make suggestions, and she will practice again.

Speaking skills increase a person's effectiveness as a leader and assist in influencing the decisions of others. An effective communicator should provide people with information, but he or she should also be able to influence others, changing their attitudes, opinions, or behaviors. Speaking in public is an art form, and it can be a beautiful thing if done well. Speaking skills are a major factor in selection of the president of the United States, chairpersons of boards, and other leaders. More often than not, a speech gives a clear statement of the speaker's beliefs and actions.[1] In short, effective public speaking gives us the ability to influence others—the essence of leadership.

PLANNING A SPEECH

As you plan a speech, consider the purpose of your speech, the audience to which you will deliver it, and the occasion that calls for it. Think like the audience. If you can **empathize** (to feel or understand as someone else does, as if you were that person) with your audience, you will be able to craft a better speech. The speaker in Figure 8–2 is good at empathizing with audiences.

Analyze the Audience

Do you select the topic first, or do you analyze the audience first? It depends on the situation. Most often, however, you should analyze the audience before you select your topic. Can you imagine the featured speaker at a Young Republicans banquet speaking on the virtues of being a Democrat, or vice versa? Once you have figured out who your audience will be, you can then select an appropriate, interesting topic.

For an FFA Public Speaking Contest, the audience is rather obvious and predictable: it consists

FIGURE 8–1 Practice does not necessarily make perfect—perfect practice makes perfect. Practicing in front of friends and getting their evaluations aids in the pursuit of perfection. *(© Amir Ridhwan/Shutterstock.com)*

FIGURE 8–2 Barack Obama shows how to get in touch with the audience as he delivers a speech. *(Courtesy of Andy Colwell, Department of Public Information Media Gallery, Pennsylvania State University.)*

| VIGNETTE 8-1 | SPEECH WRITING IS A SKILL THAT LASTS A LIFETIME |

JENNIFER DEDMAN IS AN ACCOMPLISHED MEMBER OF THE AGRIBUSINESS COMMUNITY

She built an incredible agritourism business in her home state of Tennessee. Young people come from schools all around to see her business, learn from her, and discover why she has been so successful in a niche area of agriculture. Jennifer is very humble, and she believes that much of her success can be attributed to her experiences with prepared public speaking. Whenever Jennifer talks to anyone, she always advises them to participate in public speaking events whenever the opportunity arises.

She usually explains to people that public speaking is the key to her success in agribusiness. Jennifer is certain that public speaking is the reason she was able to overcome her shyness. Public speaking is the reason she is able to market and advertise her business to others. She also maintains that her experience in writing speeches, as part of the public speaking contests she participated in, continues to help her organize her thoughts, think critically about business decisions, and feel confident in the research she has conducted. Certainly, other experiences and skills have contributed to Jennifer's success as well, but she firmly believes that public speaking is a wonderful avenue for developing life skills.

---

of agricultural education students, agricultural education teachers, and those interested in the fields of production agriculture, agribusiness, or agriscience. The rules of the FFA Prepared Public Speaking Contest state, "Participants may choose any current subject for their speeches, which is of an agricultural nature." Specifically, "[o]fficial judges of any FFA Prepared Public Speaking Career Development Event shall disqualify a participant if he or she speaks on a nonagricultural subject."[2] As you can see, both the audience and the general topic area are predetermined for FFA Public Speaking Contests.

Analyze the audience by finding out as much information as possible concerning who will be in it. In some instances, this may be impossible until minutes before the actual presentation. It may be helpful to know the number of people in the group, their ages, their interests, their knowledge about your topic, whether it is a formal or informal group, their occupations, their educational levels, sensitive subjects, the setting, the time, your place on the program, room size, seating arrangements, where speakers stand, microphone availability, and special occasions. For example, if you are speaking at a banquet to a group of agricultural education teachers, you might want to allude to how their jobs have an effect on the food that was consumed during the meal.

Be Audience Centered

"Good speakers are audience centered." They know that the primary purpose of speaking is not to display their knowledge, demonstrate their superiority, or vent their anger. The primary purpose should be to gain a desired response or to please the listener. You need not compromise your beliefs to get a favorable response from your audience. You can remain honest and true to yourself while adapting your speech and delivery to the needs of a particular audience. When analyzing your audience, keep three questions in mind:

- What are the folks like to whom I am speaking?
- What knowledge, beliefs, and behavioral changes do I want the audience to consider as a result of my speech?
- How can I best prepare my speech to help the audience experience those changes?[3]

The answers to these questions influence every decision you make along the way: selecting a topic, determining a specific purpose, settling on the main body of your speech and supporting materials, organizing the speech, and, finally, delivering the speech.

SELECT A TOPIC

First, choose a topic that interests you. Next, choose a topic about which you are knowledgeable or have an interest in becoming knowledgeable. Third, pick a topic of interest to your audience. If possible, choose an area in which you are experienced, have convictions, and are accomplished. If you feel strongly about your topic, this feeling will be conveyed in your delivery. If you have knowledge and experience, you will do a better job in responding to questions.

When searching for a topic for an FFA speech, you may wish to start with five general agricultural areas: agriscience and technology, agribusiness, agrimarketing, international agricultural relations, and agricultural communications and education. List topics that interest you in one area. Quickly jot down every word or phrase you know relating to that area. Spend no more than two minutes on each area. Then, from the list you have compiled, select the most interesting topic.

An example of brainstorming is as follows. While brainstorming, one student wrote down *tractor*. Tractors made her think of John Deere. John Deere reminded her of mowing her lawn. Her lawn-mowing thought led to thoughts of making money through her own lawn-mowing business. This thought caused her to investigate business models for the lawn care industry. After considerable research, she developed an excellent speech entitled "Lawn Care Business Models for Rookie Entrepreneurs."

That is a long way from tractors! If you started out free-associating from the word *tractor*, you would probably end up somewhere completely different. This is what brainstorming is all about. The students in Figure 8–3 are brainstorming and associating words to come up with speech topics.

Figure 8–4 lists ideas for speech topics in the five major areas: agriscience and technology, agribusiness, agrimarketing, international agricultural relations, and agricultural communications

FIGURE 8–3 Selecting a speech topic can be a challenge for some students. These students are brainstorming and doing word association to come up with a topic. *(© iStock/sturti)*

AGRISCIENCE AND TECHNOLOGY	AGRIBUSINESS	AGRIMARKETING	INTERNATIONAL AGRICULTURE RELATIONS	AGRICULTURAL COMMUNICATIONS AND EDUCATION
Genetically modified organisms (GMOs)	What is agribusiness?	Functional foods	Global food security	Positive or negative media for the agriculture industry
Cloning	How big is agribusiness?	Web-based agrimarketing	Poverty reduction	Consumer decision making about agricultural products
Organic agriculture	Impacts of vertical integration	Market alternatives for the farmer	Value of exporting agricultural products	Dangers of agricultural ignorance
Biofuels or alternative energy	Agricultural finance	Niche markets	Commodity and trading standards	Technologies bringing agricultural information to those in need
Sustainable and natural resource development	Forecasting in agribusiness	Country of Origin Labeling (COOL)	African Growth and Opportunity Act (AGOA)	Agricultural literacy of elementary youth
Food microbiology	Strategic management and leadership	Role of the government in agricultural product marketing	Agricultural trade policy	Solving the agriscience teacher shortage

FIGURE 8–4 Selecting a topic for an FFA speech can be challenging. The topics listed in this chart may help you think of other ideas for a speech.

and education. Whatever topic you choose, nearly every speech will require some additional research before you are ready to write and deliver it. Give yourself plenty of time in advance of your speaking date to do this additional preparation.

GATHER INFORMATION

Benjamin Franklin once said, "An empty bag cannot stand upright." Gathering information is of the utmost importance. Without solid material (i.e., facts, research, and personal experience), your speech will fold like Franklin's empty bag. Fortunately, there are many sources of information for researching a speech. Especially when beginning your research, or when you need an overview of a subject area, do not forget the obvious: encyclopedias and general reference books. Most libraries have a recent edition of *Encyclopaedia Britannica, Encyclopedia Americana,* or *World Book Encyclopedia.* Many of these resources also contain outlines or breakdowns of general topics.

A great place to find both basic and supporting material is in your personal books and magazines. *Farm Journal* and *Progressive Farmer* are excellent sources of general information. Trade magazines, business journals, and breed association publications also are excellent sources.

In addition to books and magazines, many research journals that used to be accessible only in university libraries are now online and available for your speech research. A first-rate research tool is Google Scholar (http://scholar.google.com). Google Scholar allows broad searches for scholarly materials: research that is credible, trustworthy, and science based.

Traditionally, most research for speeches has been conducted in a good library, with the help of good reference librarians. Library research should prove effective for you even if you know very little about using a library catalog or reference works such as the *Readers' Guide to Periodical Literature.* Librarians can show you where to look for what you need. They can also be very helpful even if you already know a great deal about using libraries and are experienced in research.[4]

Consult not only the Internet, articles, and books but also organizations and experts; you need not limit yourself to the sources mentioned here. Search far and wide for sources, but make sure they are credible sources. It is important to understand what a credible source is. A credible source is based on research or written by experts who work for research-based organizations. Just because you read information on Twitter or Wikipedia doesn't mean it is credible; it just means someone knew how to construct a sentence and publish it to the Internet. You can also do research by requesting information from organizations active in the field you will be speaking about. Persons who are experts in their fields are often glad to furnish comments or opinions when such information is professionally requested.

Government officials and their staff members or aides are often a great source of information on issues of public importance. An example is the U.S. Department of Agriculture (USDA), which operates experiment stations throughout the United States. The USDA produces a monthly magazine of related research, and it publishes hundreds of bulletins that are available at little or no cost. A list of these can be obtained from the U.S. Government Printing Office, Washington, DC, 20402. USDA information, and quite a bit of other data, is on file at the National Agriculture Library in Beltsville, Maryland. In fact, the USDA and the National Agricultural Library maintain a very informative website (http://www.nal.usda .gov) where you can request library materials and search through an archive of current news and events. Your state and local extension offices are also excellent sources of materials for your speech. If you are in doubt about which government agency's mission and purview might cover a particular subject, try calling or writing your congressional representative. His or her staff can point you in the right direction. Simple Web searches using your favorite search engine are also very useful for your information dig—but make sure that what you find is credible (and preferably verifiable) before you use it.

If the subject is controversial, you can be pretty sure there is an organization for it or against it. For example, if you have chosen to speak on the issue of exploring for oil in national parks, you can be reasonably certain that oil companies and trade organizations will be in favor of it, whereas environmental groups such as the Sierra Club will oppose it. As another example, think about tobacco products: the American Tobacco Institute is for the use of tobacco; the American Cancer Society opposes its use. The *Encyclopedia of Associations*,[5] available in most libraries, lists special-interest groups, along with their mailing addresses and other contact information. Most of these groups offer free pamphlets and literature, and may make materials available online. Regardless of the position you eventually take in your speech, be careful to present both sides of your topic when you use information from special-interest organizations.

Speakers can find **quotations** (use of words or phrases written or spoken by another person) to support or illuminate their ideas in sources such as *Bartlett's Familiar Quotations, Brewer's Dictionary of Phrase and Fable, Columbia Granger's Index to Poetry,* and the *Oxford Dictionary of Quotations.* Quotations can be particularly useful in introductions and conclusions, as they may reflect your main point in a pithy, memorable, or unusually well-turned sentence.

Several other good sources of information for agriculture-related topics include the following:

- Livestock breed associations
  Examples: American Quarter Horse Association, American Hereford Association, Holstein Association USA
- Personal interviews
  Examples: president of a civic club, farmer or rancher, agribusiness person, agricultural scientist
- Education facilities
  Examples: universities, local colleges, private research groups
- Newspapers
  Examples: *The New York Times, The Wall Street Journal,* local newspapers
- Almanacs and atlases
  Examples: *World Almanac, Rand McNally Atlas*
- Surveys and polls
  Examples: public opinion polls, suggestions for improvement
- Biographical sources
  Examples: *Who's Who in America, Dictionary of National Biography, Dictionary of American Biography*
- Microfilm indexes
  Examples: *COM (Computer Output Microfilm), Magazine Index, Business Index*
- Computer databases
  Examples: ERIC (Educational Resources Information Center), Agridata Network, AG STAT

* World Wide Web
  Examples: Google (http://www.google.com),
  Yahoo! Search (http://www.search.yahoo
  .com), Ask.com (http://www.ask.com)

Gathering materials for a speech is like gathering information for any project. For most topics, you will uncover far more sources than you can use. Skim or rapidly go through the material. If you are evaluating a magazine article, spend a minute or two finding out what it covers. If it is a book, read the table of contents carefully, look at the index, and skim pertinent chapters. You can then decide which sources you should read in full, which you should read in part, and which you should abandon. Minutes spent in such an evaluation may save you hours of reading. In Figure 8–5, a woman and her research team go straight to the source as they gather information for her speech. Interviews are sources and should be cited and referenced just like other sources.

RECORD IDEAS

When gathering material, write the important information you want to use on a note card or in a computer file. On each note card, be sure to include identifying information such as author name, date, title, and page number. If you use the note-card method, you should also make a bibliography card or entry for each source. The bibliography card includes everything just listed plus other important information (e.g., edition, city/state of publication, and publisher). What the bibliography looks like on your card, and at the end of your written-out speech, depends on the style guide you use. The National FFA Prepared Public Speaking Career Development Event requires that you use APA style, set out in *Publication Manual of the American Psychological Association*. The following gives an example of APA style for a book; other examples of the style can be found at the APA's website (http://www .apastyle.org).

Herren, R. V. (2012). *The science of animal agriculture* (4th ed.). Clifton Park, NY: Cengage Learning.

If you prefer to use a computer for note taking and reference keeping, you can type important information and reference-material source entries into a simple word-processing system (e.g., Microsoft Word), a spreadsheet (e.g., Microsoft Excel), or various reference management software packages (e.g., EndNote). Either method (note-card system or electronic files) allows easy movement of the information from one place to another.

Use a separate card or entry for each note. Take plenty of notes. Even when you are not certain you will use a particular fact or piece of evidence in your speech, record it and the source in which you discovered it.

When you are taking notes and recording the source of your information, paraphrase the findings instead of using only direct quotation. Use quotations and **statistics** (facts or data

FIGURE 8–5 Interviews are an excellent source of information. This student is interviewing an expert for her research. *(© iStock/KatarzynaBialasiewicz)*

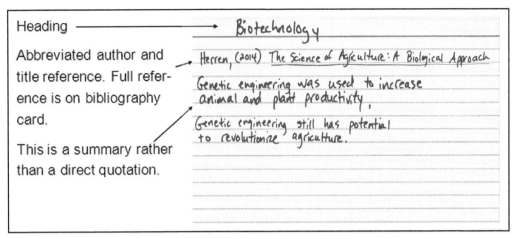

FIGURE 8–6 Book note card example.

used to present information of a numerical or empirical nature) in your speech only when they are needed to make a point. Also, maintain the context for quotations and statistics, preserving their original meaning. Although recording what others believe and have said shows that you have done your homework, the audience wants to hear *your* opinions in your own words. When you use material from others, always recast and synthesize it into your own unique expression of the point; whether spoken or written, cited or uncredited, plagiarism is still plagiarism.

Make sure the source information (including author, date, title of book or journal, title of chapter or article in an edited book or journal, publisher, page numbers, Internet address) is correct. This accurate information supports the speech and adds to its credibility. If you use direct quotations, do so sparingly and cite the page number where the quoted material appears, both in the text of your speech and in the reference list/bibliography section.

Information from Books

On each note card or computer entry, write your note; list the author, date, title, and page number of the source; and create a heading indicating the subject of the note (Figure 8–6). The subject heading is particularly important. It simplifies the task of arranging your notes when you start to organize the speech.

Information from Magazines or Journals

For magazine articles or journal articles, write your note; the subject heading; and the article title, name of the author (if one is credited), magazine name, year and date of publication, and the page number on each note card (Figure 8–7).

Source Citations

In a written report, ideas taken from other sources are identified by short citations in the text, and a full reference is given at the end of the report in a bibliography section or reference list. However, in a speech, some of these citations and identifications must be included within the context of your manuscript. This not only will help the audience evaluate the context but also will add to your credibility. Figure 8–8 gives examples of several source citations in the body of a speech.

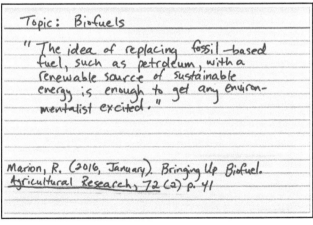

FIGURE 8–7 Magazine note card example.

In a speech on wetlands delivered to the Tennessee Farm Bureau Convention last January, Julius Johnson, Tennessee's Commissioner of Agriculture, said . . .

In an interview with Ben Jordan, retired mechanical engineer with the United States Army, Mr. Jordan stated that . . .

According to an article about the use of pesticides in last month's issue of *Progressive Farmer Magazine* . . .

In 2016, a molecular reproductive physiology researcher at the Clay Center Agricultural Research Service in Nebraska, Dr. Clay Lents, wrote . . .

FIGURE 8–8 Examples of source citations in running text.

MAKE AN OUTLINE

An outline helps you prepare the speech. In creating an outline, you actually begin to put your speech together. There are three main reasons for using outlines:

1. To help you organize and develop your ideas
2. To help you recognize the speech's strengths and weaknesses
3. To help you save time when writing the speech

While preparing the outline, you decide what you will say in the introduction, how you will organize the main points and supporting materials in the body of the speech, and what you will say in the conclusion.

Format

Use a standard set of symbols for your outline. Several styles of outlining exist, so choose one that you are comfortable with and use it consistently. In one preeminent outline style, main points are indicated by Roman numerals, major subdivisions by capital letters, minor subheadings by Arabic numerals, and further subdivisions by lowercase letters. Although you can show further breakdown of ideas, it is rarely necessary to subdivide a speech outline beyond the level shown in Figure 8–9.

Topical Outline The topical outline style uses key words or phrases to express the order of the thoughts to be presented in a speech. This topical outline can be used for the speaker's notes.

TITLE
Alternative Energy Sources
   I.   INTRODUCTION
        Importance of research
  II.   BODY
      A.   First main point
          Renewable
         1. Subpoint #1
         2. Subpoint #2
      B.   Second main point
          Produced on farm
         1. Subpoint #1
            a.
            b.
         2. Subpoint #2
            a.
            b.
 III.   CONCLUSION
        Economically and environmentally correct

FIGURE 8–9 Topical outline style.

Sentence Outline The sentence outline style uses complete sentences to express the order of the thoughts to be presented in a speech (Figure 8–10). The sentence outline is more complete and therefore usually a better outline for beginning speakers to use. Using complete sentences enables you to see (1) whether each main point actually enhances the theme or goal of your speech and (2) whether the wording makes the point you want it to make.

Parts of an Outline

Like the speech itself, an outline consists of an introduction, a body, and a conclusion. Because the **introduction** is the beginning of the speech, it should motivate the audience to listen to your speech. The introduction should get the audience's attention, state the purpose of your speech, relate the topic to the audience and describe its importance to them, and preview the main points to be covered.

TITLE

Production Agriculture

Alternative Energy Sources

 I. INTRODUCTION

   Research of new agriculture alternative fuel sources is needed for the farm economy and the future of the environment.

 II. BODY

   A. First main point

    The research begins by perfecting existing crops and resources.

    1. Subpoint #1

     a.

     b.

    2. Subpoint #2

     a.

     b.

   B. Second main point

    After perfecting the alternative energy sources, marketing will be needed.

    1. Subpoint #1

    2. Subpoint #2

 III. CONCLUSION

   The research, perfecting, and marketing will lead to a stronger farm economy.

FIGURE 8–10 Sentence outline style.

The **body**, which is the bulk of the speech, carries the central theme and main ideas. The body of the speech is, in essence, the speaker's message. Its main objective is to develop fully the theme of the presentation. It consists of the main points, subdivisions, and supporting detail. A speech usually contains from two to five main points. If there are more than five, the audience will have trouble following the speaker (and the speech will probably be too long).

The conclusion reviews or summarizes the main points of the speech or calls for action. The end of the speech is just as important as the beginning and must be allotted the necessary amount of preparation time. Your conclusion must do two things: give your audience a feeling of completion or satisfaction, in terms of your having done what you set out to do; and center attention on the whole idea of your speech rather than on any single part of it.[6] In short, it must drive home the central message of the speech.

WRITE THE SPEECH

The saying for writers, "Write the way you talk," applies even more aptly to public speaking. Your talk should represent you. You should use words that are comfortable for both you and your audience. Select words that convey your exact message.

Body

You may assume that because the introduction is the first part of the speech, that is where you should begin—but that assumption would be wrong. How are you going to write the introduction until you have put together the material that you will be introducing? Therefore, start with the body of your speech. Write the main points; arrange or order the main points in a logical sequence; and select effective language, examples, and quotations to convey those main points.

Main points are the key building blocks of your speech. They are the ideas that you want your audience to remember. The main points should be specific, vivid, and parallel. They should be of equal importance. The wording should explain exactly what you mean while arousing interest in the subject. For example, here are the main points of a speech about alternative fuels from agriculture:

Specific purpose: To inform my audience that a production agriculturalist can produce fuel in times of national crisis.

Central idea: Production agriculturalists can produce their own ethanol, methane, soy-diesel, and hydrogen.

Main points:

 I. Ethanol can be produced from corn raised on the farm.

 II. Methane can be produced from manure in dairy and feedlot facilities.

 III. Soy-diesel can be produced from soybeans raised on the farm.

 IV. Hydrogen can be produced from water with solar, wind, water, or nuclear energy as the power source.

These four points form the skeleton on which to flesh out the body of the speech. If there are four major alternative fuel types that production agriculturalists can produce, then logically there will be four main points.

Once you establish your main points, you need to decide in which order to present them. A speech can be organized in many different ways. Your goal is to find a structure that will achieve your goal and help the audience make the most sense of your material. These organizational patterns can help an audience follow your speech:

- *Time or chronological order*—follows a chronological sequence of ideas or events
- *Space order*—tells the audience that there is a special significance to the positioning of the information, such as top to bottom
- *Topical order*—emphasizes categories or divisions of a subject
- *Causal order*—emphasizes the causal relationship between the main points and the subject of the speech resulting from specific conditions
- *Reasons order*—emphasizes why you believe an audience should accept a statement as true or behave in a particular way; often arranged in order of strongest to weakest
- *Problem solution order*—the main points are written to show that:
  - there is a problem that requires a change in attitude or behavior
  - the solution you are presenting will solve the problem
  - your solution is the best way to solve that problem

Supporting materials back up the ideas in your main points. When selecting supporting materials, make sure they are directly relevant to the main points they are supposed to support. This supporting information comes from the material you gathered earlier, when planning and researching the speech. Each item of information listed under a key element heading may contain further subpoints. These subpoints show the relationship between each idea and the preceding main point. Use illustrations, examples, and stories for each point. List enough material to have both quantity and quality. Remember, it is always easier to edit than to search for new material when time is short.

Choosing effective language for a speech is like choosing the right clothes for a special occasion. Wise speakers choose their words very carefully to display their ideas most effectively. This involves using simple words, using more concrete and specific words, and restating certain words. Effective language choices enhance clarity and allow proper emphasis. Often you can do this by using figures of speech, such as simile, metaphor, personification, hyperbole, and irony. **Simile** is a figure of speech that presents a brief comparison of two basically unlike items using the word *like* or *as*. **Metaphor** is a figure of speech containing an implied comparison that omits the word *like* or *as*, in which a word or phrase ordinarily and primarily used in relation to one thing is applied to another. **Personification** is a figure of speech in which a thing, quality, or idea is represented as a person. **Hyperbole** is exaggeration for effect that is not meant to be taken literally. **Irony** is humor based on use of words in a way that suggests the opposite of their literal meaning.

> *Words are the garments with which speakers clothe their ideas.*
>
> —J. Regis O'Connor

When writing a speech, be aware of your legal and **ethical** (conforming to the standards of conduct of a given profession or group) responsibilities. A speaker must refrain from making cruel statements that may harm another's reputation (**defamation of character**). Concerning ethical issues, speakers must remember that they are personally accountable for their speech. Do not make statements without first checking the facts. A speaker must allow free choice by the audience (that is, allow them to hear both sides of an issue). Ethical speaking builds respect for the speaker and in turn displays the speaker's respect for the audience.[7]

Introduction

Once the body of the speech is ready for practice, begin working on the introduction. (Prepare two or three different introductions to determine which will be most effective.) The introduction may be one sentence or many, depending on the length of the speech. A good introduction should grab the audience's attention, create an atmosphere of credibility and goodwill, and lead directly into the body of the speech. You may have a "captive audience," but physical presence does not guarantee mental presence. The speaker's first goal is to create an opening that will earn the audience's undivided attention. Do something to gain the interest and attention of the audience:[8]

- Tell a joke.
- Ask a question.
- Tell a story.
- Use a quotation.
- Use a personal reference or anecdote.
- Create suspense.
- Give a compliment.

All these methods draw attention, but that attention may wander if the remaining introduction does not lead directly into the speech. With most speeches, the audience members experience a feeling of goodwill if they perceive the speaker as sincere and concerned for them.

The introduction must also focus the audience on the goal of the speech. To focus attention, simply tell the audience what you are going to tell them in the speech. This is the last statement of the introduction and leads into the body. This is usually the main purpose statement.[9]

Conclusion

"All's well that ends well."

—Shakespeare

Write the conclusion last. This offers the speaker one last opportunity to remind the audience of the speech content. It reinforces the audience's understanding of, or commitment to, the central idea. Do not startle the audience by ending too abruptly, nor ramble so long that they become exhausted. With the most common types of conclusions, the speaker

- Summarizes the main points
- Uses a story
- Is humorous
- Makes an appeal
- Makes an emotional impact

As with the introduction, write two or three different conclusions to find the most effective one.[10]

PRACTICE THE SPEECH

Keep the following considerations in mind when practicing your speech.

Practice Time Limits

When should you begin preparing your speech? Many students of public speaking do not allow themselves enough time for planning, research, speechwriting, and practice. Consequently, they do themselves an injustice by not leaving sufficient time to do the job they are capable of doing. It is not unusual for a student to attempt to write, memorize, practice, and deliver a speech for a contest within a week. Even an experienced, accomplished speaker would be heavily taxed to do a job of this magnitude in such a short time span.

Think in terms of the minimum time that will enable you to do your best. You should start working on the speech at least two months before the delivery date. Try to complete the necessary research and writing within a month. This allows another month before the speech date to memorize, practice, and otherwise prepare to deliver the speech. To succeed in public speaking, you must be willing to give much time to practice. Continue to practice even beyond the point when you think sufficient time has been devoted. Revise sentences so they are easier for you to say; change words if they are difficult for you to pronounce; find substitutes for words that do not adequately convey your intent.

Memorize the speech as quickly as possible so that you can spend more time in practicing delivery. Too frequently, high school students become convinced far too early in the process that they are fully prepared, and as a result do not practice enough. This overconfidence can cause you to forget your speech and thus make a dismal showing.

Practice Methods

Demosthenes, the famous Greek orator, developed his speaking ability by going off into the woods alone and standing on a rock while speaking to an imaginary audience. There are several resources available to help you practice your speech:

School Classes and Teachers Most teachers are happy to have you appear before their classes for practice. Ask them to listen to your speech privately, as well as in class, and to offer constructive criticism.

Home and Mirror Practice at home as much as possible. At times when you would otherwise be doing nothing, practice that speech! Stand in front of a mirror and observe your facial expressions and gestures. This will help you see yourself as others see you. Do you like what you see in the mirror? If not, try to do something about it.

Auditorium When opportunity permits, go into the auditorium for practice (Figure 8–11). This gives you an opportunity to become accustomed to different acoustic conditions. It also gives you the additional advantage of learning to look at greater space as you talk and to imagine encompassing a large audience.

Civic Organizations Arrange to talk to every civic organization and other group outside the school that you possibly can. This will give you a decided advantage in becoming accustomed to facing an audience and an opportunity to get over those **pangs** (sudden, sharp, and brief pains; may be physical or emotional) of stage fright that afflict most beginning speakers.

The Camera Use of videotape or digital video is perhaps one of the finest modern methods of speech rehearsal. With this device, you can record your speech and then watch and listen to yourself. This gives you an excellent opportunity to detect your mistakes and correct them. Record your speech, watch and listen to it, then rewind and record it again. When satisfied that you have it about right, ask your teacher, as well as others, to listen to it. Solicit their suggestions and criticisms, and then practice again and again.

Refer to Chapter 7 for other fun and effective ways to practice and rehearse.

Things That You Need to Practice

When you are practicing your speech, be aware of the following.

Your Smile Loosen up and smile a little for your audience! You will know when to smile at the right time in your speech; you will also know the passages that require a more serious demeanor. Practice these facial expressions just as **ardently**,

FIGURE 8–11 This student is practicing in an auditorium. It gives him an opportunity to become accustomed to different acoustic conditions. *(© iStock/Viorika Prikhodko)*

with warm or intense feeling, as you practice the words of your speech.

Gestures As soon as you have your speech memorized, work on the finer points of delivery. Practice your gestures and get used to making them gracefully and naturally. The issue of when and how much to move your hands can be difficult, but it is safe to say that the amateur public speaker should leave most of the hand gestures to the experienced speaker. Though they are often used to add emphasis or give expression, gestures must appear natural and be appropriate.

Even many experienced speakers greatly overdo the use of gestures. This is usually due to lack of early training and practice. It is much better to use no gestures than to use too many. If you use gestures too much, you will find that the audience becomes focused on your movements rather than the message in the subject matter.

Head and Eyes One of the best ways of giving that added touch of emphasis without distracting the audience is with the head and eyes. A gentle nod, shake, or backward movement of the head may have a far greater effect than you might expect.

You often hear the remark, "She talks with her eyes." When you are talking to an audience, you are talking to individuals. Do not make the deadly mistake of looking at the floor, the ceiling, or out a window while talking. Look individual audience members in the eyes, and they will feel that you are talking to them. You may find this distracting or even frightening when you first try it, but with a little practice, this feeling will disappear and you will find that you really are talking to them and enjoying the feeling.

Sincerity Practice being sincere in what you say. People who are willing to spend their time listening to a speaker deserve to hear a speech that is the result of much thought and in which the speaker demonstrates sincerity. The speaker who delivers a speech in a disinterested manner should not be surprised or insulted if members of the audience start text-messaging in the middle of the speech.[11] If you are not interested in or committed to what you are saying, why should they be? Be sincere as you write the speech, as you practice it, and especially as you deliver it.

PRESENT THE SPEECH

Chapter 7 gives a detailed description of delivery (presentation), and specifically addresses the FFA score card and divisions. Chapter 7 also discusses techniques to overcome stage fright and nervousness. This section reviews presentation techniques that are more suited to a prepared speech than to FFA Creed delivery.

The actual delivery of the speech is much easier if you have done a good job of preparation. When you are working on delivery, here are a few more things to consider: the salutation, being deliberate, using the hands, using the body, humor, dress and physical appearance, where to stand, notes, and special considerations.

Salutation

The **salutation** is beginning a speech with a greeting, either addressing or welcoming. It should be pleasant and inclusive. The exact wording depends on the particular audience. Some speakers begin simply with "Ladies and gentlemen," and then launch into the speech. In many instances, a better approach might be: "Mr. Chairperson, fellow members, distinguished guests, and friends, I shall speak to you on the subject of...."

Being Deliberate

Try to be **deliberate**—careful and intentional—from the time you rise to your feet until the time you sit down again. One of the most common mistakes is to appear to be in a hurry. This gives the impression that you are nervous. A deliberate manner of speaking helps you enunciate your words more clearly. When talking rapidly, we tend to run words together and use less-than-optimal diction. This can be a serious handicap to a public speaker (and very off-putting to the audience).

Using the Hands

What to do with your hands will likely be one of your first problems—suddenly you have no place to put them, and you may feel like they are strange, disconnected appendages with a life of their own. This need not be true, especially if you have devoted enough time to practicing gestures and posture. No one else is apt to notice your hands until you start using them in an unaccustomed or bizarre manner.

Let your motions and hands appear natural. If they want to clasp behind you, let them do that; if one of your hands wants to stray to a pocket, let it stray; if a hand wants to hold on to a lapel, let it do so. When you start hunting for places to put the hands, you are asking for trouble.

Using the Body

One of the most important tips for public speakers is to avoid swaying back and forth. This is a very common error of the beginning speaker. However, slight movements of the body coordinated with movements of the head or limbs are occasionally permissible to add emphasis. A slight step forward to indicate emphasis or a change of pace in the speech, or perhaps a slight backward step in certain instances, may be used to your advantage. However, do not walk around the stage too much; this can easily turn into nervous pacing or cause mispositioning that cuts off part of your audience.

Humor

Nothing puts an audience more at ease, especially at the beginning of a speech, than a good joke. This, of course, is not common practice in contest speeches and is not recommended as such. It is almost never advisable for the beginning speaker to try to incorporate humor into a speech. Nevertheless, at times humor may be used to gain an advantage even in a contest speech.

You may not be a good joke teller. If not, do not try to tell jokes. However, you may improve your speech by including stories, examples, and illustrations (humorous and otherwise) to help people understand the point you are discussing. Examples are word pictures; people understand pictures better than abstract ideas. The experienced speaker in Figure 8–12 appears natural as she injects humor into her speech.

Dress and Physical Appearance

The importance of physical appearance should not be underestimated. The general appearance you present can have a definite impact on your listeners. For any official FFA speech, you should be in official FFA dress.

FIGURE 8–12 You may improve your speech, as this FFA member has done, by injecting anecdotes, examples, illustrations, and humor, if appropriate. *(Courtesy of Georgia Agricultural Education.)*

Let neatness be your watchword in this area. Clothes that are clean and neatly pressed, gleaming teeth, well-groomed hair, and, for men, a clean-shaven face are seemingly small items, but they merge to help create a pleasant picture and make a positive first impression.

Where to Stand

Where should you stand while delivering your speech? Usually this decision is left entirely up to you. Sometimes, especially in contests that are conducted in large auditoriums, the speakers are encouraged to stand behind the lectern so their voices will travel well over the public address system. If there is a microphone, determine whether it is fixed in position or detachable from its stand; this will in great part determine how far and where you can move. It is also recommended that you arrive early and check the area where you are going to speak.

A good point to remember is that you should stand where the audience can hear you best. If the acoustics are poor, stand near the edge of the stage. If there is a lectern and you can adjust to it, stand behind it. (This is especially helpful for beginners because it acts as a place to put one's hands.) If there is a table, stand to either the left or the right of it; standing directly behind it creates a separation between speaker and audience that may be construed as hiding, nervousness, or aloofness.

Notes

You should know your speech so well that notes are unnecessary. However, if you have to use notes, write them large enough so that you can see them at a glance. Use a system to keep your note cards in order, such as an outline or numbers. Do not hold your notes in your hands; place them on the lectern so your hands will be free.

Special Considerations

As a speaker, you must know how to deal with a number of situations: the lectern, microphone, distractions, and following a written speech.[12]

Lectern A speaker's stand is intended to be a convenient place to lay notes or a manuscript. When a lectern is not available, notes must be held inconspicuously in the hands.

Microphone A microphone can be very helpful when speaking to a large audience or in a large space. To get the most from a microphone, however, you should practice with it, preferably in the space where you will be using it. Test it shortly before the speech. While delivering your speech, keep your mouth 10 to 12 inches from the head of the microphone at all times.

Distractions If distractions or interruptions occur during a speech, you must handle them calmly and with poise. If the interruption is minor, it is usually best to ignore it; if it creates a major disturbance, you must handle it decisively and wisely.

Following a Written Speech For contest purposes, should you follow your written speech exactly? It is up to the judges and the organization

conducting the contest. In real life, no one does— or should—follow his or her speech exactly. Daily news, daily events, introductions, previous speakers, and other things on the program may necessitate last-minute changes to make your speech more meaningful and memorable. Minor word changes during speech delivery should never be a factor in the outcome of a contest, but for some judges it is—so be sure to stay true to your manuscript.

ANSWER QUESTIONS

The rules for many public speaking contests allow up to five minutes for the judges to ask the speaker questions about his or her subject matter after the speech. Therefore, you must be well acquainted with the subject; after all, you are supposed to have studied and researched well while writing your speech. Be sure you can answer all questions completely; continue reading and researching right up to the contest date. Ask your teacher and others to prepare lists of questions that might be asked on the subject matter, and prepare answers for them.

During the questioning period, follow these guidelines:

- Be deliberate. Take time to think through the answer and organize your reply.
- Be thorough and complete. There is no scarcity of words, so use enough of them to render a complete answer. If you answer with several points, count them off on your fingers as you answer.
- Answer with confidence. A hesitating answer convinces the judges that you do not know much about the subject of the question.
- If you do not know the answer, say so without hesitation, but do not show embarrassment. Do not try to bluff your way through. If you have some knowledge of the answer, admit that your information on the question is limited; then go ahead with what you do know.
- If you did not hear or did not understand the question, ask the person to repeat or rephrase it. Some speakers, while stalling, say, "I'm glad you asked that question."

Obviously, if everyone used that phrase, the judges would know what you were doing. Be creative rather than leaving dead space while you think.

LISTEN AND EVALUATE

General Evaluation

Evaluating speeches by others not only allows you to analyze where the speech went right and where it went wrong but also gives you insight into techniques to use as you prepare your own speech. If a speech has good content, is well organized, uses language well, and is presented well, it is more likely to achieve its goal.[13]

FFA Speech Evaluation

The FFA Prepared Public Speaking Contest uses the score sheet shown in Figure 8–13. Scoring covers two areas. The first part (shown in Figure 8-13A) includes content and composition of the manuscript. You may wish to ask your English teacher to review your speech for needed corrections. The content could be reviewed by friends, professionals, family, teachers, and others who are willing to provide input. When you enter a contest, there is no excuse for not scoring near perfect on the first part if you are sufficiently prepared.

The second part, shown in Figure 8-13B, scores the presentation of the speech, including oral and nonverbal communication and response to questions. Refer also to Chapter 7, in which the FFA Creed score card is discussed, for a comprehensive explanation of the delivery section.

The points allotted to each item are listed on the judge's score sheet shown in Figure 8–13. Note that points are deducted for time if your speech is too long or too short.

NATIONAL FFA PREPARED PUBLIC SPEAKING CAREER DEVELOPMENT EVENT

The National FFA Prepared Public Speaking Career Development Event (CDE) is designed to develop agricultural leadership by encouraging member participation in agricultural public speaking activities and stimulating interest in leadership and citizenship. The national event is held in conjunction with the National FFA convention.

Event Rules

The speaker must follow these event rules:[14]

- Participation in the National FFA Prepared Public Speaking CDE is limited to one person from each state association.
- "Each participant's manuscript must be the result of his or her own efforts. It is expected that the participant will take advantage of all available training facilities at their local school in developing their speaking ability. Facts and working data may be secured from any source, but must be appropriately documented."
- Participants receive instructions from the event superintendent at the time and place shown in the current year's program for National FFA events.
- "It is highly recommended that participants be in official FFA dress as defined in the current official FFA Manual."
- "Each state with a speaker shall provide a competent individual to judge the national event." Representatives from organizations connected with agricultural education act as judges for the final round. "Three to eight competent and impartial persons will be selected to judge the event." Advisors with students competing in the speaking event may not serve as judges for the event.

Event Format

The speaker must provide the following:[15]

- A manuscript in PDF format that is double-spaced and typewritten to fit 8½ × 11-inch white bond paper, with a cover page that gives the speech title, participant's name and state, and year. The body of the manuscript must have 1-inch margins, using 12-point serif (Times New Roman, Cambria, etc.) or sans serif (Ariel, Calibri, etc.) font. Follow the most current American Psychological Association (APA) style manual for the reference list/

NATIONAL FFA
CAREER AND LEADERSHIP
DEVELOPMENT EVENTS

Manuscript Content and Composition Rubric
200 points

NAME _____ MEMBER NUMBER _____

CHAPTER _____ STATE _____

INDICATOR	Very strong evidence of skill is present 5-4 points	Moderate evidence of skill is present 3-2 points	Strong evidence of skill is not present 1-0 points	Points Earned	Weight	Total Points
Topic relevance	Topic addresses an issue facing the industry of agriculture.	Topic addresses an issue that may show some relationship to the industry of agriculture.	Topic addresses an issue that is unrelated to the industry of agriculture.		x 6	
Persuasive explanation of position on topic	Position clearly stated and ample evidence is provided.	Position is not obvious and evidence is not clearly provided.	Position is not stated and evidence is not provided.		x8	
Alternative viewpoints recognized	Identifies and counters alternative viewpoints.	Only identifies alternate viewpoints.	Does not identify alternate viewpoints.		x 4	
Logical order and unity of thought	Clearly organized and concise with strong introduction, body and conclusion layout.	Good organization with few statements out of place or lacking in clear construction.	Little to no organization is present; sometimes awkward and lacking construction.		x 4	
Spelling/grammar (sentence structure, verb agreement, etc.).	Spelling and grammar are extremely high quality with two or less errors in the document	Spelling and grammar are adequate with three to five errors in the document.	Spelling and grammar are less than adequate with six or more errors in the document.		x 7	
Quality of resources	Resources are from reputable sources.	Resources are from questionable sources.	Resources are unreliable and invalid.		x 6	
Manuscript written according to guidelines	5 points		0 points			
Double-spaced formatted to 8½" x 11" with 1" margins						
12 point serif (Times new roman, Cambria, etc.) or sans serif font (Ariel, Calibri, etc.)					x 1	
Cover page with speech title, participant's name, state and year					x 1	
APA style for references and citations					x 3	

TOTAL POINTS [   ]

FIGURE 8–13A The National FFA Prepared Public Speaking Career Development Event Score Card. *(Courtesy of the National FFA Organization.)*

NATIONAL FFA
CAREER AND LEADERSHIP
DEVELOPMENT EVENTS

Presentation and Questions Rubric

800 points

NAME   MEMBER NUMBER

CHAPTER   STATE

INDICATORS	Very strong evidence of skill is present 5-4 points	Moderate evidence of skill is present 3-2 points	Strong evidence of skill is not present 1-0 points	Points Earned	Weight	Total Score
Oral Communication and non-verbal communication						
Supporting evidence	Examples (stories, statistics, etc.) are vivid, precise and clearly explained.	Examples are usually concrete and sometimes need clarification.	Examples are sometimes confusing leaving the listeners with questions.		x 15	
Persuasive use of evidence	Exemplary use of evidence to persuade listeners.	Sufficient use of evidence to persuade listeners.	Has difficulty using evidence to persuade listeners.		x 15	
Pace	Speaks very articulately at rate that engages audience.	Speaks articulately but occasionally speaks too fast or has long unnecessary hesitations.	Speaks too slow or too fast to engage audience.		x 15	
Command of audience	Speaker uses appropriate emphasis and tone to captivate audience.	Speaker presents speech as mere repeating of facts and speech comes across as a report.	Speaker lacks enthusiasm and power to engage audience.		x 20	
Eye contact	Constantly looks at the entire audience (90 to 100 percent of the time).	Mostly looks around the audience (60 to 80 percent of the time).	Occasionally looks at someone or some groups (less than 50 percent of the time).		x 10	
Mannerisms and gestures	No nervous habits are displayed. Hand motions are expressive and used to emphasize talking points.	Sometimes exhibits nervous habits. Hands are sometimes used to express or emphasize.	Displays some nervous habits. Hands are not used to emphasize talking points; hand motions are sometimes distracting.		x 10	
Poise	Portrays confidence and composure through appropriate body language (stance, posture, facial expressions)	Maintains control most of the time; rarely loses composure.	Lacks confidence and composure.		x 15	
Response to questions						
Response to questions	Is able to respond with organized thoughts and concise answers.	Answers effectively but has to stop and think and sometimes gets off focus.	Rambles or responds before thinking.		x 20	
Knowledge of topic	Answer shows thorough knowledge of the subject and supports answer with strong evidence.	Answer shows some knowledge of the subject but lacks strong evidence.	Answer shows little knowledge of subject and lacks evidence.		x 40	
					TOTAL	

FIGURE 8–13B (continued)

bibliography section. Manuscripts are to be uploaded by the designated deadline on FFA.org. "Manuscripts not meeting these guidelines will be penalized."

- A complete and accurate reference list/bibliography that cites all sources used in writing the speech and uses the APA publication manual style. All participants in the National FFA Prepared Public Speaking CDE "should give credit to others where any direct quotes, phrases, or special dates used in the manuscript, in order not to be guilty of plagiarism."

The manuscript, prepared according to the handbook guidelines, must be uploaded online and sent to the CDE Program Manager, National FFA Center, by the posted deadline each year. "A penalty of 20 points (10% of available manuscript points) will be assessed by the judges scoring the manuscripts for any late submissions. Any manuscripts received later than the deadline will not be entered into the event and the speaker will be disqualified from speaking in the event."

Subjects

Participants may choose any current subjects of an agricultural nature for their speeches. This may include the areas of agribusiness, animal systems, plant systems, environmental services, food products and processing, natural resource systems and power, and structural and technical systems. A participant will be disqualified if he or she speaks on a nonagricultural subject.[16]

Time Limit

Each speech shall be a minimum of six minutes in length and a maximum of eight minutes. Participants are penalized one point per second on each judge's score sheet for being under six minutes or over eight minutes. No time warnings will be given. Each participant has five minutes of additional time in which to answer questions relating to his or her speech.[17]

Judging

The judging of entries is as follows:[18]

- Event officials will place the speaking order by the manuscript scores. The program chairperson introduces each participant by name in the order they were drawn. Participant may use notes while speaking, but "deductions in scoring may be made for this practice if it detracts from the effectiveness of the presentation. No props are to be used. Applause shall be withheld until all participants have spoken."

- A timekeeper(s) records the time taken by each participant to deliver his or her speech, noting undertime or overtime, if any, for which deductions are made. Timekeepers sit together.

- "Before any CDE speaking event, the judges will have any opportunity to review, judge, and score the manuscript. The manuscript will then be given to the presentation judges in order to gather questions to ask."

- During the event, the judges sit in different sections of the room in which the event is held. Using official score sheets, they score each participant on delivery. Each judge also fills out a comment card, which is given to the participant at the awards function.

- A five-minute question-and-answer period follows presentation of each speech. Each judge asks questions pertaining directly to the speaker's subject. Judges will not pose questions with multiple parts. Each participant is scored on his or her ability to answer all questions asked by all judges.

- "When all participants have finished speaking, each judge will total the score on composition and delivery for each participant. The timekeeper(s) record will be used in computing the final score for each participant. The judges' score sheets will then be submitted to event officials to determine final ratings of participants."

- Participants are ranked on the basis of the final score determined by each judge without consultation; the judges' rankings of each participant are then added. The winner is the participant whose total ranking is the lowest. Other placings are determined in the same manner.

Note that these rules may change. Go to the National FFA website (http://www.ffa.org) and read the latest CDE handbook to make sure you are following the most current rules for the Prepared Public Speaking event.

Tiebreakers

Ties are broken by counting participants' low ranks; the participant with the greatest number of low rankings is declared the winner. If a tie still exists, the event superintendent ranks the participants' responses to questions. The participant with the lowest rank from the response to questions is declared the winner. If a tie still exists, the participants' raw scores are totaled and the participant with the greatest total of raw points is declared the winner.[19]

SPEAKING TO GROUPS AND ORGANIZATIONS

Speaking at contests and in classes is formal. For informal speaking, we must still be poised, but also act naturally and be somewhat flexible. Here are suggestions for when you speak to groups and organizations:

Acknowledge Your Introduction Graciously acknowledge any introduction. If the chairperson has been funny at your expense, laugh good-naturedly along with the audience. If the chairperson has expressed real admiration for your accomplishments, do not make jokes at the chairperson's expense. The response to an introduction should not be overdone, but neither should it be completely ignored.

Show Appreciation Express pleasure for the opportunity to speak. Do so directly, sincerely, and briefly. Direct your comments to the program committee or whoever invited you to speak.

Relate to Local Interests Refer to matters of local interest and civic or organizational pride. If there is a particular event or person for which a town is famous, you can establish rapport by mentioning it. If speaking before a civic club or charitable organization, note some of the organization's accomplishments. By tactfully displaying such awareness, you compliment the audience or community.

Respond to the Mood of the Audience Try to pick up on the group's mood or feeling and capitalize on it. A statement such as, "As I listened to the reports, I sensed that this organization really wants to do something for the community," will enhance rapport. You should also make note of situations that the audience is aware of; there is almost always a little something about the room, the stage, the occasion, noises outside, or other relatively unimportant things that you can use as the basis for a humorous remark or graceful beginning.

The groups and organizations listed in Figure 8–14 are possible places to deliver a speech. Be sure that you know something about the organization before you address its members. Not only does this build rapport quickly; it may save you from making an embarrassing or awkward comment. This tailoring of your speech to the audience and occasion is part of any good speaker's preparation.

Historical society	Lions Club	Secondary school principals
State board of education	Jaycees	Rotary Club
Kiwanis Club	Optimist Club	Veterans of Foreign Wars
Chamber of Commerce	American Legion	Labor unions (AFL-CIO, etc.)
Sales and marketing executives	Century Club	Farm Bureau
Local business or industry board meetings	House of Representatives and/or Senate	Youth organizations (e.g., Girl Scouts, Boy Scouts, 4-H, etc.)
League of Women Voters	City council meetings	Local garden club (e.g., master gardeners)
FFA alumni meeting	Toastmasters	Better Business Bureau
Parent-Teacher Association	Local school board	Public school assemblies

FIGURE 8–14 These organizations are places where you could deliver speeches while in high school or after you graduate. Some of these groups are also good places to practice your speech if you are entering a speaking contest.

POINTS TO REMEMBER

A speaker's immediate purpose is to communicate ideas and feelings to the listeners. Communication is the transfer of meaning. The experienced speaker is aware of the audience's response and is able to interpret the many clues given by audience members.

The speaker's basic purpose is to relate to the listeners in order to achieve some goal. Communication is a means rather than an end in itself. The speaker communicates to listeners to get something done. It may be to share information, solve mutual problems, ease tensions, or inspire improved performance. Always determine the purpose of a speech before you give it; this will guide both your preparation and your delivery.

A skilled speaker makes an accurate analysis of the situation and adapts the speech to it. No two speech situations are ever identical, so the able speaker quickly analyzes each situation to answer the following questions: Am I prepared? Am I enthusiastic about the subject? How will the audience react to the subject? How do factors such as time, length of program, and meeting room influence the situation? Is the audience already interested in the subject, or does interest have to be created? Can I improve?

Here are a few points to remember:

- Open your speech with a sentence that secures the attention of the audience.
- End the speech in a forceful manner.
- Take appropriate pauses; do not allow yourself to run out of breath.
- Maintain good posture while speaking.
- Keep your voice well modulated; use variety in tone and pitch for emphasis and interest.
- Strive for correct pronunciation and clear enunciation.
- Do not eat a big meal immediately before speaking.
- Invite constructive criticism.
- Cultivate a sincere interest in people.
- Constantly strive to increase your vocabulary.

Speech is too powerful a tool to let it be used only by the dishonest, the manipulative, and the destructive. Orators who have a vision and a goal, such as Martin Luther King Jr., Henry Clay, and Abraham Lincoln, will always be influential in shaping people's opinions, their actions, and even their destinies. If your talents are such that they could be developed to make you a fine and skilled speaker, you as a future leader will serve both your own goals and humankind in general by putting in the work and practice required to do so.

CONCLUSION

The field of public speaking has never been overloaded with good practitioners. As in all other fields, there is always plenty of room for honest people of high principles and high ideals. To be a good public speaker and leader, you must be willing to sacrifice and ready to give of yourself, while remembering that you are serving humanity. You are challenged to study earnestly, work hard, and practice constantly to make yourself into an effective, skilled speaker. The rewards will far exceed your fondest dreams!

SUMMARY

Speeches are given to inform, to persuade, or to bring the members of an audience together. Through persistence and dedication, you can achieve your goals in the area of public speaking. Speaking skills increase a person's effectiveness and ability to influence the decisions of others. Effective public speaking is influence. Influence is leadership.

As you plan a speech, consider the purpose, the audience, and the occasion. Good speakers are audience centered. They know that the primary purpose of speech preparation is not to display their knowledge, demonstrate their superiority, or express their anger. The primary purpose should be to gain a desired response or to please the listener.

When selecting a topic, select one that interests you. Next, choose a topic about which you are knowledgeable or have an interest in becoming knowledgeable. Third, pick a topic of interest to your audience. When searching for a topic for an FFA speech, you may wish to start with general agricultural areas such as production agriculture, agribusiness, or agriscience and then drill down to a more specific topic in that area.

Start gathering information for your speech by searching the Internet, using encyclopedias and general reference works, and going through your own books and magazines. Most research for speeches is done in the library, with the help of good reference librarians. However, trade, business, and breed association magazines are also excellent sources. Consult organizations and experts as well as articles and books. Whatever or whoever your sources are, make sure they are credible and reliable.

When recording your ideas, keep track of the source details: name of the source, the page number(s), the author(s), and the publication date. Use direct quotations and statistics in your speech only when they are needed to make a point. The audience wants to hear your opinions and words, not those of another author.

An outline helps you prepare the speech and begin to put your speech together. An outline helps you organize and develop your ideas, recognize strengths and weaknesses in your argument, and save time when writing the speech. An outline has an introduction, a body, and a conclusion. Two common types of outlines are topical and sentence outlines.

In writing the speech, write the way you talk; use words that are comfortable for both you and your audience. Select words that convey your exact message. Start by listing the main points and then write the body of your speech. Once the body is completed, begin working on the introduction. Write the conclusion last.

Practice your speech. There are several resources available to help you practice, including school classes and teachers, the home and mirror, auditoriums, civic organizations, and the video camera. When you are practicing your speech, be aware of your smile, gestures, and head and eyes. Be sincere as you write the speech, as you practice it, and especially as you deliver it.

The actual delivery of the speech is much easier if you have done a good job of preparation. As you deliver your speech, consider the following: salutation; being deliberate; using the hands; using the body; humor; dress and physical appearance; where to stand; notes; use of lectern and microphone; and distractions.

The rules of many public speaking contests allow up to five minutes after the speech for judges' questioning on the subject matter. You must be well acquainted with the subject, because your speech is supposed to be well studied and researched.

If a speech has good content, and is well organized, well written, and well presented, it is more likely to achieve its goal. The FFA Prepared Public Speaking Career Development Event (CDE) assesses the content and composition of the manuscript. The delivery of the speech is also evaluated on the basis of voice, stage presence, power of expression, response to questions, and general effect.

When speaking to groups and organizations, graciously acknowledge any introduction you were given. Express pleasure for the opportunity to speak. Refer to matters of local interest. Last, respond to the mood of the audience.

 Take It to the Net

Research your speech topic on the Internet. Go to your favorite search engine. Type one of the five major topical areas (agriscience and technology, agribusiness, agrimarketing, international agricultural relations, or agricultural communications and education) into the search field. Browse the websites that come up as hits. Pick five that you think will be useful in preparing your speech and print them out.

Chapter Exercises

REVIEW QUESTIONS

1. Define the Terms to Know.
2. How is public speaking related to leadership?
3. Should you select the speech topic first or analyze the audience before selecting a speech topic?
4. Briefly explain how to gather (review) materials for a speech.
5. What are nine sources of information for preparing a speech?
6. What are three reasons for using outlines?
7. What are the three major parts of an outline?
8. Explain the speaker's ethical responsibility when writing a speech.
9. List six organizational patterns within the body of the speech.
10. What are seven things that can be done in an introduction to gain attention?
11. What do you believe would be the best practice method for you? Explain.
12. Discuss the use of notes when speaking.
13. When you are working on your delivery, what are nine things to consider?
14. Briefly discuss five practice methods.

COMPLETION

1. _____ is an uncritical, nonevaluative process of generating ideas, much like the word-association game.
2. When recording information from books on the computer or note cards, write the note, the _____, _____, _____ and the page number.
3. The most common types of conclusions are _____, _____, _____, _____ and _____.
4. Using a _____, _____ is perhaps one of the best modern methods of speech rehearsal.
5. _____, _____, _____ and _____ are four things to practice when rehearsing a speech.
6. The scoring for content and composition of a speech allots _____ points and _____ points are allowed for the delivery.

MATCHING

_____ 1. Entering fully, through imagination, into another's feelings.

_____ 2. The reference/bibliography style required by the National Prepared Public Speaking CDE.

_____ 3. An uncritical, nonevaluative method of developing speech topics.

_____ 4. A figure of speech containing an implied comparison that omits the word *like* or *as*.

_____ 5. A type of humor in which the literal meaning expressed is the opposite of the meaning intended.

_____ 6. Verbatim use of another person's words.

_____ 7. A figure of speech that presents a brief comparison of two basically unlike items using the word *like* or *as*.

_____ 8. The greeting in a speech.

_____ 9. Being slow and careful in deciding what to do.

A. simile

B. quotation

C. irony

D. brainstorming

E. APA

F. metaphor

G. empathizing

H. deliberate

I. salutation

ACTIVITIES

1. Divide a sheet of paper into three columns. Label column 1 with your major career interest, such as "art history"; label column 2 with a hobby or activity, such as "chess"; and label column 3 with a concern or issue, such as "water pollution." Working on one column at a time, brainstorm a list of at least 20 words or phrases for each column. Then, check three of the words or phrases in each column that are most compelling, of special meaning to you, or of potential interest to your audience. From this exercise, select a topic for your speech.

2. Your library probably has the *Encyclopedia of Associations* in its reference section. Using this book, find the names and addresses of three organizations you might write to for information about one of your speech topics. (If your library does not have the *Encyclopedia of Associations,* use the Internet to find organizations for this assignment.)

3. Using the library at your school, locate three books on the subject of your next speech. Following the APA format explained in this chapter, prepare a reference list/bibliography entry for each book.

4. In the *Readers' Guide to Periodical Literature,* find three articles in magazines on the subject of your next speech. Prepare a preliminary reference list/bibliography entry for each article, and track down the articles in the stacks or in the periodicals section of the library. Follow the APA format explained in this chapter.

5. List three other sources that would provide information for your next speech topic.

6. Select a speech topic and gather information for a prepared speech.

   a. *Select a speech topic.* It would probably be best to select a topic that you already know something about and that interests you. Choose from the areas of agriscience and technology, agribusiness, agrimarketing, international agricultural relations, or agricultural communications and education.

   b. *List references.* List possible references that may help you find background information to develop your speech topic.

   c. *Record information.* Read the following guidelines for recording information.

      • Record information on index cards or in a spreadsheet or word-processing system.

      • If you are in doubt about whether to take a note, take it. It is easier to discard a note you do not need than to track down information again.

      • Summarize information in your own words but without distorting the original thought.

      • Do not copy information word for word from the original unless your source concisely sums up a great deal of information and you want to incorporate that quotation into your speech.

      • If you copy something exactly, make sure to write down the page number on which the quotation appears, and make sure you enclose it in quotation marks so you will know later that the note was quoted verbatim.

      • Cite the source and date of the information you have recorded on each card. Use the format presented in this chapter.[20]

7. Write the outline for your prepared speech. Use either the topical outline style (Figure 8–9) or the sentence outline style (Figure 8–10).

8. Write the body of your speech. Use the outline you prepared in activity 7. Read through your speech and make modifications as needed.

9. Write an introduction to your speech using the outline you prepared in activity 7.

10. Write a conclusion to your speech using the outline you prepared in activity 7.

11. Practice your speech using one of the methods discussed in this chapter. Write a paragraph explaining which practice method works best for you.

12. Take part in a class discussion on the topic "What I find most difficult about speech delivery." Notice during this discussion whether many of your classmates experience the same problems as you do.

13. Collect pictures from magazines to show different gestures. Display these pictures in the classroom, and ask students to identify the ideas being communicated by each picture.

14. Watch television commercials for one night. Take notes on them, and report to the class about which commercials make the best use of gestures to convey their messages.

15. Prepare the final draft of your speech. Make sure the introduction, body, and conclusion flow together. Review the speech closely. Reword any parts that need improvement.

NOTES

1. Appreciation is extended to Dr. Joe Townsend, professor of agricultural education, Texas A&M University, for sharing notes from his "Leadership" class about speechwriting.
2. "National FFA Prepared Public Speaking Career Development Event," in *Career Development Events Handbook, 2012–2016* (Indianapolis, IN: National FFA Organization, 2006); available at http://www.ffa.org/documents/cde_prepared.pdf Note: *2017-2021 National FFA Proposed Revision,* The National FFA Organization. Indianapolis, IN. Compliments of National FFA Staff Member, Frank Saldana.
3. S. E. Lucas, *The Art of Public Speaking,* 12th ed. (New York: McGraw Hill Education, 2014).
4. E. Ehrlich and G. R. Hawes, *Speak for Success* (New York: Bantam Books, 1984), p. 148.
5. The *Encyclopedia of Associations,* published by Gale, has several volumes and special editions, such as *National Organizations of the U.S.* and *International Organizations.*
6. "Effective Communication Skills," in L. Kauer (Ed.), *Introduction to Agricultural Sales and Service,* Teacher Edition (Stillwater, OK: Curriculum and Instructional Materials Center, Oklahoma Department of Vocational and Technical Education, 1990).
7. Townsend, "Leadership" class notes.
8. Ibid.
9. Ibid.
10. Ibid.
11. E. Johnson, *Public Speaking* (Nashville, TN: Tennessee Department of Education, Division of Vocational-Technical Education, n.d.).
12. J. R. O'Connor, *Speech: Exploring Communication,* 4th ed. (Lincolnwood, IL: National Textbook Company, 1996), pp. 275–276.
13. R. F. Verderber, *Communicate!* 9th ed. (Belmont, CA: Wadsworth, 1998).
14. *National FFA Career and Leadership Development Events Handbook 2017–2021: Prepared Public Speaking Career Development Event,* pp. 2–5. Retrieved July 24, 2016 from https://www.ffa.org /SiteCollectionDocuments/cde_2017_2021_prepared_public_speaking.pdf.
15. Ibid.
16. Ibid.
17. Ibid.
18. Ibid.
19. Ibid.
20. "Effective Communication Skills."

9 FFA EXTEMPORANEOUS PUBLIC SPEAKING

Extemporaneous or impromptu speaking is a skill that takes much preparation and dedication to develop. Although you do not write or memorize a speech as with prepared public speaking, this type of speaking still entails research, practice, delivery, and answering questions in a professional manner.

Objectives

After completing this chapter, the student should be able to:

- Discuss the advantages and disadvantages of extemporaneous speaking
- Prepare and research a speech
- Develop an extemporaneous speech
- Prepare an outline of a speech
- Implement various practice procedures
- Describe procedures to follow in the speech preparation room
- Discuss the use of note cards
- Deliver an extemporaneous speech
- Explain the parts of an extemporaneous speech judge's score sheet
- Answer questions about the subject matter of the speech after delivery

Terms to Know

- extemporaneous
- impromptu
- scantily
- ambiguous
- simultaneous
- earnestness
- slovenliness
- verbose
- exaggeration
- mastery
- credence
- rudiments
- acronym
- animate
- discreet
- captivate
- empathy

Speech delivery may be classified into four types: extemporaneous, impromptu, reading from manuscript, and memorized. Eventually, you should be able to deliver all four styles. In this chapter, we study **extemporaneous** speaking, a type of speech for which the speaker prepares ideas but does not memorize exact words. It is a skill that you will use many times in your career to influence people and achieve your goals. Adrienne Gentry, an FFA member who excelled in the Extemporaneous Public Speaking Career Development Event, went on to use her skills as a state FFA officer, a leadership trainer in Eastern Europe, and a congressional intern in Washington, DC (Figure 9–1). The most widely used form of public speaking addressed in this text, extemporaneous speaking, develops the fluency needed for impromptu talks and the communicative ability needed for speaking either from manuscript or from memory.

EXTEMPORANEOUS SPEAKING

Unlike for the **impromptu** speech, in which the speaker talks "off the cuff," with no chance for preparation, for an extemporaneous speech one uses only a set of brief notes on a speaking outline developed after a limited amount of preparation time. Exactly what the speaker says is selected at the time the speech is delivered.[1]

Nevertheless, the extemporaneous speech is carefully prepared. Mark Twain said, "It usually takes me more than three weeks to prepare a good impromptu [ad-lib] speech."[2] You know exactly how you will begin the speech, how you will develop each main point, and how you will conclude the speech. It is not **scantily** (scarcely; barely sufficient) prepared. However, you do not write out the speech word for word, nor do you memorize it. For a true extemporaneous speech, then, you have ideas firmly fixed in your mind, but you do not memorize the words in which you will express those ideas.[3]

Keys to Effective Extemporaneous Speaking

More than anything else, the ability to speak *ex tempore* requires that you read, study, gather information, and memorize information. When you actually present your speech, you must remember only your organization, supporting materials, and planned mode of delivery.[4] However, an extemporaneous speech should be

FIGURE 9–1 Adrienne Gentry was an accomplished extemporaneous speaker as a high school FFA member. She has used her speaking skills to develop leaders around the world and to work as a congressional intern in Washington, DC. *(Photo by Blane Marable.)*

the product of careful planning and thorough rehearsal. The extemporaneous speaker's preparation is extensive and painstaking. To do well with an extemporaneous speech, especially in the FFA Extemporaneous Speaking Career Development Event (CDE), you will have to do a great deal of research, rehearse outlining and delivery, and practice answering questions.

Adaptability of Extemporaneous Speaking

You can adapt the final delivery of an extemporaneous speech to the occasion because you are not locked into a predetermined, memorized sequence or any exact words. You can offer details, if necessary, during delivery. If the listeners look puzzled, you can restate an idea, add an illustration, make a comparison, or define an **ambiguous** phrase that has more than one possible meaning. Extemporaneous speakers who are sensitive to audience reactions and responses vary their language during delivery. Your speech takes its final form only as you deliver it.[5]

The circumstances of delivery mandate that extemporaneous speakers in large part create their speeches as they deliver them. Sometimes, as in the impromptu mode, speakers craft the whole speech as they talk. In fact, in every impromptu speech (as well as in most conversations), creation and presentation are **simultaneous**, happening at the same time. No other artists, not even actors or musical performers, create their entire work while in front of an audience.

Advantages of Extemporaneous Speaking

The first advantage of an extemporaneous speech is flexibility: you can adapt it and your delivery of it to each new situation. For example, suppose that a group to which you are speaking includes members of another group to which you spoke previously. These audience members would recognize a joke or illustration if you used the same one. Therefore, you adapt your planned presentation, changing it to remain fresh and better suit the current situation.

New ideas will flash into your mind, inspired by the audience. Therefore, adapt your thoughts and choose your words as you go along. Though you plan and generally arrange your ideas in advance, when new illustrations come to you at the last moment,

you can include them as you speak. An extemporaneous speaker must always be ready to meet unforeseen speaking situations as they develop.

The second advantage of the extemporaneous speech is that it promotes a personal relationship and fluid interaction between the speaker and the audience. The extemporaneous speaker is not glued to a manuscript, nor is he or she trying to remember each memorized word correctly. Because the speaker is prepared (even overflowing) with subject matter, he or she can talk about it with the audience as though in personal conversation. There should be no barriers between the speaker and the audience.

Of course, you do not automatically speak better when extemporizing than when reciting a memorized speech. In fact, you will probably not speak nearly as well—but with experience, you can learn to speak better. The reasons are obvious. Neither your mind nor your eyes are on a manuscript. You are not trying to call up each memorized word. Instead, you face a live audience: as you talk, they respond to what you say. They show their response by facial expression, by nods of the head, and by their degree of attention. Their responses spur you to greater **earnestness** (sincerity or seriousness) and power. In this feedback loop, you stimulate the audience and the audience stimulates you.[6]

Disadvantages of Extemporaneous Speaking

A potential disadvantage of extemporaneous speaking is **slovenliness** (carelessness, sloppiness, or messiness). If you do not prepare carefully, you will ramble. You may also develop sloppy or careless habits of using phrases such as "very unique" or "most important." You may fall back on words that save you from really having to think: "I believe in *true Americanism* and in *justice* and *liberty.*" As used in this example, the four italicized words have no real meaning. Actually, you may have wanted to express that people have the right to free elections, but did not think fast enough to state this specifically in well-chosen words.

A careless extemporaneous speaker often relies on pet words and phrases that he or she repeats constantly. For example, each new idea may be introduced by *also* and summarized by *do you see?* The audience will notice these formulaic expressions, because they get old and boring after a very few repetitions.[7]

The unprepared or sloppy extemporaneous speaker can also be overly **verbose**, using too many words while searching for the right ones. Brevity requires exactness and thought. In other words, if you do not plan exactly, you will wander, ramble, and digress, using many unnecessary words just to fill the time.

Unless you watch yourself carefully, extemporaneous speaking may also lead you into **exaggeration**, or overstatement of the importance of a person, thing, or occurrence. In the excitement of speaking and the heat of the moment, a speaker may expand or stretch the truth: *a few* becomes *many*, *sometimes* becomes *always*, and *seldom* becomes *never*. This tendency can develop into a habit that is hard to break and dangerous to maintain. Exaggeration often comes perilously close to dishonesty; certainly, it does not preserve the accuracy your audiences deserve.

PREPARATION AND RESEARCH

You may think that **mastery** (high accomplishment and skill in a given area) is not a factor in extemporaneous speech, wherein you must collect, organize, and present your thoughts almost simultaneously. You will probably discover that you are quite wrong. Because you do not have a speech memorized—and in fact do not even know your exact subject until just before you speak—you need a great deal of information to draw on. Your best assets, therefore, are a good memory, a well-stocked information bank, and a background of reading and experience that is both broad and deep.[8] Some claim that the FFA Extemporaneous Speaking CDE is actually an information contest, and this is partially true. A speech without a factual background and a solid underpinning of subject knowledge is merely words without substance.

Preparation

Get started by investigating and analyzing your topic areas. Arrange your ideas, select material for the development of those ideas, prepare a suitable outline, and choose a method of attack. Even though you will not know your specific topic, you can still rehearse your delivery in the general thematic areas specified by the FFA. With good preparation, appropriate words will come to you during delivery of your speech. "The best extemporaneous speech develops from a solid foundation of rigorous preparation. Once sufficiently prepared, you will find the freedom, flexibility, and spontaneity needed to adapt to the shifting reactions of your listeners."[9]

Research

Notebook Start by preparing a notebook for the contest. Your notebook will become the key to your success. For this FFA CDE, topics will be drawn from the general themes of agricultural literacy and advocacy, current agricultural issues, advancing agriculture through agriculture science, current technology uses and applications in agriculture, agrimarketing and international agriculture, and food and fiber systems, so arrange your notebook by these six areas.

Read, read, read! There is no substitute for research as you prepare for the contest. Start by surfing the Internet (Figure 9–2). This is one of the best places to gather a large amount of information in a relatively short time. Print out or bookmark any information that may be important for your research process. After you have a good idea of possible topics, start investigating and scanning material from your agricultural education classroom. When you see articles on the Web and in classroom texts and magazines that relate to the six theme areas, place printouts or copies of the articles into your notebook. If you own the magazine you are researching, you may simply cut out the article and tape it into your notebook.

Insert tabs for subtopics. For example, agrimarketing and international agricultural relations is one of the six sections in your notebook. Within this section, place tabs for pertinent subtopics, such as (1) the Farm Bill and (2) international perceptions of genetically modified organisms (GMOs). The student in Figure 9–3 is preparing her notebook with sections and tabs.

Sources A huge variety of sources is available for use in researching your topics. The *Agriculture Yearbooks* are published annually by topic areas. For agrimarketing, there is another yearbook called *Marketing U.S. Agriculture*. To study up on international agriculture relations, the yearbook *U.S. Agriculture in a Global Economy* is an excellent source

FIGURE 9–2 These students are using the Internet to find information in support of their extemporaneous speeches. (© iStock/sturti)

for topics under this theme. For the other theme areas, such as advancing agriculture through agriculture science or current technology uses and applications in agriculture, you can find articles in the *Americans in Agriculture* yearbook. (If a library near you does not have these references, you can obtain them through your congressperson.)

Textbooks, magazines, research publications, and newspapers are all terrific sources of material. Many textbooks even have the same titles as the theme areas, so you are assured of finding relevant, useful information in them. Trade publications, such as breed association magazines, often address many relevant issues. *Progressive Farmer*, *Successful Farming*, and *Hoard's Dairyman* provide information in approachable language. They can be helpful in giving you background, an overview, or in-depth understanding of a particular issue. Research publications are often technical, but many are written in readable language. One such publication is *Agricultural Research*, published monthly by the U.S. Department of Agriculture. Articles in this publication report on the latest research done at the various USDA research stations throughout the country.

Newspapers provide up-to-date information. Look beyond your community and city newspapers; for example, *USA Today* has articles pertinent to many of the major theme areas. *USA Today* also produces a magazine with articles reprinted from its daily newspaper.

Many agricultural education programs have access to the AG/ED Network, which is a national online educational database. In the AG/ED Network, you will find many different agricultural lessons. Of course, there is an abundance of information on the Internet. By using search engines, you can find practically anything on the six FFA subject areas of extemporaneous speaking. As with any research, though, you must carefully judge the credibility, truthfulness, and accuracy of Internet information and sources.

An excellent online resource that will usually yield more credible findings is Google Scholar, https://scholar.google.com. It is a free search engine that returns scholarly publications (e.g., journals, books, reports, etc.). In other words, it will give you results that are based on science, not opinion or propaganda.

Use whatever resources you need to reach your objectives as long as they are within the rules. Do not forget the *Readers' Guide* and other library resources. Interviews are also appropriate for specialty information. The

FIGURE 9–3 Extemporaneous speaking in the FFA is in large part an information contest. This student is preparing a notebook and grouping information into tabbed sections. *(© iStock/Catherine Yeulet)*

rules of the FFA Extemporaneous Speaking CDE limit preparation-room sources to five items: four books and a notebook that contains no more than 100 pages.

The more research you do, the more knowledgeable you become. Knowledge leads to confidence, which makes you more able to relate to the audience. The better you relate to the audience, the better your speech will be. Finally, the more you research, the better you will be able to answer questions about the subject matter of your speech.

DEVELOPING YOUR SPEECH

You now have in your possession information and ideas from which to develop a speech. Remember, for an impromptu speech, you do not know what you will speak about; for extemporaneous

speaking, you have at least a general idea of what your topic might be. Start practicing by developing topics in each of the main theme areas.

Be sure that your information for your main points is accurate and sufficient. If necessary, provide additional facts, statistics, examples, and other materials that will lend **credence**, or believability, to your main points. Draw on your own resources, but also verify and substantiate your knowledge of a subject through additional observations, interviews, and reading.[10] As you develop your talk, devise preparation methods that will work with any topic you are assigned. Five possible methods of organizing and attacking a topic instantly are (1) six honest servants; (2) PREP; (3) past, present, and future; (4) object or visual aid; and (5) order of events.

Six Honest Servants

Rudyard Kipling, when asked how he became such a prolific writer, replied, "I have six honest servants. They've taught me all I know. Their names are who, what, when, where, why, and how." In six words, you have the **rudiments**, or basic principles, of all good speaking, writing, and reporting.[11] If you use this formula correctly and conscientiously, you should always be successful, because you will be specific, organized, concise, and interesting. Consider the following:

> Terry Young from Middle Tennessee State University set the world land speed record for a hydrogen-fueled vehicle at the Bonneville Salt Flats in Wendover, Utah, in 1992. Young and his professor, Cliff Ricketts, wanted to demonstrate the viability of hydrogen from water as a fuel and show the speed and power of an environmentally clean vehicle. This was made possible by building an engine based on the physical principle of thermodynamics.

This paragraph includes all six honest servants.

PREP

PREP is an **acronym** (a word formed from the initial letters of a phrase or title) for point, reason, example, and point.

Point: State your point.
Reason: Give a reason for your point.
Example: Furnish one example (a statistic,

comparison, incident or experience, illustration, or exhibit or demonstration) supporting your point. Any talk that does not include a concrete example will be weak and probably uninteresting as well.

Point: Restate or rephrase your point or position.[12]

Here is an example of the PREP method:

Point: Biotechnology can lead to great economic benefits in the area of ornamental horticulture.

Reason: Through plant genetics, multiple offspring can be created from one plant.

Example: Hundreds of roses can be cloned by using tissue cultures from cells from a single rose petal.

Point: Because of biotechnology, less land, labor, and resources are needed to produce equal amounts of products, leading to tremendous economic gains.

The PREP formula will be of great assistance in wise use of your time. Take any topic from the four main theme areas mentioned earlier and apply the PREP formula to develop an extemporaneous speech on that topic. You may amaze yourself with the results.

Past, Present, and Future

The past-present-future formula applies to everything you could imagine, whether **animate** (possessing or characterized by life) or inanimate. Again, the rules of extemporaneous speaking apply: be specific, brief, and organized.[13] If you apply the past-present-future formula to agricultural power technology, for example, the flow will come easily:

Past: In the early days of American agriculture, fields were cultivated with horses and mules.

Present: Today, we cultivate fields with powerful four-wheel-drive tractors equipped with air conditioners, radios, CD players, and global positioning units.

Future: What does the future hold? Already in the news we hear about no-till cultivators equipped with pesticide applicators to prevent insects, weeds, and diseases. Tractors can be remotely controlled by computers from your home. Satellites from space can give you information about the condition of your crops.

Object or Visual Aid

Using a visual aid is a modified version of the elementary-school "show and tell" activity.[14] For example, if you are assigned a topic on international agriculture, you could use your shoes to support your speech. You could make the point that leather is produced from American cattle, cured and processed in Mexico, made into shoes in Thailand, and marketed in Europe by wholesalers to retailers in the United States. You could also use an item that you have in the pocket of your FFA jacket that illustrates a point you wish to make about agriscience technology or agrimarketing.

Do not overlook a chance to use something small and **discreet** (modest, reserved) as a visual aid. Examples are bracelets, key chains bearing quotations, pens, pencils, cards, medicine bottles, credit cards, or money. Take to heart the warning about size, though. As a general rule, if you are not already using it, or if it does not fit inconspicuously into your pocket, do not use it.

Some people use an adaptation of the object/visual aid method known by the acronym **SPEECH**, which stands for

Subject
Point
Enthusiasm
Exhibit
Concise (or **C**lear)
Humor

Montgomery suggests that all you really need is an object for the exhibit; if you have that, the rest will come easily. You make a point about it concisely, clearly, and enthusiastically. Your point is clearer (and your speech is better) because you have a visual aid, something that listeners can readily relate to as you talk about it.

Order of Events

One other method of organizing your speech is to use a **time sequence** in which you discuss events in terms of the hour, day, month, or year, moving forward or backward from a certain time. A **space order** would take you from east to west, top to bottom, or front to rear. Using **causal order**, you might identify certain causes and then discuss the results or consequences that follow from them. You can use a special order devised to fit your specific purposes,[15] but make sure that whatever order you choose enables your speech to flow smoothly and logically.

OUTLINING YOUR SPEECH

An outline organizes the results of your research as you develop your speech. It charts the main points and subpoints along the course that your speech will follow. Main points support or explain your topic. Subpoints support or explain either your main points or other subpoints.

An excellent rule of thumb for any speech is "tell them times three." A fully developed outline is divided into three main sections:

1. An introduction (which attracts audience interest, initiates rapport between the speaker and the audience, and reveals the subject and sequence of the speech); tell them what you are going to tell them.
2. The body or discussion (the major part of both the outline and the speech, in which the speaker develops his or her theme); tell them.
3. The conclusion (a summary of the entire speech); tell them what you told them.[16]

Chapter 8 discusses outlines and the process of outlining in depth.

Curtis Childers, former National FFA Extemporaneous Speaking Champion and National FFA president, offers a slightly different slant on outlines (Figure 9–4). Experiment a bit with a traditional outline or the Childers-style outline, and use whichever is most suited to the way you like to work.

Attention-Getter

Make sure your attention-getter is as dynamic as possible. Tell a story; use an analogy; pose a question; find an unusual, intriguing way of pulling your audience in. Many speakers use a line from a popular song. When the hit song "Don't Blink" by Kenny Chesney first came out, many speakers alluded to it. Many stories were told about precious moments in an agriculturalist's life. The primary purpose is to **captivate** the

---

CHILDERS'S EXTEMPORANEOUS OUTLINE

   I. Childers's attention-getter

  II. Topic introduction

 III. Main points

     A.

     B.

     C.

  IV. Topic conclusion

   V. Tie-in to the attention-getter

FIGURE 9–4 Curtis Childers uses this type of outline for his extemporaneous speeches. The attention-getter, which is usually short (15 to 30 seconds), comes before the topic is introduced. He goes back to the attention-getter after the conclusion of the speech.

---

audience (hold the audience's attention) so they think, "Hey, this is going to be interesting!" and pay close attention.

Topic Introduction

The topic introduction is where most speeches begin. Introduce an extemporaneous speech just as you would a prepared speech: state the topic and tell audience members what you are going to tell them. You basically want to say (1) why this topic is important and (2) why the audience needs to know this (that is, how this information will benefit the audience personally, financially, or professionally).

State the Main Points

In Chapter 8, we stated that you need three to five main points for a speech. Three is usually preferred, and you should never try to cover more than five. The body of your speech is where you "tell them" the first time. Time will go quickly if you have subpoints for each of your main points. Support each main point with recognizable, everyday examples. For instance, if agrimarketing is the topic, you could use the Associated Milk Producers slogan, "Taste the difference," as an example. "Pork, the other white meat" was used successfully by the National Pork Producers Council; beef and poultry producers roll out new campaigns annually, such as the dairy cow telling people to "eat more chicken." Just make sure your examples, illustrations, stories, and other materials are up to date and relevant.

Whenever possible, tie the main points together with a memorable introductory phrase. For example, the phrase "Grow them, know them, show them" was used to illustrate the main points of a speech about educating the media regarding agriculture.[17] As short and simple as it is, this phrase is memorable, punchy, and leaves the audience with a positive impression. This phrase is not the topic introduction, nor is it a full introduction to the main points, but it does embody the following three main points:

1. The media need to be educated about the facts of production agriculture ("Grow them").
2. Use public relations and human relations skills to become better acquainted with members of the media ("Know them").
3. Invite the media to the farm to spend a day experiencing production agriculture ("Show them").

This device is neither necessary nor possible with every speech. However, a memorable tagline, motto, or example immediately attracts interest and attention, and certainly adds clarity and understanding if you can work it into your speech.

Topic Conclusion

As you prepare your outline, simply remember (whether you write it down or not) to link your main points together. For the conclusion, summarize the main points; that is, "tell them what you just told them."

Tie-in to the Attention-Getter

Go back to the story, object, or whatever else you used as an attention-getter, and briefly connect it to your main points. This is useful as a concluding mechanism whether or not you mentioned your attention-getter in the body of the speech.

PRACTICING YOUR EXTEMPORANEOUS SPEECH

Lloyd George, former prime minister of England, wrote:

"To trust to the inspiration of the moment"—that is the fatal phrase upon which many promising careers have been wrecked. The surest road to inspiration is preparation. I have seen many men of courage and capacity fall for lack of industry. Mastery in speech can only be reached by mastery in one's subject.[18]

Practice Builds Success

Preparation and practice are the keys to self-confidence, enthusiasm, and competence. Mastery comes from study and practice—or, in one word, **preparation**. There is just one sure cure for all our negative speech habits, weaknesses, faults, and mannerisms. Get truly excited about practice and preparation, and 99 percent of your speaking problems will disappear. Because attitude is so important for success in any kind of speaking, especially extemporaneous speaking, follow Oliver Wendell Holmes's advice: "Success is the result of mental attitude, and the right mental attitude will bring success in everything you do."[19]

Public Practice

Set up formal practice times with your agricultural education teacher, your classmates, and the speech teacher in your school. Practice as though you were actually at a contest. Draw a topic; prepare for 30 minutes, using only the five permitted references; and actually deliver an extemporaneous speech. There is no substitute for this kind of experience. In this arena, you really do learn by doing. The more you practice, the more you will be able to convert your thoughts into a coherent, interesting speech. You will learn how to connect the attention-getter and introduction to the main points, and practice will improve the flow of and transitions in your speech. The FFA member in Figure 9–5 is practicing in front of the speech teacher and others in her class.

Preplanned Introduction

It is wise to think up a few good sentences that are usable in any introduction, to ensure against a fumbling or tentative start. Prepare introductions for each of the six main theme areas. Remember, this is an extemporaneous speech and not impromptu speaking. Therefore, it is appropriate to prepare introductory thoughts even before you draw a topic.

FIGURE 9–5 By speaking in front of teachers and other adults, you gain confidence in your delivery and topic-area knowledge. *(© iStock/Mlenny)*

Preplanned Conclusion

To prevent a total collapse of your speech, you may wish to draft a few good sentences to avoid an inconclusive ending. Obviously, these sentences must be open-ended and flexible, because you do not yet know your specific topic. As with the introduction, you can develop focused thoughts for each of the six theme areas.

Practice Writing Main Points

When you draw a topic area during practice, quickly select three to five points. This is the time to prepare and outline your thoughts. Practice finding data and statistics to support your points by using your four books and notebook. Practice aloud to test your ideas. Fix the outlined main points in your mind. Develop an ear for the sound and swing of your talk.

Practice Thinking Skills

When you first practice with your outline, you are likely to find that your speech takes too much time and that you are wordy or tend to ramble. During the first run-through, your statement of the main points may be confused, and even your telling of a simple story may be hard to follow.

This is because your thinking is confused. Why does your choice of words get better after several trials? Because your thinking has improved. As you rehearse ideas extemporaneously, you literally think out loud; and as you word ideas differently, you clarify your thoughts.

Although verbosity is the most common difficulty, sometimes you may find yourself at a total loss for words, standing in uncomfortable silence as you rack your brain. After stating your idea, you may not be able to think of anything else to say. What is the trouble? Again, it is your thinking. You have an idea but you have not thought it through; you have not developed it. So, practice out loud. Try stating the thought in different ways. By talking about it, you will clarify your thinking.

Practice Choosing the Right Words

As you practice, you must also learn to choose words that will communicate your thinking to listeners. The words that best convey your thinking for yourself may or may not be the right words for an audience. Therefore, practice with your prospective audience consciously and constantly in mind, and in each practice session select and try out various words that might suit them.[20]

Specific Practice Sequence

- Organize three to five main points in your mind.
- Organize the specific and supporting material in your mind. Do not memorize it as you would a prepared speech. Picture or visualize it rather than memorizing.
- Read your written outline silently and slowly.
- Read the outline aloud.
- Put down your written outline and rehearse the speech aloud. If you forget any part, go on to the next part that you do remember. In this part of the exercise, you are trying to maintain a whole thought pattern. Do not get tangled up in details.
- Study your outline and note places where you forgot, skipped things, or got out of sequence. "Fix" your speech mentally. Then read your outline again aloud, slowly and thoughtfully.
- Put aside the written outline and rehearse the speech aloud from start to finish.
- Rehearse the speech formally 5 to 10 times (do not count the previous silent rehearsals). Start with thoughts on paper in the outline. Then, transfer the written points to mental images (thoughts in your mind). Now get ready to create the speech while standing on your feet.
- Rehearse standing. Plan your posture and actions.
- Rehearse in a room about the size of the room in which you will speak. Get the feel of your voice filling the room. Make sure everyone can hear you.[21]

Learning to speak extemporaneously is like learning to read, learning to listen, or learning to use a typewriter. It takes practice. Practice builds knowledge. Knowledge builds confidence. Confidence builds success.

PREPARATION ROOM

Participants in FFA Extemporaneous Speaking contests are given 30 minutes in a preparation room. You are permitted to take in and use only a limited number and type of resources while you prepare your speech. Other contestants will be sharing this room at the same time.

Topic Selection

Thirty minutes before you are to speak, you will be asked to "draw" a topic. According to the rules of the FFA Extemporaneous Public Speaking CDE, the contestants "draw three specific topics relating to the industry of agriculture." After the contestant selects one of those topics, all three topics are returned to the original pool before the next drawing. Obviously, choose the topic for which you are best prepared.

Resources Needed

Make sure you have your notebook and four other references, as well as a pencil and note cards. To help keep you on track with the allotted time, wear a watch or have a stopwatch with you.

Get Comfortable and Organized

Take off your FFA jacket and shoes if it helps. The speaker in Figure 9–6 is releasing tension as she prepares her speech. Once you start to prepare, eliminate everything not needed. For example, if only the agri- marketing book is pertinent to your topic, set aside the other three books. Then you will have only one book and your notebook to search through for information.

Select Three to Five Main Points

Quickly combine all relevant resources you think you may want to use for your topic. Brainstorm and write down words quickly as they come to mind. Select three to five main points and organize the speech. Use your resources for specific information and supporting materials. Use the table of contents or index to speed up your research. Some contestants mark the outer pages of their reference books, color-coding them for selected topics. This enables them to access pertinent information very quickly. Make notes on the back of your outline note cards in anticipation of possible questions.

Supply Supporting Facts

Do not just prepare a flowery presentation without any substance or grounding in reality.

FIGURE 9–6 This contestant has removed her jacket to relax as she prepares for her speech.

As you add comments and gather supporting evidence, include some facts and figures, and then transfer what you know to your listeners. In other words, let people hear what you know about the topic. Present your own thoughts and let your opinions come through. Do not just parrot someone else's ideas: you could lose your credibility and would probably be boring in the process.[22]

Manage Your Time

Remember, you have only 30 minutes in the preparation room. Spend 10 to 15 minutes searching out materials and writing down facts in your outline. You have 10 minutes of preparation time to use a provided computer for research. Spend the remaining time rehearsing your presentation. Go over in your mind at least two or three times what you are going to say. Go over and over the three to five main points. Make your presentation to the wall, a picture, or whatever is available in the preparation room.[23] Do not forget to review your attention-getter, introduction, and conclusion, along with your main points. Each presentation room will have a designated person to answer your questions.

Get Mentally Ready

Other contestants will probably be in the room with you. Do not bother them or let them distract you. Talk at a normal pace as you practice. If you cannot practice your speech aloud, go over your presentation mentally several times. Ready yourself mentally to perform. In athletics, we would say "Psych yourself up."

NOTE CARDS

The contest rules state that a participant "will be permitted to use notes while speaking, but deductions in scoring may be made for this practice if it detracts from the effectiveness of the presentation."[24] It is only reasonable to assume that most beginning speakers will need note cards (Figure 9–7). However, if you do use them, be discreet: fold the card into the palm of your hand. Refer to your notes when you need to, but do not read them verbatim. Use your note card(s) as an aid, not a crutch. Try to look at your note card(s) only as you go to your next point. Have a skeleton outline ready before you go into the preparation room. Once you begin your preparation, fill in information for the topic you drew.

PREPLANNED OUTLINE

TOPIC: New Ways of Marketing and/or Advertising Agriculture Products

I. ATTENTION GETTER

Tell Grocery Store Strike Story

II. TOPIC INTRODUCTION

Products are advertised for the 2000s to increase sales by stressing health, convenience, and packaging.

III. MAIN POINTS

   A. Products are advertised and marketed to promote health:
      1. Fat-Free   3. Low Calories
      2. Low-Fat   4. Low Cholesterol

   B. Products are marketed for the out-of-home working mother (and Father)
      1. Pre-cooked   3. Instant
      2. Microwave   4. Frozen

   C. Packaging of Products
      1. Bright Packaging   4. California Raisins
      2. Small Quantities   5. Ziploc Packaging
      3. Easy Access

IV. TOPIC CONCLUSION

As you can see, maybe the good old days are not as good after all.

V. TIE-IN ATTENTION GETTER

The Strike is over and we are glad.

FIGURE 9–7 A sample note card with a skeleton outline that has been completed in the preparation room.

DELIVERY OF YOUR SPEECH

Chapter 7 provides extensive coverage on delivery. Refer to it for a discussion of power of expression, voice, stage presence, and general effect. Chapter 8 has some additional comments about delivery in prepared speaking. The comments in this section, although they could apply to any type of speaking, are specifically geared toward extemporaneous speaking. All of your preparations culminate either directly or indirectly in the delivery of your speech through your voice, words, and body movements. Your goal is to talk *with* people, not *at* people. If you truly believe in the importance of your topic,

you will be able to transfer these convictions to your audience.

When it is your turn to speak, take your time. Begin speaking only when you are ready; your allotted time does not begin to run until you start talking. Your poise will help establish rapport with the audience.

Transferring Feelings

Because you do not follow a script with extemporaneous delivery, you can be in more contact with the audience. Through your words, voice, gestures, and mannerisms, you can respond to the listeners' reactions as they appear. Two words underline this concept: *empathy* and *rapport*.

Empathy means involving yourself, imaginatively and sympathetically, with the feelings, thoughts, or attitudes of other people. You try to see things as they see them and feel as they feel, even though you personally may not subscribe to their views.[25] At least for the length of your speech, you walk in their shoes.

When both speaker and listener are open to each other's thoughts and emotions, the resulting interaction is known as *rapport*. "Rapport is the hallmark of good speaker-audience relationships." Without rapport, your communication is always incomplete and therefore imperfect; in the worst case, it fails completely.[26]

Conversational Norm

Former FFA extemporaneous public speaker and National FFA President Curtis Childers suggests that you use a conversational norm as you speak. Because lively conversation exemplifies the best attributes of communicative delivery, the conversational norm provides the best guide to the delivery of any type of speech. The qualities of voice and action that characterize lively conversation—directness, animation, variety, and spontaneity—are appropriate to all speaking, formal as well as informal. Many people have the mistaken notion that public speech calls for an oratorical manner that is significantly and mysteriously different from their natural mode of private speech. The truth is that if you can comfortably talk easily, directly, and responsively in conversation, you may proceed with assurance in adopting the conversational norm for public speaking.

Note that this style may not work equally well or be appropriate in all parts of the country. For an FFA CDE, charismatic enthusiasm is the norm. Your FFA adviser will help you identify what the norm is for your situation and surroundings. Nevertheless, you are fairly safe in basing your approach on this method. Here are a few suggestions that will help strengthen your conversational style of delivery during your extemporaneous speech.[27]

Interest The first requisite of good delivery is an evident interest in what you are saying and in the people with whom you are talking. If you take a sincere interest in your subject and your listeners, many of the most common speech-delivery problems will vanish or never even arise.

Think Think about what you are saying while you are saying it. Audiences can spot mental and emotional "drifting" almost immediately; the speaker's manner and voice almost always betray this kind of distancing. You cannot truly communicate unless you stay in touch with what you are saying as you speak.

Respond React to your audience. Think about what you are saying while you are saying it, but think about it in relation to your audience. Every audience sends out signals. Tune into those signals to see whether they carry messages of understanding, puzzlement, interest, boredom, weariness, or disapproval. Ask yourself questions like, "Can they hear me?" "Would another example help?"

Captivate Hold your listeners' attention by talking things over with them instead of just talking *at* them. The more they feel like they are interacting with you, the better your speech will be.

Voice Use your voice to communicate meaning and feeling. The greatest benefit of the conversational norm is as a guide to the use of your voice. Listen to the ordinary conversations around you. You will hear pleasant and unpleasant voices, clear and slovenly enunciation, good and bad diction, differences in dialect, mispronunciations, and grammatical errors. However, through it all usually come remarkable meaning and feeling. Try to preserve this immediacy and relevance while using the well-honed delivery skills that you have practiced so hard.

Body Language Speak with physical animation and directness. This can best be explained by the following:

> Speaking is action in which mind and body cooperate. Bodily action supplements and reinforces words; it energizes thought; it reveals you as a person who is self-confident rather than self-conscious. A listless person fails to sustain interest for very long; an anxious, distracted person communicates his distress; an overwrought person wears us out by trying too hard. A nice balance between relaxation and tension contributes to poise and directness.[28]

Eye Contact Establish eye contact. People like personal attention. When you look audience members straight in the eye, you are taking notice of them. You are saying, "I invite you to share this information or observation." This simple action contributes to good human relationships. It also helps you adapt and react sensitively to the responses of others.

Total Group Speak to the total group. If you are talking to a large group, you cannot focus your attention on everyone at once. Simply shift your attention smoothly from one area of the audience to another. Exclude no one. Do this easily and naturally, without swinging your head back and forth. Usually there is no need to single out individuals unless you happen to be especially interested in their reactions. Even then, be careful not to make them feel uncomfortable or targeted.

NATIONAL FFA EXTEMPORANEOUS PUBLIC SPEAKING CAREER DEVELOPMENT EVENT

The National FFA Extemporaneous Public Speaking Career Development Event (CDE) is designed to develop the ability of all FFA members to express themselves on a given subject without having prepared or rehearsed its content in advance. This gives FFA members an opportunity to formulate remarks for presentation in a very limited amount of time. The event is held in connection with the National FFA convention.

Event Rules

The speaker must follow these event rules:[29]

- Participation in the National FFA Extemporaneous Speaking CDE is limited to one person from each state association.
- "It is highly recommended that participants be in official FFA dress in each event."
- Participants will receive copies of the rules and score sheet in advance of the national event.
- "Three to eight competent and impartial persons will be selected to judge the event. At least one judge should have an agricultural background. Each state with a speaker shall provide a judge for preliminary rounds of the national event. Any adviser who has a student competing in a speaking event may not serve as a judge for that respective speaking event."
- "Any participant in possession of any electronic device, not provided by the event committee, in the preparation room is subject to disqualification."

Event Format

The format of the event should be as follows:[30]

- "Event officials will randomly draw speaking order. The superintendent will announce each participant by name and in order of the drawing."
- "The selection of topics will be held 30 minutes before the event. The participants will draw three specific topics, selected at random from the pool of 18, relating to the industry of agriculture. After selecting the topic they desire to speak on, all three topics will be returned for the next drawing."
- "Eighteen topics will be prepared by the event superintendent and will include three each from the following categories:
  - Agricultural literacy and advocacy,
  - Current agricultural issues,
  - Advancing agriculture through agriculture science,
  - Current technology uses and applications in agriculture,
  - Agrimarketing and international agriculture, and
  - Food and fiber systems."
- "Participants will be admitted to the preparation room at 15-minute intervals and given exactly 30 minutes for topic selection and preparation."
- The officials in charge of the event screen the permitted reference material for the following:
  - No more than five items total (e.g., notebook, textbook, etc.)
  - Materials may be printed books, magazines, or compilations of materials collected from Internet research.
    - Students may receive 10 minutes for online research with the help of a computer that may be provided. Access to pre-prepared documents online will be prohibited (e.g., Dropbox, email, etc.)
- "To be counted as one item, a notebook or folder of collected materials may contain NO more than 100 single-sided pages or 50 pages double sided numbered consecutively" (cannot be notes or speeches prepared by the participant or notes prepared by another person for the purpose of use in this event).
  - "References should be in original format."

- Cutting and pasting into a word processor will be considered pre-prepared materials.
- "Each speech should be the result of the participant's own effort using approved reference material which the participant may bring to the preparation room. No other assistance may be provided. Participants must use the uniform note cards provided. Any notes for speaking must be made during the 30-minute preparation period. A participant will be permitted to use notes while speaking, but deductions in scoring may be made for this practice if it detracts from the effectiveness of the presentation."
- Before the event, the judges will get and review a list of all possible topics. "Each speech should be not less than four and no more than six minutes in length. An additional five minutes will be allowed for the questioner to ask questions." The room coordinator of the event will introduce the participant by name and state, and the participant may introduce his or her speech by title only. Participants are to be penalized one point per second on each judge's score sheet for being over six minutes or under four minutes. Time commences when the speaker begins talking. Speakers may use a watch to keep a record of their time. Event officials or observers will give no time warnings."
- The speaking order is drawn at random. The program chair introduces each participant by name and in order of the drawing. A participant is "permitted to use notes while speaking, but deductions in scoring may be made for this practice if it detracts from the effectiveness of the presentation. Applause shall be withheld until all participants have spoken."
- The national contest has three rounds: preliminary, semifinals, and finals. Only the final four speakers are ranked.
- Timekeepers will record the time for each participant's speech delivery, "noting undertime or overtime, if any, for which deductions should be made." The timekeepers sit together.
- During the speeches, the judges sit in different sections of the room. They score each participant's delivery using the official score sheet.

- "Each room in all rounds will have one person designated as a questioner. This individual will ask and score all questions for the event round. Questions will pertain directly to the speaker's subject. Questions containing two or more parts should be avoided. Judges will score each participant by their ability to answer. The full five minutes should be used."
- When all participants have finished speaking, each judge totals the score for each participant. "The timekeepers' record will be used in computing the final score for each participant." The judges submit their score sheets to event officials for determination of participants' final ratings. "Participants will be ranked in numerical order on the basis of the final score to be determined by each judge without consultation. The judges' ranking of each participant then will be added, and the winner will be that participant whose total ranking is the lowest. Other placings will be determined in the same manner (low rank method of selection)."

Tiebreakers

Ties are broken according to the greatest number of low ranks. The participants' low ranks are counted, and the participant with the greatest number of low ranks is declared the winner. If a tie still exists, the event superintendent ranks the participants' response to questions. The participant with the lowest ranking from the response to questions is declared the winner. If a tie still exists, the participants' raw scores are totaled and the participant with the greatest total of raw points is declared the winner.[31]

Note that these rules may change from time to time. Go to the National FFA website (http://www .ffa.org), and find the latest CDE Handbook to make sure you are following the most current rules for the extemporaneous public speaking event.

JUDGE'S SCORE SHEET

Be sure that you know how you are being evaluated. Figures 9–8A and 9–8B show the judge's score sheet, which has a summary of the possible score points.[32] Voice, stage presence, power

NATIONAL FFA
CAREER AND LEADERSHIP
DEVELOPMENT EVENTS

Extemporaneous Public Speaking Rubric
1000 points

NAME _____   MEMBER NUMBER _____

CHAPTER _____   STATE _____

INDICATORS	Very strong evidence of skill is present 5-4 points	Moderate evidence of skill is present 3-2 points	Strong evidence of skill is not present 1-0 points	Points Earned	Weight	Total Score
Oral Communication – 600 points						
Examples	• Examples are vivid, precise and clearly explained. • Examples are original, logical and relevant	• Examples are usually concrete, sometimes needs clarification. • Examples are effective, but need more originality or thought.	• Examples are abstract or not clearly defined. • Examples are sometimes confusing, leaving the listeners with questions.		X 10	
Speaking without hesitation	• Speaks very articulately without hesitation. • Never has the need for unnecessary pauses or hesitation when speaking.	• Speaks articulately, but sometimes hesitates. • Occasionally has the need for a long pause or moderate hesitation when speaking.	• Speaks articulately, but frequently hesitates. • Frequently hesitates or has long, awkward pauses while speaking.		X 10	
Tone	• Appropriate tone is consistent. • Speaks at the right pace to be clear. • Pronunciation of words is very clear and intent is apparent.	• Appropriate tone is usually consistent. • Speaks at the right pace most of the time, but shows some nervousness. • Pronunciation of words is usually clear, sometimes vague.	• Has difficulty using an appropriate tone. • Pace is too fast; nervous. • Pronunciation of words is difficult to understand; unclear.		X 10	
Being detail-oriented	• Is able to stay fully detail-oriented. • Always provides details which support the issue; is well organized.	• Is mostly good at being detail-oriented. • Usually provides details which are supportive of the issue; displays good organizational skills.	• Has difficulty being detail-oriented. • Sometimes overlooks details that could be very beneficial to the issue; lacks organization.		X 30	
Connecting and articulating facts and issues	Exemplary in connecting facts and issues and articulating how they impact the issue locally and globally. • Possesses a strong knowledge base and is able to effectively articulate information regarding related facts and current issues.	Sufficient in connecting facts and issues and articulating how they impact the issue locally and globally. • Possesses a good knowledge base and is able to, for the most part, articulate information regarding related facts and current issues.	Has difficulty with connecting facts and issues and articulating how they impact the issue locally and globally. • Possesses some knowledge base but is unable to articulate information regarding related facts and current issues.		X 30	
Speaking unrehearsed (questions & answers)	Speaks unrehearsed with comfort and ease. • Is able to speak quickly with organized thoughts and concise answers.	Speaks unrehearsed mostly with comfort and ease, but sometimes seems nervous or unsure. • Is able to speak effectively, has to stop and think and some- times gets off focus.	Shows nervousness or seems unprepared when speaking unrehearsed. • Seems to ramble or speaks before thinking.		X 30	
			Oral Communications Total			

FIGURE 9–8A and 9–8B The National FFA Extemporaneous Public Speaking Contest Score Card with an explanation of the points. *(Courtesy of the National FFA Organization.)*

Extemporaneous Public Speaking Rubric continued

INDICATORS	Very strong evidence of skill is present 5-4 points	Moderate evidence of skill is present 3-2 points	Strong evidence of skill is not present 1-0 points	Points Earned	Weight	Total Score
Non-verbal Communication – 400 points						
Attention (eye contact)	• Eye contact constantly used as an effective connection. • Constantly looks at the entire audience (90-100% of the time).	• Eye contact is mostly effective and consistent. • Mostly looks around the audience (60-80% of the time).	• Eye contact does not always allow connection with the speaker. • Occasionally looks at someone or some groups (less than 50% of the time).		X 20	
Mannerisms	• Does not have distracting mannerisms that affect effectiveness. • No nervous habits	• Sometimes has distracting mannerisms that pull from the presentation. • Sometimes exhibits nervous habits or ticks.	• Has mannerisms that pull from the effectiveness of the presentation. • Displays some nervous habits – fidgets or anxious ticks.		X 20	
Gestures	• Gestures are purposeful and effective. • Hand motions are expressive and used to emphasize talking points. • Great posture (confident) with positive body language.	• Usually uses purposeful gestures. • Hands are sometimes used to express or emphasize. • Occasionally slumps; sometimes negative body language.	• Occasionally gestures are used effectively. • Hands are not used to emphasize talking points; hand motions are sometimes distracting. • Lacks positive body language; slumps.		X 20	
Well-poised	• Is extremely well-poised. • Poised and in control at all times	• Usually is well-poised. • Poised and in control most of the time; rarely loses composure	• Isn't always well-poised. • Sometimes seems to lose composure.		X 20	

	Non-verbal Communication Total Points	
	Oral Communication Total Points	
	Time Deduction *	
	NET TOTAL POINTS	
	RANK	

*–1 point per second under 4 minutes or over 6 minutes, determined by the timekeepers

FIGURE 9–8A and 9–8B (continued)

of expression, and general effect are discussed in Chapter 7. The following are reminders relating specifically to extemporaneous speaking.

Content Related to Topic

Preparation is the key to extemporaneous speaking. This does not mean having a "canned" speech for each of the six theme areas. For each of six main areas, choose a topic. Select three to five points about the topic and add any supporting data or statistics for each point.

Knowledge of the Subject

Know your subject. Make sure your information is correct, important, and appropriate. Make sure the examples and stories you use to illustrate your main points are accurate and relevant. For example, you would not use artificial insemination as an illustration of genetic engineering.

Organization of Material

The key words are **smooth flow** and **transition**. Make sure to preserve unity of thought through the introduction, body, and conclusion. The main points should be presented in a logical order, such as time sequence or causal order. Use appropriate language and sentence structure.

Introduce the topic. Introduce the main points. Discuss the first main point and use illustrations to explain it. Discuss the second main point. Tie it in with the first point. Address the third main point. Show the connection with the first two points. Your conclusion should briefly recap the main points. If you used an attention-getter, refer to it again as you conclude your speech.

ANSWERING QUESTIONS

The only difference between answering questions for prepared public speaking (discussed in Chapter 8) and for extemporaneous speaking is the uncertainty factor. With prepared speaking, you can anticipate the questions that are likely to arise. Obviously, this is more difficult for extemporaneous speaking. First, you do not know what your exact topic will be until the event starts. Second, it is hard to predict or anticipate questions from the judges until you are in the preparation room. Nevertheless, preparation is vital to answering questions effectively. There is no substitute for being ready! Consider the following suggestions and techniques.

Ask for Questions

Once you conclude your speech, say something like, "I would now like to address a few questions." Many contest speakers just stand silently after they finish. This statement provides a smooth transition from the speech to the question-and-answer period. It also enables more accurate timekeeping.

Positive Response

For questions that have two or more potential answers, you may want to introduce your answer with "In my opinion," or "As I see it," or "My present thinking is. . . ." The judges may not agree with your opinion, but because the answer is not presented as a fact or truth (only your belief or opinion), technically your response cannot be considered incorrect. In other words, although you may not have given the judge the answer he or she wanted, you gave a positive response.

Stall Tactfully If You Do Not Know

Sometimes you will have no idea how to respond. Rather than waiting out 30 seconds of dead silence—which may seem like 30 minutes—stall

tactfully and discreetly. Say, "I'm glad you asked that question," or "Would you repeat the question, please?" This could give you the time you need to collect your thoughts and continue without a noticeable break. However, do not get in the habit of using these phrases.

What if you do not know the answer? You have two choices. You can say, "I don't know," or you can say something to the effect of, "I didn't come across that information, but I did find that. . . ." Another possible response is, "I don't know, but if you had asked. . . ."

When the senior author of this text was speaking before the energy congressional committee, he was asked a question to which he did not know the answer. He responded by saying, "I do not know the answer to that question, but I wish you had asked me this." He then asked his own question, and responded to that to the satisfaction of the congressional committee. The key is to stay in control of the situation.

Response Time

How much time should you spend answering a question? Remember, you have a five-minute question-and-answer period, and the judges are supposed to use up all of it. If you give short answers, you could field as many as six to eight questions. If you elaborate on your answers, you could get as few as two to four questions. Therefore, if a question relates to an area in which you are very knowledgeable, you may respond to it for longer than usual; just do not overdo it. If you get a hard question, give a brief response and move on to the next question.

Data and Statistics

Use data and statistics when appropriate. Your credibility is greatly enhanced when you can quote data to support your answers. The more you read, the more able you will become in doing this. In the absence of data, relevant stories that make the point (which are also a form of supporting evidence) can be very effective.

Conversational Norm

Do not forget that you are still performing. Speakers often forget the importance of voice, power of expression, stage presence, and general effect when they are answering questions. Use the same delivery style for answering questions as you do for your four- to six- minute presentation.

When the judge calls time, it seems awkward or disrespectful to just sit down or leave the lectern without some closure or response. Say softly, "Thank you," or give another appropriate response that you and your adviser have agreed on. What does the "thank you" mean, as used here? It is a short phrase that acknowledges the contest officials' effort in permitting you to develop your speaking skills and abilities.

CONCLUSION

There is little point in extemporaneous public speaking if you merely try to regurgitate a preconceived lecture. Instead, use the preparation methods and techniques in this chapter to attack each subject as a unique "conversation" with the audience. Do not panic; remember, some nervousness is a good sign of readiness. Your audience will expect nothing extraordinary from you because they know you are speaking extemporaneously. Actually, they probably will be very encouraging. If you approach your speech with poise and determination, your chances of success are exceedingly good. A well- rounded knowledge base obtained from a consistent reading program will assist you immeasurably. The FFA Extemporaneous Public Speaking CDE is a challenge: take it! This event will truly help you develop your speaking abilities.

SUMMARY

Few people ever develop the ability to speak extemporaneously and answer questions on an unknown topic with only 30 minutes of preparation. However, this skill is an important tool for the leader's objective of influencing others. Influence is leadership. Although almost anyone can learn to speak extemporaneously, if one is willing to undertake the necessary preparation and practice, not many do. The skills of verbal ability, speaking, and thinking on your feet will set you apart from the crowd, thus helping tremendously

in advancing your career once you enter the workforce.

The advantages of extemporaneous speaking are that (1) you can adapt it to unforeseen situations, (2) it promotes a personal relationship between the speaker and the audience, and (3) it leads to superior delivery. The disadvantages are that it can lead to (1) slovenliness, (2) verbosity, and (3) exaggeration.

Preparation is the key. Read! Read! Read! There is no substitute for research. Read books, articles, and Web pages, and assemble a notebook with a section for each of the theme areas. Select four other books that can be used later in the preparation room.

As you develop your speech, make sure your information is both accurate and sufficient. Draw on your own resources, but also verify and enhance your knowledge of a subject through additional observations, interviews, and reading. Develop a method of attack that will work for any topic you select. Five methods of organizing a topic are six honest servants; PREP formula; past, present, and future; an object or visual aid; and order of events. An outline organizes the results of your research as you develop your speech. One method of outlining is to have an attention-getter, topic introduction, body (three to five main points), topic conclusion, and tie-in to the attention-getter.

Practice your speech. Preparation and practice are the keys to self-confidence, enthusiasm, and competence. Set up formal practice times with teachers, friends, and parents. Give yourself only 30 minutes while you are in the preparation room. Take your five references, eliminate everything not needed, and work on selecting three to five main points. Spend 10 to 15 minutes on your outline and the remaining time rehearsing your presentation.

If you use note cards, do so discreetly. Try to glance at the card only when you go to your next point.

Once you begin, approach the lectern and reflect for a moment; be natural, be confident, and look sharp. Deliver your speech conversationally. Express your- self while exhibiting empathy and rapport. Be aware of your four- to six-minute time limit, and know how the scoring is done. Your speech will be evaluated on content related to topic, knowledge of subject, organization of material, power of expression, voice, stage presence, general effect, and responses to questions.

Answering questions is the true challenge for the extemporaneous speaker. Give clear and thorough answers. Do not forget that you are still performing. If you do not know an answer, or if you need more time to organize your thoughts, use some techniques to discreetly stall and make a transition statement; then give a positive response.

Take It to the Net

Search the Internet for information on the six theme areas (agricultural literacy and advocacy, current agricultural issues, advancing agriculture through agriculture science, current technology uses and applications in agriculture, agrimarketing and international agriculture, and food and fiber systems). Using your favorite search engine, search each of the four theme areas separately by typing them into the search field. Find as many useful websites as you can.

Chapter Exercises

REVIEW QUESTIONS

1. Define the Terms to Know.

2. What is the difference between extemporaneous speaking and impromptu speaking?

3. What are three advantages of extemporaneous speaking?

4. What are three disadvantages of extemporaneous speaking?

5. List five types of publications that could be used as resources for preparing an extemporaneous speech.

6. Is it possible or appropriate to have a preplanned introduction and preplanned conclusion before you speak on an unknown topic? Explain.

7. What are five things you should do in the preparation room once you have drawn a topic?

8. What are eight things that will help your conversational style of delivery?

9. Explain what you would do if you were asked a controversial question with two or more potential answers.

10. What can you do if you do not know the answer to a question?

11. What are the six topic areas in the FFA Extemporaneous Speaking CDE?

12. Briefly discuss five possible methods of attacking a topic instantly.

13. According to Rudy and Kipling, what are the six honest servants?

14. List the components of the acronym *PREP*, and briefly explain each one.

15. In practicing your speech, what are the 10 steps in the Specific Practice Sequence?

16. List and briefly discuss eight things that will help strengthen the conversational style of delivery in your extemporaneous speech.

COMPLETION

1. For the FFA Extemporaneous Speaking Contest, you will be asked to draw your topic _____ minutes before you are to speak.

2. Preparation and practice are the keys to self-confidence, enthusiasm, and _____.

3. Success is the result of mental attitude, and the right mental attitude will bring _____ in everything you do.

4. Practice builds knowledge. Knowledge builds confidence. Confidence builds _____.

5. Answering _____ is the true challenge for the extemporaneous speaker.

6. In the conclusion, you tell them what you just _____ them.

For the remaining questions, insert the number of points after each category on the extemporaneous public speaking judge's score sheet.

7. Content related to subject: _____ points

8. Organization of material: _____ points

9. Power of expression: _____ points

10. Voice: _____ points

11. Stage presence: _____ points

12. General effect: _____ points

13. Response to questions: _____ points

14. Total points: _____ points

MATCHING

_____ 1. Start with a story or an analogy.

_____ 2. "Tell them what you are going to tell them."

_____ 3. "Tell them."

_____ 4. "Tell them what you told them."

_____ 5. Go back to the story at the end of the speech.

_____ 6. Speaking on the subject selected.

_____ 7. Gained through research and preparation.

_____ 8. Smooth flow and transition.

A. topic conclusion

B. tie-in to the attention-getter

C. main points

D. attention-getter

E. topic introduction

F. knowledge of the subject

G. organization of material

H. content related to topic

ACTIVITIES

1. Prepare a notebook for extemporaneous speaking. Arrange the notebook into six major sections. Title the sections *Agricultural literacy and advocacy, Current agricultural issues, Advancing agriculture through agriculture science, Current technology uses and applications in agriculture, Agrimarketing and international agriculture*, and *Food and fiber systems*. Read magazines, journal articles, newspapers, pamphlets, Extension Service bulletins, credible Web articles, and other materials. Copy articles that address the six theme areas and place them in the notebook. Insert tabs within each major section so that subtopics are arranged together. For example, tab a section called "agricultural exports." Have at least five subtopics tabbed for each main section of your notebook. Find at least one article for each subtopic.

2. Select a topic or make up one within one of the theme areas mentioned in activity 1. Prepare a one- to two-minute talk using one of the following methods of organizing and addressing the topic: six honest servants; PREP formula; past, present, and future; object or visual aid; order of events.

3. Refer to the outline section in this chapter. Select a topic in one of the theme areas and follow Childers's outline style to prepare an outline using your notebook and reference books.

4. Select a person to listen to your speech based on the outline prepared in activity 3. As you practice, ask the person to list five areas in which you need to improve.

5. Deliver your speech to the class. (Since you began preparing this extemporaneous speech as you completed the previous activities, you will have had more than 30 minutes to prepare ... but you have to start somewhere.) As appropriate, duplicate contest conditions as closely as possible: your teacher will have you draw a topic, prepare for 30 minutes, and deliver a speech.

6. Answer questions about your extemporaneous speech.

NOTES

1. S. E. Lucas, *The Art of Public Speaking*, 3d ed. (New York: McGraw Hill Education).
2. Retrieved October 16, 2009, from http://www.whatquote.com/quotes/Mark-Twain/500-It-usually-takes -me-.htm.
3. W. G. Hedde, W. N. Brigance, and V. M. Powell, *The New American Speech*, rev. ed. (New York: J. B. Lippincott, 1963), p. 144.
4. W. W. Braden, *Public Speaking: The Essentials* (New York: Harper & Row, 1966), p. 78.
5. D. C. Bryant and K. R. Wallace, *Fundamentals of Public Speaking*, 4th ed. (New York: Appleton-Century-Crofts, 1969), p. 20., (5th, ed., 1976).
6. Quoted in Hedde, Brigance, and Powell, *The New American Speech*, p. 144.
7. Ibid., p. 145.
8. Braden, *Public Speaking*, p. 76.
9. J. H. McBurney and E. J. Wrage, *Guide to Good Speech*, 4th ed. (Englewood Cliffs, NJ: Prentice-Hall, 1975), p. 20.
10. R. L. Montgomery, *A Master Guide to Public Speaking* (New York: Harper & Row, 1979), p. 59.
11. Ibid., p. 59.
12. Ibid., p. 60.
13. Ibid., p. 61.
14. Ibid.
15. C. S. Carlile and A. V. Daniels, *Project Text for Public Speaking*, 6th ed. (Boston: Allyn & Bacon, 1997).
16. McBurney and Wrage, *Guide to Good Speech*, p. 26.

17. "Competitive Public Speaking: The Extemporaneous Speech Contest" [videotape] (Lubbock, TX: Creative Educational Video, 1993). Note: Curtis Childers narrates.

18. Quoted in D. Carnegie, *Public Speaking and Influencing Men in Business* (1913; reprint, Whitefish, MT: Kessinger, 2003), p. 17.

19. Quoted in Montgomery, *A Master Guide to Public Speaking*, p. 67.

20. McBurney and Wrage, *Guide to Good Speech*, p. 27.

21. Hedde, Brigance, and Powell, *The New American Speech*, pp. 146–147.

22. "Competitive Public Speaking: The Extemporaneous Speech Contest."

23. Ibid.

24. "National FFA Extemporaneous Public Speaking Career Development Event," in *Career Development Events Handbook 2012–2016* (Indianapolis, IN: National FFA Organization, 2006); available at http://www.ffa.org/

25. McBurney and Wrage, *Guide to Good Speech*, p. 25.

26. Ibid., p. 25.

27. Ibid., pp. 26–28.

28. Ibid., p. 29.

29. *National FFA Career and Leadership Development Events Handbook 2017–2021: Extemporaneous Public Speaking Career Development Event*, pp. 2–6. Retrieved July 24, 2016 from https://www.ffa.org/SiteCollectionDocuments/cde_2017_2021_extemporaneous_public_speaking.pdf

30. Ibid., pp. 2–6.

31. Ibid., pp. 2–6.

32. Ibid., pp. 2–6.

Section 3 LEADING INDIVIDUALS AND GROUPS

10

BASIC PARLIAMENTARY PROCEDURE

Because parliamentary procedure is a democratic process, its use develops important leadership skills. Specifically, parliamentary procedure requires and incorporates extemporaneous speaking, critical thinking, decision making, and many other leadership skills. Other than local, state, and national elections, the most democratic experience occurs when a person participates in a meeting according to the rules of parliamentary procedure. Everyone has equal rights to discuss every proposal, and everyone is guaranteed a vote. This is democratic leadership in action.

Objectives

After completing this chapter, the student should be able to:

- Discuss the characteristics of a chairperson
- Demonstrate the proper procedure for handling a motion
- Describe the standard characteristics of a motion
- Describe the purpose and types of voting
- Demonstrate the following motions:
  - Main motion
  - Amend
  - Previous question
  - Refer to a committee
  - Lay motion on the table
  - Take motion from the table
  - Postpone definitely
  - Postpone indefinitely
  - Suspend the rules
  - Rise to a point of order
  - Appeal from the decision of the chair
  - Division of the house (assembly)
  - Reconsider a motion
  - Recess
  - Adjourn
- Explain the proper order of business (agenda)
- Discuss common errors in and misconceptions about parliamentary procedure

Terms to Know

- parliamentary procedure
- discuss
- vote
- chairperson
- out of order
- poise
- parliamentarian
- recognition from the chair
- decorum
- precedence
- majority vote
- two-thirds vote
- general (unanimous) consent
- main motion
- debatable
- amendable
- subsidiary motions
- germane
- primary amendment
- secondary amendment
- immediately pending motion
- committee
- standing committee
- special committee
- adjourned meeting
- order of the day
- incidental motion
- violation
- tie vote
- vote by voice (viva voce)
- privileged motion

Agricultural education and the FFA have been credited with developing leadership not only by many advocates of these programs but also by objective researchers. In Chapter 1, we discussed the areas within the FFA that contribute to leadership development. One such area is **parliamentary procedure**, which is a set of rules and procedures for keeping a meeting orderly and harmonious, and guaranteeing that all persons have equal opportunity to express themselves. The basic question is: *How* does parliamentary procedure contribute to leadership development?

In Chapter 1, we discussed behavioral leadership as one of the seven categories of leadership. Behavioral leadership includes democratic, authoritarian, laissez-faire, and situational leadership. Except in certain situations, Americans tend to favor democratic leadership. If democratic leadership is the best type of leadership, how do we learn to be democratic leaders? The answer lies with parliamentary procedure.

Parliamentary procedure rules guarantee that a meeting proceeds in a purely democratic fashion. Everyone has equal rights to **discuss** (talk about a motion after being properly recognized) every proposal, and everyone has a **vote** (method by which members express approval of or opposition to a particular action or motion). The process ensures that the majority of members support the actions taken by the organization. Using correct parliamentary procedure in organization meetings is the most democratic way of doing things.[1]

The rules of parliamentary procedure are based on consideration for the rights of the majority, individual members, and absentees. Under the rules of parliamentary law, a group is free to do what it wants while ensuring the greatest measure of protection for itself and consideration for the rights of its members.[2] Outside of local, state, and national elections, the most democratic experience possible in our country occurs when a person participates in a meeting according to the rules of parliamentary procedure. This is democratic leadership in action.

Though parliamentary procedure has evolved over the years, its roots were a few basic rules instituted in the English Parliament. Because of its efficiency and ability to support democracy, other democratic groups and societies started using parliamentary procedure as well, and the rules gradually expanded and changed. Today, *Robert's Rules of Order* has become the standard code of parliamentary procedure.

This chapter introduces parliamentary procedure to anyone who has never engaged in this useful activity. After reading this chapter, you should be able to run a meeting as a capable chairperson and as a committed member of the organization in the audience. You should be able to handle a variety of privileged, subsidiary, and incidental motions.

After practicing and becoming competent with the parliamentary procedures covered in this chapter, you should be a capable, thinking, and active chairperson or member of any organization that conducts its business according to parliamentary procedure. The information presented here covers most of the parliamentary procedure that a member would encounter in most business meetings. Chapter 11 goes a step further for those desiring more advanced, in-depth knowledge of parliamentary procedure.

THE CHAIRPERSON

The presiding officer of a meeting is ordinarily called the **chairperson**. In an organized group, the chairperson's title is usually prescribed by the bylaws. In many organizations, the president is assigned the role of chairperson. The term *chair* refers to the person or leader who is actually presiding at the meeting, whether or not that person is normally the chairperson.[3]

Addressing the Chairperson

The chairperson should be addressed by an official title. In fact, you should seek **recognition from the chair** or properly attain permission to speak. If the chairperson is also the president, "Mr. or Madam President" is proper; if the person is not the president but is the chairperson, then "Mr. or Madam Chairperson" is always appropriate. Use of this designation is recommended as a standard practice to avoid the common error of saying "Mr. or Madam President" when someone other than the president is the chair. The gender of the chairperson is recognized by the use of "Mr." or "Madam."[4]

The chairperson speaks of himself or herself only in the third person. The chairperson never uses the personal pronoun "I" to refer to himself or herself in the capacity of chair. In actual parliamentary proceedings, he or she always refers to himself or herself as "the chair." For example, if the chair must rule that a motion is **out of order** (inappropriate action in parliamentary procedure), he or she should say, "The chair rules the motion out of order."

Qualities of a Good Chairperson

The chairperson should possess certain qualities that guarantee the rights of the majority, the minority, and individual members. The chairperson should maintain order and be in complete control of the assembly at all times. The chairperson should:

Be Fair Respect and hear members on both sides of the question. In fact, the chair should deliberately alternate pro and con speakers on a motion as much as possible.

Possess Good Judgment Handle each situation thoughtfully and seriously. Always keep the discussion directed to the topic under consideration.

Manifest Poise Poise is composure under pressure or stress. Demonstrate good posture and speak with a strong, clear voice; face the group with a confident manner. The chairperson's leadership position should be obvious.

Have a Working Knowledge of Parliamentary Procedure It is highly desirable for the chairperson to be well educated in parliamentary procedure, so that he or she can make quick and correct rulings. The chair should refer to the **parliamentarian** (an expert on rules governing meetings who serves as an adviser to the chairperson) as little as possible.

Be Deliberate and Tactful The chair should neither race through business nor drag matters out unnecessarily. The chair tries to maintain group harmony in all situations.[5]

The chairperson in Figure 10–1 appears to be poised and in control of the meeting.

FIGURE 10–1 A good presiding officer should be fair, possess good judgment, manifest poise, be deliberate and tactful, and have a working knowledge of parliamentary procedure. *(© wavebreakmedia/Shutterstock.com)*

PROCEDURE FOR PROPERLY HANDLING A MOTION

One of the basic elements of effectively conducted parliamentary procedure is handling motions in the correct manner. There are eight steps in receiving and disposing of a motion (Figure 10–2).

Step 1 *A member rises and addresses the chair.*

Step 2 *The member gains* **recognition from the chair.**

Step 3 *The member makes a motion before the assembly.*

Step 4 *Another member seconds the motion.*

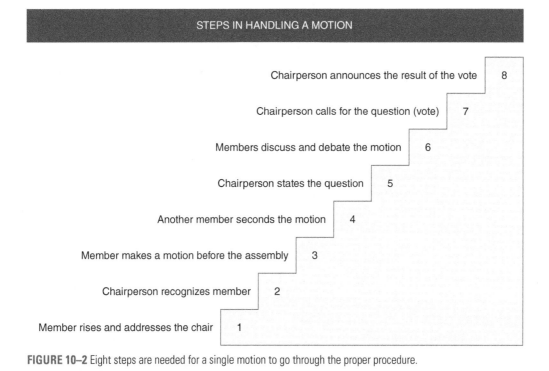

STEPS IN HANDLING A MOTION

Chairperson announces the result of the vote	8
Chairperson calls for the question (vote)	7
Members discuss and debate the motion	6
Chairperson states the question	5
Another member seconds the motion	4
Member makes a motion before the assembly	3
Chairperson recognizes member	2
Member rises and addresses the chair	1

FIGURE 10–2 Eight steps are needed for a single motion to go through the proper procedure.

Step 5 *The chairperson restates the motion.*

Step 6 *The members discuss and debate the motion.*

Step 7 *The chairperson calls for the question (vote).*

Step 8 *The chairperson lets members know vote results.[6]*

It is vital for the chairperson to know and be able to follow these steps automatically when handling a motion.

Example

1. Member rises and addresses the chair. Member stands and says, "Madam Chairperson."
2. Chairperson recognizes member. Chairperson says, "Mr. Speaker."
3. Member makes a motion by stating, "I move that . . ."
4. Another member seconds the motion by saying, "I second the motion."
5. Chairperson states the question. Chairperson says, "It is moved and seconded that . . . [repeat the motion]."

6. After the chairperson asks, "Is there any discussion?" members discuss and debate the motion. Members debate the motion while observing rules of **decorum** (the propriety of debate).
7. After discussion has apparently ceased, chairperson calls for the question (vote). Chairperson says, "If there is no further discussion, all in favor say 'Aye.' All opposed say 'Nay.'"
8. Chairperson announces the result of the vote. Chairperson states, "The motion passes," or "The motion fails."

STANDARD CHARACTERISTICS OF A MOTION

The five major classifications of motions are main, subsidiary, unclassified, incidental, and privileged. When each of these is applied to a motion as a question, it helps the chairperson keep his or her thoughts organized. Proper classification also preserves correct parliamentary procedure during a meeting. When you know the characteristics of each motion, you will know how to classify them by asking yourself certain questions:

- Does this motion take **precedence** or priority over other motions? What motions can be pending without making this motion out of order? What motions can be made and considered while this motion is pending?
- In what situations is this motion applicable? What motions are applicable to this motion, if any?
- Can this motion be considered when another motion is on the floor?
- Does this motion require a second?
- Is this motion debatable?
- Is this motion amendable?
- What kind of vote (majority, percentage rate, etc.) is required for adoption of this motion?
- Can this motion be reconsidered?[7]

Purpose of Seconding a Motion

The purpose of seconding a motion is to keep the assembly from having to waste its time on a motion that only one member wants to consider. A second shows that at least two people in the assembly are in favor of considering the motion; it does not, however, necessarily mean that they are in favor of passing the motion. By requiring a second on most motions, the assembly prevents a single member from wasting everyone's energy with time-consuming motions that do not interest anyone else.

If a motion that requires a second does not receive one, it dies from lack of support. The chairperson states, "The motion dies for lack of a second." The chairperson can then call for further business.

PURPOSE AND TYPES OF VOTING

The purpose of a vote on a motion is to see whether the assembly is in favor of or opposed to that particular motion. Most of the motions that require a vote require a majority vote. A **majority vote** is defined as a decision by more than half of the votes cast by those legally entitled to vote.

Two-Thirds Majority Vote

A two-thirds vote is required on motions that take away the rights of members, such as motions to suspend the rules, previous question, rescind, or limit debate. In figuring a **two-thirds vote**, a chairperson must remember it is a decision by at least two-thirds of those voting who are legally entitled to vote.

An easy way to determine a two-thirds vote is to double the number of negative votes. If the doubled number of negative votes is equal to or less than the number of affirmative votes, there is a two-thirds majority vote. For example, for a motion requiring a two-thirds vote to pass, if the affirmative vote is six and the negative vote is three, there is a two-thirds majority vote in the affirmative. When the negative vote is doubled, it becomes six. Because six is equal to the number of affirmative votes, it is a two-thirds majority vote. In contrast, if the doubled negative vote is more than the positive votes, it is not a two-thirds vote. For example, six positive and four negative votes are cast. The four is doubled to equal eight, and eight is more than six; hence, the motion fails.

General Consent

Another type of vote is **general (unanimous) consent**, a decision-making technique in which everyone agrees and no vote is necessary. It is useful in dealing with routine business, or for questions that are unopposed or unimportant, because it often saves quite a bit of time. For example, general consent is usually used to adopt the minutes. The chairperson says, "Are there any corrections, additions, or deletions, to the minutes? If not, the minutes are approved as read." General consent can be used either to adopt a motion without going through the eight steps discussed previously or to take action without the formality of a motion.[8]

If it is obvious that the motion is favored by all in the group, the chair may state, "If no one objects, we will adopt this motion [or "it passes"]." If even one member of the group objects, however, the consent is not unanimous, so then the chair must conduct a regular vote, first calling for the affirmative and then the negative votes.

Methods of Voting

There are five major methods of voting: voice, hand, standing, secret ballot, and roll call. For votes about which the chairperson is relatively sure or most people feel the same way, a voice vote is used. The chairperson often asks for a voice vote for an adjournment. Raising the right hand is the voting method of choice for most meetings. For larger groups or more precise voting, a standing vote is often used. Most organizations use the secret ballot. It is especially used when the voting involves electing people in order to prevent hurt feelings and causing anger among the group. After a vote is announced, the chair taps a gavel like the one in Figure 10-3 to let members know it is time to move on to the next item of business.

For local, state, and national government-elected officials, a roll-call vote is used. The votes are a matter of record. With new technology, many elected officials have vote-tallying equipment at their stations. When they press the "aye" or "nay" button, their vote is recorded and shown on a display board.

If you ever begin to think that your vote does not make a difference, consider the following:

- One vote made English rather than German the official language of America.
- One vote admitted Texas to the Union.
- One vote gave Rutherford B. Hayes the presidency.
- One vote gave Adolf Hitler leadership of the Nazi party.[9]

FIGURE 10–3 The gavel, often a symbol of parliamentary procedure or the authority of the chair, helps with management of a meeting and the handling of business. (© StudioSmart/Shutterstock.com)

MAIN MOTION (ORIGINAL)

Rationale

An original **main motion** introduces new business before the assembly. A main motion is a formal suggestion of action to be taken on an item of business, by means of group discussion and debate and ultimately a group vote for approval or rejection of the motion.

Explanation

Main motions are the foundation of parliamentary business. There can be only one main motion on the floor at a time. A main motion takes precedence over no other motion. Therefore, a main motion may be made only when no other business is pending on the floor. Because a main motion introduces new business before the assembly, a second is required to consider the motion. A main motion is **debatable** (open to discussion) and **amendable** (modifiable or changeable) and requires a majority vote to pass. If a main motion receives the majority vote required to pass, it becomes the official statement of the action taken (or to be taken) by the group. The wording of a motion should be clear, concise, complete, and in a form appropriate for the intended purpose. All official motions should be introduced with the two words "I move." Finally, no main motions are to be considered if their intent conflicts with a law, bylaw, or organizational rules. If a motion "against the law" is adopted, it is as if it were never passed.

Example

Member: *Mr. [or Madam] Chairperson.*
Chairperson: *Mr. [or Madam] Speaker.*
Member: *I move that our FFA chapter sponsor a science fair at our school.*
Chairperson: *Is there a second to the motion that our chapter sponsor a science fair?*
Another Member: *I second the motion.*
Chairperson: *It is moved and seconded that our chapter sponsor a science fair at our school. Is there any discussion?* [There may or may not be discussion.]

Chairperson: *If there is no further discussion, all in favor say "Aye." [Response.] All opposed say "No." [Response.] The motion passes [or fails].*

Note: Either way the vote goes, the chairperson then asks for further business.

MOTION TO AMEND

Rationale

The motion to amend is probably the most widely used of the **subsidiary motions** (motions that relate to some other motion on the floor and can change or alter the main motion). The purpose of the motion to amend is to modify or change the wording, and in some cases the meaning, of the motion to which it is applied. A motion to amend may also be referred to as an **amendment**. There are three ways to amend a motion:

- Insert (or add)
- Strike out (or subtract)
- Strike out and insert or substitute

Explanation

For an amendment to be considered, it must be germane to the motion to which it is being applied. A **germane** amendment is relevant to or has significance in relation to the subject of the motion to be amended. In other words, no new subject can be introduced as an amendment, especially a new subject that is irrelevant.

There are two types of amendments: primary and secondary. A **primary amendment** is one that is applied to any amendable motion, except the motion to amend. A **secondary amendment** is applied to a motion to amend. A secondary amendment is debatable; however, it is unamendable. Therefore, there may be only two motions to amend pending on a motion at a given time. For example, there is a main motion on the floor to have a science fair. There is also an amendment on the floor to add "with the FCCLA [Family, Career, and Community Leaders of America] on October 14." There is also a motion to amend the amendment on the floor to strike the words *October 14* and insert *October 13*.

The motion to amend takes precedence over a main motion and a subsidiary motion to postpone indefinitely. The motion to amend requires a second and is debatable and amendable. A secondary amendment is also debatable, but it is not amendable. The motion to amend requires a majority vote to pass; if it is adopted, the motion to which it was applied should be changed accordingly. If the motion to amend fails, the motion it was applied to remains as it was.

Note: A member's vote on an amendment does not obligate him or her to vote in a particular way on the motion to which the amendment applies.

Example

Member: *Madam Chairperson.*
Chairperson: *Madam Speaker.*
Member: *I move to amend the motion by adding the words "on October 25 at 2:00 P.M." after the word "fair."*
Chairperson: *Is there a second to the motion that we amend the main motion by adding the words "on October 25 at 2:00 P.M." after the word "fair"?*
Another Member: *I second the motion.*
Chairperson: *It has been moved and seconded that the words "on October 25 at 2:00 P.M." be added to the motion. Is there any discussion?*

Note: The chairperson takes a vote on the motion to amend and, if it passes, proceeds to attend to the main motion as amended. If the motion to amend fails, the assembly returns to discussion of the unaltered main motion.

PREVIOUS QUESTION

Rationale

The purpose of the previous-question motion is to end discussion and bring a motion to a vote. It is used when members are tired of long debates or when members feel that most of them know how they want to vote without continued discussion. It is useful when the assembly is taking more time than is really needed to consider a motion. The previous-question motion not only immediately closes debate but also stops amendments, the **immediately pending motion** (a formal proposal that is under consideration on the motion being discussed), and other pending motions, as the motion may specify. It also prevents the making of any other subsidiary motions except the higher-ranking "lay on the table."

Explanation

Consider the following example as an explanation of the previous-question motion. A main motion is pending that also has an amendment pending. During discussion of the motion to amend, a member moves the previous question. If the member specifies that the previous question is for all items of business on the floor and the previous-question motion passes, the assembly would vote immediately on the motion to amend and then on the main motion. If the member does not specify the motions to which the previous-question motion applies, the chair should interpret the previous-question motion as applying only to the immediately pending motion, which in this example is the motion to amend. The previous-question motion is commonly used by legislators to bring a motion to a vote before the opposing side can gain enough votes for its position.

The previous-question motion takes precedence over all debatable or amendable motions to which it is applied and over the subsidiary motion to limit or extend limits of debate. It yields to the subsidiary motion to lay on the table and to all privileged motions. It requires a second, is not debatable or amendable, and requires a two-thirds vote to pass.

Example

There is a main motion pending that our FFA chapter send two delegates to the National FFA Convention and pay their expenses. As soon as the motion is open for discussion, a member moves the previous question. This is done as a strategic tactic so that any opposition does not have a chance to gain enough votes for its cause.

> **Member:** *Madam Chairperson.*
> **Chairperson:** *Madam Speaker.*
> **Member:** *I move the previous question.*
> **Chairperson:** *Is there a second?*
> **Another Member:** *I second the motion.*
> **Chairperson:** *This motion is nondebatable and requires a two-thirds vote. All those in favor raise your right hand.* [Response.] *All those opposed raise your right hand.* [Response.]

Note: If the previous-question motion passes, the next business is to vote on the motion. If it does not pass, the original motion is still open for debate.

REFER TO A COMMITTEE

Rationale

The motion to refer to a committee is used on items of business that cannot or should not be handled by the group at the present time.[10] Sending business to a committee is especially helpful when the group needs more detailed study on a time-consuming or complicated matter. By referring a motion to a committee, the assembly can save time by entrusting action to those particularly qualified to act and avoiding protracted, unproductive discussion among the whole group.[11]

A **committee** is one person or a group of people elected or appointed to research and/or take action on certain matters. The committee itself is not a form of assembly.[12]

Explanation

There are three major types of committees. Two of these are in common use; the other is rare. Standing committees and special committees are commonplace in most organizations, but a **committee of the whole** is used only in special instances.

A **standing committee** has a continuing existence. Most organizations have several standing committees that carry out most of the committee work. The FFA Program of Activities consists of standing committees. Examples of standing committees are the Budget and Finance Committee, the Ethics Committee, and the Executive Committee. A **special committee** is created for a specific, special purpose, and it goes out of existence as soon as it has completed its task.

An example of a special committee of an assembly is one that has been appointed to study the possibility of hosting a celebrity basketball game fund-raiser. This item of business is obviously a one-time special purpose, so a special committee is developed to deal with it. After the committee is finished with its business, it is dissolved.

Some details may be included when the motion is made or amended during discussion:

- The number of members on the committee
- Who the committee members are to be

- When the committee should report back to the assembly
- Whether the committee has the power to make a decision (power to act)
- The duties the committee is to perform
- The type of committee (standing or special)

If a motion to refer to a committee passes, any such details not specified in the motion are then decided by the chairperson.

The motion to refer to a committee requires a second; it is debatable, amendable, and requires a majority vote to pass. The motion to refer to a committee takes precedence over a main motion, a motion to postpone indefinitely, and a motion to amend. The motion to refer to a committee is sometimes called the **motion to refer** or the **motion to commit**.

Example

During discussion on the main motion that an FFA chapter sponsor a $500 scholarship for a graduating senior who will attend college, it becomes obvious that some detailed study of this matter will be required.

Member: *Madam Chairperson.*
Chairperson: *Mr. Speaker.*
Member: *I move to refer this motion to the Scholarship Committee; that this committee have the power to act, and that it should report back at the next month's meeting.*
Chairperson: *Is there a second?*
Another Member: *I second the motion.*
Chairperson: *It has been moved and seconded that we refer the motion to the Scholarship Committee. Is there any discussion?*

Note: If the motion to refer to a committee passes, the matter goes to the standing scholarship committee. This committee has the power to act and must report back at the next meeting. After the chair instructs the secretary to inform the committee members of their duties, the floor is open for further business.

If the motion to refer to a committee fails, the assembly returns to discussion of the main motion, "That our FFA chapter sponsor a $500 scholarship for a graduating senior who will attend college." The member in Figure 10–4 is waiting his turn to discuss a motion to refer to a committee.

FIGURE 10–4 This person is waiting to be properly recognized in order to discuss the motion to refer to a committee. *(Courtesy of Greg Grieco, Department of Public Information Media Gallery, Pennsylvania State University.)*

LAY MOTION ON THE TABLE

Rationale

A motion to lay a motion on the table, when passed temporarily, lays aside an item of business so the assembly may immediately handle a more urgent matter. The motion to lay on the table does not set a time at which the question (motion) is to be resumed. Only a majority vote, through a motion to take from the table, can bring the tabled motion back before the assembly.

Explanation

The motion to lay on the table is not intended to kill the motion to which it is applied, but only to put it aside in such a way that it may easily be resumed. However, when an organization holds regular meetings at least each quarter, a motion that was laid on the table stays on the table until it is taken from the table or until the next meeting is over, after which the motion dies.[13]

A motion that has been laid on the table is still within the group's control, so no other motion on the same subject is in order as long as the original motion is still on the table. It is also out of order to move to lay a motion on the table if no other matter requiring immediate attention is evident.

The motion to lay on the table takes precedence over the main motion, over all subsidiary motions, and over any incidental motions that are pending when it is made. The motion to lay on the table requires a second, is not debatable, is not amendable, and requires a majority vote to pass.

Example

During discussion of the main motion "that our FFA chapter participate in *June Is Dairy Month*," the assembly learns that the school's beloved secretary has just been admitted to the hospital. So that the assembly may adopt a motion to send the secretary some flowers, it must first lay aside the main motion concerning *June Is Dairy Month*.

> **Member:** *Mr. Chairperson.*
> **Chairperson:** *Madam Speaker.*
> **Member:** *I move to lay on the table the main motion that our FFA chapter participate in "June Is Dairy Month."*
> **Chairperson:** *Is there a second?*
> **Another Member:** *I second the motion.*
> **Chairperson:** *It has been moved and seconded that we lay the main motion to participate in "June Is Dairy Month" on the table; this motion is nondebatable and unamendable and requires a majority vote.*

Note: A vote is taken by the chairperson; if the motion to lay on the table passes, the motion to which it applies is set aside until a motion to take from the table is made or until a time limit terminates it. If the motion to lay on the table fails, the assembly returns to discussion of the main motion concerning *June Is Dairy Month*.

TAKE MOTION FROM THE TABLE

Rationale

A motion to take from the table means to bring again before the assembly a motion or a series of adhering motions that were previously laid on the table. The motion to take from the table is not in order unless at least one item of business has been transacted since the motion was laid on the table.

A motion that has been laid on the table remains there and can be taken from the table during the same meeting or during the next session after it was laid on the table. If not taken from the table within these time limits, the tabled motion dies; however, it can be reintroduced later as a new question.[14]

Explanation

The motion to take from the table is an unclassified motion and is in order only when no other question is pending. It requires a second, is not debatable, is not amendable, and requires a majority vote to pass. If the motion to take from the table passes, the question is back on the floor along with any subsidiary motions that were pending at the time it was tabled. If it fails, the tabled motion remains tabled.

Example

A main motion was laid on the table earlier in this meeting, and at least one item of business has been transacted since that time. While no business is pending, the following business takes place.

> **Member:** *Mr. Chairperson.*
> **Chairperson:** *Mr. Speaker.*
> **Member:** *I move to take from the table the main motion that our FFA chapter participate in "June Is Dairy Month."*
> **Chairperson:** *Is there a second?*
> **Another Member:** *I second the motion.*
> **Chairperson:** *It has been moved and seconded that we take from the table the main motion that our FFA chapter participate in "June Is Dairy Month"; this motion is neither debatable nor amendable and requires a majority vote to pass. We will now proceed to vote.*

Note: If the motion passes, discussion returns to the main motion that had been tabled: namely, that the chapter participate in *June Is Dairy Month.* If the motion fails, the floor is open for further business.

MOTION TO POSTPONE DEFINITELY

Rationale

The motion to postpone definitely puts off action on a pending motion and fixes a definite time for its future consideration. A main motion can be postponed only to the next scheduled meeting or to a later time in the current meeting. If a matter of business that was postponed has to be discussed before the next regularly scheduled meeting, which had a specified time, it is necessary to provide an additional **adjourned meeting** (a continued meeting set to meet again at a certain time) and then postpone the motion to a special meeting. This special meeting time could be set in an amendment to a motion to adjourn (see Chapter 11).

Explanation

If the motion to postpone definitely passes, the question to which it is applied is taken up again at the time specified in the motion to postpone. The motion becomes the **order of the day** (an item of business on the agenda) for the time to which it is postponed.

The motion to postpone definitely requires a second, is debatable and amendable, and requires a majority vote to pass. It takes precedence over the main motion and the subsidiary motions to postpone indefinitely, amend, or refer to a committee. The motion to postpone definitely is sometimes referred to as the **motion to postpone to a certain time.**

Example

Members of an FFA chapter are discussing the main motion "that our FFA chapter volunteer as group leaders at summer camp." Because it is early in the winter and most students have not made plans for the summer, the following transactions take place.

Member: *Madam Chairperson.*

Chairperson: *Madam Speaker.*

Member: *I move to postpone the main motion until our next meeting in March.*

Chairperson: *Is there a second?*

Another Member: *I second the motion.*

Chairperson: *It has been moved and seconded that we postpone the main motion that our FFA chapter volunteer as group leaders at summer camp until our next meeting in March. Is there any discussion?*

Note: If the motion to postpone definitely passes, the main motion is treated as unfinished business in the March meeting. If it fails, the assembly returns to discussion of the main motion.

MOTION TO POSTPONE INDEFINITELY

Rationale

The purpose of a motion to postpone indefinitely is to stop the passage of a main motion without letting it come to a vote by the assembly.[15] A motion to postpone indefinitely may be proposed when it is in the best interest of the assembly not to take a position on a particular main motion. This motion comes in handy if a group needs to get rid of a poorly chosen motion that will cause problems if it passes or is rejected.[16] Only a main motion may be postponed indefinitely. If the motion to postpone indefinitely passes, it kills the main motion for the meeting. The only way a main motion that was postponed indefinitely can come up again is as a new motion at a later meeting.

Explanation

The motion to postpone indefinitely takes precedence over nothing except the main motion to which it is applied. It is the lowest ranking of the subsidiary motions. It can be applied only to a main motion and therefore can be made only while the main motion is immediately pending. It requires a second, is debatable, is not amendable, and requires a majority vote to pass. During debate of the motion to postpone indefinitely, discussion can proceed fully on the merits of the main motion.

The motion to postpone indefinitely also has another purpose. It is sometimes used strategically by members who wish to test their strength on a particular main motion, according to the vote the motion to postpone receives.

Example

During discussion of the pending main motion "that our FFA chapter sponsor a science fair," a member proposes a motion to postpone the main motion indefinitely.

Member: *Mr. Chairperson.*
Chairperson: *Mr. Speaker.*
Member: *I move to postpone indefinitely the main motion that our FFA chapter sponsor a science fair.*
Chairperson: *Is there a second?*
Another Member: *I second the motion.*
Chairperson: *It has been moved and seconded that we postpone indefinitely the main motion; this motion is debatable but unamendable. Is there any discussion?*

Note: A majority vote is taken by the chairperson. If the motion to postpone indefinitely passes, the main motion is lost for the remainder of the meeting. If the motion to postpone indefinitely fails, the assembly returns to discussion of the pending main motion "that our FFA chapter sponsor a science fair."

SUSPEND THE RULES

Rationale

The motion to suspend the rules is desirable when the best interests of the organization are served by a temporary suspension of the written rules governing its operation. A group can adopt a motion to suspend the rules provided that the proposal does not go against the organization's bylaws (or constitution); local, state, or national laws; or the fundamental principles of parliamentary procedure.

Only rules of procedure can be suspended. Rules that cannot be suspended are

- Common parliamentary law
- Rules in the organization's charter or its constitution
- Rules in the organization's bylaws, unless the bylaws have provisions that allow for suspension of the rules

Explanation

The motion to suspend the rules is an **incidental motion** (motion that arises out of other motions and takes precedence over other motions when appropriate). It requires a second, is undebatable and unamendable, and requires a two-thirds vote to pass. The motion to suspend the rules can be made whenever no motion is pending. When a motion is being discussed, the motion to suspend the rules takes precedence over any motion if it is for a purpose connected with that motion.

Example

During discussion of the main motion "that our FFA chapter send a $35 floral arrangement to the hospital for the school secretary," the following business transpires.

Member: *Madam Chairperson.*
Chairperson: *Madam Speaker.*
Member: *There is a rule in our local chapter that we may spend no money without the approval of the Budget and Finance Committees; therefore, I move to suspend this rule.*
Chairperson: *Is there a second?*
Another Member: *I second the motion.*
Chairperson: *It has been moved and seconded that we suspend the rule stating "that the Budget and Finance Committees be previously consulted before spending money" so that we can send flowers to the hospital for the school secretary.*

Note: Because the motion to suspend the rules is undebatable and unamendable, an immediate two-thirds vote is in order.

If the motion to suspend the rules receives the two-thirds vote required to pass, the rule is suspended temporarily and discussion continues on the main motion. The motion regarding rule suspension fails if it receives less than a two-thirds vote. If it fails, the chair should rule the main motion out of order, because it is in derogation of a chapter rule.

RISE TO A POINT OF ORDER

Rationale

A member may rise to a point of order to draw attention to a **violation** or misuse of the rules and insist on enforcement of proper parliamentary procedure (Figure 10–5). Ordinarily, this

FIGURE 10–5 The person standing in this photo did not need recognition to rise to a point of order. *(© Monkey Business Images/Shutterstock.com)*

motion is used by members to call attention to errors made by other members that have not been corrected by the chair in the course of a meeting. After ruling on the point of order, the chairperson gives reasons for his or her decision or confers with the parliamentarian for a ruling.

Explanation

If a point of order arises concerning business that is currently pending, the point of order must be ruled on before the business can proceed. If the chair rules incorrectly on a point of order, a member should appeal from the decision of the chair (see the following section). A point of order is used when a rule has been broken either intentionally or unintentionally. Members should also correct the chair's willful violations of accepted procedures by rising to a point of order.

These are the steps in rising to a point of order:

- A member rising to a point of order should rise without recognition and say, "Mr. or Madam Chairperson, I rise to a point of order."

Note: If the point of order requires immediate attention, the member may interrupt another speaker.

- The chairperson replies, "State your point."
- Member explains the violation.

- The chairperson then rules on the point of order.

If the ruling is incorrect or a member does not agree, the member can appeal from the decision of the chair.

Example

There is a main motion pending before the assembly that "our FFA chapter participate in *June Is Dairy Month*." During discussion, a member moves to lay the main motion on the table. The chair receives a second on the motion to lay on the table and calls for discussion. An observant member realizes that the motion to lay on the table is undebatable and unamendable; therefore, the following comments are made (without recognition).

Member: *Mr. Chairperson, I rise to a point of order.*
Chairperson: *State your point.*
Member: *Since the motion to lay on the table is undebatable and unamendable, the chair should not have called for discussion.*
Chairperson: *Thank you, Madam Speaker. Your point is well taken. The chair should not have called for discussion, and we will proceed to vote on the motion to lay on the table.*

APPEAL FROM THE DECISION OF THE CHAIR

Rationale

Any member of the assembly who does not agree with a ruling of the chair may appeal from the decision of the chair. The appeal must take place immediately following the ruling. The purpose of an appeal is to prevent the chair from improperly controlling the meeting.

Explanation

Most commonly, an appeal from the decision of the chair is made after a point of order. However, an appeal can be made after any ruling by the chair. Members should not criticize a ruling of the chairperson unless they appeal from his or her decision.

The appeal from the decision of the chair requires a second; it is debatable but not amendable. A **tie vote** (the same number for and against a motion) sustains the chair. This is an application of the rule that a motion is automatically lost in the case of a tie vote. Here, the motion is to overrule the chair. In the case of a tie vote, the motion to overrule is lost; thus, the decision of the chair is sustained. Why? Common sense recognizes that the chair would vote for himself or herself anyway, so time is saved.

Example

A member makes a main motion "that our class have a cookout." Before receiving a second, the chair calls for discussion on the main motion. An alert member notices the breach and rises to a point of order, stating, "The main motion did not receive a second." The chair hastily replies, "I am sorry, Mr. Speaker, but your point is not sustained since a main motion does not require a second." At this point, it becomes obvious that the chair is trying to push this motion through the assembly, because he rules incorrectly on the point of order. The procedure the assembly should use to keep the chair from improperly controlling the meeting is an appeal from the decision of the chair. Therefore, a member rises after the chair rules on the point of order and the following business occurs.

Member: *Mr. Chairperson, I appeal from the decision of the chair.*
Chairperson: *Is there a second?*

Note: It may be necessary for the chairperson to ask what the appeal is.

Another Member: *I second the appeal.*
Chairperson: *Those who agree with the decision of the chair, please raise your right hand. Thank you, hands down. Those who do not agree with the decision of the chair, please raise your right hand. Thank you, hands down.*

If the appeal from the decision of the chair sustains the chair, the assembly returns to discussion of the main motion "that our class have a cookout." If the appeal from the decision of the chair does not sustain the chair, the chair must go back and ask for a second on the main motion. A second must be received before the main motion can continue. Remember, a tie vote sustains the chair.

DIVISION OF THE HOUSE (ASSEMBLY)

Rationale

Whenever a member doubts the result of a vote that was taken by voice, he or she may request a revote by calling for a division of the house. A member can demand a division from the moment the negative votes have been cast until the question is stated on another motion. A request for a division is ordinarily granted by the chair as long as there is a reasonable doubt about the outcome of the previous vote. When a division of the house is called, the revote should be taken by a more accurate means, such as standing (or rising), ballot, or roll call.

Explanation

A single member cannot order a counted vote by calling for a division of the house. Only the chair or assembly can order a counted vote.

If after a **vote by voice (viva voce)**, in which the chairperson asks for verbal responses, the chair is unsure which side has won, the chairperson may initiate a division by explaining the circumstance and calling for a revote by the raising of hands or standing (Figure 10–6).

Because a single member can request it, a division of the house does not require a second and does not receive a vote. There are different ways in which a member may request a division of the house. The most common form is for a member

FIGURE 10–6 The chairperson is counting the hand votes. Other methods of voting are voice, standing, and roll call.
(© Africa Studio/Shutterstock.com)

to rise without receiving recognition from the chair and state, "Madam Chairperson, I call for a division." Other phrases that would serve the same purpose include:

Mr. Chairperson, I call for a division of the assembly.
Madam Chairperson, I call for a division of the house.
Mr. Chairperson, I doubt the vote.

Example

After a voice vote is taken, the results are unclear, so the following transpires.

Member: *Madam Chairperson, I call for a division of the house.*
Chairperson: *A division of the house is called for. We shall vote again with a more accurate means; all those in favor of the motion please stand.*

RECONSIDER A MOTION

Rationale

The purpose of the motion to reconsider is to bring back before the group a motion that has already received a vote, thus making possible a change of vote on the original motion. In a sense, you are voting to see whether you want to revote. The motion to reconsider may be made only by a person who voted on the winning or prevailing side. If a person voted against the motion and it passed, he or she was not on the prevailing side. A member may vote on the prevailing (victorious) side for the specific purpose of being in a position to move to reconsider later in the meeting.

The motion to reconsider a motion that previously lost must be made during the meeting at which the main motion was introduced, except when the main motion passed and was entered in the minutes. If so, it can be considered only in the next meeting. Otherwise, use the motion to rescind, discussed in Chapter 11.

Explanation

The motion to reconsider may be applied to all motions *except* a motion to adjourn, recess, or suspend the rules; affirmative vote to lay the motion on the table; affirmative vote to take business from the table; or a motion to reconsider. The motion to reconsider requires a second; it is debatable (if the original motion was debatable) but is not amendable. It requires a majority vote. When the chairperson receives the motion to reconsider, the first thing to do is ask the speaker whether he or she voted on the prevailing side. If so, the motion is in order.

Suppose a motion to sell magazines with the band was moved and properly seconded. After the discussion, a majority of the people voted in favor of the motion; the motion passed and was recorded in the minutes. Before the next meeting, a person who voted on the winning (prevailing) side learned that the band did not want to sell magazines with the FFA. At the next meeting, this person would gain recognition from the chairperson and say, "I move that we reconsider the vote to sell magazines with the band since they do not want to work with us." The chairperson asks, "Did you vote on the prevailing side?" The member answers yes. The chairperson directs the discussion of the motion to reconsider, takes a vote, and announces the results. Once the motion passes, it places the original motion before the assembly at the same place it occupied before a vote was taken.

Example

Chairperson: *Is there any business to be presented at this time?*
Speaker: *Madam Chairperson.*
Chairperson: *Mr. Speaker.*
Speaker: *I move to reconsider the motion to have an FFA fruit sale since we are not selling magazines.*
Chairperson: *Mr. Speaker, did you vote on the prevailing side?*
Speaker: *Yes, I did.*
Chairperson: *Then your motion is in order. Is there a second?*
Another Member: *I second the motion.*
Chairperson: *It has been properly moved and seconded to reconsider the motion to have an FFA fruit sale. Is there any discussion? If not, all in favor of reconsidering the motion say "aye."* [Response.] *All those opposed say "nay."* [Response.] *The motion is carried. Discussion is now open on the motion to sell fruit. Hearing none, we shall proceed to vote. All in favor of the motion please stand. Thank you. Please be seated. All opposed, please stand. Thank you. Please be seated.*

Mr. Secretary, the motion passed; 20 for the motion and 6 opposed. The motion carries. [Tap gavel.]

MOTION TO RECESS

Rationale

A **recess** is a short intermission in the meeting, but it does not close the meeting. After the recess, business starts at exactly the point at which it was interrupted.

Explanation

There are two forms of the motion to recess: qualified and unqualified. A motion to recess is sometimes used by members as a strategic action to hold an informal group discussion or seek information on a question during a meeting.

A **qualified motion to recess** is a **privileged motion** (has nothing to do with the pending motion, but is of such urgency and importance that it is allowed to interrupt the consideration of other questions). To be in the privileged form, a motion to recess must be made while another question is pending. In its privileged form, the motion takes precedence over all subsidiary and incidental motions and over all privileged motions except the privileged form of the motion to adjourn or to fix the time to which to adjourn. It requires a second and is undebatable; amendments must relate to the time limits of the recess. If it receives the majority vote required to pass, the recess must begin immediately.

An **unqualified motion to recess** is made while no other question is pending. It is treated as a main motion. It requires a second and is debatable and amendable. When made as a main motion, the recess moved for may begin immediately or at a future time.

Example

While a main motion is pending that the FFA chapter "hold its annual FFA Banquet on May 13," it is brought out during discussion that the school administrator must grant permission for any such functions. So that the motion may be decided on at that meeting, and to further enable the members to hold an informal discussion on the topic, a member makes a qualified motion to recess.

Member: *Mr. Chairperson.*
Chairperson: *Madam Speaker.*
Member: *I move that we take a recess for fifteen minutes.*
Chairperson: *Is there a second?*
Another Member: *I second the motion.*
Chairperson: *It has been moved and properly seconded to take a recess for fifteen minutes. Amendments are related only to time limits.*

If the motion to recess passes, the assembly takes a 15-minute recess. When the recess is concluded, the business is taken up at the point at which it was left. If the motion to recess fails, the assembly continues with discussion of the main motion without taking a recess.

If no business is on the floor, the following unqualified motion to recess can be made.

Example

Member: *Madam Chairperson.*
Chairperson: *Mr. Speaker.*
Member: *I move that we recess.*
Chairperson: *Is there a second?*
Another Member: *I second the motion.*
Chairperson: *It has been moved and seconded that we recess. This motion is open for debate and amendments if needed. Is there any discussion?*

MOTION TO ADJOURN

Rationale

The motion to adjourn is used to legally end a meeting. When a group meets again following an adjournment, its members do not take up at the point at which they left off on the agenda (as in the case of a recess). Instead, they begin a new meeting with a new agenda. Business that was not completed before the adjournment comes up on the agenda of the next meeting under "unfinished business."

Explanation

There are two types of motions to adjourn: unqualified (privileged motion) and qualified (main motion).

Unqualified (Privileged) Motion The privileged motion to adjourn is a motion to close the meeting immediately. It takes precedence over all subsidiary and privileged motions except a motion to fix the time to adjourn. The privileged motion to adjourn is made if another meeting time already exists, or if no time has been set for the existing meeting to adjourn. It requires a second. It is undebatable and unamendable and requires a majority vote to pass.

Qualified (Main) Motion When the motion to adjourn is made in its qualified (main motion) form, it is debatable and amendable as a main motion. It is used in the following situations as a main motion:

* When the motion is qualified in any way, such as a motion to adjourn at, or to, a future time
* When a time for adjournment has already been established
* When the effect of the motion to adjourn would be to dissolve the group (an example would be a last meeting of a convention)

Example

During a meeting, members are discussing an unimportant main motion. Realizing the late hour on a weeknight, a member moves to adjourn the meeting.

Member: *Madam Chairperson.*
Chairperson: *Mr. Speaker.*
Member: *I move to adjourn.*
Chairperson: *Is there a second?*
Another Member: *I second the motion.*
Chairperson: *The motion has been moved and seconded that we adjourn. This is a privileged (unqualified) motion, which makes it undebatable and unamendable. We will proceed to vote.*

If the motion passes:

Chairperson: *The meeting is adjourned.*

Note: If the motion to adjourn fails, the assembly returns to discussion of the main motion.

The following is an example of a qualified motion to adjourn that occurs when no other business is currently being discussed.

Example

Member: *Mr. Chairperson.*
Chairperson: *Madam Speaker.*
Member: *I move that we adjourn.*
Chairperson: *Is there a second?*

Another Member: *I second the motion.*
Chairperson: *Is there any discussion?*

Note: A qualified motion to adjourn is treated just like a simple main motion.

Discussion of advanced parliamentary procedure continues in Chapter 11. Rules for the National FFA Parliamentary Procedure Contest are discussed in Chapter 11 and in Appendix E.

ORDER OF BUSINESS (AGENDA)

Order of business and **agenda** are synonymous in parliamentary procedure. The agenda is an established order of business or sequence of activities. The following is an example of an agenda for a group that holds regular business sessions with no special requirements:

- Reading and approval of minutes
- Reports of officers, boards, and standing committees
- Reports of special committees
- Special orders of unfinished business and general orders
- New business

Some organizations perform ceremonies or have other special requirements for which they must customize their agendas. The FFA is such an organization, as evidenced by the following sample FFA order of business:

- Opening ceremony
- Minutes of previous meeting
- Officer reports
- Special features
- Unfinished business
- Committee reports
- New business
- Ceremonies
- Closing ceremony
- Entertainment

To maintain group harmony and organization, it is expected that the agenda will be followed at all times. If for some reason the assembly fails to conform to its agenda, the motion to call for the orders of the day may be used to bring the assembly back to its agenda. (Calling for the orders of the day is further explained in Chapter 11. Agendas are further discussed in Chapter 12.) Figure 10–7 shows a sample FFA meeting agenda.

FFA MEETING AGENDA

1. Opening ceremony.
2. Minutes of the previous meeting.
3. Officer reports.
4. Report on chapter program of activities (chairpersons of the various sections of the program of activities are called on to report plans and progress).
5. Special features (speakers, special music, and the like).
6. Unfinished business.
7. Committee reports.
   a. Standing
   b. Special
8. New business.
   a. Plateau Experiment Station 50th Anniversary
   b. State FFA Dairy Judging Contest
   c. State Fair
   d. State FFA Livestock Judging Contest
   e. Forestry Conclave
   f. FFA membership dues deadline
   g. Ag Day—Varsity Visit
   h. National FFA Convention
   i. District Parliamentary Procedure Contest
   j. National FFA Week
9. Degree and installation ceremonies (if appropriate).
10. Closing ceremony.
11. Entertainment, recreation, refreshments.

FIGURE 10–7 A sample FFA meeting agenda. The agenda should be followed at all times to maintain group harmony and organization.

COMMON ERRORS IN AND MISCONCEPTIONS ABOUT PARLIAMENTARY PROCEDURE

Sometimes a single error is perpetuated throughout an entire meeting if no one knows enough about correct parliamentary procedure to notice the mistake. Theoretically, there is a right way to perform every motion (as well as many wrong ones). In meetings in which you are a member of the assembly, you should strive not to make any errors. You should also notice errors made by others and bring them to the attention of the chair, through point of order if necessary. In the following list, we correct some common errors and misconceptions concerning parliamentary procedure:

- In presenting a motion before an assembly, a member should use the proper terminology; that is, "I move. . . ." Less skilled members might say, "I make a motion. . . ."
- In determining the outcome of a vote, the only votes counted are those cast by members who are legally entitled to vote.
- A member making a motion about something that has just been said by the chair or another member in an informal discussion during a meeting should avoid statements such as "I so move." The member himself or herself should recite the complete motion being offered.
- If members are required to be recognized by the chair before they speak, a member should wait until the chair has given him or her the floor or permission to speak before addressing the assembly.
- A member's making or seconding of a motion does not mean he or she *must* vote in favor of that motion. In a democratic society, a member may vote any way on any motion or even abstain from voting.

QUICK REFERENCE GUIDES

When you have successfully completed this chapter but still desire further knowledge and skills in the area of parliamentary procedure, go to Chapter 11. Refer to Figure 10–8 for a summary of the motions reviewed in this chapter. The chairperson may want to keep this at his or her side to use as a quick reference while conducting business or to briefly review

BASIC PARLIAMENTARY PROCEDURE MOTIONS OVERVIEW						
MOTION	RECOGNITION FROM THE CHAIR REQUIRED	SECOND REQUIRED	DEBATABLE	AMENDABLE	VOTE REQUIRED	CLASS OF MOTION[1]
Main motion	Yes	Yes	Yes	Yes	Majority	M
Amend	Yes	Yes	Yes	Yes	Majority	S
Previous question	Yes	Yes	No	No	Two-thirds	S
Refer to committee	Yes	Yes	Yes	Yes	Majority	S
Lay on the table	Yes	Yes	No	No	Majority	S
Take from the table	Yes	Yes	No	No	Majority	U
Postpone definitely	Yes	Yes	Yes	Yes	Majority	S
Postpone indefinitely	Yes	Yes	Yes	No	Majority	S
Appeal from the decision of the chair	No	Yes	Yes	No	Majority[2]	I
Division of assembly	No	No	No	No	Revote	I
Reconsider a motion	Yes	Yes	No[3]	No	Majority	U
Recess	Yes	Yes	Yes/No[4]	Yes	Majority	P
Adjourn	Yes	Yes	Yes/No[5]	Yes/No[5]	Majority	P
Suspend the rules	Yes	Yes	No	No	Two-thirds	I
Point of order	No	No	No	No	None	I

1. M = motion; S = subsidiary; I = incidental; P = privileged; U = unclassified.

2. Majority vote. However, a tie vote sustains the chairperson.

3. Yes, if original motion is debatable.

4. Recess as a main motion is debatable and amendable. Recess in privileged form is undebatable but amendable to time only.

5. Adjourn as main motion is debatable and amendable. Adjourn as a privileged motion is undebatable and unamendable.

FIGURE 10–8 An overview of the basic parliamentary procedure motions covered in this chapter. *(Courtesy of the National FFA Organization.)*

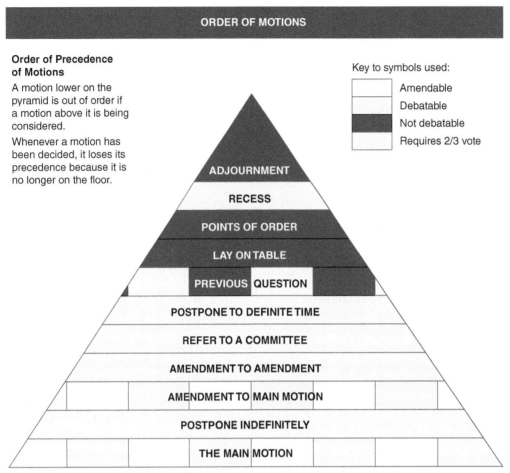

ORDER OF MOTIONS

Order of Precedence of Motions

A motion lower on the pyramid is out of order if a motion above it is being considered.

Whenever a motion has been decided, it loses its precedence because it is no longer on the floor.

Key to symbols used:
- Amendable
- Debatable
- Not debatable
- Requires 2/3 vote

ADJOURNMENT

RECESS

POINTS OF ORDER

LAY ON TABLE

PREVIOUS QUESTION

POSTPONE TO DEFINITE TIME

REFER TO A COMMITTEE

AMENDMENT TO AMENDMENT

AMENDMENT TO MAIN MOTION

POSTPONE INDEFINITELY

THE MAIN MOTION

FIGURE 10–9 The visual order of precedence of motions covered in this chapter. *(Courtesy of the National FFA Organization.)*

it before serving as chairperson in either a practice or a real situation. Figure 10–9 provides a good visual guide to the order of precedence of the motions. Figure 10–10 is a sample script showing how parliamentary procedure can be used in a meeting. For beginning practice, your teacher may wish to assign parts to you and your classmates and have class members go through the script so you can all become comfortable with parliamentary procedure.

CONCLUSION

Parliamentary procedure has always been a basic curriculum component of agricultural education and the FFA. Because parliamentary procedure is a democratic process, leadership skills are developed through its use. Parliamentary procedure rules guarantee that group action proceeds in a purely democratic fashion. The rules of parliamentary procedure are based on consideration for the rights of the majority, the minority, individual members, and absentees.

SUMMARY

The chairperson of a meeting is ordinarily called the president. The chairperson should be fair, possess good judgment, manifest poise, have a working knowledge of parliamentary procedure, be deliberate, and be tactful. The chairperson should maintain complete control of the assembly at all times.

One of the basic elements of effective parliamentary procedure is receiving and disposing of motions correctly. There are eight steps for handling a motion: (1) member rises and addresses the chair, (2) chairperson recognizes the member, (3) member makes a motion, (4) another member seconds the motion, (5) chairperson restates the motion, (6) members discuss and debate the motion, (7) chairperson calls for the vote, and (8) chairperson announces the result of the vote.

SAMPLE OF PARLIAMENTARY PROCEDURE USED IN A MEETING

Note: We assume that the opening ceremony has been completed and each officer knew their part and did a good job. It is time for the president to ask the secretary for the first order of business. Your meeting could and should proceed along this general pattern:

President: "Mr. Secretary, what is our first order of business?"

Secretary: "Minutes of the previous meeting, Madam President."

After the minutes are read,

President: "You have heard the minutes of the previous meeting. Are there any corrections? If not, the minutes stand approved as read."

(Do not say "of our LAST meeting." It sounds like there will never be another.)

President: "Thank you, Mr. Secretary.

What is our next order of business?"

Secretary: "Officer reports, Madam President."

President: "Are there any officer reports to be presented?"

Treasurer: "Madam President, I should like to present the monthly treasurer's report. We began the month with a balance of $871.35. Our receipts for the period were $786.42 and our disbursements totaled $411.19, leaving a balance of $1,246.58. As the members entered the meeting room, a copy of the month's report was distributed. Madam President, I move the acceptance of this report."

President: "Is there a second to the motion?"

Member: "I second the motion."

President: "You have heard the motion and the second. Is there any discussion? If not, all in favor say 'aye.' All opposed say 'no.' The motion is carried."

(In a chapter meeting, the secretary can easily identify the person who seconds a motion. By the way, one does not say "I second THAT." "THAT" was not the motion under consideration).

President: "Are there any other officer reports to be presented?

Reporter: "Madam President, during the past month we have published a total of 137 column inches of news and put on one five-minute radio broadcast about our summer camp. Madam President, I move acceptance of this report."

President: "Is there a second to the motion?"

Member: "I second the motion."

President: "You have heard the motion and second. Is there any discussion?"

Member: "Madam President."

President: "Julie."

Julie: "I wish to know why the article in *The FFA New Horizons* magazine was not included?"

President: "Will the reporter please respond to the question?"

Reporter: "The article was not included since it did not appear during the month covered in the report. It will be included in the next month's report."

President: "Is there any further discussion on this motion? Hearing none, all in favor of the motion say 'aye.' All opposed say 'no.' The motion is carried and the report is accepted. Are there any other officer reports to be presented? Hearing none, Mr. Secretary, what is our next order of business?"

Secretary: "Reports on Chapter Program of Activities committees."

Committee Chairperson: "Madam President, this is the report from the leadership committee. After reviewing the applications, our committee recommends that Todd Jackson, Jason Erickson, Martha Mork, and Donna Anderson be this year's finalists for the Star Greenhand award. I move the acceptance of our committee report."

President: "Is there a second to the motion?"

Member: "I think Tommy Kawalski should be included."

Member: "I rise to a point of order."

President: "State your point."

Member: "Since the motion had not received a second, discussion was not in order."

FIGURE 10–10 Scripts such as this one provide the opportunity to build your confidence as you learn the correct parliamentary procedure techniques. Your teacher can assign you a part in the script. *(Script format source: The National FFA Organization.)*

SAMPLE OF PARLIAMENTARY PROCEDURE USED IN A MEETING

President:	"Your point is well taken. We call again for a second to the motion."
Member:	"I second the motion."
President:	"The motion has received a second. Is there any discussion?"
Member:	"At our last meeting this committee was directed to select four finalists. It was agreed that their decision would be final."
President:	"Is there any further discussion? Hearing none, are you ready to vote?"
Members:	"Question—Question."
President:	"The question has been called. All in favor, signify by saying 'aye.' Those opposed say 'no.' The motion is carried."
President:	"Are there any other program activities reports to be heard? If not, Mr. Secretary, what is our next order of business?"
Secretary:	"Special features, Madam President."
President:	"Our agenda indicated that there are no special features scheduled so we shall proceed to the next order of business."
Secretary:	"Madam President, unfinished business is the next order of business."
President:	"Is there any unfinished business to come before the chapter?"
Member:	"Madam President."
President:	"Helen."
Helen:	"At the previous meeting a special committee was appointed to plan our Food for America program. What did they decide to do?"
President:	"This should come up during special committee reports. Mr. Secretary, is that not our next order of business?"
Secretary:	"Yes, Madam President. Special committee reports are next on the agenda."
President:	"Are there any special committee reports to be presented?"
Chairperson:	"Madam President, our committee recommends that our Food for America program be presented to elementary students on the first Monday of April with Tammy Koble as the coordinator. I move the acceptance of this report."
President:	"You have heard the motion. Is there a second?"
Member:	"I second the motion."
President:	"Is there any discussion? Hearing none, all in favor signify by saying 'aye.' Those opposed say 'no.' The motion is carried. Are there any other special committee reports to be presented? If not, what is the next order of business?"
Secretary:	"New business, Madam President."
President:	"Is there any new business to be presented?"
Member:	"Madam President."
President:	"Joan."
Joan:	"I move that our chapter purchase student handbooks for every first-year member in our chapter."
President:	"Is there a second to this motion?"
Member:	"I second the motion."
President:	"Is there any discussion?"
Member:	"Madam President."
President:	"Wilfred."
Wilfred:	"I move to amend the motion by substituting the words 'pay $2.00 of the cost' in place of 'purchase.' "
President:	"Is there a second?"
Member:	"I second the motion."
President:	"Is there any discussion? Hearing none, all in favor of the amendment to the motion, signify by saying 'aye.' Those opposed say 'no.' The amendment is defeated. Is there any further discussion on the original motion? Hearing none, all those in favor say 'aye.' Opposed say 'no.' The motion carried. Is there any more new business to come before the meeting? If not, what is the next order of business?"
Secretary:	"Since we have no degree or installation ceremonies, I have none, Madam President."
President:	(Begins closing ceremony.)

FIGURE 10–10 (continued)

There are five major classifications of motions: main, subsidiary, unclassified, incidental, and privileged. Once a motion is made, it receives a second and a vote. Most motions require a majority vote, but four motions require a two-thirds majority vote: suspend the rules, previous question, rescind, and limit debate. Another type of vote is general consent. There are five major methods of voting: voice, hand, standing, secret ballot, and roll call.

After the main motion, the subsidiary motions include amend, refer to committee, lay on the table, postpone definitely, postpone indefinitely, and previous question. Incidental motions include suspend the rules, point of order, appeal from decision of the chair, and division of the house (assembly). Privileged motions include recess and adjourn. Two motions are unclassified: take from the table and reconsider a motion.

The agenda is an established order of business or sequence of activities for a meeting. The agenda should be followed at all times, to maintain group harmony and organization.

 Take It to the Net

Explore parliamentary procedure on the Internet. Use your favorite search engine and look up "parliamentary procedure" or "Robert's Rules of Order." When you find a suitable website, print the first page of the site. Write a summary of the site or of an article in the website. The summary should include why you feel the website is useful.

Google parliamentary procedure on YouTube, and view students practicing parliamentary procedure. Discuss your findings with your classmates, and tell your teacher one thing that you observed that gave you a better understanding of parliamentary procedure.

Chapter Exercises

REVIEW QUESTIONS

1. Define the Terms to Know.
2. Discuss five characteristics of an ideal chairperson.
3. List the eight steps in properly handling a motion.
4. What are eight questions the chairperson should ask when a main motion is received?
5. What is the purpose of seconding a motion?
6. What are five methods of voting?
7. What is the purpose of a main motion?
8. List the order of business for an FFA meeting.
9. Identify and explain five common errors and misconceptions about parliamentary procedure.
10. What policy on which our country is built is an application of parliamentary procedure?
11. Whose rights are protected by parliamentary procedure?
12. Name four motions that do not require a second.

COMPLETION

1. The _____ is the person who is actually presiding over a meeting.
2. A _____ is necessary so the assembly does not waste time on a motion that only one person wants to discuss.
3. If there is no opposition on a question of little importance, the motion may be adopted by _____.
4. The FFA organization has a motion to _____ built into its closing ceremony.
5. A(n) _____ is the order in which business is presented in an assembly.
6. If a motion is _____, discussion on its merits is allowed.
7. Ninety members are voting, which means _____ are needed for a two-thirds majority.
8. A motion that has been tabled must be taken from the table at the _____ meeting or the _____ meeting, or it will cease to exist.
9. Motions to postpone definitely must include a _____ when the original motion will be back on the floor.
10. Every _____ has a right to express an opinion when parliamentary procedure is being used.

MATCHING

_____ 1. Recess.

_____ 2. Amend.

_____ 3. Vote required on suspend rules.

_____ 4. Vote required on postpone indefinitely.

_____ 5. Postpone indefinitely.

_____ 6. Division of assembly.

_____ 7. "I make a motion."

_____ 8. Bring back a motion that has already received a vote.

_____ 9. A member makes a parliamentary error.

_____ 10. You disagree with the ruling of the chairperson.

_____ 11. You want to end the debate now.

_____ 12. A subgroup can deal with this matter better and more efficiently.

_____ 13. Undebatable motion to put off original motion until later.

_____ 14. Short or extended break for informal discussion.

_____ 15. Add to original motion.

A. modify; change

B. motion to amend

C. incorrect terminology

D. majority vote

E. recess

F. short break

G. kills a motion

H. appeal from the decision of the chair

I. rise to point of order

J. reconsider

K. refer to a committee

L. lay on the table

M. previous question

N. revote

O. two-thirds vote

ACTIVITIES

1. Participate in the following "Parliamentary Procedure" game in class: stand with your classmates in line in the classroom as you would for a spelling bee. Your teacher names a motion. The first person must state whether or not it requires a second; the next person, whether or not it is debatable; the next person, what vote it requires; and the next person, what other motion is directly above it in the order of precedence. (For incidental motions, the correct answer is "No order of precedence.") A student missing any of these responses must be seated. The teacher continues asking questions about motions until only the winner remains standing.

2. Select one of the motions and write a script similar to the one in the Reconsider a Motion example. Use these scripts to practice parliamentary procedure. Write a script for a positive vote and a script for a negative vote. If the teacher assigns or lets students volunteer for different motions, perhaps you could create a practice script for each motion.

3. List the motions that require a two-thirds vote. Write a 100-word essay explaining why these motions require a two-thirds vote but other motions require a simple majority.

4. Write a 100-word essay on why you think some motions are debatable and others are not.

5. Write a paragraph on the circumstances under which it is advisable to table a motion.

6. Write a paragraph on the circumstances under which it is advisable to refer a question to a committee.

7. Conduct a mock meeting, with each class member taking a turn serving as chairperson. After the meeting, discuss the problems each of you had while acting as chair. The class should be divided into three groups. The chairperson and three to six class members will study the problems in the following Skills to Be Demonstrated sections. One chairperson can do all three problems with each demonstration, or different students can serve as chairperson for each problem.

Skills to Be Demonstrated (TEAM ONE)

-   Main motion
-   Amendment to the amendment
-   Previous question
-   Motion to reconsider

Problem 1

a. Member moves that each member is to have a minimum of four exhibits at the state fair.

b. Motion receives a second.

c. Motion is discussed by at least two members.

d. Member moves to amend the motion by substituting the number *five* for *four*.

e. Amendment is discussed by at least two members.

f. Member moves to amend the amendment by adding the words *at least one of which will be a crop or horticulture exhibit*.

g. Amendment to amendment, amendment, and main motion all are passed.

Problem 2

a. Member moves to reconsider action taken at the previous meeting when this motion was defeated: "Jackie moved that FFA dues must be subtracted from state fair premium money before the member receives the balance of the award." Second was by Ruth. After much discussion, motion was defeated.

b. Motion to reconsider receives a second.

c. Motion is discussed by at least two members.

d. Motion is passed.

e. Chair announces that the original motion is now on the floor in its debatable form.

f. Motion carries.

ABILITIES TO BE DEMONSTRATED	KIND OF MOTION	SECOND REQUIRED	DEBATABLE	AMENDABLE	VOTE REQUIRED
To receive and dispose of a motion and an amendment to the amendment					
a. Main motion	Main	Yes	Yes	Yes	Majority
b. Amendment	Subsidiary	Yes	Yes	Yes	Majority
c. Amendment to the amendment	Subsidiary	Yes	Yes*	No	Majority
Previous question	Subsidiary	Yes	No	No	Two-thirds
Motion to reconsider	Other	Yes	Yes	No**	Majority

*If the main motion is debatable.

**This motion is not amendable. If the motion is carried, the item to be reconsidered is sent back on the floor in its original form. If the original motion was debatable and amendable, the same situation now exists. Because it is classified as "unfinished business," this should be handled before the other motions in the demonstration.

Note: Properly, secretaries introduce orders of business. Items of business come from the floor.

Problem 3

a. Member moves that we sponsor an all-school harvest dance.

b. Motion receives a second.

c. Motion is discussed by at least three members.

d. Member moves the previous question.

e. Previous-question motion is defeated.

f. At least one member discusses the original motion.

g. Motion is defeated.

Skills to Be Demonstrated (TEAM TWO)

- Main motion
- Division of the house

- Lay on the table
- Amendment
- Previous question

Problem 1

a. Member moves that each member donate $1.00 to the FFA Foundation, Inc.

b. Motion receives a second.

c. Motion is discussed by at least three members.

d. Member moves to amend the motion by substituting the amount *$.50* for *$1.00*.

e. Amendment is seconded and discussed by two members.

f. Motion to table the main motion and amendment is presented, seconded, and passed.

Problem 2

a. Member moves that the chapter pay for all meals and lodging expenses for members of the dairy judging team and livestock judging team and the prepared public speaker participating in the state contest.

b. Motion receives a second.

c. Motion is discussed by one member.

d. Member moves to amend the main motion by striking out the word *meals* from the original motion.

e. Amendment receives a second and is discussed by three members.

f. Amendment is voted on and sounds as though it passes. A division of the house is called for, and the amendment fails.

g. Discussion returns to main motion by two members.

h. Motion for the previous question is made and seconded.

i. Previous question is voted on and passed.

j. Main motion is voted on and passed.

ABILITIES TO BE DEMONSTRATED	KIND OF MOTION	SECOND REQUIRED	DEBATABLE	AMENDABLE	VOTE REQUIRED
Main motion*	Main	Yes	Yes	Yes	Majority
Amendment	Subsidiary	Yes	Yes	Yes	Majority
Division of the house	Incidental	No	No	No	None
Lay on the table	Subsidiary	Yes	No	No	Majority
Previous question	Subsidiary	Yes	No	No	Two-thirds

*In this problem, when the main motion is voted on as a result of a previous-question passage, the main motion requires a two-thirds majority vote. Therefore, a rising vote is necessary.

Note: Properly, secretaries introduce **orders** of business. **Items** of business come from the floor.

Skills to Be Demonstrated (TEAM THREE)

- Main motion
- Amendment
- Postpone definitely
- Take from the table

Problem 1

a. Member moves that our chapter pay the expenses of one member to attend the Washington Leadership Conference.

b. Motion receives a second.

c. Motion is discussed by at least two participants.

d. Motion to amend the motion to substitute *two members* for *one member*.

e. Discussion by at least two participants.

f. Motion and amendment are disposed, with the amendment failing and the main motion passing.

Problem 2

a. Member moves that every member have at least one exhibit at the state fair.

b. Motion receives a second.

c. Discussion by at least two members.

d. Motion to postpone definitely.

e. Motion receives a second.

f. Motion passes.

Problem 3

At the previous meeting, a motion was tabled that would have had a committee appointed to select an activity that would earn enough money to pay the dues for all members.

a. Motion to take the proposal from the table.

b. Motion receives a second.

c. Motion carries.

d. Discussion on the main motion (now back on the floor in its debatable form) by at least two participants.

e. Motion carries and president appoints a three-member committee, identifying the committee chairperson.

Note: The preceding three problems appear courtesy of the National FFA Organization.

NOTES

1. D. B. Jameson, *Leadership Handbook* (New Castle, PA: LEAD, 1978), p. 31.
2. Henry M. Robert III et al., *Robert's Rules of Order Newly Revised*, 11th ed. (Da Capo Press, 2011).
3. Ibid.
4. Jameson, *Leadership Handbook*, p. 40.
5. R. E. Bender, G. S. Guiler, and R. J. Woodfin, *Mastering Parliamentary Procedure,* rev. 4th ed. (Columbus: Department of Agricultural Education, Ohio State University, 1983), p. 5.
6. Robert, *Robert's Rules of Order*, p. 47.
7. Ibid.
8. Ibid.
9. "One Vote (Your Vote) Could Make a Difference," *The Volunteer Voice* (newsletter of the Tennessee Association of Parliamentarians), February 1994, p. 12.
10. K. L. Russell, *The "How" in Parliamentary Procedure*, 6th ed. (Danville, IL: Interstate Printers & Publishers, 2000).
11. Bender, Guiler, and Woodfin, *Mastering Parliamentary Procedure*, p. 18.
12. Robert, *Robert's Rules of Order,* p. 9.
13. Ibid.
14. Ibid.
15. Bender, Guiler, and Woodfin, *Mastering Parliamentary Procedure*, p. 18.
16. Robert, *Robert's Rules of Order*, p. 63.

11

ADVANCED PARLIAMENTARY PROCEDURE

Parliamentary procedure is based on fairness, rights, and common sense. It is a useful and vital skill by which leaders put the democratic process into action. Mastery of parliamentary procedure further enhances a person's leadership ability. Using the information in this chapter, you can raise your parliamentary procedure skills to the next level.

Objectives

After completing this chapter, the student should be able to:

- Describe the duties of the chairperson
- Explain decorum (propriety) in debate
- Discuss the classification of motions
- Demonstrate each of the following motions:
  - Fix the time at which to adjourn
  - Question of privilege
  - Limit or extend limits of debate
  - Parliamentary inquiry
  - Division of the question (motion)
  - Motion to withdraw
  - Rescind
  - Call for the orders of the day
  - Object to consideration of a question
- Describe common errors in and misconceptions about parliamentary procedure
- Take official minutes
- Discuss serving as a qualified parliamentarian
- Explain how FFA parliamentary procedure contests are conducted and scored

Terms to Know

- floor
- question of privilege
- parliamentary inquiry
- previous question
- withdraw
- unstated subsidiary motion
- minutes

Parliamentary procedure is not a difficult subject to learn, but as with any broad subject, it takes time and practice to master. Once a person has a good command of basic parliamentary procedure, he or she can join a professional group focused solely on the study of parliamentary procedure. One such group is the National Association of Parliamentarians. In this type of group, a member can achieve certain levels of mastery and eventually offer his or her services as a certified parliamentarian.

This chapter is a continuation of Chapter 10. The material covered in these two chapters is by no means exhaustive. Further knowledge of parliamentary procedure can be obtained through the study of parliamentary authorities, such as *Robert's Rules of Order*.

The information contained in this chapter includes basic guidelines on useful motions and topics that are common to most meetings, such as decorum in debate, minutes, and duties of the chairperson. Several motions are summarized in this chapter, and useful tools and guidelines are presented. Examples of the advanced concepts discussed include serving as a qualified parliamentarian, parliamentary procedure in governmental bodies, and parliamentary procedure contests.

We recommend that after completing these two chapters on parliamentary procedure, students continue to investigate the subject. Parliamentary procedure is a useful, vital skill that few people truly master. The following section describes how a mastery of parliamentary procedure which may empower an FFA member and add significantly to his or her effectiveness as a leader.

DUTIES OF CHAIRPERSON

Routine Duties

The chairperson of an assembly has routine and specific duties to perform that relate to the proceedings of a business meeting. To ensure a smooth, efficient meeting, the chairperson must fulfill these duties to the best of his or her ability.

The chairperson opens the meeting by calling it to order and closes it by declaring it adjourned. It is also the chairperson's duty to handle all business according to the rules of parliamentary procedure and to expedite the business in every way compatible with the rights of members. Other routine duties are discussed in Chapter 10.

Specific Duties

Receiving Motions The chairperson, either on his or her own initiative or at the secretary's request, can require any main motions, amendments, or instructions to a committee to be in writing before the chairperson states the question. Also, if a motion is made in wording that requires clarification before it can be recorded in the minutes, it is the chairperson's job to see that the motion is put into suitable form before the question is stated.

Discussion on the Floor While a motion is open to debate, there are three important instances in which the **floor** (recognition from the chairperson of the right to speak) should be assigned to a certain person:

1. "If the member who made the motion claims the floor and has not already

---

VIGNETTE 11-1 **WHO IS THE MOST POWERFUL PERSON IN A MEETING?**

IN ANY GIVEN BUSINESS MEETING CONDUCTED UNDER THE RULES OF PARLIAMENTARY PROCEDURE, very few of the members truly understand the dynamics and specifics of sound parliamentary procedure. In some instances, an impartial person who is not a member of the group or organization conducting the meeting is asked to serve as the chairperson. In other cases, the president or any member of the group who knows a little about parliamentary procedure may serve as the chairperson.

Is the chairperson the most powerful person in a business meeting that conducts its business according to parliamentary procedure? What do you think? Perhaps this is so, but we submit to you that if knowledge is power, then the most powerful person in the meeting is the person who knows and understands parliamentary procedure the best. Just think about it for a moment. You have the power, even if you are not the chairperson, to move the pace of the meeting along, end debate, keep the assembly honest, and much more. However, this knowledge and power come with much responsibility; use your skills carefully and responsibly.

spoken on the question [(motion), that member] is entitled to be recognized in preference to other members."

2. "No one is entitled to the floor a second time in debate on the same motion on the same day as long as any other member who has not spoken on this motion desires the floor."

3. "In cases where the chair knows that persons seeking the floor have opposite opinions on the question (and the member to be recognized is not determined by [the previous two cases]), the chair should let the floor [speaker] alternate, as far as possible, between those favoring and those opposing the measure."[1]

Ending Discussion The chairperson cannot end discussion as long as any member who has not exercised the right to debate desires the floor, except by order of the group, which requires a two-thirds vote. When the discussion appears to have ended, the chairperson may ask, "Are you ready for the question?" If no one then rises to speak, the chairperson can proceed to a vote. The chairperson must always call for a negative note, no matter how nearly unanimous the affirmative vote may appear. The chairperson announces the result of the vote immediately or as soon as he or she has paused to permit response to the call for the negative vote.

DECORUM (PROPRIETY) IN DEBATE

The following practices and customs, which should be observed by all speakers and other members in a meeting, allow discussion and debate to proceed in a smooth and orderly manner. Unless otherwise stated in the bylaws or any other rule of the group, the following rules of debate apply.

Member Rights in Debate

Each member has the right to speak twice on the same motion on the same day, but may not discuss the motion a second time as long as any member who has not spoken on that motion wishes to speak. If the member who made the motion wants to discuss it and has not done so already, he or she is entitled to be recognized in preference to other members. Discussion must be relevant to the pending motion. The man in Figure 11–1 is openly discussing a motion, which is his right after receiving recognition from the chair.

FIGURE 11–1 After receiving proper recognition from the chair, a member can debate a motion. *(© iStock/Alina Solovyova-Vincent)*

Member Responsibility

Members or speakers must talk directly to the chair, remain courteous, stay objective, and refrain from including personal matters in a debate. Makers of a motion, although they can vote against it, are not allowed to speak against their own motion. While discussing a motion, members' remarks must be relevant to the pending motion. More specifically, the comments must have a bearing on whether the pending motion should be adopted.

Chairperson Rights

If the chairperson is a member of the group, he or she has the same rights to discuss a motion as any other member. However, the requirement that a chair remain impartial during a meeting will keep her from exercising those rights while governing the meeting. To participate in discussion, the presiding officer must relinquish the chair for as long as that item of business is pending.

Yielding to Chairpersons

If at any time the chairperson rises to make a ruling, give information, or make comments within privilege, any member who is speaking should be seated (or step back slightly, if standing at a microphone some distance from a seat) until the chairperson has finished talking.

Meeting Disruptions

During discussion or comments by the chairperson, and during voting, no member should be permitted to disturb the meeting by whispering, walking across the floor, or in any other way. This rule does not mean that members can never whisper to one another, or walk from one place to another in the meeting area while the meeting is in progress. However, the chairperson should make sure that such activity does not disrupt or hamper the transaction of group business.

CLASSIFICATION OF MOTIONS

All motions fall into one of five classifications: main motions, privileged motions, subsidiary motions, incidental motions, and motions that bring a motion (question) before the assembly again.

Main Motion

A main motion is the basis of all parliamentary procedure. It is the method of bringing business before the assembly for discussion and action. It can be introduced only if no other business is pending. Main motions must ordinarily be made first before any of the other classes of motions (this only makes sense, as the other classes of motions all relate to a main motion). Main motions require a second before they may be debated and require a simple majority to pass.

Privileged Motions

Privileged motions are of such urgency or importance that they are entitled to immediate discussion even though they may not relate to the original motion. They take precedence over the pending motion and all other items of business.[2] Privileged motions, in order of precedence, are as follows:

- Fix the time to which to adjourn
- Adjourn
- Recess
- Raise a question of privilege
- Call for the orders of the day

Subsidiary Motions

Subsidiary motions may be applied to another motion for the purpose of modifying it, delaying action on it, or disposing of it. Subsidiary motions are always made after the main motion to which they apply. They must be debated and voted on before the group resumes debate on the original (main) motion.[3] The subsidiary motions, in order of precedence, are as follows:

- Lay on the table
- Previous question
- Limit or extend limits of debate
- Postpone definitely
- Refer to a committee
- Amend
- Postpone indefinitely

Incidental Motions

Incidental motions either arise out of a pending motion (question), arise out of a motion that has just been passed, or relate to the business of the meeting. Incidental motions usually relate to the way business is transacted rather than to the business itself. They have no rank or hierarchy among themselves because they are incidental to the business of the assembly. Following are the commonly used incidental motions:

- Appeal from the decision of the chair
- Division of the assembly

- Object to consideration of a question
- Parliamentary inquiry
- Point of order
- Suspend the rules
- Withdraw

Motions to Bring Back

Motions to bring back return a motion to the meeting for consideration. These are sometimes referred to as **unclassified motions**. Three unclassified motions are commonly used:

- Take from the table
- Reconsider
- Rescind

Ranking Motions

The 13 ranking motions, in order of precedence, are as follows:

Privileged motions:

- Fix the time to which to adjourn
- Adjourn
- Recess
- Question of privilege
- Orders of the day

Subsidiary motions:

- Lay on the table
- Previous question

- Limit or extend limits of debate
- Postpone definitely
- Refer to a committee
- Amend
- Postpone indefinitely

Main motion:

- Main motion

A main motion may be made only when there is no business on the floor. Although they are the lowest-ranking motions, main motions are the foundation of parliamentary business. A motion to fix the time to which to adjourn can be made whenever a member can legally obtain the floor. It is the highest-ranking motion.

The ranking motions are placed in a logical sequence. Privileged motions are ranked higher than subsidiary and main motions; therefore, privileged motions take precedence (have privilege) over all other motions. Subsidiary motions cannot be made while any privileged motion is pending, because privileged motions are higher ranking. However, subsidiary motions can be made while a main motion is pending. Figure 11–2 shows the order in which motions could be made and ultimately voted on if every ranking motion could be pending at one time.

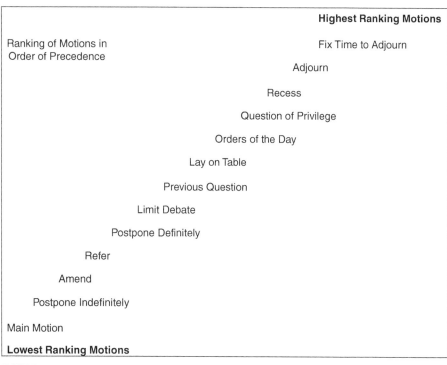

FIGURE 11–2 Motions have to be taken in a logical, orderly manner. If every ranking motion could be pending at one time, this is the order in which they would be addressed.

Once a main motion is made, each motion can be applied in succession, as long as it is higher ranking than the motion to which it is being applied. The only exception is the motion to amend. Amendment can be applied to higher-ranking motions if those higher-ranking motions are amendable. This is described, along with the characteristics of each individual motion, later in this chapter.

Once a series of motions has been made, the motions are voted on in the opposite order in which they were made. Thus, the last motion made is the first one voted on; the first one made is the last one to come to a vote. For example, a main motion is made and then a motion to postpone indefinitely, and then a motion to refer:

Main motion

Postpone indefinitely

Refer

When the motions come to a vote, the motion to refer is voted on first, then the motion to postpone indefinitely, and finally the main motion.

FIX THE TIME TO WHICH TO ADJOURN

Rationale

The purpose of the motion to fix the time to which to adjourn is to set the time for another meeting to continue the business of the present session. It has no effect on when the current, existing meeting will adjourn.

Explanation

This motion is privileged only when it is made while a motion is pending in a meeting and there is no existing provision for another meeting on the same or the next day. If a motion to fix the time to which to adjourn is made while no motion is pending, it is handled as a main motion. Therefore, it is subject to all the other rules just as if it were a main motion. In this instance, it requires a second, is debatable and amendable, and requires a majority vote to pass.

Whether the motion to fix the time to which to adjourn is a privileged or a main motion, the result of this motion is to establish an adjourned meeting. An adjourned meeting is a legal continuation of an existing meeting to another meeting. Whether the motion to fix the time to which to adjourn is privileged or main, the newly scheduled adjourned meeting must be set for a date before the next regular meeting.

If a motion to fix the time to which to adjourn is in its privileged form, it takes precedence over all other motions. It requires a second, is not debatable, and is amendable only as to time. An amendment is also undebatable. A privileged motion to fix the time to which to adjourn requires a majority vote to pass.[3]

Example

In an assembly in which a motion is pending, the following exchange takes place.

> **Member:** *Madam Chairperson.*
> **Chairperson:** *Madam Speaker.*
> **Member:** *I move that when this meeting adjourns, it will adjourn to meet at 10:00 A.M. tomorrow.*
> **Chairperson:** *Is there a second?*
> **Another Member:** *I second the motion.*
> **Chairperson:** *It has been moved and seconded that we fix the time to which to adjourn at 10:00 A.M. tomorrow. This motion is undebatable and amendable only as to time. Is there any discussion?*

If the motion to fix the time to which to adjourn passes, the assembly has set up an adjourned meeting. When the assembly adjourns the current meeting, it will adjourn to meet at 10:00 A.M. tomorrow. If the motion to fix the time to which to adjourn fails, the assembly carries on with its regular business. When it adjourns the present meeting, it will adjourn to meet at its next regularly scheduled meeting.

QUESTION OF PRIVILEGE

Rationale

A **question of privilege** is a request made by a member and granted or denied by the chairperson. A question of privilege may be made to secure immediate action on some urgent matter relating to the comfort, convenience, rights, or privileges of the group or assembly as a whole, or of a single group member. As this is a privileged motion, any member may rise to a question of privilege even when other business is before the group.[5]

There are two types of questions of privilege: those relating to the group as a whole and those relating to individuals (personal privilege).

Explanation

A question of privilege takes precedence over all other motions except the three higher-ranking privileged motions: fix the time to which to adjourn, adjourn, and recess. It does not require a second, is not debatable, is not amendable, and does not receive a vote. The chair rules on the question. However, if the chair's ruling is challenged through a motion to appeal from the decision of the chair, a vote is taken on the appeal.

Example

In an FFA meeting, a member rises and addresses the chair without awaiting recognition.

> **Member:** *Madam Chairperson, I rise to a question of privilege.*
> **Chairperson:** *State your question.*
> **Member:** *The temperature should be increased; it's cold in here.*
> **Chairperson:** *Your request is granted. Will the sentinel adjust the thermostat to make the room warmer?*

Note: The chairperson then continues with business.

LIMIT OR EXTEND LIMITS OF DEBATE

Rationale

The motion to limit or extend limits of debate either restricts the time to be devoted to debating a pending motion or removes any limitation placed on that time. Other than the previous-question motion covered in Chapter 10, the motion to limit or extend limits of debate is the only motion a group can make to exert special control over debate on a pending motion or series of pending motions.

Note: Neither a motion to limit or extend limits of debate nor a previous-question motion is allowed in committee.

Explanation

The motion to limit or extend limits of debate can be applied to any immediately pending debatable motion; to an entire series of pending debatable motions; or to any consecutive part of such a series, beginning with the immediately pending question. An order limiting or extending limits of debate is exhausted when one of the following situations occurs:

- when all of the questions on which the order was imposed have been voted on
- when questions affected by the order and not yet voted on have been either referred to a committee or postponed indefinitely
- at the conclusion of the session in which the order was adopted.[6]

The motion to limit or extend limits of debate takes precedence over all debatable motions. It yields to a previous-question motion, a motion to lay on the table, all privileged motions, and applicable incidental motions. It requires a second, is undebatable, is amendable, and requires a two-thirds vote to pass. Amendments applied to it are not debatable.

Example

A main motion is pending "that our FFA chapter sponsor a $1,000 scholarship for a high school senior." This important motion deals with a large amount of money, so the members have much to discuss. The following transactions occur:

> **Member:** *I move to extend debate to three times per person on this motion.*
> **Chairperson:** *Is there a second?*
> **Another Member:** *I second the motion.*
> **Chairperson:** *The motion has been made and seconded to extend debate to three times per person on this motion. This motion is undebatable, but amendable, and requires a two-thirds majority vote. Are there any amendments to the motion?*

If the motion to extend debate passes, a person may speak on this particular main motion up to three times. If the motion to extend debate fails, the members may offer the regular amount of discussion, which is only two times per speaker per motion. The members in Figure 11–3 are eager to discuss a motion. You can see why extending or limiting debate can be so important!

FIGURE 11–3 Parliamentary procedure is used in business sessions at the local, state, and national levels of the FFA. Without the motion to limit debate, debate could continue indefinitely on important issues, as evidenced by the eagerness of these members to speak. *(© iStock/Miroslav Georgijevic)*

PARLIAMENTARY INQUIRY

Rationale

A **parliamentary inquiry** is a question that any member may ask the chair regarding the correct usage of parliamentary procedure or the rules of the organization. Although a parliamentary inquiry usually pertains to the business that is currently pending, it is not mandatory that it do so.

The chair may answer the question or direct the parliamentarian to answer. In either case, the question is answered, so no vote is necessary. This is a useful tool when a member is in doubt as to what skills to use or what the proper parliamentary procedure is to accomplish a certain purpose.

Explanation

When making a parliamentary inquiry, a member need not obtain recognition from the chair. A parliamentary inquiry takes precedence over any motion as long as the inquiry relates to that motion. However, a parliamentary inquiry can be made at any time, even if no motion is pending. It is in order even if another motion is on the floor, if the parliamentary inquiry requires immediate attention. Because a parliamentary inquiry is usually answered by the chairperson, it does not require a second, is not debatable or amendable, and is not voted on.

The chairperson's reply to a parliamentary inquiry is not subject to appeal because it is an opinion, not a ruling. A member has the right to act contrary to this opinion.

Example

During lengthy discussion of a main motion, a member rises to a parliamentary inquiry.

> **Member:** *Mr. Chairperson, I rise to a parliamentary inquiry.*
> **Chairperson:** *Madam Speaker, please state your inquiry.*
> **Member:** *Would it be in order at this time to move the* **previous question** *[a formal motion requesting a vote and requiring two-thirds vote for passage] to end debate?*
> **Chairperson:** *Yes, a motion to move the previous question would be in order at this time.*

The assembly then returns to discussion of the pending main motion. A member may move the previous question (acting on the information received from the chair) but is not required to do so.

DIVISION OF THE QUESTION (MOTION)

Rationale

If a pending main motion contains two or more parts capable of standing as separate questions, a division of the question (motion) may be used to separate a main motion into two or more separate motions for the purpose of separate debate and voting.

Explanation

If the question (motion) is to be divided, each part should be capable of standing alone as a complete motion to receive a vote. The person who proposes dividing the motion must clearly state the manner in which the question is to be divided.

A division of the question takes precedence over the main motion and over a subsidiary motion to postpone indefinitely. It requires a second, is amendable, is not debatable, and requires a majority vote to pass.

Example

A main motion is pending before the assembly "that our FFA chapter conduct a fundraiser for

Muscular Dystrophy and we donate $200 to the Muscular Dystrophy Association." After some discussion, it appears that the majority of the FFA chapter believes it is an excellent idea to conduct the MDA fundraiser, but the chapter does not have any money in the budget at present to donate to the MDA. So, instead of voting down the entire motion, a member moves to divide the question into two parts.

> **Member:** *Madam Chairperson.*
> **Chairperson:** *Madam Speaker.*
> **Member:** *I move to divide the question to consider separately the question of conducting a Muscular Dystrophy fundraiser and the question of donating $200 to the MDA.*
> **Chairperson:** *Is there a second?*
> **Another Member:** *I second the motion.*
> **Chairperson:** *It has been made and seconded that we have a division of the question. This motion is only amendable. Is there any discussion?*

Note: The chairperson proceeds to take a majority vote.

If the motion to divide the question (motion) passes, the assembly will consider two main motions. The first motion the chapter will consider is "that our FFA chapter conduct an MDA fundraiser." The assembly will discuss and vote on this main motion without affecting the second main motion. As soon as the first main motion is disposed of, the second main motion is considered. The second motion is also treated as a separate main motion. If the motion to divide the question fails, the original main motion continues to be considered as one main motion.

MOTION TO WITHDRAW

Rationale

A member makes a motion that seems reasonable at the time. However, after brief discussion, the member who made the motion realizes the motion was inappropriate. The motion to **withdraw** a motion "enables a member who has made a motion to remove it from consideration before a vote is taken."[7]

Explanation

"Before a motion has been stated by the chair, it is the property of its mover, who can withdraw it or modify it without asking the consent of anyone." Thus, in the brief interval between the making of a motion and the time when the motion is placed before the assembly by the chair's stating it, the maker of the motion can withdraw it.[8]

After a motion has been stated by the chairperson, it becomes part of the meeting as a whole and the person who made the motion must ask the group's permission to withdraw or modify the motion. A request for permission to withdraw a motion, or motions to grant such permission, can be made at any time before voting on the question has begun, even if the motion has been amended, and even when subsidiary or incidental motions are pending.[9]

Any member can suggest that the maker of a motion ask permission to withdraw it, which the maker can do or decline to do at his or her discretion. The motion to withdraw takes precedence over any motion with which it is connected. It requires a second if the motion to which it refers has been given to the assembly and the maker of the motion wishes to withdraw it. It is not debatable or amendable, and requires a majority vote to pass if a vote is taken.

Example

During an FFA meeting, a member makes a main motion "that our group take a trip to Wyoming to go skiing." After some mostly negative discussion, the member realizes that no one really wants to go to Wyoming and decides not to press the issue.

> **Member:** *Madam Chairperson.*
> **Chairperson:** *Madam Speaker.*
> **Member:** *I move to withdraw my main motion that our group take a trip to Wyoming to go skiing.*
> **Chairperson:** *Is there any objection to the withdrawal of the main motion?* [There is no objection.] *The motion is withdrawn.*

If even one member objects, the chair should call for a second and then proceed to vote on the motion to withdraw. If the majority is in favor of withdrawing the motion, the motion to withdraw passes and the main motion is withdrawn. If there is not a majority in favor of withdrawing the motion, the main motion remains on the floor.

RESCIND

Rationale

The motion to rescind repeals or cancels a main motion that has been passed at the present or a previous meeting. As situations change, an organization often needs to change and update its plans and activities. Therefore, this motion is valuable when motions previously passed could hold a chapter back from doing new and creative things to make the organization stronger and more effective.

Explanation

There is no time limit for rescinding a motion, but the motion to rescind does not release the organization from any previous commitments. For example, if a chapter had a four-year contract for soft drinks for the concessions, which after three years proved unsatisfactory, rescinding the motion would not cancel the contract.

Rescinding has the effect of voiding a motion after the time it is rescinded, but it is not retroactive. For example, a year earlier a chapter passed a motion to provide free FFA manuals to new members. If this motion were rescinded, the effect of rescission would apply only prospectively from the time the motion to rescind passed. Those who had already received free FFA manuals would not be asked to pay for them.[10]

The motion to rescind requires a second, is debatable, and opens the main motion to debate. Any amendment must apply only to the main motion; a motion to amend requires a two-thirds vote unless previous notice was given to all members, in which case it requires only a majority vote. Also, if the motion to rescind was scheduled in the agenda before the meeting began, an amendment is out of order. The motion to rescind requires a majority vote.[11]

Example

> **Speaker:** *Mr. Chairperson.*
> **Chairperson:** *Mr. Speaker.*
> **Speaker:** *I move to rescind the motion passed at our April meeting to hold a picnic meeting in June, since so many of our members will be away on vacation.*

Chairperson: *Is there a second to the motion?*
Another Speaker: *I second the motion.*
Chairperson: *It has been properly moved and seconded to rescind the motion to hold a picnic meeting in June. Is there any discussion? Hearing none, we shall proceed to vote. Since no prior notice was given on this motion, it requires a two-thirds vote. All in favor of rescinding the motion stand. Thank you. Please be seated. All those opposing the motion, please stand. Thank you. Please be seated. The vote is unanimous and the motion is rescinded.[12]* [Tap gavel once.]

CALL FOR THE ORDERS OF THE DAY

Rationale

A call for the orders of the day is a privileged motion by which a member can require the group to follow the agenda, program, or order of business or to follow a general or special order that is to be handled at a specific time. However, two-thirds of those voting may decide that the agenda need not be followed.

Explanation

A call for the orders of the day in a regular meeting is a demand that the group adhere to its program and not talk about other things. If an inappropriate motion that does not follow the written agenda were made, a member could rise and say, "I call for the orders of the day." The chairperson would either state that the orders of the day (agenda) were being followed or put it to a vote by saying, "Will the meeting follow the orders of the day?" A two-thirds vote against following the orders of the day is required for the nonagenda motion to be considered. If less than two-thirds of the members vote against following the orders of the day, the chair must dismiss the main motion and return to the agenda. However, if two-thirds of the members vote against following the orders of the day, the main motion is in order and discussion may continue.[13]

As a motion, a call for the orders of the day takes precedence over all motions except other privileged motions or a motion to suspend the rules. It does not require a second and is not debatable

or amendable. The orders of the day can be suspended only by a two-thirds vote.[14]

Example

In the middle of an FFA meeting, a member moves a main motion that is not on the agenda. Another member does not believe that the matter should be discussed at this time and without awaiting recognition, addresses the chair.

Member: *Madam Chairperson, I call for the orders of the day.*
Chairperson: *The orders of the day are called for. ... Will the meeting follow the orders of the day? A two-thirds vote in opposition to following orders of the day is required to consider the main motion.*

Note: The chairperson proceeds with the vote. If there are less than two-thirds against following orders of the day, the main motion is dismissed. If two-thirds or more are in favor of dealing with the motion, they defeat the call for the orders of the day, whereupon the main motion becomes part of the orders of the day and may be considered immediately.

Figure 11–4 includes an agenda of a community service committee. If items that are not on this agenda come up for discussion, a member could call for the orders of the day.

AGENDA

Community Service Committee
Gallatin High School FFA Chapter
I. Call to Order
II. Roll Call
III. Reading of the Minutes
IV. Roadside Cleanup Report
V. Christmas Caroling Report
VI. Report from committee assigned to investigate FFA's Million Hour Challenge
VII. Unfinished Business
VIII. Announcements
IX. Adjournment

FIGURE 11–4 Orderly and harmonious meetings are the result when agendas are prepared and followed. If someone brings up an item of business that is not on the agenda, a member can call for the orders of the day.

OBJECT TO CONSIDERATION OF A QUESTION

Rationale

A motion objecting to the consideration of a question is used to prevent debate of matters that are not worthy of discussion, such as humorous or ridiculous motions and those in poor taste or of questionable legality.

If an original main motion is made and a member believes that it would be harmful for the motion even to be discussed in the meeting, he or she can raise an objection to consideration of the question, provided the objection motion is made before discussion has begun or any subsidiary motion has been made. The group members vote on whether the main motion should be considered. If there is a two-thirds vote against consideration, the motion is dropped.

Explanation

Because its purpose is to head off any discussion of a main motion, an objection to the consideration of the question must be stated before discussion begins on the main motion and before any subsidiary motions have been made. Only original main motions may be the subject of motions objecting to consideration of a question.[15]

An objection to the consideration of the question takes precedence over original main motions and over an **unstated subsidiary motion** (a subsidiary motion that may not have been seconded and has not been stated by the chair), except lay on the table. It does not require a second and is not debatable or amendable; a two-thirds vote against consideration sustains the objection.

Example

A ridiculous main motion is made in the assembly "that each member must give $200 a month to be in good standing with the group." As soon as this motion is made, and before discussion takes place, another member rises without recognition.

Member: *I object to the consideration of this motion.*

Chairperson: *Thank you; a motion has been made to object to the consideration of the question. This motion is undebatable, unamendable,*

and requires a two-thirds majority vote to sustain the objection.

If two-thirds of the members are opposed to consideration of the main motion, the main motion is dismissed. If less than two-thirds of the members are opposed to consideration of the main motion, the main motion will be considered.

REQUEST FOR INFORMATION

Rationale

In connection with business in a meeting, members may wish to obtain information or have something done that requires permission of the assembly. Any member can make a *request for information.*

Explanation

A request for information takes precedence over any motion with whose purpose they are connected, and can also be made at any time when no questions is pending. A motion on a request that is pending yields to all privileged motions and to other incidental motions. It can be applied in reference to any motion or parliamentary situation out of which they arise. No subsidiary motion can be applied. A request for information is in order when another person has the floor if it requires immediate attention.

A request for information does not require a second except when moved formally by the maker of the request. A motion to grant the request of another member does not require a second, since two members already wish the question to come up – the maker of the request and the maker of the motion.

It is not debatable or amendable. Also, no vote is taken on a request for information, and it is not subject to reconsideration.[16]

Example

A request for information (also called a *Point of Information*) is a request directed to the chair, or through the chair to another officer or member, for information relevant to the business at hand but not related to parliamentary procedure. During a meeting, a member made a motion to send

12 FFA members to a state FFA convention with $200.00 being allowed for lodging and meals.

> **Member:** *Madam Chairperson, I have a request for information.*
> **Chair:** *The speaker will state his question.*
> **Member:** *This motion calls for a large expenditure. Will the Treasurer state the present balance in our account?*
> **Chair:** *Mr. Treasurer, would you please give us the present balance in the budget?*

Note: The Treasurer responds. If the information is desired of a member who is speaking, the inquirer, upon rising, may use the following form instead:

> **Member:** *Madam Chairperson, will the member yield for a question?*
> **Chairperson:** *Jack, will you entertain a question?*

Note: Jack responds to the question.

COMMON ERRORS IN AND MISCONCEPTIONS ABOUT PARLIAMENTARY PROCEDURE

Many errors are made in business meetings when members attempt to use motions incorrectly. For this reason, each organization should have a chairperson who is qualified to handle every situation correctly and efficiently. Whenever the chairperson or any other member notices a violation of a rule, it should be made known to the group. Following are some common errors that occur in business meetings, with appropriate corrections:

- When asking for corrections while approving the minutes, the chair should say, "Are there any corrections to the minutes?" Members of some societies say, "Are there any additions, deletions, or corrections to the minutes?" This is redundant, as additions and deletions obviously fall into the category of corrections.
- The motion to lay on the table is used to temporarily lay aside a matter so more urgent business may be handled. Members who often use the motion to lay on the table as a tool to "kill a motion" are incorrect and should be ruled out of order.
- A motion to amend should state only how the motion is going to be changed (e.g., strike out, add), not how the whole motion would read after being changed.

TAKING OFFICIAL MINUTES

Content of Minutes

The official written record of the proceedings of a meeting is usually called the **minutes**. Ordinarily, meeting minutes should consist mainly of a record of what was done at the meeting, rather than what was said by the members. In other words, the minutes need only reflect motions introduced by the words "I move that," indicating an official action. Discussion, opinion, and debate need not be recorded unless the comments lead to an official question (motion) either approved or failed by the group. Official rulings by the chairperson, such as general consent, should also be included in the minutes.

The minutes should not be biased towards or against any actions taken. The first paragraph of the minutes should contain the following:

- Kind of meeting: regular, special, adjourned regular, or adjourned special
- Name of the group, organization, or assembly
- Date, time, and place of the meeting
- The fact that the regular chairperson and secretary were present, or the names of the persons who substituted for them if the regular officials were absent
- Whether the minutes of the previous meeting were read and approved as read, or as corrected (also, the date of that meeting is recorded if it was other than a regular business meeting)[17]
- Whether a quorum is present

Reading and Approval of the Minutes

The minutes of each meeting are normally read and approved at the beginning of the next regular meeting. Corrections, if any, and approval of the minutes are normally done by general consent. If one member objects, formal approval by voting should be done.

The reading of the minutes can be dispensed with (not carried out) at the regular time by a majority vote without debate. If this happens, a reading can be ordered at any time later in the meeting while no business is pending, again by majority vote without debate. If the minutes are not read before adjournment, the minutes must be read at the following meeting before the reading of the later minutes. A motion to dispense with the reading of the minutes is not a motion to omit the minutes.[18] If the group wants to approve the minutes without having them read, it will be necessary to suspend the rules.

If a copy of the minutes is sent to all members before the meeting, it can be presumed that the members have reviewed the minutes, and in this case the minutes are not read at the meeting unless a member so requests. Even if the method of prior notification of the minutes is used, correction and approval must still take place, and an official copy of the minutes must be kept on record as the authoritative set of minutes. If an error is found in the minutes, even years later, the minutes can be corrected by means of the motion "to amend something previously adopted," which requires a two-thirds vote. The member in Figure 11–5 is busy taking minutes.

FIGURE 11–5 Accurate minutes are a crucial element of parliamentary procedure and the democratic process. To prevent authoritarian decisions, the minutes of the previous meeting are always approved by members. *(Courtesy of Greg Grieco, Department of Public Information Media Gallery, Pennsylvania State University.)*

SERVING AS A QUALIFIED PARLIAMENTARIAN

Need for Parliamentarians

A parliamentarian may be a certified professional or an appointed member of the group. For small local meetings, such as the FFA or other youth organizations, a parliamentarian is rarely required. However, if one is needed, a member of the group could be appointed by the president to the post of parliamentarian. For larger groups, professional organizations, or corporations, the parliamentarian may be a consultant, sometimes a professional, who advises the president, other officers, and committees on matters of parliamentary procedure. A parliamentarian's role during a meeting is only advisory, because only a chair may make decisions on questions of order and answer parliamentary inquiries.

Some state or national organizations may need parliamentarians on duty at meetings to help interpret bylaws and rules or to assist and advise the board, officers, or committees as they work. Parliamentarians with these large organizations may have duties above and beyond the usual, such as assisting with planning and the order of business to be introduced.

Duties of Parliamentarians

It has often been said that the best parliamentarian in a meeting is one who is never heard (or even noticed) by the assembly, only by the chairperson. The president should confer with the parliamentarian before meetings and during recesses about potential problems. This helps to avoid frequent consultation during meetings. During the meeting, the following duties or actions of parliamentarians are appropriate:

- Give advice to the chairperson and, when requested, to any other member.
- As inconspicuously as possible, call the chair's attention to any error in the proceedings that may affect a member's substantive rights or cause other harm.
- Confer with the chair and make suggestions even if this momentarily distracts the chairperson's full attention from the business at hand.

- Recognize a developing problem and "be able to head it off with a few words to the chairperson." The parliamentarian should not wait to be asked for advice; it may be too late.[19]

The parliamentarian should sit next to the chairperson so as to be able to consult quietly and unobtrusively when needed. The parliamentarian should avoid speaking to the assembly if at all possible. The chairperson should try to avoid consulting too often with the parliamentarian, as each conference delays the conduct of business and distracts the chair's attention from the meeting. Parliamentarians are resources, not an excuse not to know parliamentary procedure. Whatever advice is given by the parliamentarian, the chairperson has the duty to make the final ruling. The chairperson always has the discretion to follow or disregard the parliamentarian's advice.

The member appointed as a parliamentarian should remember that (1) he or she does not vote on any motion, except in the case of a ballot vote; and (2) he or she does not cast a deciding vote, even if the vote would affect the result, as this would interfere with the chairperson's duty. If a member does not want to give up the right to vote, he or she should not serve as parliamentarian.

PARLIAMENTARY PROCEDURE CDE

Parliamentary procedure CDEs are a useful teaching tool for youth organizations and school clubs. Through participation in these events, students are introduced to and trained in basic and advanced parliamentary laws. In the process, the principles of democratic leadership are instilled in the participating students.

Parliamentary procedure team competitions allow students of parliamentary procedure to become more accurate and efficient while participating in business meetings. They also help students develop their leadership, research, and problem-solving skills. These events vary in format, length, and skill level from group to group and region to region.

Figure 11–6 shows an example of a script that may be useful for developing your parliamentary

---

PARLIAMENTARY PROCEDURE PRACTICE SCRIPT

Chairperson 1:   Receive and dispose of a main motion   Recess
                 Lay a motion on the table   Question of group privilege
                 Rescind a motion   To postpone definitely

1. I move that our chapter buy new jackets for the officer team.

2. I second the motion.

3. Because we do not know how much money is in the chapter funds, I move we table the motion.

4. I second the motion. (Fail the motion to table, fail motion.)

5. In our monthly meeting, we passed a motion stating that all members must wear official FFA overalls while in the shop. I move we rescind the motion.

6. I second the motion. (Pass the motion)

7. I move that we take a 20-minute recess.

8. I second the motion. (Fail the motion)

9. I move that we have a Sadie Hawkins dance added to our yearly activities.

10. I second the motion.

11. I rise to a question of group privilege. (Wait for response.) There is a draft in here. Could you please close the windows? (Comment from chair; return to item of business.)

12. Since the Recreation Committee chairperson is not present, I move that we postpone this motion until our next chapter meeting.

13. I second the motion. (Pass the motion to postpone.)

FIGURE 11–6 Until you get some experience and confidence with parliamentary procedure, your teacher may want to assign you a part in a script so that you will know what to say.

PARLIAMENTARY PROCEDURE PRACTICE SCRIPT

Chairperson 2:	Reconsider a motion	Adjourn
	Amend a motion	Rise to a point of order
	Withdraw a motion	Appeal from the decision of the chair

14. Earlier in the meeting, we referred to the Recreation Committee a motion to have a Halloween party. I move that we reconsider the motion to refer to a committee. (Give the chairperson time to ask if he voted on the prevailing side.)

Yes, Mr. Chairperson.

15. I second the motion. (Pass motion to reconsider.)

16. I move to amend the motion by substituting the words *our chapter officers* for *Recreation Committee*.

17. I second the motion. (Pass the amendment. Pass the motion to refer.)

18. I move we have a "Boots and Boxers" party on Halloween.

19. I second the motion.

20. Halloween is a time for ghosts and goblins, not boots and boxers; besides, it will be cold!

21. I move to withdraw the motion of having a "Boots and Boxers" party on Halloween. (Wait for comment.)

22. I move that we adjourn.

23. I second that motion.

24. I move to amend the motion by adding the words *until our next biweekly meeting on Thursday at 4:30.*

25. I second the motion to amend. (Pass the amendment; fail motion.)

26. I rise to a point of order. You did not get a second on the last motion. (Wait for comment.)

27. I appeal from the decision of the chair. (Wait for comment from chair.) I believe you did not get a second on the last motion.

28. I second the motion to appeal. (Fail the motion to appeal.)

Chairperson 3:	Take from the table	Postpone indefinitely
	Refer to a committee	Object to the consideration of the question
	Suspend the rules	

29. At our last meeting, we tabled the motion to buy a quarter horse **sire** for breeding purposes. Since we have found one with good conformation, I move that we take this motion from the table.

30. I second the motion. (Pass motion to take from table. Give chairperson time to state original motion.)

31. I think we need to give this more thought, so I move this be referred to a committee.

32. I second the motion. (Pass a motion to refer.)

33. I move that we buy refreshments for the meeting.

34. I second the motion.

35. The chapter rules state that we cannot spend any money not in the budget. (Wait for comment.)

36. I move that we temporarily suspend the rule.

37. I second the motion. (Pass with two-thirds vote.)

38.*I move that we serve refreshments during the meeting.

39.*I second the motion. (Fail motion.)

40. I call for division of the house. (Fail motion.)

41. I move that we sell beef jerky for our fund-raiser.

42. I move to postpone this motion indefinitely.

43. I second the motion. (Pass motion to postpone.)

44. Since our goal every year is to be the best chapter, I move that we purchase a banner for inspirational **purposes that reads** "Smash the competition."

45. I object to the consideration of the question. (Two-thirds negative vote prevents discussion.)

46. I move we adjourn.

47. I second that motion. (Pass motion.)

*Numbers 38 and 39 can be eliminated. Continue with main motion to buy refreshments.

FIGURE 11–6 (continued).

procedure skills. Use such a script in class as you learn the proper procedure for stating, receiving, and handling a motion. Most of the motions discussed in this book are included in the script. You or your teacher may want to write your own script(s) concerning motions relevant to your school or organization.

CDE Format

Many states use a CDE format similar to that of the National FFA Parliamentary Procedure event. Students must take a written test, answer oral questions, and perform a parliamentary procedure demonstration during which the secretary prepares a set of minutes. For the presentation, the contest official assigns a main motion. Before the event begins, five members (excluding the chairperson) are given a contest card with five required motions (Figure 11–7). There is no limit to the number of subsidiary, incidental, privileged, and unclassified motions to be demonstrated except that the team must demonstrate two subsidiary, two incidental, and one privileged or unclassified motion designated by the contest officials. Each team consists of six members: a chairperson and five members offering motions.[20] Figure 11–8 lists the motions a team needs to prepare for the Parliamentary Procedure Career Development Event.

MAIN MOTION: I move that the state FFA Degree recipients be sent to the National FFA Convention.

Required Motions:
- Extend Debate
- Fix Time to Which to Adjourn
- Refer to a Committee
- Appeal
- Division of the Assembly

FIGURE 11–7 A sample parliamentary procedure contest card. The motion to appeal is highlighted, which means that it is the required motion for the holder of this card.

CDE Rules

The rules of the National FFA Parliamentary Procedure CDE are available in PDF form on the National FFA website in the Career Development Event Resources section. This publication also covers the format of the written test, oral questions, presentation, presentation minutes, and instruction on minutes. Be sure to check the FFA website for the most current version of the rules.

Scoring the Official Minutes

The tabulation sheet (scoring rubric) for the minutes is Parliamentary Procedure Form 3, found in the "Parliamentary Procedure" chapter of the *National FFA Career Development Event Handbook*. The rubric and the instructions on minutes in the handbook provide the necessary guidance for developing skills in writing minutes. Form 1, in which the final copy of the minutes is written, is reproduced in Figure 11–9A. The "Tabulation Sheet for Scoring Minutes" is reproduced in Figure 11–9B.

Team Problem-Solving Activity

Teams advancing to the semifinal and final rounds will complete a team problem-solving activity in lieu of the minutes. Teams will be provided with a short parliamentary procedure scenario outlining a practical problem. The team will have 30 minutes to research the problem and write a short solution with reference to specific page and line numbers in *Robert's Rules of Order, Newly Revised*. All team members are required to provide their own copy of the most current edition of *Robert's Rules of Order, Newly Revised*.[21]

Event Scoring

A total of 1,000 points is used for the CDE. As Figure 11–10 shows, the written test is worth 150 points, oral questions are worth 135 points, and the minutes are worth 45 points. The presentation is worth 670 points. Refer to Figure 11–10B for the Team Score Sheet.

The National Parliamentary Procedure Team Score Sheet (Form 2) is also provided in the *National FFA Career Development Events Handbook*. All of the contest areas, totaling 1,000 points, are accounted for on this form, as are possible

CHART OF PERMISSIBLE MOTIONS					
MOTION	**SECOND REQUIRED**	**DEBATABLE**	**AMENDABLE**	**VOTE REQUIRED**	**RECONSIDER**
PRIVILEGED MOTIONS					
Fix the Time to Which to Adjourn	Yes	No	Yes	Majority	Yes
Adjourn	Yes	No	No	Majority	No
Recess	Yes	No	Yes	Majority	No
Raise a Question of Privilege	No	No	No	Chair Grants	No
Call for the Orders of the Day	No	No	No	No vote, Demand	No
SUBSIDIARY MOTIONS					
Lay on the Table	Yes	No	No	Majority	Neg only {3}
Previous Question	Yes	No	No	2/3	Yes
Limit or Extend Limits of Debate	Yes	No	Yes	2/3	Yes
Postpone to a Certain Time (or Definitely)	Yes	Yes	Yes	Majority	Yes
Commit or Refer	Yes	Yes	Yes	Majority	Yes
Amend	Yes	Yes (1)	Yes	Majority	Yes
Postpone Indefinitely	Yes	Yes	No	Majority	Affirm only
Main Motion	Yes	Yes	Yes	Majority	Yes
INCIDENTAL MOTIONS					
Appeal	Yes	Yes (1)	No	Majority	Yes
Division of the Assembly	No	No	No	No vote, demand	No
Division of a Question	Yes	No	Yes	Majority	No
Objection to the Consideration of a Question	No	No	No	2/3	Neg only
Parliamentary Inquiry	No	No	No	Chair answers	No
Point Of Order	No	No	No	Normally no vote Chair rules	No
Request for Information	No	No	No	No Vote. Chair responds	No
Suspend the Rules	Yes	No	No	(2)	No
Withdraw a Motion	No (3)	No	No	Majority (3)	Neg. Only
MOTIONS THAT BRING A QUESTION AGAIN BEFORE THE ASSEMBLY					
Reconsider (4)	Yes	Yes (1)	No	Majority	No
Rescind (4)	Yes	Yes	Yes	Majority with notice 2/3, or majority of entire membership (3)	Neg. Only
Take From The Table (4)	Yes	No	No	Majority	No

(1) If applied to a debatable motion

(2) Rules of Order 2/3 vote, standing rules - majority vote

(3) Refer to *Robert's Rules of Order Newly Revised*, current edition for rule (s)

(4) Refer to LDE Parliamentary Procedure event rules before using these motions in the demonstration

FIGURE 11–8 Summary of permissible motions for the National FFA Parliamentary Procedure Contest. *(Courtesy of the National FFA Organization.)*

Parliamentary Procedure   Form 1

Chapter: _____   Date: _____

State: _____   Place: _____

NATIONAL PARLIAMENTARY PROCEDURE EVENT
Official Minutes

Chair's Signature: _____   Secretary's Signature: _____

FIGURE 11–9A The final copy of the minutes for the National FFA Parliamentary Procedure Contest is written on a form such as this one. *(Courtesy of the National FFA Organization.)*

National FFA Parliamentary Procedure CDE   Form 3

TABULATION SHEET FOR SCORING MINUTES

(Preliminary Round)

State: _____

SCORING CRITERIA	POSSIBLE POINTS	POINTS EARNED
Completeness and Accuracy Name of chapter and state Date and place of meeting Minutes accurately reflect all business transacted during demonstration	15	
Format of Minutes Separate paragraph for all items Name of person making motion Name of seconder NOT included All main motions (including those withdrawn) All secondary motions (including those lost) All points of order and appeals Signed by the President and Secretary	15	
Grammer, Style and Legibility Complete sentences Correct spelling (deduction of 1 pt./error) Correct punctuation (deduction of 1 pt./error) Legibility and clarity	15	
TOTAL POINT:	45	

Comments:

_____

_____

_____

_____

_____

FIGURE 11–9B Tabulation sheet for scoring minutes. *(Courtesy of the National FFA Organization.)*

deductions. Figure 11–11 lists the 25 motions to be discussed, as well as practical uses of the motions.

Presentation

The required motions account for 100 points. An additional motion is worth 50 points. The debates are worth 300 points. The Chair's ability to preside and lead are worth 100 points. The team's presence, use of debate, and conclusions are allotted 120 points.

Note: Please remember to check the 2017–2021 National FFA Career Development Events Handbook for revisions to the Parliamentary Procedure CDE.

Phase	Breakdown of Points	Section Points	Total Points
Written Test (average of 6 members scores)			100
Presentation			750
Total of 5 members on the floor		500	
Required motion	20		
Debate (max. of 4 debates @ 20 pts. each)	60		
Additional motion	20		
Chair		100	
Ability to preside	80		
Leadership	20		
Team's General Effect		150	
Conclusions Reached by Team		50	
(Team's use of motions and debate support disposal of the main motion)			
Team Effect (Degree to which debate was convincing, logical, realistic, orderly, and efficient)	50		
Team's voice, poise, expression and appearance	50		
Oral Questions			100
Total for members' questions (6 × 12 pts)	72		
Additional clarification questions	28		
Presentation Minutes			50
Completeness and accuracy	25		
Format	10		
Grammar, style, legibility	15		
Deductions			
Deductions for parliamentary mistakes		5–20 pts/minor mistake	
Deductions for omitting assigned motion		50	
Deductions for going overtime		2 pts./second over 10:30	
TOTAL			1000

FIGURE 11–10A This form gives a breakdown of the points that can be scored in the Parliamentary Procedure CDE. *(Courtesy of the National FFA Organization.)*

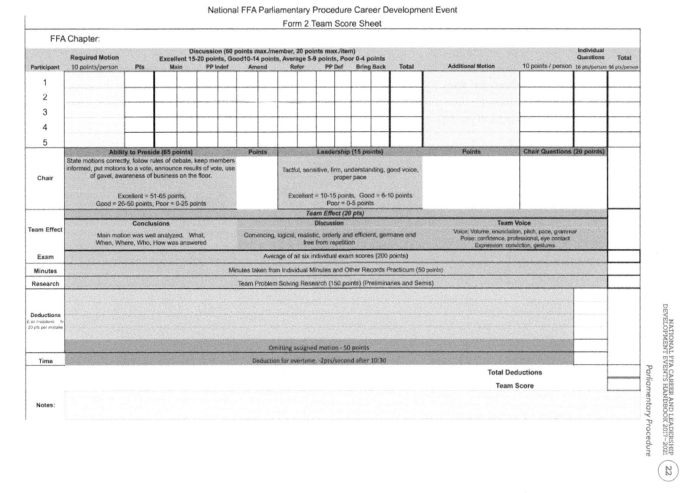

FIGURE 11–10B This form is the Team Score Sheet used to assign points for the Parliamentary Procedures CDE. *(Courtesy of the National FFA Organization.)*

CONCLUSION

Parliamentary procedure is based on fairness, rights, and common sense. It is a useful, vital skill that is necessary to put the democratic process in action. A mastery of parliamentary procedure greatly enhances your leadership potential.

SUMMARY

The chairperson of an assembly has routine, specific duties to perform that relate to the proceedings of a meeting. Ensuring a smooth, efficient meeting is the chairperson's responsibility.

Group members observe decorum (propriety) in debate to keep discussion and debate moving in a smooth and orderly manner. Decorum includes member rights in debate, member responsibility, chairperson rights, yielding to chairpersons, and avoidance of meeting disruptions.

There are five classes of motions: main motions, privileged motions, subsidiary motions, incidental motions, and motions that bring a motion (question) before the assembly again.

A main motion is the basis of all parliamentary procedure. Privileged motions are of such urgency or importance that they are entitled to immediate discussion even though they may not relate to the original motion. Privileged motions discussed in this chapter are the time to which to adjourn, question of privilege, and call for the orders of the day.

Subsidiary motions may be applied to another motion for the purpose of modifying it, delaying action on it, or disposing of it. This chapter examined the subsidiary motion to limit or extend limits of debate.

USE OF PARLIAMENTARY PROCEDURE IN A MEETING	
PARLIMENTARY PROCEDURE MOTION	PRACTICAL USE DURING MEETING
Main motion	Present business for action
Postpone indefinitely	Kill the main motion
Lay on the table	Set matter aside for a later time
Amend	Change or modify
Commit or refer to a committee	Let a subgroup attend to a matter
Previous question	Stop debate, order immediate vote
Suspend the rules	Action contrary to the rules
Withdraw a motion	Prevent a vote on a motion
Object to consideration	Oppose a foolish motion
Question of privilege	Attend to well-being of individual or the chapter members as a whole
Adjourn	Dismiss meeting
Fix the time to which to adjourn	Set a time for next meeting
Reconsider	Bring a motion back to discuss it and vote again
Parliamentary inquiry	Ask question relating to parliamentary procedure
Point of order	Call attention to parliamentary error
Appeal	Allow member to question the decision of the chair
Take from the table	Bring back an item that was laid on the table
Extend or limit debate	Establish length of time for discussion on a motion
Postpone definitely	Defer a motion to a specified time
Rescind	Cancel action taken by chapter
Division of the assembly	Question the chair's decision on a vote
Division of the question	Divide motion into two seperate motions
Call for orders of the day	Want to follow agenda
Recess	Take a break in the meeting

FIGURE 11–11 A review of the 24 motions and why they are used during meetings.

Incidental motions either arise out of a pending motion, arise with regard to a motion that has just been passed, or relate to the business of the meeting. Incidental motions discussed in this chapter are parliamentary inquiry, division of the question, motion to withdraw, and object to consideration of a question (motion).

Motions that bring a motion back before the meeting are sometimes referred to as unclassified motions. This chapter discussed the unclassified motion to rescind.

Motions are ranked in order of precedence. Privileged motions rank higher than subsidiary and main motions; therefore, privileged motions take precedence over all other motions. Subsidiary motions cannot be made while any privileged motions are on the floor. However, subsidiary motions can be made while a main motion is pending.

Parliamentary procedure contests are a useful teaching and learning tool. Teams are judged by their performance on a written test, oral questions, and a parliamentary procedure presentation, which includes participation of members, performance of the chairperson, and general effect of the team. By participating in contests, students learn the principles of democratic leadership—and they have fun doing it.

Take It to the Net

Using your favorite search engine, compile a list of five organizations that cite the importance of parliamentary procedure to their business philosophy. Write a one-page summary of how parliamentary procedure is used in these companies.

Chapter Exercises

REVIEW QUESTIONS

1. Define the Terms to Know.
2. What are three primary duties of the chairperson?
3. Briefly explain three important situations in which the floor should be assigned to a certain person.
4. Explain two ways the chairperson can end discussion of a motion.
5. List the five classes of motions.
6. What are three common errors in parliamentary procedure?
7. Name six items that should be in the first paragraph of the minutes.
8. When are minutes normally read and approved?
9. What are six duties or actions of a parliamentarian during a meeting?
10. Name four parts of the National FFA Parliamentary Procedure CDE that are awarded points.

COMPLETION

1. Unless otherwise specified, a member may speak _____ times on a motion each day.
2. Makers of a motion are not allowed to _____ their own motion.
3. To participate in discussion, the chairperson must _____ the chair for as long as that item of business is pending.
4. Whenever the chairperson is making a ruling or comments, any member who is speaking should be _____ until the chairperson is finished.
5. The motion _____ may be made to secure action relating to comfort of the meeting room or an individual member.
6. The motion _____ sets the time for another meeting to continue business of the existing session.
7. The motion _____ may be used to separate a main motion into two (or more) motions for the purpose of separate debate and voting.
8. The motion _____ is a question that any member may ask the chair regarding the correct use of parliamentary procedure.

9. The motion _____ enables a member who has made a motion to remove it from consideration before a vote is taken.

10. The motion _____ restricts the time devoted to debate or removes any time restrictions.

11. The motion _____ is used to prevent debate of matters that are not worthy of discussion.

12. The motion _____ can require a group meeting to follow the agenda.

MATCHING

Terms may be used more than once.

_____ 1. Fix the time to which to adjourn.

_____ 2. Limit or extend limits of debate.

_____ 3. Introduced if no other business is pending.

_____ 4. Reconsider.

_____ 5. Object to consideration of a question.

_____ 6. Official record of proceedings.

_____ 7. Presentation part of National Parliamentary Procedure CDE.

_____ 8. Written test of National Parliamentary Procedure CDE.

_____ 9. Minutes or Team Problem-Solving of the National Parliamentary Procedure CDE.

_____ 10. Oral questions of the National Parliamentary Procedure CDE.

_____ 11. A member wishes to remove a motion passed earlier.

A. motion to bring back

B. incidental motion

C. subsidiary motion

D. privileged motion

E. rescind

F. 150 points

G. 670 points

H. 45 points

I. minutes

J. main motion

K. 135 points

ACTIVITIES

1. Select one of the motions discussed in this chapter and become an authority on it. Present a report about it to your class and demonstrate it to the class. When questions arise during class practice, you will be called on to answer any questions about that motion. Know such things as whether the motion is privileged, incidental, or subsidiary; debatable or amendable; the type of vote required; whether a second is needed; and anything special about the motion.

2. Use class time to discuss and practice parliamentary procedure. Take turns serving as chairperson. Divide the class and compete according to the rules of your state's parliamentary procedure contest.

3. Give a demonstration of parliamentary procedure at a high school assembly or at local meetings.

4. Demonstrate the following skills with a team of five members and a chairperson:

Skills to Be Demonstrated

- Main motion
- Reconsider
- Division of the house
- Refer to a committee
- Amendment
- Postpone definitely

Problem 1

a. A member moves to build 10 picnic tables for the new commemorative park as part of the chapter's service learning program. Motion receives a second.

b. The motion is discussed by three members, one of whom moves to amend the main motion by changing the number from 10 tables to 5 tables. Amendment receives a second.

c. The amendment is voted on and passes.

d. A motion is made to refer to a committee the main motion as amended. The motion receives a second, is voted on, and passes.

JUDGING SUMMATION					
ABILITIES TO BE DEMONSTRATED	KIND OF MOTION	SECOND REQUIRED	DEBATABLE	AMENDABLE	VOTE REQUIRED
Main motion	Main	Yes	Yes	Yes	Majority
Amendment	Subsidiary	Yes	Yes	Yes	Majority
Refer to a committee*	Subsidiary	Yes	Yes	Yes	Majority
Reconsider	Other	Yes	Yes	No	Majority
Division of the house	Incidental	No	No	No	None
Postpone definitely**	Subsidiary	Yes	Yes	No	Majority

* In referring an item of business to a committee, the member so moving must stipulate the number of people on the committee, who appoints the committee, and when the committee is to report back.

** In postponing definitely, a member must state when the item of business will again come before the chapter, such as the next regular meeting, or a special meeting.

Source: The National FFA Organization.

Problem 2

a. A member moves to reconsider the item of business passed at the last meeting regarding the increase in the chapter dues from $8.00 to $9.00. Motion is seconded, voted on, and passed.

b. The chairperson announces that the motion is now on the floor in its original debatable form.

c. The motion is discussed by three members, one of whom moves to amend the original motion by substituting *$10.00* for *$9.00*. Amendment receives a second.

d. The amendment is voted on, with a close vote resulting. Division of the house is called, and the amendment passes.

e. Discussion returns to the main motion as amended, with one member moving to postpone definitely this item of business. Motion is seconded, voted on, and passes.

5. Complete the following table by stating (1) whether a second is required; (2) whether the motion is debatable or amendable; (3) whether the motion requires a two-thirds vote, simple majority vote, or no vote; and (5) whether the motion can be reconsidered. (6) Write yes, no, or the vote required in the appropriate blanks. Then, state whether the recognition is required and (7) whether a speaker may be interrupted.

KIND		SECOND REQUIRED	DEBATABLE	AMENDABLE	VOTE REQUIRED	CAN BE RECONSIDERED	RECOGNITION REQUIRED	MAY INTERRUPT A SPEAKER
Privileged	Fix the time to which to adjourn							
	Adjourn							
	Recess							
	Question of privilege							
	Call for the orders of the day							
Subsidiary	Lay on the table							
	Previous question							
	Extend or limit debate							
	Postpone definitely							
	Refer to a committee							
	Amend							
	Postpone indefinitely							
	Main motion							
Incidental	Appeal							
	Withdraw a question							
	Division of the assembly							
	Division of a question							
	Objection to the consideration of question							
	Parliamentary inquiry							
	Point of order							
	Suspend the rules							
Unclassified	Reconsider							
	Take from the table							
	Rescind							
Request for Information								

6. To perform well in a parliamentary procedure contest following the national format, a team must be able to discuss and debate motions well. Each member can earn a maximum of 60 points for debating. No more than 20 points can be earned per one recognition of the chair. Examples of quality debate can be found in the national contest rules mentioned earlier. Use this exercise to improve debate by a parliamentary procedure team. Give the team a main motion to discuss. Allow the team members one minute to think about the debate, and then begin. Call the names of the team members and have them stand and debate either for or against the main motion. Students can practice being for and against the main motion. The debate should always be directed to their chair.

NOTES

1. Henry M. Robert III et al., *Robert's Rules of Order, Newly Revised*, 11th ed. (Philadelphia, PA: Da Capo Press, 2011), p. 31.
2. Russell and K. L. Russell, *The "How" in Parliamentary Procedure*, 6th ed. (Danville, IL: Interstate Printers & Publishers, 2000), p. 67.
3. J. R. O'Connor, *Speech: Exploring Communication*, 4th ed. (Lincolnwood, IL: National Textbook Company, 1996), p. 458.
4. Robert et al., *Robert's Rules of Order, Newly Revised*, pp. 242–245.
5. R. E. Bender, G. S. Guiler, and R. J. Woodfin, *Mastering Parliamentary Procedure*, rev. 4th ed. (Columbus: Department of Agricultural Education, Ohio State University, 1983), p. 12.
6. Robert et al., *Robert's Rules of Order, Newly Revised*, pp. 195–196.
7. O'Connor, *Speech: Exploring Communication*, p. 192.
8. Robert et al., *Robert's Rules of Order, Newly Revised*, pp. 295–296.
9. Ibid., p. 296.
10. Bender, Guiler, and Woodfin, *Mastering Parliamentary Procedure*, p. 31.
11. Ibid.
12. D. B. Jameson, *Leadership Handbook* (New Castle, PA: LEAD, 1978), p. 299.
13. Bender, Guiler, and Woodfin, *Mastering Parliamentary Procedure*, p. 13.
14. Ibid., pp. 13–14.
15. Ibid., p. 22.
16. H. M. Robert, D. H. Hoffmann, and T. J. Balach, *Robert's Rules of Order, Newly Revised, 11th ed.* (Philadelphia, PA: Da Capo Press, 2011), pp. 294.
17. Ibid., pp. 468–469.
18. Ibid., pp. 473–474.
19. Ibid., p. 466.
20. *National FFA Career and Leadership Development Events Handbook 2017–2021: Parliamentary Procedure Career Development Event*, pp. 2–20. Retrieved July 24, 2016 from https://www.ffa.org/SiteCollectionDocuments/cde_2017_2021_parliamentary_procedure.pdf
21. Ibid., pp. 10–11.

12 CONDUCTING SUCCESSFUL MEETINGS

When you conduct and participate in meetings, you will find many opportunities for developing your leadership skills. The ability to conduct a meeting effectively is a critical leadership and communication skill. This skill will be valuable both in the workplace and in professional business and civic organizations. Building morale should be a key element of any meeting, along with efficiency, effectiveness, and excitement.

Objectives

After completing this chapter, the student should be able to:

- Identify leadership skills developed by conducting successful meetings
- List skills developed by being an officer
- Discuss basic meeting communication functions
- Describe the effect of meetings on the development of leader sensitivity
- Explain the characteristics of a good meeting
- Plan and prepare for a meeting
- Secure attendance at meetings
- Describe the arrangements, paraphernalia, equipment, and supplies needed in the meeting room
- Properly use committees in conducting business
- Select informative and motivational topics for meetings
- Conduct effective meetings
- Describe ways to involve group members
- Discuss problem solving and decision making by a group
- Describe characteristics of mature groups
- Discuss the responsibilities of officers and members
- Describe the FFA Program of Activities
- Explain why meetings should be evaluated

Terms to Know

- sensitive
- sensitivity
- apathy
- policy
- bylaws
- Executive Committee
- Program of Activities
- agenda
- paraphernalia
- committee
- energy cycle
- buzz groups
- role-playing
- problem solving
- groupthink

Scene 1:

"Are you going to the FFA meeting today?"

"Do I have a choice?"

"Not much, it's either that or homeroom."

"Every second and fourth Wednesday. Just like clockwork."

"One of these days, you'll figure out why."

Scene 2:

"Time for the FFA meeting."

"Okay! I'm anxious to find out what's happening next."

"The agenda says Ms. Tippens, our adviser, is talking about the spring fling."

"I've got some good information to share that I think the group will like."

"Good, let's go."

Which scenario is more familiar to you? Do you approach an FFA meeting expecting it to be a waste of time and energy? After the meeting, do you wonder what it was even for? If this is the case, you are probably feeling the pain of an inefficient meeting. When unproductive meetings occur regularly, students begin to avoid attending.

The main reason meetings fail is poor planning and organization. Even though many professionals in the workforce spend a great deal of time in meetings, research shows that most were never trained in planning, organizing, and conducting efficient, successful meetings.[1] When meetings are well managed, they are an effective communication tool. Important decisions are made, ideas are generated, and information is shared. Meetings are a critical part of building group enthusiasm, and as enthusiasm and excitement grow, the chapter benefits. The chapter also benefits as the group's ability to work together and make decisions grows.

LEADERSHIP SKILLS DEVELOPED

Fortunately, the process of conducting and participating in meetings offers many opportunities for developing leadership skills. These skills will be valuable when members graduate and participate in professional, business, and civic organizations. Members can practice public speaking,

parliamentary procedure, communication with groups, and various other leadership and personal development skills.

Interesting meetings build members' interest and participation. Arranging for special programs and encouraging involvement in committees are valuable learning experiences. People have always assembled to plan, negotiate, and delegate responsibility. Regular meetings of the FFA chapter are necessary so that activities can be accomplished and members can gain experience in group planning and functions.

Conducting meetings also gives members opportunities to learn how to lead individuals and groups, establish and maintain good public relations, and develop cooperation. Members can learn how to plan meetings, write up minutes, establish rapport, lead discussions, and follow democratic procedures. The students in Figure 12–1 are actively involved in a group project, and they are developing their cooperation skills while doing so.

All members of an organization are responsible for electing a capable leadership team (e.g., FFA officers). FFA members learn what is expected from officers, what the necessary qualifications are, and how to nominate and elect a person for positional leadership. By learning the duties of officers and how to select them, FFA members can better contribute to clubs and civic and private organizations, and as advisers to youth clubs and activities.

SKILLS DEVELOPED BY BEING AN OFFICER

As an officer, a member can practice better communication skills and further develop his or her public speaking ability. Members can learn what is expected of good officers, how to install officers, and the proper functions of a nominating committee. By understanding each officer's duties and responsibilities, as mentioned in the FFA Opening Ceremony, an officer can learn the following values of the FFA: leadership, willingness to change, cooperation, self-confidence, knowledge, attitude, record keeping, public relations, neatness, honesty, budgeting, alertness, participation, and respect. Many FFA officers are

FIGURE 12–1 These students are working on a group project, and enhancing their cooperation skills in the process. *(© iStock/meshaphoto.)*

able to attend their respective state FFA conventions and/or the National Convention, held each October in Indianapolis, Indiana. At these conventions, members and officers are required to attend assemblies and sessions in which official business is conducted.

BASIC MEETING COMMUNICATION FUNCTIONS

There are five functions that meetings perform better than any other communication technique:

Share Knowledge "Meetings provide a forum where individual information and experience can be pooled. The group revises, updates, and adds to what it knows."

Establish Common Goals "Meetings help every member of the team understand the goals and objectives of the group and how [individual efforts] will affect those objectives."

Gain Commitment Meetings foster commitment. Attendees develop a sense of responsibility for implementing and supporting the group's decisions.

Provide Group Identity "Meetings define the team. Those present belong to the team; those absent do not. Attendees develop a sense of collective identity."

Allow Team Interaction Often meetings are the only time the entire group works as a team.[2]

DEVELOPING LEADER SENSITIVITY THROUGH MEETINGS

Leaders, supervisors, and administrators trained in effective human relations are **sensitive**, or keenly aware of, the needs of others. **Sensitivity**, the ability to understand the needs of others, is a trait of the democratic leader.

Untrained leaders may be so intent on applying rules and maintaining their own positions that they are unaware of individual needs. When they appear to be ignoring individual needs, frustration within the membership results. When leaders do not acknowledge the interests of members, they may discourage participation and create **apathy** (lack of interest or concern) in the group.

Leaders who are aware of individual needs and differences in group situations recognize the importance of these differences. They can help each individual make a unique contribution to the decision-making process. A leader who is alert and sensitive to the needs and feelings of individuals accomplishes far more than a leader who is concerned only with getting a task done.

CHARACTERISTICS OF GOOD MEETINGS

Meetings held on a regular basis are important to the FFA chapter—but chapter meetings are only as interesting and effective as their leaders make them. The following are characteristics of good meetings:

- Each meeting has a specific purpose. If there is no business to be transacted, good leaders cancel meetings.
- They are carefully planned and prepared for.
- They are publicized well in advance of the meeting time.
- The agenda is posted.
- A high percentage of members attend.
- Proper parliamentary procedure is used.
- The agenda is followed.
- Meetings proceed in an orderly, businesslike manner.
- Meetings are conducted by qualified officers who are well prepared and know their duties and responsibilities.
- Participating members are well prepared and able to contribute.
- The program offers a variety of business, education, and recreation opportunities.
- The members are interested.
- Meetings are held in a suitable and comfortable place.
- Meetings are conducted by members, with minimum participation by the chapter adviser.
- Minutes are taken, posted, and distributed.[3]

Meeting Policies

Once you enter the business world, you may encounter policies regarding meetings. Why? Efficient use of time saves a company money.

Policies help personnel lead and participate in effective meetings. A **policy** is a rule, action, or norm of an organization, usually adopted for the sake of consistency, equity, expediency, and/or efficiency. Good meetings improve morale and team spirit because objectives are met. The meeting policy used at the Western Center of General Dynamics includes these guidelines:

- All meetings start and end on time, with a maximum length of 90 minutes.
- Each participant has a commitment to making the meeting successful.
- A clear objective is established for the meeting and the agenda is followed.
- Common courtesy and caring are emphasized. Issues, not people, are criticized.
- A positive orientation of "We are here to help" is established.[4]

Barriers to Successful Meetings

Unfortunately, some meetings are boring and do not meet the members' expectations. The following are potential problems that undermine morale and derail many meetings:[5]

- Poor organization
- Failure to focus on important issues
- Inconclusiveness
- Boredom
- Excessive length
- Held too frequently or not frequently enough
- No need for the meeting
- Disruptive behavior by members
- Diversion from the agenda by members with selfish motives
- Domination by officers, formal leaders, or a handful of influential members

Appropriate Time for Meetings

Most FFA chapters hold regularly scheduled meetings; the schedule is often specified in the chapter's constitution or **bylaws** (the standing rules governing the regulation of an organization's internal affairs). One FFA meeting per month throughout the school year and summer is considered the minimum. More active chapters may need to meet more than once per month to conduct their business effectively. Meetings

FIGURE 12–2 Chapter meetings will be as interesting and effective as you make them. These students are planning and preparing for a meeting.

should be scheduled well in advance. Regular chapter meeting dates and a list of programs for meetings should be planned for the full year as part of the chapter's Program of Activities. This schedule of programs should be posted where all members can access it. Major topics should be scheduled on the basis of interest and benefit to a majority of the members.

The primary challenge for most chapters is selecting an appropriate meeting time. Many schools allow less than 30 minutes for club meetings. This is not enough time for an effective meeting with a full agenda. An alternative to meeting during school hours is evening or night meetings. Although attendance tends to be somewhat lower, this time slot attracts a more serious and focused group. However, if meetings are appropriately planned to include business, educational, and recreational activities, members will attend.

A more functional approach may be a combination of day and night meetings. The day meetings could be used for announcements, discussion of upcoming activities, and short motivational

talks. The night meetings could follow the full agenda, including the minutes, treasurer's report, committee reports, old and new business, and the program.

There are a huge number of reasons why scheduling meetings can be problematic. Concentrate on the one time that works for your chapter. Remember to have the school administration approve your meeting times. The members in Figure 12–2 are working hard to plan a meeting.

PLANNING AND PREPARING FOR MEETINGS

Planning

Planning is the key to good chapter meetings. An order of business should be established and followed at each FFA meeting. The responsibility of developing the order of business for the chapter meeting falls upon the **Executive Committee**. This committee is generally composed of chapter officers and those who chair the major committees, often referred to as *standing committees*.

The president presides over the Executive Committee. The Executive Committee needs to work closely with the Conduct of Meetings and Recreation Committees.

Well-planned, regular chapter meetings are essential to maintain member interest, secure attendance, ensure efficiency, and promote the general welfare of the group. Each meeting should be a unit in the series for the year instead of a separate entity unrelated to other chapter interests and activities. Remember, the primary reason for holding meetings is to conduct the business of the organization; a majority of the business of the organization should be your effort to carry out a challenging **Program of Activities** (a document within the FFA organization that records the chapter's goals and the ways and means to achieve them). Successful meetings generally consist of the business to be transacted, the program, refreshments, and recreation.[6]

Planning your chapter meetings on a yearly schedule enhances unity and reinforces purpose. It also ensures that everything is included. Although some revisions may have to be made as the year progresses, a schedule contributes to effective chapter operation. Having a yearly schedule enables you to schedule important speakers in advance, secure needed school facilities and equipment, and alert members to their responsibilities.

Purpose

During planning, the Executive Committee should define the purpose of each meeting. A clear idea of what to do is the foundation on which everything rests: The planners should have a good idea of what they want to accomplish. Unless the meeting leader is utterly certain about the purpose of the meeting, it cannot possibly be effective. Meeting leaders should remember to ask for suggestions from group members well before a meeting. This input helps to ensure that meetings focus on the members' needs and promotes their interest in and ownership of the chapter.

Questions for Responsive and Productive Meetings

Meetings can be either great time wasters or very productive. It all depends on how well the officers plan and how responsive they are. Responsive officers know that good meetings are key to maintaining a good organization. More than anything else, a productive meeting depends on good planning and good execution. Following are some questions that can help officers plan productive and responsive meetings:

- Is this meeting really needed?
- What is the reason for holding this meeting?
- What is the goal of this meeting?

VIGNETTE 12-1 | **PERSONAL WORST LEADERSHIP EXPERIENCE**

WHAT IS THE ABSOLUTE WORST LEADERSHIP EXPERIENCE YOU HAVE EVER HAD?

Were you teaching someone or a group of people how to do something? Were you in charge of a certain project? What was the project?

For many of us, the worst leadership experience occurred when we did not plan—and that is exactly what happened to Elizabeth during the biggest meeting of the year for her sorority. She was put in charge of the banquet and, subsequently, the banquet committee for the year. Traditionally, the banquet was also the meeting where most of the business for the upcoming year was discussed.

Elizabeth made sure the food was ordered. She made sure the decorations were put up. She even sent out invitations to alumni. What she forgot to do was plan for the business portion of the banquet. She thought drafting an outline one day before the banquet would be enough. It was not, by a long shot. She seemed to be the only person in the room who knew what was going on (or what should have been going on) that night. The food was good and the room was nice, but the disorganized meeting seemed to last forever. Members discussed business items several times over because they did not use parliamentary procedure and there was no posted agenda to keep them on track. The members had to ask repeatedly for clarification about exactly what was being discussed. After about an hour, people just started to leave, and the evening was ruined.

How could Elizabeth have made the evening different? What could she have done to make the meeting run more smoothly? (You probably identified poor planning as one of the many problems she encountered!) After that night, Elizabeth vowed never to go into a meeting unprepared, ever again.

- Which officer will preside at the meeting?
- Should all members attend this meeting?
- Will parliamentary procedure be used to guide the meeting?
- Where should the meeting be held?
- How will members be informed before the meeting?
- How will information be presented during the meeting?
- When will the meeting begin?
- When will the meeting end?
- What should be done to follow up on the meeting?

Members' Preparation

Before meetings, members should

- Study what the meeting is all about and be ready to contribute to the discussion
- Make sure they have the right information to make a decision
- Prepare by thinking of the action steps it will take to bring a project to the next stage
- Represent other groups or people that will be affected by decisions made at the meeting[7]

Preparing a Meeting Agenda

Once the plans have been drawn up, an **agenda** (a list, plan, or the things to be done) should be prepared to outline and reflect the planning and preparation for the meeting. Some groups have a clause in the constitution and bylaws that no business may be discussed unless it is on the agenda.

After the Executive Committee has completed the planning, the secretary should prepare a written agenda for the meeting. The agenda should be posted in a convenient place where all members can see it. Other options include emailing the agenda to all members or posting the agenda on a social networking site used by the members (e.g., Facebook, Twitter, or even Instagram). In some cases, the agenda can be duplicated and handed out to members.

The agenda is the single most important component of meeting planning. A well-thought-out agenda distributed prior to the meeting provides participants with purpose and direction. It prepares participants and helps to create a solid

structure for the meeting. The presiding officer can ask participants to submit agenda items if he or she wants to promote group involvement and avoid surprises at the meeting. A written agenda is useful because it accomplishes the following:

- Allows participants to come to the meeting with the resources and materials necessary to make important decisions
- Provides a plan, including time limitations for each topic, that aids in keeping the meeting running smoothly
- Keeps meeting participants on course and focused with tangible reminders
- Alerts all participants who will be in attendance (e.g., special speakers, observers, and other invitees)[8]

Typical FFA Agenda

Make sure the agenda is clear and concise, but avoid making it too brief or vague. In general, keep the agenda and meeting short. The following order of business can be used in planning and conducting chapter meetings. Your chapter should feel free to modify this agenda to fit your local needs. However, you should have an order of business that is followed at all regular meetings.

1. Opening ceremony (official ceremonies should always be used for FFA meetings)
2. Minutes of the previous meeting (read by the secretary and approved by the membership)
3. Officer reports (treasurer's report at every meeting, others as needed)
4. Report on chapter Program of Activities (chairs of the various sections of the Program of Activities are called on to report plans and progress)
5. Program (it is often more appropriate to place this as the last item before the closing ceremony)
6. Unfinished business
7. Committee reports (All committee reports should be in writing. After a report has been accepted, the committee chairperson should note any changes and file his or her report with the secretary for inclusion in the minutes.)
   a. Standing committees
   b. Special committees

8. New business
9. Degree and installation ceremonies (used only when new members are initiated, when Greenhands are raised to Chapter Degree, and when officers are installed)
10. Closing ceremony
11. Entertainment, recreation, refreshments (Always adjourn the meeting before participating in recreation and before refreshments are served. This helps to keep the formal and informal portions of a meeting separate.)[9]

Item 11 in the agenda is often ignored or skipped, but it can add life and interest to every meeting. Every chapter needs a program director who arranges entertainment, recreation, and refreshments for each meeting in addition to the business or program. Figure 12–3 illustrates the major components of an FFA meeting.

(a)

(b)

(c)

FIGURE 12–3 Three components are (a) the program, © *iStock/Izabela Habur* (b) recreation, © *iStock/stock_colors* and (c) refreshments. © *iStock/Lauri Patterson*

LUCI IS A GREAT LEADER IN HER COMMUNITY

She is in charge of more than 20 employees at the local gym, and she heads up the annual Susan G. Komen Breast Cancer 3-Day for the Cure, a 60-mile walk for those who want to make a personal difference in the fight against breast cancer. Whether she is having a meeting at work or with the volunteers who help with the big walking event, Luci has learned that meetings matter.

She has found that one of the first things people look for in a meeting is the agenda. First of all, they want to know whether everything on the agenda is important to them. Luci thinks that you should not waste time on things people do not really care about. She also believes that the most important issues to the volunteers should be first or last on the agenda. This practice is very effective at getting them to pay attention and for improving their recollection of the most important issues.

Luci has also found that it is important for people to physically attend meetings. When people are at the meeting, the decisions made mean more because they participated in making those decisions in a face-to-face format, rather than being uninvolved witnesses via emails or memos. In addition, people like hearing and being motivated by their leader, and Luci is always happy to feed this desire.

Each leader will dream up even more creative ways to improve and ensure the success of meetings. Whether you use your own techniques or Luci's or a combination, just remember that what your followers think and want is one of the most important considerations as you prepare the agenda.

SECURING ATTENDANCE AT CHAPTER MEETINGS

If you feed them, they will come. This actually refers to feeding members psychologically, not physically. Although refreshments (especially pizza) have a special appeal to most FFA chapter members, securing high attendance at meetings is always a challenge. Time is precious to everyone, and people tend to make time for what they really want to do. Therefore, the adviser and Executive Committee should ask, "What is there about our meetings that makes members want to attend?" The short answer is that members usually want to attend when they have some responsibility for the success of the meeting or when the meetings are interesting and beneficial to them.[10]

Students go where the action is. If motivating business, program, and recreational activities (discussed later in the chapter) are included as a part of each meeting, students will attend. However, you can follow some logistical hints to improve attendance:

- Schedule meetings at a convenient time, as discussed earlier. Do not schedule meetings that conflict with school athletic events and other school programs.
- Develop carpools. Not only does this reduce gas usage and parking problems, but it also makes students realize that other people are counting on them to be there.
- The adviser could give a grade for attendance or award bonus points (after all, meetings are educational and develop leadership skills).
- Have members address postcard reminders to themselves, to be mailed by the secretary.
- Hold purposeful, beneficial, efficient, and motivating meetings.

MEETING ROOM ARRANGEMENTS, PARAPHERNALIA, EQUIPMENT, AND SUPPLIES

Arrangements

A well-arranged meeting room that includes all the necessary items lends dignity to FFA gatherings and creates a desirable spirit of enthusiasm and pride. The sentinel should make sure the meeting room is properly arranged, according to the official FFA manual. Officer stations should be located according to the manual and the paraphernalia prominently displayed. Figure 12–4 shows the room arrangement for an official FFA meeting.

Sentinel Duties

Before the meeting, pay attention to the meeting place, especially the ventilation, heat, and lighting. The sentinel should continue to monitor these matters during the meeting to ensure member alertness and comfort. The sentinel should make arrangements to get and set up any equipment and supplies needed during the

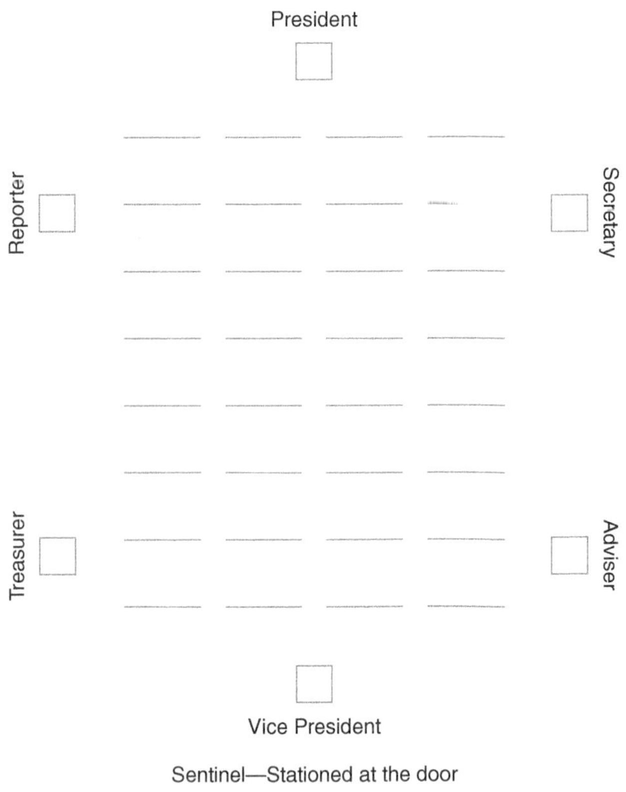

President

☐

Reporter ☐

Secretary ☐

Treasurer ☐

Adviser ☐

Vice President

☐

Sentinel—Stationed at the door

FIGURE 12–4 Seating arrangement for an official FFA meeting.

meeting, such as video, television or monitor, projector, screen, and speaker's stand. After the meeting, the sentinel should see that all paraphernalia, equipment, and supplies are returned and the meeting room is left in good order.

Official Chapter Paraphernalia, Equipment, and Supplies

The following is a list of chapter **paraphernalia** (the gavel, owl, plow, flag, secretary's book, treasurer's book, and other items used in an FFA meeting), equipment, and supplies (available from the National FFA Merchandise Center):

1 American flag
1 FFA felt banner (3 feet by 6 feet)
1 plow
1 ear of corn
1 bust of George Washington
1 owl
1 rising sun
1 flag and base (miniature)
1 shield for sentinel station
1 picture of Thomas Jefferson

1 gavel and block
1 secretary's book
1 reporter's book
1 treasurer's book
1 scrapbook
7 or more official FFA manuals
1 charter (framed)
1 official FFA flag (optional)
1 Creed (framed—optional)
1 Purposes (framed—optional)

Proper Use of the Gavel

The presiding officer should exhibit authority and leadership, become familiar with use of the gavel in chapter meetings, and use the gavel as a symbol of authority to guide members in proper procedures.[11]

2 taps of the gavel—Call meeting to order
3 taps of the gavel—Members stand
1 tap of the gavel—Members sit
1 tap of the gavel—After announcing results of a vote or decision of the chair
A sharp tap or series of sharp taps—To restore order

COMMITTEES

A **committee** is a chosen group of members with specified responsibilities. The use of committees is an important part of the FFA organization. Chapter 14 discusses the committee structure within the FFA. Committee reports are a part of most FFA meetings. Committees have their own meetings, and the committee chairperson reports to the whole-chapter group about the progress his or her committee has made on its assigned task. The chairperson makes sure that the committee members have a clear purpose, know what to do, and understand why they were selected or appointed.

Organization

The key to effectiveness for committees is for leaders to organize them properly. This includes setting goals, identifying steps needed to reach the goals, setting target dates for completion, and estimating and gathering the resources needed. Committee responsibilities should be divided among the committee members. The Committee Activity Planning Sheet helps committees organize their tasks (Figure 12–5).

Reasons for Using Committees

Involvement is the main reason for using committees. In democratic organizations, the leader delegates. Members need to feel that they are a contributing part of the group. Members also need to be involved because they offer expertise that can help meet the needs of the group. For example, Corey should be on the Public Relations Committee because he has excellent mechanical abilities, and one of the goals of the committee is to enter a float in the homecoming parade.

Committees help develop leadership by bringing out the best in people: talents are discovered and confidence is built. Members become excited when they know they are contributing members of the team. Because of members' involvement, the group's effectiveness is increased and the team becomes stronger.

COMMITTEE ACTIVITY PLANNING SHEET

Activity: FFA Week

Division: Public Relations

Members Responsible: Chuck Barstow, Betty Zetlow, and Mary Carlson

GOALS	STEPS	TARGET DATE	ESTIMATED BUDGET	EVALUATION NOTES
(What do we want to do?)	*(How are we going to meet the goals?)*	*(When?)*	*(How much?)*	*(Information for future planning)*
1. Sponsor faculty breakfast.	1. Set date and place on school calendar. Date set: _____	Sept. 15	None	
	2. Arrange to use home economics facilities.	Sept. 15	None	
	3. Have members sign up for the following jobs: room setup, greeters, cooks, servers and cleanup.	Jan. 15	None	
	4. Prepare invitations and place in teachers' school mailboxes.	Feb. 1	None	
	5. Select menu.	Feb. 15	None	
	6. Buy food.	Feb. 20	$50	
	7. Acquaint workers with jobs.	Feb. 20	None	
2. Present a five-minute radio program.	1. Discuss plans with manager of KRNT.	Oct. 15	None	
	2. Select three members for program.	Jan. 15	None	
	3. Develop script using materials from National FFA Center.	Jan. 25	$5	
	4. Review plans with KRNT program manager and set taping date. Date set: _____	Feb. 2	None	
	5. Revise script.	Feb. 8	None	
	6. Rehearse program.	Feb. 15	None	
	7. Tape program.	Feb. 20	None	

FIGURE 12–5 Committees work best when their activities are planned and organized. The Committee Activity Planning Sheet, which is part of the FFA Program of Activities, helps a committee organize its task. *(Courtesy of the National FFA Organization.)*

Responsibilities of Committee Members

The first responsibility of committee members is availability. Unless you are committed to serving, the goals will not be met. Serving includes attending meetings, contributing ideas, listening, being tolerant of other ideas, and helping to execute the committee's decisions.

The second responsibility is involvement. Do not be a spectator. Make decisions based on facts, be innovative, show initiative, accept responsibility, follow through with assignments, and be cooperative. If you do not like an idea or a person, learn to "agree to disagree."

Responsibilities of Committee Chairs

The responsibility of a committee chairperson is basically the same as that of the presiding officer of a larger meeting, such as an FFA meeting. It includes planning; preparing an agenda; collecting background information on an item of business or committee assignment; and selecting a meeting date, time, and place.

Before the committee meeting, the chairperson should supply members with appropriate materials. During the meeting, the chairperson should direct the members' discussion, keep members on task, and help them develop a sense of responsibility. Each member should be given a chance to express his or her ideas. When necessary, it may be appropriate to develop subcommittees.

After the meeting, the chairperson should compliment members for their work when they have done a good job or tactfully motivate them toward improvement if they have not. The chairperson should make sure that committee reports and minutes are complete, and that officers of the chapter or organization are kept informed as to committee progress and actions. Reports should be given to the chapter during FFA meetings as necessary. Committee reports should be recorded in the minutes.

Committee Seating Arrangement

You are probably asking what seating arrangement has to do with anything, much less committee meetings. Verbally, it has nothing to do with anything. Nonverbally, seating arrangements communicate much. In Chapters 1 and 2, we discussed three types of leadership styles: autocratic, democratic, and laissez-faire (participative). Seating arrangements nonverbally communicate a leadership style to the members. Figure 12–6 shows seating arrangements for autocratic, democratic, and participative leadership styles.

In the frequently used autocratic seating arrangement, the chairperson presides from the end of the table. This arrangement focuses attention on the leader and enables him or her to exert tight control over the agenda and meeting.

When the leader wants to encourage participation, he or she can use one of two equalizing patterns whereby the leader sits among the other attendees. This works best when the leader wants to limit her influence, promote creativity, or maintain a low profile. The leader should use a seating style that best serves the purpose and type of meeting desired.

In the democratic seating arrangement, the chairperson sits on one side of the table, leaving the ends open. Although the leader still has control, there is a feeling of openness, and it is easier for members to discuss topics with each other as opposed to directing all communication toward the leader.

In the participative, round-table arrangement, everyone's position is equal and each person is

COMMITTEE SEATING ARRANGEMENTS

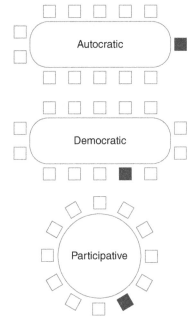

FIGURE 12–6 Seating arrangements nonverbally communicate your leadership style. Depending on the situation, the meeting or committee chairperson may want to vary seating arrangements to meet various objectives.

encouraged to participate fully. This arrangement minimizes the display and effects of status and power.

MEETINGS THAT INFORM AND MOTIVATE

Scene 1:

"What a waste of time!"

"Right? I know Calvin was presiding, but he never shut up! The one time I tried to say something, he cut me off."

"And he laughed at Nancy's idea even though it sounded great."

"What a jerk!"

Scene 2:

"I'm amazed that everyone bought our idea!"

"Especially since Bobby wanted to go the other way. Still, he really encourages everyone to say what they think."

"I can't believe the number of ideas that came up."

"Some of them might be implemented later on."

"Bobby's meetings always seem relevant and exciting; they're always worth going to!"[12]

Excitement

Are your meetings like those discussed in Scene 1 or Scene 2? We hope they are like the infinitely more positive and exciting Scene 2. Every meeting should be packed with excitement. Remember, we are in an entertainment age, and there are many possible activities competing for members' attention every moment. With so many attractive options available for how we spend our time, FFA meetings have to be exciting and appealing, as well as educational, if we want to make sure our members will show up.

Meeting Components

Every meeting should contain three key elements: business, program (education), and recreation. The business section should be organized and efficient. It should include a crisp opening ceremony, a clear and distinct reading of the minutes of the previous meeting, the treasurer's report, committee reports by the committee chairs, and proper parliamentary procedure as old and new business is discussed.

Educational (Program) Components

Each meeting should also have a program feature. Variety is the key here. The more members are involved, the more motivational and interesting the meeting will be. Programs involving members could include team development activities; or the program could simply be the chapter Creed contest, the chapter public speaking contest, a parliamentary procedure demonstration, and/or an FFA talent contest. Such programs often include invited speakers, but the persons putting the programs together should make sure that anyone invited is entertaining and motivational. Special chapter programs could include state officers, FFA alumni, or reports about national and state FFA conventions or FFA Camp that recruit members to attend (Figure 12–7).

FIGURE 12–7 FFA or 4-H camps and leadership conferences can become excellent meeting topics. If conducted well, such meetings should motivate and recruit students for these activities. *(Courtesy of University of Georgia—College of Agricultural & Environmental Sciences.)*

MONTH	PROGRAM	RECREATIONAL ACTIVITY
July	Program of Activities Report	Swimming, boating, games
August	Advisers/Officers Report	Softball (alumni-member game)
September	Initiation of Greenhands	Class flag football competition
October	Parliamentary Procedure Demonstration	Hayride and "weenie" roast
November	Chapter Creed Contest	Class volleyball competition
December	Christmas Program for Underprivileged Children	Christmas party
January	Film of Member's SAEP	Class basketball competition
February	State FFA Officer's/Chapter Public Speaking Contest	Class basketball competition
March	Careers in Agriculture	Skating party
April	State FFA Convention Report	FFA talent contest
May	FFA Camp Program	Class softball competition
June	Film on Year's Activities	Cookout and games

FIGURE 12–8 Members will attend meetings that are well planned, organized, and full of excitement. Students go where the action is. This sample schedule of FFA meetings balances educational and recreational components.

Recreational Components

Recreational activities could include a year-long athletic competition. Teams can be formed around class, year, individual classes, or another format to provide teams with a balance of talent and ability. Athletic competition could include softball, flag football, volleyball, team development activity contests, and a variety of minor sports.

Whatever the format, meetings should inform, motivate, and attract students. Again, students go where the action is. The FFA can be the group in your school "where it's happening." The list in Figure 12–8 suggests possible programs and recreational activities for FFA meetings.

The major topics included in the preceding section are but a few that you might consider in developing your own yearly schedule. Following are some additional topics that might suit your needs and interest your members:

FFA alumni
Washington Leadership Conference
Made for Excellence
Leadership roles of former FFA members
National Chapter Award Competition
Proficiency awards
High school principal (speaker)
High school guidance counselor (speaker)
College Night
National FFA Week
Service learning activities
FFA contest and awards
Election of officers
Open house/Parent's Night
State wildlife resource agency
Community service agency
Community service activities
Successful agriculturalist (speaker)
Thanksgiving program

CONDUCTING EFFECTIVE MEETINGS

FFA meetings should operate in such a way that every potential member wants to attend. If this is not the case with your chapter, go back to the drawing board and try some different approaches. However, when a meeting has been carefully planned and officers and members are well prepared, a good meeting is almost assured.

As the presiding officer, the president plays an important role in a good meeting. The presiding officer sets the mood for the meeting by being interesting, motivating, and concerned. However, a meeting should never be a one-person show; the president should encourage all members to take part and share their ideas.

Democratic Decisions

Chapters 10 and 11 provide the basic skills and understanding needed to conduct business meetings adequately using parliamentary procedure. FFA members should have a working knowledge of parliamentary procedure and should also be familiar with the chapter's constitution and bylaws.

Some presiding officers seem to believe that the speed with which they conduct business is somehow related to effectiveness. In reality, the purpose of meetings is to reach sound decisions based on facts and appropriate debate. Decisions should reflect everyone's alert thinking, contributed through thorough discussion and exploration of a matter before a decision is made final by formal group action. It should be obvious that this type of participative decision making usually results in better attitudes, higher acceptance of decisions, and improved group cohesiveness.[13]

Democratic Problem Solving

The presiding officer should focus on problems, not personalities. He or she should look at the quality of the idea, not the person who suggested it. The development of positive attitudes contributes to chapter morale and effective programs.

A good presiding officer is democratic and willing to respond to the group's wishes. He or she believes the democratic process is a way to get other members to think, speak, plan, and work together. The presiding officer is fair and impartial, reserving judgment until the opinions of others have been fully aired. A secure presiding officer perceives individual differences as a potential opportunity for the group rather than as a threat to the leader's position or the program. Following are some steps to help the presiding officer conduct meetings effectively within the framework of parliamentary procedure.

Presiding Officer Abilities

A presiding officer should work conscientiously on developing skills and procedures for the following:

- Start and end the meeting on time
- Set an atmosphere for the meeting that encourages order, discussion, and efficiency
- Prepare for the discussion of business

- Introduce each topic properly
- Keep the discussion centered on the topic
- Keep the discussion moving forward
- Maintain poise in handling the meeting
- Speak with a clear voice and loudly enough so all can hear
- Be fair with all members and show good judgment
- Give everyone a chance to speak, comment, question, or otherwise contribute
- Keep the discussion from becoming overly emotional or personality oriented
- Stimulate discussion by asking questions
- Summarize the discussion when it appears to be finished
- Formally conclude the business and announce the disposition of the topic
- Announce the next topic for discussion

The presiding officer needs to develop a sense of timing, so that he or she knows when to step in and encourage a response or contribution from a nonparticipant. The presiding officer should also know how and when to discourage someone who is monopolizing a discussion, when to ask for a summary or explanation if members seem lost or confused, and when to table a discussion if an issue cannot be resolved.[14]

Energy Cycle of Meetings

"Meeting energy is affected by attention cycles, interest in topics, complexity of topics, the number of topics to be addressed, the scheduling of those topics, and the level of participation of attendees. Every meeting has an **energy cycle** (scheduling things at certain times in a meeting in order to be the most productive) that can be managed and enhanced" by the presiding officer. Follow these guidelines for effective timing and management of the energy cycle:

- "The early part of the meeting tends to be more lively and creative than the end of it, so items requiring more imaginative ideas, mental energy, and clear heads should be addressed early in the meeting."
- Any critical items that absolutely must be addressed should be placed first on the agenda. This keeps you from getting stuck and spending too much time on matters of low importance.

- If no critical, high-priority items exist, you have the option to address first any items that can be dealt with quickly and easily; this leaves the rest of the meeting for matters that require a greater investment of time.
- "Consider reserving a controversial, high-interest item" until the end of the session. "This way, useful work can be accomplished before the topic comes up. The high interest level in that item will keep attention from lagging."
- Try grouping items of business so that people can come and go from the meeting as they are needed. Changing the composition of the group "automatically raises the energy level as fresh faces and new voices appear."
- "If the meeting will be long, with many agenda items, consider alternating working items with reporting items to avoid boredom."
- "Try to find a unifying item to end the meeting."[15] It is especially important to finish on a harmonious note if there has been contention or heated discussion about any of the agenda items.

Minutes

"If a meeting is important enough to be held, it's important enough to have the decisions and action items recorded. Meeting minutes make sure that important points and decisions are recorded accurately, without misinterpretations." Taking minutes is the responsibility of the secretary. The minutes should be posted in the classroom or on a chapter website, or typed and distributed to those who need them. The point is that all members should be able to access the minutes. Minutes should contain the following information:[16]

- Time, date, and place of the meeting
- Name of the person who presided
- The balance of the chapter's fund, from the treasurer's report
- All agenda items (and other items) discussed and all decisions reached; task assignments and persons responsible for each task
- Date, time, and place of the next meeting
- Signature of the secretary (or other person) who took and wrote up the minutes

GROUP MEMBER INVOLVEMENT

The presiding officer, leader, and FFA member have many occasions to lead groups (both large and small) in discussions. Effective leaders plan the kinds of interaction they want. If member involvement is a serious problem, ask the members themselves for ways to get the whole group involved. Open discussions by letting the members know that the presiding officer is serious about generating dialogue and new ideas.

Nothing affects the productivity and outcome of a meeting more than the attendees' participation. A meeting is held so that everyone can benefit from the combined information, wisdom, and experience of the members; otherwise, there is little point in having a meeting at all. In other words, if some members fail to contribute their ideas and thoughts, poor decisions will be made and organizational commitment to decisions will suffer. Some techniques for promoting group member involvement follow.

Questions and Questioning

Questions stimulate group interaction. They help people generate ideas and promote thinking. Questions can summarize a discussion or move a group to action. Ask specific questions of individuals to encourage them to get involved. Ask for participation and reward individual contributions both verbally and nonverbally. Treat each person courteously and acknowledge all ideas and contributions, even if they do not reflect your opinions and beliefs.[17]

Buzz Groups (Small-Group Discussion)

In **buzz groups**, an assembly is broken into small groups of six to eight members each for the purpose of generating ideas, solutions, and possibly common ground in a given amount of time. A leader and recorder are selected for each group. Often, a representative from each buzz group is asked to report on the discussion. Buzz groups have three advantages:

1. They examine as many ideas as possible and report them to the main group.
2. All members get a chance to speak.
3. More members can practice their leadership skills.

Panel Discussions

Panel discussions are a good way to present information from experts. Various forms and formats can be used for panel discussions:

Question-and-Answer Sessions These sessions give members a chance to ask the panel members questions throughout the duration of the discussion.

Symposium In a symposium, each panel member speaks on an area in which she or he has expertise. Experts may entertain questions following the panel presentations.

Problem-Solving Sessions In this type of session, panel members deal with specific problems through interaction with the audience.

Role-Playing

In **role-playing**, two or more people assume parts and act out a situation relating to an issue or problem that the chapter or group is considering. A brief summary of the role-play, or a discussion that clarifies the role-play, is often helpful. Role-play gives members a picture of how things look from another person's perspective.

Brainstorming

Just as with brainstorming ideas for a speech, in group brainstorming members state every idea that comes into their heads, without criticism or judgment from other group members. After everyone's ideas have been collected, the suggestions are all evaluated. The evaluation session usually uncovers some patterns or themes that may be helpful for the decision-making processes in a meeting, and sometimes yields fresh ideas that provide new insights into the matter or a new way to attack a problem.

Parliamentary Procedure

Effective parliamentary procedure remains the key to conducting effective meetings. As group members work together, personality types tend to become obvious. Behavior typical of certain personalities (the silent member, the talker, the wanderer, the bored one, and the arguer) can cause meetings to bog down. Effective use of parliamentary procedure can help the presiding officer to deal appropriately with each.

Hold Back Your Opinion

If you are the leader, try not to give your opinion right away. Doing so can make someone who disagrees hesitant to participate in the discussion. Hold back on showing members your thoughts as the leader if you want their honest opinions.

Listen

In most instances, it is more valuable for the leader to listen—and listen actively—than to talk. Do this through note taking, asking clarifying questions, seeking confirmation of what you are hearing, and making a genuine attempt to understand the group.

GROUP PROBLEM SOLVING AND THINKING

Problem-solving techniques can help members arrive at decisions, answers, or solutions. Effective problem solving tends to follow a pattern. In the first part of a discussion, the members investigate the nature of the problem. In this phase, the presiding officer should "encourage factual, nonevaluative discussion" of the problem. Next, identify all possible solutions, again without evaluation, so that the group has at hand as many alternatives as possible. After all solutions have been considered, "select the solution that best fits the needs of everyone involved."[18]

Group problem solving is a cooperative process in which members pool their thinking to reach a satisfactory solution. The following steps are characteristic of good problem solving and thinking:

- Reorganize, state, and comprehend the problem
- Review the background of the problem
- Keep the main problem clearly in mind as solutions develop
- Propose and apply any relevant suggestions
- Critically examine and evaluate proposed solutions
- Abandon ideas that prove false, unrealistic, or impractical
- Suspend judgment on an idea until all the facts are known
- Test all conclusions
- Agree on a plan to implement the decision

FIGURE 12–9 One sign of quality group member involvement is effective problem solving and decision making. The intensity of this speaker indicates that a good meeting is in progress. *(© Glynnis Jones/Shutterstock.com.)*

Group problem solving has several advantages over individual problem solving, for the following reasons:

- It solicits and uses the contributions of all group members.
- It stimulates, modifies, and refines individual thinking.
- It makes available different points of view and more resources.
- It appeals to collective wisdom and cooperative action.

The intensity of the member in Figure 12–9 indicates that effective problem solving and decision making are probably occurring.

GROUP MATURITY

Agree to Disagree

Good presiding officers should maintain a climate of constant inquiry in which all assumptions can be questioned—even their own. Different points of view, critical thinking, and constructive disagreement should be encouraged. Presiding officers should stimulate creativity and curb members' desires to make decisions quickly. Productive meetings often involve conflict, but the conflict should be a clash of ideas, not personalities. In other words, agree to disagree on ideas. Do not take it personally if other members do not agree with you.

Perils of "Groupthink"

In *Communicating at Work*, Alessandra and Hunsaker discuss the term **groupthink**, which is basically a tendency toward a herd mentality. Once groupthink begins, the group no longer does critical, in-depth investigation of an issue. Group members become content with easy solutions and fail to question or challenge ideas and recommendations. Groupthink is especially likely to occur in groups that meet over long periods of time with mostly the same members. Cooperation is encouraged within the FFA, but cooperation is less important than maintaining our beliefs, values, and opinions when we do not agree. Groupthink is one of the critical pitfalls of FFA meetings, even in meetings that appear to be working well.

Some of the causes of groupthink include

- **Illusion of togetherness,** or the perception that everyone agrees. "The group begins to take pride in its lack of disagreement and ability to come to rapid decisions."

- **Conformity pressures,** which cause dissent or questioning to be perceived as an improper or unhealthy attack by people who are not "team players."
- **Self-censorship,** whereby group members do not bring up negative factors or consequences of decisions and do not air misgivings about the direction the group is taking.
- **Time pressures,** which "can block deep examination of issues and make people grab at easy solutions and avoid the interpersonal processes that make constructive disagreement possible."

To avoid groupthink, the presiding officer should create an atmosphere in which members feel free to disagree. Encourage openness, "devil's advocate" hypothesizing, and real consideration of minority viewpoints. Silence should not be interpreted as agreement: ask for different ideas and solicit new views.[19]

Develop Member Sensitivity

Leaders can and should develop sensitivity to their own needs, other members' needs, and the group's needs and goals.[20] Using good communication and interpersonal skills, they can also assist members to become more sensitive in these areas.

Members who are sensitive to their own needs

- Recognize their own needs but do try not to meet them at the expense of others
- Share their ideas, feelings, and differences with the group
- Are able to change their own ideas without emotional upset
- Do not dominate or monopolize meetings
- Recognize the value of other group members, even when they dislike those members personally

Members who are sensitive to other members' needs

- Are aware of and ready to try to understand others' needs
- Encourage fulfillment of others' needs
- Encourage others to participate

Members who are sensitive to the group's needs and goals

- Are ready to take on responsibilities
- Recognize the importance of objectivity
- Are willing to abide by group decisions
- Are flexible about assuming roles recommended or requested by the group
- Put the group's goals first and their own second

OFFICER RESPONSIBILITIES

Officer Team Responsibilities

The officer team has various responsibilities regarding the conduct of successful meetings. They post the agenda in the agricultural education classroom or on the website several days before the meeting. They make sure that other members are prepared for the meeting.[21] They are responsible for guest speakers, room arrangement, refreshments, and entertainment, whether they take on these tasks themselves or oversee others' fulfillment of these responsibilities.

One of the most important officer responsibilities is performing the FFA opening and closing ceremonies. Ceremonies that impress with a display of meaning and conviction contribute substantially to a successful meeting.[22] Business-like ceremonies give the meeting status, purpose, and prestige.

All members of the officer team must

- Have a genuine interest in being a part of a leadership team
- Be able to lead by example
- Be familiar with the organization's constitution and bylaws
- Be familiar with parliamentary procedure
- Be willing to accept and delegate responsibility
- Be willing to memorize their parts in the various FFA ceremonies[23]
- Be able to work together
- Be willing to share ideas
- Be able to help satisfy members' needs
- Emphasize *we*, not *I*
- Be concerned with what is best for the group rather than themselves

The officers in Figure 12–10 are carrying out their responsibilities and making last-minute preparations before a meeting.

Specific Responsibilities

Each of the individual FFA officers has specific responsibilities.[24]

Note: The chapter officers listed in this section are required by the National FFA Constitution. Optional officers may be added at the local level, as needed and desired.

President

- Presides over all chapter meetings
- Serves as chair of the executive council
- Appoints committees
- Serves as the chapter's official representative
- Coordinates the chapter's activities and keeps in touch with the progress of each division of the Program of Activities

Vice President

- Assists the president and serves as presiding officer in the president's absence
- Coordinates all committee work
- Assumes responsibility for guest speakers and chapter meeting programs

Secretary

- Prepares and reads minutes of each meeting
- Prepares an agenda for each meeting
- Attends to chapter correspondence
- Prepares, posts, and distributes minutes
- Compiles chapter reports
- Keeps member attendance and activity records
- Issues membership cards

Treasurer

- Receives and deposits FFA funds
- Collects dues and assessments
- Helps plan public information programs
- Prepares and submits the membership roster and dues to the national organization, in cooperation with the secretary
- Maintains a neat and accurate official FFA Treasurer's Book
- Chairs the Earnings and Savings Committee
- Prepares monthly treasurer's reports for chapter meetings

FIGURE 12–10 A chapter is no better than its officers. These officers are checking last-minute details to make sure everything is in order. *(Courtesy of iStock.com)*

Reporter

- Publishes a chapter newsletter and maintains a reporter's scrapbook
- Releases news and information to local news media
- Maintains chapter social media accounts
- Helps plan public information programs
- Sends local stories to area, district, and state reporters
- Sends articles and pictures to the *FFA New Horizons* magazine and other national and regional publications
- Works with local media on radio and television appearances and FFA news

Sentinel

- Prepares the meeting room and cares for chapter equipment and supplies
- Attends the door and welcomes visitors
- Keeps the meeting room comfortable
- Takes charge of the candidates for degree ceremonies
- Assists with special features and refreshments

MEMBER RESPONSIBILITIES

Members should become familiar with the parts of the opening and closing ceremonies and with basic rules of parliamentary procedure. Members should also know what items of business are on the agenda for each meeting and be prepared to discuss them or raise questions to clarify the issues. Members should arrive at meetings on time and be alert to ways in which they can participate as informed and willing FFA members in all chapter meetings.[25]

The performance of individual FFA members determines the effectiveness of the chapter meeting. When members interact, help set goals, or take part in activities, they are engaged in positive group performance. Members should be involved in as many things as possible, as often as possible, and leaders should try to engage and use the interest and abilities of each member.

The secret of a vital and successful meeting is to involve all members so that they make the group their own. If students share participation and pride in the meeting, they will want to work hard. The best thing is that they will have fun in the process!

PROGRAM OF ACTIVITIES

The FFA Program of Activities (POA) is designed to provide FFA chapter officers and agricultural education instructors (FFA advisers) with activities that are appropriate for the chapter. "Conduct of Meetings" is one segment of the chapter POA. (The chapter POA is discussed in detail in Chapter 14.) The National FFA Merchandise Center (formerly the FFA Supply Service) has excellent materials to help plan the chapter POA. There are 12 divisions in a chapter's POA. Refer back to Figure 12–5 to see an example of the public relations POA area.

Many businesses and organizations also develop and follow a plan of action. A properly planned and developed POA enhances the effectiveness of any organization. Meetings will run more smoothly if they have a functional POA. Other benefits of a POA are that it does the following:

- Specifies what the organization intends to do and how it plans to do it
- Allows the members to set their individual plans and schedules
- Helps meet each member's needs and ensures that all members have an opportunity to participate in decision making
- Helps provide consistency in the various committees' efforts and prevents duplication or omission of duties and responsibilities
- Assists in the drafting of a viable budget
- Provides a means of measuring the group's success
- Helps maintain continuity over the years
- Provides direction for meetings[26]

MEETING EVALUATION

End-of-meeting evaluations are an easy way to determine the effectiveness of your meetings. Group members are asked to evaluate selected

MEETING EVALUATION FORM					
Rank each of the items. Please circle appropriate number.					
	Very Low	Low	Average	High	Very High
1. Physical arrangement and comfort	1	2	3	4	5
2. Orientation	1	2	3	4	5
3. Group atmosphere	1	2	3	4	5
4. Interest and motivation	1	2	3	4	5
5. Participation	1	2	3	4	5
6. Productiveness	1	2	3	4	5
7. Parliamentary procedure used	1	2	3	4	5
8. Agenda followed	1	2	3	4	5
9. Opening and closing ceremony	1	2	3	4	5

Please answer the following questions:

1. How would you rate this meeting   No good _____ Average _____ All right _____ Good _____ Excellent _____
2. What were the strong points? _____
_____
3. What were the weak points? _____
_____
4. Suggestions for improvement. _____
_____

FIGURE 12–11 Democratic leaders listen to members. Evaluations help to make sure members' needs and wishes are met. Using this evaluation form and following the suggestions you receive should improve your meetings.

aspects of meeting leadership, process, and productivity. Through these evaluations, members can formally analyze their chapter meetings.

Every group has problems from time to time. Evaluations should capture information and member reactions related to these issues: what caused the problem, how group leaders handled it, how members reacted, and so on. It is also valuable to have the evaluation measure and reflect the methods the group used for dealing with problems and uncomfortable situations, and how successful it was at resolving conflicts.

For several reasons, the questions that make up meeting evaluations should be designed to identify strong points as well as weaknesses. First, a positive approach makes members feel better about doing an evaluation; they do not feel that they are only criticizing and finding fault. Second, it is important to identify strong points so they can be emphasized as future activities are planned.27 Figure 12–11 shows a sample evaluation form.

CONCLUSION

Being able to conduct a meeting effectively is a critical leadership and communication skill. The benefits of a well-planned and well-conducted meeting are enormous: identification of solutions to problems, shared ideas and information, development of plans to which the group is committed, and strong team cohesiveness and morale. Building morale should be a key element of meetings. FFA members need to leave each meeting motivated to become or to continue to be contributing members of the team.

SUMMARY

Many opportunities for developing leadership skills arise while you are conducting and participating in meetings. These skills will be valuable after you graduate and participate in professional, business, and civic organizations.

The basic meeting communication functions are sharing knowledge, establishing common goals, gaining commitment, providing group identity, and interacting as a team. Conducting meetings in a democratic style and communicating with others help leaders develop sensitivity and allow them to foster sensitivity in others.

Good meetings result from careful planning and preparation. They should be well publicized in advance of the meeting time. The agenda should be posted, and the meeting should have a specific purpose. Meetings should be conducted by officers who are well prepared, know their duties and responsibilities, use correct parliamentary procedure, follow the agenda, and involve members. The meetings should be held in a suitable, comfortable place and include business, a program, recreation, and refreshments.

Minutes of the business portion of any meeting should be taken, posted, and distributed.

Committees are an important part of FFA meetings. The key to an effective committee is proper organization by the committee chairperson. Committee chairs should report to the group as a whole at regular intervals and when their committee's task is completed.

Democratic leaders want, invite, and encourage member involvement. If members are apathetic, ask them about ways to get the whole group involved. Some techniques for promoting group member involvement are buzz groups, panel discussions, role-playing, brainstorming, withholding of the leader's opinion, and listening. Group problem solving and thinking also increase member involvement. Successful meetings are interesting, exciting, and build member morale and commitment.

As groups and group members mature, they become more secure. Group leaders should learn to agree to disagree. If disagreement and conflict never occur, a worse thing can develop—groupthink. The presiding officer should promote an atmosphere in which members feel free to disagree.

Officers of the organization are responsible for conducting successful meetings. Chapter officers should also check to see that other members are prepared for meetings. Evaluations should be done at the end of the meetings to plan for and improve future meetings.

 Take It to the Net

Explore the Internet for information on how to conduct a successful meeting. You can look at as many sites as you want, but choose one for this assignment. Once you find an article about conducting meetings, summarize the article.

Search Terms
meeting
effective meetings
how to conduct a meeting
conducting effective meetings

Chapter Exercises

REVIEW QUESTIONS

1. List five leadership skills that can be developed by conducting successful meetings.

2. What are 14 values learned or benefits obtained by understanding each officer's duties and responsibilities mentioned in the FFA Opening Ceremony?

3. Name five basic meeting communication functions.

4. What are 15 characteristics of a good meeting?

5. What are ten barriers to successful meetings?

6. List 12 pre-meeting questions that help leaders ensure that meetings will be responsive and productive.

7. What nine items should be included in a typical meeting agenda?

8. What are five helpful hints for improving attendance at meetings?

9. What are three major components of an FFA meeting?

10. What 15 skills should the presiding officer develop?

11. List six items of information the minutes should include.

12. What are eight responsibilities of a committee chairperson?

13. Name 20 educational activities that could be used for FFA programs.

14. Briefly describe seven techniques to manage the energy cycle of a meeting.

15. List and briefly discuss eight techniques for promoting group member involvement.

16. Why can groupthink be a problem?

17. List 11 requirements of all FFA officers.

18. What are five specific responsibilities of the president?

19. What are three specific responsibilities of the vice president?

20. What are seven specific responsibilities of the secretary?

21. What are seven specific responsibilities of the treasurer?

22. What are seven specific responsibilities of the reporter?

23. What are five specific responsibilities of the sentinel?

24. What are eight benefits of a Program of Activities (POA)?

25. List nine items that are useful in evaluating a meeting.

COMPLETION

1. Leaders, supervisors, and administrators trained in effective human relations are more _____ to the needs of others.

2. Unless the meeting has a _____, the meeting will not be efficient and effective.

3. _____ is the key for using committees.

4. Every meeting should contain three elements: _____, _____, and _____.

5. In any meeting, the secretary should record _____.

6. The _____ assists the president and serves as presiding officer in the president's absence.

7. End-of-meeting _____ are an easy way to determine the effectiveness of your meetings.

8. Members should develop _____ to their own needs, other members' needs, and the group's needs and goals.

MATCHING

_____ 1. Call meeting to order.

_____ 2. Members stand.

_____ 3. Members are seated.

_____ 4. Restore order.

_____ 5. Chairperson presiding at end of table.

_____ 6. Chairperson sits on one side of table.

_____ 7. Round table arrangement.

_____ 8. Receives and deposits money.

_____ 9. Releases news and information to media.

_____ 10. Takes care of the meeting room.

A. sentinel

B. one tap of gavel

C. three taps of gavel

D. participative

E. two taps of gavel

F. autocratic

G. democratic

H. reporter

I. series of sharp gavel taps

J. treasurer

ACTIVITIES

1. Write a policy on the conduct of members at meetings.

2. Write a meeting policy (guidelines) for your FFA chapter meetings. If you need help, refer to the General Dynamics meeting guidelines discussed earlier in this chapter.

3. Prepare an agenda for a meeting. Be creative and professional in suggesting new ideas or items.

4. Develop a plan justifying and supporting three recommendations for improving attendance at chapter meetings.

5. You are appointed to a committee to plan an FFA summer recreation event. Use the Committee Activity Planning Sheet in Figure 12–5 as a guide to plan the event. Provide information for each of the categories.

6. Your teacher will present three topics for discussion. You are to lead the group to a decision on all three topics. Use parliamentary procedure, which you learned in the preceding chapters.

7. Compare the steps in problem solving with those in the scientific method.

8. Write a one-page essay on the statement "let us agree to disagree," including thoughts from a leader, members, and your personal perspective.

9. Video record a meeting. Watch and evaluate it along with other chapter members. Write at least five suggestions for improvement according to the recommendations in this chapter.

NOTES

1. T. Alessandra and P. Hunsaker, *Communicating at Work* (New York: Fireside, 1993).
2. Ibid., pp. 188–189.
3. R. E. Bender, R. E. Taylor, C. K. Hansen, and L. H. Newcomb, *The FFA and You* (Danville, IL: Interstate Printers and Publishers, 1979), p. 300.
4. Alessandra and Hunsaker, *Communicating at Work*, p. 191.
5. "Successful Meetings," *Communication* series, No. 8741-I (College Station, TX: Instructional Materials Service, Texas A&M University, 1988), p. 1.
6. *Agriculture Teacher's Manual* (Indianapolis, IN: National FFA Organization, 1998), pp. 9-3, 9-6–9-7; available at http://www.ffa.org/documents/edr_teachnbk.pdf.
7. Alessandra and Hunsaker, *Communicating at Work*, p. 195.
8. Ibid.
9. *Agriculture Teacher's Manual*, p. 9–6.
10. Bender, Taylor, Hansen, and Newcomb, *The FFA and You*, p. 308.
11. "Successful Meetings," p. 3.
12. Alessandra and Hunsaker, *Communicating at Work*, pp. 201–204.
13. Bender, Taylor, Hansen, and Newcomb, *The FFA and You*, p. 311.
14. Alessandra and Hunsaker, *Communicating at Work*, p. 206.
15. Ibid., pp. 197–198.
16. Alessandra and Hunsaker, *Communicating at Work*, pp. 208–209.
17. Ibid.
18. Ibid., p. 211.
19. Ibid., pp. 216–217.
20. *Leadership of Youth Groups* (Kansas City, MO: Farmland Industries, n.d.).
21. Bender, Taylor, Hansen, and Newcomb, *The FFA and You*, p. 303.
22. Ibid., p. 304.
23. *Official FFA Manual 2014* (Indianapolis, IN: National FFA Organization).
24. Ibid.
25. Bender, Taylor, Hansen, and Newcomb, *The FFA and You*, p. 305.
26. "Program of Activities," *Efficiency* series, No. 8742-B (College Station, TX: Instructional Materials Service, Texas A&M University, 1988), p. 1.
27. G. M. Beal, J. M. Bohlen, and J. N. Raudabaugh, *Leadership and Dynamic Group Action* (Ames, IA: Iowa State University Press, 1962), p. 290. (Latest ed., 1967).

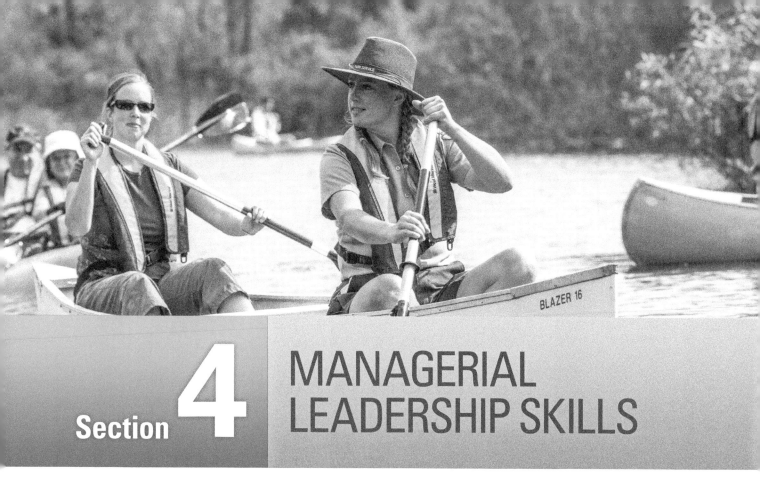

Section 4 MANAGERIAL LEADERSHIP SKILLS

13 PROBLEM SOLVING AND DECISION MAKING

Problem solving and decision making are critically important skills for any leader. Problem solving as a leader involves identifying issues standing in the way of a productive team and taking corrective action to address those issues for the good of the group. The process by which actions are selected is decision making. The way you make decisions depends on the situation, your leadership style, your personality type, and your communication style.

Objectives

After completing this chapter, the student should be able to:

- Explain the importance of problem solving and decision making
- Define the terms *problem, problem solving,* and *decision making,* and differentiate between the latter two
- List mistakes in problem solving and decision making
- List skills needed in problem solving and decision making
- Describe three decision-making styles and identify your own style
- Identify two approaches to problem solving and decision making
- List and describe the seven steps of problem solving and decision making, and use them to solve problems
- Describe three types of problems and how to solve them
- List advantages and disadvantages of group problem solving and decision making
- List methods that groups can use to solve problems and make decisions
- Identify leadership styles used in group problem solving and decision making

Terms to Know

- problem
- decision making
- alternatives
- reflexive style
- reflective style
- consistent style
- minimizing approach
- optimizing approach
- exact-reasoning problems or decisions
- creative problems or decisions
- judgment problems or decisions
- conventional method
- brainstorming method
- devil's advocate method
- Delphi method
- consensus method
- nominal group method
- synectics
- left-brain people
- right-brain people
- holistic
- autocratic leadership style
- consultative leadership style
- participative leadership style
- laissez-faire leadership style

This chapter will help you identify the skills needed to be an effective problem solver and decision maker. It will show you how to develop these skills by making you aware of your own problem-solving and decision-making style, by teaching you approaches to solving problems and making decisions, and by describing and illustrating the types of problems and decisions you may encounter. This chapter also focuses on the group problem-solving and decision-making process (Figure 13–1) and the styles leaders use in directing groups.

IMPORTANCE OF PROBLEM SOLVING AND DECISION MAKING

Daily and Life Decisions

In your personal, family, and working life, you solve problems and make decisions every day. You are probably not even aware of the many decisions you make or the problems you solve. You decide when to get up, what to wear, where to go, how to get where you are going, and when to come home. These types of decisions are relatively easy to make. Others, however, are of such importance that they require a great deal of thought. Will you attend college? If so, where? Whom will you marry? What occupation will you choose? Will you buy a house, and how will you pay for it? The choices you make regarding these "big" decisions will affect you for the rest of your life. No matter how important a decision is, you have to live with the consequences of both the good and bad decisions you make.

Problem Solving and Decision Making on the Job

The ability to make good decisions and to solve problems is a characteristic of an effective person, and a requirement for anyone who wants to become an effective leader. Consider the following:

- Leaders are often hired mainly to solve problems and make decisions.
- The activity requiring the second most time for leaders is solving problems and making decisions. It requires almost 13 percent of a leader's time.

FIGURE 13–1 These students are solving problems and making decisions. These abilities are signs of effective leadership. (© iStock/South_agency)

- Leaders can make as many as 10 decisions per hour and hundreds per day.
- Problem solving/decision making is ranked among the top six critical skills for success at the supervisory level for leaders.[1]

Obviously, effective problem solving and decision making are important if we wish to hold leadership positions in the workplace; these skills are also important to our personal financial, educational, and emotional well-being. Fortunately, the ability to solve problems and make decisions is a skill that can be learned and improved through practice. Let us begin by examining what we mean when we use the terms *problem*, *problem solving*, and *decision making*.

DEFINING PROBLEM SOLVING AND DECISION MAKING

Problems and Problem Solving

Most of us are familiar with the word *problem*, but may have trouble putting an exact definition into words. A **problem** arises when there is a perceived gap between reality and the desired state of affairs for an individual, group, or organization.

For example, if a car dealership has the goal of selling 20 cars per week but is selling only 10 cars per week, a problem exists. The dealership must do something to solve the problem. Problem solving, then, is doing whatever it takes to reach the goals of an individual, group, or organization.

Decision Making

As part of the process of problem solving, you must take a new or different course of action to correct a problem. The process by which you select this course of action is called **decision making**. At times, after careful consideration, you may find that your best decision is to do nothing, because the problem is simply too difficult to be solved in the amount of time you have to solve it. In other instances, you may decide to change your goals to eliminate the problem.[2] In our example, the dealership could simply change its goal from selling 20 cars to selling 10 cars per week, and instantly the problem no longer exists. However, if the financial success of the dealership depends on selling 20 cars per week, reducing the goal is only going to cause more problems. Therefore, some other decision(s) must be made to resolve the situation (Figure 13–2). By developing

FIGURE 13–2 Penn State President Graham B. Spanier works with and listens to others to solve problems and make decisions. *(Courtesy of Annemarie Mountz, Department of Public Information Media Gallery, Pennsylvania State University.)*

our skills in problem solving and decision making, we can learn to avoid making the same type of mistake over and over again.

MISTAKES IN PROBLEM SOLVING AND DECISION MAKING

Both groups and individuals frequently make mistakes in confronting problems or decisions, and these mistakes tend to fall into recognizable categories or types. Phipps et al. identify the following as common mistakes in attempting to solve problems or make decisions:[3]

- Poorly defining the problem, or denying that a problem even exists
- Treating the symptoms rather than the cause; attempts are made to solve or change the situation resulting from the problem, not the problem itself
- Not specifying goals and objectives, or not clearly defining them
- Not gathering enough information about possible **alternatives** (different courses of action that one might take; solutions)
- Not carefully, correctly, or adequately considering and evaluating all possible alternatives
- Allowing opinions, emotions, feelings, and self-interest to interfere with—and often override—objective, rational thought
- Jumping to unwarranted conclusions
- Becoming paralyzed and taking no action because of fear of making mistakes or making a wrong decision

SKILLS NEEDED IN PROBLEM SOLVING AND DECISION MAKING

To avoid the mistakes listed in the previous section, Phipps et al. suggest that individuals and groups must develop certain problem-solving and decision-making skills.[4] Foremost among these are the ability to

- Recognize problem situations
- Clearly distinguish the problem from the situation caused by the problem

- Clearly define goals and objectives
- Develop creative, imaginative solutions to problems
- Gather information relating to possible solutions
- Be open-minded about possible solutions offered by others
- Carefully evaluate information as a basis for accepting or rejecting solutions
- Work with others to solve problems
- Avoid jumping to unwarranted conclusions (be flexible)
- Accept the fact that mistakes may occur
- Put aside opinions, emotions, and self-interest that may interfere with objective thinking
- Understand different types of problems and techniques for solving them
- Know and use a systematic approach to problem solving and decision making[5]

How can we develop these important skills? The following sections help us answer this question.

DECISION-MAKING STYLES

Determining Your Decision-Making Style

One step toward developing the skills necessary to be a good problem solver and decision maker is to understand your decision-making style. If your current style is not conducive to making good decisions, you may want to consider ways to change it. Each person tends to favor one of three problem-solving and decision-making styles: reflexive, reflective, or consistent.[6]

Reflexive Style If you have a **reflexive style** of problem solving and decision making, you probably tend to make quick and sometimes unthinking decisions or choices of solutions. As a result, you may not take the time to consider and evaluate all possible solutions to your situation before acting. You tend to be decisive, and you are not likely to put off taking action in a problem situation. However, your tendency toward speed can result in hastily made decisions that you may later regret.

To improve your problem solving and decision making, take more time to identify possible solutions. Try to gather information concerning each possible solution, and analyze and evaluate the alternatives thoughtfully. Follow the steps in the problem-solving and decision-making process described later in this chapter.

Reflective Style If you have a **reflective style**, you take the time you feel you need to identify, analyze, and evaluate as many alternatives as possible for solving a problem or making a decision. The advantage to this is that you carefully consider your decisions and do not make them haphazardly. However, you may also take so long to make a decision that you appear indecisive or fail to act in a timely manner.

To improve your problem solving and decision making, continue to be careful, but attempt to make your decisions more quickly, and learn to accept and work with less-than-complete information. Andrew Jackson said, "Take time to deliberate; but when the time for action arrives, stop thinking and go on."[7] For many decisions,

you must take action on the basis of incomplete or even insufficient information.

Consistent Style If you use a **consistent style** as a problem solver and decision maker, you know the appropriate amount of information to consider and evaluate before making a decision, and you act within a reasonable amount of time. You do not make decisions too quickly, as do reflexives, and you do not act too slowly, as do reflectives. Your decisions are timely, reliable, and consistently sound.

APPROACHES TO PROBLEM SOLVING AND DECISION MAKING

Another way to develop your skills is to understand the approaches you can take to problem solving and decision making and learn when to use them. There are two general approaches.[8]

VIGNETTE 13-1 **SO WHAT'S YOUR DECISION-MAKING STYLE?**

IT WAS THEIR SENIOR PROM AND THE LAST TIME MANY OF THEM WOULD BE TOGETHER IN ONE PLACE

The prom committee, Reflexive Rachel, Reflective Ronnie, and Consistent Connie, felt they had planned one of the best proms ever. So far, everything about the night was going great, but trouble was on its way. About 30 minutes before the king and queen of the prom were to be presented to the class, Reflective Ronnie noticed that the students who were to be "crowned" had not yet arrived. What should the committee do? All the committee members, being the leaders that they were, huddled up and tried to come to a decision.

Reflexive Rachel was the first to speak up, and without much forethought she exclaimed, "We should just call everything off. There's nothing we can do now."

Reflective Ronnie just stood there rubbing his chin. He did not operate well in such a pressured situation. His gut told him that such an important decision should not made in haste.

Consistent Connie took a few moments to ponder the situation and consider all of the available options, but she knew an appropriate decision had to be made soon. She realized that the first thing to be done was to get someone to call the "royal" couple and make sure they were all right. As it turned out, they were fine, but had been detained by a traffic accident of which they were not a part. Connie decided that the next steps should be to go ahead and announce who the king and queen were, and then explain the situation to everyone and let them know that the king and queen were safe. She suggested that everyone continue dancing and having a good time until the king and queen arrived and the rest of the ceremonies could be held. The other prom committee members were glad Consistent Connie was there to keep the balance.

Each style is appropriate in certain circumstances. If they had noticed the problem only two minutes before the announcement, Reflexive Rachel's idea might have been the best, as there would not have been time to think things through or gather any information. If they had noticed the problem one hour before the announcement, Reflective Ronnie might have come up with an ingenious solution. However, it is usually best to aim for the calm, but assertive, style of Consistent Connie.

Which decision-making style do you have? How has it helped you or hurt you in your leadership roles?

Minimizing Approach

In the **minimizing approach**, you simply opt for the first solution available, even though this solution may not necessarily be the best. If this first solution does not work (or work well enough), you then seek out a new one. This process of trial and error continues until you find an acceptable solution. The minimizing approach is useful if the situation is an emergency or if the delay incurred while making a more considered decision could be costly.

For example, you may choose the minimizing approach when it is raining hard and water is pouring in through a leak in your roof. Your objective is to stop the leak immediately so that your property is not ruined. There is no time to develop an elaborate plan for fixing the roof permanently (even though this will be desirable in the long run); you just do what you have to do immediately.

Optimizing Approach

In the **optimizing approach**, you take the time to review many different alternatives and solutions before making a decision, in order to choose the most effective, appropriate, or helpful solution. This approach requires more time, more thought, and more detailed planning and consideration than the minimizing approach, but it tends to be more reliable. The optimizing approach is usually best (most appropriate) when you have ample time to make a decision, when you cannot easily change your decision once it is made, or when an emergency situation that may have initially required the minimizing approach has passed and you have time to find a more permanent solution.

Using the previous example, once the rain has stopped and water from the leak no longer threatens to flood your house, you may wish to use the optimizing approach to decide whether the whole roof should be replaced (or just a portion repaired), what sort of roof you want to install, and which roofing contractor will be the most cost-effective, competent, and reliable.

THE SEVEN STEPS OF PROBLEM SOLVING AND DECISION MAKING

Another way to develop your skills as a problem solver and decision maker is to use a systematic approach to the situations you face. The steps commonly used in problem solving and decision making have their roots in the work of scientists such as Bacon, Newton, Galileo, and their successors, who sought to develop a systematic approach for acquiring knowledge. The work of these individuals and others led to formulation of the scientific method.

The scientific method consists of (1) identifying a problem situation or asking a question, (2) doing background research to define the problem, (3) creating hypotheses (solutions, answers) for the problem, (4) testing your hypothesis by doing an experiment and gathering data (information), (5) analyzing data, drawing conclusions, and/or making necessary revisions to the hypotheses or creating new hypotheses to test, and (6) communicating results.[9]

The process of problem solving and decision making is very much like the scientific method. If we use a logical, systematic approach to solving problems and making decisions, as scientists do, we are more likely to make better choices. Remember, problem solving and decision making are skills that can be learned, practiced, and improved.

The steps in problem solving and decision making are presented in Figure 13–3. We examine each of these steps and use the example of the car dealership to clarify the process.

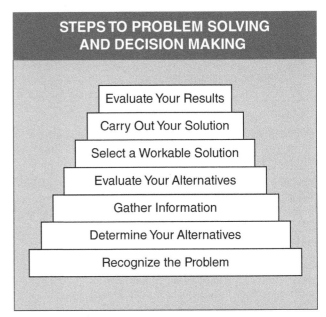

STEPS TO PROBLEM SOLVING AND DECISION MAKING

Evaluate Your Results

Carry Out Your Solution

Select a Workable Solution

Evaluate Your Alternatives

Gather Information

Determine Your Alternatives

Recognize the Problem

FIGURE 13–3 The steps in solving problems and making decisions are similar to those of the scientific method. By following these steps from the bottom up, you are likely to make good choices.

Mr. Brown owns and is the sole employee of a small car dealership. He has determined that he must sell 20 cars per week to make his dealership profitable. Currently, however, he is selling only 10 cars per week. Mr. Brown is determined to make his business successful and wants to develop a plan for doing so. He decides to use the steps for problem solving and decision making to help him.

Step 1 *Recognize and identify the problem.*

Problems are inevitable, in almost any area of life, and they must be solved, not ignored, if you are going to be succeed. The first step in solving a problem or making a decision is realizing that you *have* a problem to solve or a choice to make. You must then make sure you understand the exact nature of the problem. Ask yourself the following questions:

- What problem am I attempting to solve?
- What goal am I trying to reach?
- Do I fully understand the problem?
- Do I need to take action at all?

It may be helpful to write the problem down; you can make it more concrete and definite just by putting it into words. This allows you to focus clearly on a specific issue. Remember, not all decision making concerns negatives; you may have a choice between two or more equally good things! Regardless, understand that you must not ignore problems.

In our example, Mr. Brown realizes that if he needs to sell 20 cars to make a profit and grow a successful business, he cannot continue to sell only 10 cars per week. He must find a way to sell more cars.

Step 2 *Determine your alternatives.*

Once you have identified your specific problem, you need to determine what alternatives are available to you. *Alternatives* are the different courses of action you can take to solve your situation. There are likely to be many alternatives to any given problem, and it is usually best for you to consider each one carefully before making a decision. It may be helpful to consult with other people who are familiar with your situation, or have experienced similar situations, to determine what alternatives are available. Be aware that any alternative may directly affect a number of

people other than yourself. It may be helpful to list the alternatives on paper so that you can maintain a clear focus.

In our example, Mr. Brown lists several alternatives to his problem. We will consider three: (1) He may sell a different type (line) of car, (2) he may lower the price of his current inventory, or (3) he may hire a salesperson to help him.

Step 3 *Gather information.*

Once you have listed the alternatives, you need to gather information about each one. Look at the facts. Relying only on opinions, emotions, and intuition may lead to hasty, poorly thought-out decisions that you will later regret. While gathering information, ask yourself the following questions:

- What do I need to know about each alternative?
- What materials, information, assistance, or action will be needed to implement this alternative?
- What will this alternative cost?
- Is this alternative feasible? Is it likely to work (solve the problem)?
- What are the probable (or possible) consequences of this alternative?

Do not try to reinvent the wheel. There is a good chance that someone has already had the same problem as you and has identified at least some of the potential alternatives. The library should lead you to an abundance of data.

In our example, Mr. Brown gathers the following information. Some of it came from trade publications on car sales and dealerships that were on the Internet.

1. If he switches to selling a different type of car, he will need to secure a loan of several thousand dollars, find a manufacturer willing to sell him a new line of cars, and sell or somehow dispose of his current inventory.
2. If he lowers the price of his current inventory, he may sell more cars because he will be selling them for less than other dealerships in the area.
3. If he hires a salesperson, he will need to place a "help wanted" ad, conduct interviews, pay a salary, and provide a benefits package. He must also be able to work well with the person he chooses.

Step 4 *Evaluate your alternatives.*

Once you have gathered information regarding each alternative, you will need to evaluate (and perhaps make a written list of) the advantages and disadvantages of each, both in relation to solving the problem and in relation to one another. Consider also what new problems may be created by the adoption of each alternative.

Mr. Brown evaluates his alternatives as follows:

Sell a New Line of Cars Mr. Brown has used his full credit line at the bank, so borrowing more money is not a viable option at this time. Furthermore, establishing a relationship with a new car manufacturer is a lengthy and costly process that will only further reduce his ability to sell more cars at this time.

Lower the Price on Current Inventory Mr. Brown knows that if he reduces the price on his current inventory, he only makes his financial problems worse. To stay profitable, he would have to sell even more cars, which would be difficult because he has other responsibilities in addition to selling cars (e.g., record keeping, management, ordering, etc.). Although selling cars is of primary importance, he must attend to these other duties as well, despite the fact that they take up time and keep him away from the sales floor. Mr. Brown also sees that lowering his prices is probably not very feasible, because his current prices are already quite competitive with those of other area dealerships.

Hire a Salesperson Hiring a salesperson will cost Mr. Brown money, but he knows that most salespeople work for a small salary plus a commission (a percentage of the final sale price of each car). Thus, the salesperson's pay would depend on his or her ability to sell cars. Having a full-time salesperson would also mean that more time would be spent actually selling cars. This is important because Mr. Brown often has had to take care of managerial matters (bookkeeping, ordering, etc.) while potential customers were on the car lot. When he is unavailable and unable to help these people, he loses sales. Mr. Brown also realizes that having another salesperson will allow him time to better fulfill his administrative duties.

Step 5 *Select a workable solution.*

After you have evaluated each alternative and its possible results and consequences, choose the one that is the most practical, reasonable, and effective in solving your problem. After considering all his alternatives, Mr. Brown decides that hiring a salesperson is his best solution. He can sell more cars, attend to his administrative duties more easily, and afford to pay a commission-based salesperson.

Step 6 *Carry out your solution.*

Once you have determined your course of action, follow through. If you fail to implement your solution, the time and effort you have expended getting to this point are wasted. A solution must be put into action if it is to work.

Mr. Brown advertises for a salesperson. He interviews a number of people, explaining his goals and compensation plan to each one. He hires a salesperson who he feels will be both competent and compatible with his own personality and values.

Step 7 *Evaluate your results.*

The problem-solving and decision-making process does not end when the proposed solution is carried out. It ends when you determine that your problem has been solved, that the same problem persists, or that new problems have been created.[10] Evaluation may lead you to accept your solution as a good one, make further adjustments to improve your solution, or abandon your solution (discard your decision) and start the process over again (Figure 13–4).

In our example, several weeks after hiring the salesperson, Mr. Brown evaluates what his decision has done for him and his business. He finds that not only is he selling more cars than he needs to be profitable, but he also has fewer worries about his other duties. In addition, he gets along well with the new salesperson. Mr. Brown is satisfied with his choice of alternative solutions; that is, he is happy with the decision he made.

Overall, learning and practicing the steps of problem solving and decision making will help you become a better leader. Our example has a happy ending, but not all of your decisions will turn out so ideally. Nevertheless, as long as you carefully followed each step, you can feel confident that you made the best decision you could at the time. Do not be afraid to start the process over again if your first solution or decision does not work or does not have the desired effects. You can learn from any mistakes you make. Thomas Edison failed numerous times while trying to develop

FIGURE 13–4 This leader is consulting with all of the members of the organization to evaluate results. *(Courtesy of Gene Maylock, Department of Public Information Media Gallery, Pennsylvania State University.)*

the incandescent light. Rather than being discouraged, he focused on the wealth of knowledge he gained from his failures. He knew well what did not work, and he kept trying until he found what did work.

TYPES OF PROBLEMS AND DECISIONS

As you progress through the problem-solving and decision-making steps, you can increase your effectiveness by recognizing the types of problems you encounter and having some techniques in mind for solving them. Essentially, the types of problems or decisions you will normally encounter fall into three general categories listed here and explained below: exact-reasoning problems or decisions, creative problems or decisions, and judgment problems or decisions.[11]

Exact-Reasoning Problems or Decisions

With **exact-reasoning problems or decisions**, there is usually one definite (usually factual) answer. Mathematical concepts and calculations often resolve such problems.

For example, Tom and Kim, a newlywed couple, are considering the purchase of their first home. They must decide whether they will finance the house through a local bank or with a mortgage company. They find that the only difference in the terms and services offered by the two institutions is that the local bank charges 6 percent interest over the life of the loan, compared to the mortgage company, which charges 7 percent. Tom and Kim decide to finance their home through the local bank, because they save 1 percent on the interest charges by doing so.

The number of potential variables in these problems can make them a bit complex, but if you "do the math" properly, an exact answer is possible. For example, is it more advantageous financially to attend a four-year, out-of-state college with a higher annual tuition, or to take six years to complete the same degree program at a local college where the annual tuition is less? For this calculation, you would have to factor in estimated living expenses for each situation as well as the tuition costs. However, if finding the overall lowest cost is your objective, this remains an exact-reasoning problem for which there is a definite answer.

Creative Problems or Decisions

When you encounter **creative problems or decisions**, you usually need to draw up a plan or create some sort of design to help you come to a solution.

For example, Phil and Joe are college roommates. They each bring several things from home to put in their 15-foot by 20-foot dorm room. They must decide how to arrange the room so that all their belongings fit in a neat and efficient setup. They must arrange two beds, two desks and desk chairs, a computer, a television and stand, a small refrigerator, and a stereo. They may want to sketch a room design that is drawn to scale so they can see the possibilities and decide on optimal placement for each item. Such advance planning will save them from having to move heavy furniture more than once!

Judgment Problems or Decisions

Judgment problems or decisions present many factors and alternatives to be listed, compared, and evaluated before you can reach a decision. Judgment problems or decisions may fall into four categories: possibilities and factors, improving a situation, steps and key points, and advantages/disadvantages.

Possibilities and Factors In some cases, you will have to consider many different possibilities and factors to reach a good solution or decision. For example, John has just started a new job and wants to buy a new car. He knows his price range, he knows he wants a red car, and he knows he wants

POSSIBILITIES	FACTORS		
	IN PRICE RANGE	RED	AIR BAG
Car style 1	No	Yes	Yes
Car style 2	Yes	Yes	Yes
Car style 3	Yes	No	Yes

FIGURE 13–5 A chart of possibilities and factors can help you decide which of the three cars to buy.

a car equipped with an air bag. He goes to look at three different styles of cars and outlines the possibilities and factors for each (Figure 13–5). Based on the information in Figure 13–5, John decides to buy car style 2, because it satisfies all the criteria he considers important.

Improving a Situation This type of judgment problem involves taking a current situation and using the problem-solving techniques to amend or improve on it. For example, Mr. Gomez is not happy with the appearance of his house and yard. He is trying to find ways to make them look better. After identifying specific problems and collecting several alternatives and ideas, he develops and decides to carry out the plan shown in Figure 13–6.

Steps and Key Points This type of judgment problem requires you to progress through a series of steps while considering the key points associated with each. For example, Fred and Ginger want to purchase a new home. They have two children. Figure 13–7 shows some steps and key

CURRENT CONDITION	NEEDED IMPROVEMENTS
Exterior of house is rundown, looks outdated	• Replace aluminum siding with wooden exterior • Paint gutters and downspouts
Yard looks plain, has bare spots	• Plant shrubs along front of house • Plant grass seed on bare spots

FIGURE 13–6 Sometimes it takes only good planning to improve a situation.

STEPS	KEY POINTS
1. Location	Near good schools, stores, and work
2. Cost	Monthly payment must fit in family budget
3. Size	Need a bedroom for each child; two baths

FIGURE 13–7 Sometimes a decision, such as choosing a house to buy, requires you to follow a series of steps while considering the key points associated with each. Decisions are made only after as many criteria as possible have been met.

FACTORS TO CONSIDER	YES	NO
1. Close to home	X	
2. Good academic program	X	
3. Offers intended major		X
4. Reasonable cost	X	
5. Scholarships available	X	

FIGURE 13–8 Factors to consider can be presented as a list of yes-or-no answers or as advantages and disadvantages. One negative can override all the positives, as shown in this chart presenting factors in deciding whether to attend a certain university.

points that will guide their decision. Armed with this information, Fred and Ginger can consider only houses that meet their criteria.

Advantages/Disadvantages (Yes/No) In this type of judgment problem, you identify factors or considerations relating to the problem or decision and weigh them as positive or negative in terms of their value for solving the problem or making the decision. For example, Mary is considering whether to attend State University. She should consider the factors shown in Figure 13–8.

In some cases, no single factor is more important than any others, so you can simply pick the option with the most advantages. In other instances, one positive factor can override many negative ones. The reverse is also true: one negative factor can outweigh many positive ones in the decision-making balance. In this example, there are more positive factors than negative for Mary's attending State University; but Mary realizes that because the university does not offer her intended major, she will not benefit from attending school there.

GROUP PROBLEM SOLVING AND DECISION MAKING

Up to this point, we have discussed problem solving and decision making types and processes that apply generally to both individuals and groups. In the following section, we focus on groups and the various ways they can use the problem-solving and decision-making process. First, we consider some of the advantages and disadvantages to involving more than one person in this process.

Here are just some of the advantages of using groups in the problem-solving or decision-making process:

- More information and knowledge are available to make decisions.
- More options are likely to be produced from which to choose.
- More acceptance of a final decision by members of a group.
- More effective communication of the decision is likely.
- Better decisions are ultimately made when more than one person is involved in the decision.

Some of the disadvantages of group problem solving or decision making include the following:

- Involving groups in the process requires more time.
- It is costlier to involve a group in the decision-making process.
- Conflicts may arise among group members.
- Keeping people working cooperatively on a given task requires great skill.
- People may feel less responsibility for the solution or decision when many people are involved in the process.
- One person or a small portion of the group may dominate the process, thus causing the solution and decision to be less democratic.
- Compromise, rather than the best decision, may occur because of group member indecisiveness.
- Groupthink may occur. The desire to be accepted by the group may cause individual members to conform to group decisions rather than thinking critically or questioning objectionable and dubious solutions or

VIGNETTE 13-2 CRITICAL THINKING IN DECISION MAKING

CRITICAL THINKING IS A REASONED, PURPOSIVE, AND INTROSPECTIVE approach to solving problems or addressing questions with incomplete evidence and information and for which an incontrovertible solution is unlikely.

Three agricultural education professors constructed this definition of critical thinking.[12] Critical thinking is a skill that all of us should strive for as leaders, especially because we frequently make decisions that affect others. Let's analyze this definition and see what it really says about using critical thinking in leadership development.

Critical thinking is

a reasoned,—Before we make decisions that affect our school, clubs, communities, or families, we should reason things out. This means that we should not take action or make decisions before we have looked at all sides of an issue, listened to everyone's stories, evaluated the facts and assumptions, and examined our own biases. Rational, logical thought processes should predominate here, without undue influence from emotions, guesses, or unsupported opinions.

purposive,—When we are leaders, we have to keep our purpose at the forefront at all times. It may be easy to preach our goals or "sell" our visions to others, but we also have to keep those visions and goals uppermost in our own minds when we are solving problems and making decisions.

and introspective approach to solving problems—The answer is not always easy or obvious. Sometimes, no matter how well we reason it out, while keeping the purpose firmly in mind, the answer is still not black or white. When this is the case, we have to be **introspective**, which simply means we have to step back and think about the situation and our place in the situation. We have to reflect on how the decision may affect us, how similar decisions have affected us in the past, and how we really feel about each alternative.

or addressing questions with incomplete evidence and information—The purpose of reasoning and introspection lies in the fact that many decisions and problems simply come down to us. You may not be able to find documentation of how someone else handled a certain problem, or how he or she made a decision about the problem, or even whether anyone has faced a similar problem. Many times, all the responsibility falls on your shoulders, and you as a leader must make a sound, well-thought-out decision without much guidance or factual support.

and for which an incontrovertible solution is unlikely.—*Incontrovertible* is a big word that means "undisputable." Simply put, this means that there is not always a clear, black-or-white, "right" solution or answer. As a leader, you have to use critical thinking to come up with the best possible alternative or decision.

What kinds of problems have you faced this week that did not really have an **incontrovertible solution**? What kinds of decisions have you made lately that you had to reason through because you wanted to ensure that you made the right choice? As leaders, we have the opportunity to think critically every day, but do we?[13]

decisions. When conformity to the group negatively affects the problem-solving or decision-making process, the result is groupthink[14] (Figure 13–9).

Methods

As a group attempts to solve problems and make decisions, it may follow the seven steps for problem solving and decision making outlined earlier in this chapter. As the group progresses through the steps, it can choose from several methods to foster varying levels of group participation or to generate alternatives for consideration. Key to the effectiveness of nearly all these methods is the ideal that all group members participate in the problem-solving or decision-making process and that the group approach be open and nonthreatening to individual members.

FIGURE 13–9 A disadvantage of group decision making is groupthink. The student in the front of this picture may be more interested in conforming to the group than in thinking critically or questioning decisions. *(Courtesy of Georgia Agricultural Education.)*

FIGURE 13–10 Voting is the conventional method of making a group decision. (© iStock/AndreyPopov)

Conventional Method In the **conventional method**, there is group discussion, but the discussion is typically dominated by one or a very few individuals. After discussion of possible solutions, a vote is taken on a single solution (Figure 13–10). If a majority of the group votes for the proposed solution, it is put into place. A potential threat with this method is that if the vote is very close (51 to 49 percent, for example), many of the group members do not favor or are actively opposed to that particular solution. This may affect the cohesiveness of the group or cast doubt on the validity of the solution.[15]

Brainstorming Method Brainstorming is a group activity where everyone has free rein to suggest as many possible solutions to a problem as first comes to mind. The **brainstorming method** is most effective for discovering numerous possible solutions to a problem. All members of the group have an equal voice and may offer any alternative, regardless of how unrealistic or unreasonable, to solve a given problem. The group members are encouraged to build on one another's ideas. The alternatives are not judged until all suggestions have been made.

Devil's Advocate Method In the **devil's advocate method**, a person may propose a solution he or she does not really support, just to make others think and react. This method requires the individual to explain and defend his or her position before the group, which in turn raises all the reasons the individual's idea will not work. The group must be careful to judge the ideas presented rather than the person presenting them. Similarly, the person playing devil's advocate must be mature enough to withstand criticism of his or her ideas, and remember that the ideas are what is being judged. The purpose of this method is to refine possible solutions to problems so that they will be feasible and effective when put into action.[16]

Delphi Method The **Delphi method** polls a group through a series of anonymous questionnaires. After a first round of opinion questionnaires is completed, the opinions are analyzed and the best ideas are resubmitted to the group for a second round. Several rounds may be necessary before the group reaches a position that is acceptable to all or nearly all members.[17]

Consensus Method In the **consensus method** (not to be confused with general consent), the group comes to substantial agreement on a solution. The consensus method is especially effective when time is not a pressing consideration, when the decision is very important, and/or when there is an overriding concern for the unity of the group.

In the consensus method, members freely submit their own ideas to the group. The group reviews all ideas and then focuses discussion on the ones it feels are the most important or viable. The discussion itself, rather than a formal vote or ranking, determines how important each idea is. Based on the discussion and the reasons proffered in support of each idea, a decision is made that is acceptable to a substantial portion of the group. It is important to realize that reaching consensus does not mean reaching a unanimous (everyone in favor) decision. Rather, a consensus decision is accepted by a large portion of the group, even if there are still a few members who would prefer a different solution.[18]

Nominal Group Method The **nominal group method** is a process of generating and evaluating alternatives through a structured voting method. Initially, this is a group interaction in name only. The group does meet, but individuals formulate their own ideas (solutions) in writing without discussing them with other members. All solutions are presented to each group member for review. The proposed solutions are discussed only if there is a need for clarification. The group then chooses, by vote, the top five ideas. After the top five ideas are selected, they are discussed, and the reasoning behind each is presented. Another vote is held, and the solution that receives the most votes in this round is the one accepted for implementation (Figure 13–11).[19]

Synectics **Synectics** is a group problem-solving process of generating creative alternatives through role-playing and fantasizing. Synectics uses analogies to stimulate mental images. A common use of synectics is to project oneself into the essence of the problem. For example, if an agricultural engineer is trying to develop a more efficient combine, she might imagine herself as wheat going through the internal components of the combine. By actually seeing yourself as wheat, you can visualize what it would take to strip the stem from the chaff and other parts, leaving only the grain.

FIGURE 13–11 These students are using the nominal group method. They are discussing and clarifying the top five ideas before the final vote. *(Courtesy of Georgia Agricultural Education.)*

Prefabricated potato chips were developed by a synectic group. The company wanted to package and compress potato chips without breaking them. The group eventually drew an analogy to leaves, which can be compressed without damage as long as they are wet. They tried it, and prefabricated potato chips became a successful commercial product.

"When Nolan Bushnell wanted to develop a new concept in family dining, he began by discussing general leisure activities. Bushnell then moved toward leisure activities having to do with eating out." The idea of a restaurant-electronic game complex where families could play games and purchase pizza and hamburgers evolved and then became a viable business.[20] Creativity of this kind has become so important to organizations that many of them are beginning to look into left-brain and right-brain thinking.

Left-Brain, Right-Brain Thinking

In recent years, much attention has been focused on creativity and brain function. Most people are either left-brain dominant or right-brain dominant. This dominance also dictates the way people do things.

Left-brain people tend to be very logical, rational, detailed, active, and objectives oriented. They tend to prefer routine tasks or jobs that require precision, detail, or repetition. They like to solve problems piece by piece, using a sequential, logical approach. People with left-brain dominance tend to be more analytical but less creative and innovative than right-brain-dominant people. They are comparable to the introverted melancholy personality type discussed in Chapter 2.

Right-brain people are more spontaneous, emotional, **holistic** (emphasizing the importance of the whole and the interdependence of its parts), physical (nonverbal), and visual in their approach. They like jobs without repetition or routines and enjoy work that requires them to generate ideas. They like to solve problems by looking at the entire matter and approaching the solution through intuition and insight.[21] Right-brain people are comparable to the extroverted, choleric, or sanguine personality type discussed in Chapter 2.

Figure 13–12 compares right-brain and left-brain leadership styles, personality types, communication styles, and communication forcefulness.

COMPARISON OF RIGHT-BRAIN AND LEFT-BRAIN DOMINANT CHARACTERISTICS

PROBLEM-SOLVING/ DECISION-MAKING CATEGORY	LEADERSHIP STYLE		PERSONALITY TYPE	DISC™ BEHAVIORAL STYLE	COMMUNICATION STYLES		COMMUNICATION FORCEFULNESS
Right-brain dominant characteristics	Authoritarian	Situational Leadership	Choleric	Dominant	Directors	Situational Communication	Aggressive
	Democratic		Sanguine	Influencing	Socializers		Assertive
Left-brain dominant characteristics			Melancholy	Cautious	Thinkers		
	Laissez-faire		Phlegmatic	Steady	Relaters		Passive

FIGURE 13–12 Just as no one is 100 percent of any personality type, there is not a 100 percent correlation between any combination of leadership, personality, and communication style. Sometimes, as in situational leadership, we must vary our problem-solving and decision-making styles as well as our leadership and communication styles to meet the needs and objectives of the moment.

With this information, you can almost predict the decision-making style that correlates with each leadership style. As we discuss each of these styles, we focus on how people with that style make decisions and situations in which each style may be used appropriately. We also take one example and show how each style would arrive at a solution or decision.

LEADERSHIP STYLES AND GROUP DECISION MAKING

Group leaders, whether consciously or unconsciously, have leadership styles. The leader of a group may be someone who has been appointed as a chairperson of a committee; someone who has been selected by the group to serve as leader; someone who simply takes charge of the group; or a supervisor, foreman, manager, teacher, or administrator. How a leader involves the group in the problem-solving or decision-making process reveals his or her leadership style. (The various leadership styles are discussed at length in Chapter 1.) Good leaders are aware of their style and can vary it according to the situation. You too should be aware of your leadership style as you are put into situations in which you may become the leader of a group or the supervisor of employees. The four leadership styles in group problem solving and decision making are autocratic, consultative, participative, and laissez-faire.

Consider this workplace scenario.

Mark, a supervisor, has been given the opportunity to promote one employee from his group to a foreman's position. Let us see how Mark could use different leadership styles in deciding who should get the promotion.

A leader using the **autocratic leadership style** makes the decision independent of the group. After the decision is made, the leader informs the group of the choice and may offer an explanation of why the decision was made the way it was. The autocratic style may be effective if (1) there is no time to consult the group, (2) the leader has enough information available to make the decision alone, (3) the group is willing to accept the leader's independent decision and put it into action, or (4) the group is unwilling to make (or is incapable of making) the decision.[22]

In our example, using the autocratic style, Mark would not consult anyone; instead, he would decide who gets the promotion and inform his superiors of his choice. He would then tell his employees who is getting the promotion and why. This type of leader probably has a choleric personality type.

In the **consultative leadership style**, the leader goes to individual group members seeking additional information that will help him or her solve the problem or make the decision. The leader then solves the problem or makes the decision. Before putting the plan into action, the leader explains the solution or decision and its supporting rationale to the group. Questions and discussion may be allowed, but the decision stands. The consultative style is effective if (1) a leader has to make a decision quickly, (2) a leader needs more information to make a decision, (3) a group is not sure if they should go along with a leader's decision, or (4) a group is willing and somewhat capable of giving a leader useful information.[23]

In our example, Mark has only some of the information that will help him decide whom to promote. Using the consultative approach, he would go to various employees and sources seeking more information. When he felt he had sufficient information, Mark would choose someone and inform the rest of the employees of his decision and his reasons for it. He might allow questions and discussion from the other employees even though his decision was final. This leader more than likely has a sanguine personality type.

In the **participative leadership style**, the leader has a tentative solution or decision in mind but goes to the group seeking its ideas and opinions before making a final choice. The leader is open to change based on the group input. Another way to use this style is for the leader to present the problem to be solved or decision to be made to the group and ask for suggestions. Their suggestions become the basis for the leader's ultimate decision. Again, after announcing the decision, the leader explains the reasoning behind the decision to the group. The participative style is effective when (1) time to make a decision is abundant, (2) a leader has limited information for making a decision, (3) a group seems unaccepting of a decision because they had limited input, or (4) a group

is highly competent and willing to participate or is directly affected by the decision.[24] This leader probably has a melancholy personality type.

In our example, Mark would go to the employees and explain that he can promote only one person. He would ask the group to provide him with suggestions and rationales for who should be promoted. When his information is complete, Mark would name the employee to be promoted and explain the reasons for his choice.

In the **laissez-faire leadership style**, the leader presents the problem or decision to be made to the group; the group, not the leader, solves the problem or makes the decision. The leader may even act as a group member. The laissez-faire style is effective when (1) there is ample time for a decision to be made, (2) the leader has limited information to make a decision, (3) the group has to have a voice in the decision-making process, or (4) the competency level of the group is outstanding. Note that of the four styles, laissez-faire is the only one that allows the group to make the final decision.[25]

In our example, using the laissez-faire style, Mark would tell the group members that they need to decide who is to be promoted. Mark would then become a group member. The group would hold a discussion and decide who gets the promotion (Figure 13–13).

FIGURE 13–13 This group is using the laissez-faire decision-making method, in which the group decides who will lead the group activity. (© iStock/boggy22)

CONCLUSION

Problem solving and decision making are some of a leader's most important skills. The way we make decisions depends on the situation, our leadership style, our personality type, and our communication style. However, the important attributes in this area are the abilities to make decisions and solve problems.

SUMMARY

A problem arises when there is a difference between what is actually happening and what the individual or group wants to happen. Problem solving is the process of taking corrective action to bring about the conditions that the individual or group desires. The process by which a new course of action is selected is called decision making.

Problem solving and decision making are skills that can be learned, practiced, and improved. People and groups tend to have certain problem-solving and decision- making styles. Reflexive-style problem solvers and decision makers make decisions quickly, without a great deal of thought. Those with a reflective style take a great deal of time to formulate alternatives, evaluate those alternatives, and choose among them. People with a consistent style make careful, reliable, well-founded decisions in a reasonable amount of time.

You can approach problem solving and decision making in two ways. The minimizing approach adopts the first, but not necessarily the best, solution to resolve the situation. The optimizing approach reviews many different solutions to a situation before choosing the best one available.

When you choose the optimizing approach, you may wish to use the seven-step process for problem solving and decision making: (1) recognize the problem, (2) determine alternatives, (3) gather information, (4) evaluate alternatives, (5) select a workable solution, (6) carry out the solution, and (7) evaluate the results. You may use these steps to find solutions to problems that have one exact answer, problems that usually involve creating a plan or design, and judgment problems that call for techniques such as weighing possibilities and factors, trying to improve the current situation, determining sequences of steps and key points, and determining advantages/disadvantages.

There are both advantages and disadvantages to group problem solving and decision making. A group may choose to solve a problem or make a decision through a number of different methods that use varying degrees of group participation. Some of these are the conventional method, brainstorming method, devil's advocate method, Delphi method, consensus method, nominal group method, and synectics method. Leaders of groups may have distinctive styles of solving problems and making decisions. The style they use depends in part on how involved and capable the group members are, and it may vary depending on the situation. These styles, beginning with the least amount of group participation and progressing to the most, are autocratic, consultative, participative, and laissez-faire.

 Take It to the Net

Explore problem solving and decision making on the Internet. Using the following search terms, try to find information you feel would be worthwhile sharing with your teacher and classmates.

Search Terms

problem solving

problem-solving techniques

creative problem solving

decision making

decision-making process

Chapter Exercises

REVIEW QUESTIONS

1. Define the Terms to Know.
2. List four reasons why being an effective problem solver and decision maker is important.
3. List eight mistakes individuals or groups make when attempting to solve problems or make decisions.
4. List 13 skills individuals and groups need to develop to solve problems and make decisions effectively.
5. List and briefly describe three problem-solving and decision-making styles.
6. Distinguish between the minimizing and optimizing approaches to problem solving and decision making. When might you use each?
7. List and briefly describe the seven steps for problem solving and decision making.
8. List and briefly describe three types of problems or decisions.
9. List five advantages and eight disadvantages of group problem solving and decision making.
10. List and briefly describe seven methods a group can use in solving problems and making decisions.
11. List and briefly describe four leadership styles used in group problem solving and decision making.

COMPLETION

1. The _____ decision maker makes quick decisions without taking time to consider and evaluate all possible alternatives.
2. The _____ decision maker evaluates all possible alternatives in a reasonable amount of time.
3. The _____ decision maker considers all possible alternatives but often takes an unreasonable amount of time to make a decision.
4. In the _____ style of group leadership, the leader may have a solution in mind but can be swayed by group input.
5. The _____ style of group leadership allows the group to make decisions.
6. In the _____ style of group leadership, the leader independently makes a decision and then informs the group of the choice.
7. In the _____ style of group leadership, the leader seeks only additional information from group members without encouraging group discussion.
8. _____ types of decisions involve listing possibilities and factors, improving a situation, listing steps and key points, or listing advantages and disadvantages.
9. _____ types of problems have an exact mathematical solution.
10. _____ types of problems require an artistic design or a plan to solve.
11. _____-brain people tend to be very logical, rational, detailed, active, and objectives oriented.
12. _____-brain people are spontaneous, emotional, holistic, nonverbal, and visual in their approach to decision making.

MATCHING

_____ 1. Arises when there is a difference between what is actually happening and what a person or group wants to happen.

_____ 2. The different courses of action you can take to solve a problem.

_____ 3. Doing whatever it takes to reach the goals of an individual, group, or organization.

_____ 4. The process by which a new or different course of action is selected to correct a problem.

_____ 5. Choosing the first available, but not necessarily the best, solution to solve a problem or make a decision.

_____ 6. Method of suggesting as many alternative solutions as possible to a problem without judging the value of the alternatives.

_____ 7. Process of discussing possible solutions to a problem and then taking a majority vote to decide the course of action.

_____ 8. Reviewing many solutions to a problem before choosing the best one to implement.

_____ 9. Results in a substantial portion of a group agreeing on a solution.

_____ 10. Uses a series of anonymous polls to solve problems or make decisions.

A. alternatives

B. brainstorming method

C. optimizing approach

D. Delphi method

E. problem solving

F. consensus method

G. problem

H. minimizing approach

I. conventional method

J. decision making

ACTIVITIES

1. Give an example of a problem you face now. Alternatively, you may choose to create a hypothetical problem situation like the example used in this chapter.

2. Follow the seven steps of problem solving and decision making to solve the problem you listed in activity 1. Be sure to label each step. If you wish, share your thought process with a classmate.

3. Write an example of each of the following problem or decision types. Exchange your examples with a classmate and have him or her design solutions to each.

   Each person should be able to show or explain how the solution was reached.

   a. Exact reasoning

   b. Creative

   c. Judgment

4. Write an example of a situation in which you would use the minimizing approach to problem solving or decision making, and one in which you would use the optimizing approach. Read one of your examples to the class. Have the class decide which approach you are demonstrating.

5. Your class needs to raise $500 more than it already has to purchase a computer. Have the class divide into three groups. The first group should attempt to come to a solution on how to raise the money by using the conventional method. The second group should use the consensus method. The third group should use the nominal group method. Report the findings of each group to the whole class.

6. Consider the following situation: a teacher is determining the rules for his or her class. Explain how the teacher would do this using an autocratic leadership style, a consultative leadership style, a participative leadership style, and a laissez-faire leadership style.

7. Have the class divide into four groups. Take the situation given in activity 6 and have each group role-play one of the group leadership styles.

NOTES

1. R. N. Lussier, *Human Relations in Organizations: Applications and Skill Building* (Boston: McGraw-Hill Higher Education, 2012).
2. Ibid.
3. L. J. Phipps, E. W. Osborne, J. E. Dyer, and A. L. Ball, *Handbook on Agricultural Education in Public Schools,* 6th ed. (Clifton Park, NY: Delmar Cengage Learning, 2008), Ch. 23.
4. Ibid., pp. 118–119.
5. Ibid.
6. R. Griffin, *Fundamentals of Management,* 8th ed. (Boston, MA: Cengage Learning, 2016).
7. Quoted in BrainyQuote. Retrieved August 3, 2016 from http://www.brainyquote.com/quotes/quotes/a/andrewjack163004.html.
8. "Decision Making," *Efficiency* series, No. 8742-C (College Station, TX: Instructional Materials Service, Texas A&M University, 1988), pp. 1–2.
9. Science Buddies, *Steps of the Scientific Method,* retrieved August 2, 2016 from http://www.sciencebuddies.org/science-fair-projects/project_scientific_method.shtml.
10. "Problem Solving," *Efficiency* series, No. 8742-D (College Station, TX: Instructional Materials Service, Texas A&M University, 1988), p. 2.
11. Ibid., p. 2.
12. R. Rudd, M. Baker, and T. Hoover, "Undergraduate Agriculture Student Learning Styles and Critical Thinking Abilities: Is There a Relationship?" *Journal of Agricultural Education* 41(3):2–12 (2000).
13. Ibid.
14. R. Griffin, *Fundamentals of Management,* 8th ed., p. 117.
15. "Decision Making," p. 4.
16. J. Lombardo, *The Devil's Advocate: Impacts on Group Decision Making,* Video course retrieved August 3, 2016 from http://study.com/academy/lesson/the-devils-advocate-impacts-on-groups-decision-making.html#lesson.
17. R. Griffin, *Fundamentals of Management,* 8th ed., p. 116.
18. "Decision Making," pp. 4–5.
19. Ibid., p. 5.
20. R. N. Lussier, *Human Relations in Organizations: Applications and Skill Building* (Boston: McGraw-Hill Higher Education, 1999, pp. 332–333).
21. R. M. Hodgetts and K. W. Hegar, *Modern Human Relations at Work,* 11th ed. (Fort Worth, TX: South-Western Cengage Learning).
22. K. Cherry, *What is Autocratic Leadership?* Verywell: Psychology, June 22, 2016, retrieved August 3, 2016 from https://www.verywell.com/what-is-autocratic-leadership-2795314.
23. Lussier, *Human Relations in Organizations.*
24. Ibid.
25. Ibid.

14 GOAL SETTING

Goals are difference makers. You must have goals: without them, it is like trying to reach an unknown, unspecified destination or to come back from a place you have never been. You must have definite, precise, clear, written goals if you are to realize your full potential in life, both personally and as a leader. If you believe it, you can achieve it.

Objectives

After completing this chapter, the student should be able to:

- Discuss the reasons for and importance of having goals
- Describe the benefits of having goals
- Explain why people do not set goals
- Explain how to set goals
- Discuss the principles of setting goals
- Describe the steps in goal setting
- Discuss types and kinds of goals
- Set immediate, medium-range, and long-range goals
- Set goals the FFA way

Terms to Know

- resources
- values
- short-term goals
- long-term goals
- criteria
- SMART
- ways and means
- talent area
- tangible
- momentum
- inertia
- immediate goal
- worst-case scenario
- psychic income

People do not plan to fail; they simply fail to plan. Prior planning prevents poor performance. The ability to set goals (the ends toward which effort is directed) and make plans is a skill we all need to develop. If we are to be successful and achieve what we want in life, setting goals and making plans are a necessity. To realize even a fraction of our potential, we need to learn how to set and achieve goals that will move us beyond where we are now. "No goals, no glory."

REASONS FOR HAVING GOALS

Provide a Target

Only 3 percent of Americans today have defined goals, and that 3 percent outperforms the other 97 percent. The 3 percent who have defined goals can direct their energies to their specific ends, whereas 97 percent waver back and forth, undirected and unfocused and therefore ineffective. Those in that small goal-oriented population are more likely to reach their goals because they have something to strive for; in contrast, those in the 97 percent of the U.S. population that does not have defined goals cannot possibly reach any target, because they have nothing at which to aim.

Provide a Destination

Having goals provides us with a destination to move toward. A person without a goal is like a raft without a rudder or a sail, drifting aimlessly. A person with goals is like a ship with a rudder, sails, a captain with a map, a compass, and a port of destination.

Provide Purpose and Meaning

We can achieve greater satisfaction if we strive to accomplish something that is important to us. Our minds tend to work like radar. A built-in "guidance system" keeps us striving for and connecting our aims to our goals. Because of this guidance system, we can accomplish almost any goal we set for ourselves, as long as the goal is clear and we are persistent. In the rest of this chapter, you will discover ways to reach your goals once you have set them.

Help Focus Our Energies on Productivity

Defined goals can help us plan our time and focus our energies, thoughts, actions, and behaviors purposefully and productively. They allow us to make responsible decisions on a daily basis. If your goal is to come home from school and play on your phone until bedtime, you will achieve it. If your goal is to be a great leader as a father or mother some day, you will probably achieve that too. The same can be said of goals you may have related to your money, physical health, or relationships. The only thing stopping you is your drive to achieve the goals you set (Figure 14–1).

Turn Activities into Accomplishments

Activity should not be confused with accomplishment. A wheel that is stuck in the mud may spin just as fast as the one on a clean, well-paved highway, but it does not get anywhere. People who lack direction or goals may work very hard and expend a lot of energy, but still accomplish very little. When we have goals, we can accomplish much more, because goals give us the ability to clarify, plan, focus, motivate, organize, set levels of achievement, formalize our intentions, and evaluate our progress.

Help Individual Success, Self-Concept, and Evaluation

Other reasons for setting goals vary for each individual, but may include these:

Direction All people need direction in their lives. Without direction, we are traveling through life without a road map or highway signs. It is easy to get lost or go down the wrong road. With goals, it is easier to reach our desired destination because we have directions.

Success What does it mean to be successful in life? It means attaining *our* goals—not others' goals that are imposed on us, but our own goals. Success is a key element in setting goals. In reality, it is the basis for goals. When you finally attain one of your goals, relax and enjoy it. However, do not stop working. You are merely at a higher plateau than you were before. There are other heights to be attained. If you stop, you may become bored and unhappy.

FIGURE 14–1 This student and his adviser have set yield and financial goals for the student's Supervised Agricultural Experience (SAE).

Self-Concept When we are headed in the right direction and have become successful, we will develop a feeling of self-worth. This obviously makes us feel good about ourselves; that is, it enhances our self- concept. Studies have shown that people who are happy about living and feel good about themselves live longer than people who are unhappy and think poorly of themselves. The more successes we have, the more positive our self-concept. Therefore, focus your energy on a positive ending.

Evaluation Setting goals gives us a standard against which to measure our progress in life. When we set goals, we can more easily determine and evaluate our level of accomplishment. Research has shown that the harder or more challenging the goal, the better the resulting performance. However, there are dangers in setting goals too high or too low. If a goal is set too low, you may mistakenly think that you have reached your full potential when you attain it. A goal that is set unrealistically high may simply intimidate and discourage you—sometimes so badly that you stop working toward it. The best goal should be high enough to present a challenge but not so high that it is totally out of reach.[1]

Did you ever wonder why the uprights on a football field are called *goal* posts or why the "0" yard line is called the *goal* line? Did you ever wonder why the metal hoop shown in Figure 14–2 is called the basketball *goal*? Just as in sports, you have to have goals in the game of life. Without goals, we never know the score. We never know whether we are on target or successful. In sports, we would not even attempt to play the game without the goals, yet most people attempt to play the game of life without goals, never knowing what the score is or whether they are winning. Goals give life a reason; as long as life has something valuable as its objective, life itself will be valuable and cherished.

BENEFITS OF HAVING GOALS

Responsible, responsive people have a way of getting their lives sorted out, deciding what really matters, and directing all their creative energies toward the goals they set for themselves. Setting

FIGURE 14–2 You could not play basketball without the goals, but most people play the game of life without goals. (© iStock/PhilAugustavo)

goals is a proven method for stimulating high achievement and produces several definite benefits. John R. Noe, in *People Power*, discusses the following five benefits:

Concentration of Effort Goals enable us to focus all our energy and work in a specific direction. If you can keep from getting distracted, you'll be amazed at how much you can achieve. You possess incredible intelligence and energy, and all you need to do is focus those superpowers with goals. Concentrated power enables us to do much more with our time and resources, because we do not dissipate our energy with uncoordinated, unfocused efforts.

Making the Most of Time Setting goals enables us to use our time more efficiently. Each of us has only 1,440 minutes each day, to waste, spend, or invest. Goals give you the option of spending your time on matters about which you care the most

and which have the highest priorities. Remember, you can't do everything in the time you have, so choose your pursuits wisely.

Letting Others Know How to Help "High achievers soon learn that they can reach their full potential only with the help of" many others. Very often, people would be willing to help us if they only knew what we needed and wanted to accomplish. When you determine and publicize your goals, other people have a reference point by which to focus their assistance efforts (Figure 14–3).[2]

Maintaining Enthusiasm Goals can keep your energy level and enthusiasm high. Those around you will also be motivated. If you are working on goals that you truly care about and believe are important, then maintaining enthusiasm will come naturally. Enthusiasm gives you the energy you need to achieve your goals, but more importantly, your followers will feed off it. If you can't lead with enthusiasm, you may need to re-evaluate the goals you have set and make sure they are goals about which you can get excited.

Monitoring Progress Monitor your progress through goal setting. A target has a focal point. If you shoot and miss, you know how to monitor and adjust. If you hit the target in the middle, you can make it more challenging by lengthening the target or by making it smaller. The same is true with goal setting.[3]

FIGURE 14–3 Former President Clinton would not have reached his goal of being U.S. president if he had not had people helping him with everything from campaign strategy to the details of tour schedules. *(Courtesy of Andy Colwell, Department of Public Information Media Gallery, Pennsylvania State University.)*

Goals Make Mental Laws Work for You

Tracy's *Maximum Achievement* lists the following nine mental laws:

Law of Control You feel positive about yourself in direct relation to "how much you feel in control of your life." Goal setting puts you in control of the direction of your life, your activities, your relationships, and the organizations in which you lead. When goals are clear and well planned, you eliminate the lack of control that causes fear and insecurity.

Law of Cause and Effect "For every effect in your life there is a specific cause." Goals are causes: Health, happiness, freedom and prosperity are effects. You sow goals and you reap results. Goals begin as thoughts, or causes, and manifest themselves as conditions or effects. The primary cause of success in life is the ability to set and achieve goals."

In the workplace, we work to achieve either our own goals or another person's goals. The best work of all is when you are achieving your own goals by helping others to achieve theirs.

Law of Belief You achieve your goals by taking actions that are consistent with your beliefs. If you really believe something will happen, then it is more likely to happen. "This is the foundation of faith and self-confidence."

Law of Expectations This mental law calls on us to confidently expect that our goals will be achieved. Leaders who do not realize their goals have probably had negative expectations, and the opposite is true for leaders who consistently reach their goals; they have positive expectations.

Law of Attraction When you put a primary focus on your goals, you begin to attract people to your life with similar goals. When you apply principles such as hard work, consistency, truth, and appreciation for diversity to your goals and activities, you attract followers and team members who value and apply the same principles—making accomplishment of your goals more feasible.

Law of Correspondence Your environment is a reflection of your thought life. If you are focused on thoughts, goals and plans to achieve the things that are important to you, your environment will start to reveal situations and opportunities for your thoughts, goals, and plans to be realized. In short, you become what you think.

Law of Subconscious Activity "Whatever thoughts you hold in your conscious mind, your subconscious mind works to bring into your reality.... [Y]our subconscious mind is dedicated to making your words and actions fit a pattern consistent with what you really want to achieve."

Law of Concentration Whatever we focus on grows. As people think, so they become. "The more you dwell upon, reflect upon, and think about the things you want and how you can attain them," the more able you become to create and recognize chances to attain your goals.

Law of Substitution "[Y]ou can substitute a positive thought for a negative one." "Whenever something goes wrong, *think* about your goals. Whenever you have a bad day, *think* about your goals.... It is impossible to think about your goals continually without being optimistic and highly motivated."

Tracy states that "when you begin using all these mental laws" in relation to "a clearly defined purpose to which you are totally committed, you become an unstoppable powerhouse of mental and physical energy that will not be denied. With clear, specific goals, you develop and use all your mental powers. You then accomplish more in a few years than most people accomplish in a lifetime."[4]

Tapping Our Resources

Each of us has untapped **resources** (something that can be drawn on for aid) within us, just as nature does. Once humanity learns to tap into or further perfect the available resources afforded by the sun, water, wind, the atom, ocean waves, ions, magnetism, and other as-yet undiscovered energy sources, great things will begin to happen (Figure 14–4). Great things happen in our lives once we set goals and tap the resources within us. In his "I Can" course, Zig Ziglar says that "if we want something badly enough, we must make it our definite goal. When we go after it as if we can't fail, many things will happen to help make certain we won't."[5] However, setting a goal but not tapping the natural resources within is a significant problem for some people.

FIGURE 14-4 This solar panel will tap the sun for energy. Great things can happen once you tap the resources within you. *(Courtesy of Greg Grieco, Department of Public Information Media Gallery, Pennsylvania State University.)*

WHY PEOPLE DO NOT SET GOALS

Some people are talkers instead of doers. They are not willing to do what is necessary to be more successful and improve their lives. You can tell what people really believe by their actions, not their words. It is what you do that counts. Your behavior expresses your true **values** (social principles, goals, or standards held or accepted by an individual, class, or society) and beliefs. One person who takes action is worth a hundred talkers who do nothing.

Until people accept that they are fully responsible for their lives and for almost everything that happens to them, they will not take the first step toward goal setting. Such people use all their

creative energies making excuses for their failure to progress.

People who are mentally and emotionally negative are not the kind of people who confidently and optimistically set **short-term goals** (goals that are set to happen immediately or very soon) and **long-term goals** (goals that are set to be reached two or more years in the future). People raised in a negative environment do not have attitudes that allow them to set goals; unless an attitude adjustment occurs, they will continue to complain as life happens to them. However, if leaders create an atmosphere that is positive and enjoyable, they can convince followers to set all types of goals (Figure 14-5).

Our environment affects our actions, attitudes, and values. If our parents do not have goals, we may grow up without even knowing that there are such things as goals. It is a fact that 80 percent of the people around us are going nowhere. It is only natural for us to go with them. Brian Tracy said, "If people knew that all their hopes and dreams and plans, all their aspirations and ambitions, are dependent upon their ability and their willingness to set goals—if people realized how important goals are to a happy, successful life—I think far more people would have goals than do today."[6] Someone who truly masters goal setting will probably be very successful.

Goal setting is more important to our happiness than almost any other single subject we could ever learn, yet, until recently, schools have

FIGURE 14-5 This agriscience student seems to be enjoying her time in the lab. With such a pleasant attitude, she ought to be able to set goals with ease. *(Courtesy of University of Georgia—College of Agricultural & Environmental Sciences.)*

not given us even an hour's worth of instruction in goal setting. People can even get college and graduate degrees without having had any instruction in goal setting.

People can be very negative and cruel. Many do not praise us for our successes, but they certainly criticize us for our failures. Children are smart; they learn how to get along and fit in. Eventually, a child who is constantly criticized or discouraged stops coming up with new ideas, new dreams, or new goals. Effective goal setters eventually learn not to broadcast their goals. To avoid criticism and ridicule, keep your goals confidential. Do not tell anyone except the few people close to you whose help you will need in the pursuit of your goals. If you need encouragement, encourage others, and they will in turn encourage you.

The biggest roadblock to success is the fear of failure. We learn this early in life when our parents do not let us try things. They say things like "if you can't do it right, don't do it at all." Little do they realize that their children would probably eventually get it right if they were permitted to fail until they learned how to get it right. The negative attitude of the parents becomes engrained in a child and does more to squelch hope and kill ambition than anything else.

People fear failure, but would they fear it as much if they understand the role it has in achieving success? You cannot succeed if you don't fail. The greatest accomplishments in history have also been the great failures. Only those who do nothing make no mistakes—and they never achieve anything, either. Babe Ruth was not only the home-run king; he was also the strikeout king. Thomas Edison is remembered for his success in creating the lightbulb, but it took more than 11,000 failed experiments before he got it right. Edison viewed his failed experiments as learning opportunities. Likewise, you will be more successful if you try to do more things. If you try more, your probability of eventual success will be higher. It's that simple.

"Great successes are almost always preceded by many failures. It is the lessons learned from the failures that make the ultimate successes possible" (Figure 14–6).[7] Look on temporary setbacks as signs saying, "STOP! Go this way instead." Great leaders never call these setbacks *failure* or *defeat*. Instead, they use phrases such as "valuable learning experiences" or "temporary glitches." You can learn to overcome the fear of failure by being absolutely clear about your goals

FIGURE 14–6 Members of this leadership team experienced many failures before they finally achieved a breakthrough that would make a positive change in their organization. (© iStock/skynesher)

and accepting that temporary setbacks and obstacles are the price you pay to achieve any great success in life.

HOW TO SET GOALS

"Set goals that are clear, challenging, and obtainable."[8] Make sure the goals are measurable. Set immediate goals, goals for the medium term, and goals for the long term. When you reach one goal, take a moment to congratulate yourself and cherish the moment. Then move on to the next goal with confidence, energized by what you have already accomplished.

Your goals must be both realistic and attainable. You cannot accomplish everything right now. Set interim and shorter-term goals that will each get you a step further toward attainment of your ultimate goal. For example, a person who has the goal of finishing college might become discouraged rather quickly without short-term goals. Those short-term goals might include making it through the week, until the next holiday, or until the end of the semester or quarter.

Make sure your goals are specific. If you don't have goals that are specific, you can get off course quickly and never develop a sense of urgency. Each goal should include a challenging time limit.

"A goal is a dream with a deadline."
—Harvey Mackay[9]

Make sure to put your goals in writing. Written goals are part of disciplined motivation. **Discipline** is the ongoing process of bringing ourselves under control. Without disciplined motivation, we lose our ability to deal with the future. We begin to function based on feelings, merely reacting to situations and living in a manner that dulls our purpose and clouds our vision. Goals that are written down act as a concrete vision that guides us toward fulfillment, because they are always there to refer to.

Set goals and then work hard to reach them. You may not always reach your goals in the timeframe you wanted, but the important thing

is to keep planning for and pursuing your ultimate goal. For more success, prioritize your goals. Make a top ten list of things you want to accomplish as an FFA member, student, spouse, parent, or leader in your place of business. Then rank those ten items by priority, and post the finished list where you can see it. This will keep you focused on what is most important and also help you formulate clear, specific goals.

PRINCIPLES OF SETTING GOALS

Criteria for Goals

Consider the following **criteria** (standards, rules, measures, or tests by which something can be judged) for the goals you consider:

- Set goals that get you excited and motivate you. If you do this, you'll have the energy you need to meet those goals. Also, if you are motivated by a goal, you'll know it's *your* goal and not someone else's.
- Set goals that are **S**pecific, **M**easurable, **A**ttainable, **R**elevant, and **T**imebound (SMART). SMART goals are described in detail below.
- Write your goals down. This way, you'll be able to more easily reflect on them to make sure they are SMART, right, legal, fair, or moral.
- Make an action plan for your goals. One of the best ways to make sure you are committed is to develop an action plan that you can follow.
- Stick with it! This should be easy if you are motivated and have written down your goals along with an action plan, but sometimes you still need the emotional maturity to keep going. Be strong and stick with it.[10]

Consider Benefits of Goals

A goal should yield some benefit once you attain it. If you cannot identify the benefits you will reap from your goals, you would do better to choose other goals. Any worthwhile goal should offer at least the following general benefits:

- Give you more purpose and direction in life
- Help you make better decisions

- Allow you to be more organized and effective
- Spur you to do more for yourself and others
- Make you feel more fulfilled, enthusiastic, and motivated
- Stimulate you to accomplish more[11]

Use the SMART Principle in Goal Setting

Goals must be meaningful, you must believe in them, and you must have a positive attitude toward achieving them. There are five ways of determining whether your goals are SMART (acronym for Specific, Measurable, Attainable, Relevant, and Timebound) goals. Goals should be

Specific—Keep goals clear, definite, and concise.

Measurable—Set standards and use appropriate metrics so you can judge your progress and know when you have reached a goal.

Attainable—Only goals that are realistic and practical can be accomplished. This is true even of high-aiming "stretch" goals.

Relevant—Goals should serve your purpose or the purpose of your group and further what you are trying to accomplish.

Timebound—When added to measurement, record keeping will let you know whether you are on course toward reaching your goals.[12]

SMART goals represent a formula that works. Have you ever written goals using this formula? You will be surprised how it can help you with personal and team activities that are important to you. Following are some examples of SMART goals. See if you can identify the S, M, A, R, and T in each one.

- Attend State FFA Convention as a CDE competitor before graduating as a high school student in order to further develop leadership skills and personal contacts.
- Achieve 100 percent membership as an FFA chapter by Christmas in order meet the Chapter Development Quality Standard of Recruitment.
- Increase the kidding percentage of our school goatherd from 120 percent to 150 percent for the coming school year in order to make sure more students have a goat to show next summer.

STEPS IN GOAL SETTING

Goals are dreams that we are willing to pursue. To reach our goals and fulfill our dreams, we must follow certain steps. Accomplishing goals is much like climbing the steps of a ladder: to get to the top, you have to climb each rung (Figure 14–7).

Our thoughts can become reality. We become and accomplish what we think. The more intense our thoughts, the quicker our goals will become reality. "There is a direct relationship between how clearly you can see your goal as accomplished, on the inside, and how rapidly it appears on the outside."[13] The following steps will help provide the **ways and means** (a step-by-step procedure detailing how goals will be accomplished) to get from where you are to wherever you want to go.

STEPS IN GOAL SETTING
19. Begin and never quit
18. Visualize your goals
17. Develop a support system
16. Consider concerns with your goal-setting
15. Work toward your goals daily
14. Take advantage of your momentum
13. Review goals often
12. Set specifi c goals with a deadline
11. Analyze where you are beginning
10. Identify the benefi ts of reaching each goal
9. Write down your goals
8. Determine your major purpose in life
7. Set goals that maintain balance in your life
6. Choose appropriate goals
5. Set big goals
4. Set goals in your talent area
3. Develop belief
2. Develop desire
1. Set goals that align with your values

FIGURE 14–7 Achieving goals is like climbing a ladder: You start at the bottom and follow certain steps to get where you want to go.

Step 1 *Set goals that align with your values.*

If you are to reach your full potential, your goals and values should be in harmony with each other. Your values represent your deepest convictions about what is right and wrong, what is good and bad, and what is important and meaningful to you. High achievement and high self-esteem result only when your goals, your values, and your actions are compatible.

Step 2 *Develop desire.*

Almost every decision you make is emotionally based in fear or desire. A stronger emotion overcomes a weaker emotion. If you focus on your desires, write them down, and make plans to achieve them, your desires will become so strong that they overwhelm your fears.[14]

Step 3 *Develop belief.*

It is crucial for you to develop a belief that achieving your goal will happen. When you believe something, it is non-negotiable. When you develop a belief that you will accomplish your goal, the small setbacks won't sting as badly; the steps in the action plan are more plausible; and your commitment is at its highest. If you believe it, you can achieve it.

Step 4 *Set goals in your talent area.*

Goal setting is necessary for success. However, set realistic goals that you believe you can achieve. If you are 5'1" tall and weigh 105 pounds, it is not reasonable to set a goal of becoming an offensive lineman in the National Football League. You could, however, set the goal of riding a Triple Crown winner. You cannot fool your subconscious mind into supporting a dream that has no basis in actuality. "Stretch" goals are laudable; completely unrealistic ones are merely fantasies.

Each of us has the capacity to be excellent at something. You can achieve your full potential by finding your **talent area** (natural ability or predilection to learn or do something) and then putting your whole heart into developing your talents in that area (Figure 14–8).

FIGURE 14–8 Make sure the goals you set are within your area of talent. This FFA member is pursuing a goal in her talent area, singing. She is performing at the FFA Convention. *(Courtesy of Georgia Agricultural Education.)*

Do not overlook the talents within you. When considering a major in college, for example, do not discount the skills and abilities you may have already developed in a particular field. You may be more successful if you pursue educational goals related to a field in which you have already attained knowledge and experience. College may not be able to prepare you in a new field of study to the level that you have already reached in another field.

Start working toward your goals early. A student may choose to go to college with the goal of becoming a veterinarian. This is an admirable goal, but may not be attainable unless the student had the opportunity to learn enough science and math in high school. College entrance scores must reflect success in high school and indicate that the student has the ability to pass the rigorous college courses required for veterinary school. If the student sets this goal as a freshman in high school, she can choose and concentrate on courses that will help her gain the proper knowledge and skills.

Make a plan to secure needed skills and abilities. You must learn what you need to know so you can accomplish what you want to accomplish. Make a list of all the information, skills, abilities, and experiences that you need, and make a plan to acquire them. Do not overlook the talents that can be developed to help you reach your fullest potential and happiness.

"Do what you can, with what you have, right where you are."

—Theodore Roosevelt[15]

Step 5 *Set big goals.*

Big goals are set by "big people." It takes big goals to create the excitement necessary for accomplishment. There is no thrill in being average or copying others. The excitement comes when you do your best, which you can do only with the proper goals. Consider the following analogy:

> Take a bar of iron and use it for a doorstop, and it's worth a dollar. Manufacture horseshoes from that iron, and they're worth about fifty dollars. Take the same bar of iron, remove the impurities, refine it into fine steel and manufacture it into mainsprings for precision watches, and it's worth a quarter of a million dollars.[16]

The way you see life largely determines what you get out of it. The way you see the bar of iron makes the difference, and the way you see yourself and your future makes the difference.

Step 6 *Choose appropriate goals.*

Appropriate goals are challenging, but reachable for you and your followers if you do the following:

- work as hard as you can on achieving your goals,
- work with enthusiasm and inspire enthusiasm in the teams and people you lead, and
- work on goals that are relevant to an overall purpose.

Step 7 *Set goals that maintain balance in your life.*

Brian Tracy suggests that you need a variety of goals, in each of seven critical areas of life, to perform at your best. He suggests developing goals in the following areas:

- Family and personal
- Physical and health
- Mental and intellectual
- Study and personal development
- Career and work
- Financial and material
- Spiritual

Having this balance allows you to work toward your goals all the time and achieve self-fulfillment in all these areas.[17]

Step 8 *Determine your major purpose in life.*

Defining a primary, central purpose is terribly difficult for many people. You have to ask yourself, "Which *one* goal would make me truly successful and happy when I accomplish it?" Is it a goal for your FFA chapter? Is it a goal for your family? Maybe someday it will be a goal for the business you run. Choosing a primary goal representing the major purpose in your life will help you stay focused and make the kind of progress you will need to be successful.

Step 9 *Write down your goals.*

When you commit a goal to writing, you make it more **tangible** (having actual form and substance). One of the most powerful of all methods for instilling a goal into your subconscious mind is to write it out clearly, concisely, in detail, exactly as you would like to see it in reality. Decide what is right before you decide what is possible. Make the description of your goal perfect and ideal in every respect.

As you write it, include *what* you want, *when* you want it, *why* you want it, and *where* you are starting. Make a list of the obstacles you must overcome, the information you will require, and the people whose help you will need.[18] There are many ways to do this. Figure 14–9 shows how Ben Franklin planned for personal achievement.

Step 10 *Identify the benefits of reaching each goal.*

To ensure that your goal is worth the effort, make a list of all the ways you will benefit from achieving it. When you have significant reasons for achieving your goals, you increase your intensity. If your reasons are important enough, your belief solid enough, and your desire intense enough, you will make your goal more attainable.

Step 11 *Analyze where you are beginning.*

The clearer you are about where you are starting from and where you are headed, the more likely it is that you will end up where you want to be. This gives you a baseline against which to measure your progress. For example, if you decide to lose weight and you set a goal of losing 20 pounds, you must first weigh yourself now.

SIX STEPS IN BEN FRANKLIN'S FORMULA FOR PERSONAL ACHIEVEMENT

1. Select 13 characteristics (principles, attributes, virtues) that you would like to acquire or develop.

2. Using 3" × 5" index cards, prepare a "pocket reminder" for each characteristic. Write a brief summary of the characteristic, indicating what it means and what you want to develop.

3. Concentrate on one characteristic every day for one whole week. Carry the reminder card with you and refer to it frequently as you try to develop and display that characteristic.

4. Start on your second characteristic the next week. Let your subconscious mind take over for the characteristic that you have worked on for the past week.

5. Continue working on one characteristic each week until you have completed the entire series of 13. Then start over. This method allows you to complete the series four times in a year.

6. After a year, or when you feel you have acquired a desired characteristic, substitute a new characteristic that you wish to acquire or develop in its place.

FIGURE 14–9 To achieve your goals, you must have a step-by-step prioritized plan. Here is one of many plans you could choose. *Source: R. I. Carter, Leadership and Personal Development. St. Paul, MN: Hobar Publications, 1990.*

Step 12 *Set specific goals with a deadline.*

It is important to set deadlines for your overall goal and deadlines for specific tasks that need to be accomplished along the way to reach your goals. If you have the attitude that you will accomplish a goal whenever it's convenient or when you have time, you will probably never find it convenient or have enough time. If you state specific goals that include definite deadlines, you are much more likely to achieve them.

Step 13 *Review goals often.*

In addition to writing down your goals, keep those goals constantly in view, displayed where you and everyone else can see them often. You will have a better chance of achieving your goals if they are visible and on your mind all the time. Part of your continuous review includes realizing that your plan will not be perfect and that you may have to change some things to reach your goal. Successful people make detailed plans, but they are not afraid to alter them as the need arises.

Step 14 *Take advantage of your momentum.*

When you have reached one goal, take advantage of the enthusiasm generated by this success and your **momentum** (the quantity of motion of a moving object, equal to the product of its mass and its velocity) and start attacking another goal right away. The law of **inertia** (tendency of matter to remain at rest if at rest or, if moving, to keep moving in the same direction) also affects people: waiting too long between reaching an intermediate goal and beginning work on another goal can allow enthusiasm to dissipate and momentum to be lost.[20]

Step 15 *Work toward your goals daily.*

If you expect to accomplish your objectives, you must work *toward your objectives every day.* Good students do not wait until the night before to study for an exam. They study daily. Our daily objective should be to improve on yesterday.

Step 16 *Consider concerns with your goal setting.*

Goals can become negatives if you do not take certain precautions. First, realize that you are the determining factor in achieving your goals and that luck is not involved. Second, a goal can become an obstacle if it is too big. For example, to accumulate a million dollars in one year is an unrealistic goal. However, to accumulate a million dollars by retirement is very realistic. Third, your goal can be a concern if it is outside your talent area, especially if you set it to please someone else.[21] For example, you set the goal of going to law school because your parents want you to become a lawyer, even though you really want to become a master mechanic and own an auto repair shop.

Prepare for any anticipated roadblocks. Be aware of the obstacles but do not dwell on them—simply deal with the problems. A roadblock for a weight-loss program might be a love of sweets; for climbing a mountain, it could be poor physical condition; for getting a college degree, it could be a lack of self-discipline; for becoming an elected official, it could be heavy social demands. "Wherever great success is possible, great obstacles exist. In fact, obstacles are the flipside of success and achievement. If there are no obstacles between you and your goal, it probably is not a goal at all, merely an activity."[22]

VIGNETTE 14-1	GOALS MAY CHANGE

JUST BECAUSE WE SET MEASURABLE, specific, and perfectly planned goals does not mean that we are locked into those goals. Goals may change. The reason they may change is because we change, especially as leaders. Often, goals are a direct reflection of our attitudes and values, and when our attitudes and values change (for whatever reason), our goals will probably change, too. As you read about Nancy in the following paragraphs, think about times when your goals have changed.

Nancy was a teacher and the school softball coach. She was not married, and she devoted most of her time to teaching and coaching. She followed all the rules of goal setting, and they worked. All of her goals revolved around work. Two of her goals were to (1) be the best teacher in school every year, regardless of the time and effort needed to do so; and (2) take her girls to the state championship every year, regardless of how hard it was on her and the team. She was named teacher of the year. Her girls won the state softball championship two years in a row. Nancy was recognized as one of the best teachers in the school and one of the best coaches in the state.

Then Nancy got married and started having children. She developed a whole new set of goals and tacked them onto the ones she already had. She was determined to be as good at home as she was at work—but pretty soon realized that this was killing her. No matter what she tried, she could not meet all of her goals.

Should Nancy drop coaching and stop trying to be so good at teaching? Should she quit trying to do a good job with her new family? Of course she did not need to stop working hard and striving to achieve. She just needed to change her goals a bit to reflect the changes in her life and her current situation. How could she change her goals? What new or different goals should she set? How should her new goals affect or interact with the old goals?

Step 17 *Develop a support system.*

Find people who will support you and your goals. This could be your family, a friend or coworker, or certain organizations. To accomplish many of your goals, you will have to get help and cooperation from many people. Make a list of all the people whose assistance you will require.

Step 18 *Visualize your goals.*

In Chapter 4, we discussed the importance of vision and leadership. The same is true for goal setting. Picture your goal as if it were already achieved. Keep this picture in your mind. By doing this, you increase your desire and belief that your goal is achievable. Why? Your subconscious mind is activated by pictures. There are stories of golfers and basketball players who improved their scores or shot percentages merely by visualizing themselves doing so.

Step 19 *Begin and never quit.*

Once your plan is complete, start immediately. Take one small step at a time. Zig Ziglar makes two excellent comments on reaching goals: "[W]hat you get by reaching your goals is not nearly as important as what you become by reaching them."[23] You become a winner. "If you expect to change and improve your circumstances, you must change and improve yourself"—because you must be something before you can do anything.[24] Reaching goals makes us something. Among other things, we prove that we are not quitters. Strive toward your goals with persistence and determination. Simple logic says that you will eventually be successful if you never quit (Figure 14–10).

TYPES OF GOALS

How Do You Begin?

The journey of a thousand miles begins with a single step, but sometimes that first step is the most difficult. Reaching the end of the journey (your goal) is great, but it is the journey itself, the planning and the struggling, that gives you a feeling of confidence and accomplishment.

Imagine yourself living in a valley surrounded by mountains. You can climb any one you choose, but you are not even sure you want to climb a mountain at all. When you finally do select a mountain, make sure that you are prepared with training and the right equipment. Ask yourself, "Is this the right mountain for me?" You may not

DON'T QUIT

When things go wrong, as they sometimes will,

When the road you're trudging seems all uphill,

When the funds are low and the debts are high,

And you want to smile but you have to sigh,

When care is pressing you down a bit

Rest if you must, but don't you quit.

For life is queer with its twists and turns,

As every one of us sometimes learns,

And many a failure turns about,

When he might have won if he'd stuck it out.

Success is just failure turned inside out,

The silver tint of the clouds of doubt.

And you never can tell how close you are,

It may be near when it seems so far.

So stick to the fi ght when you're hardest hit,

It's when things seem worst that *you must not quit!*

—*Anonymous*

FIGURE 14–10 Many people believe that persistence is more important than intelligence in achieving goals. One thing is certain: you will not achieve your goals if you quit trying.

be able to make a final, definite, once-and-for-all choice. So, while you continue to climb and move ahead and put the best you have into the effort, you stay flexible. Even if you realize that you are on the wrong mountain, you will have learned things during the climb that will help you evaluate, plan, and prepare for changes. You might

make a complete change to a different mountain that fits your needs and suits your talents better, or you might look at where you are and decide that although the mountain you are on is not the wrong one, you need to find a different path to reach where you want to go. The important thing, whether or not you choose the right mountain or the right path the first time, is to begin.

Identify Your Goals

We have talked about all the reasons for setting goals. Now, let us select our goals. You may not need any help. You may already have more goals than you think you can accomplish in two lifetimes! However, those of you who need help coming up with goals could consider the following seven questions:

1. *What are your five most important values?* This question helps you clarify what is truly important to you. Once you have identified the five most important things, rank them from one to five, with one being the most important.
2. *What are your three most important goals in life right now?* Answer this question, in writing, within 30 seconds. When you work within this time limit, your subconscious mind quickly identifies the goals that really matter to you.
3. *What would you do, how would you spend your time, if you learned today that you had only six months to live?* This helps you define what is truly important to you.

VIGNETTE 14-2 | **CELEBRATE ACCOMPLISHMENTS**

AS IMPORTANT AS GOALS ARE, it is even more important to celebrate the accomplishment of those goals. Whenever we reach a personal goal, we should celebrate by rewarding ourselves. Whenever leaders and their followers reach a team goal, they should have a party! And when specific persons reach a goal, you should publicly recognize them and celebrate their accomplishments.

Are we saying you should "party down" and have a good time at work? Yes, we are! People respond to being recognized. Many times celebrating met goals and hard work is the fuel that keeps people working so hard. Celebrations foster an atmosphere of cooperation and collaboration.

As a leader, you need to schedule time to celebrate the attainments of individuals and the team or work unit that you lead. You also need to make an effort to spontaneously celebrate followers' accomplishment of goals. Both types of celebration are very effective in developing trust, rapport, and interpersonal relationships. We encourage you to start partying and celebrating success. You will be amazed at the effects!

Note: Celebrating accomplishments is one of Kouzes and Posner's 10 commitments under the leadership practice *"Encouraging the Heart."*[25]

4. *What would you do if you won a million dollars in cash, tax free, in the lottery tomorrow?* Think about the choices you would make if you had all the time and money you needed. Consider the things you would do differently if you could choose freely.

5. *What have you always wanted to do but have been afraid to attempt?* This helps you realize what constraints your fears have been imposing on what you really want to do.

6. *What do you most enjoy doing? What gives you the greatest feeling of self-esteem and personal satisfaction?* You will always be happiest doing what you most love—that which makes you feel the most alive and fulfilled.

7. *What one great thing would you dare to dream and do if you* knew *you could not fail?* Imagine that you are magically guaranteed that you will be successful in one thing you attempt. What specific goal would you set?[26]

Examples of Goals

Goals can include developing better test-taking ability; graduating; getting a job; maintaining good personal health; channeling stress productively; saying no to drugs, tobacco, and alcohol; improving grades; and improving relationships at home. A group of high school students made a list of all the things they would like to do if they could do anything. A few of their choices are listed here. Would any of these be on your list?

- Graduate
- Make a million dollars
- Get married
- Travel around the world
- Be a teacher
- Be a marine biologist
- Climb a mountain
- Own a classic car
- Skydive
- Be a doctor
- Drive an 18-wheeler
- Own my own store
- Have my own band
- Build houses
- Live in another country
- Publish a story
- Act
- Learn to ride a horse
- Go somewhere on an airplane
- Learn to swim

To begin the goal-setting process, brainstorm. Do not stop to think about it. Write down on a sheet of paper all the things you would like to do if nothing were standing in your way.

Categories of Goals

You can have goals in eight general areas: physical, mental, personal values/spirituality, family, school, social, financial, and career. The following list contains examples of goal statements for each of these eight categories. Using the goal statements, develop a chart with seven steps that include identifying the goal, listing its benefits, listing the obstacles to overcome, identifying support groups, listing the skills and knowledge required to attain it, developing a plan of action, and setting a deadline for achievement. (Activity 6 at the end of this chapter provides a sample chart.)

Physical

- Walk 30 minutes every day.
- Use our health club membership regularly.

Mental

- Read two inspiring books per month.
- Discover inspired ideas as I concentrate and meditate daily.

Personal Values/Spirituality

- Spend 30 minutes each morning reading an inspirational book.
- Give back to your community by volunteering 30 minutes per week.

Family

- Spend five hours per week helping my parents.
- Spend Friday nights together.

School

- Accept more responsibility at school.
- Strive for excellence in my classes.

Social

- Visit the senior citizens' home one weekend per month.
- Open our home for neighborhood gatherings one evening each month.

Financial

- Save $50 per month.
- Earn $400 per month with our part-time family business.

Career

- Take courses to prepare me for my career goal.
- Travel to get experience with my career goal.[27]

Set goals for yourself in each of the eight areas. Refer to activity 6 at the end of this chapter.

Timelines for Goals

In order for goals to be meaningful and useful, one has to "divide and conquer." In order for a student to complete college, they cannot set only a long-term four-year goal to graduate. They must set short-range and medium-range goals. A short-term goal may be getting through the week. A medium-range goal may be getting through a semester. A further explanation of type goal follows:

Immediate (Short-Range) Goals An **immediate goal** is a goal that is set to happen within a day or week. It may be as small as washing dishes right after a meal. Another one may be obtaining a grade of B on a chemistry assignment. A short-range goal can be a step toward accomplishment of a medium- or long-range goal, or it can be an independent, stand-alone objective. Washing the dishes immediately after each meal may be a totally independent goal. Getting a B on a chemistry assignment may be part of receiving a B in the chemistry class. You act to reach an immediate or short-range goal within the next few days or week.

Medium-Range Goals A medium-range goal is similar to the short-range goal except for the accomplishment time frame. Medium range is longer, usually 6 to 12 months. Medium-range goals may be independent, may include several short-range goals, or may be intermediate steps necessary to achieve a long-range goal.

Long-Range Goals Setting a long-range goal involves planning where you want to be months, years, or even decades from now. Short- and medium-range goals are usually parts or elements of long-range goals. For example, a high school student has a long-range goal of becoming a doctor. His medium-range goals may include finishing high school, college, and medical school. His short-range goals may include getting a B on the upcoming chemistry test or finishing his English paper by next weekend.

Prioritize Your Goals

Prioritizing is an important step in planning. Evaluate your goals and rank them in each category as well as by timing (immediate, medium range, and long range). Start work on accomplishing the top-priority goals by putting yourself in a favorable environment. For example, if being a state FFA officer is one of your goals, you may attend every possible state FFA function just to get exposure and experience. Develop sub-objectives for each prioritized goal. Also, develop alternatives: a plan B, a plan C, or a **worst-case scenario** (the least desirable thing that could possibly happen in a given situation) for each goal. Then, make a commitment to your goal, take action to achieve it, and periodically review the outcome.

Meeting and Even Exceeding Your Goals

A goal generates commitment. Our identities are shaped by what we commit to. You can acquire and maintain the drive and desire to succeed if you do the following:

- Find a position in which you can use your talents and abilities
- Make sure you receive a "payoff" or **psychic income**, praise or positive things that happen to us that give us a psychological boost or make us feel good (Some refer to this as planning small wins)
- Focus on the future
- Become a "doer" who takes action while focusing on results

Stay focused on your goals. Do not let others interfere with or rearrange your priorities. Remain committed to the plans, actions, and directions you have identified as necessary for achieving your goals, and use these as a guide in daily decision making. Remember that activity does not equate to accomplishment: never confuse efforts with results.

Perseverance Is Important

Goal setting is a continuing process, not a one-time practice. You must pursue your goals on a daily basis and maintain your momentum. Your psychological self responds to Newton's physical principles of inertia and momentum, which state that a body in motion tends to remain in motion unless acted on by an outside force. It takes a large amount of energy to get a body from a resting position to a state of motion; it takes a much smaller amount of energy to keep it in motion.

Nothing succeeds like success. Do something every day to keep moving toward your goals. Maintain your momentum, keep up the pressure, and develop a mind-set of becoming a high-momentum, goal-setting, goal-achieving person.

GOAL SETTING THE FFA WAY

FFA Program of Activities

During most of its existence, the FFA organization has had a procedure that provides a tremendous mechanism for goal setting: the FFA Program of Activities. Programs are developed at the local, state, and national levels within the organization. Forms and resources for developing a Program of Activities are available to everyone at the National FFA website.[28] The format presented for developing goals for your FFA chapter can also be used for developing your personal goals.

Learning how to prepare and implement an FFA Program of Activities teaches us how to be goal setters. The FFA also helps develop leadership: when we learn how to set and achieve goals, we also learn to become leaders. When we achieve our goals, we become successful. Only you, not others, can determine your success.

The FFA Program of Activities is a written plan, developed and published annually, of all activities that the chapter wishes to accomplish during the school year. It serves the chapter much as a road map serves a traveler. It is a guide that identifies the activities necessary to make the FFA chapter meet the personal and occupational needs of its membership.[29]

The Program of Activities is organized according to three primary divisions: Student Development, Chapter Development, and Community Development. Student Development is the division for activities that improve the life skills of FFA members. Chapter Development activities are geared toward encouraging students to work together. The Community Development division includes activities through which FFA members work with other groups to make their community a better place to live and work. Each division consists of five program areas around which chapter goals may be identified, executed, and evaluated. The program areas are listed under the primary divisions here:

- Student Development
  - Leadership
  - Healthy lifestyles
  - Supervised agricultural experience
  - Scholarship
  - Agricultural career skills
- Chapter Development
  - Chapter recruitment
  - Financial
  - Public relations
  - Leadership
  - Support group
- Community Development
  - Economic
  - Environmental
  - Human resources
  - Citizenship
  - Agricultural awareness

For any organization to be effective, it must plan its activities. A plan gives a sense of direction and an outline for action; this is how it helps an organization reach its goals. Without a plan, the group's activities become random and haphazard, and movement toward achievement of the goals will be very limited because the activities are no longer focused or targeted.

Developing a Program of Activities is part of the leadership training and educational process essential to preparing for a selected occupation. A Program of Activities fosters cooperative spirit and develops individual leadership talent and ability among FFA members by assigning them duties and responsibilities for developing and conducting planned activities. Perhaps the most important benefit of a well-planned Program of Activities is that it provides a means of evaluating and improving chapter activities each year.[30]

A well-planned Program of Activities is written so that it provides definite, measurable, and understandable answers to the following questions:

1. What is going to be done?
2. Who is going to be involved?

3. When are you going to do it?
4. How many are going to do it?
5. How are you going to do it?
6. How much is it going to cost?[31]

A chapter can use one of several formats for developing its Program of Activities. It is best to choose one standard format and stick with it. Regardless of the planning format and/or forms used, a Program of Activities should include the major sections that guide the planning process. These are listed and described in Figure 14–11.

Program of Life Goals

The Program of Life Goals chart, shown in Figure 14–12, is adapted from the Program of Activities concept and follows the same basic format or process as an FFA Program of Activities. Review the Steps in Goal Setting section of this chapter, and you will see the similarities to the FFA Program of Activities. This chart is a planning tool for the personal goals you set in life. The eight goal-setting areas of life are included: physical, mental, financial, family, school, social, personal values/spirituality, and career. Instead of "Amount Budgeted" from the Program of Activities, the Program of Life Goals chart uses "Obstacles, or Price You Will Have to Pay"; "Support System" substitutes for "Members Responsible." As you can see, all other sections are basically the same.

Plan of Action

Figure 14–13 shows a sample of two goals from the leadership program area of the FFA Program of Activities. All sections that can be found in various Program of Activities forms are included, with actual data for each.

PROGRAM OF ACTIVITIES	
Division/Area	Three divisions and/or as many as 15 program areas help organize planning and may frame the committee structure.
Activity	An activity is any function that the chapter conducts during the year to meet the stated objectives in each division of the Program of Activities. Be specifi c. Tell exactly what is to be done. *Example:* "Improve the scholarship of all FFA members."
Goals	Every activity should have one or more goals. Each goal should be written in specifi c terms so that the committee can determine if the goal was accomplished. Be specifi c. Tell exactly what is to be done. Avoid generalizations. Use positive words such as *prepare, build, sponsor, use, submit.* Avoid nonspecific verbs such as *encourage, urge,* or assist. *Example:* "Have 50 percent of the chapter members raise their overall grade point average by two-tenths of a point."
Activities	These are the methods or the steps that the chapter plans to use in accomplishing its goals. The actual number of steps will depend on the complexity of the goals. 1. Devote one meeting a year to scholarship with a speaker. 2. Present scholarship certifi cates to the top three students in each class at the chapter award banquet. 3. In cooperation with the school principal, select and provide a scholarship trophy to the member who improves his or her grade point average the most during the year.
Amount Budgeted	Show the amount of money the chapter plans to spend on the ways and means of conducting each activity. In cases in which no cost is involved, this column should be left blank.
Members Responsible	The members responsible are the individuals who will coordinate each goal. Do not include all students who will conduct or participate in the activities.
Date to Be Completed	Projected completion dates are an important part in planning the Program of Activities. A deadline for completing each of the ways and means should be established and adhered to as nearly as possible.
Accomplishments	As each activity progresses and is completed, make comments that will be useful in making that activity more effective should it be conducted another year.

FIGURE 14–11 The FFA Program of Activities format offers an excellent example of how to achieve organizational, professional, and personal goals.
(Courtesy of the National FFA Organization.)

PROGRAM OF LIFE GOALS	
Areas	Physical Mental Financial Family School Social Personal Values Career
Activity (Objective)	There may be several activities within each area. For career, for example, summer employment, travel, specialized courses, or training may be activities within each area.
Specific Goal	A specific goal can then be set for each activity. Write it down. Be specific. Be realistic.
Activities	Develop a step-by-step prioritized plan. Begin and never quit.
Obstacles, or Price You Will Have to Pay	Identify skills or knowledge required. Identify obstacles. Identify sacrifices.
Support System	Identify people who will support you. Visualize your goals. Identify benefits of these goals.
Deadline to Complete Goal	Set a specific deadline.
Accomplishments/Evaluation	Compare to where you began. Monitor and adjust.

FIGURE 14–12 The Program of Life Goals format is adapted from the FFA Program of Activities format. See Figure 14–14 for an example of a completed plan of action.

PROGRAM OF ACTIVITY PLAN OF ACTION

PROGRAM AREA: Leadership

OBJECTIVE: Each FFA member will take part in leadership development activities sponsored by the FFA chapter and/or by the State and National FFA organizations.

GOALS	STEPS	AMOUNT BUDGETED	DATE TO BE COMPLETED	ACCOMPLISHMENTS AND/OR EVALUATION
Provide every member with the opportunity to improve his or her speaking ability.	1a. All FFA members write and give a five-minute speech as an Ag-Ed class assignment.		January 5	
	1b. Video each member's speech and replay for constructive criticism.		January 20	
	1c. Select one member from each class to present his or her speech to local service clubs, radio, etc.		February 15	
	1d. Invite three community leaders to assist in selecting a chapter public-speaking winner to participate above the chapter level.	$10.00	March 15	
	1e. Present local chapter winner with FFA Foundation Congratulations booklet and award medal at chapter banquet.		May 12	
	1f. Publicize in school paper and local newspaper.		May 15	
Provide committee work experience for all members.	2a. Through member choice and/or appointment, have each member serve on one of the fifteen Program of Activities standing committees.		September 15	
	2b. Provide each member with a copy of the chapter Program of Activities.		September 15	
	2c. All committees and/or subcommittees meet at least once a month.		Second Wednesday	
	2d. Have a different committee member bring and present the committee report to the Executive Committee and chapter meeting each month.		Each month	

FIGURE 14–13 The program area shown here is leadership. By using this format, you can identify chapter goals, the steps needed to achieve them, and the budget necessary to accomplish the goal. Note the spaces left for projected completion date and evaluation comments. *(Courtesy of the National FFA Organization.)*

PROGRAM OF LIFE GOALS PLAN OF ACTION					

AREA: Career

ACTIVITY: Obtain work experience to help enhance employment opportunity in an agribusiness after college.

SPECIFIC GOAL	WAYS AND MEANS (PRIORITIZED STEPS)	OBSTACLES	SUPPORT SYSTEM	DEADLINE	ACCOMPLISHMENT/ EVALUATION
Secure a part-time job with Tractor Supply Company (TSC) while in high school	Prepare myself in school Dress appropriately Secure an application Request an interview Type and return application Get the job Work hard and practice good employability skills Learn all I can about the business	Scheduling hours to continue school extracurricular activities	Parents Agriculture teacher Friends	January of my junior year in high school	
Secure full-time summer employment while in college and continue part-time during school year	Be an excellent part-time worker Assume more responsibilities as experience is gained	Maintain good grades while working during school year	TSC manager College advisor Friends Parents	Fall of my freshman year in college	

FIGURE 14–14 This format, adapted from the FFA Program of Activities, is an excellent way to develop a plan of action for your program of life goals. Career is just one of eight areas in which to plan life goals.

Career is an example with two goals from one of the eight areas of life (Figure 14–14). Each section provides an example of how you can use this format to plan your personal goals. A complete plan would run to approximately eight pages—a page for each area of your life. By using the eight-page, eight-area format, you could have three to five goals for each area, or a total of 20 to 40 goals. Numbers are not important. The important thing is that this gives you a methodical process to use in setting and achieving your goals. Activity 8 at the end of this chapter gives you the opportunity to do this.

CONCLUSION

The importance of setting goals can be summed up by saying that if you have goals, you know what you want and are likely to achieve success. If you do not have goals, you will not know what you want to succeed at, so you are less likely to achieve success.

Just as your organization, chapter, or club needs a plan of action, people need a plan of action for their lives. With a plan of action, they can succeed. By using the Program of Activities format to set personal goals, you use a proven model. You can succeed!

SUMMARY

The ability to set goals and make plans is a skill that each of us should develop. If we are to reach our full potential, we must learn to set goals.

There are several reasons for having goals. Goals provide a target, a destination, purpose, and meaning. Goals help focus our energies on productivity and accomplishment. They help individuals succeed, develop self-control, and evaluate their accomplishments.

Goals let you know whether you are on target. Goals benefit you by helping you concentrate your efforts, make the most of your time, let other people know how to assist, keep your enthusiasm, and monitor your progress. Goals make mental laws work for you: the laws of control, cause and effect, belief, expectations, attraction, correspondence, subconscious activity, concentration, and substitution. Goals also help you find and use your resources.

People who set goals are serious about being successful. They have accepted responsibility for their lives. Some people do not realize how important goals are. They may not know how to set goals, may be afraid of rejection, or may be afraid of failure.

Goal setting is an important and necessary part of taking control of your own destiny. Goals should be clear, challenging, obtainable, and realistic. In the course of goal setting, you will have to identify skills or knowledge required to reach your goals and develop a step-by-step prioritized plan of action. Avoid setting goals that are too high or too low, and never adopt a goal just to please someone else. Write down your goals, make them specific, and prioritize them. Then, plan and pursue your goals.

There are 19 steps for setting goals: set goals that align with your values; develop desire; develop belief; set goals in your talent area; set big goals; choose appropriate goals; set goals that maintain balance in your life; determine your major purpose in life; write down your goals; identify the benefits of reaching each goal; analyze where you are beginning; set specific goals with a deadline; review goals often; take advantage of your momentum; work toward your goals daily; consider concerns with your goal setting; develop a support system; visualize your goals; and begin and never quit.

There are several types of goals. Questions to help you identify your goals include asking what your values are, what you want right now, what you would do if you had only six months to live, what you would do if there were no time or financial constraints, what you have always wanted to do but been afraid to attempt, what gives you the greatest feeling of self-esteem and satisfaction, and what you would really want to do if you knew that you could not fail.

Categories of goals include physical, mental, personal values/spirituality, family, school, social, financial, and career. You may set several kinds of goals within each category. For best success, develop immediate, medium-range, and long-range goals in each category.

The FFA uses the Program of Activities to accomplish the goals of the FFA chapter. You can easily adapt the time-tested Program of Activities format for a personal life goals plan to use in setting your personal goals. A Program of Life Goals Plan of Action includes area, activity, specific goal, prioritized steps, obstacles, support system, deadline, and accomplishments or evaluation.

Take It to the Net

Explore goal setting on the Internet. Using your favorite search engine (Google, Yahoo!, Goodsearch, or another), and search terms such as *goals, goal setting, achievement,* and *personal improvement,* identify a website that interests you. Print the first page of the site. Summarize the site and tell the class why you find it interesting.

Chapter Exercises

REVIEW QUESTIONS

1. What are six reasons for having goals?

2. Compare the physical goals in basketball, football, and hockey with the mental goals of life.

3. What are five benefits of goals, according to John R. Noe?

4. What are five reasons people do not set goals?

5. What are six benefits of attaining goals?

6. List five key words to remember when setting goals.

7. Name eight principles to consider when setting goals.

8. Give three instances in which goals can be negative.

9. List 19 steps to use in setting goals.

10. What are seven questions to help you select your goals?

11. What planning tool does the FFA use to accomplish chapter goals?

12. When we have goals, what are eight things that we accomplish?

13. What are five criteria for setting goals?

14. What are eight areas or categories in which we should have goals?

COMPLETION

1. People do not plan to fail; they simply fail to _____.

2. It is just as difficult to reach a destination you do not have as it is to come back from a place you have _____.

3. Goals written down become a written _____.

4. SMART is an acronym for S_____, M_____, A_____, R_____, T_____.

5. Theodore Roosevelt said, "Do what you can, with what you _____, right where you _____"

6. Big goals are set by _____ people.

7. If you believe it, you can _____ it.

8. "What you get by reaching your goals is not nearly as important as what you _____ by reaching them."

9. Prior planning prevents _____ performance.

10. If we are to be successful and achieve what we want in life, setting _____ and making _____ are a necessity.

11. No goals, no _____.

12. Only _____ percent of Americans today have defined goals.

13. If you have no defined goals, you cannot possibly reach any target because you have nothing at which to _____.

14. The best goal should be _____ enough to present a challenge, and not so _____ that it is totally out of reach.

15. Setting goals is a proven method for stimulating _____ achievement and produces several definite benefits.

16. You sow goals and you _____ results.

17. One person who talks _____ is worth a hundred talkers who do nothing.

18. Someone who truly masters _____ will probably be very successful.

19. Goal setting is more important to our _____ than almost any other subject we could ever learn.

20. To avoid criticism and ridicule, keep your goals _____.

21. The biggest roadblock to success is the fear of _____.

22. Set goals that are _____, _____, and _____.

23. A goal should yield some _____ once you attain it.

24. We become and accomplish what we _____.

25. One thing is certain; you will not achieve your goals if you _____ trying.

26. Stay _____ on your goals.

MATCHING

_____ 1. Feeling positive about yourself to the degree that you determine your own destiny.

_____ 2. For every effect in your life there is a specific cause.

_____ 3. What you dwell on grows.

_____ 4. When you put a primary focus on your goals, people with similar goals are drawn to you.

_____ 5. Let a positive thought take the place of a negative thought.

_____ 6. Anticipating that everything that will happen will move you toward your goals.

_____ 7. You become what you think.

_____ 8. What you think about with your conscious mind, your subconscious will bring into reality.

_____ 9. Self-confidence and faith that you will achieve your goals by taking appropriate actions.

A. law of belief

B. law of substitution

C. law of correspondence

D. law of control

E. law of subconscious activity

F. law of concentration

G. law of expectations

H. law of attraction

I. law of cause and effect

ACTIVITIES

1. This chapter discussed several benefits of goals. Select the one that most applies to you, and write a paragraph or short essay supporting your choice and explaining your reasoning.

2. Mental laws were briefly discussed in this chapter as they relate to goals. Select and write a paragraph or short essay on one of the mental laws that causes you to think about yourself and how you react in certain goal-setting situations.

3. Read the section on why people do not set goals. Write a paragraph or short essay about the excuse of which you are most guilty.

4. Write the names of five people (friends, parents, teacher, boss, etc.) who would wholeheartedly support you if you shared your goals with them.

5. Brian Tracy suggests that you answer seven questions to help you come up with your goals. Answer each of the following questions:

   a. What are your five most important values in life?

   b. What are your three most important goals in life right now?

   c. What would you do, how would you spend your time, if you learned that you had only six months to live?

   d. What would you do if you won a million dollars cash, tax-free, in the lottery tomorrow?

   e. What have you always wanted to do but been afraid to attempt?

   f. What do you most enjoy doing? What gives you the greatest feeling of self-esteem and personal satisfaction?

   g. What one great thing would you dare to dream and do if you *knew* you could not fail?

6. Complete the accompanying "Goal-Setting Chart" by writing goals within each of the eight areas. Do not limit yourself to only five for each. Add more if you wish.

GOAL-SETTING CHART

On a separate sheet of paper, write three to five goals for each of the categories.

Physical	*Evaluation**	**Financial**	*Evaluation*
1.	10 _____	1.	10 _____
2.	9 _____	2.	9 _____
3.	8 _____	3.	8 _____
4.	7 _____	4.	7 _____
5.	6 _____	5.	6 _____
	5 _____		5 _____
	4 _____		4 _____
	3 _____		3 _____
	2 _____		2 _____
	1 _____		1 _____

GOAL-SETTING CHART			
Personal Values	*Evaluation*	**Social**	*Evaluation*
1.	10 _____	1.	10 _____
2.	9 _____	2.	9 _____
3.	8 _____	3.	8 _____
4.	7 _____	4.	7 _____
5.	6 _____	5.	6 _____
	5 _____		5 _____
	4 _____		4 _____
	3 _____		3 _____
	2 _____		2 _____
	1 _____		1 _____
Mental	*Evaluation**	**Family**	*Evaluation*
1.	10 _____	1.	10 _____
2.	9 _____	2.	9 _____
3.	8 _____	3.	8 _____
4.	7 _____	4.	7 _____
5.	6 _____	5.	6 _____
	5 _____		5 _____
	4 _____		4 _____
	3 _____		3 _____
	2 _____		2 _____
	1 _____		1 _____
School	*Evaluation**	**Career**	*Evaluation*
1.	10 _____	1.	10 _____
2.	9 _____	2.	9 _____
3.	8 _____	3.	8 _____
4.	7 _____	4.	7 _____
5.	6 _____	5.	6 _____
	5 _____		5 _____
	4 _____		4 _____
	3 _____		3 _____
	2 _____		2 _____
	1 _____		1 _____

*The numbers 1 to 10 in each category represent an "area" for achievement in your life. Rate your proficiency by placing an "X" next to the number that best states where you are today (1 is poor and 10 is excellent).

7. Select a goal from one of the areas in the "Goal-Setting Chart" used in activity 6 and break it down into long-range, medium-range, and short-range goals, using the following format.

SETTING LONG-RANGE, MEDIUM-RANGE, AND SHORT-RANGE GOALS

1. On a separate piece of paper, write one long-range goal. This should be something that will take you more than a year to do. (For example, become a state FFA officer in two years.)

2. Name one medium-range goal. This should be something you can do in a few weeks or a few months (less than a year). (For example, make the parliamentary procedure team.)

3. Write down two short-range (immediate) goals. These are things you can do in a day or a week at the most. (For example, make at least a B+ on my agricultural education test tomorrow; attend the state fair and see all the agriculture exhibits.)

8. Develop a "Program of Life Goals Plan of Action." Study Figures 14–12 and 14–14. How do they compare to the FFA Program of Activities format in Figures 14–11 and 14–13? Complete the accompanying "Program of Life Goals Plan of Action" using at least one goal in each of the areas. (Note: Your teacher may want you to do a plan of action for each of your goals.)

PROGRAM OF LIFE GOALS PLAN OF ACTION

Area

Activity

SPECIFIC GOAL	PRIORITIZED STEPS	OBSTACLES	SUPPORT SYSTEM	DEADLINE	ACCOMPLISHMENTS/ EVALUATION

NOTES

1. "Personal Goals," *Efficiency* series, No. 8742-E (College Station, TX: Instructional Materials Service, Texas A&M University, 1988), p. 1.
2. J. R. Noe, *People Power* (Nashville, TN: Oliver Nelson, 1986), pp. 145–146).
3. Ibid., p. 146.
4. B. Tracy, *Maximum Achievement,* (New York: Simon & Schuster, 2011), Google Play e-book, p. 143–147.
5. Zig Ziglar, *The "I Can" Course Learner's Manual* (Carrollton, TX: The Zig Ziglar Corp., 1989).
6. Tracy, *Maximum Achievement,* p. 148.
7. Ibid., p. 151.

8. S. R. Levine and M. A. Crom (Dale Carnegie & Associates, Inc.), *The Leader in You: How to Win Friends, Influence People, and Succeed in a Changing World* (New York: Simon & Schuster, 1993), p. 177.

9. Quoted in T. Boone, "The Science of Leadership," Professionalization of Exercise Physiology, (8) 5, May 2005, para 23, retrieved Aug. 11, 2016 from https://www.asep.org/asep/asep/SCIENCEofLEADERSHIP.html

10. Adapted from Mind Tools YouTube Channel, retrieved Aug. 11, 2016 from https://www.mindtools.com/pages/article/newHTE_90.htm

11. "Personal Goals," p. 2.

12. Ibid.

13. Tracy, *Maximum Achievement*, p. 160.

14. Ibid.

15. Quoted in Brainy Quote website, retrieved Aug. 11, 2016 from http://www.brainyquote.com/quotes/quotes/t/theodorero100965.html.

16. Ziglar, *The "I Can" Course Learner's Manual*, p. 176.

17. Tracy, *Maximum Achievement*.

18. Ibid., p. 167.

19. Noe, *People Power*.

20. Ibid., pp. 153–154.

21. Ziglar, *The "I Can" Course Learner's Manual*, p. 178.

22. Tracy, *Maximum Achievement*, p. 167.

23. Zig Ziglar, *See You at the Top*, 25th anniversary rev. ed. (Gretna, LA: Pelican Publishing, 2010), p. 207.

24. Ibid., p. 173.

25. J. M. Kouzes and B. Z. Posner, *The Leadership Challenge: How to Get Extraordinary Things Done in Organizations* (San Francisco: Jossey-Bass, 1987), pp. 259–276. (5th ed., 2012).

26. "Goal-setting: Self-Esteem in Action" (Lubbock, TX: Creative Educational Video, 1989).

27. Tracy, *Maximum Achievement*, pp. 153–155.

28. http://www.ffa.org (accessed November 3, 2016).

29. National FFA Association, "Chapter Management and Resources"; available at http://www.ffa.org/index.cfm?method=c_aged.chapters#poa.

30. Ibid.

31. Ibid.

15 TIME MANAGEMENT

Leaders who manage their time well are more productive leaders. Time management involves setting goals, balancing the various parts of your life, setting priorities, and exercising self-discipline.

Objectives

After completing this chapter, the student should be able to:

- Discuss the four levels of time management
- Differentiate between urgent and important
- Explain ultimate time management for quality leaders
- Set and prioritize personal and professional goals
- Discuss the importance of balancing the major categories of personal and professional time
- Analyze your use of time with a daily log
- Describe time wasters
- Manage time more effectively and efficiently

Terms to Know

- time management
- prioritization
- urgent
- important
- matrix
- coherence
- portability
- personal time
- professional time
- socializing
- daily log
- Pareto's principle
- procrastination
- qualifier
- prime time
- accountability

Did you ever get to the end of a busy day, when you were tired and looking forward to going to bed, and then realize that you had a five-page assignment due for history class the following morning? Where did the time go? Looking back over your day, you think, "There just wasn't enough time to get everything done!" You might know those who seem to be involved in everything—student council, basketball team, band, academic Olympics, a local youth group—and yet are on the honor roll every time report cards come out! How do those people do it?

People who seem to have an endless supply of energy do not have any more time to get things done than you do; they have just learned to become good managers of their time. **Time management** is planning how to control your time so that you can do the things you need and want to do. Good managers of time have the same 24 hours per day as everyone else. The only difference is that they have taken the time to decide how to control what they will do in that 24-hour period. Time is your most valuable personal resource; use it wisely, because it cannot be replaced or recovered.

FOUR LEVELS OF TIME MANAGEMENT

People are always looking for ways to do things more efficiently. Time management is no exception. Each generation, or level, of time management builds on the one before it and helps us gain greater control of our lives. Each level is important, but the more effective leaders operate at the fourth level. The four levels or generations are as follows.[1]

1. **Notes and checklists** are the first level of time management. Keeping track of what we have to do can help us meet and manage the many demands placed on our time and energy.
2. **Calendars and appointment books,** the second level, help us schedule events and activities in the future. This also alerts us to possible overloading and conflicts.
3. Third-generation time management includes everything in the preceding levels, but adds the important concepts of **prioritization** (arranging or dealing with in order of importance), clarification of values, comparison of the worth of various activities based on our values, and goal setting (long range, medium range, and immediate). This level also includes making a plan to accomplish specific goals and objectives on a daily basis.
4. The fourth generation of time management consists of **managing ourselves** rather than just managing time. Rather than focusing on things and deadlines, fourth-level expectations focus on improving personal and professional results and relationships—perhaps over a meal, as shown in Figure 15–1.

FIGURE 15–1 Fourth-generation time management focuses on preserving and enhancing relationships. These friends have decided that taking time to eat together is an important activity. *(© iStock/monkeybusinessimages)*

URGENT VERSUS IMPORTANT

Urgent

Activities can be described as either urgent or important. **Urgent** matters require immediate attention. Urgent matters are usually visible, right in front of us, and demanding action. Often they are pleasant and easy to do—but frequently they are unimportant. For example, three phone lines are ringing, demanding immediate (urgent) attention, but one is for a lunch appointment with a friend, the second is to schedule a golf game, and the third is a call asking the time of the charity softball game next month. Urgent matters do not help us reach our goals and are usually time wasters.

Important

Important matters have great value, significance, or consequence and have to do with results. If something is important, it contributes to attainment of your goals. We react to urgent matters, but important matters that are not urgent require more initiative and planning. We must act to take advantage of opportunity and make things happen. If we do not have a clear idea of what is important—of the results we want in our lives—we are easily distracted and channeled into responding to the urgent. An example of a matter that is both *urgent* and *important* is completing a scholarship application in time to meet a deadline. An example of an important matter is daily exercise, such as walking two miles. It is not *urgent:* no one is in a crisis; there is no pressing problem or deadline to be met; it does not scream to be done right away. However, it is *important* if your long-range goal of maintaining good health is to be met.

Time-Management Matrix

In *The 7 Habits of Highly Effective People*, Stephen Covey presents the time-management box, or **matrix**, a rectangular array or network of intersections (Figure 15–2).

Area 1: Get It Done—No Debate These matters are both urgent and important, being significant problems that require immediate attention (crises). Leaders who operate in area 1 are crisis managers, problem-minded people, and deadline-driven producers. Leaders who focus on area 1 allow the matters in this area to dominate them.[2] Leading out of urgency is not good—but it is often unavoidable. People who operate in area 1 too much need to plan so that they manage their time instead of allowing time to manage them.

Area 2: Relevant in the Long Run This is the area from which real leaders operate. This area is all about activities that are not urgent, but important. Activities such as maintaining friendships or your health, strategic planning, focusing on prevention of problems, or preparing for the future. This area is full of activities that we know we need to do, but that get pushed to the side for lack of time. Effective leaders who operate in area 2 are opportunity minded rather than

TIME-MANAGEMENT MATRIX		
	URGENT	**NOT URGENT**
Important	1. Get it done—no debate • Crises • Pressing problems • Deadline-driven projects	2. Relevant in the long run • Relationship building • Recognizing new opportunities • Planning, recreation
Not Important	3. Nagging time wasters that require attention • Interruptions, some calls, mail, reports, meetings, most e-mail	4. Pleasant busywork to stay organized • Trivia • Some mail, some e-mail • Some phone calls, some answering-machine messages

FIGURE 15–2 The four areas of the time-management matrix. Area 1 activities are important and urgent, such as meeting deadlines. Area 2 activities are important but not urgent, such as exercising for health reasons. Area 3 activities are urgent but unimportant (e.g., meetings, most email). Area 4 activities are neither urgent nor important.

problem minded. "They feed opportunities and starve problems. They think preventively." The result is vision, perspective, balance, discipline, control, and few crises.[3]

Area 3: Nagging Time Wasters That Require Attention Some leaders spend a great deal of time on urgent, but not important, area 3 activities, thinking that they are operating in area 1. They spend most of their time reacting to things that are urgent and assuming that these things are also important. However, the urgency of these matters is often based on the priorities and expectations of others. Some call this "majoring in the minors." Examples of area 3 activities are any interruptions, some phone calls, mail, reports, meetings, and most email that do not help accomplish any of your long-term goals.

Area 4: Pleasant Busywork to Stay Organized Area 4 provides a place of escape or relief for leaders who are besieged by problems all day, every day. They turn to the unimportant, not urgent activities of area 4: trivia and pleasant activities such as reading through brochures or intriguing junk mail. Although leaders should avoid this area as much as possible, there are times when you need to unwind to release stress, especially between major projects.[4]

Reality versus Ideal

Most leaders should spend most of their time in area 2, on important matters that will lead to professional and personal growth. In fact, most leaders spend most of their time in area 1, dealing with crises and deadlines; considerable time in area 3 with nagging time wasters that require attention; and the rest of their time in area 4, mentally regrouping while doing trivial tasks. No matter how hard you try, it is probably impossible to keep from spending some time in each area. However, leaders should strive to increase their time in area 2 where they will be most productive.

Make a Positive Difference in Your Life

Consider this question in relation to the time-management matrix. What one thing could you do (that you aren't doing now) that would make a tremendous positive difference in your personal or professional life if done on a regular basis? More than likely, your answer would fall into area 2—something that is obviously important but not urgent. Because they are not urgent, such things often do not get done.

Whether you are a high school student, a university student, an assembly-line worker, or a

---

VIGNETTE 15-1 | **RELEVANT IN THE VERY LONG RUN**

TIME MANAGEMENT IS IMPORTANT IN ALL ASPECTS OF OUR LIVES

For John, a man who formerly weighed 300 pounds, time management was his key to losing more than 100 pounds. One day he just decided that taking the extra time to exercise was very important. John was not at death's door, so it was not urgent, but it was very important that he lose weight. John began operating in area 2 of Covey's time-management matrix.

Every day, instead of going to lunch and taking a break, John hit the track. First he started walking, and then he started running a mile or two. The next thing he knew, John was running five to seven miles at lunchtime. He was losing weight and feeling great. His goal of losing weight and becoming healthy was very important to him. Because he was patient, John was able to get to a very desirable weight.

Because John was healthier than he had ever been in his life, and because many of his friends, family, and coworkers were supporting him, he chose to commit more time to running. He decided to train for a marathon. That's right, a 26.2-mile marathon. Talk about relevant in the long run! John committed hours each day and weekend to his goal, and the rewards were great. John finished not one but three marathons before he decided to rest his legs.

Time management was very important to John's life, but it was also very important to many others. The local newspaper did an article on his success, and friends, family, and many others were so inspired by John's success that they made commitments to improving their own quality of life, by dedicating themselves to accomplish goals of their own.

John is a leader not only because he has his priorities in order and is a good time manager but also because he is able to influence many other people. We encourage you to use your time wisely. Make a commitment to dedicate daily time to something productive that will change your life. Find something that is relevant in the long run. You are already a leader, so you should not have any trouble deciding on a goal of this nature.

professional, the same principles apply. Those who operate in area 2 increase their effectiveness dramatically. Crises and problems shrink to manageable proportions because you are thinking ahead, planning, and taking preventive action to keep situations from developing into crises in the first place. You take control of your time instead of letting time take control of you.

ULTIMATE EFFECTIVE TIME MANAGEMENT FOR LEADERS

Leaders put first things first; then they proceed to organize and execute around those things. The best leaders and time managers (operating in area 2) need to meet six important criteria, according to Covey:

- **Coherence.** This is an orderly or logical relation of parts that affords comprehension or recognition. There is harmony among your roles and categories, goals and objectives, things to do, priorities, and scheduling.
- **Balance.** Keep balance in your life so that you do not neglect the important areas of health, family, school or professional preparation, and community.
- **Focus.** Stay focused on important issues so that you are preventing rather than prioritizing crises.
- **People-centeredness.** A good leader deals with people effectively and humanely.
- **Flexibility.** Keep some flexibility in your schedule to allow for the unexpected.
- **Portability.** Carry a portable organizer that is able to be moved easily.[5]

SETTING AND PRIORITIZING PERSONAL AND PROFESSIONAL GOALS

We discussed goals in the previous chapter, but it is hard not to mention goals again in relation to time management. We cannot manage our time correctly and efficiently if we do not know what our goals are. Without goals, we cannot know whether our activities are furthering our interests.

Goals Provide Direction for Time Management

What do you want to do with your life in terms of short-range goals—the next 24 hours, the next week, the next month? How about your long-range goals—a year, 5 years, or 10 years from now? The words *goal* and *objective* are often used interchangeably, but a *goal* is an end that you strive to attain or reach in the long term. An *objective* is a specific end that can be reached in the short term. You may achieve several objectives in the process of attaining your long-range goal. For instance, you have to get certain grades in particular courses to maintain a grade point average that will earn you a high school diploma. You must work to achieve a good grade in each class as part of reaching the goal of obtaining a high school diploma.

Goals are necessary to give direction to your life. Without them, it would be difficult, if not impossible, to manage your time. Consider the statements on goals in Figure 15–3.

Periodically Adjust Goals and Time Commitments

Keep in mind that to manage your time you will have to update your goals occasionally. Former long-range goals may become short-range goals, and short-range goals may disappear from your list as you attain them. People also change their minds. In high school a student may wish to go on to college and become a veterinarian. After attending college for a year, though, she might get interested in another field, such as biotechnology. It is perfectly acceptable to change your mind and your direction, as long as you realize

---

GOALS AND THE FOUR BASIC QUESTIONS

- Goals develop **knowledge**, which answers the question *what?*
- Purpose develops **understanding**, which answers the question *why?*
- Planning and strategy develop **wisdom**, which answers the question *how?*
- Priorities develop **timing**, which answers the question *when?*

FIGURE 15–3 Goals are the beginning of the journey to your destination. Along the way, you must gain knowledge, plan, and determine your priorities, which means managing time efficiently.

that the change will require you to set new goals and revise how you spend and manage your time. A new goal may mean spending additional time on education or preparation for a career.

BALANCING PERSONAL AND PROFESSIONAL TIME

Many of the categories of **personal time** (free or leisure time) and **professional time** (time spent at or relating to school or employment) overlap. They are all parts of the whole and are interdependent. To be the best time manager possible, you must take care of your whole self. In other words, time management does not apply only to our professional lives; it applies to and affects our personal lives as well. We must manage our time in all areas to meet our goals in life. Time spent on various activities can be divided into four major categories: health, family, community (personal time), and school or employment (professional time). Each category is important. If any one of them is neglected, there will be adverse effects in the other categories.

Health

Taking care of yourself takes up much time, but it is time well spent. No health, no work, no life. You will get more done if your body is healthy, and a healthy body improves your mental, emotional, and spiritual well-being. The way you take care of your health can add years to your life or steal them away.

Sleep Most people need a good eight hours of sleep in every 24-hour period to function at their best. This amount of time varies from person to person and often changes as you get older. No matter what your norm is, when you skip or lose sleep, your immune system does not have time to recharge, which makes you more prone to infection and illness.

Diet People who eat a well-balanced diet and maintain their proper weight experience fewer energy "burnouts." When you eat better, you feel better, work better, and are better able to achieve your time-management objectives. The old "Garbage in, garbage out" saying applies aptly to diet. Substances such as caffeine can adversely affect your long-term performance; caffeine is a quick stimulant that can eventually act as a depressant. It is best not to overindulge in any one type of food. Instead, eat a balanced and varied diet (Figure 15–4).[6]

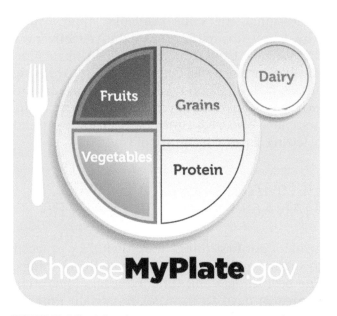

FIGURE 15–4 Good diet gives you more energy to meet your time management objectives. The MyPlate figure is a great reminder of how to eat a balanced diet.
Source: USDA's Center for Nutrition Policy and Promotion, www.choosemyplate.gov

Exercise The American Medical Association has said that you can dramatically reduce your risk of death due to heart disease, cancer, and other diseases with only moderate exercise, such as 30 minutes of brisk walking, or some other aerobic exercise, at least every other day.[7] Adequate exercise, a balanced diet, and enough sleep will give you the energy and stamina needed to finish what you start and accomplish your daily goals.

Recreation Good health does not pertain merely to the physical; it also includes your mental, emotional, and spiritual self. You need to set aside time for recreation, not only for the body but also for the mind. Sports, games, reading a good book, watching a play, listening to music, or engaging in the arts, special interests, and hobbies all play an important part in achieving good health. A relaxed body leads to a relaxed mind, which can make your use of time more productive.

Socializing Spending time with friends and others who have common interests is important for developing interpersonal skills. Skills in **socializing** (learning to get along with others) can help you as you develop a career and new interests.

Spiritual Needs Many researchers have found that people who spend a certain amount of time attending to their spiritual needs tend to have long, healthy, and happy lives. Morals, ethics, and positive lifestyles, often influenced by various faiths, are important for happy, well-rounded perspectives on life. The amount of time spent on spiritual matters and activities depends on the needs of the individual person.

Family

Time spent with your family is also important. A doctor once said, "I have never heard a dying man say that he wished he had spent more time at the office." For most of us, when all is said and done, families are the most important thing. Families instill a sense of belonging and have always been an essential part of human survival. Older people teach younger members of a family how to take care of themselves and how to be part of a strong group.

Responsibility and cooperation are two important skills that can be developed in the family setting (Figure 15–5). Many families divide up the chores of cooking, cleaning, and home maintenance. These are skills that high school youth need for independence when they leave their homes to start their own lives as adults.

Community

We have obligations as community members. It is part of the American dream to make our communities better places to live. Well-rounded individuals participate in groups and activities whose goals include community improvement. This may mean being a part of a neighborhood watch, helping to establish and maintain parks and libraries, or working with youth groups. Charitable work for nonprofit organizations is another means of improving your community. What if no one budgeted time in this area? The friends in Figure 15–6 are contributing their time through a community service project to clean up trash around the shore.

Employment

Employment is also an important time category. Aside from meeting our need for income, employment can greatly enhance personal growth and development. The luckiest people are those who have a job doing something they enjoy. It is best to choose a career that you enjoy and find interesting. Much of the helpful time-management advice in this chapter centers around the workplace.

FIGURE 15–5 This family enjoys working together on their fruit and vegetable farm. Families are closer when they have similar goals and objectives. *(© Chinaview/Shutterstock.com)*

FIGURE 15–6 Being a contributing member of your community should be a goal for all of us. Our communities will be better places to live if we all contribute. *(© iStock.com/Jason Doiy)*

ANALYZING YOUR USE OF TIME WITH A DAILY LOG

Before you decide on changes to improve your use of time, you may want to find out where your time is actually going. Seeing how you have used your time in the past helps to make planning your future use of time easier.

Keep a Daily Log

If you keep a **daily log** (diary; record and schedule of activities and appointments throughout the day) of everything you do for at least one week, you can then analyze how you have been managing (or mismanaging) your time and make the necessary changes. Take this log with you everywhere you go. Mark down in 15-minute intervals how you spend your time. Do not wait until the end of the day to record in this log; record activities as you do them. For example, "6 A.M., woke up, took shower, and dressed; 6:15, ate breakfast; 6:30, caught school bus; 6:45, at school, studied for test first period." Continue making brief notes every 15 minutes until you go to sleep that night. Continue this daily log for at least one

week. You may want to use a log sheet similar to the one in Figure 15–7. Using an electronic calendar system (such as Google Calendar) is also an excellent way to keep track of your time if you prefer to incorporate technology into your time-management analysis.

Pareto's Principle

You may be surprised to find out where all your time went! You might have had the best of intentions to get an outline done for a research paper one night, only to find that 10:00 P.M. came and went and you had been on the phone since 7:00. Watch out for the time wasters that can keep you from doing what is important for reaching your goals. You will probably find that 80 percent of the things you do are low-priority tasks. The "80–20" rule, or **Pareto's principle**, applies to many different situations in life. The 20 percent of tasks that are high priority (important rather than urgent) can generate 80 percent of the results you need to reach your goals.[8]

Pareto's principle is also called the **time-value ratio**. We need to put first things first. Is what you are doing worth the time it is taking to do

DAILY LOG SHEET	
Day_____	Date_____
6:00	2:00
6:15	2:15
6:30	2:30
6:45	2:45
7:00	3:00
7:15	3:15
7:30	3:30
7:45	3:45
8:00	4:00
8:15	4:15
8:30	4:30
8:45	4:45
9:00	5:00
9:15	5:15
9:30	5:30
9:45	5:45
10:00	6:00
10:15	6:15
10:30	6:30
10:45	6:45
11:00	7:00
11:15	7:15
11:30	7:30
11:45	7:45
12:00	8:00
12:15	8:15
12:30	8:30
12:45	8:45
1:00	9:00
1:15	9:15
1:30	9:30
1:45	9:45
	10:00

FIGURE 15–7 Completing a daily log such as this one will help you analyze how you are spending your time.

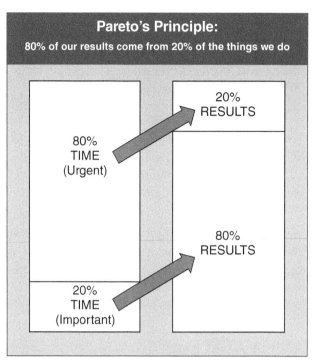

FIGURE 15–8 Urgent things do not lead to achieving your goals. Something is important when it leads to achievement of your goals. Twenty percent of your tasks (the high-priority ones) can generate 80 percent of the results you need to reach your goals.

and effective (doing the right task). Figure 15–8 is a chart illustrating Pareto's principle.

Analyzing Your Time Logs

After keeping time logs for a week, you can analyze them by taking the following actions:

- Review the time logs to determine how much time you are spending on your primary responsibilities. How do you spend most of your time?
- Identify areas in which you are spending too much time.
- Identify areas in which you are not spending enough time.
- Identify major interruptions that keep you from doing what you want to get done. How can you eliminate these interruptions?
- Identify tasks you are performing that you do not have to be involved with. If you are a manager, look for nonmanagement tasks. To whom can you delegate these tasks?
- How much time is controlled by others? How much time do you actually control? How can you gain more control of your own time?

it? You increase the return on your investment of time and energy if you work on the important things. Something is important when it leads to the achievement of your goals. We have to learn to be both efficient (doing the task the right way)

- Look for crisis situations. Were they caused by something you did or did not do? Do you frequently encounter recurring crises? How can you eliminate recurring crises?
- Track habits, patterns, and tendencies. Do they help or hinder your job accomplishment? How can you change them to work to your advantage?
- List your three to five biggest time wasters. What can you do to eliminate them?
- Determine how you can manage your time more efficiently.[9]

You can then use this daily log to help you better plan your weekly and daily activities (discussed later in the chapter). First, let us examine some of the common time wasters that occasionally hinder and entrap us.

TIME WASTERS

Procrastination

It is one thing to identify a time waster and an entirely different matter to get rid of it. One of the worst and most widespread habits in time usage is **procrastination** (putting off doing something until a later time). Procrastination often takes the form of doing a low-priority (and often enjoyable) task instead of tackling a high-priority, difficult, or boring task. If you are avoiding doing something you do not need to do, no problem. However, if the task you are avoiding is one you need to complete to reach an important and necessary goal, procrastination can really hold you back.

Lack of Understanding and Direction The hardest part of any job is getting started. Similarly, once you can get beyond procrastination, you will have learned much and become able to accomplish a great deal. Why do people procrastinate? Many people procrastinate because they do not understand what they are supposed to do or because they do not really know how to attack what seems to be a huge job.

These seemingly undesirable tasks are often imposed on the procrastinator by an outside party. A common example for students is a major research paper for one of their classes. Often these are long-term projects that require many hours of work over several weeks. To make this task less intimidating, the procrastinator needs to get specific directions from the instructor on how to complete the task. A large project is usually best approached by breaking it down into smaller tasks that can each be completed in one sitting or work session. These smaller tasks might include selecting a topic, doing library research, compiling a bibliography, drafting an introduction and a conclusion, outlining the research paper, writing paragraphs to go under the various subheadings in the outline, and writing footnotes or endnotes.

Identify Short-Term Rewards for Accomplishing Tasks Another aid in eliminating procrastination is to sit down and list the rewards for finishing each stage of the research paper on time and the consequences for not finishing or even doing each stage on time. B. F. Skinner, a famous psychologist, found out through original laboratory research that positive reinforcement or rewards for desirable behavior (finishing each stage of the paper on time) worked better than suffering punishment for or the adverse consequences of undesirable behavior (not finishing each stage of the research paper on time). Promising yourself that you can surf the Web for 30 minutes if you first complete all of the library research for your paper is much more likely to change your behavior in a positive manner than not eating dessert that night because you did not do your homework.

Identify Long-Term Rewards for Accomplishing Tasks If you do well on your research paper, you will receive a good grade on the assignment, which will help you to get a good grade in the course. If you learn to make good grades in one class, that could lead to making good grades in other subjects. If you graduate with a high overall grade point average, you are more likely to receive scholarships to attend the college of your choice. If you attend college and do well in your studies, this could help you get started in a career you enjoy and make a desirable salary. Conversely, if you do not learn how to complete papers well and on time, the ultimate consequence may be that you do not get the job you wanted. The habit of procrastination can sabotage your efforts and accomplishments in the workplace, too, so do your best to kick the habit early!

Characteristics of a Procrastinator Procrastinators always think they have more time tomorrow. When faced with a difficult task, procrastinators wait for a better time. They do not start a project unless they can finish it. When procrastinators are facing an important but difficult project, they spend their time doing other things rather than getting started. Procrastinators tend to be perfectionists. They also have trouble with priorities and lack self-confidence; they do not plan; and they overcommit themselves.

Overcoming Procrastination To overcome procrastination, do the following:

- Do it now. At the very least, start it now.
- Take small steps; break tasks into doable chunks.
- Do the easy parts first.
- Tell someone else about your commitment to the task and the deadline.
- Whenever possible, find and use time-savers; reduce the amount of time it takes to complete a task.
- Plan to do work at a time when you are at your best and least fatigued; know when you tend to be the most productive and schedule accordingly.
- Reward yourself.

The Telephone and Cell Phone

Another big time waster is your cell phone. You might say, "But I didn't text him; he sent me a text!" Let family and friends know that you must spend so many hours a week doing homework and taking care of other responsibilities. Try setting aside a specific time each day for studying, and ask people ahead of time not to interrupt you during that time. With tact, you can handle this appropriately, and your request will be respected.

Terminating Conversations When you have more important things you need to be doing, there are several things you can say—politely, of course—to terminate phone conversations. Refer to Figure 15–9 for help on ending conversations.

Another way to minimize your time on the phone is to stand while you are talking; this helps you be aware of how long you have been talking so you can keep the call brief. You can use a clock, the timer on the microwave oven, or any other timer; set it for no more than 15 minutes. When you hear the timer go off, it is time to say goodbye.

POLITE CONVERSATION TERMINATORS

"I can't talk right now . . ."

"Can I get back to you?"

"I have company right now, so I need to call you back later."

"I have to go now . . ."

"I can't talk much longer . . ."

"I was just on my way out the door . . ."

"I know you're busy, so I'll let you get back to what you were doing."

"Well, I'm going to have to let you go . . ."

"Right now is my study time, but I can call you back later."

FIGURE 15–9 Sometimes we just do not have time to talk on the telephone, but we do not want to be rude. These tips should help you end telephone conversations effectively but tactfully.

If you cannot afford the time even to tell people that you will call back later, let them leave a message. (If they do not leave a message, they probably did not really need to talk to you in the first place.) Listen to the message to make sure it is not an emergency. If it is not an emergency, call back when you have scheduled time to talk or text.

Positive Phone Use Your phone can also be used as a positive, time-saving tool. For example, you are not able to remember how to solve a certain math problem, so you call a friend for help. Instead of having to go to your friend's house and consult in person, a quick phone call saves you both time.

It has even been suggested that some businesspeople should disable the ringer on their phones. This emphasizes that a phone is for their convenience for outgoing calls—it is not an instrument with which to be disturbed. The phone is a tool for your use; do not become a slave to its every signal! Consider the following sample phone policy.

A Manager's Guide for Handling Incoming Calls

Incoming calls

- Log your calls according to their priority.
- Redirect calls to others better suited to handle them. "Coach" callers to do it themselves.

- Automate a callback system.
- Gather name, number, best and worst times to return the call, and the nature of the call.

Outgoing calls

- Keep the numbers you need to call often logged in your phone.
- For group or conference calls, good times are usually after lunch, mid-morning, or midafternoon. This obviously depends on your work (e.g., sales).
- Have materials at hand, with the agenda for conversation outlined.
- Ask, "Are you free to talk?" Don't assume that now is a good time for the other person to talk; remember the power of public relations and diplomacy.

Alternative communication systems

- Voice mail
- Phone message center
- Pager or answering service[10]
- Email
- Social networking (e.g., Facebook, Twitter, Instagram, Snapchat)

The Internet

The Internet has changed the world. Some people believe that you can do anything on the Internet. If this is true, then it stands to reason that we can waste time there as well. Useful, informational websites are plentiful, but what starts as learning could turn into a "time bomb." On the Internet, you can make purchases, do research, conduct business, email, and talk (chat). You can watch television shows, view movies, and network with your friends. Although many Internet capabilities and opportunities are very useful and capable of helping us meet our goals, we have to be careful to keep our focus and purposes in mind. It is extremely easy to get sidetracked while using the Internet.

Social Networking

Social networking sites have become a preferred form of communication for many individuals. There are now several different ways to communicate, but people like these sites because they are also a form of entertainment. It is easy to spend many hours finding old and new friends and learning all about them by reading their profiles, viewing the pictures they have posted, and reading about the causes they support. You may also dedicate a surprising amount of time to updating and maintaining your own pages. As with the phone, it is important to manage your time wisely with regard to social networking sites. Limit the time you spend on social networking sites, but do not dismiss them as complete wastes of time. They could actually help you be more efficient in finding people, getting assistance, or communicating quickly.

Television

Television is another media form that can help or harm you. If you have attained all your higher-priority goals for the day, you might watch TV as a form of recreation or socialization. Television can be entertaining and educational; it can also turn you into a "couch potato" who risks poor health by not exercising and numb your brain and critical faculties by broadcasting only what the advertisers and TV executives choose to present. Studies report that teenagers typically watch between 20 and 50 hours of TV per week. If you are spending several hours a day watching TV, you may need to reassess how your time is spent and get back to a schedule that is balanced among the five major categories of personal and professional time.

Inability to Say No

If you think you cannot get everything done that you are supposed to because you have too much to do, that is probably true. Good time managers keep a portable calendar/planner with them and record all their obligations. If you find that you have too many things to do, you need to back up and regroup. Reprioritize your goals. Do what is most important to you; do what you need to achieve to reach *your* goals, not someone else's. You also have to realize that there are only so many hours in a day; there will always be times when someone wants you to do something and you will have to tell that person no. When you say no, say it early, be willing to take a risk, and understand the benefits of saying no.

Say No in a Positive Way Refusals can be made in a positive, nonthreatening way that prevents hurt feelings and angry misunderstandings. You need to be honest when people ask you to help them and you just do not have the time to do it.

For instance, you are asked to work on a committee to decorate for an upcoming school party, but you have already obligated yourself to fund-raising for that party and practicing for an upcoming debate contest. You can say no politely with, "I'm flattered you want me to help you, but I'm already committed to doing two other things." Most people will understand and respect your honesty and integrity. Realize that you may upset the requester far more if you take on too much and end up dropping the ball later on.

Use Qualifiers You can also say no with a **qualifier**, which is a word or phrase that qualifies, limits, or modifies something. An example is, "No, I can't help you this time with the fall dance, but I can help you later with the spring dance." When you must refuse, you do not want to offend anyone or burn your bridges behind you. Learning how to deal with situations such as these in high school can help you deal with similar situations when you start working full time. There may come a time when you need to call on those same people to help you with something. People are more inclined to agree to help you when you have said yes and helped them in the past. This reciprocal assistance is more than just "calling in markers"; rather, it is part of interpersonal networking with people whom you know and trust and who know and trust you.

Saying No and Following Through Do not say yes when you know you cannot or will not follow through on the commitment. If you do, you will lose your credibility. If a task or job is worth doing, it is worth doing to the best of your ability. You cannot devote the proper attention to each task or priority if you simply have too many of them. If someone becomes upset or hurt by your polite refusal, you can point out that you would much rather disappoint them a little bit now rather than wreck their project by being unable to fulfill your commitment later on, when it is too late to find someone else or make other arrangements. Do not let people pressure you into overextending yourself. The stress incurred by trying to do too many things at once is not worth it.[11]

Other Time Wasters

There are many other time wasters out there (and almost anything can become a time waster if you let it). Some of the most common are

- Allowing work to expand to fit the time available
- Inertia
- Unclear goals and objectives
- Lack of priorities
- Over-commitment
- Fear of failure
- Lack of organization
- Poor communication
- Failure to listen
- Lack of delegation
- Perfectionism
- Haste (if you do not take time to get it right initially, where will you find the time to do it over?)
- Socializing
- Complaining about having too much to do instead of doing something about it
- Daydreaming
- Committees
- Crises
- Junk mail
- Unnecessary correspondence
- Email
- Chat rooms
- Instant messaging/texting
- Web surfing

You can control almost all of these time wasters if you commit to doing so and exercise some self-discipline. The next section discusses ways to take charge of your time and your actions.

MANAGING TIME MORE EFFECTIVELY AND EFFICIENTLY

There is never enough time to do all the things we want and need to do. We must choose between competing demands on our time. Because we make these choices, we are ultimately responsible for how we spend our time. Procrastination can take the form of filling your time unnecessarily. Some small tasks need not be completed to perfection, and some need not be done at all. Learn to recognize when you are overcompleting tasks (i.e., spending too much time on them). You may use far too much time for far too little result.[12] Manage time both effectively (by doing the right task) and efficiently (by doing the task the right way).

We have already discussed the importance of plans, goals, and objectives. To manage time well, we must also make and keep schedules, adapt daily, construct to-do lists, set priorities, and delegate.

Scheduling

Look at the week ahead with your goals in mind, and schedule time to achieve them. Having identified roles and categories and set goals and objectives, you can translate each goal and objective to a specific day of the week, either as a "thing to do" or, even better, as a specific time or appointment. Refer to Figure 15–10 for an example of a scheduling calendar with categories, things to do, and priorities. In this sample, spots are also available for people/relationship goals, creativity/ideas, and greatest accomplishment of the day.

Scheduling Weekly If your goal is to stay in shape through exercise, you may want to set aside 30 minutes every day (or 3.5 hours per week) to accomplish that goal. There are some goals that you may be able to accomplish only during work hours and some that you can do only on weekends. Be sure to schedule other things that are important in your life, such as reading, family time, and entertainment. As you will see, there are advantages to organizing your week.

Scheduling Daily Successful leaders also have daily schedules. At the end of each day, you should complete your schedule for the next day. This should take 15 minutes or less. Fill in time slots that are not already scheduled. Schedule your goals and objectives of the week and your "things-to-do" list each day. Allow enough time to do each task. Many leaders find that they get accurate estimates of the time it will take to perform a nonroutine task by doubling an initial estimate.[13]

Schedule High-Priority Items during Your Prime Time Prime time refers to the hours of the day when we get the most done. For many people, this is early morning. However, some of us are slow starters or night people who perform better later in the day. Determine your prime time, and schedule the tasks that require your full attention during that time. Do routine things outside of your prime-time

WEEKLY PLANNING, ORGANIZING, SCHEDULING CALENDAR

ROLES OR CATEGORIES	WEEKLY GOALS/ OBJECTIVES	WEEK OF: WEEKLY PRIORITIES	SUNDAY	MONDAY	TUESDAY	WEDNESDAY	THURSDAY	FRIDAY	SATURDAY
Professional or School						TODAY'S THINGS TO DO AND PRIORITIES			
Family									
Community									
Health									
My Time									
Meaningful People/Relationship Goals									
Creativity/ Ideas									

SHEDULE OF APPOINTMENTS/COMMITMENTS

	SUNDAY	MONDAY	TUESDAY	WEDNESDAY	THURSDAY	FRIDAY	SATURDAY
	8	8	8	8	8	8	8
	9	9	9	9	9	9	9
	10	10	10	10	10	10	10
	11	11	11	11	11	11	11
	12	12	12	12	12	12	12
	1	1	1	1	1	1	1
	2	2	2	2	2	2	2
	3	3	3	3	3	3	3
	4	4	4	4	4	4	4
	5	5	5	5	5	5	5
	6	6	6	6	6	6	6
	Evening	Evening	Evening	Evening	Evening	Evening	Evening
	7	7	7	7	7	7	7
	8	8	8	8	8	8	8
	9	9	9	9	9	9	9
	10	10	10	10	10	10	10

MY GREATEST ACCOMPLISHMENT TODAY

FIGURE 15–10 Many types of time-management calendars are available. This sample is useful because it provides space for goals and objectives for professional or school, family, community, health, personal time, meaningful people or relationship goals, and creativity or ideas.

hours.[14] For example, set a specific time when people can call you (or you can call them).

Schedule a Time for Unexpected Events Regardless of how well you plan, unforeseen things will come up. Therefore, expect the unexpected. Do not do an unscheduled task before a scheduled task without prioritizing it first. If you are working on a high-priority item, and a medium-priority item is brought to you, let the new matter wait. As discussed earlier, often the so-called urgent things can wait. Of course, if your superior says, "Do it," you are best advised to do it regardless of what your time-management plan says!

Daily Adapting

Because the best leader organizes by weeks, daily planning becomes more a process of daily adapting: prioritizing activities and responding meaningfully to unanticipated events, relationships, and experiences.[15]

Using Things-to-Do Lists to Reach Daily Goals

In essence, a things-to-do list is a list of objectives. If we reach our objectives, we reach our goals. If we reach our goals, we are successful. Being successful makes us happy, and if we are happy, we feel good about ourselves. Preparing and completing items on things-to-do lists give you a sense of accomplishment that affects your attitude. Lists help you stay focused and finish the things that you need to do daily.

To use a things-to-do list effectively, record each activity that must be done and assign it a priority (Figure 15–11). Remember that priorities often change during the day because of unexpected events that must be added to your list. Start with the high-priority (H) activities by performing the most important one. When it is completed, mark it as finished and select the next, until you have completed all your high-priority activities. Then do the same with the medium (M) priorities, then the low (L) priorities. Make sure that you update your priorities. Deadlines usually drive one's priorities: with time, low priorities often move up to become high priorities.[16]

Each day, you can either write up a new to-do list or simply update an electronic list if you are keeping it on a computer or smart phone. Yesterday's low priorities, if not completed, become the next day's medium or high priorities. The important thing to remember is to leave some flexibility in the schedule. Failing to account for unforeseen circumstances and occurrences can land you facing the hard truths of Murphy's Law: (1) nothing is as simple as it seems; (2) everything takes longer than it should; and (3) if anything can go wrong, it will. Good planners are rarely tripped up by Murphy's Law.[17]

Setting Priorities

At any given time we are faced with many different tasks. One thing that separates successful from unsuccessful people is their ability to do the important things (priorities) first and the less important things later. However, to do what is really important, you must know the priority for each item and work according to those assignments.

Priorities should be determined according to your major responsibilities. To determine priorities, a good leader should answer the following questions:

1. Must I be personally involved because of my unique knowledge or skills?
2. Does the task fall within my primary area of responsibility?
3. When is the deadline? Is quick action needed? Should I work on this activity right away, or can it wait? Time is relative: In one situation, months or even a year may be considered quick action, but in another situation only a matter of minutes may constitute quick action.[18]
4. Can I delegate this?

Delegation

We accomplish all that we do through delegation—either to time or to other people. "If we delegate to time, we think *efficiency*. If we delegate to other people, we think *effectiveness*. Many people refuse to delegate to others because they feel it takes too much time and effort, and they could do the job better

THINGS TO DO BY PRIORITIES	Delegate	High (H)	Medium (M)	Low (L)	Deadline
DATE:					
ACTIVITY					

FIGURE 15–11 In reality, a things-to-do list is a list of objectives. By delegating or assigning priorities to each, we reach those objectives more efficiently.

themselves."[19] Also, some people do not delegate because they have no experience in delegating work, they have no confidence in their subordinates, or they are perfectionists. Some perceived leaders are threatened by the possibility of someone else doing it better, because they believe this would make them look bad or incompetent.

Why Delegate? Besides helping others to develop, delegation is the key to the leader's sanity. You cannot do it all yourself. The more you try, the greater the pressure and tension become, and the less effective you will be. The late J. C. Penney was quoted as saying that the wisest decision he ever made was to "let go" after realizing that he could not do it all by himself any longer. That decision enabled the development and growth of hundreds of stores and thousands of people.[20] Extensive delegation also frees up valuable time for you to concentrate on your goals. Get into the habit of asking yourself whether what you are doing could be handled by someone else.

Delegate with Deadlines "'Work expands to fill the time available for its completion.' In other words, people tend to stretch out jobs as long as they can." For example, if you have three hours to complete a task, you will take three hours to do it, even if you could have been done in two hours or less. To head off procrastination and delay, never give "open-ended" directives to your group members; always specify a reasonable completion date. "Be sure to keep deadlines realistic. If your personnel keep missing them for reasons beyond their control, they will soon start ignoring them completely."

Delegating with deadlines benefits both the leader and those to whom tasks are assigned. Leaders who delegate well enable themselves to concentrate on their own tasks while avoiding interference or undue pressure.[21]

Two Types of Delegation Stephen Covey, in *The 7 Habits of Highly Effective People*, states that there are basically two kinds of delegation: "gofer" delegation and stewardship delegation.

Gofer delegation means "go for" this item or go for those items or do this for me or do that for the team and let me know when you have completed my simple task(s). This sort of delegation is not a good practice because the micromanaging leader does not allow creativity, imagination, self-direction, or growth.

Stewardship delegation is more appropriate for leaders in most cases. It is based on appreciation of the self-awareness, imagination, conscience, and free will of other people. It is focused on "results instead of methods. It gives people a choice of method and makes them responsible for results. . . . Stewardship delegation involves clear, up-front mutual understanding and commitment regarding expectations in five areas":

Desired Results "Create a clear, mutual understanding of what has to be accomplished, focusing on *what*, not *how*; on *results*, not *methods*."

Guidelines "Identify the parameters within which the individual should operate. These should be as few as possible to avoid methods delegation, but they should include any formidable restrictions."

Resources "Identify the human, financial, technical, or organizational resources the person can draw on to accomplish the desired results."

Accountability Accountability is being answerable or capable of being explained. "Set up the standards of performance that will be used in evaluating the results and the specific times when reporting and evaluation will take place."

Consequences "Specify what will happen, both good and bad, as a result of the evaluation. This could include such things as financial rewards, psychic rewards, different job assignments, and natural consequences tied into the overall mission of an organization."[22]

Figure 15–12 illustrates a clear and concise way to handle delegation. When leaders delegate, they should clarify what degree of initiative is expected. Clear instructions will save time and embarrassment. If you are not clear and consistent about delegation, those who answer to you may hesitate to take the appropriate initiative for fear of being criticized. If you use this form, there should be no excuses based on lack of communication.

DELEGATING AND CLARIFYING DEGREE OF INITIATIVE EXPECTED

From: _____

To: _____

Re: _____

❏ 1. Look into this problem. Give me all the facts. I will decide what to do.

❏ 2. Let me know the alternatives available with the pros and cons of each. I will decide which to select.

❏ 3. Recommend a course of action for my approval.

❏ 4. Let me know what you intend to do. Delay action until I approve.

❏ 5. Let me know what you intend to do. Do it unless I say not to.

❏ 6. Take action. Let me know what you did. Let me know how it turns out.

❏ 7. Take action. Communicate with me only if your action is unsuccessful.

❏ 8. Take action. No further communication with me is necessary.

Comments: _____

_____

_____

Signature   Date

FIGURE 15–12 When leaders delegate, they should clarify what degree of initiative is expected. The use of this form in delegating should remove all doubts and prevent miscommunication and misunderstandings.

CONCLUSION

Leaders who manage their time well are more productive leaders. Time management involves setting goals, balancing the various parts of your life, setting priorities, and exercising self-discipline. By failing to plan, we plan to fail. If we do not know where we are going, we will not know when we get there. If we do not have objectives to accomplish during the day, we can work all day without doing anything that moves us toward reaching our goals.

SUMMARY

Time management consists of planning how to control your time so that you can do the things you need and want to do. Good leaders thoughtfully decide what they are going to do in a 24-hour period. Time is your most valuable personal resource. Use it wisely, because it cannot be replaced.

Time management has evolved through four generations or levels. The first generation promotes use of notes and checklists. The second generation uses calendars and appointment books. The third generation includes assigning priorities, clarifying values, comparing the worth of activities based on one's values, and setting goals and objectives. The fourth generation is concerned with managing ourselves through managing time; it includes enhancing relationships and meeting human needs as well as personal and professional goals and objectives.

Activities may be urgent, important, or (occasionally) both. Urgent matters require immediate action, but they do not move you toward accomplishing any of your goals. Important matters have to do with results; they contribute to reaching your goals and objectives. We often react to urgent matters, but should preserve as much time as possible for the truly important matters that require more initiative and planning.

Ultimate time management for good leaders requires putting first things first. The ultimate leader and time manager uses six important criteria as a guide to scheduling: coherence, balance, focus, people-centeredness, flexibility, and portability. Goals provide direction for time management: We cannot manage time correctly, effectively, and efficiently if we do not know what our goals are. Periodically, you will have to adjust and update your goals and time commitments.

You must balance the major categories of personal time and professional time. Time is spent on various activities that can be divided into four major categories: health, family, community (personal time), and school or employment (professional time). Each category is important. Health includes sleep, diet, exercise, recreation, socializing, and spiritual pursuits. Family should always be a high priority; likewise, we want to make our communities better places to live. We are most concerned with productivity in the employment category.

Analyze your use of time with a daily log. When you know how you have been spending your time, you can begin to change your priorities to emphasize high-priority tasks and activities. Pareto's principle, or the 80-20 rule, states that 20 percent of tasks generate 80 percent of the results you need to reach your goals. Therefore, identify that 20 percent as high priority and work on them first.

There are many time wasters: procrastination, the telephone, the Internet, social networking, television, the inability to say no, unexpected occurrences, lack of planning, and emotional conflicts. Some other time wasters include making work expand to fit the time available, unclear goals and objectives, disorganization, over-commitment, lack of organization, and perfectionism.

There is never enough time to do everything we want and need to do. Therefore, we must manage our time effectively and efficiently. This includes using planning, organizing, and scheduling calendars; making things-to-do lists; setting priorities; and delegating. There are two types of delegation: gofer and stewardship. Gofer delegation imposes specific instructions and methods and does not enhance self-worth. Stewardship delegation is based on self-awareness and an appreciation of the imagination, conscience, and free will of others. Tools of stewardship delegation include desired results, guidelines, resources, accountability, and consequences.

 Take It to the Net

Explore time management on the Internet. Browse the articles you find and pick one that you feel is useful. Print out the article and summarize it. In your summary, describe why you feel the information in the article is useful. Following are some search terms that may help in your quest for articles.

Search Terms

time management

effectively managing time

time-management skills

online time-management tools

time-management tips

Chapter Exercises

REVIEW QUESTIONS

1. Define the Terms to Know.

2. Name at least two types of things included in each of the four levels or generations of time management.

3. Within the time-management matrix (Figure 15–2), which area deserves the most attention? Explain why.

4. Briefly explain what the other three areas in the matrix represent.

5. What six important criteria must be met for ultimate time management?

6. What is the relationship between goals and time management?

7. What are the four major categories in which you spend your time?

8. What does diet have to do with time management?

9. List eight ways to overcome procrastination.

10. List four major time wasters.

11. What are 10 ways to deal with phone calls once you are in a leadership or management position?

12. What five areas does stewardship delegation involve?

13. List two helpful time-management tips for school and college students.

14. In determining priorities, what four questions should you answer?

15. What seven major categories should a scheduling calendar accommodate?

16. Without goals, it would be nearly impossible to manage your time. What are four basic questions of goals as they relate to time management?

17. What are 10 actions to take in analyzing your time logs?

18. What are nine ways to politely terminate phone conversations?

19. What are three hard truths of Murphy's Law?

COMPLETION

1. _____ is one of a leader's most valuable resources.

2. Effective leaders are not problem minded, they are _____ minded.

3. Health, family, and community are types of _____ time, whereas school and employment are types of _____ time.

4. The "80-20" rule or Pareto's principle can be stated thus: The _____ percent of tasks that are high priority generate _____ percent of the results you need to reach your goals.

5. B. F. Skinner found that _____ _____, or rewards for desirable behavior, work better than negative consequences as a tool for changing behavior.

6. Work expands to fill the _____ available for its completion.

7. _____ delegation means "go for" this item or go for those items or do this for me or do that for the team and let me know when you have completed my simple task(s)."

8. _____ delegation is based on appreciation of the self-awareness, imagination, and free will of other people.

9. Schedule goals and objectives for the _____ and make a things-to-do list each _____.

10. An _____ matter appears to require your immediate attention, even though it has nothing to do with achievement of your goals.

11. People need to plan so that they manage their _____ instead of allowing _____ to manage them.

12. Taking care of you takes up much time, but it is time well spent. No health, no work, no _____.

13. The luckiest people are those who have a job doing something they _____.

14. It is one thing to identify a time waster and an entirely different matter to get _____ of it.

15. The hardest part of any job is _____ started.

16. If we reach our goals, we are successful. Being successful makes us happy, and if we are happy, we feel _____ about ourselves.

17. Time management consists of planning how to control your time so that you can do the _____ you need and want to do.

18. _____ provide direction for time management.

MATCHING

_____ 1. Schedule your high-priority items.

_____ 2. Could cause you to be a "couch potato."

_____ 3. Can cost time because you do not take time.

_____ 4. More time tomorrow.

_____ 5. Time-value ratio.

_____ 6. "I was just on my way out the door."

_____ 7. Putting first things first.

_____ 8. "I can do it better."

_____ 9. Daily objectives.

_____ 10. "I'm flattered you want me to help you, but I'm already booked."

A. procrastination

B. lack of planning

C. prime time

D. telephone etiquette

E. saying no

F. things-to-do list

G. television

H. lack of delegation

I. setting priorities

J. Pareto's principle

ACTIVITIES

1. What one thing could you do (that you are not doing now) that would make a tremendous positive difference in your personal or professional life if you did it on a regular basis? Explain why you are not doing it.

2. Before you start a daily log for one week, take a few minutes to record everything you think you do in a typical day or week and estimate how much time you spend on each activity. After you keep a daily log for one week, record how much time you actually spent on each activity. Then answer the following questions:

   • On what activities did you spend more time than you had estimated?

   • On what activities did you spend less time than you had estimated?

   • How could you use this information to manage your time?

3. Your teacher can provide you with a daily log similar to the one in Figure 15–7. Complete the daily log. Record in 15-minute intervals how you spent your time. Record activities as you do them. If you do this every day for a week, you will have a much more accurate picture of your current use of time.

4. Using the daily log from activity 3, complete it as if you had done all the things you should do (whether you want to or not). Record your ideal day from the minute you wake up to the minute you fall asleep. Make sure to include activities from each of the four categories of personal and professional time.

5. List your own personal five biggest time wasters and what steps you will take to eliminate them.

6. Make a "to-do" list of goals you want to work on and achieve tomorrow. Be sure to prioritize them. You may want to use a form similar to Figure 15–11 or an online tool.

7. Your teacher can provide copies of a weekly planning/organizing/scheduling calendar similar to the one in Figure 15–10. Complete one of these for the week. Write a report and share with the class what you learned by completing and following the weekly or daily calendar.

8. Select five things you learned that will best help you manage your time in the future. Write a paragraph explaining each and supporting your choices.

NOTES

1. S. R. Covey, *The 7 Habits of Highly Effective People* (Miami: Mango Media, 2015), pp. 198–202.
2. Ibid., p. 192.
3. Ibid., p. 193.
4. Ibid.
5. Ibid., pp. 201–205.
6. M. E. Douglass and D. N. Douglass, *Manage Your Time, Your Work, Yourself* (New York: AMACOM, 1993), pp. 142–143.
7. E. C. Bliss, *Getting Things Done: The ABCs of Time Management* (New York: Charles Scribner's Sons, 1991), p. 30.
8. "How to Get Things Done" (Shawnee Mission, KS: Fred Pryor Seminars, 1997), p. 5.
9. R. N. Lussier, *Human Relations in Organizations: Applications and Skill Building* (New York: McGraw-Hill Education, 2017, 10th Ed), p. 91–93.
10. "How to Handle Multiple Priorities" (Shawnee Mission, KS: Fred Pryor Seminars, 1989), p. 45.
11. J. L. Yager, *Creative Time Management for the New Millennium*, 2nd ed. (Stamford, CT: Hannacroix Creek Books, 2008), pp. 16–18.
12. "How to Handle Multiple Priorities," p. 40.
13. Lussier, *Human Relations in Organizations*, p. 95.
14. Ibid.
15. Covey, *The 7 Habits of Highly Effective People*, p. 210.
16. Lussier, *Human Relations in Organizations*, p. 93–94.
17. Douglass and Douglass, *Manage Your Time, Your Work, Yourself*, p. 52.
18. Lussier, *Human Relations in Organizations*, p. 93.
19. Covey, *The 7 Habits of Highly Effective People*, p. 215.
20. Ibid.
21. F. A. Manske Jr., *Secrets of Effective Leadership* (Columbia, TN: Leadership Education and Development, Inc., 1990), pp. 84–85.
22. Covey, *The 7 Habits of Highly Effective People*, pp. 219–222.

16 MOTIVATING OTHERS

Leaders of groups and organizations must learn to motivate others. The major goals of most companies and organizations are productivity and efficiency. When a leader is able to motivate his or her followers to reach (or exceed) the productivity and efficiency standards, the leader has met a major objective.

Objectives

After completing this chapter, the student should be able to:

- Explain motivation
- Describe the characteristics of motives
- Discuss various theories of motivation
- Explain four types of reinforcement
- Name and describe the types of reinforcement scheduling
- Discuss positive reinforcement
- Explain the various forms of positive reinforcement

Terms to Know

- motivation
- need
- behavior
- internal motivation
- self-motivated
- motives
- external motivation
- content theories
- process theories
- hierarchy
- Esteem needs
- E.R.G.
- hygienes
- motivators
- recognition
- responsibility
- advancement
- learned needs theory
- equity
- relevant others
- inequity
- operant conditioning
- reinforcement
- positive reinforcement
- avoidance reinforcement
- negative reinforcement
- extinction
- punishment
- reinforcement scheduling
- fixed interval schedule
- variable interval schedule
- stroking
- fixed ratio schedule
- variable ratio schedule
- physical strokes
- verbal strokes
- stroke deficit
- confirmation behaviors
- praise
- positive written communication
- misconceptions

A primary responsibility of leaders of groups and organizations is to motivate others. The major goals of most companies and organizations are productivity and efficiency. Motivation is for leaders at any level and one of the major keys to being successful.

MOTIVATION

Have you ever wondered what makes a person do what he or she does? Why does a student strive for good grades? Why does an athlete give her all for the team? Why is it that some workers excel and others do not? The best answer to all these questions is simply motivation. Leaders must learn how to motivate followers, through the exercise of good "people skills" and the use of nonmanipulative, sincere methods and techniques.

Motivation in Everyday Life

Motivation is a key component of the classroom and the workplace. As a matter of fact, it is a key component of life. We do homework because of motivation. We go to jobs every day because of motivation. We go to college because of motivation.

We take part in religious activities because of motivation. *Motivation* comes from the word *motive*, meaning the reason for doing something or behaving in a certain way. Much of what motivates us depends on our goals, as discussed in Chapter 14. Our goals arise from our needs and desires.

As a leader, you must be aware of what makes people tick. Needs, motives, and motivation are a big part of why people act the way they do. Why do you like some of your teachers better than others? The answer probably lies in their ability to motivate you. Good teachers understand your needs and motives. Therefore, they tend to be dynamic and exciting in the classroom, inspiring students to reach their full potential (Figure 16–1).

Leaders in the workforce also must motivate people. The major goal of most companies and organizations is productivity, so a leader's primary objective is to motivate his or her followers to achieve the company's or organization's goals.

Leadership is a process in which an individual influences group members to work toward the group's goals. Good leaders build positive relationships among members by using motivation. They understand that the group's success depends on the leader's ability to use influence effectively.

FIGURE 16–1 This aerobics instructor is encouraging his students to keep working hard. Such behavior will motivate these participants to improve even more. *(© iStock/Christopher Futcher)*

Good leaders rely more on influence than on authority derived from power. When you think of great leaders, perhaps you think of actual people, such as Washington, Lincoln, Churchill, or Napoleon. Influence combines enthusiasm, excitement, charisma, and wisdom. When leaders use influence, they appeal to a person's values, skills, and knowledge. Keeping followers motivated and enthusiastic is basic to effective leadership. The way a leader uses power determines whether his or her influence will motivate followers.

The FFA awards program is based on motivation. Our motive is the attainment of awards: the Greenhand degree, chapter FFA degree, state FFA degree, and American FFA degree. There are many proficiency awards, several speaking events, and many contests (discussed in previous chapters). One of the reasons that the FFA is so successful is that it meets the needs that motivate students.

Motivation Defined

Motivation is an individual's desire to demonstrate constructive behavior and reflects a person's willingness to expend effort; it makes us do what we do. For example, Sarah is hungry (a **need**, a physiological or psychological requirement for well-being) and thus has a drive or a motive to find food. When she finally eats (**behavior**, the actions or reactions of persons or things under specified circumstances), her need for food is satisfied. The hunger she felt was her motivation to get something to eat.

Types of Motivation

Internal motivation is an inner force or power that spurs a person to want to achieve a goal or stimulates the person to action. Internal motivation deals strictly with the individual and the internal rewards that he or she receives. Feelings of responsibility, personal growth, and the satisfaction of achievement are some internal rewards.[1] A person with strong internal motivation is sometimes referred to as **self-motivated**.

External motivation, which is the second type of motivation, comes from an outer force or power that causes a person to want to achieve a goal or urges the person to action. External motivation can come from incentives that encourage a person to behave in a specific way, or from rewards, such as plaques or ribbons, given by others in recognition of performance or approved behavior (Figure 16–2). Whereas internal motivation has to do with the individual and inner rewards, external motivation involves rewards offered by outside sources for achieving a certain goal.[2]

FIGURE 16–2 Even ribbons and belt buckles can be a source of motivation for some. Awards meet the need for achievement and recognition. *(© iStock/BrandyTaylor)*

If you are the individual or leader who is responsible for motivating those around you, you must recognize people, include them, encourage them, and train them in the desired behavior. You must make them feel valued. You must show them that they are trusted, respected, and cared for. Above all, you must motivate them to excel for themselves and for the organization.

CHARACTERISTICS OF MOTIVES

The motivators that cause us to act in certain ways are often referred to as **motives**. To know how to motivate someone, we must understand five important characteristics of motives.

Motives Are Individualistic

Each person has a set of needs (or motives), which may be completely different from that of another person. One student may be motivated to "ace" a test by the need to earn a scholarship for college. Another student taking the same final exam may be motivated by the chance to make the honor roll. Each student's motives are different, although their actions are the same. Sometimes, when we do not understand another individual's actions, we can gain some insights by discovering the motives underlying and driving those actions.

Motives Can Change

During a person's lifetime, motives can change. What motivated us as children may no longer motivate us as adults. For instance, early in one boy's academic career, he may study for a test just so he can get out of the class. Later, he may study so that he can win a good scholarship for college. In college, he begins to study so that he can graduate with honors and receive good job offers. Though this is the same student, his motives for studying continually changed throughout his academic career.

Motives Can Be Unconscious

Often, we are not aware of or do not fully understand the inner forces and needs that shape our behaviors. Those inner motives may be feelings of inadequacy, loneliness, or lack of acceptance by peers. Such feelings and perceptions may motivate people to strive for top awards or higher positions in an organization. They may also motivate people to stop trying, withdraw, or become angry. One of a leader's responsibilities may be to help followers channel their responses to these unconscious motives in productive, self-affirming ways.

Motives Can Be Inferred

In most cases, we notice people's behavior, but we can only guess what motivates them to act the way they do. The best way to learn what motivates another person is through discussion. There is no more practical method to obtain information about motives than by simply asking, "What motivated you to act in that way?"

Motives Are Hierarchical

Motives come in various strengths. The stronger the motive, the more likely a person is to achieve a goal. A student who works part time to earn money for a trip during spring break knows she must also make car payments, so she is motivated to work by the fact that if she does not keep up her car payments she will lose the car. For her, the ability to drive to the beach for spring break outweighs the prospect of spending her free time with friends. Behavior, therefore, is ruled by those motives or needs we deem most important.[3]

Here is where the issue of conflict of values arises. You have a new job; your brother is having a bachelor party; you have an FFA parliamentary procedure contest; your coach has scheduled an extra, unannounced practice—and they are all on the same night. What do you do? My choice would be different from yours. Your values determine your motives and thus your choices. A person cannot do everything in life; life necessarily involves choices. By following the problem-solving and decision-making process outlined in Chapter 13, you can come to a satisfactory decision. Regardless, your decision will depend on your motives.

THEORIES OF MOTIVATION

Several theories of motivation exist, but they all can be classed as one of two types: either content theories or process theories. **Content theories** are psychological suppositions that focus attention on the factors within an individual that cause the individual to behave in a particular manner. Basically, a content theory looks at individual needs that motivate behavior. **Process theories**, in contrast, are psychological approaches that examine how behavior is motivated. For a process theory to be effective, it must identify the processes surrounding the behavior. The first content theory we discuss is the hierarchy of needs theory.

Maslow's Hierarchy of Needs

Abraham Maslow's timeless theory believed that all behavior is based on the needs of the individual. He arranged these needs in a **hierarchy**, a ranked order that denotes the relative importance of each item. An individual begins at the lowest level of need and cannot move to the next higher level until the needs on the previous level have been met (satisfied). Human needs, as determined by Maslow, are (1) physiological needs, (2) needs for safety and security, (3) social and belonging needs, and the need for (4) self-esteem and (5) self-actualization (Figure 16–3).[4]

Physiological Needs We all have basic survival needs, such as air, water, food, shelter, clothing, and sleep. All these needs must be satisfied before a person expends any effort on meeting the next level of needs. For example, you cannot teach someone who is hungry. Thus, the need for school lunch programs, and in some cases, breakfast programs.

Safety and Security Needs Once physiological needs have been satisfied, we become aware of the need for freedom from the threat of harm. People yearn for a feeling of order, safety, and predictability. Students in a school building need to feel secure and safe from personal harm. Families need to feel protected from the violence that plagues our streets. Workers need to feel assured that their working conditions are safe. Only when people feel secure can they focus on the third level of needs. Firefighting squads and police departments, such as the police officers shown in Figure 16–4, provide many people with a sense of safety and security.

Social and Belonging Needs People have social needs, too. We need to feel that we

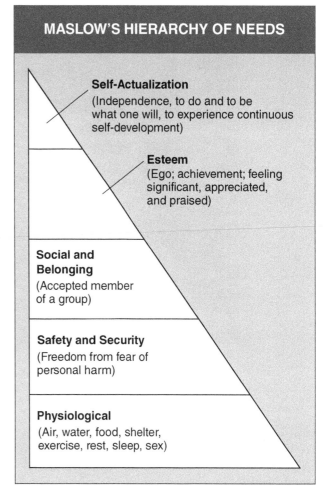

FIGURE 16–3 Basic physiological needs have to be met before we become particularly concerned about psychological needs. We do not worry about self-concept if we have no food or water. *(Adapted from Abraham H. Maslow, Motivation and Personality, New York: Harper & Row, 1954.)*

belong to a group; and we also need friendship, interaction with others, and love. These needs affect our mental health and well-being, whereas the previous two levels of needs dealt with physical health and well-being. In an attempt to fulfill some of these mental health needs, an individual may join a team or an organization in which he or she can satisfy the needs for belonging, friendship, and interpersonal interaction.

In high school, the FFA may meet some social and belonging needs. In college, fraternities, sororities, and friends can meet these needs. Earlier in life, the 4-H Club, Boy or Girl Scouts, sports teams, family, and a variety of other groups can meet social needs.

FIGURE 16–4 Police officers help people feel secure and safe from personal harm. *(Courtesy of Annemarie Mountz, Department of Public Information Media Gallery, Pennsylvania State University.)*

Self-Esteem Individuals need to feel good about themselves. How a person feels about himself is referred to as *self-esteem*. Individuals also need others to recognize their achievements. **Esteem needs** include a person's need for self-respect as well as respect, recognition, and affirmation from others. Status symbols, personal titles, and awards that denote achievement can help meet esteem needs. Self-esteem can be affected by words of praise and appreciation from peers and others. Self-concept and self-esteem are discussed further in Chapter 18.

Self-Actualization The need for self-actualization is often the most difficult need to understand and satisfy. Self-actualization represents the maximum use of an individual's talents, abilities, and skills. A self-actualized person has realized his or her fullest potential as a human being. Each person's search for self-actualization is different because of the uniqueness of each person. Many people find it difficult to satisfy their self-actualization needs because they are limited by circumstances. If you fulfill this need, you no longer have any challenges to be met. Perhaps this is the reason so many people never completely satisfy this need.

E.R.G. THEORY

A second classical content theory of needs was proposed by C. P. Alderfer. **E.R.G.** stands for three types of needs: existence, relatedness, and growth. Alderfer defined them as follows:[5]

- **Existence needs** are material and are satisfied by environmental factors such as food, water, pay, benefits, and working conditions.
- **Relatedness needs** involve relationships with "significant others," such as coworkers, superiors, subordinates, family, and friends.
- **Growth needs** involve the desire for unique personal development. They are met by developing whatever abilities and capabilities are important to the individual.

Alderfer's theory differs from Maslow's in three important dimensions:

1. Alderfer proposed three need categories in contrast to Maslow's five.
2. Alderfer arranged his needs along a continuum, as opposed to a hierarchy. Existence needs are the most concrete and growth needs the most abstract.

3. Alderfer allowed for movement back and forth on the continuum, in contrast to Maslow's perception that people move steadily up the needs hierarchy. A person who becomes frustrated or unable to satisfy higher needs would regress to fulfilling lower needs.

Expectancy Theory

The third content theory was introduced by Vroom. V. H. Vroom pushed expectancy theory into the arena of motivation research. According to Vroom's cognitive theory, each person is assumed to be a rational decision maker who will expend effort on activities that lead to desired rewards. The theory has five major parts: outcomes, valence, instrumentality, expectancy, and force.[6]

- **Job outcomes** are the things an organization can provide, such as pay, promotions, benefits, and vacation time. Theoretically, there is no limit to the number of outcomes. They are usually thought of as equivalent to rewards.

- **Valence** refers to one's feelings about job outcomes and is usually defined in terms of attractiveness or anticipated satisfaction.
- **Instrumentality** is the perceived degree of relationship between performance and outcome attainment. It exists in a person's mind. If a person thought that a pay increase depended solely on performance, the instrumentality associated with that outcome would be very high.
- **Expectancy** is the perceived relationship between effort and performance. In some situations, there may not be any relationship between how hard you try and how well you do. In others, there may be a very clear, direct relationship: the harder you try, the better you do.
- **Force** is the amount of effort or pressure on or within someone. The larger the force, the greater the motivation.[7]

Expectancy theory explains why some jobs seem to create high or low motivation. On assembly lines, group performance level is determined by the speed of the line. No matter how hard one

---

VIGNETTE 16-1 YOU CAN BE KNOCKED DOWN THE HIERARCHY

MASLOW'S HIERARCHY OF NEEDS IS ONE OF THE BEST-KNOWN THEORIES OF MOTIVATION, but few writers ever bring up the fact that you can go back down the hierarchy. As a leader of teams, you need to recognize when someone who formerly was satisfied is no longer content. Consider the following scenario.

Veronica was a beautiful young woman who lived in a comfortable home and drove a nice car. Because she felt so blessed and wanted to give back to the community, she began volunteering at a local shelter for the homeless. Her supervisor, Brenda, thought Veronica was the greatest, and thus felt that she did not need to motivate Veronica to do a great job. Veronica was not thirsty, hungry, or in need of shelter. She had safety and security in her life. She was well connected socially, and her self-esteem was strong and accurate. The only thing left for Veronica was self-actualization. Every day she tried to do something good for people at the shelter, just because she knew it was the right thing to do. Veronica was one of the best volunteers Brenda had ever worked with.

Things took a turn for the worse when Veronica was in a car wreck one weekend. Although she survived the crash, Veronica ended up having to use a wheelchair for mobility. She was healthy otherwise, but was just not the same mentally. Her parents thought that working at the shelter again would help her, but it did not. Brenda had to spend an extraordinary amount of time and energy coaching and motivating Veronica to undertake the routine tasks she had done so well before her accident. However, with consistent coaching and motivation from Brenda, Veronica gradually recovered her old self.

Besides the wreck, what happened to Veronica to disrupt her motivation for work at the shelter? Think back to Maslow's hierarchy of needs. After the accident, Veronica's physiological needs were in danger, she did not feel secure, her sense of belonging was in jeopardy, and her self-esteem was at an all-time low. In the blink of an eye, Veronica moved down to the lowest level of the hierarchy, and it took a long time for her to work through her problems and climb up again.

This happens with surprising frequency to both leaders and followers. Not all situations are as serious as Veronica's, of course, but both individual and team performance can be negatively affected when someone's needs are not being met. As a leader, you have to recognize when someone's situation has changed and build people up instead of yelling at or reprimanding them during their roughest times. Have you ever tumbled down the hierarchy and suddenly not been as effective? Have you ever known someone to whom this has happened? What could you do to help someone climb the needs ladder?

works, one cannot produce any more until the next object comes down the line. Thus, there is no relationship between effort and performance. However, salespeople who are paid on commission realize that the harder they try (the more sales calls they make), the better their performance is (sales made). Expectancy theory is very good at analyzing the components of motivation.

Herzberg's Two-Factor Theory

The fourth content theory was proposed by Herzberg. Frederick Herzberg and his associates interviewed accountants and engineers and asked them to describe situations in which they were satisfied, or motivated, and dissatisfied, or unmotivated. Using the data from these interviews, Herzberg concluded that the opposite of *satisfaction* is not *dissatisfaction*, as had been traditionally believed, and argued that removing dissatisfying characteristics from a job does not necessarily make the job satisfying.

Herzberg classified needs into two categories. He called the lower-level needs **hygienes** (conditions and practices that promote or preserve health or psychological well-being). They include salary, job security, working conditions, relationships, status, company procedures, and quality of supervision. When these conditions are not present or are inadequate, employees become dissatisfied; however, neither the presence nor the abundance of these conditions guarantees motivation. Herzberg called the second category, of higher-level needs, **motivators**. They include achievement, recognition, responsibility, advancement, the work itself, and the possibility of growth. When these conditions are present, so are the strong levels of motivation that usually result in good job performance. With motivators, amount does matter.

Once hygiene factors are adequate, employees can be motivated through their jobs. The best way to motivate employees is to build challenge and opportunity for achievement into the job.[8]

| VIGNETTE 16-2 | YOU MEAN IT'S NOT ALL ABOUT MONEY? |

CAN YOU BELIEVE IT?

Research actually shows that money is not a motivator! Now, we hope you apply critical thinking when you read that, but research actually supports the notion that you need more than money to motivate people to do a good job. Money is a hygiene factor, which means that if it is not there, workers can become upset and dissatisfied. But, according to Herzberg, money is not what makes people work harder, do a better job, or go the extra mile to make an organization successful. So, what could be more important than money?

Achievement is one thing people are looking for. If people constantly feel as though they are meeting their goals and accomplishing worthwhile things, they are motivated to work harder. That sense of accomplishment is something they can take with them when they go home.

Giving people **recognition** is the best investment you can make as a leader. People claim that recognition is not important—but do not believe that for one second. It is the fuel that keeps many people's fire lit. Recognize people in private, in public, and on paper. Some researchers go a step farther and say that you should not just recognize people; you should celebrate specific accomplishments as a form of recognition. Have a party!

People also want and appreciate **responsibility**. They want to own something or to be in control of something for which they are responsible. Especially when accompanied by a sense of accomplishment, responsibility is a fantastic motivator.

People are also motivated by the opportunity for **advancement**. This only makes sense: would you be motivated to keep working as hard as you could for someone who had no intention of promoting you at some point? Advancement could come in many forms: position, job title, pay grade, and so on. Whatever their specific motivators, people need to know they are *not* stuck in the same mundane spot with no hope of improving their situation.

People are motivated by the work itself. They have a need to work. If you are the leader, you have to plan and coordinate so that your followers have work to do and that the work means something to them. Because of poor planning and poor allocation of human resources, many volunteer organizations, teams, or places of business waste people's time, and work is frequently left undone.

Even if you do not entirely buy the idea that things other than money are more important in motivating people, as a leader in any situation, you should keep in mind the motivators discussed here. Achievement, recognition, responsibility, advancement, and the work itself are variables that should not be left out as you lead and motivate your followers, team and group members, and coworkers.

Learned Needs Theory

Another content theory of motivation is the learned needs theory developed by McClelland. **Learned needs theory** states that people can acquire needs from society itself; when such a need is strong enough in a person, the person is motivated to a behavior that will satisfy the need. According to McClelland, an individual learns what his needs are by interacting with his environment. The needs essential to this theory are the needs for achievement, power, and affiliation. Once individuals understand their needs, their behavior, personality, and performance will be affected.

Achievement The need for achievement is the need to excel and to strive to succeed. People with a high need for achievement are usually goal oriented. They seek challenges and take moderate risks. They also desire personal responsibility for solving various problems. People with a low need for achievement do not perform as well as people with a high need in challenging, competitive, or nonroutine situations. People with a strong achievement need tend to enjoy entrepreneurial positions.

Power The need for power is often associated with a need to make others behave in ways that they would not behave otherwise. People with a great need for power want to control situations and other people. They like competition, but only if they can win—they hate losing. They also do not mind confrontation with others. People with a strong need for power seek positions of authority and personal status.

Affiliation The need for affiliation is the desire for interpersonal relationships and friendships. People with a high need for affiliation enjoy social activities and want to be liked by others. They also tend to join organizations and groups, to enhance their feeling of belonging. These individuals seek positions or situations in which they can help and teach others. The students in Figure 16–5 are meeting their need for affiliation by choosing to eat lunch together.[9]

Learned Needs Theory and Personality Types Each individual has varying degrees of each need. The degree is based on what a person has learned by dealing with his or her environment. One of the three needs is usually dominant in each of us and, thus, is a primary motivator of our behavior.

You may see a correlation between the personality types discussed in Chapter 2 and the learned needs theory. Sanguine personality types tend to have a high achievement need. Choleric personality types

FIGURE 16–5 These young people are meeting their need for affiliation by enjoying a meal together at a leadership training camp. (© iStock/kali9)

have the need for power and control. Melancholy personality types like affiliation but do not have a particularly great need for power or achievement.

Equity Theory

Process theories, as stated earlier, describe how behavior is motivated. The first process theory of motivation to be discussed is the equity theory.

The equity theory was developed by J. Stacy Adams. It is based on the concept that all individuals want to be treated fairly. **Equity** is an individual's belief that she is being treated fairly in relation to her relevant others. (**Relevant others** may be peers, coworkers, or a group of people with whom an individual can compare herself.) **Inequity** is an individual's belief that he is being treated unfairly in relation to his relevant others. Adams contended that people are motivated to seek social equity in the rewards they receive for the performance they give.

Individuals compare their inputs (what they contribute to the situation) and their outcomes (what they receive in return for their input) to that of others. Inequity usually occurs when the perceived input–outcome ratio is out of balance.[10] For example, a student does an extra-credit project and receives five extra points on his final grade. If that student put hours of work into his project but finds out that another student received five extra-credit points for 30 minutes of work, the hardworking student will feel a sense of inequity. Therefore, the next time extra credit is offered, the student's behavior will be motivated (and perhaps not in a positive way) by the inequity he initially felt. A perceived inequity causes tension that motivates a person to try to balance the input–outcome ratio or to make a situation equitable.

There are several ways to reduce a perceived inequity so as to balance the input–outcome ratio. Here are the most common methods:[11]

Change Inputs Reduce your level of effort.

Change Outcomes Request more rewards, such as pay, vacation time, or extra-credit points.

Change Self-Perception Change the way you think of your own inputs. You may rethink exactly how much you actually put into a situation.

Change Perception of Others Change your view of others' inputs, outcomes, or overall situations. One student may begin to believe that another student's project was indeed better than his own.

Change Comparisons Change the group to which you compare yourself. A student may focus his or her comparison on another student who worked as hard but received only three extra-credit points.

Abandon the Situation Choose simply to leave the situation or not to participate next time. You may not do extra credit the next semester so that you do not have to experience feelings of inequity.

As a leader, you should always be aware of your followers' perceptions and feelings of inequity. Seek to remove the unintended inequities or, at the very least, explain the reason for them. As a team member, focus on your individual situation and try not to worry about comparing your productivity and performance to that of others. Doing so may actually hinder your performance, and it will most certainly waste valuable time and brainpower that could be applied to accomplishing your goals.

Reinforcement Theory

Another process motivation theory is reinforcement theory. In its basic assumptions, developed by psychologist B. F. Skinner, reinforcement theory states that behavior is motivated and controlled through the use of rewards. Reinforcement theory proposes Skinner's **operant conditioning** or behaviorism as a way to modify an individual's behavior through the appropriate use of immediate rewards or punishments.[12] This theory is also concerned with maintaining a behavior over time; it contends that rewarded behavior will be repeated, whereas unrewarded behavior will not be repeated (Figure 16–6). Two important aspects of controlling behavior are the type of reinforcement and the schedule of the reinforcement.

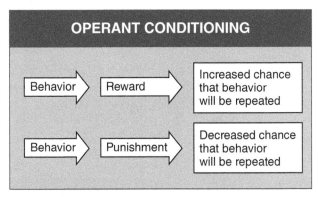

FIGURE 16–6 Psychologist B. F. Skinner theorized that a person's behavior is motivated by and controlled through the use of rewards and punishments.

TYPES OF REINFORCEMENT

There are four types of **reinforcement** (the motivation and control of behavior by the use of rewards): positive reinforcement, avoidance reinforcement, extinction, and punishment.[13]

Positive Reinforcement

Positive reinforcement is a method of using pleasant consequences or rewards to encourage performance and repetition of a desired behavior. For example, after a student studies several hours for a test and receives an A on the test, it is more likely that those study habits will occur again, because of the reward of the good grade. Positive reinforcement is the best motivator of people. Positive reinforcement also works for animals. The child and the dog in Figure 16–7 will continue to treat each other appropriately as long as positive reinforcement continues.

Avoidance Reinforcement

Avoidance reinforcement is often called **negative reinforcement**. Avoidance reinforcement entails the threat of punishment but no actual punishment; by performing or exhibiting the desired behavior, an individual can avoid a negative consequence. For example, a student hurries to get to class on time in order to avoid

FIGURE 16–7 If the dog and/or the child is given praise (positive reinforcement), each will continue to treat the other appropriately. If either or both of them are scolded, the positive behavior will eventually cease.

a reprimand by the teacher or a visit to the principal's office. The student has learned what to do to avoid the unpleasant consequences of an undesirable behavior. This is the basic premise of avoidance reinforcement.

Extinction

Extinction involves the removal of a positive reward. Extinction occurs when one tries to eliminate an undesirable behavior by withholding rewards when the behavior occurs. A student who is late for class misses the teacher's reward for promptness that day. By withholding the reward, the teacher can eventually eliminate the chronic tardiness (the tardiness becomes extinct). However, extinction can also work on good, desirable behaviors that are not rewarded. A student who works hard during class, but whose hard work goes unnoticed, may soon stop working as diligently. Can you imagine receiving the chapter, state, or American FFA degree with no awards, no banquet, and no State Convention or National Convention to attend? Without these awards or some other type of recognition, students would no longer do the work that goes into earning these awards—that behavior would become extinct.

Punishment

Punishment provides a negative consequence for an undesirable behavior. It is used after the behavior has occurred. Punishment may reduce the undesirable behavior, at least temporarily, but it may also stimulate or increase other unwanted behaviors. A student who is publicly reprimanded for harassing his teacher may stop harassing the teacher but begin disrupting the class in other ways. Punishment is the least effective way of modifying human behavior, but in some circumstances it may be the only way to handle an irrational person.

REINFORCEMENT SCHEDULING

The second consideration in the use of reinforcement is when to reward behavior. The timing of rewards or punishments is known as **reinforcement scheduling**. Reinforcement scheduling has two classifications: continuous and intermittent.

Continuous Reinforcement

With continuous reinforcement, the individual is rewarded every time the desired behavior occurs. Continuous reinforcement is better at sustaining the desired behavior, but it is not always as practical as using intermittent reinforcement.

Intermittent Reinforcement

With intermittent reinforcement, the positive consequence occurs after the passage of time or after an output. An interval schedule is based on the amount of time between rewards. A ratio schedule is based on the number of outputs between rewards. Intermittent reinforcement has four different scheduling alternatives from which to choose.[14]

1. **Fixed Interval Schedule** A **fixed interval schedule** rewards behavior at regular, set times. An employee is rewarded with a weekly paycheck for performing a desired behavior. This alternative can lead to irregular behavior and often rapid extinction of the desired behavior once the rewards stop.

2. **Variable Interval Schedule** A **variable interval schedule** rewards a desired behavior after a varying period of time. An employee is rewarded with an unexpected promotion for the desired performance. This alternative leads to moderately high and stable performance and slow extinction of the desired behavior.

3. **Fixed Ratio Schedule** A **fixed ratio schedule** rewards the behavior after a predetermined number of actual outputs. A student receives five extra-credit points after bringing her completed homework to class three times. This alternative quickly leads to very high and stable performance and slow extinction of the desired behavior.

4. **Variable Ratio Schedule** A **variable ratio schedule** rewards a desired behavior after a varying number of outputs. A student receives an unexpected award for turning in his completed homework. This alternative leads to very high performance and slow extinction of the desired behavior. This is often the most powerful method of reinforcement, because individuals maintain the desired behavior despite possibly long intervals between reinforcing rewards.

POSITIVE REINFORCEMENT

Positive reinforcement is the best motivator of people. It is also a good method of improving interpersonal relationships. People who understand and use positive reinforcement are more likely to be successful in their careers and in their relationships. Also, by understanding that all humans need some kind of positive feedback, we can better understand ourselves and our behavior.

Positive Feedback

Individuals need and desire some type of appreciation or positive feedback (positive reinforcement). Included in Maslow's hierarchy are the needs for security, belonging, and self-esteem. It would seem impossible to fulfill these needs without some type of positive feedback.

Security. An individual might never feel secure without someone recognizing his accomplishments and truly making him feel secure about what he is doing.

Need to belong. An individual would never feel that he really belonged until others showed him somehow that he was, in fact, a fully accepted member of the group.

Self-esteem. Satisfaction of the need for self-esteem definitely requires positive reinforcement. The student in Figure 16–8 apparently is not lacking in self-esteem. It would be difficult to have a very high level of self-esteem without some positive, external recognition of one's accomplishments.

Best Motivator of a Person's Behavior

B. F. Skinner's work examined the importance of reinforcement and reinforcement scheduling. He, too, theorized that positive reinforcement was the best motivator. According to the theorists, then, the best way to make a person perform well is to recognize and acknowledge his or her accomplishments. By having his or her work praised, the person's need for reinforcement is satisfied, and he or she will likely repeat the praised behavior.

This is basic knowledge for every leader. Democratic leaders lead by positive reinforcement. They also use the theory Y assumption that people do not have to be told everything to do and

FIGURE 16–8 Positive feedback meets people's needs for security, sense of belonging, and self-esteem. These needs appear to have been met for this student. *(Courtesy of Georgia Agricultural Education.)*

how to do it. People respond best to praise, not to ridicule or micromanagement.

Strokes

Eric Berne's work also identifies the need for positive reinforcement, through what he called *strokes*. **Stroking** was the term Berne used to describe various ways in which a person may recognize another person's behavior. **Physical strokes** are any recognition that occurs physically, such as an approving smile, an arm squeeze, or a pat on the back. **Verbal strokes** are any recognition that occurs verbally, such as words of thanks, gratitude, appreciation, or praise.

Berne argued that stroking is important for both physical and mental well-being. A person who does not get enough recognition from others has a **stroke deficit**. A person with stroke deficit

may actually ask for strokes; this is typically called "fishing for compliments." It is important for people to acquire positive feedback so that they feel worthwhile, needed, and valued. People who suffer from stroke deficit may damage their relationships by criticizing or undermining the work of others;[15] because they are so insecure, they put others down in an attempt to lift themselves up.

In extreme cases, people with stroke deficit may become overly critical of themselves, exaggerating any faults in their work or their personality. With the use of positive reinforcement, individuals who have a stroke deficit may overcome their constant hunger for attention and approval. Leaders should be aware of this problem and compliment workers or followers whenever possible. Just a little attention can bring a big reward in terms of increased productivity.

FORMS OF POSITIVE REINFORCEMENT

Positive reinforcement comes in various forms to be used at the appropriate times. No matter which form is used, we must stress that positive reinforcement has to be used continually for it to be effective. Effective use of positive reinforcement basically means varying the form of the reinforcement. If you always praise someone with the same words, your praise may appear rehearsed and therefore not genuine. You can usually avoid this pitfall by being specific about what you are praising and noting some detail about what was particularly good or well done.

Confirmation Behaviors

Confirmation behaviors, as Evelyn Sieburg calls them, are positive behaviors that enhance the recipient's feelings of self-worth[16] The six confirmation behaviors include orientation, praise, courtesy, active listening, positive written communication, and performance reviews.

Orientation A proper orientation consists of an adequate introduction to an organization's procedures, policies, and environment. An individual who is well acquainted with his or her surroundings is more likely to perform well and to feel like an important part of the organization.

Praise The least difficult and most powerful form of recognition to give an individual is praise. To **praise** means to express approval of a behavior and to commend its worth. The use of praise as reinforcement ensures repetition of the desired behavior. It also reassures the individual that his or her behavior is noticed and respected.

In *The One Minute Manager*, Blanchard and Johnson stated that praising works well when you

- give praise immediately;
- tell the person quite specifically what he or she did right;
- tell the person "how good you feel about what [the person] did right, and how it helps";
- encourage similar performances in the future; and
- make it known that you believe in them and want them to be successful.[17]

Courtesy Another confirmation behavior that recognizes people's need for acknowledgment is courtesy. Courtesy is simply respecting others and being polite, helpful, and considerate. It also means showing each individual that he or she is important.

Active Listening Active listening consists of letting the speaker know that you were indeed listening, by interpreting and giving feedback on what the speaker just said. This shows the speaker that you do care about what he or she has to say and makes the speaker feel important.

Positive Written Communication Notes and letters that express positive recognition can strongly reinforce behavior. **Positive written communication** refers to any written materials that express appreciation for a well-performed behavior (Figure 16–9). Positive written communication lets the recipient know that the writer of the letter cares enough to take time to express his or her consideration and gratitude. This form of confirmation behavior can lift an individual's spirits, raise his or her self-esteem, and increase the likelihood that the desired behavior will occur again.

Performance Reviews A timely performance review by a higher-up is an insightful way of reinforcing desired employee behavior. Instead of criticizing the employee's performance, an effective review focuses on the positive aspects and recognizes and rewards them. This meaningful recognition motivates the individual to strive for

Dear Carlos,

What a great job you did organizing the food drive for those laid off from the factory. Many people have told me how impressed they were with your compassion and dedication. I totally agree. Your leadership ability continues to emerge daily.

Keep up the good work.

Sincerely,

Juan Clemente, Principal

Grove City High

FIGURE 16–9 Positive letters or thank-you notes such as this one will raise self-esteem and cause the behavior to be repeated.

more positive feedback in future performance reviews; thus, it reinforces the desired behaviors.

Barriers to Positive Reinforcement

Although recognition is easy to give, it is often avoided or neglected because of certain barriers. People like to hear how well they are doing, and they appreciate receiving recognition for a job well done. Why, then, is positive reinforcement so often lacking?[18]

Preoccupation with Self Self-preoccupation interferes with an individual's ability to acknowledge the performance of others. Preoccupation with self can make a person focus on himself or herself to the exclusion of others who deserve attention. A healthy ability to take care of your own business should not interfere with your relationships with others. As a leader, you should be able to set goals and follow your dreams without completely ignoring other people.

Misconceptions Misconceptions are false or mistaken ideas or thoughts that arise from lack of information. Some common misconceptions about positive reinforcement are that it takes power away from the giver; that receivers of positive feedback will just want more rewards; and that individuals do not deserve any reward beyond what they work for, such as a grade or a paycheck. Others believe that once a person receives positive reinforcement, he or she will not work as hard. (In reality, exactly the opposite is true.) These misconceptions can be damaging to relationships and to the motivation to perform the desired behavior. By understanding how positive reinforcement works, leaders can overcome this obstacle.

"Too-Busy" Syndrome Some people feel that they are too busy to recognize the work of others or even to review their performance. To overcome this barrier, individuals must put time aside to plan for reviews and recognition. Just a few minutes is all that is necessary to offer verbal praise or to write a note letting someone know that her behavior is respected and acknowledged. Even a short note can work wonders (Figure 16–10). Good leaders find the time.

Fear of Sounding Rehearsed Sometimes it is difficult to know what to say or do to recognize behavior appropriately. When an individual says or does the same thing over and over again, it seems rehearsed and insincere, and eventually it will fail to reinforce the desired behavior. There are several verbal and nonverbal ways to show appreciation and approval. For example, you could say, "Great thinking," or "Keep up the good work." You could also make direct eye contact, pat someone on the back, or simply smile to let someone know you approve of his or her work. Some actions demonstrate approval, such as asking for advice, displaying a person's work, or publicly recognizing a person's work. Any of

these methods, along with a large variety of others, can help a leader give positive feedback to group members and coworkers.

Lack of Role Models Some leaders have never encountered an individual who gives effective positive reinforcement. They have no idea how to begin using positive reinforcement because they lack role models. This barrier can be overcome by simply doing it: start almost anywhere and give some kind of positive reinforcement. If you are just beginning this practice, you may feel clumsy or awkward about it at first, but few people will even notice if your praise is specific and sincere. Positive attitudes are contagious. By becoming a role model, an individual ensures that positive reinforcement will continue to be a powerful source of motivation.

Environment Motivation flourishes only in a positive, supportive environment. People who work together should be courteous, respectful of one other, and actively involved in maintaining the positive environment. Positive reinforcement will soon disappear in an atmosphere of negativity and dismal attitudes.

As a leader, make the changes necessary to build a positive work environment. A positive work environment includes, in order of importance, interesting work, full appreciation of efforts, involvement, good pay, job security, promotion and growth, good working conditions, loyalty to employees, help with personal problems, and tactful discipline. Good leaders have control over most of these elements.

CONCLUSION

Motivation comes from the word *motive*. It is a key component of the classroom and the workplace. Our needs result in a drive or a motive to perform certain behaviors. For example, hunger (need) is the motive for us to eat (behavior).

Positive reinforcement is the best way to motivate people. It is also a good method of improving interpersonal relationships. People who understand and use positive reinforcement are more likely to be successful in their careers and in their relationships. When we know that all humans need some kind of positive feedback, we can better understand ourselves and our behavior.

Nadine,

You did a great job on this project. I appreciate your hard work and attention to details. You are really an asset to the company.

Thanks again,

SCR

FIGURE 16–10 It does not take much time to write a note like this. The results can be amazing.

SUMMARY

Motivation is a key component of the classroom, work, and life. Motivation is the individual's desire to demonstrate constructive behavior and reflects a person's willingness to expend effort. There are two types of motivation: internal motivation, which is an inner force or power that drives a person to want to achieve a goal; and external motivation, which is any outer force or power that stimulates a person to want to achieve a goal.

There are five important characteristics of motives: motives are individualistic, motives often change, motives can be unconscious, motives can be inferred, and motives are hierarchical.

Theories about motivation come in two types: content and process theories. Content theories focus attention on the factors within an individual that cause the individual to behave in a particular manner. Process theories examine how behavior is motivated. Theories of motivation include Maslow's hierarchy of needs, E.R.G. (existence, relations, growth) theory, expectancy theory, Herzberg's two-factor theory, learned needs theory, equity theory, and reinforcement theory.

There are four types of reinforcement: positive reinforcement, avoidance reinforcement, extinction, and punishment. Positive reinforcement is the use of pleasant consequences or rewards to encourage performance and repetition of a desired behavior. Avoidance learning is often called negative reinforcement. Extinction is achieved by the removal of a positive reward. Punishment provides a negative consequence for an undesirable behavior.

The timing of rewards or punishments is known as reinforcement scheduling. With continuous reinforcement, the individual is rewarded every time the desired behavior occurs. With intermittent reinforcement, the positive consequence occurs after a passage of time or after an output. Four types of intermittent scheduling are fixed interval schedule, variable interval schedule, fixed ratio schedule, and variable ratio schedule. Any of these may be effective in a given situation.

Positive reinforcement is the best motivator. It is also a good method of improving interpersonal relationships. Individuals yearn for some type of appreciation or positive feedback. Closely related to positive feedback are the needs for security, belonging, and self-esteem. Eric Berne's work also recognized the need for positive reinforcement, which he called strokes. Stroking describes the various ways in which a person may recognize another person's behavior. There are physical strokes and verbal strokes; a lack of strokes is called a stroke deficit.

Confirmation behaviors are positive behaviors that affect feelings of self-worth in the receiver of the confirmation. Six confirmation behaviors are orientation, praise, courtesy, active listening, positive written communication, and performance reviews.

There are also six barriers to positive reinforcement. These are preoccupation with self, misconceptions, "too-busy" syndrome, fear of sounding rehearsed, lack of role models, and a negative environment.

 Take It to the Net

Explore motivation on the Internet. With just a little searching you will find information on motivation, including quotes, articles, and anecdotes. Using your favorite Web browser or search engine, identify some information you find motivating, whether it be a quote or an inspirational story. Print out or write down what you found and share it with the class. If you are having problems finding what you are looking for or want more information, use the following search terms.

Search Terms

motivation

motivational quotes

motivational theories

inspirational

Chapter Exercises

REVIEW QUESTIONS

1. Define the Terms to Know.
2. Explain the relationship between (a) motive and motivation, and between (b) need and behavior.
3. What is the difference between internal and external motivation?
4. What are five characteristics of motives?
5. What are the five levels of Maslow's hierarchy of needs?
6. How does the E.R.G. motivation theory differ from Maslow's motivation theory?
7. What are the five major parts of the expectancy theory of motivation?
8. Explain the difference between hygiene factors and motivators in the Herzberg two-factor theory of motivation.
9. What three essential needs does the learned needs theory identify?
10. List the methods used to reduce a perceived inequity.
11. Name the four types of reinforcement.
12. Compare avoidance reinforcement and punishment in reinforcement theory.
13. Explain the four types of intermittent reinforcement scheduling and their effects on behavior.
14. What are six forms of confirmation behaviors?
15. What are six barriers to positive reinforcement? Give one example of each and how they can be overcome.
16. Why do you like some of your teachers better than others?
17. Research actually shows that money is not the greatest motivator at work, so what five things could be more important than money?
18. In the "One Minute Manager," Blanchard and Johnson stated praise works well when you do what five things?

COMPLETION

1. By using a _____, you can recognize the particular behavior by smiling or by patting the person on the back.
2. _____ is any letter or note that expresses appreciation for a job well done.
3. Orientation, praise, and active listening are examples of _____.
4. _____ comes from the word *motive*.
5. _____ represents the maximum use of an individual's talents, abilities, and skills, according to Maslow.
6. _____ is the best motivator.
7. _____ are false ideas or thoughts that people form because of a lack of information.

8. _____ may be the least difficult and most powerful form of recognition to give.

9. _____ is the removal of a positive reward.

10. A positive _____ reinforces and supports positive, productive behaviors.

11. _____ is for leaders at any level, and one of the major keys to be successful.

12. Much of what motivates us depends on our _____.

13. The major goal of most companies and organizations is productivity, so a leader's primary objective is to _____ his or her followers to achieve the company's or organization's goals.

14. The FFA awards program is based on _____.

15. E.R.G. stands for _____, _____, _____.

16. The best way to motivate employees is to build _____ and _____ for achievement into the job.

17. People with a strong _____ need tend to enjoy entrepreneurial positions.

18. People with a strong need for _____ seek positions of authority and personal status.

19. Sanguine personality types tend to have a high _____ need.

20. Choleric personality types have the need for _____ and _____.

21. Melancholy personality types like _____ but do not have a particulary great need for power achievement.

22. People who understand and use _____ are more likely to be successful in their careers and in their relationships.

23. It is important for people to acquire positive feedback so that they feel worthwhile, _____, and _____.

24. When people are insecure, they put others down in an attempt to _____ themselves up.

25. Positive attitudes are _____.

26. _____ _____ will soon disappear in an atmosphere of negativity and dismal attitudes.

MATCHING

_____ 1. To express approval of a behavior.

_____ 2. Outside forces that help a person achieve a goal.

_____ 3. Belief that one is being treated fairly in relation to others.

_____ 4. Removal of a positive reward.

_____ 5. Undesirable consequence for an undesirable behavior.

_____ 6. Any recognition that occurs verbally.

A. verbal stroke

B. equity

C. extinction

D. praise

E. punishment

F. external motivation

ACTIVITIES

1. Select two people you know well. On a piece of paper, write three factors that you believe would motivate each of them. Then ask the same two people what would motivate them to do their best. Compare their answers to yours. Did you accurately judge their motives? How could knowing what motivates others be helpful to you?

2. Analyze Maslow's hierarchy of needs. At which level are you presently? Explain your answer.

3. Demonstrate the use of each of the following reinforcement tools for reinforcing a behavior:

   a. Avoidance reinforcement

   b. Extinction

   c. Punishment

   d. Positive reinforcement

4. Refer to the needs associated with the learned needs theory. For each need, write a paragraph describing a person in whom that need is dominant.

5. Review the confirmation behaviors. Practice each one on a classmate. Then reverse roles.

NOTES

1. B. L. Reece, *Human Relations in Organizations: Interpersonal and Organizational Applications,* 12th ed. (Mason, OH: South-Wester, Cengage Learning, 2014), p. 148.
2. Ibid., p. 81.
3. C. Berryman-Fink, *The Manager's Desk Reference* (New York: AMACOM, 2nd ed., 1996), pp. 156–157.
4. A. H. Maslow, *Motivation and Personality* (New York: Harper & Row, 1954).
5. C. P. Alderfer, *Existence, Relatedness, and Growth: Human Needs in Organizational Settings* (New York: Free Press, 1972), esp. pp. 31–44.
6. V. H. Vroom, *Work and Motivation* (San Francisco: Jossey-Bass, 1995).
7. Ibid., pp. 17–23.
8. R. N. Lussier, *Human Relations in Organizations: Applications and Skill Building* (New York: McGraw-Hill Education, 2017, 10th Ed), p. 239.
9. Ibid., p. 240–242.
10. J. Stacy Adams, "Toward an Understanding of Inequity," *Journal of Abnormal and Social Psychology* 67(5): 422–436 (November 1963).
11. "Equity Theory of Motivation, Management Study Guide (MSG)," retrieved August 30, 2016 from http://managementstudyguide.com/equity-theory-motivation.htm.
12. B. F. Skinner, *Science and Human Behavior* (New York: Macmillan, 1953).
13. R. N. Lussier, *Human Relations in Organizations*, pp. 244–245.
14. Ibid., pp. 245.
15. E. Berne, *Games People Play: The Basic Handbook of Transactional Analysis* Old Saybrook, CT: Tantor eBooks, 2011), pp. 6–7.
16. E. Sieburg, "Confirming and Disconfirming Organizational Communication," in *Communication in Organizations*, edited by J. L. Owen, P. A. Page, and G. L. Zimmerman (St. Paul, MN: West, 1976), pp. 129–149, at pp. 130–134.
17. K. Blanchard and S. Johnson, *The New One Minute Manager* (New York: Harper Collins Publishing, 2015), p. 19.
18. B. Reece, *Human Relations in Organizations*, pp. 217–219.

17 CONFLICT RESOLUTION

Conflict *is what occurs when two or more groups or individuals have or think they have different goals, values, beliefs, or ideas about the way things should be. Conflict resolution, which consists of getting opposing groups of people or individuals to agree on a solution or course of action with which everyone wins, is one of the tools that every leader should have in the toolbox. Conflicts arise between people every day, and a leader skilled in conflict resolution can hold things together.*

Objectives

After completing this chapter, the student should be able to:

- Define conflict resolution
- Explain the relationship between conflict resolution and leadership
- Discuss ways of assessing conflict
- Describe the styles with which people approach conflict
- Successfully resolve conflict using some of the tools outlined in this chapter
- Discuss how to manage anger

Terms to Know

- conflict
- conflict resolution
- dimensions
- perceptions
- emotions
- structure
- culture
- avoider
- accommodator
- compromiser
- competitor
- collaborator
- dialogue
- negotiation
- distributive negotiation
- integrative negotiation
- collaborate
- mediation

t is hard to move or influence others toward a group goal if they do not get along. Just think about the kinds of problems that could have been avoided in your life if someone—maybe you—had known a little bit about conflict resolution.

In this chapter, you will see why conflict resolution is important to leadership. You will also learn how to assess conflict and identify its dimensions and causes. You will learn what people need when a conflict arises, as well as how different people approach conflict. Finally, you will be presented with some ideas and tools for successfully resolving conflict.

CONFLICT RESOLUTION AND LEADERSHIP

Conflict resolution may be defined as the tools and processes used by a leader to assess conflict and solve disputes and problems that exist between individuals and groups. There are other definitions of conflict resolution, but you will notice that our definition has four key parts: tools and processes, the leader, assessing conflict, and solving disputes or problems. These four parts are the basis for conflict resolution, and once you understand their roles you can master conflict resolution.

Leadership

You have learned so far that leaders have different personality types, they need to have good communication skills, and they have to know how to work well with others. What is unique about conflict resolution is that it takes all of the leadership skills you have learned about to this point to be successful. You will eventually encounter conflict in any situation in which you are leading. The question then becomes, "How will I handle this conflict?" Good leaders know about conflict resolution, and it helps them continue to fulfill their leadership duties. Leaders who choose to leave something as important as conflict resolution to chance or ignorance may be doomed to failure.

Imagine that because of your leadership training and experience you are hired as the supervisor of the local furniture company. During the first month you are doing great. Everyone is getting along, sales are high, and you are really enjoying your new job as the supervisor of the company. Then one day, you hear two salespersons arguing over the commission on a particular sale. You try to smooth over the situation and diffuse the hard feelings, and appear to have been successful, but you start to notice that neither of them is productive anymore. They will not communicate, but even worse, they will not sell furniture (Figure 17–1). What do

FIGURE 17–1 These individuals seem to be unable to understand each other. Leaders need tools to calm situations like this. *(© iStock/mediaphotos)*

you do? Solving this problem is your responsibility. We hope that this chapter will give you the tools you need to handle situations like this as you lead in your school, community, and workplace.

ASSESSMENT OF CONFLICT

Good public speakers analyze the audience before they begin speaking. To understand conflict resolution and be able to implement it effectively, you must be able to analyze or assess the conflict in the same way you would analyze an audience before giving a speech. How could you possibly be able to resolve a conflict if you did not know how bad it was, who was involved, or even what caused it? This section introduces you to two different types of assessment for conflict resolution: dimensions and causes. These two types of assessment set the stage for successful conflict resolution.

Dimensions of Conflict

Trying to answer the question "what is a conflict?" is sometimes harder than giving a simple definition. According to Mayer, conflict may be viewed as occurring in several **dimensions**, or types; he identified these dimensions as perception, feeling, and action. Seeing conflict as a perception, as a feeling, and as an action can help us analyze the complex nature of a conflict.[1]

Conflict as a Perception As a set of **perceptions** (that is, how we view the problem), "conflict is a belief or understanding that one's own needs, interests, wants, or values are incompatible with someone else's."[2] For example, if you and your best friend were both leaders of two different youth clubs in your school, and you both wanted the school auditorium for your end-of-year banquet on the same night, then your wants are getting in the way of your friendship, the progress of both your clubs, and a practical solution to the problem. Consider another example: you believe that homework is a useless waste of time, and your teacher believes that you will learn the material if you do homework. In this case your belief, which is really just a perception, is incompatible with your teacher's belief.

Conflict as a Feeling Conflict "involves an emotional reaction to a situation or interaction that signals a disagreement of some kind."[3] Emotions such as fear, sadness, bitterness, anger, or hopelessness indicate the presence of conflict regarding another person or situation. For the most part, if we feel that we are in conflict, then we are. Many times people get very emotional about a situation, and when leaders try to diffuse the conflict and discover the source of the problem, no one can really pinpoint the reason for the heightened emotions in the first place. All they know is that they are upset. It does not always take two (or more) to have a conflict: if conflict is a feeling, then even if only one person has that feeling, there is a conflict, although the other person(s) may not be aware of the problem.

Conflict as an Action "Conflict also consists of the actions that we take to express our feelings, articulate our perceptions, and get our needs met in a way that has the potential for interfering with someone else's ability to get his or her needs met."[4] These actions may take the form of a direct attempt to make something happen, despite the fact that success may come at someone else's expense. A conflict action could be an exercise of power; it could be violent or destructive. In contrast, conflict as an action could be constructive, friendly, or helpful. Like feelings, actions can be two-way or one-way. If we both complain to the principal about not having the auditorium on the night we want to have our event, our action conflict is two-way. If you feel that this book is not helping you learn about leadership and you complain to your teacher, but we (the authors) never know about your complaints, there is still a conflict—reciprocity is not required for a conflict to arise. In any case, the purpose of conflict action is to bring to light the fact that there is a conflict or to get one's needs met.

It is important that we understand the dimensions of a conflict before we try to resolve it. If we try to deal with the feelings without knowing what the perception of the conflict is, it will be very hard to help our followers settle their problems. If we try to solve only the conflict action without attending to the other dimensions, our attempted resolution could also fail. All three dimensions do not always exist together, but we should be aware of their possible coexistence and be aware of the dimension with which we are dealing. When we have assessed the dimension

---

VIGNETTE 17-1 CONFLICT CAN BE GOOD

MANY PEOPLE BELIEVE THAT CONFLICT can actually be a very good thing if you know how to manage it. Steven Kerr, the Chief Learning Officer for General Electric Company (GE), outlined the following steps for managing conflict:

1 *"Acknowledge its importance.* Don't worry that your people are upset with each other—they're supposed to be."

If you are a leader who supports change, conflict will occur.

2. *"Do not tolerate personal attacks."*

There are rules about respecting and disagreeing with others. As a leader, you should teach your followers and other leaders about such rules and always model them. Lead by example.

3. *"Provide support."*

When you know that an upcoming meeting involves people with varying opinions and ideas about how things should be handled, and that the discussion could become heated, it is a good idea to use facilitators or mediators to avoid difficult situations and guide participants through rough patches.[5]

Good conflict management can be very valuable to leaders who encourage people to voice their opinions. If handled correctly, conflict can stimulate a very empowering atmosphere. Do you think you could work in an environment that encourages a healthy amount of dialogue and conflict? Can you think of a conflict that turned ugly that might possibly have turned out to be positive had it been managed correctly?

---

of the conflict, then we can address the causes of conflict.

Causes of Conflict Conflict always has a source (that is, a cause). The possible sources or causes of conflict may be nearly infinite, but in this section we try to establish a basic outline for understanding some of the broad categories into which most conflict sources or causes fall. First of all, we should note that there are almost as many theories about the causes of conflict as there are leadership theories. According to one book written by a team of authors, the many theories about the causes of conflict break down as follows:

- Principled negotiation theory
- Community relations theory
- Human needs theory
- Identity theory
- Intercultural miscommunication theory
- Conflict transformation theory

Each of these theories makes certain assumptions about the causes of conflict and suggests certain goals for resolving conflict. The theory that we consider the most practical and easiest to understand is the human needs theory. This theory "assumes that deep-rooted conflict is caused by unmet or frustrated basic human needs—physical, psychological, and social."[6]

The needs of the people who are in conflict are at the center of the many different causes underlying conflict. Figure 17–2 shows the major sources or causes of conflict. Notice that "Needs" are in the center of the diagram like the hub of a wheel. The outside components of the wheel are communication, emotions, structure, values, interests, procedures, relationships, power, culture, and history.[7]

Communication Communication can be a cause of conflict because many of us are not very good at it most of the time. We are sure you can think of instances in which you and someone else did not really disagree, but you thought you did because of poor communication. We cannot always assume that our communication is perfect, and we should realize that others' communication may be inaccurate or incomplete as well.

A number of factors can contribute to communication problems. Many times, people from different cultures have trouble communicating because the symbols they use are different. Males and females can have trouble communicating because of their different needs and ways of viewing the world. The same thing could happen between the diverse personality types. Can you imagine a choleric personality having communication problems with a sanguine personality? Although communication is a serious cause of conflict, it can also be

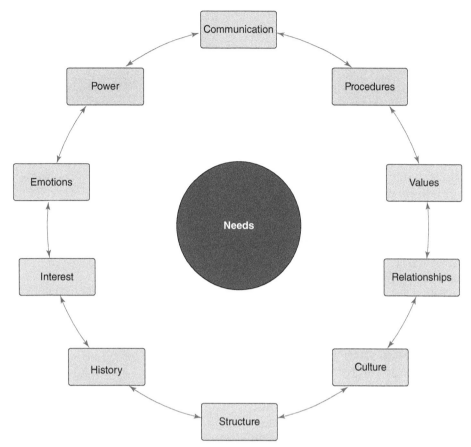

FIGURE 17–2 This diagram represents the complexity of conflict resolution. At the center of the diagram are the needs to be met before a conflict can be resolved. Surrounding those needs are the major sources of conflict. If there is a misunderstanding or need in any of the surrounding parts of the wheel, then a significant conflict is present and should be dealt with accordingly. *(Adapted from Bernard Mayer, The Dynamics of Conflict Resolution: A Practitioner's Guide [San Francisco: Jossey-Bass, 2000], pp. 4–5.)*

one of the greatest assets in reaching solutions, as discussed later in this chapter.

Emotions "Emotions are the energy that fuel conflict."[8] A small conflict can become a large one—very quickly—when emotions take over. We cannot hide from emotions or keep them from becoming involved. They are always present in our lives, especially when we consider our needs. As a leader, you have to be and act emotionally mature, which means that you do your best to look at the facts and data logically and rationally while removing yourself emotionally from situations.

Emotional issues arise from underlying psychological needs that those involved in conflict perceive to be at risk. The psychological needs that usually spark our emotions are

- Power—the need to be able to influence others

- Approval—the need to be liked and respected by others
- Inclusion—the need to be accepted as a social-group member
- Justice—the need to be treated fairly and equitably
- Identity—needs for "autonomy, self-esteem, and affirmation of our personal values"[9]

When any of these psychological needs is threatened, our emotions are engaged. Although it might seem that this involvement or elevation could only worsen conflict, our emotions can also be used to diffuse and resolve conflict. When our needs are not threatened, we can use our emotions in a productive manner to help our followers.

Structure Another major source of conflict is the structure of an interaction between people or of the situation in which they are reacting.

The **structure** is the external framework of a problem and could "involve available resources, decision-making procedures, time constraints, communication procedures, and physical settings." Structure issues may get in the way even when the people in conflict are close to agreement or to finding a solution to their conflict.[10]

What if you were in a business meeting where two opposing groups, both of which agreed that a certain item should be purchased, could not agree on how it should be purchased? One group wants to finance the item, whereas the other group wants to generate the income from sales and donations. Which of the structure elements would you say is at work in this scenario? The lack of resources, which is an element of structure, seems to be the only real problem in this particular conflict.

Values Values are beliefs that each person has about what is right or wrong. These beliefs can be a serious source of conflict. Sometimes values can create the conflicts that are the hardest to solve, because of how personally people take certain issues. Over the years, conflicts between people's beliefs about right and wrong (on matters such as religion, politics, equal rights, statehood, and the like) have escalated to become the cause of wars within and among nations and people.

Differences of opinion may lead to everyday conflicts as well. If you were the president of an environmental protection organization and your friend was the head of a timber production company, your ideas about right and wrong in forestry management could lead to a conflict before each organization's needs have been met. Can you think of other leadership situations in which you might find yourself in a conflict and what you believe to be right or wrong appears to be standing in the way of progress?

Interests Specific wants or desires are the most common sources of disagreement. Resolution of disputes between people, businesses, or any combination of various parties usually involves solutions that will satisfy the needs and interests of all parties—but to achieve resolution, we must keep working until acceptable solutions are found.[11]

Interests are a unique conflict cause that leaders must be aware of, because they can prevent some conflicts by paying attention to the interests of

FIGURE 17–3 This student leader has prevented conflict from arising by identifying interests that he and his team members have in common. *(Courtesy of University of Georgia—College of Agricultural & Environmental Sciences.)*

their followers. Charismatic leaders should also be great at convincing followers to "buy into" the leaders' way of thinking or their interests. Some of the best leaders have developed effective ways of "getting everyone on the same page" for the good of their collective purpose. The student in Figure 17–3 has convinced his team that they have a common set of interests as they partner on the group project and avoid conflict.

Procedures People accept election results when they believe that the election process was fair, and they submit to a court's decision because the legal procedures that produced that decision followed a predictable and understood process. However, when people disagree about procedures—that is, the way in which problems are solved, decisions are made, or conflicts are resolved—conflict frequently arises.

Procedures serve as another excellent way for leaders to avoid conflict, quashing it before it ever starts. You can organize people and situations, as well as the processes they perform, in a way that minimizes the potential for conflict. Parliamentary procedure, for example, is a conflict-limiting process for conducting meetings and business that works primarily by establishing majority rule and respect for the rights and opinions of everyone. The meeting taking place in Figure 17–4 looks to be productive and conflict-free. Trying to conduct a meeting without parliamentary procedure, or at least some agreed-upon, accepted procedure, is just asking for conflict to arise.

FIGURE 17–4 The meeting taking place in this image is calm, productive, and without conflict. Parliamentary procedure is a process that limits and channels conflict. Can you think of other processes that limit conflict?
(© iStock/Christopher Futcher)

Relationships This text has already discussed the value of important relationships. Relationships with followers and other leaders can be the disease or the cure, but what about poor relationships? What kind of power do they have? "People may resist cooperating if they do not trust others, do not feel respected by others, do not believe that the other person is honest, or do not feel" that they are being listened to.[12] Conversely, when we develop solid relationships in which we trust, respect, and listen to each other, the potential for conflict is greatly reduced (Figure 17–5).

Power Although power is involved in some way in every conflict, power is also a source of conflict in itself. Bernard Mayer called power "the currency of conflict."[13] Misuse or abuse of power can cause conflict to arise or escalate very quickly. Suppose you were just elected as the president of the FFA. You do not want to let the outstanding reputation of your chapter fade, so you start shouting orders and exerting your newfound power wherever and whenever you can. Do you think conflict is going to find you? Of course it will.

Power is a double-edged sword, though. It can also be a great source of help in resolving conflict.

Sometimes a person in a position of power is able to make decisions that resolve a conflict more quickly than someone without power or authority ever could. As a leader, do not hesitate to use your influence to make decisions if those decisions will resolve conflict in a way that is helpful and acceptable to everyone involved.

History Conflict, its causes, and potential solutions cannot be realized without understanding the background and history of the conflict. Specifically, the history of the people involved in the conflict can provide a wealth of information that will be helpful in successful conflict resolution. Mayer states that history provides the momentum for the development of conflict. Studying the background allows for slowing the momentum of developing conflicts and for understanding volatile situations and personalities.

Culture Culture refers to socially transmitted behavior patterns, beliefs, and all other products of human work and thought patterns of a society, community, or population. It has been defined as "the particular practices and values common to a population living in a given setting."[14] What this means is that people who live

FIGURE 17–5 The supervisor on the left is listening to the concerns of the employee at this farm. The employee wanted to feed the dairy calves in a different manner than the supervisor had taught her; and after she explained her reasons for the proposed change, which were well thought out, the supervisor agreed to her suggestions with some concessions. *(Courtesy of University of Georgia—College of Agricultural & Environmental Sciences.)*

in certain situations, and are brought up by certain groups, are expected to act in certain ways. These ways—their values, traditions, mores, and so on—are their culture. When a leader's actions go against someone's cultural norms, conflict can ensue, because people usually act in accordance with their culture so as not to be rejected by their friends and family. It is our responsibility to be aware of the differences in the cultures of the people with whom we work. Through sensitivity to these issues, we could identify and eliminate situations in which our followers might have to compromise their cultural beliefs.

CONFLICT STYLES

Do you remember the four personality types, leadership styles, and communication styles? Well, people also approach conflict with different styles. The style used not only relates to personality type but also depends on the particular problem and how much a person is affected by the conflict. As a leader, you can use different conflict and conflict-resolution styles to help you diffuse almost any situation. The five styles discussed in this section represent the ways in which people most commonly approach conflict.[15]

Avoider

The **avoider** denies the very existence of a conflict, changes the subject, ducks discussion, and is noncommittal. This style is most effective when the issue is not terribly important, there is no apparent way to achieve goals, the complexity of the situation prevents resolutions, or there is a chance of physical violence. The avoider is not interested in confrontation of any kind and will give up or concede his position on the issue just to avoid conflict.

Accommodator

The **accommodator** sacrifices his or her interests and concerns while helping others to achieve their interests. "This style is effective in situations in which there is not much chance of achieving one's own interests, when the outcome is not important, or when one party believes that satisfying his own interest will in some way alter or damage the relationship."[16] Accommodators are different from the avoiders in that they have a concern for others, whereas the avoider's major concern is for himself or herself.

Compromiser

The **compromiser** tries to get all parties to work together to settle for partial satisfaction of their interests. This style is effective in situations "that require quick resolution of issues, when other parties resist" cooperation, "when complete achievement of goals is not important," and no hard feelings will be generated by settling for less than what is desired or expected.[17] This style calls for a leader to be very sensitive to the interpersonal signs (particularly nonverbal communication) displayed by the people who are resisting cooperation.

Competitor

The **competitor** conflict style is "characterized by aggressive, self-focused, forcing, verbally assertive, and uncooperative behavior" that is intended to achieve one's own interests at the expense of others. This style is effective when decisions must be made quickly or when options are limited. It is a useful style when nothing is risked by pushing or when other parties are resisting cooperation. However, this style should be used only when there is no concern about damaging the relationship between the parties.[18]

Collaborator

The **collaborator** is an active listener who focuses on the issues. This person is a good communicator who tries to meet interests and solve concerns of everyone involved in a conflict. This style is effective in situations in which the parties hold approximately equal power, when all parties value their relationship and wish to continue it, and when the parties are willing to cooperate. When you use this style, there must be enough time and motivation to come up with a workable solution for everyone.

Conflict Styles and Personality Types

Did you see any similarities between these styles and the personality types? Which of the conflict styles do you think the sanguine personality type

would naturally choose and work best with? Which style do you think an autocratic leader would use? By now you should be able to spot the relationships among styles. If not, the sanguine personality type would accommodator; the choleric personality type would bet the competitor; the melancholy personality type would e the compromiser; and the phlegmatic, the collaborator. We hope that your knowledge of the styles used to approach conflict will help you to actually resolve conflict by assessing the conflict and using negotiation and/or mediation.

RESOLVING CONFLICT

You should already be familiar with the first step involved in resolving conflict, which is assessment. If you can determine how much power one person or group has over another, it becomes easier to begin to solve disputes or change situations so that they will lead to answers. The same can be said of knowing what cultural issues you are dealing with, the intensity of the conflict, and even the personality or conflict style of the persons involved. All of these things will help you decide whether to use confidence building, dialogue facilitation, negotiation, or mediation.

Confidence Building

Confidence building involves rebuilding and enhancing mutual trust and confidence between two conflicting parties. Sometimes confidence building is the answer by itself, simply because the only reason for the conflict was that one party felt oppressed by the perceived power of the other party.

Confidence building can also be the backbone of other conflict-resolution skills. If two individuals in an organization are arguing over a committee decision, confidence building can be the first move. It is preferred as the first resolution attempt because decisions are not easily made when anger is present, or when those involved do not perceive one another as trustworthy.

Dialogue Facilitation

Dialogue is when two or more people engage in a conversation from which everyone learns something. When a leader has the ability to facilitate dialogue, he or she can get conflicting parties to communicate directly in order to learn something and come to a win-win agreement.[19] In literature, *monologue* is when only one character speaks; "dialogue occurs when characters speak with one another."[20] Dialogue as a form of communication is a great tool for conflict resolution. It is a form of communication in which participants engage in each of the following activities:

Exploring the Rules of the Communication This is important because many times people do not know how much or how little they should share in pursuit of a resolution. People need to feel that it is safe and acceptable to discuss the problems at hand.

Exploring Contexts of Meaning This refers to identifying each person's frame of reference, or the different perspectives about what a good solution looks like, based on how that person sees the issue.

Exploring Differences This discovery task simply involves making sure that everyone involved understands all sides of the situation (even if they do not agree with others' positions). At the least, the true bases of the conflict tend to be clarified.

Exploring Common Ground This effort is important because—believe it or not—all parties to a conflict almost always share some similar interests or concerns. When you find common ground, you are on your way to resolving the conflict.

To be good at facilitating dialogue, you have to be a good communicator. To be a good communicator, you have to be a good listener. Here is a list of items good communicators and dialogue facilitators listen for as they seek to resolve conflict:

- **Lived experience**—What personal experiences is the speaker sharing? What is it like to be involved in the things she or he has been through?
- **Stories told**—What do the stories tell you about the speaker? What are the speaker's ideas about beliefs, morals, and the issue at hand?
- **Values**—What is important to the speaker?
- **Contexts**—What is the speaker's frame of reference?
- **Rules**—What do certain things mean to the persons involved? How do they understand or interpret various words and actions?
- **Differences and levels of difference**—Do certain things take on more or less

---

VIGNETTE 17-2   SAVING FACE

HAVE YOU EVER HEARD the term *saving face*? If you have not yet, you will. It is usually the first thing people do when conflict arises. Saving face means preserving your pride and/or status (or perceiving that you have done so). You still win. This desire is also why conflict resolution is such a balancing act. You have to arrive at a workable solution to the conflict while at the same time keeping everyone's pride intact.

Jason had a conflict with Jill at their law firm. Jill was Jason's superior, but he knew a good client when he saw one, so he insisted they take on a new client he had identified. Jill told him "no" once, but Jason was persistent. He was pressuring Jill about the new client in the middle of the office one day, and Jill started yelling at Jason and telling him why he was not ready for such a client. Immediately, Jason was demoralized. Everyone in the office saw him get reprimanded. Jason had "lost face" through that very public incident, and he quit right then and there.

When someone loses face, his or her available options are limited. In this case, Jason decided to quit the law firm. Many times, the person who loses face will fight on to the end, all the while knowing that he or she cannot win. Acting out by continuing to struggle is that person's way of winning or trying to restore some of his or her "face."

In the case of Jason and Jill, how could the argument and Jason's resignation have been prevented? If Jill had allowed Jason to save face by discussing the matter behind closed doors, might the situation and result have been different?

Have you ever felt as though you were trying to save face, or have you ever seen someone trying to save face? Just remember that no one likes to be embarrassed or humiliated; everyone seems to retain a certain amount of pride that he or she would like to keep intact. As a leader who knows how to manage conflict, you should allow all parties involved to save face.

importance in different situations and contexts?

- **Common ground and levels of common ground**—Where are the areas of agreement?[21]

Negotiation

Negotiation is a tool of conflict resolution that simply helps people to work together (Figure 17–6). When a leader finds that followers have a problem with each other, with the leader, and/or with other leaders, negotiation is usually the first step. Most leaders use one of two basic types of negotiation. Those with authoritarian leadership styles more commonly use the first type, distributive negotiation. Integrative negotiation is the approach used by democratic leaders.

Distributive Negotiation **Distributive negotiation** is a bargaining form of negotiation in which the negotiators elevate their own self-interests over any collaborative answers to conflict. People who engage in distributive negotiation are referred to as *bargainers*. Bargainers

attempt to attain their goals at the expense of others and often use threats and demands to coerce others.

Integrative Negotiation **Integrative negotiation** "is a constructive, problem-solving process, the goal of which is to maximize [the] interests of both parties while protecting the relationship."[22] This is basically the philosophy of democratic leadership, whereby everyone's voice is heard, considered, and appreciated. A leader who is able to engage in democratic or integrative negotiation wants mutual gain for everyone. This is accomplished through creativity, innovative "thinking outside the box," and flexibility on the part of all involved in the conflict.

Imagine that you are the manager of a successful software company, and another company accuses you of stealing its ideas. You know that the accusation is baseless, because you are the only person who would know about the product it claims your company stole. Which type of negotiation would be most helpful? Would it be distributive, integrative, or neither? The answer would depend on a lot of things, but it could be neither. (We will explain why later in

FIGURE 17–6 Negotiation can take place anywhere. These leaders from different countries work hard to understand each other's point of view before resolving a conflict related to a best course of action for an intercultural leadership development activity. *(© iStock/KatarazynaBialasiewicz)*

this chapter.) First, though, let us examine the steps of the negotiation process for democratic leadership.

The Negotiation Process The process for integrative negotiation depends on a number of things, the first of which is how willing each party is to **collaborate**, or work together to reach a decision that brings mutual gain. The steps of the negotiation process are as follows:

1. Get the parties' commitment to negotiate.
2. Discuss the cost and consequences of not settling. Talking at the beginning about what will or may happen if the parties cannot reach an agreement helps reduce the fear of losing.
3. "Begin with an agreement on a definition of the problem." This allows everyone to focus on solving the same problem and avoids working at cross-purposes.
4. Identify the interests, concerns, and goals of everyone involved.
5. "Discuss the most important issues first."
6. Explore alternatives and trade-offs without requiring commitment. This is where creativity and thinking outside the box can lead to answers no one ever considered before.
7. Devise solutions that "maximize gains for all parties."
8. Evaluate the costs and benefits of each solution. Choose the best (most acceptable) one.
9. "Agree on a plan for implementation."
10. "Formalize the agreement."[23]

Following this process should lead to successful conflict resolution; at worst, the parties can agree to disagree, to bring in a mediator, and to maintain positive relationships.

Mediation

Either the parties have tried the negotiation process without success, or the conflict was such that the parties could not sit in a room together and effectively and rationally discuss the issues. This appears to create an impasse, but there is still mediation. **Mediation** is a process by which a third party intervenes in a conflict or disagreement to help clarify the problem and resolution options. In negotiation, a leader is usually involved in the conflict; in mediation, a leader can act as the neutral, third-party mediator only if she or he is not a part of the conflict.

To mediate effectively, a leader should work under the following principles outlined by Fisher et al. in *Working with Conflict*:

- Mediators need to become involved with and attached to all sides. One might think that the mediator should remain completely uninvolved and utterly neutral, but mediation is actually more effective when the mediator cares for and empathizes with all those involved.
- All sides must voluntarily agree to participate in the process and must accept the particular mediator, granting him or her authority to conduct the process.
- Mediators must be willing to work with all sides.
- Mediation is not really concerned with the best solution overall, but rather with finding a solution to which all parties can agree.
- Mediators guide and control the mediation process, but must avoid trying to direct the content of discussions.
- Answers and solutions must come from the parties to the conflict, not the mediator.[24]

A mediator who has been trained in leadership, as many of you have, should have no problem abiding by these principles. As a mediator, you are a facilitator, a coach, a teacher, a resource, a catalyst, and someone who helps people in conflict stay connected to reality. Keep all of these things in mind as you lead people through the mediation process.

The Mediation Process The mediation process bears a strong resemblance to the format of a speech or letter.

1. Make introductions. In this very important phase, the mediator explains what mediation is and what the mediator's role is in the process. He or she then introduces the parties and establishes and explains the rules by which the mediation will be governed. At this time, all parties agree on an agenda for the mediation.
2. Share information. In this phase, the context of the problem is reviewed. Both parties tell their stories, one at a time,

and interests are identified, shared, and clarified.

3. Generate options. Through brainstorming or other methods, possible answers are discussed, as well as the costs and benefits of each solution. With the guidance of the mediator, the parties then choose the best option (the one they prefer most).
4. Write up a draft agreement. It is best to do this collaboratively so that all parties have ownership of the final agreement.
5. Decide how to implement the agreement.
6. Set plans for follow-up evaluation of the implementation and results.[25]

Mediation is a tool for conflict resolution that any good leader can use. A good leader understands the personalities of others. A good leader understands that different people are motivated by different things. A good leader listens, teaches, and empowers others. A good mediator does the same.

MANAGING ANGER

Anger, especially as the result of unresolved conflict, is a powerful emotion. Think about the last time you were angry. What did you do? Did you tell the person you were angry, and why you were angry? Did you snap at the person about something else? Or did you keep your feelings to yourself? People express and react to anger in different ways.

Expressing Anger Directly

People often express their anger directly. If someone annoys you, you tell him or her so, or you glare at that person, or take any number of other inappropriate (and often ineffective) actions to display your anger. Obviously, a direct expression of anger can range from assertiveness to aggression to violence. How people express anger directly depends on their personalities and the extent to which they have been provoked. People with negative self-belief often have an underlying attitude of hostility that is easily triggered by even minor events. Others who are more secure in their self-belief and self-image can express anger more calmly without being aggressive.[26]

Expressing Anger Indirectly

Anger may also be expressed indirectly. For example, instead of confronting the person with whom you are angry, you direct your anger at a third party who is less threatening or more available.

There are many situations in which it is inappropriate to express anger at the person with whom you are angry. Suppose you have just started your own landscaping business and one of your clients keeps changing his mind about what he wants you to do. You are angry because he is wasting your time, but expressing your anger directly will cause you to lose a customer. In this case, your anger may find an outlet when you snap at one of your friends or yell at your dog.

Internalizing Anger

A third way to deal with anger is to keep it bottled up inside. Many people consider outward expressions of anger to be threatening or rude, so they internalize the anger and keep it covered up. Unfortunately, the normal result of internalizing anger is a growing resentment. Because you do not express your anger, there is no way for the conflict to be resolved, and the anger only intensifies. Internalized anger can cause stress and harm your emotional and physical health.

Controlling Anger

You can minimize the destructiveness of anger by controlling it properly. There are several approaches you can take:

- Never say or do anything immediately. It is usually best to calm down and give yourself a chance to think. The old trick of counting to 10 may help.
- Figure out why you are angry. Sometimes the cause of the anger is something you can easily change or avoid.
- Channel your anger into physical exercise. Even a walk can relieve the tensions of anger.
- Use relaxation techniques such as deep breathing to calm yourself.

Once your anger is under control, you can try to resolve the conflict that caused it. The energy created by your anger can be channeled into solving the problem. Here are some suggestions:

- Commit yourself to resolving the problem that caused the conflict. Do not just decide to keep the peace by remaining silent.
- Ask yourself what you hope to achieve by resolving the conflict. Is it critical to "get your way," or is your relationship with the other person more important? Your objective will influence how you settle the conflict.
- Make sure you and the person or party with whom you have a conflict are in agreement about what the actual conflict is. Ask questions and really listen. You may be surprised. Some conflicts are the result of misunderstanding.
- Be assertive, not aggressive. Remember that the other person has rights and feelings, too.
- Try to stick to the facts. Make sure you understand the distinction between facts and feelings. The more you can keep emotion out of your discussion of the issue, the better your chances are for resolving the conflict.[27]

At first, you may find it difficult to control your anger and approach conflicts in a more thoughtful, rational way. With practice, though, you will become more comfortable in dealing with conflict. You may find that effectively resolving conflict is a way to learn more about yourself and to grow, as well as to improve the quality of your relationships with the people around you.

CONCLUSION

Any people or groups who remain together for any length of time are sure to disagree about something. The conflict may be over what time a dating couple wants to go to the movies or whether the FFA should invite the mayor to the banquet. If disagreements cannot be settled, people become frustrated and angry. This scenario is so common that you may think conflict is always a negative experience. Yet, if handled properly, conflict can have healthy and productive results.

According to Aristotle, the ancient Greek philosopher, "Anybody can become angry—that is easy, but to be angry with the right person and to the right degree and at the right time and for the right purpose, and in the right way—that is not within everybody's power and is not easy."[28]

SUMMARY

Conflict occurs when two or more groups or individuals have or think they have different goals, values, beliefs, or ideas about the way things should be. Conflict resolution consists of the tools and processes used by a leader to assess conflict and resolve disputes and problems.

You assess a conflict in much the same ways as you would analyze an audience before speaking. There are two different types of assessment for conflict resolution: dimensions and causes. The dimensions of conflict resolution include conflict as a perception, conflict as a feeling, and conflict as an action. The causes of conflict include communication, emotions, structure, values, interests, procedures, relationships, power, history, and culture.

As a leader, you can use different styles of approaching conflict and conflict resolution to help you diffuse almost any situation. The five styles of dealing with conflict are avoider, accommodator, compromiser, competitor, and collaborator.

If you can determine how much power one person or group has over another, it becomes easier to resolve a conflict between them. Four ways to resolve conflicts are confidence building, dialogue facilitation, negotiation, and mediation.

Anger, which results from unresolved conflict, is a powerful emotion. People handle anger in different ways: expressing it directly, expressing it indirectly, internalizing it, and/or controlling it. At first, it may be difficult to control your anger, but you will find that doing so allows you to approach conflicts in a more thoughtful and rational manner and leads to a better chance of conflict resolution.

Take It to the Net

Explore conflict resolution on the Internet. Use your favorite search engine and identify policies and procedure used by different groups (businesses, youth clubs, etc.) to resolve conflict.

1. This chapter listed six different theories about the causes of conflict, but discussed only one of them. Go to the Internet and look up the other theories mentioned in this chapter. Outline the basics of each theory.

Chapter Exercises

REVIEW QUESTIONS

1. Define the Terms to Know.
2. Explain why conflict resolution is important to leadership development.
3. List and describe the dimensions of conflict resolution.
4. List and briefly describe the causes of conflict.
5. Explain human needs theory.
6. List and briefly describe the psychological needs that spark our emotions.
7. Name the conflict styles and explain how each deals with conflict.
8. Describe five tools of conflict resolution.
9. Compare and contrast distributive and integrative negotiation.
10. Outline the steps of the negotiation process.
11. Describe the principles of mediation in this chapter.
12. List the steps of the mediation process.
13. What are four ways to manage anger?
14. List and describe five psychological needs that usually spark our emotions.
15. List and describe why dialogue as a form of communication is a great tool for conflict resolution.
16. List seven items that good communicators and dialogue facilitators listen for as they seek to resolve conflict.

COMPLETION

1. _____ occurs when two or more groups or individuals have or think they have different goals, values, beliefs, or ideas about the way things should be.

2. _____ needs theory assumes that conflict is caused by unmet or frustrated basic human needs, such as physical, psychological, and social needs.

3. _____ building involves rebuilding and enhancing mutual trust and confidence among the parties to a conflict.

4. _____ negotiation is a constructive, problem-solving type of negotiation.

5. _____ negotiation is a type of negotiation in which the bargainer considers only himself or herself.

6. Although _____ is a serious cause of conflict, it can also be one of the greatest assets in reaching solutions.

7. A small conflict can become a large one very quickly when _____ take over.

8. It is our responsibility to be aware of the differences in the _____ of the people with whom we work.

9. When you find _____ _____, you are on your way to resolving conflict.

10. To be good at facilitating dialogue, you have to be a good _____.

11. A _____ is a facilitator, coach, teacher, resource, catalyst, and someone who helps people in conflict stay connected to reality.

MATCHING

Terms may be used more than once.

_____ 1. Types of conflict.

_____ 2. How we see the conflict.

_____ 3. The energy that fuels conflict.

_____ 4. The external framework of a problem.

_____ 5. A person's beliefs about the right and wrong of an issue.

_____ 6. The "currency of conflict," according to Mayer.

_____ 7. The particular interests and values common to a population living in a given setting.

_____ 8. When two or more people are engaged in a conversation.

_____ 9. Conflict-resolution tools that help people work together.

_____ 10. Necessary tool when a third person must intervene in a conflict.

A. mediation

B. negotiation

C. dialogue

D. culture

E. power

F. values

G. structure

H. emotions

I. perceptions

J. dimensions

ACTIVITIES

1. Look at the diagram representing the complexity of conflict. Draw a picture representing solutions to conflict.

2. Develop a scenario of conflict in which there is a problem between two people or groups, and then detail how each of the conflict styles would deal with the conflict.

3. Think back to a conflict you have had as a leader. Describe the structure of the conflict and how you could have handled the conflict differently.

4. Explain how you would use your values and culture to handle conflict.

5. Match each conflict style with the personality type to which you think it is most similar. Remember, there are five conflict styles, so you may have more than one for each of the personality types.

6. Consider the following situation: Two students want the lead role in the school play, and both would be good. Divide the class into five groups, with each group representing one of the conflict styles. Have each group act out how its style would solve the problem.

7. Consider the same scenario as in activity 6, but divide students into groups of three. There is one mediator in each group. Have two of the students adamantly argue their positions while the third uses the steps of mediation to solve the conflict.

NOTES

1. B. Mayer, *The Dynamics of Conflict Resolution: A Practitioner's Guide*, Kindle ed. (San Francisco: Jossey-Bass, 2010).
2. Ibid., p. 132.
3. Ibid., p. 135.
4. Ibid., p. 142.
5. S. Kerr, "GE's Collective Genius," in *Leader to Leader: Enduring Insights on Leadership from the Drucker Foundation's Award-Winning Journal*, edited by F. Hesselbein and P. Cohen (San Francisco: Jossey-Bass, 1999), pp. 227–235, at p. 231.
6. S. Fisher, J. Ludin, S. Williams, D. I. Abdi, R. Smith, and S. Williams, *Working with Conflict: Skills and Strategies for Action* (New York: Zed Books, 2000), p. 8.
7. Mayer, *The Dynamics of Conflict Resolution*.
8. Ibid., p. 198.
9. D. Dana, *Conflict Resolution: Mediation Tools for Everyday Worklife* (New York: McGraw-Hill, 2001), p. 109.
10. Mayer, *The Dynamics of Conflict Resolution*, p. 226.
11. M. W. Isenhart and M. Spangle, *Collaborative Approaches to Resolving Conflict* (Thousand Oaks, CA: Sage, 2000), p. 14.
12. Ibid., p. 15.
13. Mayer, *The Dynamics of Conflict Resolution*, p. 50.
14. Fisher et al., *Working with Conflict*, p. 41.
15. See also S. Evans and S. S. Cohen, *Hot Buttons: How to Resolve Conflict and Cool Everyone Down* (New York: Cliff Street Books, 2000), pp. 61, 70–79; L. Eisaguirre, *The Power of a Good Fight: How to Embrace Conflict to Drive Productivity, Creativity, and Innovation* (Indianapolis, IN: Alpha Books, 2002), pp. 50–59.
16. Isenhart and Spangle, *Collaborative Approaches to Resolving Conflict*, p. 26.
17. Ibid., pp. 26–27.
18. Ibid., p. 27.

19. Fisher et al., *Working with Conflict*, p. 95.
20. S. W. Littlejohn and K. Domenici, *Engaging Communication in Conflict: Systemic Practice* (Thousand Oaks, CA: Sage, 2001), p. 25.
21. Ibid., pp. 37–38.
22. Isenhart and Spangle, *Collaborative Approaches to Resolving Conflict*, pp. 45–47.
23. Ibid., p. 48.
24. Fisher et al., *Working with Conflict*, pp. 117–118.
25. Isenhart and Spangle, *Collaborative Approaches to Resolving Conflict*, p. 79.
26. R. K. Throop and M. B. Castellucci, *Reaching Your Potential: Personal and Professional Development*, 4th ed. (Clifton park, NY: Wadsworth, Cengage Learning, 2011), pp. 242.
27. Ibid.
28. Retrieved August 31, 2016 from http://www.brainyquote.com/quotes/authors/a/aristotle.html.

Section 5 PERSONAL DEVELOPMENT

18 SELF-CONCEPT

Positive self-concept can help leaders and followers who are striving for a specific goal get there in efficient and enjoyable ways. Unfortunately, a poor self-concept can limit not only group success but also self-improvement efforts and personal achievement. Leaders make it their mission to maintain a positive self-concept for themselves and to foster it in their followers.

Objectives

After completing this chapter, the student should be able to:

- Discuss the importance of self-concept
- Discuss the ingredients of self-concept
- Discuss the factors that affect the development of self-concept
- Develop a positive self-concept
- Describe the characteristics of people with a positive self-concept
- Explain how leaders can raise the self-concept of others

Terms to Know

- self-concept
- conceit
- self-determination
- self-responsibility
- resilient
- fear
- doubt
- self-fulfilling prophecy
- anxiety
- surface analysis
- desire
- action
- Pygmalion

"An individual's self-concept is the core of his personality. It affects every aspect of human behavior: the ability to learn, the capacity to grow and change, the choice of friends, mates and careers. It's no exaggeration to say that a strong, positive self-image is the best possible preparation for success in life."

—Dr. Joyce Brothers[1]

An effective leader possesses a wide variety of skills, as you have learned in previous chapters. The effective leader must use his or her abilities and styles to master these skills and put them to use in achieving personal goals. The leader also encourages and motivates others to achieve their own goals as well as the goals of a group or organization.

Where do these abilities come from? How can you gain these abilities? In this chapter you will learn that **self-concept** (the way one perceives and feels about oneself) is a key ingredient in attaining the abilities and skills of an effective leader. You will learn the definition of positive self-concept, why it is important, what affects a person's self-concept, and what you can do to improve your self-concept. Becoming an effective leader begins with self, and that self is important and valuable. There is only one you. You are unique! You can make a difference!

To have a positive self-concept is to respect yourself and appreciate who you are. Even though you are aware of your weaknesses, as well as your strengths, when you have a positive self-concept you believe in yourself. A person with a strong self-concept is able to admit mistakes and accept setbacks without shame, guilt, or blame. As you develop a positive self-concept, you will no longer feel that you constantly need to prove yourself. You will be able to handle tasks without fear of failure or defeat. Relationships with others will become stronger; a person with a positive self-concept accepts and understands the weaknesses and mistakes of others. Because of your positive outlook and enthusiasm, people will want to be around you (Figure 18–1).

FIGURE 18–1 A person with a positive self-concept is pleasant, emotionally secure, and content, like the person shown here. *(© iStock/michaeljung)*

IMPORTANCE OF SELF-CONCEPT

When your self-concept improves, your performance improves. Having a positive self-concept is the single most important factor in achieving success. If you do not believe in yourself, how can you expect to gain the confidence of your peers? To operate in this world, you must have this type of belief and confidence. For example, if you apply for a job that you do not think you can do, an employer is unlikely to hire you. To gain the respect of others, leaders must be able to prove themselves through a positive self-concept.

Self-acceptance is one thing that none of us can live without. If we cannot accept ourselves for what we are, we lessen our chances of being happy, productive, and successful. This does not mean that we as individuals should stop striving for improvement; on the contrary, it means that we should be able to recognize our faults and weaknesses so we can do something about them. For example, if we cannot accept the fact that we are not good spellers, we may never become good spellers, because we will not spend the extra time and effort to achieve this goal. Self-acceptance involves identifying and understanding our strengths and weaknesses, setting goals for improvement, and realizing that a positive self-concept gives us the ability to reach our goals and become successful.

Individuals do not have to become phonies to develop a positive self-concept. Sometimes, a poor self-image will cause you to put up a false front, trying to convince others that everything is going well or to blame others for your present situation. As an individual strives to build self-esteem, he or she will also develop the potential for leadership. A positive self-concept yields many rewards:[2]

- Stronger sense of self
- More confidence
- Trust in one's own ideas, skills, and knowledge
- Ability to identify and take advantage of opportunities
- Ability to learn from mistakes, so that you can improve and do better next time
- Increased courage, endurance, perseverance, and fortitude
- Ability to overcome and push past fears and obstacles
- More dynamic and interesting personality
- Enhanced social approval
- Ability to focus on bigger, more important goals
- Emotional security
- Ability to control your direction, instead of letting life just happen to you
- Ability to accept and celebrate success, and to use each success as a stepping-stone to further achievement
- Positive personal interactions and relationships

Positive Self-Concept versus Conceit

There is a big difference between a positive self-concept and an overly inflated ego; no one likes an arrogant person! So, what is the difference between healthy self-concept and conceit? **Conceit** is an attitude of grossly overestimating your worth. Although those with good self-esteem also feel good about themselves, the difference is the overestimation. Conceited people like to boast and brag about how great they are, often pointing out, in the most annoying way, how superior they are to others or simply gossiping about others' shortcomings to make their own greatness seem more obvious.

Quite often such braggarts and blowhards actually have a very poor self-concept and are using these behaviors to try to make themselves feel better. Sometimes conceited persons are really seeking affirmation or confirmation of their own worth from others, because they lack any true confidence or belief in themselves. However, there are arrogant, conceited persons who really do think that they are the greatest, holding an extremely inflated view of their own accomplishments and abilities.

In contrast, people with a healthy, positive self-concept are secure enough that they do not need to boast and remind people of their greatness. Such people let their actions and accomplishments speak for them; it is usually others who take public notice of the worth of individuals who have a strong self-concept. Although they readily celebrate successes, individuals with a positive self-concept do not do so at the expense of others. They acknowledge others' efforts and participation in any victories, giving credit where credit is due.

Related Terms

At first glance, self-concept is like many terms: it may or may not have a specific meaning to you. Figure 18–2 lists similar terms. Confusing? Which word means what? Obviously, not all of these words have exactly the same definition, but they do relate to a common idea or concept. This chapter uses the term *self-concept* because it best expresses the general theme underlying the definitions of all the words in Figure 18–2. You will first learn a general definition that is clarified by the use of an analogy. You will then learn the individual elements and characteristics that combine to create self-concept.

TERMS ASSOCIATED WITH SELF-CONCEPT	
self-actualization	self-image
self-acceptance	self-importance
self-adjustment	self-improvement
self-affected	self-knowledge
self-asserting	self-perception
self-awareness	self-pride
self-belief	self-realization
self-concern	self-recognition
self-confidence	self-reflection
self-content	self-regard
self-determination	self-respect
self-discipline	self-responsibility
self-esteem	self-scrutiny
self-expression	self-trust
self-fulfillment	self-worth

FIGURE 18–2 Many other terms are used to convey the idea that we call self-concept. Whatever term we use, the fact remains: the more we value ourselves, the happier we are.

INGREDIENTS OF SELF-CONCEPT

A positive self-concept can be achieved only through the understanding of its ingredients: self-esteem, self-image, self-confidence, self-determination, and self-responsibility. Each component is important in creating a positive self-concept. Just like the different ingredients that are needed to make a cake, each ingredient in self-concept has its own role.

Self-Esteem

Self-esteem is an internal feeling. How do you feel about yourself? Self-esteem relates to how much you accept yourself and how you perceive your worth and value as a human being.

Self-esteem is the core or beginning of a positive self-concept. Low self-esteem tends to suppress the ability and desire to explore and take on the challenges of life.

Low self-esteem does not totally disable an individual. Many people who have a low opinion of themselves have jobs, get married, have children, buy homes, and so on. The difference is within: they are generally unhappy or dissatisfied with life and usually constantly fight feelings

---

VIGNETTE 18-1 A TALE OF TWO INTERVIEWEES

SELF-CONCEPT IS IMPORTANT TO LEADERS because their behavior influences and directly affects many other people. When does our self-concept directly affect what we do? We would argue that your self-concept affects you personally every day. Just ask Tina and Ken about the importance of self-confidence.

Tina and Ken were from the same high school. They were both starting their freshman year at the local community college, and they both had applied for a part-time job with an agribusiness in the community, where the work experience would be almost as valuable as their school work. Tina and Ken had similar education. They were both very bright and hardworking, and both of them had always dreamed about owning their own business, so this part-time opportunity seemed perfect for them both.

Ken interviewed on Tuesday, and Tina interviewed on Thursday. Tina got the call the following Monday that she would be the newest employee with the agribusiness firm. What made the difference? They were so similar! Ken decided to call the firm's human resources office and politely ask what he could improve upon for further interviews. He was told that the deciding factor was his apparent lack of confidence in his abilities. Confused, Ken asked how he could be confident when he had never done a job like the one he had applied for. The folks at the agribusiness firm told him that self-confidence was about believing you could figure out how to do the job, not already knowing how to do it. Tina was hired because she had confidence that she would be able to figure out certain responsibilities and quickly learn what she needed to learn. She was hired because her positive portrayal of herself was contagious. The personnel office felt like she could take constructive criticism, and they were not so sure Ken could.

This tale of two interviewers should be enough to motivate you to start working on improving your self-concept and self-esteem. Your feelings about yourself are something that you will never be able to hide. How can other people feel good about the work you could do if you yourself do not feel good about your competence and capabilities?

---

of hopelessness, inability, and despair, much like the person in Figure 18–3.

Enjoying life is what life is all about, but the enjoyment begins with you. You must feel good about the person within you. If you have low self-esteem, you will struggle through life. If you have positive self-esteem, you can use it as a driving force and fully realize the pleasure of life.

FIGURE 18–3 A person with low self-esteem is generally unhappy or dissatisfied with life. Friends can help such a person overcome discouragement and despair and start working toward building a positive self-concept. *(© iStock/juanmonino)*

Nathaniel Branden, in *How to Raise Your Self-Esteem*, made the following comment:

> Apart from problems that are biological in origin, I cannot think of a single psychological difficulty—from anxiety and depression, to fear of intimacy or of success, to alcohol or drug abuse, to underachievement at school or at work, to spouse battering or child molestation, to sexual dysfunctions or emotional immaturity, to suicide or crimes of violence—that is not traceable to poor self-esteem. Of all the judgments we pass, none is as important as the one we pass on ourselves. Positive self-esteem is a cardinal requirement of a fulfilling life.[3]

Self-Image

Self-image (the idea, concept, or mental picture one has of oneself) relates to accepting yourself and presenting yourself in such a manner that you confidently display that self-acceptance. Outward physical appearance is not a measure of a person's worth, just as money is not the measure of a person's success.

John Foppe was born without arms. As he grew up, he had to deal with the pain of looking different. He had to endure the thoughtless and cruel remarks of others. Although life was not easy, John accepted his appearance and developed a positive self-image. Today, John is a very successful lecturer and leads a happy and rewarding life. John has a positive self-concept that has enabled him to overcome obstacles and achieve success. His self-image is a part of that overall self-concept.

Self-Confidence

Self-confidence is being secure in your ability to take on new tasks and develop new skills. Self-confidence is knowing that you can accomplish tasks that are presented to you. People with a high level of self-confidence are eager to apply the abilities they know they possess. Confidence also means that you do not allow fear to dominate your decisions or prevent you from pursuing new opportunities or trying new things. Possessing confidence allows a person to grow and expand his or her horizons. It is amazing what can be done with self-confidence and persistence. Consider the accomplishments of these "late bloomers," who exhibited self-confidence and persisted on their course despite the negativism and criticism they encountered on the way:[4]

- Beethoven's music teacher said, "As a composer, he is hopeless."

- Isaac Newton's work in elementary school was rather poor.
- Einstein did not speak until the age of four, and he could not read until age seven.
- Edison's teacher told him he was unable to learn.
- F. W. Woolworth's employer refused to allow him to wait on customers because he "didn't have enough sense."
- Louis Pasteur was given a rating of "mediocre" in chemistry at Royal College.
- Admiral Byrd was deemed "unfit for service" before his pole-exploration expeditions.
- Winston Churchill failed sixth grade.
- Walt Disney was fired by a newspaper editor because he had "no good ideas."
- Henry Ford was evaluated as "showing no promise."

Your confidence level can greatly determine how much success you enjoy in life. Have you ever said, "I just don't think I can do that"? That statement indicates a low confidence level and is based on either fear or lack of motivation. Self-confidence is a state of mind that exists regardless of ability or prior experience. Many people possess adequate skill, but lack the confidence to use that skill. Many people with low self-esteem have been "taught" that they are not good enough, so they never take chances. Other people with low self-confidence have experienced failure and do not want to experience failure again. In short, they are afraid, and they allow that fear to steer their decision-making process.

Skill level increases when accompanied by a feeling of self-confidence. Most people develop certain skills that become almost second nature to them, so that they perform those skills with little conscious thought about what they are doing. Remember when you learned to ride a bicycle? At first you had to concentrate on certain fundamental skills, and it was not easy. As your skill level increased, so did your confidence. After a while, you rode your bike and never thought about how you were doing it. Most people possess a high skill level in areas in which they feel confident. The combination of experience, practice, learning, and success also builds confidence that you can successfully develop other skills as well.

Self-confidence helps overcome fear. The true test of self-confidence comes when you are confronted with a situation in which your skill level is not high, or you have little experience. Will you be able to learn new skills and complete unfamiliar tasks? A high level of self-confidence allows you to confront such challenges and overcome the fear of failure and rejection.

Self-confidence also includes an understanding of your limitations and the extent of your abilities. Confidence is a positive trait, but you must not allow yourself to become overconfident. When a person becomes overconfident, he or she is more vulnerable to failure. Take driving a car, for example. As you first learn to drive, you build confidence in your ability to perform that task. If you reach the point of overconfidence, you are more likely to take risks. You then become an unsafe driver because you create an environment for failure. Be confident, but recognize your limitations; respect your ability to learn and improve, but allow for the time and experience needed to do so.

Self-Determination

Self-determination deals with inner motivation to achieve goals. Motivation is a force that drives you to seek out and accept new challenges and to explore different areas of life. You will always have certain external sources of motivation: through words or actions, parents, teachers, friends, and others may provide motivation that stimulates you to achieve and grow. The obvious rewards of a situation may motivate you to action. Money and prestige are examples of outward motivation. Everyone is influenced, to a degree, by outside motivation, but all people possess inward motivation or self-determination. Having drive and determination from within is what allows you to push yourself continually beyond normal levels of accomplishment. Self-determination helps you overcome doubt and insecurity.

To be self-determined is to be in charge of your own fate. Nobody can be in total control of all situations and events, and many things will happen to you that you have little control over, but a self-determined person has the ability to work through those obstacles. This ability allows you to remain on a positive course. You will experience failure and disappointment, but do not let that deter you from moving forward. Who better to determine your fate in life than yourself?

Self-Responsibility

Self-responsibility is the ability to accept the consequences of any effort, good or bad. This characteristic is difficult to acquire because society places such a high value on success. Thomas Edison invented the lightbulb, but his ultimate success was preceded by tremendous failure. He accepted the consequences of his failures but did not allow those failures to prevent him from reaching his goal. You will occasionally fail, and how you use those failures will in part determine what direction your life will go. Failure is not always a negative: you can use failure to evaluate yourself, learn, and find other options. This is a positive way to use mistakes. Thomas Edison evaluated his failures, corrected mistakes, and eventually achieved success.

You need to be **resilient**, bouncing back after a setback or failure. You need the ability to accept the consequences of any decision, action, or venture, regardless of failure or success. In the course of making choices, you must try to foresee results and expect the unexpected. Avoid placing blame on others. Unfortunately, finger-pointing and blaming others seem to be a growing trend in our society; people seem less and less willing to accept responsibility for their own actions. Blaming others is a way of avoiding or shifting responsibility. You must develop the internal strength to acknowledge both successes and failures, mistakes and achievements, as the results of your own actions and behavior.

Self-responsibility also includes the ability to recognize when you need help. Do not hesitate to seek assistance when you need it, but do not develop a habit of always depending on others to do it for you. Seeking help from others is not a weakness; dependence is.

FACTORS THAT AFFECT THE DEVELOPMENT OF SELF-CONCEPT

Factors that affect the development of an individual's self-concept can be classified as chronological, external, or internal. These three areas are closely connected and interrelated. Many things that affect your self-concept originate from your environment, an external factor. All external factors affect how you feel, think, and react. First, we examine the chronological factors, then the external factors. Last, we examine the internal factors. Figure 18–4 shows the relationship among chronological, external, and internal factors.

Chronological Factors

You are not born knowing who you are or what you will become. You acquire your image of yourself over time by discovering and using your natural abilities and by constantly receiving messages about yourself from the people around you. During childhood, your self-concept is developed mostly from your parents' reactions to and influence on you. As you grow toward adolescence and adulthood, teachers, coaches, friends, and others also exert influence on your self-concept.

Childhood The self-concept formed in childhood lays the foundation for your attitude toward work, your future success, your personal abilities, and

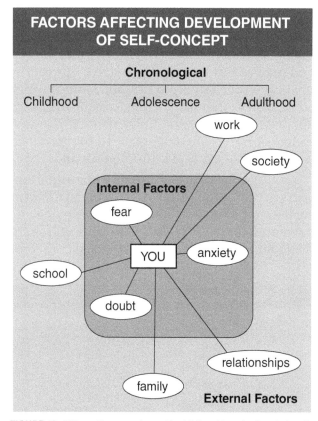

FIGURE 18–4 The self-concept formed in childhood lays the foundation. As you get older, you start comparing yourself to other people. By the time you reach adulthood, your self-concept has been influenced by many internal and external factors.

the roles you play. Your family is the first source of information about yourself. Every person with whom a child comes into contact leaves a mark. Parents do not actually teach their children self-concept, but they shape it with positive and negative messages,[5] as shown in Figure 18–5.

As a child, you accepted these messages as truths and recorded them in your memory and your subconscious mind. You most likely changed your behavior and attitudes in accordance with this ever-changing mental picture of and beliefs about yourself. Your subconscious mind gradually developed a self-concept that you came to believe was real and true, regardless of how accurate or distorted it actually was.

Adolescence As you get older, you start comparing yourself to others. Typically, you become less happy with who you are. You may wish you were more like others whom you perceive as better. During this stage, some people begin to use put-downs as an equalizer. Often, teenagers criticize or ridicule others in an attempt to reduce insecurities about themselves and their negative self-concept.[6]

Today, the media play a strong role in how adolescents perceive themselves. Television, movies, and magazines constantly present unrealistic images, particularly of physical appearance, which adolescents often use as measures of their own attributes and lifestyles. Unfortunately, it is easy to feel deficient in comparison to the multimillion-dollar fantasy that is the media world.

Your teenage years (12 to 18) are important years in your self-concept development. Changes such as not relying on your parents as much and becoming more independent in your search for personal success have a huge impact on your self-concept. You must also deal with physical changes; relationships with peer groups; an emerging, often confusing identity and sense of self; the loss of childhood; and the assumption of some adult responsibilities. In fact, many people never move beyond the self-concept they had of themselves while in high school, and throughout adulthood they continue to compare themselves to others.

Adulthood When you reach adulthood, your self-concept has been formed by those in your environment and from all your past experiences. You may compare yourself to others, as you did during adolescence, or you may focus on your own inner sense of self-worth. Adults tend to define themselves in one of three ways, each of which can influence self-concept. We may define ourselves in a variety of ways:

In terms of the things we possess This is the most primitive source of self-concept. We buy material things (such as cars, clothes, houses, and land) to enhance our self-concept. People who define themselves in terms of what they own may have difficulty deciding how much is enough, and may spend their lives attempting to acquire more and better material possessions, subconsciously feeling that these things make them better or worthier persons.

In terms of what we do for a living Too often our self-concept depends on our job titles. If we allow external entities, such as a corporation, a school or university, the media, counselors, friends, or our parents, to select a job or career for us, we are just following someone else's plan. This can happen easily if you do not set personal goals; you end up adopting others' ideas about what you should do and be. People who have been pushed

NEGATIVE	POSITIVE
Bad boy! Bad girl!	You're great!
You're so lazy!	You can do anything!
You'll never learn!	You're a fast learner!
What's wrong with you?	Next time you'll do better.
Why can't you be more like . . . ?	I like you just the way you are.
It's all your fault.	I know you did your best.

FIGURE 18–5 As children we were told, "Sticks and stones may break my bones, but words will never hurt me." This is simply not true. Words play a big part in psychological development, and negative words can deeply damage self-concept.

into jobs they do not like may make lots of money but hate going to work, even though their job or position seems desirable to others.

In terms of internal value system and emotional makeup Emmett Miller, in *The Healing Power of Happiness*, says this is the healthiest way for people to identify and define themselves. Miller also points out that if you do not give yourself credit for and build your self-worth through accomplishments and excellence in areas of life other than your job and material possessions, you have nothing on which to ground your identity if you lose your job or possessions.[7]

It is important to learn how to protect your self-concept against those who criticize you unhelpfully or hurtfully; it is equally important to listen to those who encourage and challenge you. This kind of learning can help you determine the difference between what is helpful and what is destructive, what is correct and what is incorrect.

External Factors

Family, relationships, school, work, and society all impact your self-concept. The people and situations that confront you daily have either a positive or negative effect on you. Learning more about these external factors will give you a better understanding of what you face and how to deal with it. People have a better chance of resolving problems when they fully understand the nature and extent of those problems; dealing with the unknown is always difficult.

Family The years from birth to age 18 are considered the formative years. During these years, the family/home environment is highly influential, and thus plays a leading role, in the development of a person's self-concept. The family environment satisfies the very basic human needs of survival, safety, and security. A child must depend on family for food, shelter, and clothing, of course, but beyond these basic needs, the family should be a place in which a child experiences love and acceptance, praise, and constructive criticism.

Praise children for small displays of competence, beginning at an early age. Arrange "success experiences" for young children so that they come to expect positive results. Avoid negative comparisons among siblings. Let children hear you praise others for behaviors they themselves demonstrate. Pass on to a child compliments you hear from others, and give positive support to children with a smile, a pat on the back, or any other appropriate gesture. Only through this sort of feedback can a child establish and internalize certain values, such as respect, responsibility, discipline, and cooperation, as well as the ability to distinguish between right and wrong.

Relationships The people with whom you come into contact outside of the family constitute your network of past, present, and future relationships. These people exert influence on you and either consciously or unconsciously make certain demands and expectations of you. Some people you may encounter only briefly: a camp counselor, a guest speaker at some public function, or even someone you meet in a department store. Today's world is mobile, and people do not always stay in one spot for long. However, the length of contact time is not always proportional to the impact others have on you.

Other influential relationships are the ones that last and remain a real part of your everyday life. The friends you choose reflect your level of self-concept. More than likely, your friends are individuals with whom you have many things in common. You and your friends probably have common likes and dislikes, enjoy common activities, and share common interests. Friendships can act as a mirror of yourself—not necessarily your physical image, but an image of your feelings, thoughts, likes, and dislikes. This is not to imply that friends always share a common interest in everything. You might enjoy watching a basketball game with certain friends and going to movies in the company of others. The key in your relationships with friends is that you tend to build friendships with people who share your level of self-concept.

As you age, your relationships will expand beyond those with immediate friends. Eventually you will branch off into the area of love, marriage, and possibly the beginning of your own family. This area of relationships can be extremely rewarding. Mixed with the rewards are new responsibilities and demands that will test, revise, and build you and your self-concept.

School During childhood, you expand your environment to more than just the home. You will spend a major part of the next 12 to 16 years of your life in school. This new environment offers an entirely new set of challenges to the individual and his or her self-concept. Along with those challenges and demands, school offers students a chance to develop and use skills that would be difficult to acquire in the more limited home environment.

School can be an exciting place, filled with new experiences, opportunities, challenges, and demands. It can awaken your mind in ways you never dreamed possible. The opportunity exists to discover a new and wonderful you that you may never have dreamed existed. The benefits you reap and the discoveries you make are your responsibility. Your teachers provide the opportunity, but you must take that opportunity and turn it into reality. The responsibility is on your shoulders, and the ways in which you meet it will have profound effects on you later on in life, as well as now. You can decide to accept challenges or allow them to pass you by. Have you ever passed up the chance to go somewhere because at the time it did not interest you? Have you ever regretted that choice when others returned from that event with great stories about how wonderful it was? The school environment is an excellent place to begin to build

your self-concept. Get involved—do not allow learning opportunities to pass without taking advantage of them.

Peer pressure can be intense, and the demands on you to follow the crowd—going along to get along—may be extreme. Following is not always bad; just be careful of whom and why you follow. Even during these formative, changeable years, the best course to follow is that of being a leader. Leaders are found in the forefront, not lost in the crowd. The group members in Figure 18–6 will have an influence on each other's self-concept.

Work When the time comes to leave school, you will be thrust into yet another environment: the world of work. Companies compete for people who possess a high self-concept because such people have the tools necessary to be effective employees. If you are a good employee, a company profits from your contributions. The stronger your self-concept, the more tools you possess. The more tools you possess, the more in demand you will be, which gives you many choices to select from. Finally, the more choices you have, the better your chances of finding a job that you enjoy and do well. The students in Figure 18–7 will become leaders in their respective careers of

FIGURE 18–6 Although these students are from vastly different cultural and ethnic backgrounds, and live thousands of miles apart, their kind actions toward one another help build their positive self-concept. *(© iStock.com/Rawpixel Ltd.)*

FIGURE 18–7 Developing your self-concept involves family, relationships, school, work, and society. The students in this photo seemed to have well-developed and positive self-concepts so far, but they must make a choice to stay positive and remain alert for opportunities to reinforce the positive aspects of life, given the many different factors in life that can affect them. *(©Monkey Business Images/Shutterstock.com)*

choice. Will you have a positive self-concept that leads you to career success?

Internal Factors

The world around you affects how you think, feel, and act. These external factors interact with and influence the internal factors that you must also deal with, such as fear, doubt, and anxiety.

Fear is overwhelming anticipation or awareness of danger. In relation to self-concept and interpersonal relations, fear most commonly centers on rejection, humiliation or embarrassment, and failure. It takes a lot of determination to overcome these fears. People protect themselves from what they fear through avoidance. Employees pass up advancement opportunities because they are afraid that they will not be able to succeed. People pass up opportunities for relationships because they fear rejection and emotional pain. Fear is a basic human emotion that serves a useful purpose: it forces us to be cautious. However, the complete avoidance of fear-inducing situations indicates a low self-concept and severe doubts about self-worth. Our emotions should assist us, not control us.

Doubt is a state of questioning your ability to learn, think creatively, make decisions, accomplish, and succeed. Too many people simply say, "I can't do that," and stop right there—the subconscious mind makes sure that their actions support their beliefs. This is often referred to as **self-fulfilling prophecy**: you tend to behave in a way that both reflects and supports what you believe about yourself. When you say, "I can't," your subconscious mind will always prove you right. The phrase you should use instead is, "I question my ability to do that." This at least leaves open the possibility that you can indeed do whatever it is. When you are faced with a totally new situation, or circumstances over which you have no control, you *may* not be able to succeed—but that is no reason not to give it a try.

Anxiety is having an uncomfortable feeling or uneasiness about a situation or event. It may be closely connected with fear of failure or embarrassment. The phrase "having butterflies in the stomach" defines anxiety in a way you can probably relate to. Of the three negative internal factors affecting self-concept, anxiety is the easiest to overcome.

DEVELOPING A POSITIVE SELF-CONCEPT

Five areas contribute to developing a positive self-concept. Think of these five areas as a set of steps. To reach the top you must go up each step, one at a time (Figure 18–8). At the top of the steps is the desired destination, a positive self-concept.

Step 1   *Restoring and Nurturing.*

The first step toward developing your self-concept is to accept yourself as you are now. The past cannot be changed, but the future is determined by how you think and act. The practices presented here have been developed to help people restore and nurture a healthy self-concept:

- Identify and accept your limitations
- Learn to accept others
- Make a list of your greatest talents
- Make a list of your positive qualities and occasionally review it
- Make decisions
- Stop procrastinating
- Develop expertise in some area
- Find a mentor
- Avoid **surface analysis** (judging the actual appearance rather than the inner nature) of yourself and of others
- Dress as though you are already successful
- Continue to learn and be observant
- Seize opportunities to learn new skills
- Use positive self-talk

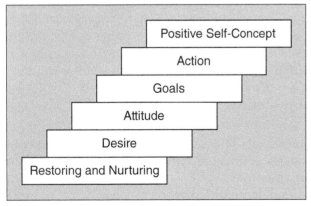

FIGURE 18–8 There are five steps to achieving a positive self-concept. Once you have secured your footing on one step, you are ready to climb to the next.

- Tackle the things you fear
- Choose your friends and associates carefully
- Learn from failures
- Forgive yourself as you would others
- Do quality work; then compliment yourself
- Go the extra mile to do your best
- Believe in yourself
- Make a victory list of your past successes
- Speak up and share your views
- Smile and compliment others
- Stand up for others
- Do good for others
- Join an organization that requires your involvement
- Look people in the eyes when speaking
- Never give anyone permission to make you feel inferior
- Finish every job you start
- Change enemies into friends
- Practice good manners
- Learn to love to read

Step 2   *Desire.*

You must begin your development of a positive self-concept with a genuine desire to change. **Desire** is a state of mind, a longing or hope for something. Have you ever wanted something so much that you could not imagine what life would be like without it? The desire to possess is strong in humans. Advertising is such a big industry because companies must strive to create a desire for their products in consumers. Examine the benefits of a positive self-concept and decide for yourself whether you possess a genuine desire to change. When you make that decision, realize that some doubts and anxiety are quite normal. Change breeds uncertainty as well as opportunity; it can make you anxious and fearful if you allow it to do so.

Step 3   *Attitude.*

When you have completed the second step, you will need to create the proper environment to allow your desire to continue to grow. Attitude is a state of mind; it includes feelings, beliefs, and outlook, and it can be either positive or negative. Previously, we stated that your attitude determines your altitude. If your attitude is positive, you will be able to maintain your level of desire. If your attitude is negative, your desire will diminish. What usually happens to a child who is made to take piano lessons when he or she would rather be doing something else? The majority of the time, the child develops a negative attitude and ends up having little desire to play the piano at all.

Be aware of the attitudes of those with whom you associate. A positive attitude may turn sour when it encounters or mixes with a negative one; it is very easy simply to take on another person's attitude. Remember, though, that people who try to discourage you are usually insecure about themselves. Surround yourself with people who display a positive, optimistic outlook. People with positive attitudes are a source of encouragement.

You determine your attitude. Being positive creates an environment for growth and helps you maintain a clear vision of where you want to go and what you want to be. A negative attitude is like driving a car in dense fog: you are unsure of what lies ahead, you have to make snap judgments when a situation suddenly appears, and you get where you are going very slowly. A positive attitude is like driving on a clear day: you can see obstacles coming and are able to plan and react in time to avoid problems. You also can enjoy the scenery around you as you travel!

Step 4 *Goals.*

There is a relationship between your goals and your self-concept. Being equipped with desire and proper attitude is good, but now you must develop some direction or purpose. This is accomplished through the establishment of goals. Use grades as an example. If you set a goal of getting all As, you either reach or do not reach that goal

when you receive your report card. In between is the effort put forth toward attaining your goal. If your goal is reached, your self-concept is enhanced.

Goals are usually categorized into short, medium, and long-range. A short-range goal may be to make an A on a six-week test. A medium-range goal may be to maintain an A average for the semester. A long-range goal may be to graduate from school with a 3.0 grade point average. The experience of success in achieving short-range goals enhances your self-concept, and this spurs you on to set and attain higher medium- and long-range goals.

Goal setting is best done through an organized process. Sit down with pen and paper and write out a step-by-step procedure so that you have a map or guidelines to keep you on track. Figure 18–9 is a good method, or you can refer to Chapter 14 for other models. Organization is a key part of being able to set and reach goals.

Step 5 *Action.*

You have now reached the last step. **Action** represents a state of motion, either physical or mental. Without action, all you have are good intentions. Have you ever set out to do something and failed to accomplish it? Maybe you wanted to learn to play the guitar. You had the desire to learn to play. You had a positive attitude and were confident you had the ability. You even set a goal of learning to play. What's left? Action. You physically have to sit down and practice. You must put your good intentions into motion. Failure to practice leaves you with only good intentions.

PROCEDURE FOR SETTING GOALS

- **Statement of Goal**—A single-sentence statement that fully defines the goal.
- **Establish Time Frame**—Categorize the goal as short, medium, or long range. Set a completion date.
- **Make an Agenda**—List all the steps you must take to reach the goal; list possible obstacles and measures and the steps you will take to avoid or overcome them. This is the "how-to" part of the process.
- **Be Realistic**—Try to set goals that are ambitious but attainable. If you have a long-range goal, it might be beneficial to set some shorter-range subgoals to be accomplished along the way.
- **Evaluate**—Take time to check your progress periodically. A more efficient method is to set evaluation points/dates in the agenda when you will assess progress.
- **Adjust**—Be flexible after an evaluation. If the process you designed for reaching your goals is not working, alter some aspect of it, such as time frame or steps involved. Change whatever is necessary to get you back on track and moving forward; do not allow yourself to hit a point of stagnation.
- **Reward**—When you reach your goal, celebrate your accomplishment in some way. Reward yourself for a job well done.

FIGURE 18–9 Goals give direction and allow you to organize and set priorities in your life. A life void of goals is a life without direction; you cannot reach a destination that you do not have. Achieving goals enhances your self-concept.

Action is by far the hardest step to take because it requires energy, both physical and mental. It necessitates discipline to keep from veering off track. It requires determination to keep from getting discouraged. It calls for persistence to try again when faced with setbacks and disappointment. It requires patience when things do not happen as fast as you want. When you take action, all the internal and external factors that affect self-concept come into play.

Be prepared at all times to overcome the negative and maintain the positive. Remember that change is not magical: you cannot just wish it and make it so. Achieving a positive self-concept is hard, but the benefits far outweigh the effort of the work needed to do so.

You should constantly take on new challenges and be open to new ideas. Everybody has certain talents and natural abilities. Build on those and then begin to expand yourself and see whether you enjoy the new you. Keep remembering that self-concept is a natural part of everyone's existence. The key difference between individuals is whether their self-concept is positive or negative. A positive self-concept enables you to take full advantage of the exciting possibilities and opportunities life has to offer. You will undoubtedly encounter some difficulties during the development of a positive, strong self-concept, but the reward can be a full, happy, and successful life.

A positive self-concept equips you with qualities and abilities that can be used in your education, career, and society. All aspects of your public life are in one of these three areas (education, career, society), and a positive self-concept is key in each area. Becoming a leader requires qualities such as innovativeness, creativity, determination, responsibility, and confidence. These traits will enable you to be successful in life—successful by your own standards. Success measured *by your standards* is an important distinction, because the world has an abundance of standards and guidelines. You may agree with and accept some of society's standards, but make sure the decisions and evaluations of success are yours and not those of society or other individuals.

Self-Concept and the Workplace

Advances in technology and science as well as fast-moving social changes are creating both tremendous progress and tremendous problems. Today's world is more complex, challenging, and competitive than ever. Today, organizations need workers with the confidence to tackle assignments requiring a higher level of knowledge and skill. In your workplace, you will also need higher levels of independence, self-reliance, self-trust, and initiative. In other words, you are going to need a healthy self-concept to be successful in your career. If you can develop a healthy self-concept, it will have an impact on the people you work with, the people you lead, and the people you follow.

Increases Productivity In the workplace, those with a positive self-concept are inclined to form nurturing, nourishing relationships. These workers tend to do more than what is strictly required on the job. They are receptive to new experiences and new people, able and eager to take on responsibility and make decisions.[8] Such people contribute to the well-being and productivity of the workplace.

Consequences of Low Self-Concept Employees with low self-concept manifest many of the following characteristics at work:[9]

- Negativity
- Fear of taking action (fear of being wrong)
- Fear of asking for help (fear of showing weakness or inability)
- Blaming others for failures (lack of self-responsibility)
- Lying
- High absenteeism
- Rebelliousness
- Clock-watching

Reason for Hostility People with a low self-concept can feel hostile, show a lack of respect for others, and attempt to retaliate against others to save face in difficult situations. Hostility toward others is a natural outcome of a low self-concept. Our emotional system is controlled by a balancing mechanism. When our self-concept is threatened, the mechanism is thrown out of balance and we start to feel hostile and anxious.[10]

Consequences of Low Self-Concept in Leaders A person in a position of leadership who has a low self-concept is not likely to treat group members, coworkers, or peers fairly. Leaders with a low self-concept can decrease the efficiency and productivity of a group because they tend to exercise less initiative, hesitate to accept responsibility or make decisions on their own, ask fewer questions, and take longer to learn procedures.[11] To bolster their self-esteem or reduce their insecurity, they may also downplay or denigrate group members' efforts, contributions, and ideas; these so-called leaders are often the same ones who always claim all the credit and always blame others for failure to achieve group goals.

Conditions Needed to Empower Self-Concept in the Workplace A workplace where the leaders and coworkers operate at a high level of self-concept would be a workplace of extraordinarily empowered (self-fulfilled) workers (Figure 18–10). The following conditions allow leaders to help their coworkers reach their full potential:[12]

- People feel challenged. Assign tasks and projects that excite and inspire them while exercising and stretching their abilities.
- People feel recognized. Acknowledge and celebrate individual talents and achievements and reward extraordinary contributions (both financially and otherwise, whenever possible).
- People are given constructive feedback. Let group members know how they could improve their performance, using approaches that stress positives and build on their strengths. Avoid critical negativism and never disparage people as individuals.
- People see that innovation is expected. Draw out their opinions, invite brainstorming, and stress that new and usable ideas are desirable and welcome.
- People are given clear-cut rules and guidelines. Give your followers a structure appropriate to their job descriptions and make sure they know what is expected of them.
- "People see that their rewards for successes are far greater than any penalties for failures." Encourage people to take risks and express themselves by minimizing the consequences of honest mistakes.
- People are treated fairly, equitably, and justly. Make your organization a consistent, trustworthy place in which to function.
- People believe in the value of what they produce and are proud of their accomplishments. In this positive environment, "they perceive the result of their efforts as genuinely useful, they perceive their work as worth doing."

FIGURE 18–10 A workplace where the leaders and coworkers operate at a high level of self-concept is one of extraordinarily empowered (self-fulfilled) workers, like those shown here. *(Courtesy of University of Georgia—College of Agricultural & Environmental Sciences.)*

Workers Perform to Expectations Workers tend to behave and perform in a way that aligns with what they believe about themselves; this is the self-fulfilling prophecy again. Your career successes and failures are directly related to the expectations you hold about your future. However, people can also be greatly influenced by the expectations of others. The Pygmalion effect sometimes causes people to become what others expect them to become.

Robert Rosenthal, a Harvard University professor, developed a theory based on a Greek legend about **Pygmalion**, the sculptor who carved and then fell in love with a statue of the goddess Aphrodite. The goddess was so moved by his devotion that she brought the statue to life. Pygmalion saw the statue as real; thus, she became real. From this story, Rosenthal and others formulated what has become known as the expectation theory ("you get what you expect").

When students or workers are expected to do great things, they do great things.[13] Their self-concept is enhanced; thus, they believe in themselves. When people believe in their ability and have faith, they can accomplish great things.

Self-Concept in School and the Social Arena

You spend most of your time in school, which is part of the social arena. As a member of the social arena, you will find yourself in numerous situations that force you to make choices. When confronted with many of society's demands, you will experience doubt, fear, despair, rejection, confusion, and more. People with a low self-concept try to avoid these feelings and emotions, sometimes by avoiding the problematic situations (which is impossible), and sometimes by refusing to take responsibility for their choices and actions. A positive self-concept makes a person much better able to face society's demands by making sound decisions and judgments.

People with a positive self-concept tend to be outgoing. People with a positive self-concept tend to be receptive to new experiences and new people. They "are willing to tolerate differences in others. They tend to go out of their way to greet

FIGURE 18–11 People with positive self-concepts are outgoing and enthusiastic, like these two individuals. *(© iStock.com/Uber Images)*

people and meet their needs. Generally, people with positive self-concepts have more friends than do people with negative self-concepts."[14] They are more willing to express emotions and share ideas with others, as the people in Figure 18–11 are doing.

People with a positive self-concept are humble. A positive self-concept does not include being arrogant and boastful. If you know any people like that, you probably realize that their behavior could be improved by a change in attitude. People do not like others to flaunt their superiority, whether real or imagined.[17]

People with a negative self-concept often have trouble relating to others and engaging with groups. They tend to be pessimistic and distrustful, and group members may wrongfully perceive them to be unfriendly. Many people with a negative self-concept feel that others are out to get them. They are not able to accept a compliment without looking for a hidden agenda or an ulterior motive.

A positive self-concept helps you cope. A positive self-concept does not shield you from difficult situations; rather, it provides a method to deal with them. This is similar to health: a healthy immune system does not guarantee that a person will never become ill. It does make the person less vulnerable to disease and better equipped to overcome it. The same is true of self-concept. Life's difficulties will cause you a degree of pain and anxiety, but you will rebound faster if you have a positive self-concept.

VIGNETTE
18-2 | THE PYGMALION EFFECT

STRONG LEADERS TEND TO HAVE HIGH EXPEC-TATIONS for themselves and for their followers. These expectations are very powerful, because we often wind up seeing what we expect rather than what is actually occurring. Researchers refer to this phenomenon as the self-fulfilling prophecy or the Pygmalion effect.

In Greek mythology, the sculptor Pygmalion carved a statue of a beautiful woman, fell in love with the statue, and brought it to life by the strength of his perceptions, belief, and emotions. According to leadership experts James Kouzes and Barry Posner, leaders play "Pygmalion-like roles in developing people."[15] Research has shown many times over that people act in ways that are consistent with our expectations of them.

Whatever leaders expect from followers is usually what they will get. If a leader feels like a team is unproductive and incompetent, that team will probably never achieve and produce at the appropriate level. In contrast, if a leader believes that a team is a strong one, then the team will be great and achieve impressively.

It is not enough just to feel or believe that your followers will be successful. Leaders have to put their expectations into action. To build self-confidence in people, leaders must tell employees or followers exactly what the expectations are. Leaders also need to support, recognize, reward, encourage, and model high expectations to improve self-confidence. This improved self-confidence is what makes group members feel empowered; it is what gives them strength to innovate, make decisions, take the initiative, and take risks. When followers begin to feel self-confident, a leader knows that he or she has "brought them to life," just as Pygmalion did with the statue he loved. If you have confidence in and set high expectations for your followers, and give them the support they need to accomplish their task, you will be amazed at how they "come to life."[16]

If self-concept is so important, why do so many people have a poor one? Several realities contribute to poor self-concept:[18]

- We live in a society in which negativity and ridicule are not only acceptable but also often encouraged. The popular use of the one-liner put-down to "zing" another person is just one example.
- We "confuse failure on a project with failure in life."
- We confuse being uneducated or having an untrained memory with being unintelligent, when in reality we are not spending enough time studying or practicing skills.
- We tend to unrealistically and unfairly compare our experience with others' experience, and conclude that we are not successful because someone else appears to have been more successful.
- We confuse experience with ability and intelligence.
- We compare our worst features to someone else's best features, rather than concentrating on developing and strengthening our own best features or abilities so that we can reach our goals in life.
- We "set standards of perfection that are unrealistic and unreachable" rather than concentrating on what we can do and have already achieved.

When we have a poor self-concept, we may exhibit the characteristics and behaviors in Figure 18–12.

Critical and jealous nature	Too much emphasis on material things
Involvement in gossip	Lack of genuine friends
Improper reaction to criticism	Senseless and erratic actions
Improper reaction to laughter	Excuses to justify failure
An uncomfortable feeling when alone	Spur-of-the-moment, impossible promises
An "I don't care" attitude	Rebellion against authority
Breakdown in decency	Foolish and impulsive actions

FIGURE 18–12 Many of these behaviors and characteristics are the result of not being secure within ourselves. As we become more secure and our self-concept improves, many of these negative behaviors will diminish.

CHARACTERISTICS OF PEOPLE WITH A POSITIVE SELF-CONCEPT

Once your self-concept is attained, improved, or enhanced, you will exhibit many positive characteristics. There are six characteristics of people with a high self-concept:[19]

Future Oriented People with a positive self-concept look forward to the future and do not overstress past mistakes and failures. These people believe that every experience can teach you something if you are willing to learn.

Able to Cope with Problems People with a positive self-concept can handle problems and disappointments. Successful people don't get down when they have problems. They simply maintain a good attitude and work out the problem.

Able to Acknowledge, Control, and Deal with Emotions People cannot help the way they feel, but they can control the way they act. Those with a positive self-concept are able to keep their emotions under control. They are able to establish and maintain healthy relationships because of their positive self-concept.

Able to Help Others and Accept Help Secure people in leadership roles surround themselves with good followers; often these coworkers are good enough that they are capable of doing the leader's job. In helping others, people benefit themselves as well.

Able to Accept People as Unique, Talented Individuals People with a positive self-concept "learn to accept others for who they are and what they can do." Acceptance of others is a good indication that you have a positive self-concept.

Exhibit a Variety of Self-Confident Behaviors People with a positive self-concept know how to accept a compliment. They can say "Thank you" without feeling that a reciprocal action is necessary (Figure 18–13). They can laugh at themselves, and they do not correct or ridicule others. They feel free to express opinions that may differ from those of other group members. They can be by themselves without feeling lonely or isolated. They can talk about themselves without bragging or showing off. Finally, they stay true to their own values and ethics.

Other Characteristics of People with a Positive Self-Concept

- Your facial expression, demeanor, and ways of speaking and moving naturally communicate your pleasure in being alive.

FIGURE 18–13 People with a positive self-concept accept compliments appropriately. *(© iStock/zoranm)*

- You are open to criticism and readily acknowledge mistakes because your self-concept is not bound up with an image of perfection.
- Your speech and movement seem easy and spontaneous because you are not in conflict with yourself.
- There is "harmony between what you say and do and how you look, sound, and move."
- You are open to and curious about new ideas, experiences, and possibilities; for you, existence is an adventure.
- If feelings of anxiety or insecurity arise, they are not likely to intimidate or overwhelm you, because you know how to manage them and rise above them.
- You enjoy the humorous aspects of life in yourself and in others.
- You are flexible in responding to challenges and maintain "a spirit of inventiveness and even playfulness, since you trust your mind and do not see life as doom or defeat."
- You are comfortable with assertive (though not belligerent) behavior; you are able and willing to defend and advocate for yourself.
- You maintain your harmony and dignity in stressful situations, because feeling centered comes naturally to you.[20]

Ways to Present a Positive Self-Concept

When you have developed a positive self-concept, you will naturally reflect it in your behavior. Most of the behaviors stemming from good self-esteem are natural nonverbal responses; you may need to do some work to cultivate others. Here are just a few suggestions:

- Smile often, every day.
- Dress neatly and appropriately for the occasion.
- Be polite and considerate of others.
- Be an active listener.
- Take pride in your work.
- Be independent and make wise decisions.
- Make the best possible situation out of everything you do and hold a positive mental attitude.

You may even see noticeable changes at a physical level as you gain a positive self-concept:

- Your eyes may well become more alert, bright, and lively.
- Your face will at some point become more relaxed and (barring illness) will tend to exhibit natural color and good skin vibrancy.
- Your chin will probably be held more naturally and more in alignment with your body.
- Your jaw will tend to become more relaxed.
- Your shoulders typically will become more relaxed yet erect.
- Your hands will tend to be relaxed, graceful, and quiet.
- Your arms will tend to hang in a relaxed, natural way.
- Your posture will tend to be relaxed, erect, and well balanced.
- Your walk will tend to be purposeful (without being aggressive or intimidating).
- Your voice will tend to be modulated with an intensity appropriate to the situation, and with clear pronunciation."[21]

HOW LEADERS CAN RAISE THE SELF-CONCEPT OF OTHERS

For leaders to bring out the best in people, they must relate to other people appropriately. In large part, this includes contributing to a positive self-concept in others. In doing this, the leader stimulates active and creative participation that allows for innovation. Nathaniel Branden suggests several things that you as a leader can do to raise the self-concept of group members or coworkers:[22]

- "Work on your own self-esteem: commit yourself to raising the level of consciousness, responsibility, and integrity you bring to your work and your dealings with people."
- Create opportunities for people to practice self-responsibility. Allow and encourage them to show initiative, suggest ideas, try new methods and new tasks, and generally expand the range of their abilities.
- Give reasons for rules and guidelines when the rationale is not self-evident. When you cannot grant a request, explain why; do not merely give instructions and directions.

- Admit your mistakes and apologize for shortcomings and mistakes in interpersonal relations, such as being unfair or snapping at someone. You will not lose face, injure your dignity, or endanger your position to admit having taken an action you now regret.
- Allow people to fail, to make mistakes, and to admit that they do not know something. People who are constantly afraid of the consequences of error or ignorance may stop taking risks and even stop trying altogether; they may never exercise their creativity, and they may lie about or try to cover up their mistakes.
- "Describe undesirable behavior without blaming: let someone know if his or her behavior is unacceptable, point out its consequences, communicate what kind of behavior you want instead, and omit character assassination."
- When someone does a great job or makes a good decision, of course praise the accomplishments—and also follow up to discover how and why the person was successful. If everyone learns what circumstances made the achievement possible, it is much more likely that you and your coworkers will be able to re-create those favorable situations and the resulting accomplishments in the future.
- "Praise in public and correct in private: acknowledge achievements in the hearing of as many people as possible while letting a person absorb corrections in the safety of privacy."
- When someone's behavior becomes problematic, avoid imposing corrections or solutions. Instead, ask the responsible party to come up with a solution. This method encourages self-responsibility, self-assertiveness, and heightened awareness.
- "Give your people the resources, information, and authority to do what you have asked them to do." The leader's job consists in large part of providing what group members need to be successful, whether it is training, financial resources, or guidance. If people are assigned responsibility for a task, they must also be given the authority needed to carry out the task. Doing otherwise is a sure invitation to failure and destruction of morale.

- Whenever possible, match tasks and objectives to individual strengths, dispositions, and interests. Build on people's strengths and let them do what they enjoy most and are best at.
- Give people as much control over their work as possible. "If you want to promote autonomy, excitement, and a strong commitment to goals, empower" your group members and coworkers.
- Aim high and expect excellence; assign tasks and projects that stretch your coworkers' known abilities.
- Help others to view problems as opportunities and challenges.
- Support talented non-team individuals and use them appropriately. There will always be some mavericks who work best on their own, even while pursuing achievement of group or team goals. Respect their individuality and give them whatever assistance you can to enhance their chances of success.
- Give written praise, commendation, and appreciation to high achievers (and encourage other leaders to do the same). When you prove that the organization values its members' *minds*, people are motivated to push "the limits of what they feel capable of achieving."

The person with a positive self-concept is one whose lower-level needs have been met, or are being met, and who is thus self-motivated—motivated from within by the need for continued self-development and the fulfillment of his or her potential as a human being.[23] Such people accept themselves because they know their strengths and build on them. They know they are human, and therefore less than perfect, so they take advantage of constructive criticism and suggestions. They value mistakes and failures for the lessons they can teach. People with a positive self-concept are realistic; they are aware of their own and others' capabilities and can cope well with pressure because they know what can be done.

People with a positive self-concept are tolerant of uncertainty. They are not afraid of new ideas and conditions. They are willing to innovate and function within new parameters. They accept other people and are open with them.

They work wholeheartedly with new personnel. People with a positive self-concept are not afraid to make independent decisions. They are committed to their work. They go beyond the call of duty because they believe in what they are doing. They are willing to work overtime if needed. People with a positive self-concept are appreciative and grateful. They have a genuine interest in others and in the infinite richness of life. They are appreciative and grateful to be, to feel, to know, to do, to create, and to become.

CONCLUSION

What can people with a positive self-concept become? Just about anything they want to be! A positive self-concept helps you become energetic, enthusiastic, positive, happy, hopeful, good, kind, fair, understanding, loving, forgiving, friendly, helpful, generous, caring, sensitive, nice, likable, loved, healthy, humble, obedient, appreciative, polite, cooperative, trusting, considerate, communicative, patient, disciplined, organized, consistent, ethical, persistent, sincere, creative, knowledgeable, confident, capable, talented, intelligent, loyal, honest, truthful, trustworthy, dependable, goal directed, independent, self-controlled, motivated, determined, and hardworking.[24]

SUMMARY

Self-concept is not a single ability. Rather, it is the result of several factors that affect how you feel about yourself, how you think about and perceive yourself, and how you act. It includes self-esteem, self-image, self-confidence, self-determination, and self-responsibility. These factors establish the level of worth and acceptance that you possess in the effort called life. Each factor is an important ingredient in self-concept, for without each the product that is the unique you remains unfinished.

Self-concept is an important part of any individual's life. Success and happiness are the ultimate goals of all people, but those goals are attained only through effort. The person who sits on the bench all the time is missing out. You, as an active player in the game of life, must get off the bench and participate to actually experience what life has to offer and take advantage of all the opportunities that come your way. A positive self-concept does not mean that your life will be totally void of disappointment. You may encounter many unpleasant situations, and you may experience disappointment, discouragement, fear, doubt, and anxiety. A positive self-concept equips you with the ability to overcome the inevitable and inescapable negative aspects of life. It helps you accept new challenges and find success in family, school, work, society, and relationships.

Several external and internal factors affect your self-concept. External factors include family, relationships, school, work, and society. Internal factors are fear, doubt, and anxiety.

There are five steps in developing a positive self-concept: restoring and nurturing, desire, attitude, goals, and action. Restoring and nurturing your self-concept includes such things as identifying and accepting your limitations, learning to accept others, making decisions, developing expertise, dressing as though you are successful, continuing to learn, being observant, tackling the things you fear, choosing your friends and associates carefully, learning from failures, and maintaining or altering your physical appearance or condition.

Many causes contribute to a poor self-image: a negative environment, ridicule, confusing failure on a project with failure in life, an untrained memory or uneducated brain, comparing one's worst features to others' best features, and setting unrealistic and unreachable standards of perfection.

Once your self-concept is attained, improved, or enhanced, you will exhibit many positive characteristics. These include being future oriented; able to cope with life's problems; able to deal with emotions; able to help others and accept help; able to accept people as unique, talented individuals; and showing a variety of self-confident behaviors.

Good leaders can raise the self-concept of others in many ways. Once people have achieved a positive self-concept, they accept themselves and are realistic, tolerant of uncertainty, and committed to work. They are appreciative and grateful to be, to feel, to know, to do, to create, and to become.

Take It to the Net

Explore self-concept on the Internet. Go to a search engine. Type your search terms in the search field (the ones in the following list are a good starting point). Browse various Web sites and find something related to this chapter. Print what you find and prepare to share it with the class.

Search Terms

self-concept

self-assessment

self-esteem

self-image

self-concept assessment

Chapter Exercises

REVIEW QUESTIONS

1. Define the Terms to Know.

2. List 15 rewards of a positive self-concept.

3. Differentiate between a positive self-concept and conceit.

4. List five ingredients of a positive self-concept.

5. Give 10 examples of people who were "late bloomers."

6. List four things that are critical in an adolescent's development of self-concept.

7. In what three ways do adults tend to define themselves (set their self-concept)?

8. What are five external factors that affect self-concept?

9. What are three internal factors that affect self-concept?

10. What are the five steps to achieving a positive self-concept?

11. List 33 things you can do to restore and nurture a healthy self-concept.

12. What are two main reasons that self-concept is important in the workplace?

13. List nine characteristics of workers with a low self-concept.

14. Explain how a low self-concept can lead to hostility.

15. What are six causes of a poor self-concept in the school or social arena?

16. List 10 characteristics exhibited by a person with a poor self-concept.

17. List 16 characteristics of people with a positive self-concept.

18. What are seven ways to present a positive self-concept?

19. What are 10 physical changes that may occur as you gain a positive self-concept?

20. What are 16 things leaders can do to raise the self-concept of others?

21. What are eight conditions that allow leaders to help their coworkers reach their full potential?

22. What are seven realities that contribute to a poor self-concept?

COMPLETION

1. Having a positive _____ is the single most important factor in becoming a success.

2. If you do not _____ in yourself, how can you expect to gain the confidence of your peers?

3. If we cannot _____ ourselves for what we are, we lessen our chances of being happy, productive, or successful.

4. An increasing trend in our society is people's inability to _____ responsibility for their own actions.

5. In the social arena, people with a positive self-concept are more _____ and _____; in this area, a positive self-concept helps you _____.

6. We often confuse experience with _____ and/or _____.

7. We often confuse failure on a project with failure in _____.

8. We confuse having an untrained memory or being uneducated with _____.

9. The ages of _____ to _____ are among the most critical in developing your self-concept.

10. It's no exaggeration to say that a strong, positive _____ _____ is the best preparation for success in life.

11. _____ _____ involves identifying and understanding our strengths and weaknesses, setting goals for improvement, and realizing that a positive self-concept gives us the ability to reach our goals and become successful.

12. Having _____ and _____ from within is what allows you to push yourself continually beyond normal levels of accomplishment.

13. Seeking help from others is not a weakness; _____ is.

14. The self-concept formed in childhood lays the foundation for your attitude toward _____, your future _____ and _____ abilities.

15. Teenagers criticize or ridicule others in an attempt to reduce _____ about themselves and their negative self-concept.

16. A _____ attitude is like driving a car in a dense fog: you are unsure what lies ahead.

17. A _____ attitude is like driving on a clear day: you can see obstacles coming and are able to plan and react in a time to avoid problems.

18. There is a relationship between your goals and your _____.

19. The experience of success in achieving short-range goals enhances your self-concept, and this spurs you on to set and attain higher _____ and _____ goals.

20. When workers are expected to do great things, they do _____ things.

21. People with a negative self-concept often have trouble _____ to others.

22. People cannot help the way they feel, but they can control the way they _____.

23. Praise in public and _____ in private.

MATCHING

_____	1. Music teacher said, "as a composer, he is hopeless."	A. Walt Disney
_____	2. Rating of "mediocre" in chemistry at Royal College.	B. Louis Pasteur
_____	3. Mysterious power of expectations.	C. Albert Einstein
_____	4. Fired by a newspaper editor because he had "no good ideas."	D. Thomas Edison
_____	5. You perform the way others believe you can.	E. Winston Churchill
_____	6. Greek sculptor in a legend on which Rosenthal based expectancy theory.	F. Ludwig van Beethoven
_____	7. Failed the sixth grade.	G. expectations
_____	8. Critical and jealous nature.	H. self-fulfilling prophecy
_____	9. Teacher told him he was unable to learn.	I. Pygmalion
_____	10. Could not speak until age four and could not read until age seven.	J. poor self-concept

ACTIVITIES

1. Several illustrations of "late bloomers" were given in this chapter. Identify a late bloomer whom you know (friend or celebrity). Explain why you believe that person has achieved success.

2. Select the external factor that you believe had the most impact on your self-concept. Write a paragraph explaining why.

3. List five things you will do when you become a parent to enhance the self-concept of your children.

4. Step 1 in developing a healthy self-concept, entitled Restoring and Nurturing a Healthy Self-Concept, listed several practices. Write a statement beginning with "I will" for 10 of the practices, describing how you will personally improve your self-concept.

5. Write three good things or positive characteristics about each member of your class. Read these out loud to the class. Write down the positive things that were said about you.

6. Suppose you were given a supervisory or leadership position in the workplace. Read the sections Conditions Needed to Empower Self-Concept in the Workplace, and How Leaders Can Raise the Self-Concept of Others. List 10 practices you would use to enhance the self-concept of workers. Share these with your classmates and get their responses.

7. Select a friend; identify five negative self-concept characteristics exhibited by that individual, and write suggestions on how he or she could improve self-concept.

   Note: Do not share these suggestions unless you do so under the supervision of your teacher and handle them carefully, as constructive suggestions for improvement. The purpose of this activity is to enhance a positive self-concept, not lower a self-concept.

8. Read the section called How Leaders Can Raise the Self-Concept of Others. Select five things that would be easy for you to do as a leader, and select five that would be a challenge to you. Read these in class and get your classmates' reactions.

9. Think of a past experience when you had to deal with someone who had an unhealthy self-concept. Answer the following questions.

   a. How was the poor self-concept manifested?

   b. How did it affect your relationship?

   c. Do you enjoy being around that person?

   d. What steps can (could) you follow to enhance the relationship?

10. Suppose your local newspaper is doing a story on each graduating senior. It plans to print a comprehensive review of you. Write the story as you would like it to appear in the paper. List your positive characteristics and accomplishments. Remember, you must occasionally "toot your own horn"; usually nobody else will toot it for you.

NOTES

1. Retrieved August 31, 2016 from http://thinkexist.com/.
2. "Self-concept," *Personal Development* series, No. 8736-A (College Station, TX: Instructional Materials Service, Texas A&M University, 1988).
3. N. Branden, *How to Raise Your Self-Esteem*, Google Play ed. (New York: Bantam Books, 2011), p. 14.
4. Z. Ziglar, *The "I Can" Course Teacher's Guide* (Carrollton, TX: The Zig Ziglar Corp., 2000), p. 37.
5. B. L. Reece, *Human Relations in Organizations: Interpersonal and Organizational Applications,* 12th ed. (Mason, OH: South-Wester, Cengage Learning, 2014), p. 78.
6. Ibid., p. 98.
7. Ibid, pp. 79–80.
8. Reece, *Effective Human Relations in Organizations*, p. 16.
9. J. Canfield, "Self-Esteem in the Workplace," *Self-Esteem Newsletter*, Fall 1990, p. 1.
10. R. Baumeister, *Self-esteem: The Puzzle of Low Self-Regard* (New York: Springer Science and Business Media, 2013), p. 171.
11. Reece, *Effective Human Relations in Organizations*.
12. N. Branden, *The Six Pillars of Self-esteem*, Kindle ed. (New York: Bantam Books, 2011), pp. 4660–4689.
13. U. Boser, M. Wilhelm, and R. Hanna. Center for American Progress. *The Power of the Pygmalion Effect.* Retrieved August 31, 2016 from http://files.eric.ed.gov/fulltext/ED564606.pdf.
14. J. M. Kouzes and B. Z. Posner, *The Leadership Challenge: How to Get Extraordinary Things Done in Organizations*, 5th ed. (San Francisco: Jossey-Bass, 2012), p. 276.
15. Ibid.
16. R. N. Lussier, *Human Relations in Organizations: Applications and Skill Building*, 10th ed. (New York: McGraw-Hill Education, 2017), p. 68.
17. Ibid.
18. Z. Ziglar, *See You at the Top* (Gretna, LA: Pelican Publishing, 2010), pp. 69.
19. Reece, *Effective Human Relations in Organizations*, pp. 81–82.
20. Branden, *How to Raise Your Self-Esteem*, pp. 168.
21. N. Branden, *The Six Pillars of Self-esteem*, pp. 911–916.
22. Ibid. pp. 4695–4784.
23. Appreciation is extended to Dr. Joe Townsend for sharing notes with me from his "Leadership" class at Texas A&M University, from which this material is adapted.
24. Ziglar, *The "I Can" Course Teacher's Guide*, p. 120.

19 ATTITUDES

Your attitude determines your altitude, or how high you go in life. We choose the type of attitudes we have. If you think you can, you can; if you think you cannot, you cannot. Attitude is the reason underlying a large proportion of our attainment of success, accomplishments, promotions, good grades, happiness, and the many other enjoyable things of life.

Objectives

After completing this chapter, the student should be able to:

- Explain the importance of attitude
- Discuss the types of attitudes
- Describe how attitudes are formed
- Discuss how attitudes affect behavior and human relations
- Describe the effect of attitudes on relationships
- Explain how positive attitudes can help you personally
- Compare and contrast attitudes and skills
- Explain how to change attitudes
- List and describe attitudes valued by employers
- Describe the effect of attitude on career success

Terms to Know

- attitude
- aptitude
- cognitive attitudes
- affective attitudes
- behavioral attitudes
- peer groups
- reference groups
- role model
- condescend
- tunnel vision
- psychosomatic
- symbiotic

Attitude affects our lives. William James said that we can alter our lives by altering our attitude. We can choose to have a positive attitude, or we can choose to have a negative attitude. Positive attitudes will have positive results because they are contagious. We cannot guarantee that only good things will happen to us, but we can control our attitudes when challenges and difficulties arise. With a positive attitude, you work to find the positive aspects and lessons in problematic situations and circumstances. The old saying is true: When life gives you lemons, make lemonade!

If you think you can, you can. If you think you cannot, you cannot. In general, we really control our own destinies. We can achieve whatever we want to if we are willing to pay the price in persistence, dedication, hard work, and sacrifice. Whether we are successful depends in large part on our attitude.

There are many definitions of and approaches to attitudes. The focus of this chapter is attitude as a positive outlook on other people, your job, and your surroundings (Figure 19–1). Ralph Waldo Emerson said, "What lies behind you and what lies before you pale in significance when compared to what lies within you."[1]

IMPORTANCE OF ATTITUDE

Attitude has an impact in our social and family lives as well as in the work place. We have observed that most people lose their jobs because of attitudes rather than job incompetence.

What Is an Attitude?

An **attitude** "is a strong belief or feeling toward people, things, and situations." All of us have positive or negative attitudes about life, human relations, work, school, and everything else. Attitudes are not changed easily. Our friends and those around us usually know how we feel about things, because our attitudes are indicated by our behavior. For example, if you make a disrespectful gesture behind someone's back, any onlookers will assume that you have a negative attitude toward that person.[2]

Attitude and Self-Concept

Attitude is usually related to self-concept. Those with a low self-concept often exhibit attitudes that are not based on the way things really are, but rather on their own feelings of inadequacy or insecurity. The way we feel about ourselves affects every other attitude we express, consciously or

FIGURE 19–1 This is a successful young showman with a positive attitude. Your attitude determines your altitude at any age. *(Courtesy of Dr. Alanna Vaught, MTSU.)*

involuntarily. If we do not like ourselves, everything we see and do will be affected by that feeling.[3]

Attitude versus Aptitude

Attitudes have a powerful influence on your life, future, and career. In reality, our attitudes hold us back more than our talents, gifts, or **aptitude** (quickness in learning and understanding) do. Zig Ziglar, American author and motivational speaker, encourages people with the following quote: "Your attitude, not your aptitude, will determine your altitude in life."[4] People who possess a positive mental attitude and an optimistic view of life are more apt to achieve personal and economic success. Such success does not happen by chance: people with a positive view of life are more apt to develop a life plan that includes a series of goals toward which they work daily.

Professionals, educators, businesspeople, parents, and students all share the opinion that attitude is the dominant factor in success. A study by researchers at Harvard University showed that 85 percent of the reasons for success, accomplishments, promotions, good grades, happiness, and many of the other good things in life were our attitudes, and only 15 percent depended on our knowledge, skills, or "technical expertise."[5]

Attitude and Control

The more control people feel they have over their lives, the more likely they are to be optimistic. Individuals who have a negative mind-set often expect the worst and believe that their efforts will not make any difference; at best, they simply hope that somehow things will work out for them and then let life "happen." These negative people usually have no long-range goals and no definite future plans. They often feel out of control because they have made little attempt to influence the future.[6]

Attitudes Can Greatly Influence Mental and Physical Health

In *Love, Medicine and Miracles*, Bernie Siegel stresses the strong connection between a patient's mental attitude and his or her ability to heal. Siegel says that when a doctor can instill some measure of hope in the patient's mind, the healing process begins. Siegel actually observed this as a doctor, so as part of his treatment plan, Siegel attempts to give patients a sense of control over their own destinies. He wants his patients to believe in the future and know that they can influence their own healing.[7]

Other people also believe that there is a relationship between good health and good thoughts. Although certain segments of our society have always maintained this, a growing number of health care professionals and specialists are taking the position that a positive mental attitude and outlook about the future can be potent treatments for a variety of mental and physical health problems. These people argue that our thoughts are the primary causes of the effects or conditions in our lives. To change our future lives, we must change our present attitudes.[8]

Attitude and Job Satisfaction

Several factors affect our attitude toward our jobs and the satisfaction we derive from them. These include attitudes toward the company's benefits, promotions, coworkers, supervision, safety, and the work itself. Job satisfaction is a concern to employers because management knows that there is a direct link between attitudes and productivity. Employees who do not like their jobs are more likely to be late or absent from work, become unproductive, or quit. A positive attitude toward work—in this case, job satisfaction—can reduce tardiness, absenteeism, and employee turnover.[9]

Attitude and Organizations

People shape their perception of you by what they see and hear. They interpret your attitudes through your behavior. How you feel about something is usually no secret to friends and schoolmates.

Attitudes are a powerful force in any organization. For better or worse, members will attract people with similar attitudes to the organization. If your FFA chapter does not hold regular meetings, does not participate in contests, and has committees that do not work, the dysfunction and lack of commitment are probably reflecting and conveying poor attitudes on the part of your adviser and officers, as well as their perception about the value of the FFA. An attitude of confidence, enthusiasm, and value can pave the way for improved meetings and greater participation from the officers and members. A sincere improvement effort by the chapter officers, filtered through attitudes of confidence, trust, and hope, results in the type of organization that everyone desires. These positive attitudes will attract new members who have similar attitudes.

Attitudes and Research

Attitudes and their effect on relationships, health, and intelligence are backed up by research. Although many studies have been done, one of the most notable was reported by psychologist Carl Rogers, in a paper titled "The Characteristics of a Helping Relationship."

In Rogers's study, children of accepting parents and of rejecting parents were compared on various traits. The children of accepting, democratic parents developed better attitudes. Children of parents with warm, caring attitudes showed higher intelligence, "more originality, more emotional security and control, and less excitability than children from other types of homes." By the time they reached school, they were popular, friendly, and nonaggressive leaders.

Children of rejecting parents showed slightly slower intellectual development, relatively poor use of the abilities they did possess, and some lack of originality. They were "emotionally unstable, rebellious, aggressive, and quarrelsome." The children of parents with other attitudes tended to fall between the characteristics of the two groups discussed.[10] What kind of attitudes do you think the parents of the young girl in Figure 19–2 possess?

Attitudes and Individual Success

Several stories have been told of winners who succeeded because of their positive attitudes. These included Lincoln, with his several defeats before winning the presidency; and Roger Bannister, who broke the four-minute-mile barrier. Glenn Cunningham was told he would never walk again after a near-fatal accident he had while starting a fire, but because of his attitude and determination, he became a world-class runner.

In fact, attitude relates to individual success in many ways:

- Academic achievement is directly related to an individual's attitude toward school and authority figures.

FIGURE 19–2 This young girl seems to be happy, healthy, and full of energy. What kind of attitude do you think her parents and/or guardians have?

- Job performance and promotion are directly related to individual attitudes toward work, supervisors, work ethic, and self.
- Athletic abilities are affected as much by attitude as by talent.
- Success in family relationships is affected by attitude.
- Responsible citizenship is directly related to an individual's attitude toward authority figures, neighbors, and self-worth.
- Personality, success in human relations, and sound judgment are based on positive attitudes.

TYPES OF ATTITUDE

We do not see things as they really are; rather, we view things through perceptual filters, or preconceptions based on the way we think or believe things are. Perception is also influenced by one's attitudes and understanding of others. Attitudes are a combination of an individual's feelings, thoughts, beliefs, and values. To put it simply, an attitude may be considered a way of thinking, feeling, and behaving.

Attitude has three components or elements. The first is the cognitive component: a person's thoughts, ideas, and beliefs. The second is the affective component: how a person feels or what emotions the person experiences. The last is the behavioral component: the tendency to act consistent with one's attitude.[11]

Cognitive Attitudes

Cognitive attitudes are the set of values and beliefs that an individual has toward a person, object, or event. For example, a fellow student tells you, "I don't like my teacher. He has it in for me." This cognitive attitude is the belief/perception that the teacher is unfair.

Affective Attitudes

Affective attitudes are the emotions attached to a person, object, or event. Happiness, anger, and disappointment are affective attitudes. When our FFA team loses in a regional competition, the sadness and disappointment we feel are a result of our affective attitudes.

Behavioral Attitudes

Behavioral attitudes are the tendency to act in a particular way toward a person, object, or event; our actions reflect and align with our other attitudes.[12] For example, when a student learns that an undesirable teacher is being transferred, the student smiles. When your parliamentary procedure team or public speaking contestant wins a contest, you are excited and applaud vigorously. If your agriculture classes are enjoyable, you are likely to participate more often.

HOW ATTITUDES ARE FORMED

We form attitudes primarily through our experiences. As we grow up, we interact with parents, family, teachers, and friends, among others. All these people play a part in teaching us what is right and wrong and how to act.

When we first encounter new people or a new situation, we usually have not had time to form an attitude. This is when we are most open and impressionable. Before entering a new situation, we often ask others who have had experience in such situations about it. Their responses begin the development of our own attitudes, even before the encounter.[13] Reece and Brandt point out that attitudes are learned from the numerous circumstances of each person's life, such as socialization, peer and reference groups, rewards and punishment, role-model identification, and cultural influences. A brief explanation of each follows.

VIGNETTE 19-1	ATTITUDE MAKES ALL THE DIFFERENCE

TRACY, WHO IS NOW THE PRESIDENT OF HER SORORITY and the Pan-Hellenic council at the university she attends, trains leaders as a service opportunity. She loves teaching young females how to do well in school and about being leaders in their respective organizations. In her eyes, attitude is the key to success. You see, Tracy knows better than anyone how important attitude is to being successful, being respected, and being a leader.

Tracy tells her story at the beginning of all of her leadership training sessions:

> I had just graduated from high school, and I, like every other high school graduate, was very happy about finishing school, but I was even more thrilled than most because I felt high school was a very negative experience for me. I did not do well in school. My grades were mediocre at best, and I did not choose to participate in any extracurricular activities. I had only a handful of friends and detested each day that I had to endure in high school. When I was lucky enough to get into college after having done so poorly in high school, I was convinced to make this experience better than high school. I realized that my problems in high school were brought about by my reactions and my attitude, not my ability. I decided from day one at this university to react positively whenever possible and to have a great attitude in all situations.

Tracy usually goes on to tell all of the great things she has experienced in college and to describe leadership more specifically. For her, it all changed when she changed her attitude. People were drawn to her positive nature. More opportunities came her way. She took more risks. She developed confidence, and she felt unstoppable in all aspects of her life. She completely changed her life by changing her attitude. Tracy hopes to change other people's attitudes and subsequently their lives. That is why she conducts the leadership seminars, and that is why they are titled "Attitude Makes the Difference."

FIGURE 19–3 This little girl is forming attitudes from daily interactions with her father. *(© iStock/Portra)*

Socialization

Young children are both extremely impressionable and amazingly observant. They interact most often with parents, teachers, and friends, and thus absorb, believe, and internalize what these authoritative and familiar figures say and do (Figure 19–3). For example, children who observe their parents recycling, using public transportation instead of cars to get to work, and saving electricity may develop a strong concern for protection of the environment; this value becomes part of their attitude toward the world around them.

Socialization occurs with both positive and negative attitudes and behaviors. The environment in which a child is raised is a powerful influence on the child's development. A child often does not have the opportunity to observe or compare other environments; the way he or she is raised is the norm for that child. Consequently, a child who grows up in an atmosphere of racism, sexism, or violence will develop and internalize those attitudes just as quickly and thoroughly as a child who grows up in a loving, tolerant environment will "grow" an attitude of courtesy and compassion. Values, beliefs, and attitudes are taught by example!

Peer and Reference Groups

As children begin to break away from their parents, they associate more closely with children their own age. These **peer groups** (groups of people who share rank, class, or age) can often be stronger influences than parents as children become young adults. With the passing of years, **reference groups** replace peer groups as sources of attitude formation and influence for young adults. The reference groups, such as college fraternities or sororities, may act as a point of comparison and a source of information for individuals. The FFA is a reference group for many. Reference groups can and should have a positive influence on positive attitudes. If yours do not, seek out different and better reference groups.

Rewards and Punishment

Authority figures generally encourage (reward) some attitudes and discourage (punish) others. It is natural for a child to want to maximize rewards and minimize punishments. A student who is praised for participating and asking questions in class is likely to repeat the behavior. Adults continue to have their attitudes shaped by rewards and punishment at work and in interpersonal relationships.

Role-Model Identification

Young people often achieve their goals of increasing status or popularity by identification with a **role model**, someone they admire who has an influence on them and whom they seek to emulate. These role models, whether they are parents or media stars, have a tremendous influence on developing attitudes. Television, movies, and the Internet also influence people's selection of role models. Why? By the time most students graduate from high school, they will have spent 50 percent more time in front of a television than in the classroom or with parents, family, or friends.

Cultural Influences

Culture consists of the values, beliefs, concepts, knowledge, and a broad range of behaviors that are acceptable within and a basic part of a specific society.[14] Every country and even sections within a country have different cultures. Although people may strive to define themselves as individuals within a culture, in large part we become what we grow up with. Most Catholics remain Catholics; most Protestants remain Protestants; most Hindus remain Hindus. We tend to maintain the basic cultural values, mores, and practices that we were raised with.

As discussed in Chapter 1, the FFA has its own symbols and culture. Generally speaking, agricultural education and the FFA represent a set of beliefs, customs, symbols, values, and norms that binds members of the organization together. FFA culture has a strong influence on member attitude.

HOW ATTITUDE AFFECTS BEHAVIOR

Attitudes affect your everyday behavior, but they are not the only cause of behavior, nor are they always a reliable predictor of behavior. Attitudes are complex. Even if you have a negative attitude toward your job, you may still work hard because of other attitudes, such as a positive attitude toward your boss, your peers, or promotion. The attitudes listed in Figure 19–4 affect behavior.

Attitude and Positive Thinking

Every thought you have has some effect. People get sick more easily when they are emotionally upset or depressed. Do not program your mind with bad thoughts, or you will end up with bad reasoning and poor attitudes. Remember, even when you cannot do anything about a situation, you can do a lot about your attitude toward the situation. When you or your followers meet situations that are not ideal, perhaps even sad or painful, work to improve your own and your followers' attitude by thinking and speaking positive thoughts. The athletes in Figure 19–5 could get through what looks to be a tough time by focusing on the positives of what lies ahead rather than focusing on the troubles of their current situation.

You Reap What You Sow

Presenting a positive attitude and conducting yourself in a positive manner will cause others to treat you with respect. The psychological mind can be compared to the physiology of plants. If you plant corn, you will get corn, not soybeans. Whatever a person sows, he or she will reap. I once had a student come to my office and start making negative comments about one of his former teachers. After a few minutes, I stopped him and said, "Ronnie, look at these two blocks on the wall. Every time you say something negative about somebody, you get a mark in this block. Every time you say something good about somebody, you get a mark in this other block. Ronnie,

Moody	Neat	Jealous	Afraid	Honest
Open-minded	Thrifty	Popular	Kind	Happy
Unreasonable	Even-tempered	Shy	Modest	Friendly
Demanding	Dependable	Clumsy	Proud	Sad
Sensitive	Angry	Show-off	Lazy	Serious

FIGURE 19–4 Our attitudes affect our behavior. Which of these attitudes do you possess?

FIGURE 19–5 These athletes seem to have trouble on the track. If they could assume a positive thinking about what could lie ahead rather than focusing on their current situation, they will be more successful eventually. *(© iStock/Steve Debenport)*

you reap what you sow. For every mark in the negative block, somebody says something negative about you. For every mark in the good block, somebody says something good about you."

Attitude and Motivation

Our attitudes can motivate ourselves and others. Attitudes can be shaped to be positive. Over time, you can condition and train your mind so that you will automatically respond positively to any negative situations you encounter in life. Psychologist David McClelland of Harvard University found that you can change your attitude and self-motivation by consciously changing the way you think about yourself and your circumstances. Needs create goals. The pursuit of goals results from motives. Our motives are what spur us to action.

Some say you cannot motivate others. Nonsense! The old saying, "You can lead a horse to water, but you can't make him drink," is really not true. Give the horse some salt and watch it drink. The salt, creating a need for water, motivates the horse to drink. Positive attitudes can motivate us to positive thoughts and actions.

Winners Find a Way

Winners find a way. Others find an excuse. Be willing to grow into greatness. A small acorn grows into a great oak tree. Seeds of greatness can grow within our lives. It takes work and discipline. It takes proper nurturing. It takes time and may not happen overnight.

Constantly keep the goals toward which you are working in view. Avoid complaining. Winners never **condescend**, or do something they regard as beneath their dignity; rather, they lift others around them to a higher level through encouragement and assistance. Help others attain success, and you will help yourself.

Enthusiasm Is an Attitude

Most people let conditions control their attitude instead of using their attitude to control conditions. If things are going well, their attitude is upbeat. If things are going badly, their attitude is poor and depressed. It is best to build a solidly positive attitude, so that when things are bad or the going gets rough, your attitude still stays good.

Enthusiasm can make a difference in your attitude. How you respond to failure and mistakes is one of the most important decisions you make. With enthusiasm, failure does not mean that nothing has been accomplished. It simply means you figured out another way something does not work. There is always the opportunity to learn something. With enthusiasm, what is within you will always be bigger than whatever is around you.

Attitudes and Failure

We all experience failure and make mistakes. In fact, successful people have more failure in their lives than average people do. Anyone who is currently achieving anything in life is simultaneously risking failure. Only those who do not expect anything are never disappointed. Only those who never try avoid failure completely. However, it is always better to fail in doing something than to excel in doing nothing. A flawed diamond is more valuable than a perfect brick. People who have no failures also have few victories. Everybody gets knocked down, but those with a good attitude get back up fast.[15]

ATTITUDES AND RELATIONSHIPS

A positive attitude or a change in attitude can dramatically improve your relationships with others. We must look for the good in others. People will build positive relationships if they live by the saying "If you can't say something good about someone, don't say anything at all." Our attitudes will be enhanced when we show sincere love, care, and concern for others. Like it or not, you live within (and depend on) a society in which you constantly interact with other people. The more you understand others, work cooperatively with others, and show respect toward others, the better your own life will be.

Find Good in Others

As a child, I used to go to a hill on our farm and start shouting to hear echoes. I would shout, "You are great." The echo would shout back, "You are great." I would shout, "You are greater," and the echo would return with, "You are greater." Life is an echo: what you give comes back. Regardless of who you are or the type of occupation you have, if you are looking to find the good in others, you are looking for the way to reap the most rewards in all areas of life, including attitude enhancement. Adopt the Golden Rule—"Do unto others as you would have them do unto you"—as a way of life. Find the good in others, and they will find the good in you.

Sculpting Special Abilities

As a teacher, I visualize every student as a block of wood. Within that block of wood are special traits or talents. If I keep carving, I will eventually discover that special trait; that is, I will not only find the good, but I will also find the special traits or talents that each student has. Then, I will help direct and cultivate those talents. The student's attitude and self-esteem begin to develop.

The film *Rain Man*, starring Dustin Hoffman and Tom Cruise, told a story of an autistic man who was practically and socially dysfunctional, but who nevertheless possessed a mathematical ability so great that he was banned from the casinos in Las Vegas. There is a genius in each of us that can be carved into a masterpiece sculpture.

When you mine gold, you have to move several tons of dirt to get an ounce of the precious metal—but we do not mine for dirt; we mine for gold. Do not look for the negatives; look for the "gold" or positive aspects of a person or situation. The harder you look, the more positives you will find.

Praise and Success

Few people are motivated or inspired to work harder or perform better with constant criticism. Most people respond better to positive reinforcement including praise and success.

Compliments A sincere compliment is one of the most effective attitude adjusters and motivational methods there is. Some say compliments are just hot air; but hot air can make you fly high or, like the air in automobile tires, can help ease you along the highway.

Tunnel Vision Praise and success can break us out of **tunnel vision**, a narrow perspective.[16] Many students believe that they are A students, and this attitude *makes* them A students. The same is true of many B and C students: they become what they believe they are.

Success Breeds Success Once students experience success, they like it. Success breeds success. One ounce of praise and successful experiences is worth more than a pound of criticism. Unfortunately, many people do not realize this. As a high school teacher, I occasionally used a technique called the "Shotgun Approach to Teaching" (see Appendix A). This technique involved introducing the unit, teaching it, reviewing, and giving a test all within one class period. By teaching short segments of information, using several principles of learning, and reviewing extensively, I helped more than 80 percent of the class to get As. When the students who were not accustomed to making good grades got As, they experienced a positive attitude shift. The success broke them out of their tunnel vision and they saw themselves as better students. Therefore, they *became* better students, because they enjoyed the taste of success, wanted to experience it again, and knew that it was possible.

Expectations

Most people live up to what others expect of them. This is true in all aspects of life, whether one is a student, an athlete, or in the workplace.

People Rise to the Expectations of Others At the beginning of the school year, a new teacher looked up IQ scores in her students' files. She had two classes. She was amazed at the difference in the intelligence level of the two classes. For the class with IQ scores in the 120 range and up, she taught challenging classes at a high ability level. The students responded according to her expectations. For the class with IQ scores in the 90 range and lower, she taught at a slower rate, being sure not to cover the material too quickly. She did not give much homework and practically spoon-fed these students. This class also responded as she expected.

An interesting thing happened later in the year. While the teacher was looking for something in one of the student files, she discovered that she had inadvertently gotten locker numbers rather than IQ scores. In reality, there was absolutely no difference in the intelligence levels of the two classes. The difference in performance was the direct result of the difference in the teacher's expectations. In short, the teacher treated the two classes differently because she perceived them differently, and the different treatment yielded different results. From the students' point of view, they simply performed the way the teacher expected them to perform. The way you see others is the way you treat them, and the way you treat them often determines the way they act, react, think about themselves, and become.

Expectations and Attitudes Can Affect Health The famous Harvard psychologist, William James, wrote, "The greatest discovery of my generation is that human beings, by changing the inner attitudes of their minds, can change the outer aspects of their lives." Physician Bernie S. Siegel said, "Years of experience have taught me that cancer and indeed nearly all diseases are **psychosomatic** [bodily symptoms experienced as a result of mental conflict or upset]."[17] "The body knows only what the mind tells it.... If one has taken part in getting sick, one can also take part in getting well."[18]

Belief in Others

Our attitude of belief in others can push individuals toward greatness. Many of you have had teachers, parents, friends, or somebody else who has given you the spark of belief that motivated you. I challenge you to do the same for others.

When Helen Keller was given England's highest award as a foreigner, Queen Victoria asked her, "How do you account for your remarkable accomplishments in life? How do you explain the fact that even though you were both blind and deaf, you were able to accomplish so much?" Without a moment's hesitation, Keller replied that had it not been for Anne Sullivan, the name "Helen Keller" would have remained unknown. Anne Sullivan was Keller's teacher, who believed that the child could learn and dedicated herself to teaching Helen to communicate. Because of Sullivan's persistence and belief, Keller learned the alphabet by touch, how to connect words with objects, how to read and write in Braille, and how to type.

FIGURE 19–6 Canada geese demonstrate the value of cooperation. The "V" formation fights the headwind, creating a partial vacuum that allows geese to fly 72 percent farther than they would if each flew alone. *(© iStock/Merrimon)*

Attitude of Cooperation

Canada geese instinctively know the value of cooperation. The geese fly in "V" formation and regularly change the lead goose. Why? The lead goose, in fighting the headwind, helps create a partial vacuum for the geese flying behind it (Figure 19–6). In wind-tunnel tests, scientists have discovered that the geese as a group can fly 72 percent farther than an individual goose can fly. Perhaps people could fly higher, farther, and faster by cooperating with, instead of fighting and competing against, those around them! As you practice an attitude of cooperation, remember that people pay more attention to what you do than to what you say.

A POSITIVE ATTITUDE CAN HELP YOU

Attitude during Job Interviews

Attitude, especially attitude toward work, is critical during the job interview. Employers look for signs of how you will work with others and deal with problems. You communicate attitude as well as words, information, and ideas during an interview.

Achieving Competence in Attitude

People can become competent in attitude just as they can become competent in a technical skill. However, attitude competence is far more difficult to measure. Nevertheless, people respond in a positive way when you have a good attitude and in a negative way when you do not. Even though attitude cannot be exactly measured, it is always being observed.

Positive Attitude Leads to a Brighter Future

People who concentrate on good attitude get the best jobs and eventually rise to the top in most organizations. All organizations are built around people, and when you build healthy relationships with your fellow workers and supervisors, you open doors that would otherwise have stay closed. For better or for worse, other people have an enormous amount of control over your job future, so the better your attitude toward them, the better things will go for you.

Better Attitudes Help You Become a Better Leader

With a good attitude, not only will you become a better supervisor or leader, you will become one sooner. Working on your attitude now, along with your willingness to work, will strongly influence your progress later.

Relationship between Attitude and Learning

There is a definite relationship between attitude and learning. When a mind is open (free of blocks, fears, prejudices), it will more readily accept and retain new data and ideas. If a student dreads taking a particular course, the chances of success are reduced because the fear becomes a barrier to learning. However, through an "attitude adjustment" in which the student adopts a positive mind-set and positive self-talk ("I know I can learn this material"), the block can be partially eliminated. Buoyed by the positive attitude, the student can talk with the teacher, find a peer mentor, or take other action to enable and enhance learning in this potentially troublesome area. Learning becomes easier because the student's attitude toward learning is more positive.[19]

Would you say that the attitude of the teacher in Figure 19–7 could enhance a student's learning?

Connection between Attitude and Personality

There is a **symbiotic** relationship between attitude and personality in that there is a close association that benefits each. Personality is generally considered to be the combination of special physical and mental characteristics through which you communicate a unique image to others. When you have a positive attitude, the image communicated is at its best. Attitude is the only characteristic that blends all your other traits—both physical and mental—together into a more attractive image. A positive attitude can overcome negative physical and mental characteristics (remember Helen Keller). When you are positive, your good features shine and your weaker aspects seem less important and less obvious.

ATTITUDE OR SKILL?

What are the characteristics of successful people? If you and a friend made lists of the qualities of the most successful person you know, your lists would be similar. It does not matter who or what this person is (an employer, employee, neighbor, friend, etc.); the qualities both you and your friend perceive in him or her are likely to be the same.

FIGURE 19–7 This teacher and student appear to have good rapport. Learning is much easier when the student's attitude toward learning is positive. (© iStock/Steve Debenport)

alert	energetic	patient
assertive	enthusiastic	personable
caring	faithful	practical
committed	friendly	prepared
common-sensical	goal oriented	prompt
communicative	hardworking	responsible
compassionate	helpful	self-respecting
confident	honest	sensitive
consistent	humble	supportive
cooperative	humorous	teachable
creative	knowledgeable	thoughtful
decisive	loving	trusting
dedicated	loyal	understanding
dependable	motivated	good listener
disciplined	optimistic	good moral character
empathetic	organized	integrity
positive mental attitude		

FIGURE 19–8 Although these are desirable qualities or characteristics for success, most are not learned in a formal academic setting. Could it be true that attitude is more important than aptitude?

Several desirable qualities or characteristics for success that would probably be on your list are shown in Figure 19–8. Study the qualities listed in this figure and decide which are attitudes and which are skills If your answers are consistent with those of most people, you probably identified most of these characteristics as attitudes. Besides knowledgeable and creative, and two or three that could be a combination of attitude and skill, most are attitudes.

Attitudes Are as Important as Aptitudes

Ask any group what it takes to make people successful, and 90 to 95 percent of the answers will be in the form of an attitude. However, we constantly hear reports about our academic weaknesses compared to those in other countries. Billions of dollars are spent on improving academic performance, but almost nothing in our educational system is dedicated to teaching and improving attitudes. Yes, academics are very important, but our attitudes are at least as important as our aptitudes.

Learning Qualities of Successful People

Can the qualities or attitudes of successful people be learned? Yes. Unfortunately, they are not typically taught in public schools. However, because you are reading this, you are being exposed to the proper attitudes for success. Former Texas Commissioner of Education William Kirby reported that in Japan, from kindergarten through high school, students receive one hour of instruction each day on the importance of honesty, character, integrity, hard work, positive mental attitude, enthusiasm, responsibility, thrift, free enterprise, patriotism, and respect for authority.[20]

It seems that most of these attitudes are learned at home, in the military, or in faith-based organizations. At one time, out of all the chief executive officers of Fortune 500 companies, 175 were former Marines. The motto of the Marine Corps is *"Semper Fidelis* [Always Faithful]"; and the Marines are strong on discipline, commitment to excellence, loyalty, and accepting responsibility for whatever task they are assigned. Additionally, many of our nation's presidents served in the military in some capacity before becoming commander in chief. Apparently, discipline, commitment, and loyalty are desirable qualities that enhance the achievement of success in top positions.[21] Attitude is important in life regardless of what you are doing or plan to do.

CHANGING ATTITUDES

Have you ever heard the expression "A fool never changes?" If we are wrong or have a negative attitude, it is in our best interest to change and seek improvement.

Changing Your Attitude

All that a man achieves and all that he fails to achieve is the direct result of his own thoughts.... A man's weakness and strength...are brought about by himself and not by another; and they can only be altered by himself, never by another.... As he thinks, so is he; as he continues to think, so he remains.[22]

You can control your attitude and thereby change your outlook on and approach to life. Usually, we cannot control our environment, but we can control our responses to external forces; simply put, we can change our own attitudes. We can choose to be optimistic or pessimistic. We can look for the positive and be happier and get more out of life. The following suggestions can help you change negative attitudes:[23]

Be Aware of Your Attitudes Research has found that people with the personality characteristic of optimism have higher levels of job satisfaction. Consciously try to maintain a positive attitude. As mentioned earlier, if a situation gives you lemons, make lemonade.

Think for Yourself Do not simply accept or copy the attitudes of others; decide for yourself what attitude you will develop. Know why you are choosing the attitude and be able to defend your reasoning.

Do Not Harbor Negative Attitudes There are few—if any—benefits to negative attitudes. A negative attitude, like holding a grudge, can only hurt

---

VIGNETTE 19-2 | **HOW TO BUILD AN ATTITUDE**

DID YOU KNOW THAT AN ATTITUDE is built in much the same way a house or car is? In *The Winning Attitude*, John Maxwell describes five foundational truths about the construction of a winning attitude.[24] Each of these truths can have profound effects on our own lives and the lives of others around us.

1. *A child's formative years are the most important for instilling the right attitude.* Most researchers have found that early child development in a positive setting is the primary determinant of a child's future success. Just as a new house must start with a perfectly aligned foundation, the early stages of attitude development must be as close to perfect as possible. This applies to leaders as well. If you are teaching employees or FFA members how to do a job, train them right the first time, while they are still new to the task. If they are not trained properly, or if they learn it wrong initially, it is much more difficult for them to unlearn it and learn it again. The formative phase of any situation is when a leader should attempt to instill an attitude that is in line with the team or organizational vision.

2. *The growth of an attitude never stops.* According to Maxwell, "[o]ur attitudes are formed by our experiences and how we choose to react to them. Therefore, as long as we live, we are forming, changing or reinforcing attitudes." When things change and problems come, train yourself to react in a positive manner whenever possible, so that you will continue to form and grow a good attitude.

3. *"The more our attitude grows on the same foundation, the more solid it becomes."* We all live by a set of certain beliefs and values, especially as leaders. If we align our good attitude growth with our fundamental beliefs and values, our beliefs and values will become stronger along with our positive attitude.

4. *Over time, many builders help construct our attitudes.* Just as a house-building project requires many specialists, so does attitude building. Many people play many different roles in our lives and in the construction of our attitudes. Take advantage of their wisdom and allow people to give to you—and do not forget to thank them for their influence.

5. *"There is no such thing as a perfect, flawless attitude.* Houses and cars get old, go out of style, and need updates; so do attitudes. We have to maintain our winning attitude. If you stop after the first four truths, your attitude will begin to sour, no matter how good it once was."

your relations with other people, and eventually it will hurt you.

Keep an Open Mind Listen to other people and use their input to support and enhance your positive attitudes.

Alter Your Thinking Letting go of attitudes that are no longer appropriate can be difficult, but it is part of changing your thinking for the better. Evaluate whether your attitudes toward people and situations are still valid, or whether a negative attitude is really warranted.

Change Your Behavior One of the most difficult approaches to dealing with an undesirable attitude is to attempt to change the accompanying behavior. If you are persistent and determined in changing your thinking, though, the behavior flowing from that thinking will change too.

Change Your Ideas and Beliefs In many cases, an undesirable attitude is based on insufficient or misleading information. Simply becoming aware of new facts is very likely to help you modify your attitude.

Change Feelings The most promising approach for dealing with feelings and emotions in others involves listening. True active listening can be extremely difficult. It requires understanding, sympathy and empathy, and being accepting and non-threatening so that the other individual feels free to express his or her feelings, problems, and attitudes.

Change the Situation When possible, change the situation that is the source of unfavorable attitudes, such as working conditions. Try to surround yourself with positive, optimistic, encouraging people.

It can be next to impossible to change others—but we can always work on changing ourselves. To adopt and maintain a positive attitude, we can smile, say something pleasant every hour, change negative statements to positive statements, view a negative problem in a positive light, and keep a mental picture of the kind of people we want to be. We can change our attitudes if we have high expectations of ourselves.

Changing Followers' Attitudes

It is hard to change your own attitudes, but it is even harder to change other people's attitudes. Nonetheless, it can be done. The following hints can help leaders change the attitudes of their followers:[25]

Give Followers Feedback Followers must become aware of their negative attitudes before they can even begin to change them. The leader must talk to the follower about the negativity and show that the attitude has negative consequences for both the individual and the group. The leader should suggest an alternative attitude.

Accentuate Positive Conditions Followers tend to have positive attitudes about the things they do well. Make conditions as pleasant as possible; make sure followers have all the necessary resources and preparation to do a good job. To the greatest extent possible, assign tasks according to followers' strengths, talents, and interests.

Provide Consequences If an activity or event has positive consequences, it will probably be repeated. Similarly, people tend to avoid actions and situations that have negative consequences. Encourage and reward followers who have positive attitudes. Try to avoid the development and spread of negative attitudes.

Be a Positive Role Model If you, the leader, have a positive attitude, your followers are more apt to have positive attitudes (Figure 19–9).

ATTITUDES VALUED BY EMPLOYERS

Several attitudes valued by employers have been discussed indirectly throughout this chapter. The following characteristics relate more specifically to the workplace.

Willingness to Assume Self-Leadership

Self-leadership emphasizes self-sufficiency. People need to be leaders of themselves. If you have self-leadership, you can find the drive to meet and handle challenges regardless of their scope. You need to set your own goals and monitor your own progress toward those goals. Assuming self-leadership also means you know how to motivate yourself. Employers place a great value on people with an internal drive to perform.

Willingness to Learn

If you are self-motivated, you will also take it upon yourself to learn as much as you can to be

FIGURE 19–9 The positive attitude of the supervisor causes a positive attitude in the employee, which will enhance the workplace. *(© iStock/DragonImages)*

the best you can be in your job. Many employers agree that employees who are willing to learn—especially those who bring a good foundation of basic education and learning skills—are worth more, because the employer need not dedicate as much time and money to training them.

Willingness to Be a Team Player

Because employers are increasingly using teams to get things done, team players are in ever-greater demand. Team players work well with others in building products, innovating and solving problems, and making decisions. Being a team player also means being dependable, remaining loyal to the brand of an organization, and lifting everyone up around you with a positive attitude. Leadership skills and positive attitudes are tremendous assets in teamwork.[26]

Concern for Health and Wellness

Many employers provide programs on health and wellness. The employees who participate in these programs usually take fewer sick days and have a higher level of energy. A positive attitude and respect for your own well-being enhance both physical and mental health and thus your ability to achieve in your job.

Enthusiasm for Life and Work

Employees who show enthusiasm for life and work are more likely to be a positive influence on their coworkers. Enthusiasm is contagious. People with enthusiasm about life and work look for the bright spots and can find good in almost everything (Figure 19–10). They avoid negative people whenever possible and refuse to be persuaded or dragged down by negative thinkers who see only problems and never solutions.

ATTITUDE AND CAREER SUCCESS

Some people "light up" a room when they enter. Others, "put out the light" when they enter. Make sure your attitude projects the "light". Your attitude will carry over into the workplace and play a major role in how successful you are.

Self-Evaluation of Attitudes

Robert Lussier, in *Human Relations in Organizations*, compiled the following positive job attitude statements.[27] How do you rate on this list? Each

FIGURE 19–10 This worker appears to be enthusiastic about her job. Individuals who have an enthusiastic attitude are productive, and people enjoy working with them. *(Courtesy of University of Georgia—College of Agricultural & Environmental Sciences.)*

of these attitudes would enhance your career success.

- I smile and am friendly and courteous to everyone at work.
- I make positive, rather than negative, comments at work.
- When my boss asks me to do extra work, I accept it cheerfully.
- I avoid making excuses, passing the buck, or blaming others when things go wrong.
- I am an active self-starter at getting work done.
- I avoid spreading rumors and gossip among employees.
- I am a team player willing to make personal sacrifices for the good of the work group.
- I accept criticism gracefully and make the necessary changes.
- I lift coworkers' spirits and bring them up emotionally.

Positive Attitudes Are Essential for Career Success

Attitude will impact your career success. In fact, a negative attitude will sabotage your career, and a positive attitude will lift it up. A positive attitude is essential to career success, for the following reasons:[28]

Energy When you think positively, you have more energy, motivation, productivity, and alertness. A positive attitude lets your inner enthusiasm surface freely.

First Impressions The initial impression you make on people is extremely important in employment. If you display a positive attitude, coworkers will pick up a friendly, warm signal and be attracted to you. If they do not sense such an attitude, they may ignore or try to avoid you.

Productivity of Others Increases An employee who is positive makes other employees more productive. "Attitudes are caught more than they are taught." People with positive attitudes are more successful because they have (1) the ability to shrug off bad news and not give up, (2) a willingness to take risks, (3) a desire to assume personal control of events rather than just allowing things to happen, and (4) a willingness to set ambitious goals and pursue them.[29]

Positive and Productive Work Environment When you show a positive attitude, coworkers enjoy your presence more, thus creating a more productive and positive work environment all around. This makes your job more interesting and exciting because you are in the middle of things and not on the outside complaining or whining.

Promotion Opportunities Your future success will depend on the kind of attitude you express toward authority figures. Your mental attitude is constantly being read and judged even though you may think it does not show. When you consistently project a positive attitude, you are more likely to be considered for special assignments and promotions.

CONCLUSION

A positive attitude will have positive results because attitudes are contagious. Emerson said, "When a happy person comes into the room, it is as if another candle has been lit." If you think you can, you can; if you think you cannot, you cannot. Why? Your attitude determines your altitude. We can alter our lives by altering our attitudes because, among other reasons, your attitude is just as important as your aptitude.

You cannot control every situation you encounter, but you can control your response to the situation. Furthermore, although you cannot always create ideal situations for yourself, you can tailor your attitude to make better situations and outcomes more likely. Most successes, accomplishments, promotions, good grades, happiness, and many of the other good things of life depend on or are based on attitude.

SUMMARY

Attitudes affect our lives. We choose the type of attitudes we have. Your attitude determines your altitude, or how high you go in life. If you think you can, you can; if you think you cannot, you cannot. Your beliefs and thinking—that is, your attitude—determine your behavior and actions.

An attitude is a predominant belief or feeling about people, things, and situations. Cultivating a positive attitude is important because it gives you more control of your life, influences your mental and physical health, increases job satisfaction, increases the strength of your organization, and helps you accomplish individual success.

There are three types of attitude: cognitive, affective, and behavioral. Attitudes are formed by socialization, peer and reference groups, rewards and punishment, role-model identification, and cultural influence. Positive attitudes cause us to be more motivated, find a way to achieve and win, be more enthusiastic, overcome failure, and learn from mistakes.

A positive attitude can dramatically enhance your relationships by helping you find the good in others, bringing out the best in others, praising others, creating high expectations for and belief in others, and enhancing cooperation.

Positive attitudes can help you personally: during job interviews, by leading to a brighter future, by helping you become a better leader, by increasing your learning ability, and by enhancing your personality.

Ninety percent of the qualities of successful people are actually attitudes. Attitudes can be learned, practiced, and developed.

We can change our attitudes if we become aware of those attitudes, think for ourselves, do not harbor negative thoughts, keep an open mind, alter our thinking, change our behavior, change our ideas and beliefs, change our feelings, and change our situations. Leaders can help change their followers' attitudes by giving followers feedback, accentuating positive conditions, providing consequences, and being positive role models.

Attitudes valued by employers include willingness to assume self-leadership, willingness to learn, willingness to be a team player, concern for health and well-being, and enthusiasm for life and work. Positive attitudes that are essential for career success include being energetic, creating good first impressions, increasing productivity of others, and contributing to a positive and productive work environment.

Take It to the Net

Using your favorite search engine, explore attitudes on the Internet. Some search terms are listed here to start your thinking process. Browse various websites and find something of interest related to attitudes and leadership. Print what you find and prepare to share it with the class.

Search Terms

attitude types

attitudes

improving attitudes

positive attitudes

Chapter Exercises

REVIEW QUESTIONS

1. Define the Terms to Know.
2. What are four broad areas in which attitude is important?
3. In what five specific ways can a good attitude contribute to individual success?
4. Give an example of a research finding that positive attitudes really do make a difference.
5. Name and explain the three types of attitudes.
6. Describe five ways in which attitudes are formed.
7. What are some behaviors that a positive attitude can affect?
8. What are five things we can do to improve relationships?
9. What are five ways attitudes can help you personally?
10. Name seven ways to change your attitudes.
11. Name four ways a leader can change the attitude of followers.
12. What are five attitudes valued by employers?
13. Give five reasons why positive attitudes are essential to career success.
14. According to John Maxwell, what are five foundational truths about the construction of a winning attitude?
15. What are nine positive job attitude statements that will enhance your career success?
16. What are four reasons that people with a positive attitude are more successful?

COMPLETION

1. We can alter our lives by altering our _____.
2. Positive attitudes will have positive results because they are _____.
3. When lemon situations arise, make _____ out of them.
4. If you think you can, you can. If you think you cannot, _____.
5. We can become just about whatever we want to in life, if we are willing to pay the price of _____, _____, and _____.
6. What lies behind you and what lies before you pale in significance when compared to what lies _____.
7. Your attitude is more important than your _____.
8. Your attitude determines your _____.
9. You reap what you _____.
10. It is better to fail in doing something than to excel in doing _____.
11. Everybody gets knocked down, but those with a good _____ get back up fast.
12. Visualize yourself as a block of wood; within that block of wood is a _____.

13. Do unto others as you would have them _____ unto you.

14. Find the good in others, and they will find the _____ in you.

15. You can have everything in life you want if you will just help enough people get _____.

16. People can fly higher, farther, and faster by _____ with, instead of fighting against, those around them.

17. People pay more attention to what you _____ than to what you say.

18. Most people lose their jobs because of _____ rather than job incompetence.

19. People who possess a positive mental attitude and an optimistic view of life are more apt to achieve _____ and _____ success.

20. Attitudes are a combination of an individual's feelings, _____, _____, and _____.

21. If you can't say something good about someone, _____ say anything at all.

22. One ounce of praise and successful experiences is worth more than a _____ of criticism.

23. Our attitude of belief in others can push individuals toward _____.

24. With a great attitude, not only will you become a better supervisor or leader, you will become one _____.

25. Academics are very important, but our _____ are at least as important as our aptitudes.

26. You cannot control every situation you encounter, but you can control your _____ to the situation.

27. Attitudes are _____ more than they are taught.

MATCHING

_____	1. Find the good in others.	A. negative attitudes
_____	2. One of the most effective attitude adjusters and motivational methods in existence.	B. never fail
		C. praise and success
_____	3. Bad deeds to others produce bad deeds to you.	D. skills
_____	4. Those who never try.	E. high expectations
_____	5. Less important than attitude in keeping most jobs.	F. ounce of praise
_____	6. People with personality characteristics of optimism have higher levels of this.	G. sculpture
		H. job satisfaction
_____	7. Like holding a grudge, these can only hurt your human relations.	I. they will find the good in you
_____	8. One within every block of wood.	J. you reap what you sow
_____	9. Worth more than a pound of criticism.	
_____	10. Students perform better when this is present.	

ACTIVITIES

1. This chapter describes five ways in which attitudes are formed. Select the one that you believe has had the most effect on your life, and write 100 words (approximately one page) explaining why.

2. Refer to the list of attitudes and skills (90 percent are attitudes) in Figure 19–8. Select the attitudes that you personally need to work on to be successful. Write a brief action plan for each.

3. Pretend that you are an employer and you want to hire the best people possible. Identify the 10 most important attitudes and skills on your priority list. Defend each of these selections.

4. You have been asked to give a speech on attitudes. Using the information in this chapter, write a five- to seven-minute speech and present it to the class. Add your own stories or other information. (You may wish to refer to Chapters 7 and 8 on writing and presenting speeches.)

5. You are the chief executive officer of a large corporation. Prepare a worker evaluation form to be used by the supervisor in each department.

6. Think of a role model in your life who has shaped many of your attitudes. Identify and explain attitudes that this person helped to form.

7. Including the stories, sayings, facts, and other information from this chapter, list five things you learned from this chapter. Share these with the class.

NOTES

1. Retrieved September 3, 2016 from http://www.brainyquote.com/quotes/quotes/r/ralphwaldo386697.html.
2. R. N. Lussier, *Human Relations in Organizations: Applications and Skill Building*, 10th ed. (New York: McGraw-Hill Education, 2017), p. 60.
3. L. H. Lamberton and L. *Minor, Human Relations: Strategies for Success*, 2nd ed. (Chicago: Irwin/Mirror Press, 2001), pp. 63–64.
4. Retrieved September 4, 2016 from http://www.brainyquote.com/quotes/quotes/z/zigziglar381975.html
5. Z. Ziglar, found at http://quote.robertgenn.com/getquotes.php?catid=22.
6. B. L. Reece and R. Brandt, *Effective Human Relations in Organizations*, 8th ed. (Boston: Houghton Mifflin, 2005), pp. 101–102, 158–159.
7. B. S. Siegel, *Love, Medicine and Miracles*, Reissue ed. (New York: William Morrow Paperbacks, 2011).
8. Ibid.
9. J. Kagan, *Psychology: An Introduction*, 9th ed. (New York: Harcourt Brace Jovanovich, 2003).
10. C. R. Rogers, "The Characteristics of a Helping Relationship," in *The Helping Relationship Sourcebook*, edited by D. L. Avila, A. W. Combs, and W. W. Purkey (Boston: Allyn & Bacon, 1977), pp. 2–18, at pp. 4–5.
11. Retrieved from web-based video September 4, 2016 from http://study.com/academy/lesson/types-of-attitudes-in-the-workplace-cognitive-affective-behavioral-components.html.
12. Ibid.
13. Lussier, *Human Relations in Organizations*, p. 60.
14. W. M. Pride and O. C. Ferrell, *Marketing: Concepts and Strategies*, 18th ed. (Boston: Cengage Learning, 2016).
15. Ziglar, *The "I Can" Achievers Course*, pp. 49–83; Ziglar, *See You at the Top*, pp. 98, 316.
16. Ibid., p. 375.
17. Siegel, *Love, Medicine and Miracles*, p. 1831.
18. Ibid., p. 1835.
19. S. O'Neal and E. N. Chapman, *Your Attitude Is Showing*, 12th ed. (Englewood Cliffs, NJ: Pearson Prentice-Hall, 2008).

20. Z. Ziglar, *Raising Positive Kids in a Negative World* (New York: Ballantine Books, 1996), pp. 42–43.

21. Ibid.

22. J. Allen, *As a Man Thinketh* (FV Editions, 2016), p. 21.

23. Lussier, *Human Relations in Organizations,* p. 63.

24. J. C. Maxwell, *The Winning Attitude* (Nashville, TN: Thomas Nelson Publishers, 2000), pp. 53–57.

25. Lussier, *Human Relations in Organizations*, p. 64.

26. *What Do Employers Want from Their Employees?* Retrieved September 5, 2016 from: http://www.americasjobexchange.com/career-advice/what-employers-want.

27. Lussier, *Human Relations in Organizations*, p. 11.

28. Chapman, *Your Attitude Is Showing.*

29. R. Sasson, *The Power of Positive Attitude Can Change Your Life*, retrieved September 5, 2016 from http://www.successconsciousness.com/positive_attitude.htm.

20 ETHICS IN THE WORKPLACE

It is important for leaders to understand ethics, values, and cultural diversity. If we do not understand others' values and culture, we probably will not be able to achieve greatness as members of the same team. A strong sense of ethics is essential for any leader. Our ability to hold jobs, build strong relationships, and even function in society or the workplace depends on our ethics.

Objectives

After completing this chapter, the student should be able to:

- Explain values in our society
- Describe terminal and instrumental values
- Explain the development of values
- Determine personal values
- Discuss basic societal values
- Discuss workplace ethics
- Make sound ethical decisions
- Discuss the importance of workplace etiquette
- Discuss workplace etiquette relating to telephones and Internet communication
- Discuss meeting etiquette

Terms to Know

- belief
- truth
- philosophy
- terminal values
- instrumental values
- media
- prudence
- temperance
- justice
- vitality
- fidelity
- integrity
- etiquette
- ethics
- value conflict

Why is it important for leaders to understand values, ethics, and cultural diversity? For one thing, if we do not understand others' values and cultures, we probably will not get along very well with other people. We most likely will not be able to hold jobs, build strong relationships with other human beings, or even function in society. It is important for us, as beings who must live together, to understand what is expected of us. We must know how to treat each other, take care of ourselves, and find meaning in our lives.

People are not born with an internal set of values. Values and ethics must be taught, just as culture is taught. "Problems occur between people of different cultures primarily because people tend to assume that their own cultural norms are the right way to do things." They mistakenly assume, often without realizing it, that specific behaviors practiced and promoted in their own cultures are universally valued and accepted.[1]

In the past, the teaching of values and ethics has been the responsibility of parents and religious institutions. Increased rates of violence, teen pregnancy, teen suicide, and vandalism, however, indicate that this society's traditional values are not being passed on. The acceptance and application of values and ethics are necessary for the harmonious functioning of society. Without values and ethics, individuals' purpose and direction in life can become unfocused or lost, and without them individuals (along with society) begin to degenerate and self-destruct.

The task of teaching values, ethics, and cultural diversity belongs to everyone. People of good character are not all going to come down on the same side of every political or social issue, but they will stand together on the same side of the truth.

VALUES IN OUR SOCIETY

Values Defined

Values are the principles and beliefs that you consider important. Interacting with motivation, your values influence your decisions and actions, which is why it is important for you to become aware of your values.[2] To say that something is a value is to imply that at least one other person holds to the same principle or belief and that others must be able to recognize it as a possible principle for regulating behavior. *Value* in this sense includes belief, but a value is more than just what one believes. Values are social elements, principles, or standards acknowledged and accepted by a large number of people, usually over a long period of time. Several values are being exhibited in Figure 20–1.

Values and Human Behavior

Values are especially important for understanding human behavior. Conflicts on the job, as well as conflicts between leaders and followers, often arise from a difference in values. When you do not get along very well with another person, you may want to investigate how your basic values differ.[3]

We also need to recognize and understand international differences in values. Many people in the United States, especially those who have not had extensive contact with persons from other countries, tend to assume that mainstream American values are shared by all and to think that "our" system and beliefs are correct. Although it is a good way, it is far from the only way: people from other countries also tend to think that their national and cultural values are the best, "right," or "correct." In the workplace, you will deal with people from many different cultures, ethnic groups, political systems, and belief systems. Be aware of differences in values and cultural differences, never make assumptions, and always be courteous and willing to adapt.[4]

Differences between Beliefs, Truths, and Values

Although the terms *belief, truth,* and *value* have different meanings, they are related. An individual's beliefs are normally based on that individual's perception of accepted truths. If an accepted truth is disproved, an individual's belief may change (importantly, it may also remain unchanged even in the face of clear evidence to the contrary). Individual values are usually the product of beliefs and truths. Therefore, individual values remain constant as long as the individual's beliefs and

FIGURE 20–1 These cadets are showing their values of honoring the flag and their country, honoring tradition, and respecting authority. *(© iStock/Mauricio Umana.)*

perceptions of truth remain the same. Consider the following definitions:

Belief is the state of believing that certain things are true or real; a belief is an individual's conviction or acceptance that something is true or real.

Truth is the quality or state of being correct, accurate, or true; agreement with a standard, a rule, or an established or verified fact or principle.

Values can be defined as the social principles, goals, or standards held or accepted by an individual, class, or society.

Influence of Philosophy on Values

Leaders must be aware of different philosophical beliefs and values even if they do not agree with them. Respect for individual differences enhances cooperation and understanding within a group.

Different values arise from different philosophies. Basically, everyone believes his or her own philosophy is the truth. **Philosophy** is a system of

values by which one lives; also, it is inquiry into the nature of things based on logical reasoning rather than empirical investigation. Philosophy has great value in our complicated world. Many people have no real foundational beliefs that they consciously use to guide their actions. Philosophy can provide people with a reasoned framework within which to think and interpret the world around us. By accepting a particular philosophy, people gain direction and standards, which affect their values and their actions. For example, those who have a stoic philosophy strive to master their emotions. Epicureans seek happiness through pleasure. Rationalists attempt to gain knowledge through logic and reason. Each philosophy leads to a particular way of thinking and behaving.

Values usually fall into one of the following six categories: theoretical, economic, aesthetic, social, political, or religious:

Theoretical Values Individuals with theoretical values seek to discover objective truths. They observe what happens and try to systematize their empirical knowledge.

Economic Values People with economic values are concerned with useful, practical matters, such as the production of goods and services and the accumulation of wealth; their concern with utility and application of knowledge may conflict with less tangible social or artistic goals.

Aesthetic Values The aesthetic person places the highest value on beauty, harmony, and artistic expression; the aesthete often opposes people who try to repress or constrain individuality, personal freedom, and original thought.

Social Values People whose philosophy is based on social values appreciate and love other people. Kindness, compassion, and selflessness are very important social values.

Political Values The political person is mainly interested in power. He or she seeks leadership, influence, and fame.

Religious Values Religious values are founded in beliefs, traditions, or texts, and generally center on a relationship with a divine being, other humans, or another higher reality. "Religious values define what people expect of themselves and of others based on the beliefs common to the religions they practice."[5]

The values a person adopts often align with his or her personality type, and frequently are reflected in the person's choice of profession. For example, scientists tend to hold theoretical values; businesspeople may have strong economic values; artists often express aesthetic values; social workers strongly espouse social values; and politicians have strong political values. Generally, each person's core values may fall into two or three of these categories.[6]

Your basic values may not change much after your high school years, but some may become less important. As you mature, your priorities change because your needs and goals change. As our needs are met, our desires change. For example, after high school, getting an education may be more important than earning money. However, once you graduate, marry, and get a full-time job, your family, achievement of material success, and acceptance by coworkers may dominate your values. Therefore, we often have somewhat different values at different stages of life.

TERMINAL AND INSTRUMENTAL VALUES

Most of us know the things we want to strive for in order to secure happiness and security when we get old or retired. Most also know the things which make us happy whether it is hunting, shopping, sports, or just hanging out with friends. We are different due to our terminal and instrumental values.

Terminal Values

Some values, such as security, self-respect, and spiritual growth, may remain the same throughout a person's life. Milton Rokeach, in *The Nature of Human Values*, calls these **terminal values**, representing goals you strive to accomplish before you die.[7] Some terminal values include

- Comfortable, prosperous, stimulating life
- Sense of lasting contribution and accomplishment
- Equal opportunity
- Family security; loved ones taken care of
- Freedom of choice; independence
- Enjoyable, leisurely life
- Social respect and admiration
- An exciting, active life
- Peace
- Happiness and contentment
- Inner harmony
- True friendship
- Wisdom
- Self-respect[8]

Instrumental Values

Instrumental values reflect the way you prefer to behave (ways to get there). Each individual determines which values are instrumental and which are terminal.[9] Some examples of instrumental values are

- Ambition and hard work
- Capability, competence, and effectiveness
- Cheerfulness
- Creativity
- Courage
- Independence, self-reliance, self-control
- Love, affection

- Respect, obedience
- Forgiveness
- Responsibility, dependability
- Neatness, tidiness
- Politeness, courtesy, being well-mannered
- Open-mindedness
- Helpfulness
- Honesty
- Imagination
- Intellectual ability
- Logic, reason, rationality[10]

DEVELOPMENT OF VALUES

People are not born with a set of values. As children, we watch how our parents, grandparents, and other relatives and caregivers act, and these observations form the basis of our early value system. Values internalized early in our life motivate us and become the foundation of a value system that carries us through life. Positive experiences and values early in life help us develop a sense of self-worth and a positive self-concept that aids in our success. Later on, our values are shaped as we interact with and observe people from different backgrounds, racial groups, and cultural groups.

We need to recognize that each group has a part to play in today's society. We must accept others and work harmoniously together, but we can do this only if we have developed good values early in life.

A child learns the values of the culture in which he lives. Parents, teachers, television, radio, electronic devices, and everything and everyone in their environment teach values to children. It is impossible for any of us to avoid becoming part of the culture in which we live. As you can see, the development of values follows approximately the same course as the development of knowledge or manners. Several factors influence the development of values: family, school, historical belief system, peers, and the media. The most important factor is the family.

Family Influence

Parents pass on to their offspring the set of values they have accumulated. This transmission process begins at birth and occurs indirectly, by example, at least as much as by direct instruction. Parents indirectly display their values, and thus influence their children, by their behavior in different situations (Figure 20–2). For example, when children hear their parents arguing, they

FIGURE 20–2 Parents pass the values that they have accumulated on to their offspring. This mother and child already share a set of values, which has been built by parental direction and observation by the child. *(© wavebreakmedia/Shutterstock.com)*

may conclude that arguing is how people resolve differences of opinion. When children see their parents help each other with household chores, they learn the value of cooperation. Values are taught and transmitted primarily by actions, behavior, and example; children absorb and believe what they observe and experience. There is no escaping parental influence on children, no matter how little time children actually spend with their parents.

The School

At one time, values were directly taught in school, stressing such things as differentiating between right and wrong, honesty, and other mainstream cultural values. Teachers were a great influence as students learned cooperation, teamwork, the importance of meeting deadlines, following directions, etiquette, sharing, and how to use their minds. However, many in the diverse U.S. society did not want specific values being expressed or taught to their children in public schools; they felt that public schools should be "value neutral." Over

time, it has become apparent that many children do not seem to learn appropriate values at home, and the public is now asking that schools give students some type of guidance in this area. Many private schools took on the teaching of values, and now some public schools have taken on some of that responsibility as well.[11]

Of course, it is impossible to avoid the expression of values completely. Young people spend almost as much time with their teachers, coaches, sponsors, and friends as they do with their parents or caregivers. All of these people influence students both by their direct instruction and by example. Teachers, coaches, and sponsors can demonstrate how to work together, how to accomplish goals, and how to meet deadlines, for instance. The teacher in Figure 20–3 is indirectly communicating values as he instructs his students.

Historical Values

Some value systems exhort people not to steal, lie, or kill; to avoid greed and respect authority; and

FIGURE 20–3 Teachers, coaches, and other educators instill values directly and indirectly as they work with students daily. (© Air Images/Shutterstock.com)

FIGURE 20–4 These young leaders are learning the value of participating in their community. *(© wavebreakmedia/Shutterstock.com)*

to be truthful. Historically, values are developed and tested over time in a culture. As a result of intellect, religion, and experience, these values become principles for a given culture as a pattern for ethical behavior.

Peer Group Influence

Peer groups also influence values. For example, if you want to become a member of a particular club, you may have to accept the values promoted by that club and espoused by its members. This would be a direct influence on your own values. If a young person spends time with friends who are active in civic organizations, he or she may develop the value that helping the community is a good thing and also become an active participant. The students in Figure 20–4 are developing values as they undertake a community service activity.

Media

Media may have a great influence on values.[12] **Media** include any form of communication that reaches a very large audience, such as the websites, social media,, television, newspapers, radio, podcasts, magazines, and outdoor advertisements. It is not hard to see the influence that mass media have had on fashion, music, and

recreational choices today. Figure 20–5 shows that many factors contribute to the development of a person's values.

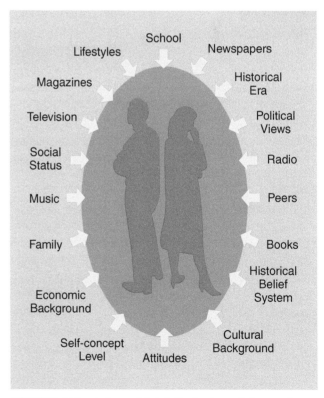

FIGURE 20–5 Many factors contribute to the development of a person's value system.

DETERMINING YOUR PERSONAL VALUES

Sometimes we pay lip service to values that we feel we should have, but inwardly we are not sure how strongly we really hold those values. Sometimes, you may find that your values are compromised when you are around certain people or in certain situations; if this happens, these may not really be your personal values.[13] It is good not only list out the values you believe you possess, but to take stock and make sure you truly possess those values. Answer the following questions about any value you consider important:[14]

1. Is it truly my value or did I pick it up from someone else?

2. Is it a means to an end or an end in itself? For instance, if you value financial security because you feel it leads to freedom to serve others, then freedom to serve others is actually the most important value.
3. Do your actions show your value?
4. When were you the happiest or most excited? Whatever you were doing at that moment is probably representative of what you value most.
5. What do you regret the most? This is another way to identify your key values.

The workplace values in the next section can be subjected to these questions. Answering these questions will help you separate your real values—the ones that truly are strong in your life—from those that you merely think might be good.

VIGNETTE 20-1 **WHY YOUR VALUE SYSTEM IS IMPORTANT**

THE OLA FFA CHAPTER DECIDED TO UNDERTAKE A SUMMER SERVICE ACTIVITY

At the May meeting, before school was out for the summer, the chapter finally had to come to a specific decision about the project. After some debate, the chapter agreed to help the community retirement center in some way. Deciding exactly what to do for the retirement center was another story. This particular FFA chapter had many leaders from different backgrounds and with different interests, so deciding what they wanted to do for the retirement center was a difficult task.

Ted, who was president of the Ola FFA chapter, had a **political value system**. He admitted that he wanted to do a summer service project because no other organizations at the school had done so yet. He thought the summer activity would help the chapter with recruitment and gain community recognition.

Then there was Ben, who had a **theoretical value system**. He wanted to do something that would really make a difference. He did not care how hard the project would be or how long it would take. He just wanted it to be special for the residents of the retirement center.

Maria, who had **economic values**, wanted a summer service project that would have a product at the end. She needed to see, at the end of the summer, the product of her labor, and she wanted the center residents to feel like there was an obvious difference when the project was complete. She advocated for a service project that would generate funds for the center, because she felt that raising more money for the residents would be a definite indicator of a positive outcome.

The **aesthetic value system** of Shawn and Rachel caused them to believe the project had to make the retirement center look better or the residents feel better. They thought Christine felt as they did, but she just wanted everyone to get along. Christine, who had a **religious value system**, was quite bothered that a decision had not yet been made, but she kept emphasizing that the project would be successful as long as everyone worked together and learned from each other.

Social values were what Jim and Rick were trying to get the club to understand. They explained that it was important to consult with the residents and center management to plan a project that was truly helpful. Like Ben, they felt that the project needed to really mean something, but Jim and Rick thought it needed to mean something to the people who lived there.

Ted was amazed by the diversity of ideas among the Ola FFA members. He decided to refer the decision to a committee that would plan a project that would satisfy the value system of each of the members who had spoken up at the FFA meeting. The committee finally chose a rather large project; the project was a success, but more importantly, most of the members planned to continue their service at the retirement center by volunteering some of their time and talents each month.

What do you think would have been a good project for the retirement center? What kind of value system do you think you possess?

BASIC SOCIETAL VALUES

Certain values are widely held in society and the workplace. These values are usually associated with people who are trying to do what is best. Remember, values, unlike truths, are not absolute; everyone's values are not the same. The general, somewhat abstract, values discussed in this section may not be part of your value system.

Honesty

Not only is "honesty the best policy," it is better than all policy. To be honest is to be real, genuine, and authentic. Honesty expresses both self-respect and respect for others. Honesty fills lives with openness, reliability, and candor. Mark Twain said, "If you tell the truth, you don't have to remember anything."[15] Honesty comes in many forms:

- Telling the truth
- An honest day's work, or giving the employer fair work for the pay
- Fair actions, such as treating people in an honest and fair manner
- Expression of true feelings, such as making sure your compliments are genuine

Developing an Honest Character To develop an honest character, the individual must first think about how he or she will react to a situation. The more the individual reacts with honesty, the more it becomes a way of life. One develops an honest character by following this progression:

- Thoughts becoming actions
- Actions becoming deeds
- Deeds becoming habits
- Habits becoming a way of life

Dishonesty Dishonesty is a willful perversion of truth, usually for personal advantage or gain. It comes from the intent to mislead, deceive, cheat, or defraud. Dishonesty takes many forms:

- Cheating on a test
- White lies
- "Stretching" the truth, even just a little
- Exaggerating
- Shoplifting
- Lying
- Making a false call in sports
- Withholding part of the truth
- Double-dealing
- Misleading, misdirection, or misrepresentation
- Mischievous deceit

Prudence

Prudence simply means common sense. It means thinking about consequences and possibilities before you act. Being prudent is the opposite of being foolish or thoughtless. A prudent person uses all of his or her intellectual ability, but that does not mean you have to have a high IQ score to be prudent. You do, however, have to think about the potential results of your behavior before you act.

Temperance

Temperance means allowing just the right amount, at the right time, of any pleasure. **Pleasure** means any activity that makes you happy for the moment, including eating, sleeping, exercising, reading, working, and playing. Temperance does not mean abstinence. Temperate people are those who have control over the pleasure in their lives; desires and overindulgence do not control them.

Justice

Justice includes everything we usually consider fair and equitable. It encompasses honesty, compromise, truthfulness, and promise keeping. It is easy to see why justice is a value. No one respects or trusts a person who is dishonest, tells lies, is ungenerous, or breaks promises.

Courage

Courage includes standing up for what you believe when others are putting pressure on you to do otherwise or to act contrary to your values. The mark of a mature, secure person is not doing what everybody else is doing, but instead doing what he or she believes to be right. With courage, you confront a dangerous situation head-on, when necessary. It takes courage to speak your mind when you know you will be scorned for doing so. Courage means being tough enough to do what is right, even when you know that the majority will oppose you.

Courage does not mean that you are fearless. Having some fear is healthy and safe; caution encourages thought before action. Aristotle said that courage is a settled disposition to feel an appropriate degree of fear and confidence in challenging situations. Courage, said Plato, is knowing what to fear. Courage is acting bravely when we do not really feel brave.

It is often difficult to maintain the "courage of your convictions." Peer groups can exert extreme pressure on people to conform, and it requires a great deal of fortitude to take a path different from that of your peers.

Perseverance

In the fable of the tortoise and the hare, slow and steady wins the race. Perseverance is an essential characteristic of high-level leaders. Often those who could have done good things fail because they hesitated, faltered, wavered, or just did not stick with it. Combined with practical intelligence and other good qualities, in the right context, perseverance is an essential ingredient of life (Figures 20–6 and 20–7).[16]

Loyalty

Loyalty is willingness to support a person or a cause. Our loyalties are demonstrated by consistency or steadfastness in our attachments, which reflect the kind of person we have chosen to become. To be a loyal worker, citizen, or friend means to care seriously about the wellbeing of one's employer, country, or associate, and

---

It Can Be Done

The man who misses all the fun
Is he who says, "It can't be done."
In solemn pride he stands aloof
And greets each venture with reproof.
Had he the power he'd efface
The history of the human race;
We'd have no radio or motor cars,
No streets lit by electric stars;
No telegraph nor telephone,
We'd linger in the age of stone.
The world would sleep if things were run
By men who say, "It can't be done."

—Unknown

FIGURE 20–7 Some believe that vision and persistence are as important as intelligence. Many things have been accomplished by persistent people in spite of other people telling them "it can't be done."

---

to express that conviction and affiliation in your religious, political, and professional life. Loyalty, like courage, is most evident when we find ourselves in stressful situations. "Real loyalty endures inconvenience, withstands temptation, and does not cringe under assault."[17]

Commitment

Commitment is a pledge or promise to do something. In today's world, commitment is an important quality that is necessary for success in many aspects of life. In a marriage, both husband and wife must be committed to their union; a lack of commitment can cause the marriage to founder or fail. In the workplace, commitment to quality and to performing a job correctly is essential in producing goods or services that can compete in today's markets. A lack of commitment to the job creates poor-quality products and morale problems among employees. Commitment means doing what you say you are going to do and following through.

Pride

It feels great to accomplish a job or task with such quality that you are proud for others to see it. Although pride may be manifested negatively, such as when an individual becomes boastful or arrogant, a certain amount of pride is healthy. If you do not perceive something as worthwhile,

---

Try, Try Again

'Tis a lesson you should heed,
  Try, try again;
If at first you don't succeed,
  Try, try again;
Then your courage should appear,
For, if you will persevere,
You will conquer, never fear;
  Try, try again.

—Unknown

FIGURE 20–6 Most goals can be accomplished through perseverance and persistence.

FIGURE 20–8 Pride is a basic societal value. The farmer in this photo surveys a job well done in this hay field. *(© iStock.com/ seanfboggs)*

there is no reason to make an effort regarding it. A lack of pride in one's job can result in goods or services that are below standard. Lack of pride in one's school can result in lackluster academic and athletic performance, vandalism of buildings and property, and behavioral problems. A lack of pride in oneself can cause poor hygiene and grooming, poor job performance, poor school performance, and problems in relationships with people. Pride in work and a job well done are basic values. The farmer in Figure 20–8 appears to be proud of and satisfied with his work.

Conviction

Conviction may be defined as belief or acceptance that certain things are true or real. A person must stand for what he or she believes or risk being swayed by every notion that comes along. A person who has convictions is dependable because people know where he or she stands. Standing up for your beliefs makes you honest because you have something to believe in. Standing up for what you believe helps to create personal happiness and satisfaction.

Responsibility

To be responsible is to be accountable. Many believe that one of the biggest problems in society today is people's reluctance or refusal to take responsibility for their own actions. In fact, everything ever done, throughout history, was done by somebody; that is, someone exercised some power to do it. Aristotle was among the first to point out that we become the people we are by the decisions we make. So grow up, make good decisions, and take responsibility for your decisions and their consequences.

Self-Discipline

Lack of control over one's temper, appetites, urges, and passions causes much of the unhappiness and personal distress we see around us. When self-discipline fails, we often think, "Why didn't I just say no?" When it comes to self-discipline—whether developing it, practicing it, or exercising it—you are your own teacher, trainer, coach, and judge. Achieving your goals requires a great deal of self-discipline; always remember, "If it's going to be, it's up to me."

Compassion

Compassion is a form of love for or empathy with other people, their inner lives, their emotions, and their external circumstances. Compassion also involves fellowship, sharing, and supportive companionship for friends when they need you. Although former President Richard Nixon made mistakes, Americans were touched when at his funeral members of both political parties showed compassion and remembered the great things he had done.

Humility

Humility is the opposite of bragging. It is perfectly fine to know that you have done a good job, to take pride in it, and to celebrate your success—once. It is not a very good idea to go around telling everyone, over and over again, about your accomplishment and how wonderful you are for having done it. Show-offs and braggarts are not humble people. Even those who accomplish great things should remember and give credit to all the other people who assisted in the achievement of a goal. Sometimes, as in a job résumé, it is appropriate to tell about your accomplishments (accurately and matter-of-factly), but the intent in those situations is to prove that your experiences and achievements make you more able to take on new challenges. The general guideline should be not to become full of yourself and repeatedly recount your accomplishments, and not to boast about how great and clever you are.

Vitality

Vitality is physical or mental strength or energy. This does not mean everyone has to be a bodybuilder or a physicist. Rather, it means that you use all your physical and mental abilities as often as possible, as much as possible. Being vital is the opposite of being lazy.

Dedication

Dedication is similar to loyalty except that it is more of an inner motivation. It involves prioritizing and focusing on the object of your concern. You can be dedicated to anything you do. People are dedicated to their jobs, to taking care of other people, to being healthy, to learning to play an instrument, to an organization, or to doing a good job at everything they attempt.

Assertiveness

Assertive people are confident about their abilities, but not pushy or boastful. Assertive does not mean aggressive. Assertiveness means you are willing to step forward and make a statement because you believe it to be true, whether or not anyone else agrees with you. Assertive people find a way to get a job done even in the face of many obstacles.

Fidelity

Fidelity is about being loyal or dedicated, with honest and firm attachment. In today's society, though, it has come to specifically signify being faithful in marriage.

Forgiveness

Forgiveness as a societal value means that you not only forgive people for mistakes they have made, but you also forget about those mistakes and do not let them affect your relationships with or perceptions of people. It means wiping the slate clean and starting all over again. Do not forget to forgive yourself also.

Integrity

Integrity refers to striving to live up to a code of rules or moral values. When we say that someone has integrity, we usually mean that he or she possesses, in some degree, all the values we have been discussing and behaves in accordance with them.

Respect for Authority

Showing respect for authority involves doing what we are told to do by the appropriate person. It may be your parent, guardian, teacher, or boss. When we do as we are told, we gain respect. Relationships are better, we learn more, we stay out of trouble, and we perform with greater effectiveness.

Etiquette

Some of the values that facilitate good relationships between people are sometimes called rules of etiquette, discussed later in this chapter. **Etiquette** consists of the manners, practices, and customs prescribed by a culture or society. Qualities such as courtesy, civility, displaying good manners, and knowing what behavior is considered proper and appropriate constitute the valuable social attribute called etiquette.

Work

"Work is applied effort; it is whatever we put ourselves into, whatever we expend our energy on for the sake of accomplishing or achieving something." A person who places a high value on work wants to do a job to the best of his or her ability. He or she takes pride in completing tasks correctly and in a timely manner. The most satisfying work involves efforts toward achieving goals that we believe are worthy expressions of our talent and character. Those who have not experienced the joy of a job well done have missed something very important.[18]

Appreciating Cultural Diversity

People tend to assume—wrongly—that the way things are done in their own culture is the right way and should apply universally. In fact, each culture has its own customs and preferred behaviors. Understanding the uniqueness of another culture can enhance our appreciation for that value system. In working with people, though, a leader needs to acknowledge and respect the uniqueness of each individual, who may or may not completely conform to the dictates of his or her culture. If you don't treat people as individuals, if you make sweeping generalizations based on what you think you know about a particular race, class, or ethnic group, you usually end up doing more harm than good (Figure 20–9).

Other Human Relationship Values

The qualities that spring from values can build strong ties with other people. Characteristics such as cooperation, consideration, dependability, generosity, kindness, patience, respect, tolerance, keeping promises, and defense of others are usually quite evident in good interpersonal relationships. If a person is lacking in any of these qualities, his or her relationships with others are likely to suffer.

FIGURE 20–9 Even though these professionals are from diverse cultures, their differences do not matter as they work together and strive for common goals. (© iStock/monkeybusinessimages)

WORKPLACE ETHICS

Ethics, as distinguished from values, refers to the rules of conduct that reflect the character and sentiment of a community or group. Kickbacks and payoffs may be acceptable practices in one part of the world but are seen as unethical elsewhere.[19] Ethics is very important to businesses (Figure 20–10). Approximately one-third of corporations that have more than 100 employees provide some type of ethics training.[20] As a result of recent scrutiny of and debate about corporate ethics, many businesses and organizations have decided to put their ethics in writing. More than 90 percent of Fortune 1000 companies have a written code of ethics, although the effectiveness of such codes in ensuring good behavior and adherence to the code is questionable.[21]

Complexity of Ethical Behavior

Many ethical issues are complex and cannot even be considered (much less decided) in absolute terms, which explains why so many companies now offer ethics training. Employees participate in in-depth discussions on topics such as methods of gathering information about competitors, fairness in hiring, and accepting gifts and entertainment from customers.[22] How do we know what is right? Consider the following:

- You accidentally broke a piece of merchandise and returned it to the seller claiming that it was damaged before you bought it.
- In completing her income tax return, a woman routinely overstates the value of contributions and claims vacation expenses as deductible business expenses.
- You copy a computer software program so that you can complete assignments on your own computer rather than enduring the inconvenience of using a computer lab at school or at work.[23]
- A businessman who travels extensively feels cheated that telephone calls are not reimbursed travel expenses. He therefore overstates car mileage to cover the telephone calls.
- An investment broker writes an enthusiastic letter to a client stating that she has achieved a 24 percent growth in her investment, without mentioning that the stock market as a whole experienced a 36 percent growth in the same period.

FIGURE 20–10 These business leaders have just closed a business transaction with a handshake. This type of business deal requires an assumption that all parties are ethical. *(© iStock/Christopher Futcher)*

- To increase profits in its service sector, a company increases the complexity of a product so that repairs have to be done by an authorized service department.[24]
- A company quotes a price for a job, knowing that the time actually needed to complete the job will be significantly less than estimated.

Depending on your value system, some of these examples may be blatantly wrong and some may not be wrong. Your values are the foundation of your ethical choices. Even minor compromises that go against your values can gradually weaken this foundation.

MAKING ETHICAL DECISIONS

Is there a way to determine what is a good ethical decision? Is there a way to determine what our values are and how they apply to a specific issue? Many methods are available to help individuals and businesses analyze ethical issues. One way of determining whether an action is ethical is the Pagano Model. According to this model, you must answer the following six questions honestly to make the correct ethical choice:

1. Is the proposed action legal?
2. What are the benefits and costs to the people involved?
3. Would you want this action to be a universal standard, appropriate for and applicable to everyone?
4. Does the action pass the light-of-day test? That is, if your action appeared on television or the front page of a newspaper, or if others learned about it, would you be proud?
5. Does the action pass the Golden Rule test? That is, would you want the same to happen to you?
6. Does the action pass the ventilation test (being aired out as a hypothetical)? Ask the opinion of a wise friend who has no stake in the outcome. Does this friend believe that the action is ethical?[25]

Even though a decision may be legal, it could be unethical. For example, many food products were labeled "lite" or "low-fat" when in fact most of the calories in the food were derived from fat. For certain people, this misinformation could cause illness or even death. James E. Perrella, executive vice president of Ingersoll Rand Company, said, "Good ethics, simply, is good business. Good ethics will attract investors. Good ethics will attract good employees. . . . You can do what's right. Not because of conduct codes. Not because of rules or laws. But because you know what is right."[26]

THE IMPORTANCE OF WORKPLACE ETIQUETTE

Proper etiquette is absolutely necessary to success in the workplace. It is often assumed that people learn etiquette at home or through experience or observation elsewhere. Unfortunately, many people do not learn much etiquette, or other human relations skills, at all.

Because they always occur between people, workplace interactions are largely social interactions. Graceless behavior that does not conform to the expected etiquette in a social situation might cause some embarrassment, some loss of social status, some lessening of others' regard for you, or exclusion from future events. However, not following the proper etiquette during business interactions can cause loss of an assignment, a promotion, or even a job; if the unfortunate interaction is with a client or customer, you may lose a sale, wreck a deal, or lose the customer entirely. Rightly or wrongly, people often equate your behavior and conformity to etiquette (or lack thereof) with your ability (or inability) to do a job. Therefore, it is in your best interest to become thoroughly familiar with the expected, appropriate behavior.[27]

BUSINESS PHONE, CELL PHONE, AND EMAIL ETIQUETTE

Picture this: you're responding to an urgent work email when your business or home phone rings, and as soon as you answer it and start talking your cell phone rings from someone returning an emergency call you made earlier. What is the proper etiquette in these situations? Consider the following as you determine the most appropriate etiquette.

The Office Telephone

A first impression, whether it is of the organization or of you personally, sends a powerful message and image to the caller. Always answer the telephone in a professional manner; never answer with just "Hello" or by simply stating your phone or extension number.[28] Using a nickname when you answer the phone is unprofessional and lacks authority.

Never immediately place a caller on hold. If you must put a caller on hold, ask for their permission to do so. They may not be able to wait, or there may be an emergency. It is also best not to leave someone on hold for more than 30 or 40 seconds. If you can't talk within the 30- to 40-second timeframe, simply get their name and number and then call them back as soon as possible.

When you do have a caller on your office phone, be sure to give him or her your full attention. Refrain from eating, drinking, or engaging in other personal activities that might sound annoying to the person on the other end of the line. If you get lots of calls at once, handle each call before moving to the next one, again giving each caller in turn your full attention. If you call a wrong number, be sure to apologize, and if you answer or make a call using a speaker that can be heard by others, be sure to make those on the other end of the line aware of that. Here are some other guidelines for telephone etiquette:[29]

- Answer the phone as soon as possible.
- Answer with your name and the name of the person or company for whom you work.
- Return calls as soon as possible, ideally in less than a day.
- Refrain from sharing personal information about coworkers.
- When you call someone, ask whether she has time to talk; if not, ask for a more convenient time at which you may call back.

Smartphones

When calling someone on a smartphone, be as brief as possible, even if this means calling back later on a landline/wired phone. If you have a smartphone provided by your company, it is your ethical responsibility to minimize your time by restricting calls to clients and company business. In any event, you should not talk on a cell phone while walking down the street; in a theater, restaurant, or classroom; in a meeting; or in other public places. Go somewhere private to make a cellular phone call, where you will not disturb others. For your own safety and that of others, avoid placing calls while you are driving—and *never* try to text-message while you are driving. In some areas, cell phone use while driving is illegal. Phones should be kept in the vibrating mode at all times when you are at work. Even on vibrate, try not to look at your phone in meetings or around friends and associates so that you can give them your full attention.

Email

Do you get too many emails? Do you get more information than you need to know?

Do you get urgent messages that are not actually important? All of us can answer yes to these questions. Email etiquette is very important. Consider the following:

Be Brief The convenience and immediacy of email has made it very popular. However, email is best for short messages. If you have to send a long message, send it as an attachment.

Be Aware of Incompatible Software Be sure that your recipients have the software needed to view your attachment. For example, do not attach a document in Mac format if your recipients all use PCs and Windows; do not send a PowerPoint file if you are not sure that your recipient has the PowerPoint program. Even within an operating system family, document formats can be incompatible (e.g., docx and.doc files); if you do not know, ask your recipient what format is best for an attachment.

Be Professional Do not put anything in an email message that you are not comfortable having everyone know. Avoid sending or receiving personal messages through your work system. Employers have the legal right to monitor (read) email messages sent from their systems, and many do so; emails are discoverable as evidence and are frequently made part of the record in legal proceedings. It is possible to retrieve a message that you thought you had deleted.[30]

Do Not Use All Capital Letters Sending messages and words in capital letters is the equivalent of shouting at the recipient. Be courteous.

Set the Return Receipt Function Do not assume that everyone in the organization either has email or reads it with the frequency you do. To confirm that someone has received a message, use the "Return Receipt" function, which alerts you when the recipient opens your message.

Use Special Functions Sparingly Don't cry wolf. If all your messages are identified as "High Priority," the recipients of your messages will soon ignore them all. Do not send copies of emails to anyone who is not directly involved or does not have a need to know. Carbon (cc) and blind copies (bcc) of your emails are not proof of your hard work. They simply make Email Mountain higher than it needs to be. Make sure any email you send is necessary to the person or persons who will be receiving it.

MEETING ETIQUETTE

Proper meeting etiquette calls for the chairperson and members of the meeting to arrive prepared and on time. If you are late, apologize but do not give an excuse. Excuses for tardiness don't usually carry much credibility, and they just delay the meeting further.

The Chairperson Should Arrive Early

The person who called the meeting (normally the chairperson) should arrive early enough to check the room arrangements and to make sure that everything is ready, that any equipment needed is in place and is working properly, and that any refreshments have been delivered. The more complex the equipment being used, the earlier the chairperson should arrive, especially if he or she is unfamiliar with operation of the equipment.

Leaving Early and Phone Calls

If some people can be present for only part of the meeting, the chairperson should arrange the agenda so that these people can present their material early in the meeting and then leave.[31]

Appropriate arrangements should be made for those who must come in late to a meeting that is already in progress.

For reasons of efficiency and common courtesy, meetings should take precedence over phone calls. Cell phones should be turned off during meetings. A member who absolutely must take a call should leave the meeting room while doing so.

After the Meeting

After the meeting, the leader is responsible for making sure the room is restored to the condition in which it was found. The chairperson should also ensure that the minutes are distributed to all attendees and to any group member who was unable to attend; the secretary may be tasked with the actual distribution, but the chairperson should assist as needed with an up-to-date contact list and reminders. Refer to Chapter 12, Conducting Successful Meetings, for more information on proper etiquette and conduct of meetings.

CONCLUSION

You may experience many kinds of value conflicts in your personal life and on the job. A **value conflict** is a problem that arises when one is torn between two competing priorities or value systems. As a leader, you may be divided between loyalty to coworkers and loyalty to those with authority over you. As a worker, you may experience conflict between fulfilling family obligations and doing what you need to do to be successful at work. Your values determine the choices you will make. How you resolve these value conflicts, make ethical choices, and conduct yourself will greatly affect your attitude toward yourself and your career.[32]

SUMMARY

Values are the principles or standards for behavior accepted by a large number of people, usually over a long period of time. Values are what make it possible for human beings to live together, develop strong relationships, and accomplish goals. Values are especially important to understanding human behavior. Individual values are usually the product of beliefs and truths.

Philosophy provides us with a reasoned framework within which to think and interpret the world around us. Leaders must be aware of different philosophical beliefs and values. People have different values because their philosophies are different. Philosophical values fall into six categories: theoretical, economic, aesthetic, social, political, and religious. Terminal values represent goals you strive to accomplish before you die. Instrumental values reflect the way you prefer to behave.

Values are developed by a number of influences. The main influence on the development of one's values is the family, but schools, historical belief systems, peers, and the media all influence people to some extent. Some influences are detrimental, whereas some are quite positive and productive. Sometimes we are not really sure how important our values are. Taking the Rath test can help you find out.

Certain values are appreciated in most segments of society and the workplace, including honesty, prudence, temperance, justice, courage, perseverance, loyalty, commitment, pride, conviction, responsibility, self-discipline, compassion, humility, vitality, dedication, assertiveness, fidelity, forgiveness, integrity, respect for authority, etiquette, work, and appreciating cultural diversity.

Human relationship values include cooperation, consideration, dependability, generosity, kindness, patience, respect, tolerance, keeping promises, and defense of others.

Ethics, as distinguished from values, constitutes the rules of conduct that reflect the character and sentiment of a community or group. Many ethical issues are complex and cannot be decided in terms of absolute right or wrong. Even though a decision may be legal, it could be unethical. The Pagano Model poses six questions to help you make ethical choices.

Being a person with high values and strong ethics is important for several reasons. Our behavior influences other people; influence is leadership; and we are judged by our actions every day. What other people see in us, they may try to imitate, especially if they are younger and are searching for guidance and role models.

Proper etiquette is critical to success in the workplace. The phone should be used in a professional manner, for both taking and making calls. Users of cell phones should remember courtesy at all times, limiting their calls to private areas and selected times. Neither phones nor pagers should be allowed to interrupt a meeting. Email messages should be brief and professional. Proper meeting etiquette calls for the chairperson to arrive prepared and on time; to conduct the meeting for maximum efficiency; and to make sure that the meeting room is returned to its proper condition.

 Take It to the Net

Explore workplace ethics on the Internet. Choose your favorite search engine and browse the Web by typing in some of the following search terms or key words found in this chapter. Choose one site that you feel contains the most useful information on workplace ethics. Write a summary of the site and in your summary include what was useful about the site.

Search Terms

ethics

business ethics

workplace ethics

ethics in the workplace

ethical

Chapter Exercises

REVIEW QUESTIONS

1. Define the Terms to Know.
2. Name and explain six categories in which values can be evaluated.
3. List 10 examples of terminal values.
4. List 10 examples of instrumental values.
5. List five factors that influence the development of values.
6. List seven historical values or belief systems.
7. What are six questions that you can ask yourself to determine the strength of your personal values?
8. List 10 basic societal values you strongly agree with.
9. Name four forms of honesty.
10. Name 10 forms of dishonesty.
11. Name eight human relation values and characteristics that can help build strong ties with other people.
12. Give three examples of questionable ethical practices.
13. What six questions can you ask yourself to help determine the correct ethical decision?
14. Explain the importance of understanding and appreciating cultural diversity.
15. Why is etiquette important in the workplace?
16. List five proper uses of the office (business) telephone.
17. List four rules about the proper use of smart phones.
18. List six rules about the proper use of email.
19. List three rules of good meeting etiquette.
20. List 18 factors that contribute to the development of a person's value system.

COMPLETION

1. Several factors influence the development of values, but the most important factor is the _____.
2. Values are especially important to understanding _____ behavior.
3. We have different values in society because our _____ are different.
4. If it is going to be, it is up to _____.
5. _____ includes standing up for what you believe when others are putting pressure on you to do something contrary to your values.

6. A thought becomes an action, an action becomes a deed, a deed becomes a habit, and a habit becomes _____.

7. Mark Twain said, "If you tell the truth, you don't have to _____ anything."

8. _____ is a form of love for or empathy with other people.

9. With practical intelligence, in the right context, and occurring in the right combination with other values, _____ and persistence are essential ingredients of life.

10. Even though a decision may be legal, it could be _____.

11. People are not born with an internal set of _____.

12. The acceptance and application of _____ and _____ are necessary for the harmonious functioning of society.

13. _____ are especially important for understanding human behavior.

14. Scientists tend to hold _____ values; businesspeople may have strong _____ values; artists often express _____ values; social workers strong espouse _____ values; and politicians have strong _____ values.

15. _____ people are those who have control over the pleasures in their lives; desires and overindulgence do not control them.

16. Most goals can be accomplished through _____ and _____.

17. Real _____ endures inconvenience, withstands temptation, and does not cringe under assault.

18. _____ means doing what you say you are going to do and following through on the promise.

19. A lack of _____ in one's job can result in goods or services that are below standards.

20. A person who has _____ is dependable because people know where he or she stands.

21. Many believe that one of the biggest problems in society today is people's reluctance or refusal to take _____ for their own actions.

22. Those who have not experienced the joy of a _____ well done have missed something very important.

23. Individual _____ are usually the product of beliefs and truths.

24. Proper etiquette is critical to _____ in the workplace.

MATCHING

_____ 1. Happy with accomplishing a job or task with quality.

_____ 2. Strong belief or acceptance that certain things are true or real.

_____ 3. Opposite of bragging.

_____ 4. Forgetting about mistakes of others and not letting past mistakes affect relationships.

_____ 5. Confident of your abilities, but not pushy.

_____ 6. Inner motivation.

_____ 7. Unblemished or near-perfect ethical behavior.

_____ 8. Pledge or promise to do something.

_____ 9. Willingness to support someone or some cause.

_____ 10. Answerable or accountable for our actions.

_____ 11. Besides being needed for economic livelihood, it can build self-esteem.

A. responsibility

B. humility

C. work ethic

D. dedication

E. integrity

F. commitment

G. forgiveness

H. pride

I. conviction

J. loyalty

K. assertiveness

ACTIVITIES

1. Review the six categories of values: theoretical, economic, aesthetic, social, political, and religious. Write a short paragraph on how you view (or the importance you assign to) each of the six areas.

2. Rank the five factors that have influenced the development of your values. Write a paragraph explaining each.

3. Write down as many other values or ethical behaviors as you can think of that were not mentioned in this chapter. Be prepared to explain why you selected them.

4. Make a list of five famous people, and give one reason for selecting each as a model of integrity.

5. Develop a visual representation (e.g., a Venn Diagram) of your concept of the "The Perfect Person." Include as many values as you can from this chapter in your visual diagram. You may add others.

6. The section called Complexity of Ethical Behavior listed seven situations that might call for ethical judgments. Select three and write a paragraph (or more if needed) on how you view the issue. Be prepared to share your ideas with the class.

7. Interview a leader you know about an ethical issue that they have faced or wondered about. You can develop specific questions ahead of time or allow the interview to be more organic. Ask the leader if you can record the interview. Whether you record the interview or not, however, be sure to take good notes. Review your notes and/or recorded interview and develop a generic "Top Five" list of solutions that could have assisted the leader in handling the ethical issue.

8. Review this chapter and add any other characteristics that you consider important personal values or ethical behavior. Complete the following chart with 25 values entitled, "These Things I Value." Identify which group influenced these values.

These Things I Value

VALUE	HOME/FAMILY	FRIENDS	SCHOOL	HISTORIC BELIEF SYSTEM	MEDIA	OTHER
1.						
2.						
3.						
4.						
5.						
6.						
7.						
8.						
9.						
10.						
11.						
12.						
13.						
14.						
15.						
16.						
17.						
18.						
19.						
20.						
21.						
22.						
23.						
24.						
25.						

NOTES

1. C. M. Lehman and D. D. DuFrene, *Business Communications*, 16th ed. (Mason, OH: Cengage Learning, 2011), p. 25.
2. J. J. Littrell, J. H. Lorenz, and H. T. Smith, *School to Career*, 10th ed. (Tinley Park, IL: Goodheart-Wilcox, 2014).
3. L. H. Lamberton and L. Minor-Evans, *Human Relations: Strategies for Success*, 4th ed. (McGraw Hill Education, 2009).
4. Ibid., p. 72.
5. D. Kraft. *Examples of Religious Values*, retrieved September 5, 2016 from http://peopleof.oureverydaylife .com/examples-religious-values-9409.html.
6. R. M. Hodgetts, *Modern Human Relations at Work*, 8th ed. (Chicago: Harcourt Brace College Publishers, 2001).
7. M. Rokeach, *The Nature of Human Values* (New York: Free Press, 1973), pp. 7–8.
8. Ibid., p. 119.
9. Ibid., pp. 8–12, 326.
10. Ibid., p. 119.
11. S. L. Nazario, "Schoolteachers Say It's Wrongheaded to Try to Teach Students What's Right," *Wall Street Journal*, April 6, 1990, p. B1.
12. B. L. Reece, *Effective Human Relations in Organizations*, 12th ed. (Mason, OH: South Western Cengage Learning, 2014), p. 101.
13. Lamberton and Minor, *Human Relations*.
14. J. D. Meier. *How to Find Your Values*, retrieved September 5, 2016 from: http://sourcesofinsight.com /finding-your-values/.
15. Retrieved September 4, 2016 from: http://www.twainquotes.com/Truth.html.
16. W. J. Bennett, *The Book of Virtues* (New York: Simon & Schuster, 2010).
17. Ibid., p. 665.
18. Ibid., pp. 347–348.
19. Reece, *Effective Human Relations in Organizations*, p. 104.
20. B. L. Thompson, "Ethics Training Enters the Real World," *Training: The Magazine of Human Resources Development* 27(10):82–91, at p. 84 (October 1990).
21. B. Weisendanger, "Doing the Right Thing," *Sales & Marketing Management* 143(3):82–83, at p. 83 (March 1991).
22. Reece, *Effective Human Relations in Organizations*, pp. 110.
23. Lehman and DuFrene, *Business Communications*, p. 19.
24. Ibid.
25. Ibid., p. 22.
26. Ibid., p. 18.
27. M. W. Drafke and S. Kossen, *The Human Side of Organizations*, 10th ed. (Reading, MA: Addison-Wesley, 2008).
28. L. H. Chaney and J. St. C. Martin, *The Essential Guide to Business Etiquette* (Westport, CT: Praeger, 2007), pp. 49–50, 51; M. Y. Stewart and M. Faux, *Executive Etiquette in the New Workplace* (New York: St. Martin's Press, 1994), pp. 85, 86.
29. Drafke and Kossen, *The Human Side of Organizations*.
30. Ibid.
31. S. Fox and P. Cunningham, *Business Etiquette for Dummies* (New York: Hungry Minds, 2008).
32. Reece, *Effective Human Relations in Organizations*, p. 104.

Section 6 TRANSITION TO WORK SKILLS

21 SELECTING A CAREER AND FINDING A JOB

Career selection is important because you will spend the majority of your life working in the career you choose. This makes finding the right job one of the most important activities you will ever undertake. If you do not spend the majority of your life doing something you enjoy, you will probably not feel that you have really lived. It is better to have tried and failed than never to have tried at all. If you follow your dream to a career that presents the best-case scenario for what you really want to do, at the very least you can live your life without regretting that you did not give it your best shot.

Objectives

After completing this chapter, the student should be able to:

- Explain the reasons why people work
- Differentiate between work, occupation, job, and career
- Discuss career planning
- Explain how to find jobs that match your personal characteristics
- Describe careers in production agriculture, agribusiness, and agriscience
- Explain the diversity of agricultural jobs and the types of agricultural education job placement
- Describe the resources available to help find a job

Terms to Know

- career planning
- work
- occupation
- job
- career
- job-related skills
- self-management (adaptive) skills
- transferable skills
- gross national product (GNP)
- value-added
- immediate job placement
- postponed job placement
- avocational (part-time) job placement
- supplementary (tributary) job placement
- job lead
- network
- classified ad

magine that you could do anything you wanted to do. Forget the necessary qualifications, abilities, or money needed to do it: just suppose that you could have any career you wanted, anywhere. What would that job or career be? Visualize yourself in that job or career, and hold that desirable image in your mind.

When we are young, we usually have many ideas about what we want to be, even if we change those ideas frequently. As we grow older, we may lose sight of our ultimate dreams because we fail to plan for the desired career. As immediate needs and responsibilities arise, we tend to drift into occupations more by chance rather than by choice. We may take any job that is available when we need one, rather than intentionally searching out and choosing jobs that will keep us moving toward our preferred occupation and career.

In an earlier chapter we discussed goals and planning. People do not plan to fail, they just fail to plan. Begin the process of creating your career plan by closing your eyes and imagining your dream job. Picture yourself at work in as much detail as possible. What are you doing? Where do you work? How much do you earn? Once you have visualized your ideal job, ask yourself how you can make part or all of this dream job become reality.

Career planning concerns mapping out your life mission, or what you see as the meaning of your life. If you do your best to achieve this dream, you will feel that you have really lived, even if you try and fail to achieve exactly what you envisioned. After studying this chapter, you will be better able to find your dream job, just as the grape grower in Figure 21–1 has. Loving what you do for a living can make for a rewarding life.

WHY PEOPLE WORK

The first and most obvious reason people work is to make money, but work is about so much more. Many people work because of the problems they get to solve, the goals they get to achieve, or the people they get to help. These people actually think work is fun. Others work because they like being in charge—in charge of other workers, in charge of their life, or in charge of their finances. Some people work because it gives them an opportunity to be around other people.[1]

People's views about work vary greatly. Although not all people value and enjoy their work, most people feel that work is an important part of a well-rounded life. Most people generally like what

FIGURE 21–1 This grape grower seems to love the career she has chosen. What is the dream job that could make you smile like this woman? *(© iStock/ATELIER CREATION PHOTO)*

they do, and many studies have shown that most people would work (in some capacity) even if they did not have to. This view of work is not limited to adults. Young people's interest in learning about and preparing for work has never been greater.

People work for many different reasons, and the reasons vary from individual to individual. Following are some of the primary reasons for working:[2,3]

Income The primary reason people work is to earn money. Income is required to meet basic needs, such as food, shelter, clothing, medical care, and other necessities of modern life. Of course, most people also want an income that covers more than the basics; money is also used to purchase goods and services that provide comfort, enjoyment, and security.

Social Satisfaction Most people enjoy engaging with other people. Working gives people a chance to be with others, to make friends, and to work toward common goals with others who share their values. Engaging as a team at work can be very rewarding. In the work environment, people can give and receive understanding and acceptance, and interact with others on a daily basis.

Positive Feelings People get satisfaction from their work. They enjoy it. For instance, your work may give you a sense of accomplishment: think of how you feel when you finish a school project or a difficult task. Working can also give you a feeling of self-worth, when you know that other people will pay for your skills, products, or services.

Freedom Freedom, autonomy, and flexibility are things that most people crave. Opportunities to create, build, and make decisions that have lasting impacts on society make employees want to get up and come to work each and every day. Can you think of occupations that would give you the good feelings that come from freedom on the job?

Personal Development Many people have a drive to improve themselves and to learn and experience new things. Work can provide numerous opportunities to learn and grow, to set goals and reach them, to be challenged and win. Work and the experience gained through work are often great teachers.

Purpose We all want to do something meaningful. A job gives us a chance to make a positive difference every day. What do you want your purpose to be? You can find meaning and purpose in any job you are in, but if you start planning now, you have a much greater chance of ending up in a job that helps you fulfill your purpose (Figure 21–2).

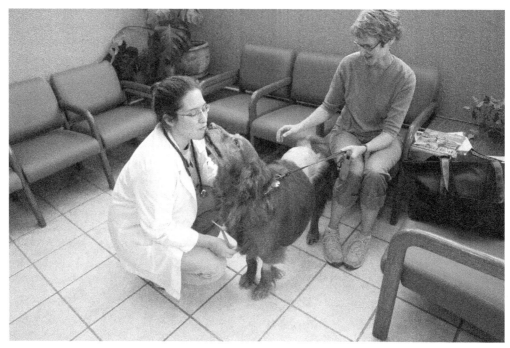

FIGURE 21–2 This worker finds her purpose in working as a veterinary technician helping animals stay healthy.
(Courtesy of University of Georgia—College of Agricultural & Environmental Sciences.)

Other reasons people work are security, success, happiness, and the esteem of peers and family.

Far too many people never follow their own dreams. Their occupational choices are often shaped by what other people think they should do. They work because other people want them to:

"You *should* go to work."
"You *should* go to college."
"You *should* serve your country."
"You *should* earn a lot of money."
"You *should* be a lawyer [or doctor, or teacher, or farmer] because your parent or grandparent was."

Many people spend their whole lives doing what others believe they should do rather than what they really want to do. They are overwhelmed and tyrannized by the "shoulds" of their parents, teachers, friends, social class, and other role models.[4]

TERMINOLOGY

The terms *work, occupation,* and *job* are often used interchangeably. They are similar and interrelated, but there are important definitional differences and distinctions.

Work

Work can be defined as activity, directed toward a purpose or goal, that produces something of value to oneself and/or society. For example, work can provide you with money and a sense of accomplishment. Work by a social worker, teacher, or nurse provides benefits to society. Work is not necessarily performed for wages. For example, service or community projects are often undertaken free of charge by volunteers.

Occupation

All occupations carry out work. An **occupation** comprises a group of similar or related tasks that a person performs for pay; normally, all of these activities and tasks are part of achieving or performing a larger objective or goal. For example, data entry, filing, maintaining records, placing telephone calls, and scheduling meetings are tasks performed by one in the occupation of

secretary. Carpenter, sales clerk, attorney, truck driver, and chef are examples of common occupations that involve groups of related tasks.[5]

Most occupations require specific knowledge and skills, which may be learned on the job or in various kinds of educational and training programs. A person can work at a number of different jobs within the somewhat broader classification of an occupation. For example, the occupation of *pilot* applies equally to those who fly crop dusters, jumbo passenger jets, and sightseeing helicopters. Any occupation may offer many specialty niches and different jobs.

Job

A **job** is a (usually) paid position at a specific place or in a specific setting to do regular work. A job can be in an office, store, factory, farm, or mine. For example, a nurse (an occupation) may have a job in a doctor's office, clinic, hospital, home, school, factory, or nursing home.[6]

Difference between Job and Career

A *job* is a specific duty, task, or chore performed for pay. It involves responsibilities based on an agreement between a worker and an employer about performance, hours, and other matters. A **career** is a series of jobs that is pursued more or less in sequence to achieve the ultimate occupation desired by the individual. Basically, a career is something you really want to do—and be—for the rest of your life.

CAREER PLANNING

Choosing a career is a serious matter. The more time you spend on career planning, the easier it will be to find the career that suits you best. No one knows for certain all the events that will occur in the future, of course, but you can anticipate changes. When you anticipate, you use your imagination to predict or forecast what might happen in the future. From these potential results or consequences, you can plan what you might need to do to make them occur.

Importance

Your career will likely be one of your most important life activities. You will spend many years in one job or another, as well as in the training required for your career. So much time will be demanded by your career that the rest of your life may revolve around it, much the way it revolves around school now. The kind of career you have will affect how much time you have for other activities, the types of people you meet, the friends you make, and the kind of lifestyle you can afford.

The amount of time you will spend working is a good reason to give careful consideration to your career. A person who works full time for 40 years, averaging 40 hours per week for 50 weeks per year, will spend 80,000 hours at work. Compare that with the 17,000 hours you will have spent in school by the time you graduate from high school!

Factors to Consider When Selecting a Career

The many factors to consider when selecting a career include or relate to nearly every aspect of one's life. Some major factors to consider are the following:[7]

Standard of Living How much money do you need to make? You will want to pursue a career that generates enough income to allow you to live the way you want to live. People's ideas of what actually constitutes a comfortable lifestyle vary tremendously, and thus their views of suitable careers vary as well. For example, some people will trade off leisure time for a career that yields high income; others will take a lower income for the opportunity to live and work in a desirable location. The decisions regarding this consideration depend entirely on the individual's needs and desires.

Opportunities Some careers have more opportunities (more jobs) than others. Consider statistics such as how many jobs go unfilled, the percentage of graduates in a certain field who are hired, or the number of unqualified workers that are hired in an area. Agriculture is a field with lots of opportunities. Every year there are more jobs available than there are qualified people to take those jobs.

Personal Contact Some people want careers in which there is a large amount of personal interaction with a great variety of individuals, whether they are customers, other business-people, or coworkers. Others do not want to deal with other people much. How well you can communicate and interact with others is a big factor in career success. When selecting a career, weigh the amount of personal contact required against the amount you desire to undertake.

Formal Education Required You will need at least a high school diploma for almost any job and certainly for a reasonable career path. Many careers and occupations also require graduation from a technical school, a college (two- or four-year degree), and, in many cases these days, graduate school.

Training and Experience Required For many careers, you will need appropriate training and experience in the field as well as formal education. You can gain experience through part-time work while you are still in school, through internships, or in entry-level (full-time) jobs. Another excellent way to get practical, hands-on experience, along with knowledge and skills, is through a supervised agricultural experience program in agricultural education. If qualifying for a job in the career you really want to pursue will take more years than you are willing to dedicate, then you may need to reassess.

Locations of Employment Many people choose the place they want to live and then select jobs and careers according to what is available in that location. Someone who is either unwilling or unable to move to a new location may have limited options both for career choice and for promotions. It's also good to think about location in terms of inside or outside, in the city or in the country, or in a small business versus a large corporation. What is the location in which you would be most happy?

Interests It is important to choose a career that is well suited to your own interests and capabilities, so that you can be happy in your work. Sometimes people take a job primarily because they will make more money at it. As they say, money doesn't buy happiness. If you are not sure what interests you have, take an online assessment or go see a career counselor. Both options have proven to be successful for people who want to make sure they choose a career about which they can get excited.

Working Conditions The fact is that much of your adult life will be spent on the job. Therefore, the probable working conditions should be an important factor in your career choice. Will you be operating dangerous machinery? Handling dangerous chemicals? Breathing pollutants? Working long hours or being on call for long periods? Will you have to do a lot of traveling? All these questions merit careful consideration.

Work–Life Balance There is more to life than working. You need to spend time with your family and do things that you love to do besides work. Many companies have policies concerning vacations and time off. Self-employed people, such as farmers and others who own their own businesses, can set their own work schedules, though they must accommodate customers and deal with many other constraints on their use of time. Entrepreneurs, managers, and some types of health care workers may spend more than the typical 40 hours per week at work. Professionals such as caregivers and attorneys may not be able to "leave work at work," especially when they are on call. The time you need to devote to your career should align with your goals and desires for other aspects of life, notably health and family.

Security during Retirement Will your job provide you with a pension during retirement? If not, will your income be high enough while you are working that you can save money for retirement? Other benefits, such as health insurance, sick leave, and disability insurance, should also be considered.

Health In many instances, the type of work you do will directly affect your health. This includes physical, mental, and emotional health. Will you have to push, pull, and lift heavy objects every day until you retire? Will you be able to remember everything you have to remember to be good at your job? Will you be stressed out for the majority of your life? These are important questions that guide not only your career choice, but also your well-being. Your job should bring you satisfaction and enhance your self-concept. If it does not do this, you probably will not be happy in that career.

Steps in Choosing a Career

A career chosen early in life may be no better than a choice made later in life. However, an early choice, even if it changes later, can give some direction to your life and your studies. If you know where you want to go, you can plan how to get

there. Here are some logical steps to follow when choosing a career or revising your career choice.[8]

Step 1 *Consider your interests, abilities, talents, and values.*

Look carefully at what you like to do and what you are best at (feel most comfortable with). What careers line up with your values system? Some careers are for you and some you should leave alone: you will be miserable if your career doesn't line up with your skill-set and worldview. Be honest with yourself about your strong and weak points. Do not forget your personality type, communication style, and learning style; for best results, all these tendencies and preferences should correlate with or contribute positively to the job or career you select. State or national FFA career shows are an excellent place for agricultural education students to explore potential careers (Figure 21–3).

Step 2 *Narrow the field of jobs.*

After you have evaluated your interests and abilities, you can begin to narrow your choices by making a list of options. In most areas, several groups of related occupations will match your interests and abilities. Do some research and figure out which ones are causes for the most excitement. Your first choice may change, but it can be a starting point for planning a career. If your strength is psychomotor skills, for example, work as a mechanic, carpenter, or building contractor could be your career choice.

Step 3 *Study the occupational requirements.*

As you research, look carefully at the requirements, working conditions, job outlook (how many job opportunities will there be), annual earnings, promotional opportunities, and other characteristics of an occupation or job, and become familiar with all the facts. You may even have some friends or family members in a profession that interests you. Interview them. Poor career choices are more likely if you do not have adequate information or if you misunderstand the facts.

Step 4 *Plan for alternatives.*

Your first choice or career attempt might not go as planned or suit you as well as you thought. Try to

FIGURE 21–3 Use every chance you have to explore career opportunities. Take advantage of high school career days, field trips to local businesses, and state and national FFA career shows. *(© Hadrian/Shutterstock.com)*

anticipate a viable second occupation option and make a backup plan. All occupations have related occupations. Determine the occupations that relate to your first choice, and be prepared to enter one of those as an alternate if the first choice does not turn out as you expected or hoped.

Step 5 *Plan ahead and set goals for your career.*

Plan ahead so that you are ready when the time comes for you to go to work. If the job requires special training, set a goal to obtain that training. Important information that should be part of a plan for career preparation could include taking a special course in high school, obtaining part-time work experience while you are still in high school, and budgeting or making yourself eligible for additional education beyond high school.

Step 6 *Be willing to pay the price for success.*

Educational planning is an important part of career preparation. Many students decide once they enter college that they want to be veterinarians, doctors, or engineers—but that decision is better made in the eighth grade or even earlier. For such a career, you will want to take as much biology, chemistry, and

math in high school as possible. Do you want to be a sportswriter or announcer? If so, you will need to work on oral or written communication skills. You can take speech courses, English courses, and journalism courses, as well as practicing public speaking in agricultural education and the FFA. Figure out what it will take to get where you want to go, and start moving toward that goal as early as possible.

Step 7 *Get work experience.*

Work experience is also important in career preparation. Direct work experience in the occupation you desire is best, but you may have to select another job because of insurance requirements, legal requirements, or job availability. If, for example, you cannot get a job working as an agricultural mechanic, you could gain valuable experience through a related job, such as working for an auto mechanic. This job would allow you to work with motors and do engine maintenance and repair.

It is also important to note that internships are a great way to ensure that you are selecting the right career. While you are in college, look for as many of those internships as possible, especially if you are uncertain about which direction you want your career to take. Not only does the firsthand experience give you important decision-making knowledge, but if you apply and demonstrate the principles learned in this book, the companies you intern for will be begging to hire you.

CHOOSING A JOB

Once you have a career path, you can determine whether it is the right choice for you. Many of today's workers don't even know how they ended up doing what they do. They didn't have a plan. Planning and careful evaluation can make the difference between working at something you love and just earning a paycheck. About half of your life, once you graduate from school, will be spent at work. Don't you want to get it right?

Are you looking for a job that offers good security or one that comes with high risk and lots of excitement? Perhaps you do not favor either extreme and would prefer something in between. What you prefer is one thing; what you get depends on how wisely you choose a career. The career you pick is the number-one influence on the kind of life you will have. It will be a major factor in determining the following:

- How much money you will make
- How often you get paid
- How much you see your family
- Who is your boss
- Where you will live
- What you will drive
- If you will have time for your hobbies and special interests

Self-Assessment

The unique qualities of an individual should be matched to a career that utilizes those qualities. A career is more fulfilling and enjoyable if there is a good match between an individual's interests, skills, and values and those associated with a given career or occupation. The United States Department of Labor makes it easy for you. They have online assessments that can help you to determine your interests, skills, and values and make suggestions about potential jobs you should consider.[9]

Evaluate your values and philosophy of life. Career counselors should candidly discuss strengths and weaknesses with you; you need honest evaluations and will eventually appreciate them. You need to know what areas of your personality to work on; that is, what areas would benefit from development. For any job, you must judge your strengths and weaknesses, evaluate potential benefit to both you and the employer, and decide whether this job moves you further toward achieving your personal goals.

Finding Your Special Abilities

Everyone is good at something. In fact, most people are good at many things, although they seldom give themselves credit for all of them. You may take for granted many things you do well that others would find hard or even impossible to do. We all have special talents and abilities.

| VIGNETTE 21-1 | PATIENCE, PATIENCE, PATIENCE |

EVEN AS YOU FOLLOW ALL OF THE STEPS TO FINDING A CAREER and the right job in that career, you must be patient. Tricia is the perfect example of someone who got into trouble because of impatience. She had just finished four years of school as a business major. She wanted a job very badly and felt that she had to get one rather quickly—so that is what she did: she took the first job she was offered, and she did not have to wait long to get it. Unfortunately, the job was only vaguely related to the career she wanted; it was also incredibly demanding and unsuited to her particular abilities. It was not what Tricia wanted to do at all. Tricia ended up wasting three years with a company where she was not happy.

We believe that you should love your job, occupation, and career. Many times, the key to finding the right job is patience. When Tricia moved to a new town, she was determined to get a job that was right for her. She found several opportunities for work soon after she moved, but none of them felt right, and this time she did not want to take just any old job because the offer was there. About two months after she moved, the perfect job in her career area became available. It felt like the right place to be, so Tricia interviewed for the job, got it, and is very happy in her current situation.

The moral of the story is patience, patience, patience. Finding good jobs takes time, for several reasons, so be patient and wait for the job that you know is right for you in the long run.

Knowing what you do best is important when you are deciding what kind of work is right for you. It makes a lot of sense to do the things you do best. If you do, you will probably be more successful. Your self-concept is enhanced and happiness is more attainable when you concentrate on the things you do best. It is also important to do things that you enjoy doing. If you enjoy what you do *and* you are good at it, your life will be satisfying and productive.

Know Your Skills

Knowing what you are good at and then letting others know what you're good at are key to landing the job you want. A very common interview approach involves an applicant explaining to a potential employer why they should hire them. If you aren't able to answer that question, you are not going to get the job.

Knowing what you are good at is important in selecting a job. It also helps you decide what type of career you will enjoy and do well in. When most people think of skills, they envision job-related skills, such as knowing how to use a spreadsheet or operate a miter saw. However, everyone also has other types of skills that are equally important for success on a job. Two of the most important ones are self-management skills and transferable skills, which we will discuss below.[10]

Job-related skills are the capabilities, abilities, characteristics, and knowledge (skills) you need to perform a specific job. An agricultural mechanic, for example, needs to know how to service and repair engines and machinery. A carpenter must know how to use various tools and be familiar with a variety of construction methods, among the many tasks related to that job. These job-related skills can be learned in career and technical classes in high school, postsecondary schools, apprenticeship programs, and other places.

Self-management (adaptive) skills are the skills that are required to be a good worker, such as diligence, honesty, enthusiasm, and ability to learn. They are often thought of as personality or personal characteristics. They help a person to adapt to or get along in a new situation. They are some of the most important things to bring

up in an interview. For example, honesty and enthusiasm are traits employers look for in any worker. Key skills are skills that employers find particularly important. They often will not hire a person who does not have most of these critical skills:

- Accepts supervision
- Excellent attendance record
- Punctual
- Hardworking
- Gets along with coworkers (team player)
- Productive
- Gets things done in a timely manner
- Organized
- Self-motivated

Transferable skills are skills you can use in many different jobs, though the importance of each skill may vary according to the job application. You literally transfer these skills from one job to another (possibly very different) job. Communicating well and writing clearly, for instance, are valuable skills that you can use in almost any job.

It is important that you know and identify the skills you possess. Most potential employees think job-related skills are their greatest asset. Although job skills are important, employers consistently list various transferable skills as some of the most sought-after skills when searching for employees. Therefore, you need to list and assess your transferable skills (Figure 21–4).

Even if you do not have a specific job title, you must know the type of things you want to do and are good at before you start your job search. This means defining the specific job you are looking for rather than a job in general. If you already have a good idea of the type of job you want, select the skills that will be most important and help you most in that job, and then emphasize those skills on your résumé and in job interviews to increase your chances of getting the job you want.

Occupational Groups

Most jobs fit into more than one category. In *How to Choose the Right Career*, Louise Schrank divided occupations into six groups. Each group attracts different kinds of people and involves a different kind of work.[11]

TRANSFERABLE SKILLS			
KEY SKILLS (CRITICAL)			
instructing others	meeting deadlines	organize/manage projects	written communication
managing money, budget	meeting the public	public speaking	skills
managing people	negotiating		

OTHER TRANSFERABLE SKILLS			
WORKING WITH THINGS		**WORKING WITH WORDS AND IDEAS**	
assemble things	observe/inspect	articulate	inventive
build things	operate tools, machines	communicate verbally	library research
construct/repair/build	repair things	correspond with others	logical
drive, operate vehicles	use complex equipment	create new ideas	public speaking
good with hands		design	remember information
		edit	write clearly
		ingenious	
WORKING WITH DATA		**LEADERSHIP**	
analyze data	evaluate	arrange social functions	mediate problems
audit records	investigate	competitive	motivate people
budget	keep financial records	decisive	negotiate agreements
calculate/compute	locate information	delegate	planning
check for accuracy	manage money	direct others	results-oriented
classify things	observe/inspect	explain things to others	risk-taker
compare	record facts	influence others	run meetings
compile	research	initiate new tasks	self-confident
count	synthesize	make decisions	self-motivated
detail-oriented	take inventory	manage or direct others	solve problems
WORKING WITH PEOPLE		**CREATIVE/ARTISTIC**	
administer	patient	artistic	expressive
advise	perceptive	dance, body movement	perform, act
care for	persuade	drawing, art	present artistic idea
confront others	pleasant		
counsel people	sensitive		
demonstrate	sociable		
diplomatic	supervise		
help others	tactful		
instruct	teaching		
interview people	tolerant		
kind	tough		
listen	trusting		
negotiate	understanding		
outgoing			

FIGURE 21–4 Job-related skills are very important, but these transferable skills are also important to employers. In the workplace, these skills translate into productivity, whether the job is in sales or human development.

Body Workers These people enjoy physical activity, manual tasks, or heavy work requiring strength and endurance. Satisfaction and a sense of achievement come from seeing the work completed; body workers enjoy seeing concrete, tangible results. They may need special clothing for their work and may be dirty or physically tired at the end of the day. Body workers often work with objects, machines, plants, or animals, and their jobs frequently take them outdoors. They may be production agriculturalists, athletes, physical instructors, or blue-collar workers. Examples of body-worker jobs are mechanic, coach, forest ranger, lumberjack, carpenter, truck driver, bricklayer, bulldozer operator, and horticulturist. These people may also work in skilled technical or service jobs. Body workers are typically practical, rugged, athletic, healthy, and aggressive.

Data Detailers These people use numbers or words in their work in very exact, precise ways. They know that attention to detail is important, and they highly value error-free work. They often have good clerical and mathematical abilities. They are often white-collar or office workers. Data detailers are best in jobs involving clerical or numerical tasks, such as banking, bookkeeping, data processing, accounting, information technology, and insurance. They are usually good at following instructions and attending carefully to detail work.

Persuaders Persuaders like to work and talk with people and enjoy convincing others to see things their way. Their success is measured by how well they influence others. Persuaders often work in sales, law, or politics. In management or sales positions, these workers persuade people to perform some kind of action, such as buying goods or services. They like to lead or manage for organizational or economic gain.

Service Workers Service workers find satisfaction in helping others. They often find rewarding jobs in schools, hospitals, and social service agencies, as teachers, nurses, or counselors. Working in education, health care, or social welfare, they teach, heal, or help people. Volunteer coordinators, waiters, instructors, health care workers, and tour guides are part of this group. They like to inform, enlighten, educate, develop, or cure people.

Creative Artists People who express themselves through music, dance, drama, writing, or other art forms are the creative artists of this world. Because demand and pay for artistic products and services is often low, many artists can only afford to work at their chosen jobs part time. They may earn all or part of their income by working in related areas in which their creativity, knowledge, and skill are used in at least part of the work. These people work with words, music, or art in a creative way. Actors, musicians, composers, authors, painters, landscapers, floral designers, and sculptors are in this group.

Investigators These people like to observe, learn, investigate, analyze, and evaluate. Investigators enjoy asking *why* and *how* questions in their work. They work with scientific or technical information, often applying it to new situations. They may work in basic scientific research or analysis as well as applied sciences. By performing scientific or laboratory work, investigators research how the world is put together and how to solve problems.

You may notice that these six categories of work align fairly closely with the personality types discussed in Chapter 2. Personality type also relates to communication style, leadership style, and learning style. Figure 21–5 shows how occupational groups correlate with personality type and learning style.

OCCUPATIONAL GROUP	USDA CATEGORIES	PERSONALITY TYPE	LEARNING STYLE
Body workers	Agricultural production specialist	Sanguine/choleric	Convergers/accommodators
Data detail	Managers and financial specialist	Melancholy	Assimilators
Persuaders	Agricultural marketing, merchandising, & sales	Sanguine/choleric	Convergers/accommodators
Service workers	Social service professionals	Phlegmatic	Divergers
Creative artists	Education and communication specialist	Melancholy	All four styles
Investigators	Scientists, engineers, & related specialists	Melancholy	Assimilators

FIGURE 21–5 Your choice of career should correlate with your personality type and learning style. Otherwise, you may choose the wrong occupation and wonder why you are not happy with your job.

CAREERS IN PRODUCTION AGRICULTURE, AGRIBUSINESS, AND AGRISCIENCE

A massive grant-funded study was conducted in 2015 to determine the job outlook for agriculture, and their findings were positive if you're interested in that industry. The researchers found that in the future, "U.S. college graduates will find good employment opportunities if they have expertise in food, agriculture, renewable natural resources, or the environment.... [By] 2020, we expect to see 57,900 average annual openings for graduates with bachelor's or higher degrees in those areas." Some people say one out of every five jobs are related to agriculture, and some say one out of every three. Either way, the agriculture career opportunities are plentiful, from Management and Business, Science and Engineering, and Food and Biomaterials, to Education, Communication, and Government Services, there is a place for you to apply your leadership skills.[12]

The Bases of Jobs in Agriculture To understand the role of agriculture/agribusiness in the U.S. economy, we need to reference the yardstick that measures the value of goods and services America produces in a year, the **gross national product (GNP)**. Agriculture accounts for 17 percent of the GNP and provides more than 20 percent of all the jobs in the United States. Two percent of the GNP comes from firms or people who sell goods and services to farmers and ranchers. Thirteen percent of the GNP comes from related industries that purchase food and fiber from farmers and ranchers and then process and package it, increasing the monetary worth, into the **value-added** products sold to consumers. This group of businesses and industries, which ranges from ice cream makers to textile mills to breakfast food makers and a host of others, makes up the economic segment called agribusiness and agriscience.

According to the National FFA Organization's new AgExplorer Career Finder tool, one in every three people work in agriculture worldwide. The job titles in Agribusiness, Animal Science, Biotechnology, Environmental Service, Food Products and Processing, Natural Resources, Plant Science, Power, Structural and Technical Systems, and Agricultural Education are almost endless. All these occupations work together to provide food, fiber, and energy in a variety of formats for the planet's growing population.[13]

Almost any career in which you may be interested can be applied to agriculture. Take engineering, for example. Today, farmers are leveling fields with lasers to decrease erosion, using global positioning systems to determine optimum land use and crop distribution, and using robotic equipment to do dangerous or repetitive jobs. If progress is to continue, agriculture needs the best and brightest young minds working to solve tomorrow's agricultural engineering challenges.

The ever-increasing world population creates a greater demand for food and fiber. It also creates a growing demand for qualified people in the agricultural industry. Almost 40 percent of today's professional jobs in agriculture are filled by graduates who do not have agriculture degrees.[14] Agriculture is changing rapidly, making it an exciting, challenging field in which to work. Whether you are interested in business, computers, mechanics, science, or communications, America's largest industry has many career opportunities and will likely spawn new careers that have not yet even been imagined. Most agriculture-related jobs fall into one of the following four categories:

Food and Biomaterials Production Specialists The production sector—people working with plants and animals—produces commodities to be processed for consumers. Production agriculture supplies many raw products, such as beef, pork, poultry, grains, fruits, vegetables, ornamental horticulture, timber, and even specialty crops such as wine grapes, organic produce of all kinds, energy crops, and products used to produce pharmaceuticals. The popularity of the local food movement and increased regulations for farmers who sell direct will create even more job opportunities for regulators and others who help farmers meet compliance issues.

Scientists, Engineers, and Related Specialists Agriscience, with its related occupations of engineering, biochemistry, genetics, and physiology, is the fastest-growing area in agriculture. This is agriculture's cutting edge, and it's never been more needed as consumers demand abundant, safe, and nutritious food at a time when water and land are scarcer than ever before. If you are interested in applying scientific principles to practical situations as a veterinarian, a computer programmer, a food-animal nutritionist, or a wildlife biologist, this may be the area that best fits your career aspirations.

Communication, Education, and Government Specialists The agricultural industry today needs to tell its story and advocate not only for its cause but also for its existence. If agriculture wants its story to be told correctly, then we will continue to need reporters, producers, and media experts who can help the public understand agricultural science, not propaganda. In all parts of the globe, education is viewed as one of the primary tools for reducing hunger and poverty, but not enough people are entering the field. Even in the United States, there continues to be a shortage of qualified secondary agriculture teachers. If you like working with people to help them improve their lives, like Extension agents do every day, this may be the sector for you. If you are interested in sharing the news and helping people interpret it correctly, a career in education and communications may suit your abilities and talents perfectly.

Management and Business Specialists This area processes, markets, and distributes agricultural products to consumers. It provides equipment and materials for production. This area is sometimes referred to as "gate to plate" because it takes the raw product from the production sector and makes it into products that are usable by consumers. There is much demand for agricultural products today. Consumers expect to walk into supermarkets and find the shelves bursting with choices. If you are interested in sales and helping people acquire the goods and services they need, a career in agribusiness or agricultural marketing could be a great fit for you.[15]

Distribution of Jobs for Graduates Potential careers for each of the agricultural categories are shown in Figure 21–6. The percentage of employment in the six agriculture categories is also shown, along with the surplus of graduates in each category. Careers in agriculture have diversified far beyond farming. Agriculture graduates are filling jobs you may not even recognize as agricultural.

AGRICULTURAL EDUCATION JOB PLACEMENT

There is often confusion on the placement rate of students coming from agricultural education programs. Many well-educated professionals do not understand the diversity of occupations in agriculture. They still see agriculture as "cows, sows, and plows" or "seeds, weeds, and feeds." They perceive the agricultural industry as employing only a very small percentage of the American workforce, whereas, as you just learned, it really accounts for more than 20 percent. Even surveys that request placement rates of students going into agriculture do not reflect the total diversity of occupations that were selected because of training received in agricultural education. To help you understand the huge range of possible job placements, we briefly discuss the following four areas: immediate job placement, postponed job placement, avocational (part-time) job placement, and supplementary (tributary) job placement.

Immediate Job Placement

Many students enter the job market immediately after high school graduation. This is called **immediate job placement**. The jobs they take may be directly related to production agriculture, agribusiness, or agriscience. Conversely, they may find jobs where they use skills learned through the agricultural education program, even though those jobs may not relate directly to agriculture. For example, indirect placement occurs when a student who received instruction in agricultural mechanics becomes a surveyor, plumber, bricklayer, or small-engine repair mechanic. Obviously, such a job would not be counted as an agricultural placement. Direct placement in agriculture could include the job of production agriculturalist, greenhouse worker, nursery worker, florist, agricultural mechanic, or produce worker. Many researchers and surveyors (especially those outside the field of agriculture) limit their study scope to these areas, but due to the diversity, business, and science of agriculture, such counts of immediate job placement reflect only a small percentage of the true employment placement of agriculture students.

AGRICULTURAL PRODUCTION SPECIALISTS	SCIENTISTS, ENGINEERS, AND RELATED SPECIALISTS	COMMUNICATION AND EDUCATION SPECIALISTS	SOCIAL SERVICES PROFESSIONALS	MARKETING, MERCHANDISING, AND SALES REPRESENTATIVES	MANAGERS AND FINANCIAL SPECIALISTS
7.5%	28.8%	7.6%	9.7%	32.4%	14.9%
CAREERS	CAREERS	CAREERS	CAREERS	CAREERS	CAREERS
Aquaculturalist	Agricultural engineer	College teacher	Career counselor	Account executive	Accountant
Farmer	Animal scientist	Computer software designer	Caseworker	Advertising manager	Appraiser
Farm manager	Biochemist	Computer systems analyst	Community development specialist	Commodity broker	Auditor
Feedlot manager	Cell biologist	Conference manager	Conservation officer	Consumer information manager	Banker
Forest resources manager	Entomologist	Cooperative extension agent	Consumer counselor	Export sales manager	Business manager
Fruit and vegetable grower	Environmental scientist	Editor	Dietitian	Food broker	Consultant
Greenhouse manager	Food engineer	Education specialist	Food inspector	Forest products merchandiser	Contract manager
Nursery products grower	Food scientist	High school teacher	Labor relations specialist	Grain merchandiser	Credit analyst
Rancher	Forest scientist	Illustrator	Naturalist	Insurance agent	Customer service manager
Turf producer	Geneticist	Information specialist	Nutrition counselor	Landscape contractor	Economist
Viticulturist	Landscape architect	Information systems analyst	Outdoor recreation specialist	Market analyst	Financial analyst
Wildlife manager	Microbiologist	Journalist	Park manager	Marketing manager	Food service manager
Surplus of Grads: +10.6%	Molecular biologist	Personnel development specialist	Peace Corps representative	Purchasing manager	Government program manager
	Natural resources scientist	Public relations representative	Population control	Real estate broker	Grants manager
	Nutritionist	Radio/TV broadcaster	Regional planner	Sales representative	Human resource Development manager
	Pathologist	Training manager	Regulatory agent	Technical service representative	Insurance agency manager
	Physiologist	**Surplus of Grads: 127.1%**	Rural sociologist	**Shortage of Grads: −18.2%**	Insurance risk manager
	Plant scientist		Youth program director		Landscape manager
	Quality assurance specialist		**Shortage of Grads: −12.7%**		Policy analyst
	Rangeland scientist				Resources and development manager
	Research technician				Retail manager
	Resource economist				Wholesale manager
	Soil scientist				**Shortage of Grads: −15.4%**
	Statistician				
	Toxicologist				
	Veterinarian				
	Waste management specialist				
	Water quality specialist				
	Weed scientist				
	Shortage of Grads: 215.3%				

FIGURE 21–6 Six categories of agricultural careers, with examples of occupations, percentage of total agricultural employment they represent, and graduate shortage in each category. *(Source: USDA)*

Postponed Job Placement

Many careers in agriculture are available only after a person has received advanced education or training beyond high school; hence, these individuals receive **postponed job placement**. The advanced, postsecondary training may be done at a two-year technical school, through a four-year college or university education, or through other experience. A problem for some states is that there are no two-year postsecondary programs for agriculture students except for university transfer (community college) students. Other opportunities for advanced training after high school are apprenticeships, on-the-job training, and service in the armed forces.

Some occupations require a college degree. Generally, a bachelor's degree requires four years of college course work. Many legal, professional, and engineering occupations require a college education. For example, to become an agriculture/agriscience teacher requires a four-year degree. The quality and type of programs offered vary from college to college. If you decide to attend a postsecondary institution, select a university or college that has a strong program in your area of interest. Size, location, and cost are other things you will need to consider. Your school guidance counselor can help you collect information on colleges and universities from the many different sources available.

Community colleges offer college-level courses for a two-year degree or credit transfer to a four-year educational institution. Community colleges offer training in more than 60 occupational areas, such as business, food management, and drafting, among many others. The credit you earn may be transferred to a four-year college. A major advantage of community colleges is that they are much less expensive than four-year colleges (both state-funded and private). In some states, community colleges and/or other types of two-year public schools are free for students with the right experiences (e.g., good grades, community service hours, etc.). Your school guidance counselor and a host of sites on the Internet can help you identify community colleges in your area and the programs they offer.

Technical schools provide work-related programs for high school students and young adults. They offer a variety of programs that train students in the skills they need to get a job. Auto mechanics, building trades, and electronics can all be learned in technical schools. Technical schools are relatively inexpensive to attend. Your school guidance counselor can also help you identify technical schools in your region that offer programs in which you might be interested.

An apprenticeship program combines classroom instruction with on-the-job training. You learn a job by working with an expert in a particular trade or occupation. Plumbing, electrical trades, and carpentry commonly offer apprenticeship programs, which last for two to five years. You earn money as you learn, but the pay is low until you finish the program. Labor unions also offer apprenticeship programs. To apply, check with a local union office. Competition is stiff for spots in apprenticeship programs.

When you start a new job, you will probably work with a more experienced employee who will help you learn the job. This is called on-the-job training. On-the-job training may take anywhere from a few days to a few years to complete. Usually, your pay is low until you complete the training program. Your state employment service can provide you with information about the type of on-the-job training you want. Company personnel offices may also tell you about opportunities available within the company.

The military services provide specialized training in many occupations and job areas of interest. If you think you would enjoy military life, service in the armed forces can be a good way to access free occupational training. New recruits must enlist for a set amount of time, usually two years, and you earn while you learn. If you want advanced training but cannot afford it otherwise, performing military service may be the answer. The disadvantage of military service is that you cannot quit or drop out if you do not like it; also, you may not be assigned to exactly the job or area in which you desire training. You can get information on military training from the local recruiting office of any branch of the armed forces.

Postponed placement is necessary for many careers. A few examples are agricultural education teachers, extension (county) agents/faculty, ASCS or soil conservationists, Farm Credit Services employees, Farm Bureau Insurance agents, veterinarians, and landscape architects.

As a future worker, you must begin to prepare yourself for the world of work. Before you make a career choice, make sure you know what advanced education, training, and skills you will need. Be prepared to spend the time and money necessary to get that training. Do not commit yourself to a career if you are not willing to train for it. Ask yourself the following questions:

- What education/training do I need for my chosen occupation?
- Where can I get the education/training needed for my chosen occupation?
- How much time and money does the education/training require?
- Am I willing to commit the time and money necessary to receive this education/training?
- How will I pay for the education/training?

Avocational (Part-Time) Job Placement

Many students use the skills learned in agricultural education in **avocational (part-time) job placement** rather than a full-time occupation. This choice is often based on the economic reality that although money can be made in production agriculture, it may not be possible to make a living from certain full-time production agriculture enterprises.

A student who learns skills in agricultural education may make an economically based decision to use those skills in an avocation (part-time work) rather than a vocational (full-time) career. Some examples of avocational agricultural enterprises are beef cattle production, swine operations, strawberry growing, nursery production, and aquaculture. It is likely that employment surveys miss a high percentage of this type of enterprise: a high school graduate who completes a placement survey may not indicate agricultural employment as an avocation if he or she has a full-time nonagricultural job.

Supplementary (Tributary) Job Placement

Some people work in areas that have no relation to agriculture, but obtained their jobs because of the leadership and personal development skills or other skills developed through a quality agricultural education program. This is called **supplementary (tributary) job placement**. These transferable skills became assets that enabled them to pursue nonagricultural careers that may offer greater financial rewards than agricultural-sector jobs. Some examples of such nonagricultural careers are attorneys, legislators, county officials, managers, supervisors, and salespeople.

FINDING A JOB

Once you have selected the career you want, it is time to find a job—preferably one that is in your chosen field and will be a step forward in pursuit of your ultimate career goal. People who lack job-seeking skills often let a job choose them instead of consciously choosing a job. They may work at a job they dislike or for which they are overqualified. If you are aware of all available job opportunities and possible job openings, you will be able to pick the one that is right for you. The more sources you use in your job search, the better your chances of finding openings for jobs you really want. Knowing where to look can help you find a job that is both satisfying and rewarding.

Jobs do not just fall into your lap; you have to hunt for them. There are many sources of information about job openings. Information about a job opening is called a **job lead**. If you can gather many job leads, you have a good chance of getting a job you want. Sources of job leads include personal networking, direct employer contact, online and newspaper classified ads (e.g., Craigslist), social networking (e.g., Facebook, Twitter, LinkedIn), employment websites (e.g., AgCareers.com, usajobs.gov), state employment services, private employment agencies, government jobs, in-school resources, trade magazines, professional associations, résumé mailings, internships and volunteering. The individual in Figure 21–7 is working hard to find job leads online. This is a popular method, and possibly a good place to start, but face-to-face communication is a more effective approach (Figure 21–8).

Networking

One of the more effective approaches in finding a job is networking. Most jobs today are obtained through networking. A **network** is an informal group of people and contacts that can be used to help one get a job. As a job seeker, your network

FIGURE 21–7 Searching for a job online is a crucial part of the job search process. It is important to set aside dedicated time to focus on your online research. *(© Odua Images/Shutterstock.com)*

FIGURE 21–8 There are several ways to find a job, but face-to-face networking and direct employer contact are by far the most effective. *(© iStock/Steve Debenport)*

is made up of all the people who can help you—and the people they know. Networking is the process you use in contacting these people. You may be surprised at how many people you can meet this way. The people in your network can do one of three things: hire you, give you job leads and information about companies, or refer you to others who can provide this kind of help.

The following is a list of potential people in your network. You may think of others to add to this list:

- Friends
- Present and former teachers
- Relatives
- Neighbors
- Former employers
- Your parents' friends
- Classmates
- Former coworkers
- Former classmates
- Members of sports groups
- Members of social clubs
- Your friends' parents
- People who sell you things
- People who provide you with services
- Members of a professional organization you belong to (or could quickly join)[16]

Each of these people is a contact for you. Obviously, some lists and some people on those lists will be more helpful than others, but almost any one of them could give you a job lead. Start with your friends and relatives. Call them up and tell them you are looking for a job and need their help. Be as clear as possible about what you are looking for and what skills and qualifications you have. It is possible that they will know of a job opening that is just right for you. If so, get the details and follow through. More likely, however, they will not. Therefore, you should ask each person three questions:[17]

- Do you know of any openings for a person with my skills? [If the answer is no, then ask:]
- Do you know of someone else who might know of such an opening? [If the person does, get that name and ask for another one. If the person does not, then ask:]
- Do you know of anyone who might know of someone else who might? [Another way to ask this is, "Do you know someone who knows lots of people?" If all else fails, this will usually get you a name.]

Each time you think of a new potential contact, you build your network infinitely. Whether you are building your network by word of mouth or by using social media platforms such as LinkedIn, you will eventually get hired. You just have to work at it.

There will always be jobs, for a variety of reasons. People retire, quit, move to another place, pass away, continue their education, or completely switch careers. However, 70 percent of all job openings are never advertised: supervisors or department heads often hire from among their own current employees, or hire friends of those employees. Therefore, anyone who becomes aware of these openings, usually through networking, has a great opportunity to snag a position that most job-seekers will never hear about. Those who wait for a job to post online may be too late.[18]

When it comes down to it, employers simply do not like to advertise. When employers put an ad on the company website, they then have to spend a huge amount of time sifting through résumés and interviewing all sorts of unknown people. Employers, in large part, do not have trained interviewers on staff, and the people who end up having to take on this task do not enjoy it. Networking can help you rise to the top. The experts agree: networking usually fills most jobs.

Direct Employer Contact

The second most effective way to find a job is by talking directly to employers. Networking is still important here. Keep in close contact with your network, and if they give you a lead make every attempt to go and meet the employer in person. Sometimes employers will only talk to you if they have actually advertised a position online (e.g., Groovejobs.com) or via some other method. If a job is advertised online, and you see something that fits, be sure to follow all the instructions the employer has placed in the announcement (e.g., "No phone calls, please") for contacting them.

For those unadvertised job openings you now know exist, it's up to you to make a list of places you'd like to work and then start doing the research to make a direct contact. Figure 21–9 lists types of jobs that are suitable for young workers. After you have an idea of the types of jobs you might want to pursue, just pick up your smartphone and start searching for names of business, phone numbers, and the names of people who work there.

Company human resources or personnel offices can provide information on job openings in that company, but you may have to dig deeper. You may telephone, email, or visit a potential employer to determine whether job openings exist in the company. When you finally go to meet a real person, dress for success. Be prepared, and bring anything along that you might need if they were to give you an interview (e.g., résumé). Check bulletin boards outside the personnel office for posted jobs, visit

Cashiers

Retail salesclerks

Machine operators, assemblers, inspectors, and tenders

Stock handlers and baggers

Cooks

Child-care workers

Janitors and maids

Food counter workers

Food preparation workers

Laborers (includes construction)

Waiters and waitresses

Secretaries and typists

Dining room attendants

Construction trade workers (includes carpenters, painters, electricians, and so on)

Receptionists and information clerks

Truck drivers

File and library clerks

Farm and nursery workers, gardeners, and grounds keepers

Stock and shipping clerks

Freight and materials movers

Garage and service station attendants

Nurse's aides, orderlies, and attendants

General office clerks

Bookkeepers and financial record clerks

Computer operators

FIGURE 21–9 Twenty-five occupation categories suited for young workers.

with other employees if possible, and ask them if they know of upcoming opportunities. Additionally, companies usually post job vacancies on their websites. Be sure to check there too.

And as we discussed before, if you meet with someone, do everything possible to make a good impression. Let all that FFA experience shine through. Look the employer in the eye, be polite, and ask them for an interview. If they give you one on the spot, you will be prepared. If they can't interview you at the moment, ask if you can schedule a time for one. After whatever kind of meeting you have, be sure to follow up soon afterward with an email thanking them for their time and any information or tips that they gave you.

Classified Ads

Newspapers and Web versions of newspapers post job openings and help-wanted ads in their **classified**

ad (want-ad) section. Your chances of getting a job this way range from only 5 to 13 percent (depending on the source), but you can learn from the ads what kinds of jobs are available and what skills are needed for various jobs. If you see a job opening you are interested in, answer the ad quickly. Many people will apply for the same job. If you wait too long, the job will have already been filled. Four types of help-wanted ads are shown in Figure 21–10.

The **open ad** describes the job requirements, identifies the employer, and tells you how to apply for the job. This is the best type of ad. The **catch-type ad** emphasizes a good salary but fails to mention the qualifications and skills needed for employment. The "catch" is that the job may involve door-to-door selling or telephone sales. The **agency spot ad** does not include the name of any particular employer. This ad is used by private agencies to advertise employment available only through the agency. In a **blind ad**, the name, address, and telephone number of the employer are omitted. This type of ad allows the employer to screen applications carefully and contact only qualified applicants for personal interviews. The list of want-ad terms and abbreviations in Figure 21–11 will help you read and understand want ads.

OPEN AD	AGENCY SPOT AD
PART-TIME CLERK TYPIST. Approximately 20 hours per week, including some evenings and Saturdays. Experience with typing, filing, telephone reception and work with public desirable. Must be dependable and able to work non-regular hours. $4.85 per hour. Applications accepted until Tuesday, Jan. 15, 20— at 5 P. M. to Mary Campbell, Alton Public Library, 405 W. Main, Alton, MA 43331. 473-6235	**Draftsman $18,000 & Up Fee Paid** Male or female, at least 2 years' experience. Call us or bring in your resume to compare your experience with our company requirements. JOLEN EMPLOYMENT AGENCY 17 Plaza Offices, P.O. Box 531 Tucker, TX 95313 635-4792

CATCH-TYPE AD	BLIND AD
Earn $100 to $500 Write for details P.O. Box 113 Sunnyvale, CA 75391	CLERICAL Local manufacturer has immediate part-time clerical position open. Involves heavy computer entry. 4 hours a day, 5 days a week, prefer afternoons. Reply to: CLERICAL P.O. Box 75A Union Station Green Hills, NY 10112

FIGURE 21–10 Four common types of help-wanted ads.

adv.	advancement		sal.	salary
aft.	after		sec.	secretary
A.M.	morning		ext.	extension (some telephones have an extension number)
appl.	applicant/application			
appt.	appointment		F.T.	full-time
asst.	assistant (helper)		ftr.	future
ben.	benefits		gd.	good
bet.	between		gen.	general
bgn.	begin/beginning		grad.	graduate
bldg.	building		hosp.	hospital
bus.	business		hqtrs.	headquarters
cert.	certified/certificate		hr.	hour
clk.	clerk		hrs.	hours
co.	company		hrly.	hourly
coll.	college		H.S.	high school
comm.	commission (pay based on sales)		hvy.	heavy
cond.	conditions		immed.	immediate
const.	construction		incl.	including
corp.	corporation (usually a larger company)		ind.	industrial
dept.	department		jr.	junior (beginner or assistant)
dir.	director		lic.	license
div.	division (part of a company)		lt.	light (a little)
D.O.T.	Dictionary of Occupational Titles		mach.	machine
D.O.E.	depending on experience		maint.	maintenance
elec.	electric		manuf.	manufacturing (making things)
empl.	employment		max.	maximum
E.O.E.	Equal Opportunity Employer		mech.	mechanic/mechanical
eqpt.	equipment		med.	medical
etc.	and so on		M/F	male/female
eves.	evenings		mfg.	manufacturing
exc.	excellent		mgr.	manager
exp.	experience		min.	minimum
nego.	negotiable		mo.	month
nec.	necessary		sh.	shorthand
of.	office		sr.	senior
opp.	opportunity		temp.	temporary
pd.	paid		trnee.	trainee (beginner)
P.M.	afternoon/evening		typ.	typing/typist
pos.	position		U-W	underwriter (insurance salesperson)
pref.	preferred		w.	with
P.T.	part-time		wk.	week or work
qual.	qualified/qualifications		wkr.	worker
refs.	references		wpm.	words per minute
req.	require/requirements		yr.	year

FIGURE 21–11 Common want-ad terms and abbreviations.

Online Employment Sites

Just about every business, of any size, has a website now, and these businesses often list their potential vacancies under links labeled *personnel, human resources, opportunities,* or *vacancies.* The Internet also offers search tools that function in much the same ways as the classified ads. With Web search tools, you can cast a wide net looking for matches with your specific set of skills. Unfortunately, nowadays almost everyone who is looking for a job trolls the Internet for leads; it is difficult to find any listings or vacancies that thousands of people across the country do not also have access to. It is like looking at classified ads on a national scale. The success rate is quite low when you generically search the Web. For greater success, search on online employment sites or use the more powerful online tools provided by social media.

One example relating to jobs in agriculture is AgCareers.com, the leading online job board for agriculture, food, and biotechnology. In AgCareers.com you can search for careers, post a résumé, set up a profile, and look for internships in the agriculture industry. They even have tools like salary surveys that tell you what you should be making in a particular position based on your experiences, and workshops and webinars to help you in your job search.

LinkedIn

Social media is a powerful tool for networking and making direct contacts not only with friends, but with potential employers. LinkedIn is a social networking tool designed specifically with job searching and networking in mind. It's a tool used by professionals to build professional identities online, discover professional opportunities, and connect with other professionals, and "has over thirty-five million members in over 140 industries."[19] You can use LinkedIn to let everyone know you are looking for a job with a simple status update. Ask people you've worked with to provide a recommendation on your job-related and leadership skills. You can also message your "connections" and ask them about "secret job requirements" and opportunities that are upcoming.[20]

Twitter

Even though LinkedIn was designed for professionals, it's not the only option for your job search with social media. Twitter is used for everything from news to conversation, but it can also help you land a job. With Twitter you can learn from your peers and build awareness of yourself as an expert, which can lead to job opportunities. Follow industries and organizations that you find most interesting, and build credibility by creating meaningful content as a member of the Twitter community. Don't be too serious, though: Twitter is also a venue where potential employers can experience and appreciate your personality and passions. Experts also warn against making your Twitter activity all about you: put others first, share helpful content, and engage with others via chats, conversations, and hashtags.[21]

You can learn a great deal from social media and various Internet sites about which companies are posting jobs, salary ranges, skill requirements, and so on. Most employers have an online presence, so it makes sense that you should give yourself one too, both to find work opportunities and to allow yourself to be found. The Take It to the Net activities at the end of this chapter should help you further explore Internet job-search possibilities.

State Employment Service

Most large cities offer or host public employment agency services. Established under federal and state laws and supported by taxes, public employment agencies exist to help people find jobs. Agencies are usually known by the name of the state where the office is located, such as the Illinois Department of Employment Security. To complete an application form, go to the employment office nearest you. After you complete the application, you will be interviewed to determine your interests and skills. If there is a job opening that seems right for you, you will be notified, and the employment office will give you a letter of introduction for a personal interview. This service is free. However, only about 5 percent of all job seekers find jobs through public agencies.

Private Employment Agencies

Because they are not tax supported, private employment agencies charge a fee for their placement services. When you apply to a private agency, you must sign a contract agreeing to pay the agency a fee if the agency helps you find a job. Often this is a percentage of your first year's salary. In some cases, your new employer will pay the fee, but this is usually done only for high-paying professional jobs.

Be very careful if you decide to use a private agency. Be sure to read the contract thoroughly before you sign. Be especially wary of paying them for research you could do on your own.[22] The sad fact is that there are many rip-off artists out there, preying on the desperate and those who have not educated themselves about job hunting. Even legitimate services have a relatively small success rate, so it is usually best to keep control of your own destiny and conduct your own search using proven methods with reasonable odds of success.

Government Jobs

The U.S. government is the largest employer in the nation. If you are interested in working for the government, there are special agencies you should contact, such as city civil service commissions, state civil service centers, and federal job information centers. Addresses and telephone numbers of these organizations are listed in the telephone book under the heading "Government." Agencies are grouped by type: city, county, state, or federal government. Ask your librarian for

---

VIGNETTE 21-2 **EMPLOYERS WILL WANT YOU!**

JAMES WAS A "4.0" STUDENT WHO HAD JUST GRADUATED FROM A VERY PRESTIGIOUS INSTITUTION

He wanted a career in journalism, so he began applying for jobs even before he graduated. James had never felt that leadership development was important, so he declined to participate in extracurricular activities of any kind and failed to take any courses in leadership development. He thought employers cared only about his 4.0 grade point average.

After several months of searching, James was completely dejected. He had not done well in any of the few interviews he had gotten, and the interviewers kept quizzing him about things like teamwork, buzz groups, and interpersonal relations. They asked about communication skills, motivating people, and getting people to respond. He did not understand why such things were so important or why he was having such trouble finding a job. Why didn't anyone want him?

Research has shown that employers are interested in people who are leaders. They like smart people who have good résumés and experience, but they also want their new hires to possess leadership skills. When the National Association of Colleges and Employers asked employers what skills and qualities they look for in new hires, they responded with the following list:

1. Leadership skills
2. Ability to work in a team
3. Written communication skills
4. Problem-solving skills
5. Strong work ethic
6. Analytical/quantitative skills
7. Technical skills
8. Verbal communication skills
9. Initiative
10. Computer skills
11. Flexibility/adaptability
12. Interpersonal skills
13. Detail-oriented
14. Organizational ability
15. Strategic planning skills
16. Friendly/outgoing personality
17. Entrepreneurial skills/risk-taker
18. Tactfulness
19. Creativity[23]

Do the items in this skills list look familiar? They should if you have studied any other parts of this book. This is a list of leadership skills. Make no mistake: you must also do well in the important content areas of math, science, agriculture, and English. However, *leadership* is the area that will set you apart with employers when you start hunting for a job.

additional information about government agencies and civil service jobs.

In-School Resources

Three sources of job leads are available in many secondary schools: your teachers, the school's guidance office, and the job placement office or career center.

Teachers Teachers and cooperative/work experience coordinators are good sources of job leads. Your own agriculture teacher or coordinator is probably already helping you. Be sure to tell other teachers that you are looking for a job. Ask for their suggestions and make them part of your networking effort.

Guidance Office Most schools also have a guidance office or guidance counselor. It is common for local employers to contact counselors when looking for new workers. The counselor usually keeps a list of job openings or posts them on a bulletin board. Tell the counselor that you are looking for a job and ask to see any information he or she has on available job openings.

Job Placement Office or Career Center Not all schools have these, but for those that do, interested students generally register with the office. They may receive job counseling and other services. Job counselors help to match students with job openings and make referrals for interviews (that is, they send students to employers who are hiring).

Trade Journals and Magazines

Trade journals and magazines sometimes publish job advertisements similar to those found in newspapers. These job advertisements may be for jobs in other cities and states. You should be willing to move to another location if you consider these jobs. A big advantage of these ads is that they are already aimed toward people with your skills and interests, because of the specialized, targeted nature of these publications.

Professional Associations

Many professions have special publications for people who work in that field. They are often a good source of information, and some list job openings. Local branches of national organizations sometimes list job openings, too. They are worth checking into. Also, joining a professional association opens up new contacts for your networking efforts, as well as keeping you current with developments and trends in your field of interest.

Mailing Résumés

Unsolicited résumés sent to no one in particular will most likely not result in any action at all. You might get lucky, but expect a 5 percent or lower response rate and even fewer interviews. It is almost always better to contact the employer in person. Once you have scheduled an interview, *then* send your résumé, addressing it (if possible) to the person with whom you will be interviewing.

Volunteering

If you lack experience or are not getting job offers, volunteer to work for free: offer your services for a short time to show an employer what you can do. Let the employer know that you understand that this temporary, however. We (the authors) personally know of people who have taken this route to enter the television, radio, and coaching professions. The employer has nothing to lose, and you have everything to gain.

Internships

An internship is a temporary position, usually while you are still in school, where you learn on the job. An employer agrees to allow you to work for them and to teach you about the organization at the same time. Sometimes an internship is a paid position and sometimes it is unpaid, but an internship experience always valuable. Internships often lead to jobs for a variety of reasons. But most importantly, if you do a good job, then you're a sure thing for an employer—you've removed all the guesswork about hiring you. Internships can lead to a permanent job either with the company you interned with or at a similar organization.

CONCLUSION

You will face stiff competition for jobs. Once you decide what you want and what your abilities are, do not hold back. Be self-confident. As soon as you learn about a job lead, follow through with quick action. The early applicant often gets the job. If you do not get the job immediately, call or go back a few days later. Let the employer know that you are really interested in the job.

The qualities of motivation and persistence, so important in finding a job, are also qualities that make you a good employee. Employers recognize this. By continuing to demonstrate motivation and persistence in pursuing a job lead, you increase the chance of such behavior being rewarded with a job offer.[24] Remember, your attitude determines your altitude.

Finding job leads may seem to take a long time. It does indeed take a lot of time to contact family and friends, search online job boards, and identify other leads. Once you find a good job lead, however, things can speed up very quickly. The next step is to apply for the job. You will need to complete job application forms, prepare a résumé, and get ready to interview.

SUMMARY

Selecting and finding a job is one of the most important decisions of your life. Working at a paid job enables you to establish your independence and define your own life. People also work for money, social contact and interaction with other people, positive feelings, freedom, personal development, meaning and purpose, self-expression, security, success, happiness, and to please peers and family.

The terms *work*, *occupation*, and *job* are often used interchangeably. However, *work* can be defined as activity directed toward a purpose or goal that produces something of value to oneself and/or to society. An *occupation* is a group of similar tasks that a person performs for pay. A *job* is a paid position at a specific place or setting. A *career* is a series of jobs pursued more or less in sequence to achieve the ultimate occupation desired and targeted by the individual.

Choosing a career is a serious step. The more time you spend on career planning, the easier it will be to find a career that suits you best. The kind of career you select will determine the types of people you meet, the friends you make, and the kind of lifestyle you can afford. Factors to consider when selecting a career include standard of living, personal contact, formal education required, practical experience required, locations of employment, whether you will like or dislike your work, working conditions, the amount of leisure time available, security during retirement, and health and happiness. Steps in choosing a career include considering your interests, abilities, and other characteristics; narrowing the field of jobs; studying the requirements of the job; planning for alternative occupations; planning any career preparation; being willing to pay the price of success; and getting work experience.

Once you have a career path, you can determine whether that career is right for you. In matching jobs to your personal characteristics, you must do a self-assessment; find your special abilities; and know your skills, including job-related, self-management (adaptive), and transferable skills. Most jobs fit into one or a combination of six occupational groups or categories: body workers, data detailers, persuaders, service workers, creative artists, and investigators.

Agriculture is the largest single industry in the United States. More than 20 percent of America's workforce is employed in some phase of agricultural industry. There are more than 8,000 job titles in agriculture. Agricultural careers are divided into four categories: food and biomaterials specialists; scientists, engineers, and related specialists; communication, education, and government specialists; and management and business specialists.

Students coming from agricultural education programs may take immediate job placement, postponed job placement, avocational (part-time) job placement, or supplementary (tributary) job placement. For further training after high school, consider colleges and universities, community colleges, vocational schools, apprenticeships, on-the-job training, and military service in the armed forces.

Once you have selected a career you want, you must find a job in that field. Sources of job leads include networking, direct employer contact, online job boards, volunteering and internships, mostly. As soon as you learn about a job lead, act on it quickly. Be motivated and persistent. Remember, your attitude determines your altitude.

Take It to the Net

The United States Department of Labor has an amazing website that can help you identify which of your interests, skills, and values are associated with certain career opportunities. Go to http://www.careeronestop.org/explorecareers/assessments/ and complete one of the online Interests, Skills, or Work Values assessments. Study the findings and see if they provide you with any insight on what you might want to do as a career. Or, if you knew what you wanted to do, did the assessment confirm those plans or amend them in any way?

Chapter Exercises

REVIEW QUESTIONS

1. Define the Terms to Know.

2. What are 11 reasons people work?

3. Explain the importance of career planning.

4. List 10 factors to consider when selecting a career.

5. List seven steps in choosing a career.

6. The career you select will be a major factor in influencing the kind of life you will have in what seven ways?

7. Name the six occupational groups.

8. Name the four categories of agricultural careers.

9. List the percentage of employment in each of the six agricultural categories, along with the corresponding surplus or shortage percentage.

10. Into what four categories may agricultural education students be placed?

11. What are six sources of advanced training after high school?

12. Name eight agricultural careers for which people postpone job placement to get advanced training.

13. What five questions should you ask yourself when planning for advanced training?

14. What are five examples of avocational (part-time) placement?

15. What are six examples of supplementary (tributary) job placement?

16. Name nine sources of job leads.

17. List 16 people or groups that could be a part of your job-hunting network.

18. If a contact in your network does not know of a job lead, what three follow-up questions should you ask?

19. What are four methods of establishing direct contact with employers?

20. What are three skill areas that are important for success on the job?

21. What are the key skills that employers are looking for?

22. What are 25 types of jobs that are suitable for young workers?

23. When the National Association of Colleges and Employees asked employers what skills and qualities they look for in new hires, what were the top 10?

24. What are three in-school resources in secondary schools available for job leads?

COMPLETION

1. _____ skills can be learned in classes in high school, postsecondary schools, apprenticeship programs, and other places.

2. _____ skills are often defined as personality or personal characteristics and help a person to get along in a new situation.

3. _____ skills are skills you can use in many different jobs.

4. One survey of employers found that _____ percent of the people they interviewed could not explain their own skills.

5. Agriculture provides nearly _____ person in every three with a job.

6. Agriculture accounts for _____ percent of the gross national product.

7. There are _____ people working in agribusiness for every farmer.

8. There are more than _____ job titles in agriculture.

9. Almost _____ percent of today's professional jobs in agriculture go unfilled because there are more jobs than people who understand agriculture.

10. The avocational (part-time) job placement concept is based on the economic reality that although money can be made in production agriculture, a _____ may not be possible for certain individuals in some full-time production enterprises.

11. Finding the right _____ is one of the most important activities you will ever undertake.

12. People who are active, happy, and engaged in their work tend to _____ better.

13. Many people spend their whole lives doing what others believe they should do rather than what they _____ _____ to do.

14. Basically, a career is something you _____ _____ to do and be for the rest of your life.

15. Your job should provide you with a _____ during retirement, and your income should be high enough while you are working that you can _____ money for retirement.

16. Other benefits should include _____, insurance, sick leave, and _____ insurance.

17. Many times, they key to finding the right job is _____.

18. If you enjoy what you do and you are good at it, your life will be satisfying and _____.

19. _____ is the largest single industry in the United States.

20. Many well-educated professionals do not understand the diversity of occupation in agriculture, thus they see agriculture as "cows, sows, and _____," or "seeds, weeds, and _____."

21. _____ percent of all jobs are never advertised.

22. You can learn a great deal from social media and various Internet sites about what is available in terms of jobs, but because companies are posting _____, _____, and skill requirements.

23. For getting a job, _____ is by far the best method.

MATCHING

_____ 1. Love to work and talk with people and enjoy convincing others to see things their way.

_____ 2. Find job satisfaction in helping others.

_____ 3. Use numbers or words in their work in very exact ways.

_____ 4. People who express themselves through music, dance, drama, or writing.

_____ 5. Work in scientific research or analysis as well as applied sciences.

_____ 6. Enjoy physical activity; work with machines, plants, and animals.

_____ 7. Cannot quit or drop out of training if you do not like it.

_____ 8. Learn skills by working with a more experienced worker.

_____ 9. Labor unions as well as schools offer these programs.

_____ 10. After successfully completing four years of study, you receive a bachelor's degree.

_____ 11. Offer two years of college-level courses.

_____ 12. Train students in a formal setting in the skills they need to get a job.

A. body workers

B. data detailers

C. persuaders

D. service workers

E. creative artists

F. investigators

G. apprenticeships

H. college degree

I. career-technical schools

J. armed services

K. community colleges

L. on-the-job training

ACTIVITIES

1. Besides money, select three reasons why you would work, and briefly explain or justify each.

2. Study the factors to consider when selecting a career. Then, write a 200- to 300-word essay entitled, "The Type of Job I Want." Include or address the career-selection factors as you write the essay. Be prepared to present your essay to the class.

3. Review the "Transferable Skills" in Figure 21–4. Select 10 skills that you feel you possess and could bring out during an interview as your strengths. Select 10 skills that you feel could use improvement. Write a short statement about each improvable skill and tell how you can improve it.

4. This chapter discussed six occupational categories. Select an occupational category and five specific jobs that might be of interest to you. Explain some general characteristics of that category that attract you to that career field.

   Note: These characteristics are explained in the Occupational Groups section of this chapter.

5. Write down 20 jobs that might be of interest to you. Match the 20 jobs you selected with the six agricultural categories in Figure 21–6.

6. Evaluate your present situation with respect to the type of career you are considering. Identify the type of advanced training you may need for that job and justify your selection.

7. Compile a list of people who could become a part of your job-search network. Make sure your list has at least 10 names.

8. Locate the open ad in Figure 21–10 and answer the following questions:

   a. What is the title of the position?

   b. What are the qualifications for the job?

   c. Who is the prospective employer?

   d. What is the phone number?

   e. Is the job full time or part time?

   f. How much money could you make weekly?

   g. Would you have to work on weekends?

9. Using any of the 11 sources of job leads, compile a list of five potential employers. These potential employers can be for part-time employment while in school or college, full-time employment after high school graduation, or full-time employment after further advanced training. Include the employer's name, address, phone number, type of work, working hours, and estimate of salary.

NOTES

1. B. Schwartz, *Why We Work* (New York: Simon & Schuster, 2015).
2. K. Nakao, "Top Five Reasons People Love Their Job and How You Can Love Yours Too," *Mashable*, November 24, 2013, retrieved September 19, 2016 from http://mashable.com.
3. Schwartz, *Why We Work.*
4. L. W. Schrank, *How to Choose the Right Career* (Lincolnwood, IL: VGM Career Horizons, 1991), p. 4.
5. L. J. Bailey, *The Job Ahead: A Job Search Worktext* (Albany, NY: Delmar Cengage Learning, 1992), p. 3.
6. D. Lennon, "What's the Difference Between a Job and a Career?" *JobMonkeyBlog*, August 8, 2016, retrieved September 19, 2016 from http://www.jobmonkey.com/.
7. S. Melone, "Factors to Consider in Choosing a Career," *eHow*, n. d. Retrieved September 19, 2016 from http://www.ehow.com/info_7918313_factors-consider-choosing-career.html.
8. "7 Steps to Take Before Choosing a Career," *Career Profiles*, retrieved September 19, 2016 from http://www.careerprofiles.info/choosing-a-career-steps.html.

9. United States Department of Labor. Self-assessments, *careeronestop*, retrieved September 19, 2016 from http://www.careeronestop.org/explorecareers/.
10. M. Farr, *100 Fastest-growing Careers*, 11th ed., iBook (Indianapolis, IN, Jist Publishing, 2010), pp. 1634.
11. Schrank, *How to Choose the Right Career*, pp. 11, 48–57.
12. D. Goecker, E. Smith, J. Fernandez, R. Ali, and R. Theller, "Employment Opportunities for College Graduates," *United States Department of Agriculture*, p. 1.
13. "National FFA Organization, Discovery Education, & AgCareers.com," retrieved September 19, 2016 from https://www.agexplorer.com.
14. D. Goecker, "Employment Opportunities for College Graduates," p. 2.
15. Ibid, p. 2.
16. M. Farr, *100 Fastest-growing Careers,* p. 1673.
17. Ibid., pp. 1675.
18. P. Bolton, "Why the Best Jobs are Never Advertised and How to Find Them," *Career Shifters*, n. d., retrieved September 19, 2016 from http://www.careershifters.org/expert-advice/why-the-best-jobs-are-never-advertised-and-how-to-find-them.
19. G. Kawasaki, "Ten Ways to Use LinkedIn to Find a Job," retrieved September 19, 2016 from www.guykawasaki.com.
20. Ibid.
21. E. Hartwig, "How to Effectively Use Twitter as a Job Search Resource," *Mashable,* Feb. 9, 2010, retrieved September 19, 2016 from www.mashable.com/2013/02/09/twitter-job-search/.
22. K. M. Dawson and S. N. Dawson, *Job Search: The Total System*, 3rd ed. (Houston, TX: Total Career Resources, 2008).
23. National Association of Colleges and Employers, *The Skills/Qualities Employers Want in New College Graduate Hires*, November 18, 2014, retrieved September 19, 2016 from https://www.naceweb.org/about-us/press/class-2015-skills-qualities-employers-want.aspx.
24. Bailey, *The Job Ahead*, p. 9.

22

GETTING THE JOB: RÉSUMÉS, APPLICATIONS, AND INTERVIEWS

By properly promoting yourself through your résumé, job application, and interview, you will increase your chances of getting a job. The qualities of motivation and persistence, which are so important in finding a job, are also qualities that make you a good employee, and employers recognize this.

Objectives

After completing this chapter, the student should be able to:

- Prepare a résumé
- Write a letter of application
- Complete a job application form
- Prepare for an interview
- Interview for a job
- Explain the steps to take after a job interview
- Discuss how to accept or reject a job

Terms to Know

- résumé
- interview
- personal management skills
- teamwork skills
- applicants
- references
- portfolios
- letter of application
- application form
- personnel office
- not applicable (N/A)
- interviewee
- interviewer
- hypothetical
- follow-up letter

Once you have selected, searched for, and located the type of job you want, it is time to get it. In all probability, you will face stiff competition for any job. Be positive and assertive. As soon as you learn about a job lead, follow through with quick action: Get the first available interview. If you do not get the job immediately, call or go back a few days later. Let the employer know that you are very interested in the job. The key is to be assertive but not so aggressive that your potential employer perceives you as obnoxious.

By demonstrating continuous motivation and persistence in pursuing a job lead, you increase your chances of getting a job offer. Besides motivation and persistence, you should have a good résumé, fill out applications appropriately, and make a positive impression during the interview. The purpose of this chapter is to develop your skills in these three areas to further enhance your chance of getting the job you want.

THE RÉSUMÉ

A well-prepared résumé is instrumental in getting a job. The résumé may not get you the job, but a bad résumé will keep you from proceeding to the next step, which is usually the job interview.

Purpose

It is estimated that between the ages of 18 and 48, a person living in the United States will change jobs 11.7 times.[1] Therefore, it makes sense to learn what is involved in obtaining the job you want and how to exit gracefully when it is time to change jobs.

First, you identify the personal characteristics, skills, education, and aptitudes that you bring to the workplace. You then decide what occupations interest you most, best match your skill set, and fit into your overall career plan. Once you have examined your capabilities, checked into employment opportunities, and set some goals for yourself, it is time to put all this information together. Getting a job consists of doing all these things and then following these efforts with a résumé, job application, and an interview.

A **résumé** is a one- or two-page description of a job applicant that lists his or her educational background, experiences, skills, and qualifications. It is the best tool for capturing and presenting all this information in a neat, concise format. In a single descriptive page, a résumé illustrates who you are, what you can do, and where you want to go. Remember that the employer uses résumés to screen job applicants (hoping to cut down the number quickly), whereas the prospective employee seeks to highlight why he or she should be considered. The process of constructing a résumé begins by compiling as much information as you can about yourself in an exceptionally concise manner. The student in Figure 22–1 is completing a résumé as a requirement for one of her classes. Her friend is helping her check for errors.

Tips for Using a Résumé

A good résumé will help you get an **interview**, a meeting between an employer and a job applicant, for the purpose of deciding whether the applicant will be hired. Use your résumé in the following ways:

- Make sure your résumé is being seen by
  - giving it to every relevant employer you meet;
  - uploading it to online job boards;
  - distributing it to your network of contacts;
  - giving to your references;
  - giving it to your instructors and mentors; and
  - uploading it to online networking sites such as LinkedIn.
- Make sure people, not places, receive your résumé.
- Follow directions and send your résumé to exactly who asked for it.
- Always write a cover letter to accompany your résumé.
- Never send out a mass mailing of your résumé.
- Include the job application form when you send a résumé and cover letter.
- Follow up after you send your résumé.[2]

Résumé Basics

Typically, a résumé has only seconds to catch the eye of a potential employer. (Remember, the employer is probably going through tons of paper, trying quickly to screen out most of the hundreds of applicants for this job.) The résumé must hold

FIGURE 22–1 If you have completed a résumé, you will be ready when you get a job interview. A résumé tells who you are, what you can do, and where you want to go. *(© iStock.com)*

the attention of the person who is scanning it and then sell the person it represents—you.

How can you ensure an employer will actually read your résumé? Each employer is different, but following some basic guidelines will give your résumé a better chance to market your unique talents and experiences that make you the right person to hire for a particular job.

The following guidelines will help you prepare a successful résumé:

Write It Yourself Look at examples of good résumés, but do not copy them exactly. There are lots of résumé templates online, but they are usually for people with more experience than a high school student. If you do not write your résumé personally, your résumé will not sound like you and may not stress what you want it to. You need to be comfortable—and intimately familiar—with your résumé.[3]

Make Every Word Count Limit your résumé length. If you're applying for an entry-level position as a new graduate, one page should do, but if you are more experienced with advanced degrees

you may need more space. Keep it short by being succinct, not by leaving out significant experiences that make you unique. After you have a first draft, edit it at least two more times. You want to present only the most relevant, important points, using short phrases or sentences that feature strong, active words.

Stress Your Accomplishments A résumé is no place to be humble or shy. Present your qualifications and talents honestly, in a positive manner.

Be Specific Give facts and numbers whenever possible. Instead of saying that you are a good stock clerk, say, "Devised method that reduced stocking time by 15 percent," or "Unload and reshelf an average of seventy-five 35-pound cases per hour." Instead of saying that you work well with other people, say, "I supervised and trained five people in shipping and receiving and increased overall productivity by 20 percent."

Keep It Lively Use action verbs and short sentences. Avoid negative words or phrases of any kind. Emphasize accomplishments and results, strengths and talents.

Make It Error-Free Ask at least one other person to proofread your résumé, looking for grammar and spelling errors. (Never depend on spell-check software to do this!) Check each word again before you print it; then print out a test copy and check it yet again before you send it to a prospective employer. Errors (even tiny ones) are an invitation to the employer to toss your résumé in the trash, because you have already demonstrated that you are sloppy and do not pay attention to details.

Make It Look Good Appearance, as you know, makes a lasting impression. Design your résumé so that it is pleasing to the eye and easy to read. Use a good-quality printer and have the résumé copied or printed on good-quality paper (for those instances where you submit a paper copy). Always remember that your résumé is advertising and selling *you*; make it reflect well on you.

Given these rules, you should be able to very quickly put together a simple, error-free résumé that will do what it needs to, which is get read by a prospective employer. Do not interrupt or delay your job search with endless polishing and improvement of your résumé; remember that it is a tool that does one part of the task and allows you to concentrate your efforts on the rest of the job search. The best approach is to draw up an acceptable version first, then actively look for a job. If you feel you need to, you can always work on improving your résumé on weekends when you are not busy making those all-important direct contacts.

WRITING THE RÉSUMÉ

There are many kinds of résumés and many ways to prepare them. This section gives you the basics on the major parts that should be included in any résumé.

Personal data should be at the top of the résumé: your name, address, home telephone number (and cell phone number, if you have one), and email address. Be sure this information is correct, because it is the only way the employer has to contact you.

The job objective states the type of job you want. This lets the employer see quickly whether it matches any job openings in the company. Unless you are applying for a particular, specific job, you may want to use a broad objective so that you will be considered for openings in related jobs. You

may slightly revise your basic résumé to match the objective with the job opening for which you are applying or the company to which you are applying.

Skills

Technical, teamwork, leadership, personal management, and employability skills should be listed in the "skills" section of your résumé. List the skills you already have that can be used in the job you are seeking. Be specific and brief. Remember to list all your skills; some of your skills may seem intangible but might be just what is needed to make your résumé stand out from all the other applicants.

Personal Management Skills Personal management skills are intrapersonal skills required to be a good employee, such as personality, attitude, honesty, enthusiasm, and an ability to meet deadlines. Perhaps you are the type of person who easily learns new skills and can identify and suggest new ways to do a job faster or more easily. If you work well without supervision, list that as a skill. Some people have the ability to organize; others are strong in imagination or initiative. Whatever your personal management skills are, you need to make a potential employer aware of them. Market yourself. Other examples of personal management skills or assets are

- Being dependable
- Being able to follow written and verbal instructions and directions
- Making suggestions
- Being organized
- Paying attention to details

Avoid using the word *I*; your résumé is all about you, and the employer/reader already knows that. Use action words. Words such as *experienced, recognized, advised,* and *operated* sound energetic and dynamic. Especially when combined with the facts-and-figures strategy, they convey a sense of expertise without arrogance. The good résumé uses action words throughout, such as[4]

- Accomplished
- Achieved
- Completed
- Constructed
- Consulted
- Coordinated
- Delivered
- Demonstrated

- Designed
- Developed
- Displayed
- Edited
- Formatted
- Identified
- Improved
- Increased
- Maintained
- Managed
- Negotiated
- Operated
- Planned
- Processed
- Reduced
- Revised

Expect to rewrite your résumé several times to get the results you want. Because the document must be short and concise, it is important to think through every part of it and make each word contribute to your self-marketing effort. Put a single idea into a simple phrase; then select the most appropriate words to convey your message.

Teamwork Skills Whether they state it specifically or not, employers are always looking for people with **teamwork skills**, the ability to work well with others in a group situation.

Even if you have never held a paying job, you can prove your teamwork skills to potential employers by showing them that you have actively participated in a group or team such as a sports team, yearbook or school newspaper staff, community group, or performing arts group. The cheerleaders in Figure 22–2 are planning and practicing last-minute details before they compete as a team. A record of participating in any group activity shows that you can follow a group's rules, listen to and communicate with other group members, and interact successfully with them to achieve a goal. Other teamwork skills include adaptability, being a team player, and being able to compromise.

Leadership Skills Include leadership skills if you possess them and if they are relevant to the job for which you are applying. Leadership is a quality that all employers look for in a potential employee. However, they also want someone who can either lead or follow, depending on what is needed, and who listens to other group members and superiors.

Education

List in reverse chronological order all the schools you have attended. This means you should begin with the last school attended and work backward

FIGURE 22–2 Being part of a competitive sports team or performance group is a great way to develop teamwork skills. These cheerleaders are learning to adapt, follow rules, and compromise. *(© iStock.com/Andrew Rich)*

in time. Write the full name of the school, indicating only the city and state in which the school is located; do not include the street address. List the years you attended the school, eliminating months or days of the month.

Programs/Courses List any program in which you were enrolled and any courses you took that specifically qualify you for the job. Be sure to include any school subjects that relate to your job objective and will help you perform effectively in the job. Include special skills you have developed and other educational experiences you have had, including cooperative education, on-the-job training, and internship programs.

Honors If you received any special honors or recognition, including academic honors, club memberships, and offices held, include that information.

Note: People who worked with you in these areas are prime candidates for your networking efforts and possibly for giving references.

Documents Include the type of document you received for completing each part of your education or training. The document could be a diploma, degree, certificate, or license. Remember to include the year in which you received the credential.

Work Experience

Work experience indicates that you have participated as a productive member of the labor force. Include all work experience, especially work directly related to your career objective. Include entrepreneurship and volunteer experience as well as paying jobs. List your work experience in reverse chronological order, with the most recent job first. Include the following information.

Place of Employment List the name of each company or firm, along with the city and state where it is located. Omit the street address.

Dates State the time periods you worked, from year to year. Do not include months or days of the month.

Job Title Include the job title for the job you held at each place of employment. Every job title is important, even if it is dishwasher or cashier. The job title shows that you were responsible for a particular position.

Description of Duties Write a *brief* description of your duties at each job. State the main tasks, using action verbs, but do not explain every detail. The students in Figure 22–3 are gaining valuable work experience with biotechnology crops.

FIGURE 22–3 Work experience is one of the most important elements in your résumé. These students will list the work experience they have gained in plant genetics on their résumés, along with the employer's name. *(Courtesy of University of Georgia—College of Agricultural & Environmental Sciences.)*

Achievements List any special accomplishments or honors you received at a particular job (for example, "Employee of the Month," "Best Trainee," attendance awards). Be sure to include the year you received the honor.

Personal

In this section of your résumé, list personal interests that will help you stand out from other **applicants** (those who apply for a job) with similar qualifications. By reading this section, a possible employer can see you as a total person. Include personal qualities, leisure-time pursuits, and other interests—but evaluate each item for its selling points and significance to the job you are trying to get. It is important to know your personal strengths and weaknesses, although you would never include weaknesses on a résumé. Be as objective as you can; an honest appraisal may even help you determine where your primary interests lie and thus save you time, energy, and frustration. Decide which interests and activities are pertinent to the job you are seeking or that indicate useful transferable skills and qualities. Here are some examples of personal characteristics:

- Adaptability
- Good manners
- Self-confidence
- Good temperament
- Spirit of cooperation
- Tact
- Assertive attitude
- Sensitivity
- Cheerfulness
- Commitment

References

The experts disagree on whether you should include **references** (persons listed on a résumé who may be contacted by a prospective employer to inquire about an applicant's qualifications) in your résumé. Do whatever you feel is the most appropriate for your region of the country, the employers to which you are applying, or the job for which you are applying. You may decide to list references on some résumés for a certain job and not list references on other résumés. Always get permission from individuals before listing them as sources of references.

Figure 22–4 is an example of a good résumé for a high school graduate. Although it takes up only one page, it provides a wealth of information for a potential employer. This person chose to list his references.

e-Résumés

Many major companies and recruiters are loading résumés into searchable databases so that applicants' skills can be matched with the specific job skills required for the job the company is trying

EDDIE MOSS
Route 3, Box 728
Watertown, TN 37185
(555) 123-4567

To demonstrate the skills and knowledge needed to work in the industrial welding field and in construction with an emphasis on safety.

SKILLS

SMAW welding	Use hand grinder	Rigging
MIG/TIG welding	Use bench grinder	Lay out materials
Oxyacetylene cutting and welding	Read blueprints	Drive skidder
Brazing	Estimate/select materials	Operate farm equipment
Operate drill press	Plumbing	Install electrical outlets
Operate metal lathe	Painting	Forming concrete
Operate gouger	Bricklaying	Build Web sites
Operate band saw	Roof/frame houses	

EDUCATION
Watertown Central High School, Watertown, TN: graduated 2016. Completed Advanced Industrial Welding program, Humphreys County Vocational Center, Watertown, TN (2013–2016). Member of VICA (2013–2014). Parliamentarian (2013–2014). Outstanding Welder (2013–2016). Completed Agriculture I, III, and IV. Member of Future Farmers of America (2013–2016). 1st Place Arc Welding, FFA District Competition, 2010. Member of National Technical Honor Society, 2015–2016.

WORK EXPERIENCE
Ironworker apprentice, HEC Steel Company, Cumberland City, TN (7/13 to present). Duties: tying rods, layout work, reading blueprints, rigging, cutting.

Service station attendant, Amoco Service Station, Watertown, TN (2015–2016). Duties: pump gas, change oil, change and repair tires.

Subcontractor, Wear-Ever Roofing, Watertown, TN (Summer 2016). Duties: roofing and framing.

Logger, James Norfleet Logging, Watertown, TN (Summer 2013). Duties: drive skidder and drag cables.

PERSONAL
Enjoy working ... dependable ... honest ... good physical stamina ... competent with tools of craft ... take pride in my work ... willing to work overtime ... follow directions ... willing to learn new skills
Hobbies include hunting; Member of Humphreys County Deer Hunters Club

REFERENCES

Oliver Dale,		Ray Hamphill,
Welding Instructor	Aubrey E. Booker	Athletic Director
Humphreys County		Watertown Central
Vocational Center	Wear-Ever Roofing	High School
1327 Highway 70 West	Highway 13 North	Highway 70 West
Watertown, TN 37185	Watertown, TN 37185	Watertown, TN 37185
(555) 123-7000	(555) 123-6000	(555) 123-9000

FIGURE 22–4 Sample résumé of a high school graduate, stressing skills, education, work experience, and personal interests and characteristics. This résumé includes references.

to fill. When you apply, your information may be entered into a database, or an electronic résumé may be translated into a format that is easily searchable and retrievable.

To enable retrieval of promising résumés, these systems usually scan for key words (particularly nouns) describing the applicant's skills and achievements. For an accounting manager position, for example, the computer may be programmed to search for words such as *supervisor*, *manager*, and *BS, Accounting*. For a salesperson, *exceeded quota* and *will travel* might be key words. Given the number of companies that are now using digital résumé screening and retrieval, it only makes sense to include these types of key words in your résumé. You must assume the company you are applying to conducts this type of analysis, so go ahead and add a summary of key words at the top of your résumé (under your name and address), labeled "Key Word Index."[5]

Portfolios

Many employers request portfolios. **Portfolios** are visual résumés used by people in creative or artistic fields to showcase their work. Physical portfolios are often oversized briefcases containing pictures, photographs, articles, illustrations, and other creative products. Portfolios are usually accompanied by some printed information. Certain fields, such as the landscaping industry, are more receptive to portfolios than others.

Electronic portfolios are becoming increasingly popular. You can create a website with your résumé, pictures, and even video clips; any potential employers can access your site whenever they wish. You can also save this information to a disk or flash drive for easy transportation to and from interviews. You may need to use special hardware and software to produce these materials, but the benefits of an electronic portfolio far outweigh the costs.

Variety of Résumés

Make your résumé work for you. Take a tried-and-true format that seems close to what you want, and add or delete categories as appropriate. For example, if you received many honors and awards in high school, create your own category entitled "Special Honors and Awards." If you participated in several leadership events, such as parliamentary

procedure or public speaking, create a section called "Leadership." If it shows your strengths to your advantage, or highlights the special value you could bring to an employer, by all means create a unique heading or section in your résumé.

Sample Résumés

The following résumés will assist you in presenting your résumé in an organized, attractive visual form that is eye-catching and easy to read. If you have little or no full-time experience, they may also give you ideas about how to include information about your summer jobs, part-time jobs, extracurricular activities, special training, and job interests. The following sample résumés describe

- A recent high school graduate (Figure 22–5)
- A person with considerable experience (Figure 22–6)
- A person reentering the job market (Figure 22–7)
- A job seeker who has yet to graduate from high school (Figure 22–8)

JOAN SMITH
123 Main Street, City, State, Zip
Home (555) 123-4567, Work (555) 123-9876

OBJECTIVE: Seeking full-time office position involving data entry and/or computer operations.

EDUCATION: Big City High School, Big City, MO
Certificate: Computer /data entry (June, 2016)
Related course work:
 —Introduction to computer literature
 —Business accounting
 —Records management
 —Computer operations, mainframe
 —Computer operations, personal computers
 —Office machines

CAPABILITIES: —Accurate and efficient worker
 —Work well without supervision
 —Pay attention to details
 —Computer experience: Word, Excel, HTML
 —Typing 65 WPM, alpha and numerical filing
 —Ten-key calculator operation

EXPERIENCE: Office worker, Ace Apple Barn, Small Town, RI (2013–2016)
 —Worked in records department.
 —Processed work orders.
 —Checked inventory.

Customer service assistant, Jane's Wireless, Summersville, CA (2013)
 —Answer calls for message center
 —Responded to recorded messages

FIGURE 22–5 Sample résumé of a recent high school graduate, including objective, education, capabilities, and experience.

BILL ROBERTS
507 Main Street, Some City, IL 12345
Home (555) 123-4567, Cell (555) 234-5678

OBJECTIVE: Seeking full-time position as Building Supervisor and/or Foreman

EDUCATION: University of Michigan, Bachelor of Science, Drafting Major.
Seminars Completed
"Managing Workers Today"
"Insight Seminars/Training for Construction"
"Advanced Constructors—Framing"

EXPERIENCE: General Builders, Detroit, MI
Duties: Framing, roofing, trimming,
—Overseeing subcontractors, plumbers, electricians, masons, painters
—Overseeing up to 30 laborers
—Arranged for purchase and delivery of all supplies and equipment
—Conducted training of many apprentice carpenters
—Residential and commercial experience

CAPABILITIES: —Attention to detail
—Excellent problem-solving ability
—Mechanically inclined

MILITARY: U.S. Army,
Second Lieutenant
Honorable Discharge

REFERENCES: Available on request.

FIGURE 22–6 Sample résumé of a person with considerable experience.

JUDITH G. SINCLAIR
445 Main Street, Berkeley, CA 12345
Home (555) 123-4567, Cell (555) 234-8901

OBJECTIVE: To obtain a position as managing editor of newspaper

EDUCATION: University of California, Los Angeles, CA (2004)
School of Journalism
Attended graduate course work (37 hours completed)
B.A. double major in Journalism, English

CAPABILITIES: —Well organized and responsible
—Excellent verbal and written communication skills
—Excellent leadership qualities
—Specialize in problem solving
—Detail-oriented and deadline-conscious

EXPERIENCE: Assistant managing editor, Los Angeles Times (2012–2016)
—Managed 10 reporters in city department
—Supervised layout of paper
—Established modern methods of conducting business

Reporter, Long Beach Daily (2004–2011)
—Managed the city desk
—Set up new procedures
—Wrote under own byline
—Researched leads on stories

REFERENCES: Available on request.

FIGURE 22–7 Sample résumé of a person reentering the job market.

GEORGE CUNNINGHAM
111 Main Street, Atlanta, GA 12345
Home (555) 123-4567

OBJECTIVE: Seeking a full-time position as stock clerk.

EDUCATION: Essex High School, Atlanta, GA
—Attended through eleventh grade
(Sept. 2014 through June 2016)
—Presently studying for GED

CAPABILITIES: —Follow written instructions and directions
—Follow verbal instructions and directions
—Attend school/work daily and on time
—Demonstrate self-control
—Pay attention to details
—Can stock 75 cases per hour

EXPERIENCE: Piggly Wiggly Supermarket, 123 Elm Avenue, Atlanta, GA
part-time stock clerk, Summer 2013

FIGURE 22–8 Sample résumé of a person with no high school diploma.

Use these examples to develop appropriate ideas and approaches in creating your own résumé.

LETTER OF APPLICATION

Another way to act on a job lead is to write a **letter of application**, a written communication used to apply for a job. You might do this at the suggestion of another person or in response to an ad online. A letter of application is often known as a **cover letter** when it is accompanied by a résumé. A combination cover letter and letter of application should have four parts or paragraphs (Figure 22–9):[6]

Reason for Writing In the first paragraph, you should explain your reason for writing. Name the job for which you are applying. Also, tell how you learned about the job. This introduction is strengthened when you can mention the name of a specific person who recommended that you apply.

Highlight Your Qualifications Use the second paragraph to briefly point out your strongest qualifications. Be factual and honest, while phrasing everything positively. Do not boast or exaggerate. Employers look carefully at this paragraph.

Call Attention to Your Fit The third paragraph invites the employer to review your résumé for how well you fit the job for which you are applying. Use this section to hammer home how well

6543 Maple Road
Cambden, OH 12345
April 6, 2020

Mr. Donald Young
Service Manager
Smith Auto Sales, Inc.
274 Main Street
Cambden, OH 12345

Dear Mr. Young:

I learned from one of your employees, Mr. Ken Jenkins, that you plan to hire a new mechanic in a few weeks. I would like to apply for the position.

I have the training and experience to do the job. For the last two years, I have worked at Goodman's Tire and Auto Center. I primarily do tune-ups, general engine repair, front wheel alignments, and wheel and brake work. I am satisfied with my present job. However, I would like to work for a new car dealership where I can better use my diagnostic and mechanical abilities. I hold a state inspection license and own my own tools.

I have enclosed a copy of my résumé that provides further details about my background. I could be available for employment following a two-week notice to my present employer.

May I have an appointment to discuss the job with you? I can be reached after 4:00 p.m. at (555) 123-4567. I would appreciate being considered for the job.

Sincerely,

Ronald Fisher

FIGURE 22–9 A combination cover letter and letter of application should have four parts or paragraphs: reason for writing, point out qualifications, call attention to résumé, and ask for appointment.

certain experiences (listed in the résumé) will help you accomplish specific job responsibilities for which they are looking.

Say Thanks and Ask for an Appointment In the last paragraph, ask for an appointment or interview. Tell how you can be contacted. Close the letter with a courteous comment and thank-you.

Be Brief and Specific Notice that the sample letter in Figure 22–9 is short and to the point. The purpose of this letter is to attract and hold the reader's interest. It should not attempt to give facts and details that are better stated in a résumé and/or job interview. The letter and résumé should let the employer know that you are qualified for the job and make the employer want to invite you for an interview.

Appearance

For both your résumé and your letter of application, appearance is critical. Any materials you send to a potential employer should have a neat, error-free, professional appearance. Use a white,

VIGNETTE 22-1 YOU MAKE THE CALL

IF YOU CONTINUE IN THE WINNING WAYS OF LEADERSHIP, you may find yourself in a managerial position where you have to "make the call" on new hires. You may have to skim through hundreds of résumés, choose the best ones for further review, and interview numerous individuals. How well do you think you could do? A tale of two people searching for the same accounting job follows. As you read about Mike and Coral, try to decide which person you would be more likely to hire.

Mike and Coral graduated from the same university with the same degree. Coral had a somewhat higher grade point average, but Mike participated in more groups and extracurricular activities during his time in school. The first thing they both had to do was to send a résumé to the accounting firm in which they were interested in working. Mike was very nervous and uncertain about writing the résumé, so he hired his sister to "help" him write it. In fact, she ended up writing most of the four-page résumé on her own. Mike took a quick look at it and said, "That's great." Coral, being a melancholy personality type, put her résumé together herself. She worked meticulously to make every word count as she presented her experience as strongly and succinctly as possible. She proofread the document several times before she decided it was finished.

Mike, the sanguine personality type, did produce his own cover letter, as his sister refused to write that for him. He drafted a two-page cover letter that did not really get to the point until the last two paragraphs. He read it again before he submitted it with his résumé. Coral was able to keep her cover letter to one page with four pithy paragraphs: one outlined her reason for writing; one paragraph pointed out her qualifications; one paragraph called attention to the attached résumé; the last paragraph specifically asked for an appointment.

About three weeks after the accounting firm received these cover letters and résumés, the company cut off acceptance of résumés. Mike's and Coral's résumés were two out of more than 100 the firm's personnel department had to wade through to decide who would even get an interview.

If it were up to you to make the call, who would get an interview and who would not? Do you think you could tell that Mike did not really write his own résumé? Do you think you could tell who believed in quality over quantity? What do you think would happen to each of the people in this scenario?

high-quality 8½-inch by 11-inch bond paper with matching envelope. Well-prepared documents are indicative of your ability to be prepared, your maturity and professionalism, the amount of effort you put into a project, and other traits an employer may value. [7]

You should not expect the first draft of your résumé and application letter to be perfect. Put in enough time to create an application that gives a positive impression; revise your letter and résumé until they say exactly what you need them to say—and make sure the mechanics of grammar and spelling are perfect. If your résumé doesn't look good, it will not get a second look by the employer. It has to look professional and polished before they will study it to see how great you are. [8]

Helpful Hints

Consider the following helpful hints for a letter of application:

Have a Good Intro The first 20 words are the most important; they should grab the reader's interest and attract attention to the rest of the letter. Don't bore the reader with the "My name is Fred Armistead, and I am applying for the sales position" comment. Make them interested by opening with a question or quote that represents your work philosophy. "Are you looking for a salesperson who has met their sales goals 100% of the time so far in their career?" This is a more effective way to open a cover letter.

Emphasize Your Value to Employer Tell your story in terms of the contribution you can make to the employer. In similar positions, what accomplishments have you had that are evidence of your value? List examples.

Be Definite about What You Want If you want an interview, ask for it. If you are interested in working for that organization, say so. Give clear reasons why the company should consider you.

Use Simple, Direct Language and Correct Grammar Employers don't want to read a cover letter, really; they want to get to the résumé. If you make the cover letter too long or complicated, or include grammatical errors, they will just throw your whole packet in the trash. Mistakes (even small typos) are simply not acceptable. Period. [9]

Don't Repeat Your Résumé What kind of message are you sending if your cover letter is just a paragraph version of your résumé? Repeating your résumé may be seen as a lack of experience, creativity, or even genuine interest in the position.

Send It to Someone by Name Get the name of the person who is most likely to supervise you. Call first to get an interview. Then send your letter and résumé. [10]

Get It Right Make sure you spell the recipient's name correctly and use the correct title. Accuracy may not attract any unusual attention, but any errors will stand out and instantly give a negative impression.

Reflect the Company's Culture A professional, informal style is usually best. However, do your research on the culture of the organization. For example, if you are applying for a position with an agricultural economist, your letter of application should have examples of your success that are quantifiable with numbers. If you are applying for marketing firm known for creativity and being social, then your cover letter should be more casual and informal.

Target Your Letter Typical reasons for sending a cover letter include (first and foremost) preparing an employer for an interview, responding to an ad, and following up after a phone call or interview. Each of these letters will be different. [11]

Follow Up Remember that contacting an employer directly is much more effective than sending a letter or an unsolicited résumé. Do not expect a letter to get you many interviews. They are best used to follow up after you have already contacted the employer.

COMPLETING A JOB APPLICATION FORM

The most common procedure for screening individuals for available jobs is to ask applicants (potential employees) to fill out a job **application form**, a general form that applicants fill out when seeking employment. Some application forms are in paper format, but it is more than likely to be something you fill out online. Any time an employer puts out a call for applications, there could be hundreds of applicants. The information provided on the form helps employers sort out the persons who are best qualified for the job. Often, the application forms are viewed before cover letters and résumés to shorten the list of applicants. After select cover letters and résumés are reviewed, a smaller number of people are invited to interview.

Reasons Job Application Forms Are Eliminated

Sometimes, even well-qualified job seekers do not get past the first cut. Again, the employer's first strategy is to eliminate as many applicants as possible, quickly and cost-effectively. In a busy **personnel office** (the department in a company that handles all staffing issues), your application could be eliminated because it is messy, because you did not list enough experience, because you don't have the right degree, because you didn't follow directions, or because your last job paid more than the one for which you are applying. Applicants are rejected for many reasons; you may be able to do the job for which you applied, but you may never get a chance to prove that.

There is no standard form for a job application. Most employers design their own forms so that they get the information they find most helpful in filling their job openings. By having the same information on each application, the employer can quickly and easily compare applicants' qualifications.

Tips for Filling Out a Job Application Form

Application forms differ from company to company (Figure 22–10). Although these forms may appear simple and straightforward, it is easy to make a mistake. If you keep a copy of your résumé or a personal data sheet with you, you will be less likely to forget important information or make errors. The following tips are helpful for filling out a job application form:[12]

- Read the entire form carefully, before you even begin to write or type in answers.
- Read, study, and understand the instructions so you will know what information to provide, and where. If you do not observe the instructions, you will give the impression that you cannot follow directions.

- If you are filling out the form by hand, it is best to print rather than using cursive or script. Be sure to sign your name wherever the form indicates that your signature is required. Use your correct legal name, not a nickname.
- A certain amount of space is allowed for each answer. Do not try to cram too much information into a space.
- Make sure your answers are in the right place.
- Neatness is a must. If you erase or cross through answers, the employer may form the perception or impression that you are a careless or sloppy person. If you can obtain two application forms, you can use one as a draft or practice copy, and then transfer your answers—neatly!—onto the second form. It is a good idea to make two or more drafts so that your final copy is perfect.
- Answer all questions. If an item does not apply to you, insert "N/A," which means **not applicable**, or draw a short line through the blank. If you simply leave a blank, the employer might think you ignored or forgot to answer that question.
- Be honest. Never give false information. You should present yourself in a positive light, but do not lie or exaggerate.
- Name the specific job for which you are applying. Although you may be willing to accept any job, never put "Anything" in this blank. Indicate that you are interested in and qualified for a specific job.
- After you have completed the draft form, check it carefully for mistakes, omissions, and possible additions. Misspelled words and crossed-out responses give a poor impression of your ability and your attention to detail. Use a dictionary to check your spelling if you are not absolutely certain. Transfer your answers carefully, accurately, and neatly to the final form.

A job application form that is neatly and accurately completed may be the determining factor in getting an interview. Just by looking at the job application form, the employer gains an impression of the applicant: Neat? Messy? Thorough in completing the assignment? Does not follow instructions? Careless? Thoughtful? As with the résumé and cover letter, the impression you make through your job application form will affect your job search.

APPLICATION FOR EMPLOYMENT

PERSONAL INFORMATION

DATE OF APPLICATION: _Jan. 15, 2020_

Name: _Fisher_ _Ronald_ _R._
     Last   First   Middle

Address: _6428 Valley Rd_ _Camden, OH_ _67423_
        Street   (Apt)   City/State   Zip

Alternate Address: _NA_ _NA_ _NA_
        Street   City/State   Zip

Contact Information: _(770)222-4539_ _(770) 351-4432_ _rrfisher@yahoo.com_
                    Home Telephone   Mobile Telephone   Email

How did you learn about our company?

POSITION SOUGHT: _Mechanical Apprentice_   Available Start Date: _Jan. 15, 2020_

Desired Pay Range: _$10.00/Hr_   Are you currently employed? _Yes_
                  Hourly or Salary

EDUCATION

	Name and Location	Graduate? – Degree?	Major / Subjects of Study
High School	Camden High School Camden, OH	Yes / H.S. Diploma	Agricultural Mechanics
College or University	NA	NA	NA
Specialized Training, Trade School, etc...	NATEF Certified	Yes	Electrical and Electronic Systems
Other Education	NA	NA	NA

Please list your areas of highest proficiency, special skills or other items that may contribute to your abilities in performing the above mentioned position.

NATEF Certified in Electrical and Electronic Systems; Experience in Engine, Transmission, and Brake repair; Member of State Winning FFA Ag Mechanics Team, High Individual in Small Engine Repair

FIGURE 22–10 Sample of a correctly completed job application form.

PREVIOUS EXPERIENCE

Please list beginning from most recent!

Dates Employed	Company Name	Location	Role/Title
1/16/16 – present	Goodman's Tire	Camden, OH	Technician

Job notes, tasks performed and reason for leaving:

Change oil, Change tires, Greet Customers, Clean shop
I am still employed at Goodman's.

Dates Employed	Company Name	Location	Role/Title
8/2/15 – 12/20/15	Hunter's Auto Repair	Camden, OH	Apprentice

Job notes, tasks performed and reason for leaving:

Clean shop, Assist technicians with various tasks (change tires,
change oil, remove transmissions)
I left Hunter's because I wanted to perform more key functions in mechanics.

Dates Employed	Company Name	Location	Role/Title
8-11-11 to 8/1/15	Fisher Farms	Camden, OH	Mechanic

Job notes, tasks performed and reason for leaving:

Service garden and farm equipment (tiller, tractors, implements);
Assist with hay harvest (mowing, raking, moving round bales);
I left full time work with the family business to earn more money.

Dates Employed	Company Name	Location	Role/Title
NA	NA	NA	NA

Job notes, tasks performed and reason for leaving:

NA

FIGURE 22–10 (continued)

Personal Data Sheet

A personal data sheet (Figure 22–11) is useful when you must complete a job application form. Take it with you and refer to it when filling out an application so that you give complete and accurate information (e.g., names of previous employers, employment dates, addresses, contact information, and references). This data sheet is also helpful in developing your résumé.[13]

Sample Application Forms

To understand how to complete applications, study the two applications in Figure 22–12 and Figure 22–13. Figure 22–12 is an application for hourly employment. Figure 22–13 is a typical form used in hiring carpenters, custodians, cooks, drivers, nurses, secretaries, shipping clerks, and various other civil service classifications.

PERSONAL DATA SHEET
IDENTIFICATION

Name _____ Soc. Sec # _____
Address _____ Zip _____
Birth Date _____ Birth Place _____
Telephone (_____) _____ Height _____ Weight _____
Hobbies/Interests _____
Honors/Awards/Offices _____
Sports/Activities _____
Other _____

EMPLOYMENT HISTORY
(Start with present or most recent employer.)

1. Company _____ Telephone (___) _____
Address _____
Employed from Mo. _____ Yr. _____ / to Mo./Yr. _____
Supervisor _____
Position/Title _____
Last Wage _____ Reason for Leaving _____

2. Company _____ Telephone (___) _____
Address _____
Employed from Mo. _____ Yr. _____ / to Mo./Yr. _____
Supervisor _____
Position/Title _____
Last Wage _____ Reason for Leaving _____

3. Company _____ Telephone (___) _____
Address _____
Employed from Mo. _____ Yr. _____ / to Mo./Yr. _____
Supervisor _____
Position/Title _____
Last Wage _____ Reason for Leaving _____

EDUCATIONAL BACKGROUND

School Name and Address
Dates Attended: From: _____ To: _____

Elementary:

Junior High:

High School:

Courses of Study _____ Rank _____ GPA _____
Favorite Subject(s) _____

REFERENCES

1. Name _____ Title _____
Address _____ Zip _____
Relationship _____ Telephone (__) _____

2. Name _____ Title _____
Address _____ Zip _____
Relationship _____ Telephone (__) _____

3. Name _____ Title _____
Address _____ Zip _____
Relationship _____ Telephone (__) _____

FIGURE 22–11 Sample personal data sheet.

APPLICATION FOR HOURLY EMPLOYMENT

Name _____ S.S.# _____
(First)   (Middle)   (Last)
Address _____ Phone Number _____
(Street) (City) (State) (Zip)   (Area Code)   (Number)
Are you at least 18 years old? ____Yes____No   If No, Birthdate _____
Do you have the legal right to remain and work in the United States? ____Yes____No
Have you ever been employed by this company? ___ Yes___No   If Yes, explain _____
What prompted you to apply for work here?
___Company___Image___Agency___Friend___Relative___Newspaper___Other_____

PERSON TO BE NOTIFIED IN CASE OF EMERGENCY
Name _____ Home _____
Address _____ Work Phone _____
Have you ever been convicted of a felony? ___ Yes___No   If Yes, explain _____
(Conviction will not necessarily void employment?) _____
Personal Interest _____

Job related organizations, clubs, professional societies _____
(Omit those which indicate sex, race, religion, creed, color, national origin, ancestry, and or age)
Do you have any medical problems that would interfere with or detract from your ability to perform
your job?_____ Yes ____ No If yes explain _____
Is any member of your family (spouse, parent, etc.) employed in this industry?
____ Yes ____ No If yes explain _____

	Name and location of school	Dates attended From To	Circle Highest Year Completed	Major & Minor Fields of study	Degree(s) or Diploma
High School			9 10 11 12		
Technical/ Vocational School					
College/ University			1 2 3 4		
Others					
Honors Received					

		SUNDAY	MONDAY	TUESDAY	WEDNESDAY	THURSDAY	FRIDAY	SATURDAY
WHAT HOURS ARE YOU AVAILABLE FOR WORK?	FROM TO							

Type of Schedule Desired _____ Part Time _____ Full Time
Do you plan to work elsewhere or attend school while employed? _____
Do you have any obligations which would affect working as scheduled? _____
How soon after accepting an offer would you be able to start working? _____

Employer		Phone	From	To
Address	City, State, Zip		Position	
Responsibilities			Supervisor's Name	
			Starting Salary / Wages	
Reason for leaving		May we contact?	Final Salary / Wages	

Employer		Phone	From	To
Address	City, State, Zip		Position	
Responsibilities			Supervisor's Name	
			Starting Salary / Wages	
Reason for leaving		May we contact?	Final Salary / Wages	

Employer		Phone	From	To
Address	City, State, Zip		Position	
Responsibilities			Supervisor's Name	
			Starting Salary / Wages	
Reason for leaving		May we contact?	Final Salary / Wages	

U.S. Military Service
Veteran of U.S. Military Service? ___ Yes ___ No If yes, what branch? _____
Dates of Service? _____ Rating or rank achieved? _____
Duties and Responsibilities? _____

I represent that the information in this application is correct to the best of my knowledge and understand that any misstatement or omission of information may be grounds for dismissal. I authorize the references listed above to give you any and all information that they may have. I hereby release all parties from all liability from any statements or information provided.

In consideration of my employment, I agree to conform to all policies of this company, and that my employment can be terminated "at will," with or without cause, and with or without notice, at any time, at the option of either the Company or me. I understand that no manager, supervisor or any representative other than the President has authority to enter into any agreement for employment. I further understand and specifically acknowledge that any agreement for employment, other than "at will," must be in writing and signed by the President and me.

This application was designed to comply with Federal Civil Rights Act, Title VII, The Age Discrimination Act of 1967, and State Fair Employment Practice Laws. The company therefore gives assurance that no question answered is or will be used to discriminate adversely in matters of race, creed, color, national origin, ethnic background, age, or sex.

Signature _____ Date _____
685281 9/88 REV.

FIGURE 22–12 Sample application for hourly employment.

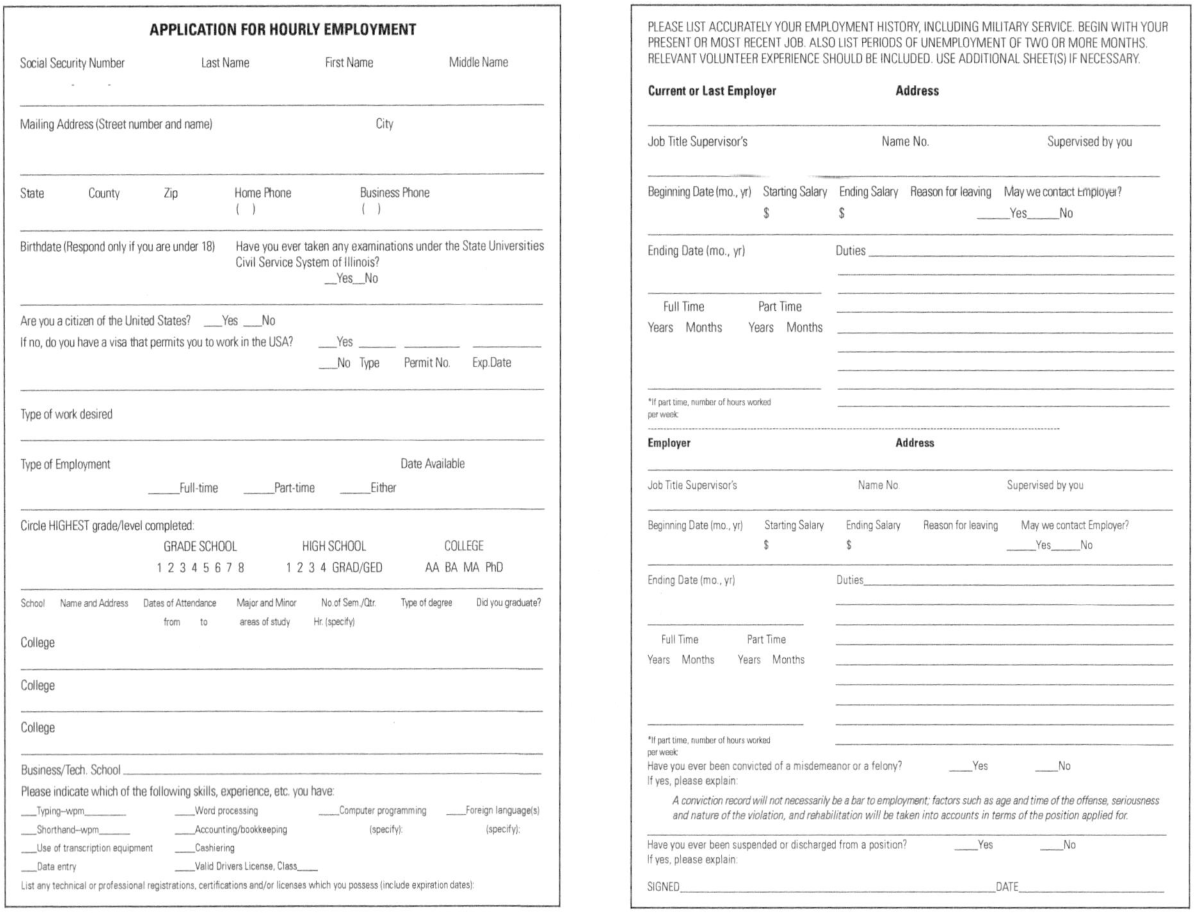

APPLICATION FOR HOURLY EMPLOYMENT

Social Security Number — Last Name — First Name — Middle Name

Mailing Address (Street number and name) — City

State — County — Zip — Home Phone ( ) — Business Phone ( )

Birthdate (Respond only if you are under 18) — Have you ever taken any examinations under the State Universities Civil Service System of Illinois? __Yes__No

Are you a citizen of the United States? ___Yes ___No

If no, do you have a visa that permits you to work in the USA? ___Yes ___No Type — Permit No. — Exp.Date

Type of work desired

Type of Employment — Date Available

_____Full-time _____Part-time _____Either

Circle HIGHEST grade/level completed:

GRADE SCHOOL 1 2 3 4 5 6 7 8 — HIGH SCHOOL 1 2 3 4 GRAD/GED — COLLEGE AA BA MA PhD

School	Name and Address	Dates of Attendance from to	Major and Minor areas of study	No.of Sem./Qtr. Hr. (specify)	Type of degree	Did you graduate?
College						
College						
College						

Business/Tech. School ____________

Please indicate which of the following skills, experience, etc. you have:

___Typing–wpm________ ____Word processing ____Computer programming (specify): ____Foreign language(s) (specify):

___Shorthand–wpm______ ____Accounting/bookkeeping

___Use of transcription equipment ____Cashiering

___Data entry ____Valid Drivers License, Class____

List any technical or professional registrations, certifications and/or licenses which you possess (include expiration dates):

PLEASE LIST ACCURATELY YOUR EMPLOYMENT HISTORY, INCLUDING MILITARY SERVICE. BEGIN WITH YOUR PRESENT OR MOST RECENT JOB. ALSO LIST PERIODS OF UNEMPLOYMENT OF TWO OR MORE MONTHS. RELEVANT VOLUNTEER EXPERIENCE SHOULD BE INCLUDED. USE ADDITIONAL SHEET(S) IF NECESSARY.

Current or Last Employer — **Address**

Job Title Supervisor's — Name No. — Supervised by you

Beginning Date (mo., yr) — Starting Salary $ — Ending Salary $ — Reason for leaving — May we contact Employer? _____Yes_____No

Ending Date (mo., yr) — Duties ____________

Full Time Years Months — Part Time Years Months

*If part time, number of hours worked per week:

Employer — **Address**

Job Title Supervisor's — Name No. — Supervised by you

Beginning Date (mo., yr) — Starting Salary $ — Ending Salary $ — Reason for leaving — May we contact Employer? _____Yes_____No

Ending Date (mo., yr) — Duties ____________

Full Time Years Months — Part Time Years Months

*If part time, number of hours worked per week:

Have you ever been convicted of a misdemeanor or a felony? ____Yes ____No

If yes, please explain:

A conviction record will not necessarily be a bar to employment; factors such as age and time of the offense, seriousness and nature of the violation, and rehabilitation will be taken into accounts in terms of the position applied for.

Have you ever been suspended or discharged from a position? ____Yes ____No

If yes, please explain:

SIGNED____________DATE____________

FIGURE 22–13 Typical application form used in hiring for civil service jobs.

PREPARING FOR AN INTERVIEW

In the job search process, an interview is a form of communication between someone looking for a job (interviewee) and the person or people /employees (interviewers) who may hire, supervise, or simply work alongside the interviewee. By being granted an interview, you have already achieved some degree of success in the job search. Someone in the company liked what he or she saw on your résumé or application and decided to find out more about you. In the interview, you present yourself as a person, as well as verifying and expanding on the information contained in your résumé or job application. This is your chance—possibly your last chance—to sell yourself to this employer. You have to be at least as good in person as you appeared to be on paper.

Purpose of the Interview

While the prospective employer is forming his or her perceptions of you, the interview also provides you with a chance to do some further investigation of the specific job and the company. You need to know whether the job, your potential coworkers, the company's policies, and the work environment meet your expectations. This information allows you to determine whether you really want to work for the organization. Remember, your needs and purposes are as valid as the interviewer's.

Neither the person being interviewed (the **interviewee**) nor the **interviewer** (the person conducting an interview) knows the other, and both parties are hoping for a positive experience. During the interview, a determination will be made about whether each of you wants the relationship to continue. If the interviewer is interested in hiring you, they will usually start trying to sell you on the job and/or initiate discussion on next steps. If the interviewer is not interested, expect small talk on everything from sports to the weather.[14]

Purpose of Practicing for Interviews

You will hardly ever have an opportunity to meet your interviewer before the interview. Considering that interviewers look for the best and brightest candidate, you need to at least meet, and preferably exceed, their expectations of whom they want to do the job. Because the interview is such an important step in getting the job, you should prepare carefully for it. The more you prepare, the better your chances of making a good impression on the employer.

Before you start practicing for the interview, ask yourself how you can best accomplish this task. Your résumé and application present a professional image that must be replicated in the interview. You need to know the company, understand how you can be a great asset to the company, and be ready to explain why you would be the best choice.[15] An excellent tool to study good interviewing skills is

YouTube, where you can find good examples of successful interviews that will help you practice.

Practice Interview Skills

An interview for an entry-level job lasts about 15 to 30 minutes. A little nervousness about a job interview is perfectly normal. To reduce stress and help build your confidence, you may want to role-play some practice interviews. As noted, something as important as a job interview deserves advance preparation. You would not take a driver's license exam without practicing your driving skills, would you? Similarly, you would not try to deliver a speech without research and practice. In fact, many of the skills you developed for public speaking are applicable and transferable to job interviews. Use what you've learned about communicating effectively and employ those skills for your interview.

You may be able to set up an interview situation in a classroom. Arrange a desk and a couple of chairs as you might find them in an office. The instructor or a fellow student can play the role of interviewer. Take turns playing the interviewee being interviewed for a **hypothetical** (a made-up situation used to convey a point or illuminate an issue) job, as shown in Figure 22–14. Try to make

FIGURE 22–14 Practicing your interviewing skills will help you respond more easily. These students are practicing the roles of interviewer and interviewee. *(© iStock.com/James Tutor)*

the interview as realistic as possible. If you have access to the equipment, videotape the interviews. It can be very instructive to see yourself participating in an interview.

By practicing the interview, you will become more aware of what is involved in thinking about a question and answering it out loud. It can be a valuable learning experience to discover, for example, how much you stumble and hesitate. Do not try to memorize answers, but do practice until you can respond easily. Specifically, practice responding to interview questions in a way that will help the interviewer see how you can help them or the company be successful.

Practice Interview Questions

Think of questions the employer might ask. There are some questions that are asked at almost every interview. Study these questions before going to an interview, prepare some material relating to them, and be sure you can answer any such questions clearly and positively. You can make a good impression by expressing yourself well.

Practice answering these questions so that you will feel more confident in the actual interview. Try preparing and then rehearsing responses to the following commonly asked interview questions:[16]

What Can You Tell Me about Yourself? Emphasize aspects of yourself that are particularly relevant to this job or that would be assets to the prospective employer.

Why Did You Leave Your Last Position? Concentrate on your desire for new challenges. Be relentlessly positive; never make negative comments about a former employer or your past performance.

What Are Your Strengths? This is the time to present your strong qualities. Do not hesitate to share these with potential employers, but do so in a matter-of-fact manner that is not boastful. Try to relate these strengths to the specific job for which you are interviewing.

What Are Your Weaknesses? To answer this question, try framing strengths as if they were weaknesses—"I tend to work too hard" or "I tend to be a perfectionist"—instead of volunteering that your work is sloppy or you often show up late.

What Were Your Major Accomplishments in Each of Your Jobs? Prepare an answer to this question in advance, highlighting achievements that relate to the job you are interviewing for. Refer to your personal data sheet if you need to (although it is better to have these important accomplishments memorized).

Why Are There Gaps in Your Work History? Tell the truth here, casting it in the best possible light, but make sure employers understand that a gap is not a red flag with regard to your capabilities: "I needed some time to reevaluate my career direction and priorities," or "I wanted to look around carefully before taking a new job in order to find the best possible match."

Why Should I Hire You? With this question, the interviewer essentially invites you to repeat any job requirements that were discussed earlier, and to make the case for how well your skills, qualifications, and experience fit. Reflect on your skills and accomplishments; match them with your understanding of the job responsibilities as well as company goals and values: "I am a highly motivated self-starter who has always taken the initiative to develop new ideas and programs." For example, if you have described yourself as a very aggressive person who really wants to take the market by storm, you could answer this question with: "Given my history of accomplishments and working style, I think we would make an excellent team."

How Do You Handle Stress and Pressure? Don't say, "I never get stressed." Everyone gets stressed. Potential employers want to know how you respond when tough times come—and they *will* come. The best approach is to give examples of real situations. Just don't give examples of stressful times that you brought on yourself, such as procrastinating or being overworked because you couldn't trust others to help you with a project.

Do You Prefer to Work Independently or On a Team? Try to portray yourself as a team player, as much as possible. Your leadership and human relations skills should be of great value in answering this type of question.

Describe Your Work Personality Be positive and honest, of course. Once again, highlight your specific strengths and abilities that relate directly to the job you want.

Which Job Did You Like the Least? Why? Do not dwell on money here; few jobs pay as well as we would like. Instead, try to pick an experience that will show in a roundabout way what you do want: "I liked working at XYZ Corporation the least because I was unable to utilize my strengths in Informational Technology, and there seemed to be no room for advancement."

Which Job Did You Like the Most? Why? By hearing what you valued in the job that you liked the most, the interviewer can assess your potential "fit" with the potential employer.

If You Could Do Things Differently, What Would You Change? Emphasize something you did well, and discuss how it could have been even better.

Have You Ever Been Fired? Why? Remember to keep your answer short and to the point, and to frame your answer as positively as possible while still being honest. Do not try to whitewash the truth; employers will see through this. If you made mistakes, take responsibility for them, but demonstrate that you learned something from these mistakes, so that you will not repeat them.

What Do You Want to Be Doing Five Years from Now? Make sure your answer reflects both some desire for career growth and an intention to remain with the company. Employers do not want people who will stagnate or decide to leave after a year.

Learn about the Company or the Job Find out as much as possible about the company and the job *before* your interview. Most companies have a website where you can learn an enormous amount about the organization and discover what further information is available from the company itself. The company website should have information on its mission statement, history, and products and/or services. LinkedIn company profiles, social media sites, and Google and Google News are also excellent tools that can assist with research on a potential employer. Researching your potential employer with diverse digital resources may also help you get an idea about the culture of the organization.[17]

If you know people who might have information about the company, ask them about it. Personal contacts can give you inside information. For instance, you might find out about such things as the working conditions or the turnover rate of personnel. If the potential employer's business place is a public facility, such as a restaurant or retail store, you can gain firsthand information. Visit the establishment to get a feel for the atmosphere. You can observe the type of work, see how it is being done, and perhaps have the opportunity to ask employees a few questions.

Do not forget the library. Information is available in several directories that describe corporations by name. A librarian can help you find such references. Gather information on products or services produced, years in operation, growth rate, standing in the industry, and so on. If you cannot find much information about a specific company, find out something about the company's type of industry. For example, if you are going to interview for a job with a property management firm, you would want to find out at least what services such firms typically provide.

When you finish gathering information about the company, write up a list of further questions that you would like to ask about the job or the company. For example, you might ask the following questions:

- Why did this job become vacant?
- Will any more training be required?
- What are the normal working hours?
- Who will my supervisor be if I get the job?

It is generally best to avoid asking about salary or benefits. If it is not public information, you can inquire about your compensation package when the job offer is extended to you.

Collect Materials Needed for the Interview

Well before you leave for the interview, collect the materials you will need. You will need a copy of your résumé to leave with the employer. (You can also use your résumé as a reference if you are asked to complete a job application form.) You will also need a copy of your references, samples of projects, background information, and documentation of your accomplishments.[18] Be sure you have a pen and pencil for the job application form, a personal data sheet, and a list of questions you want to ask. It may be best to carry your papers in a folder or briefcase so that they are neatly arranged and quickly accessible.

Appearance Matters!

The first impression an employer forms of you is created by your appearance. Attire can depend on each company's situation or work culture, but grooming and dress inevitably influence the interviewer's final decision. A very carefully chosen interviewing "uniform," worn by a well-groomed job applicant, is very important. In general, be conservative, clean, well-groomed (neatly cut hair, brushed teeth, clean hands with trimmed nails), and be appropriately dressed in clean, well-maintained clothing that fits you well.[19]

Check Last-Minute Details

Input the date, time, and place of the interview on your phone, but also write this information on a card as a backup. Check and then double-check the information. Drive to the address of where your interview will take place a few days ahead of time so you know exactly where the company is located, how long it takes to get there, where you can park, and which door you should use. If more than a week goes by between the time you made the appointment and the actual interview, email or call to confirm the interview.

Clear your schedule so that you do not feel rushed or nervous about making it to your next activity. Plan to arrive at the interviewer's office 5 to 10 minutes ahead of schedule. Introduce yourself, state why you are there, and give the name of person with whom you are scheduled to meet. There is no need to bring a friend or family member along. You need to be perceived as independent.[20] If you had a parent or friend drive you to the interview, ask that person to wait outside or to come back to pick you up when your interview is over.

You may have to wait a short time in a designated area. During that time, you should relax, read, or look over your list of questions and any research you did about the company and interviewer. Make a last-minute restroom break and check your appearance in a mirror one last time. Be pleasant to others in the reception area, especially the receptionist, secretary, or administrative assistant. Interviewers often ask for the receptionist's opinion of job applicants. Lastly, turn your phone off and insert one last breath mint.[21]

INTERVIEWING FOR THE JOB

Interviews are the gateways to jobs. Almost no one (with the possible exception of the boss's children) can get a job offer without undergoing an interview. Even when your paper credentials are stellar, and your references give you great recommendations and glowing support, the person who is doing the hiring will want to meet you face to face and talk with you in person before he or she decides whether to offer you a job. "People hire people, not résumés."[22]

Link with the Interviewer

First impressions are everything when you meet the interviewer. When you meet your interviewer, be friendly, give a firm handshake, and know that how you sell yourself in the first minute will move you to the top or bottom of the list of potential hires. Say simply, "Hi, I'm Jane Doe. It's nice to meet you." You should also use the interviewer's name at the first opportunity, and pronounce it correctly. Wait for the interviewer to offer you a seat, but if that doesn't happen by the time they start the conversation, choose a chair across from the interviewer and sit. Whether the interviewer begins with casual talk or gets straight down to business, maintain eye contact, and stay alert and interested.[23]

There are many things you could do during the first few minutes of an interview. Following are four suggestions from experienced interviewers:

Allow Things to Happen Relax. Do not feel you have to start a serious interview immediately—but remember that your "performance" begins when you walk in the prospective employer's door.

Smile Look happy to be there and to meet the interviewer.

Use the Interviewer's Name Be formal. Use "Mr. Smith" or "Ms. Sharpe" unless you are asked to use another name.

Ask Some Opening Questions The interviewer will probably start the conversation off, but if not, be sure to have some prepared questions with which you could lead off the interview. For example, you could mention a certain aspect of the job announcement that excited you about this job and ask for further clarification about that aspect. You could also let the interviewer know you have researched their mission statement and ask how that is integrated into the advertised roles and responsibilities of the position.

Body Language

As in any interpersonal interchange, how you say something is often as meaningful as what you say. The importance of body language (nonverbal communication) cannot be overemphasized. An astute interviewer gains many clues about your personality from what you do with your body and how you move; these nonverbal messages and signals reveal how you feel about yourself and others. Everything you learned in Chapters 6 through 9 about communication and public speaking holds true in the interview situation, which resembles extemporaneous speaking in many ways. Following are five ways to project a positive image:

Posture The way you stand and sit can say volumes about your confidence, preparedness, and comfort level. If you lean back or slump in a chair, you may look too relaxed or even careless. A poised body that leans slightly forward conveys a relaxed but highly interested attitude to the interviewer. Gestures should be natural and never overdone; do not flap your hands and arms around. Maintain eye contact with the interviewer, demonstrating sincere interest and engagement, and keep your breathing regular. Your entire posture should project a high level of confidence, as shown in Figure 22–15.

Voice Even if you are nervous, try to speak in a normal, pleasant voice, neither too soft nor too loud. Keep a tone of optimism, enthusiasm, and confidence in your voice. Applicants who feel at ease during an interview usually answer questions without much hesitation, but also do not rush into their responses. Make sure you listen actively to the interviewer's questions and pause briefly to think about your answers if you need to. Speak at a normal rate; resist the nervous urge to talk fast and say too much.

Eye Contact In our culture, people who do not look a speaker in the eyes are usually perceived as shy, insecure, or even dishonest. Although you should not stare fixedly, you appear more confident when you look at the interviewer's eyes while you listen or speak. Try to make the interview as much like a normal (if somewhat intense) conversation as possible.

Distracting Habits You may have nervous habits you are not even aware of. For example, do you play with your hair or clothing? Say something like "You know?" or "Umm" over and over? Pay attention when you practice interviewing, and do your utmost to get rid of these habits. Most interviewers find such habits annoying. If the applicant blinks rapidly and repeatedly, speaks with a hand over the mouth, constantly shifts in the

FIGURE 22–15 The interviewee appears relaxed and self-confident. She has positive nonverbal messages, as indicated by her facial expression. *(© iStock.com)*

chair, rubs the chin, has a blank or glazed look, or avoids eye contact, the interviewer will conclude that the applicant is very uncomfortable and may think that he or she is not telling the truth.

Video The best way to see yourself as others do is to have someone videotape you while you role-play an interview. You can do it yourself with a smartphone or tablet camera. Find out how others see you and try to change any negative behaviors. Your friends and relatives can help raise your awareness of any annoying habits that could distract or put off an interviewer.

Effective Communication

Let the interviewer set the tone and pace of the interview. Adjust yourself to the style of the interviewer. For example, if the interviewer is serious and businesslike, your style should be similar. If the interviewer is cheerful and outgoing, you may need to brighten up a little. Research indicates that it is important to establish some similarity with the interviewer: it helps them to like you.[24]

Communication skills are important for every step of the job search, but never more so than in the job interview. Be sure to listen carefully and speak clearly. Answer each question briefly, but do not give one-word or one-line answers. If you think that the interviewer has not understood your answer or that you have not made yourself clear, try again. Stay on the topic until you are sure that the interviewer has understood your message.

After each question from an interviewer, begin your answer only after the interviewer is completely finished speaking. Interrupting or jumping in too quickly will give the impression that you are rude. You may also miss the real question or important information within the question. Listening to the interviewer is as important as speaking thoughtfully and clearly. The ability to listen shows your attentiveness and reflects your interest in the job. If you find that you don't understand something the interviewer has said, there is no harm in asking her to clarify.

Answering Questions

If you have practiced answering questions during your preparation for the interview, you should be fine. However, realize that interviewers can and may ask almost anything. They are looking for potential problems. They also want to be convinced that you have the skills and experience to do a good job and that your personality will be a good fit in the work environment. Part of your skill set is the ability to communicate about yourself. In one survey, employers said that more than 90 percent of the people they interviewed for jobs could not answer difficult questions. In addition, more than 80 percent could not adequately describe or explain the skills they had or how their abilities qualified them for the job.[25] With proper practice, this should not be a problem for you!

Asking Questions

An interview involves two-way communication. Of course, the interviewer will ask you questions, but he or she will also expect you to ask questions. Part of the interviewer's impression of you is created by the questions you pose. When you ask questions, you show employers that you are assertive, that you know how to gather information, and that you want to make a good decision. Make a list of questions that you couldn't answer when you were researching the company. Otherwise, you may seem unprepared.[26]

Do not be in a hurry to ask questions. Wait until the interviewer invites them. A pause in the conversation once the interview is well under way may be a good time for you to ask a few questions. Be careful, though, not to interrupt the interviewer or appear to be trying to take over control of the conversation. If you have not been given a specific opportunity to ask questions, request a chance to do so before the interview ends.[27]

To land a job, you will need more specific information and a better feel for how this company operates, so don't be afraid to ask. Do not, however, ask questions that indicate you haven't done your homework (e.g., "What are the job requirements?" or "What products do you make?"). "Avoid questions about salary, required overtime, and benefits that imply you are interested more in money and the effort required than in the contribution you can make."[28]

Good questions can help demonstrate your interest in the business and the job. Also, the interviewer's responses could give you clues to what the company is looking for, so that you can shape your answers during the rest of the interview and any later interviews for the same position. The following questions are good starters for this phase of an interview:

- "How would you describe the ideal employee for this position?"
- "What do you think is the best background for this job?"
- "What is the typical day like in this job?"
- "What type of people would I be working with (peers) and for (supervisors)?"
- "What advice do you wish you had been given when you were starting out?"
- "Why do you continue to work for the company?"
- "Would you describe the initial training program for people in this position?"
- "How much value does your company place on advanced degrees?"[29]

Concluding the Interview

Try to get a feeling for when the interview has run its course. The interviewer may stand or simply say, "Well, I think I have enough information about you for now." To help bring an interview to its conclusion, you can ask, "Do you have any other questions I can answer?"

When the interview is closing, be sure to shake the interviewer's hand, thank them for the opportunity to meet, express your genuine interest in the position, and actually ask for the job. Many job applicants fail to ask for the job before the interview ends. This is a big mistake. If you want the job, say so. Say something like, "I know I can do the work, Mr. Morgan, and I would like to have the job."

An interviewer seldom either makes a job offer or rejects an applicant at the conclusion of an interview. In most companies, the interviewer meets with all the applicants before making a final decision, so that they can all be compared and weighed against each other. In some cases, the interviewer's role is only to evaluate and make recommendations to another person who makes the actual employment decision. Nevertheless, make the end of the interview work to your advantage:[30]

- Accept the handshake.
- Thank the interviewer by name for the opportunity to meet.
- Express interest in the job and ask for the position if you want it.
- State that you look forward to hearing from the company.
- Say good-bye.

If you do learn at this point that the company cannot use you, ask about other employers who might need a person with your skills. Then thank the interviewer, shake hands, and leave. On the way out, thank the secretary or receptionist.

AFTER THE INTERVIEW

Take the following steps after each interview. They will help you prepare for your next interview and ensure that you leave a positive impression with each prospective employer.

Step 1 *Evaluate your performance.*

Take a few minutes to review what you learned in the interview and assess your performance. Could you have answered some questions better? Did the interviewer ask questions you were not expecting? Write down what you learned, whom you met, and how you will improve on your answers in future interviews so you will be ready if you have another interview.[31]

Step 2 *Immediately send a follow-up email.*

Within 24 hours of your interview, write a brief follow-up email or **follow-up letter** thanking the potential employer or interviewer for the interview (Figure 22–16). Timing is crucial for this critical follow-up step. You want to impress them with this contact before they make their decision on whom to hire. Send a follow-up email or letter to each person you met, especially if they gave you a business card. Keep the email or letter fairly short and personal. Mention again that you are interested in the job, review key discussion points in the interview, and briefly recap your job-related skills and experience. This follow-up contact will help the interviewer remember you and may be the deciding factor in whether you get a job offer.[32]

Step 3 *Send a thank-you note.*

Even if the interviewer tells you that you will not be hired (or you inform the interviewer that you are no longer interested in the job), consider sending a handwritten note thanking the interviewer for the

123 Mainline Drive
Henson, TN 12345-1001

September 12, 2020

Ms. Linda Odum
HOME BUILDING SUPPLIES
1603 Main Street
Henson, TN 12345-2002

Dear Ms. Odum:

Thank you for the interview yesterday for the job as stock clerk. I appreciate the time you took to show me through the warehouse.

My experience as a carpenter's helper has made me familiar with most of the home building supplies your company manufactures. In fact, we use your products whenever possible.

Thank you again for the interview. It is my hope that you will find me the most qualified and enthusiastic candidate for the job as stock clerk.

Sincerely,

John Potter

FIGURE 22–16 Sample follow-up letter to send after an interview.

interview. Sending a thank-you note is a simple but rare act of appreciation. Such notes also have a practical benefit: people who receive them will remember you. Employers rarely get thank-you notes. As with the follow-up email or letter, your note should be brief and timely and have a personal tone. A thank-you note will not get you a job you are not qualified for, but it will impress people. When a job opens up, they will remember you.[33]

After completing these steps, try to relax. Continue to pursue other job leads while you wait for a decision from the company you interviewed with. If you do not hear from the company within the time period specified (or a reasonable amount of time, if no time limit was mentioned), get in touch. Good follow-up demonstrates your continued interest and perseverance.

National FFA Employment Skills Career Development Event

Every year, the National FFA Organization holds an Employment Skills Career Development Event. It includes the cover letter, résumé, job application, interview, and follow-up correspondence. See Appendix D for the rules of the event and score cards.

VIGNETTE 22-2   SHE HAD A GREAT INTERVIEW

HAVING A GREAT INTERVIEW IS IMPORTANT, but there is more to a great interview than just hitting it off with people. Angelea was very good at communicating, and she knew how to relate to other people, so it seemed as though she was heading for a great interview with the state attorney's office. She had contacted one of the administrative assistants before the interview was set up, so when she got to the office, there was some confusion, and some of the interviewers were not quite ready for her. Angelea did not bring copies of her résumé because she had already mailed it.

The seemingly disorganized interviewers had lost part of her mailed résumé, so they did not have anything to refer to during the interview. However, Angelea thought they probably would not have needed it, because she made a great impression. The atmosphere was comfortable; she had people laughing and talking about their families. Occasionally she was stumped when they inquired about her knowledge of the state attorney's office, but no one seemed to mind or even really notice.

After the interview, Angelea shook hands all around and left the office convinced that she had the job. She did not send a follow-up letter of any kind because she felt that the interview had been so positive that all she had to do was wait. As it turned out, Angelea did get a call—informing her that she would not be getting a job offer. The job went to someone who was nowhere near as exciting, personable, or fun as Angelea.

The person they hired was Dale, who interviewed after Angelea. He made sure everyone knew when he was going to be there for the interview. Dale brought extra copies of his résumé and sent a follow-up letter as soon as he could. The interviewers remembered him. As much as they liked Angelea personally, Dale's organization and persistence made him too impressive to pass up.

What could Angelea have done differently? What else could she have done to make sure he got the job? Did the attorneys make the right decision?

ACCEPTING OR REJECTING A JOB

Only rarely will you be offered a job during the interview. Usually the employer makes the decision later, after all the suitable candidates have been interviewed. Employers like to be able to compare applicants before making a final selection.

Evaluating Job Offers

In all companies, employees come and go and new jobs open up. If you are good enough to have been invited for an interview, you are qualified for a job there. Do not get discouraged. Whether at that company or somewhere else, a job will open up for you. So, what do you do when you finally get a job offer?

The first rule of thumb is never to say yes immediately. Express your thanks and interest, but take a day or two to evaluate the offer and make sure it really is the best thing for you. Do the positives outweigh the negatives? Be sure to think qualitatively as well as quantitatively. For example, opportunities to do challenging work for good pay may outweigh the disadvantages of a long commute. Perhaps the starting salary is a little low, but the opportunities for advancement and growth are great. Try to evaluate whether the long-term benefits outweigh the short-term difficulties. When all this is done, ask yourself: "Do I want *this* job?" Only you can determine the right answer to this question![34]

Discuss Conditions of Employment

Job offers usually come by telephone. This gives the employer and the applicant a chance to discuss the details of the job offer. If you did not discuss the conditions of employment during the interview, now is the time to ask about them. Terms of employment include such things as salary, working hours, benefits, and any number of other items depending upon the type of job. Unless you've already talked salary, and you stated you were agreeable to a certain salary, it usually pays off to negotiate terms. Lots of times employers are expecting you to ask for a little more salary or make some other requests. When terms have been agreed upon, you should ask for the offer and terms in writing. An email is fine. You will want to know when you are to start the new job and if there is anything special that you should bring or be prepared to do the first day. It is fine to ask for some time to think about the offer, especially if you are still not sure you want to accept it.[35]

What to Do if You Get Two Job Offers

It is possible that you will be considered for a job by two (or more) different employers at the same time. Suppose that you have been interviewed for jobs at both a department store and a restaurant. The department store offers you a job and you accept it. Once you have nailed down the specifics of your new department-store job, you should phone the restaurant and let the manager there know that you have taken another job. Try to avoid burning any bridges: be sure to send a follow-up notice to the employer whose offer you rejected, thanking the company for considering you and sending your regrets that you were unable to join the team there at this time.

Rejecting a Job

What if a company offers you a job you do not want? Turn it down politely and briefly explain your reasons for the refusal. No matter what your real reasons are, do not criticize or say anything negative about the employer. Again, avoid burning any bridges: you never know if or when you may want a job with that company in the future.

Dealing with Rejection

Not all interviews result in job offers. In fact, most of them do not. Therefore, you must learn to deal with rejection. Being disappointed is normal, but do not take it too personally or get angry with the employer. Learn what you can from the experience and move on.

Do not neglect to send a thank-you letter. By accepting rejection gracefully, you keep alive your chances for a future job. For example, suppose that the person chosen for the job turns it down? You may be next in line for it.

CONCLUSION

By properly selling yourself through your résumé, job application, and interview, you will eventually find a job. Then you will be leaving, or at least spending less time in, the familiar world of the classroom. The changes you will experience may be scary at first. Remember what it was like going from junior high to high school? A similar experience awaits you now. You are going from the known into the unknown. This can be exciting and frightening at the same time. You are going from high school into the world of work. By taking the time now to learn what to expect, you can prepare yourself for a smooth transition into your new role as a worker.

SUMMARY

Once you have selected and located the type of job you want, it is time to get it. The qualities of motivation and persistence, which are extremely important in finding a job, are also qualities that make you a good employee, and employers recognize this. To get a job, you will need to conduct an organized campaign that includes networking, direct employer contact, writing résumés, filling out job applications, and doing well in interviews.

A good résumé will help get you an interview. Write it yourself, make every word count, make it error-free, make it look good, stress your accomplishments, be specific, and keep it lively.

The résumé should include personal data, job objective, skills (personal management, teamwork, leadership, academic, and technical), education, work experience, and references (optional). Many résumés become part of a database that can match your skills with specific jobs. There are many formats and organizational schemes for résumés; choose one that works for you.

Another way to act on a job lead is to write a letter of application. A letter of application is called a cover letter when it is accompanied by a résumé. A letter of application should have four parts: your reason for writing, pointing out qualifications, calling attention to your résumé, and

asking for an appointment. In developing your letter of application, be brief and specific; make it look good; be assertive; use simple, direct language and correct grammar; let your letter reflect your individuality; send it to someone specific, using the correct name and title; be clear about what you want; be friendly; target your letter to the job you are interested in; and follow up.

Job application forms are often used to screen potential employees. Employers will quickly eliminate applicants who submit messy or incomplete forms, do not list enough experience, or show that their last job paid more than one for which they are applying. Carefully follow the instructions and directions as you complete a job application form. Refer to a previously completed personal data sheet when filling out an application so that your responses are accurate, neat, and complete.

By being granted an interview, you have already achieved some success in your job search. An interview is an opportunity to sell yourself to the employer and gives you a chance to further investigate the specific job and prospective employer. To do your best in an interview, you must prepare and practice. There are common questions you can always expect during an interview. Develop responses to these questions and practice answering them before the interview. Learn about the company, attend to your appearance, collect materials needed for the interview, and check last-minute details before the actual interview takes place.

When the interview begins, connect with the interviewer, smile, use the interviewer's name, and ask some opening questions. Be aware of the messages your body language is sending, including posture, voice, eye contact, and habits. Communicate effectively, answer questions, and ask intelligent questions. Conclude the interview by accepting a handshake and thanking the interviewer by name for the opportunity to meet. Express interest in the job, and even ask for the job if you want it. Let the interviewer know you look forward to hearing from her, and say good-bye. After the interview, evaluate your performance and write a follow-up email and/or a thank-you note. Evaluate any job offer and either accept it or reject it gracefully.

Take It to the Net

Explore résumés and interviewing on the Internet. The Internet has a multitude of sites that can help you with résumé writing and preparing for an interview. Use your favorite search engine and browse for sites. Choose a couple that you find useful and summarize them. For additional information on résumés and interviews, use the following search terms:

Search Terms

résumés

interviewing

writing a résumé

preparing for an interview

interview questions

Chapter Exercises

REVIEW QUESTIONS

1. Define the Terms to Know.
2. List eight basic guidelines for preparing a successful résumé.
3. What are the seven major parts of a résumé?
4. Give 10 examples of personal management skills.
5. Explain what each of the four paragraphs in a letter of application should include.
6. What are 11 helpful hints for writing a good letter of application?
7. What are three reasons your job application may be eliminated by an employer?
8. What is the purpose of an interview?
9. What are 15 questions that you will probably be asked in an interview?
10. What are five things that you need to take to an interview?
11. What are four things you could do during the first few minutes of an interview?
12. What four major areas should you give special attention to during the actual interview?
13. List seven questions an applicant could ask an interviewer.
14. What are five things to do at the conclusion of an interview?
15. What should you do if you get two job offers?
16. Explain the procedure for evaluating a job offer.
17. What are three steps you must complete in acquiring a job?
18. List 24 action verbs that could potentially be used in a résumé.
19. Name three steps that should be followed after each interview.

COMPLETION

1. At best, a résumé will help you get an _____.

2. In preparing your résumé and letter of application, _____ is critical.

3. On a job application form, you should write _____ if a question does not apply to you.

4. Before you write any answers, _____ the job application carefully.

5. Study the _____ so you will know what information to provide on the job application form.

6. It is a good idea to _____ the information on a job application form.

7. Do not try to fit too much _____ into a space on the job application form.

8. Every _____ on a job application form should be filled in.

9. Find out as much as possible about the _____ and the _____ before your interview.

10. If more than a _____ goes by between the time you made the appointment and the actual interview, call to confirm the interview.

11. Plan to arrive at the interviewer's office _____ to _____ minutes ahead of schedule.

12. As soon as you learn about a job lead, follow through with _____ action and get the first available _____.

13. By demonstrating continuous _____ and _____ in pursuing a job lead, you can increase your chances of receiving a job offer.

14. A résumé is nothing more than an _____, selling your services and time to a potential employer.

15. Employers are impressed by applications that say, "I _____ enough to do my very _____."

16. A job application form that is neatly and accurately completed may be the _____ _____ in securing an interview.

17. In the interview, you need to _____ good, know what the _____ involves, understand how you can be an _____ to the company, and be ready to explain why you are the _____ choice.

18. How you say something is often as meaningful as _____ you say.

19. A poised body that leans _____ _____ conveys a relaxed but highly interested attitude to the interviewer.

20. During the interview, keep a tone of _____, _____, and _____ in your voice.

21. _____ to the interviewer is as important as speaking thoughtfully and clearly.

22. In one survey, employers said that more than _____ percent of interviewers could not describe or explain the skills they had or how their abilities qualified them for the job.

23. Good _____ can help demonstrate your interest in the business and the job.

24. Many job applicants fail to ask for the _____ before the interview ends.

25. A letter of application is called a _____ _____ when it is accompanied by a résumé.

MATCHING

_____ 1. Same as cover letter.

_____ 2. Same as nonverbal communication.

_____ 3. Lack of it during an interview can cause suspicion.

_____ 4. Speaks with hand over mouth.

_____ 5. Use when trying to improve interview techniques.

_____ 6. Should project a high level of confidence.

_____ 7. Complimenting during beginning of interview.

_____ 8. Purpose is to obtain an interview with a prospective employer.

_____ 9. Written to the interviewer thanking him or her for the interview.

_____ 10. Gives the applicant an opportunity to present detailed information about himself or herself.

A. videotape

B. linking with interviewer

C. eye contact

D. follow-up letter

E. letter of application

F. posture

G. distracting habit

H. cover letter

I. résumé

J. body language

ACTIVITIES

1. Write a résumé for a job that you would like to have right now using accurate facts about yourself. Type or have the résumé typed or printed with a quality printer. Be neat and include all information that might help you get a job. Use Figure 22–4 as an example.

2. Find a professional in the industry of your choice who would be willing to assess your résumé for you. Take your résumé to them, and then sit down with them and ask them what you could do to improve it. Afterwards, prepare a presentation to share what you learned with your class.

3. Select a help-wanted ad for a job in your occupational area from the classified ad section of the local paper. Write a letter of application to accompany the résumé you prepared in activity 1. Use the example in Figure 22–9 as a guide.

4. A personal data sheet is useful in completing a job application and in developing a résumé. It is also helpful to take the personal data sheet with you when filling out an application, so that you do not omit any pertinent information. Complete a form provided by your teacher (or use a copy of Figure 22–11) for each item that applies to you.

5. Figure 22–12 is a form used for hiring entry-level workers for hourly employment. Complete the form provided by your teacher by filling in each item that applies to you.

6. Figure 22–13 is a typical form used in hiring for various civil service classifications. Complete the form provided by your teacher by filling in each item that applies to you.

7. Suppose you are an employer and you have 35 applicants for one position. Would you interview all applicants for the job? If not, how would you decide which individuals to interview?

8. If you were an employer, what questions would you ask an applicant during an interview?

9. Describe how to convince an employer to hire you.

10. Ask three other classmates to volunteer to participate as a group. Select one person in the group to play the role of interviewer; other group members will play the roles of job applicants. Everyone in class should carefully observe the interviews. Following the interviews, discuss with the class the behavior shown in the interviews, noting both positive features and behaviors that could be improved. Part of the evaluation is appropriate dress and grooming.

11. Write a follow-up letter thanking the employer for an interview. Use any form you wish, or follow the format of Figure 22–16.

NOTES

1. Bureau of Labor Statistics, *Number of Jobs Held, Labor Market, Activity, and Earnings Growth Among the Youngest Baby Boomers: Results from a Longitudinal Study* (Washington, DC: United States Department of Labor, 2015), retrieved September 24, 2016 from http://www.bls.gov/news.release/pdf/nlsoy.pdf.
2. "How to Use a Resume," CAREERWise Education, n. d., retrieved September 24, 2016 from: https://www.careerwise.mnscu.edu/jobs/resumeuses.html.
3. K. M. Dawson and S. N. Dawson, *Job Search: The Total System,* 4th ed. (Total Career Resources, 2015), Loc. 748.
4. Ibid., Loc. 859.
5. A. Doyle, "Resume Keywords and Tips for Using Them," *the balance,* 2016, retrieved September 25, 2016 from https://www.thebalance.com/resum-keywords-and-tips-for-using-them-2063331.
6. A. Doyle, "How to Write a Cover Letter," *the balance,* 2016, retrieved September 25, 2016 from https://www.thebalance.com/how-to-write-a-cover-letter-2060169.
7. K. M. Dawson and S. N. Dawson, *Job Search: The Total System,* Loc. 803.
8. A. Doyle, "How to Create a Professional Resume," *the balance,* 2016, retrieved September 26, 2016 from https://www.thebalance.com/how-to-create-a-professional-resume-2063237.
9. Ashe-Edmunds, "Cover Letter Hints, *Houston Chronicle,* n. d., retrieved September 26, 2016 from http://work.chron.com/cover-letter-hints-1552.html.
10. L. Zhang, "The 3 Rules of Addressing Your Cover Letter," *themuse* (2017), retrieved March 3, 2017 from https://www.themuse.com/advice/the-3-rules-of-addressing-your-cover-letter
11. S. Porges, "6 Secrets to Writing a Great Cover Letter," *Forbes* (2012), retrieved September 26, 2016 from http://www.forbes.com/sites/sethporges/2012/.
12. C. Conlan, "How to Write the Perfect Cover *Letter,"* *Monster,* n. d., retrieved September 26, 2016 from http://www.monster.com/career-advice/.
13. J. M. Kelly-Plate and R. Volz-Patton, *Career Skills* (Encino, CA: Glencoe/McGraw-Hill, 2006), pp. 132–134.
14. Job Applications, *CAREERwise Education,* n. d., retrieved September 26, 2016 from https://www.careerwise.mnscu.edu/jobs/jobapplications.html.
15. The Three-Step Interview Process, *CollegeGrad,* 2016, retrieved September 26, 2016 from https://collegegrad.com.

16. "10 Steps to Interview Success," *GrooveJob,* n. d., retrieved September 26, 2016 from www.groovejob.com.

17. A. Doyle, "Job Interview Questions and Answers: How to Answer the Most Frequently Used Interview Questions," *the balance,* 2016, retrieved September 26, 2016 from https://www.thebalance.com /job-interview-questions-and-answers-2061204.

18. A. Doyle, "Tips for Researching Companies before Job Interviews," *the balance,* 2015, retrieved September 26, 2016 from https://www.thebalance.com/tips-for-researching-companies-before-job-interviews-2061319.

19. Dawson and Dawson, *Job Search,* Loc. 2403.

20. Ibid., Loc. 2886.

21. L. Quast, "8 Tips to Navigate the Interview Day with the Least Amount of Stress," *Forbes: Leadership*, 2013, retrieved September 27, 2016 from www.forbes.com.

22. F. Danzo, *People Hire People, Not Resumes* (New York: Sterling & Ross Publishers, 2009), Loc. 649.

23. C. M. Lehman and D. D. DuFrene, *Business Communication*, 16th ed. (Mason, OH: South-Western, Cengage Learning), pp. 529–530.

24. E. Barker, "How to Ace a Job Interview: 7 Research-Backed Tips," *Time,* 2014, retrieved September 27, 2016 from http://time.com/.

25. J. M. Farr, *How to Get a Job Now: Six Easy Steps to Getting a Better Job* (Indianapolis, IN: JIST Works, 1997), p. 87.

26. Lehman and DuFrene, *Business Communication,* p. 537.

27. L. J. Bailey, *The Job Ahead: A Job Search Worktext* (Albany, NY: Delmar Cengage Learning, 1992), p. 70.

28. Lehman and DuFrene, *Business Communication*, p. 537.

29. Ibid., pp. 537–538.

30. Ibid., p. 538.

31. Dawson and Dawson, *Job Search*, Loc. 2966.

32. J. Goudreau, "After the Job Interview: Five Crucial Steps to Seal the Deal," *Forbes,* 2012, retrieved September 28, 2016 from www.forbes.com/sites/jennagoudreau/2012.

33. Ibid.

34. A. S. Hirsch, *VGM's Careers Checklist: 89 Proven Checklists to Help You Plan Your Career and Get Great Jobs* (Lincolnwood, IL: VGM Career Horizons, 1991), p. 167.

35. A. Green, "5 Things to Do When You Get a Job Offer," *US News: Money,* 2010, retrieved September 28, 2016 from http://money.usnews.com/money/.

23

EMPLOYABILITY SKILLS: SUCCEEDING IN THE JOB

You've selected a career. You've found a job and been hired. Now you want to be able to succeed in the job. This chapter helps you learn the human relations skills that employers want so your job can be secure and you can move up within your place of employment.

Objectives

After completing this chapter, the student should be able to:

- Identify skills that employers want
- Discuss personal management skills that employees need
- Discuss teamwork skills that employees need
- Discuss academic and technical skills that employees need
- Discuss the employability characteristics of successful workers in the modern workplace
- Identify and describe employer and employee responsibilities
- Explain the importance of responding to authority
- Explain the role and importance of ethics in the workplace
- Describe how to get job promotions
- Outline a complaint and appeal process
- Demonstrate knowledge of personal and occupational safety practices in the workplace
- Explain the proper procedure for leaving a job

Terms to Know

- employability skills
- academic skills
- technical skills
- insubordination
- memorandum
- team
- competence
- occupation-related skills
- technical knowledge
- cooperative skills
- gossip
- compromise
- dependability
- trustworthiness
- ambitious
- diligence
- capability
- commitment
- authority
- ethics
- boundaries
- resignation
- pride

Technology is changing rapidly. The workplace is changing, too, so the skills that employees must have to keep up and to succeed are also changing. One thing that has not changed, however, is the fact that employers are looking for employees with leadership ability, a good work ethic, and human relations skills.

You are developing job skills each day of your life. The personal qualities and skills you develop today go with you into the workplace. The kind of person you are today will affect the kind of employee you become. Employers look for employees who have certain qualities and skills. **Employability skills** "are those basic skills necessary for getting, keeping, and doing well on a job."[1] Although **academic skills** (skills and subject-matter material learned in a formal school situation, such as science, English, and math) and **technical skills** (psychomotor skills, such as typing, welding, or drafting; generally learned through high school course work and/or on- the-job training) are important, employers primarily want good, honest, hardworking people with positive attitudes. Developing the qualities and skills prized by employers will help you to be a success on the job.

Skills that employers want can be divided into three major categories. Personal management skills, teamwork skills, and academic and technical skills are the foundational employability skills. Personal management skills include dependability, responsibility, ability to set and accomplish goals, ability to make decisions, honesty, and self-control. Teamwork skills include organizing, planning, listening, sharing, and flexibility. Academic and technical skills include communicating, planning, understanding, problem solving, and competency in a chosen occupation.[2] This chapter addresses these topics and others relating to the employability skills that you must learn to obtain and maintain meaningful and productive employment. The intensity of the workers in Figure 23–1 is a good indication that they possess many of the employability skills just mentioned.

SKILLS EMPLOYERS WANT

Some high school students find work immediately after graduation, but many others take much longer to get hired. Why? Although their educational degrees and attainments may be the same, other

FIGURE 23–1 Employability skills include hard work, dependability, sharing, listening, making decisions, communicating, and competing in a chosen occupation. *(© iStock/Figure8Photos)*

factors come into play during the hiring process. Every time employers are asked about the most important skills and characteristics they look for in new employees, they answer with items such as good communication, positive attitude, dependability, punctuality, responsibility, professionalism, politeness, and self-confidence. Employers seek individuals who have a vision of where they are going and who set goals for themselves. In the latest top ten list, nearly all of the skills desired by employers were leadership skills:

- Teamwork skills
- Decision-making and problem-solving skills
- Verbal communication skills
- Planning, organizing, and prioritizing skills
- The ability to obtain and process information
- Data analysis skills
- Technical knowledge related to the job
- Software and computer skills

- Written communication skills
- The ability to sell to and influence others[3]

An experienced interviewer noted that "too many applicants put the emphasis on educational credentials and ignore the qualities we seek with keenest interest." According to him, employers look for the following qualities in résumés and during interviews, in approximately this order:

- Maturity and stability
- Loyalty
- Honesty and sincerity
- Diligence and reliability
- Sense of humor (ability to form positive interpersonal relationships)
- Experience
- Good appearance (personal grooming and dress)
- Well-organized résumé
- Education
- Hobbies and personal interests[4]

The following isw a partial listing of occupation-related skills and personal attributes that employers look for when they hire people. An interviewer observes or searches for nearly all of these factors, in one way or another.

Achiever—Has an internal drive to get things done; energetic, competitive, ambitious.

Activated—Present in the moment and able to give useful, intelligent input.

Anticipative—Able to predict consequences.

Attitude—Positivity; you cannot succeed unless you think you can! Your attitude largely determines the outcome of any endeavor.

Commanding—Shows an ability to take charge and speak out with authority; tends to adopt a definite stance.

Committed—Is dedicated to assisting coworkers to make a success of the organization; intent on achieving both personal and organizational goals.

Competitive—Has the desire to achieve and rise to the top.

Courageous—Remains determined in the face of resistance; able to carry out difficult assignments.

Credible—Defines self as professional; takes pride in quality and accomplishment.

Dedicated—Has commitment to a vision that empowers self and others.

Dependable—Can be relied on and trusted in all situations.

Developer—Desires to help others mature and succeed; takes satisfaction from each increment and indication of growth.

Disciplined—Able to structure own time and environment; able to organize for increased efficiency.

Driven—Has a desire for accomplishment, even in the face of setbacks and discouragement.

Empathetic—Understands other people's positions and needs; able to recognize how others think and feel.

Ethical—Lives by a set of principles and clear standards of right and wrong; makes choices and decisions according to what is morally correct.

Focused—Able to choose and maintain a direction or goal; can identify key priorities, target attention appropriately, and stay on track.

Ideational—Able to understand and explain events; acts as a problem solver and source of innovative ideas.

Knowledgeable—Has clear and certain mental perceptions and understanding of processes and problems. You must know something about what you are doing!

Loyal—Is devoted to a person, group, or cause.

Organized—Has a system for establishing goals and meeting objectives within a given time frame.

Responsible—Takes ownership of personal behavior, especially work, and encourages others to feel responsible for their own work; able to see that directions and rules are followed.

Self-confident—Belief in own ability; not afraid to make judgments; desire to convince others of one's ability.

Team player—Fits within and works well with a group of employees who set out to accomplish similar goals; willing to do what is necessary for organizational success.

Values-based—Believes in and adheres to high principles, ideals, and standards of conduct.

If you adhere to the highest standards of business and professional conduct, present

GENERAL SKILLS THAT EVERY STUDENT SHOULD HAVE

Personal Management Skills

- Attend school/work daily and on time
- Meet school/work deadlines
- Develop career plans
- Know personal strengths and weaknesses
- Demonstrate self-control
- Pay attention to details
- Follow written instructions and directions
- Follow verbal instructions and directions
- Work without supervision
- Learn new skills
- Identify and suggest new ways to get the job done

Teamwork Skills

- Actively participate in a group
- Know the group's rules and values
- Listen to other group members
- Express ideas to other group members
- Be sensitive to the group members' ideas and views
- Be willing to compromise if necessary to accomplish the goal
- Be a leader or a follower to accomplish the goal
- Work in changing settings and with people of differing backgrounds

Academic and Technical Skills

- Read and understand written materials
- Understand charts and graphs
- Understand basic math
- Use mathematics to solve problems
- Use research and library skills
- Use specialized knowledge and skills to get the job done
- Use tools and equipment
- Speak in the language in which business is conducted
- Write in the language in which business is conducted
- Use scientific method to solve problems

FIGURE 23–2 Personal management skills, teamwork skills, academic skills, and general skills every student should have for jobs at all levels. *(Used with permission of the Michigan Educational Assessment Program.)*

yourself as accurate and honest, and can convince the employer that you will conduct yourself in this manner as an employee, you will probably get a job offer if the employer is doing any new hiring. At least you will be referred to someone who is looking for an employee with your skills. On the job, you have the opportunity to demonstrate all the employability skills discussed in this chapter.

More and more, employers are looking for employees who have not only specific or "technical" skills for a job, but other skills as well. The Michigan Employability Skills Task Force—a group that included leaders from business, labor, government, and education—specified the general skills that every student should have for jobs at all levels, not just for entry-level jobs. These skills, shown in Figure 23–2, along with their subcategories, are personal management skills, teamwork skills, and academic and technical skills.[5]

Today, employers are stressing these skills for the simple reason that they are important in getting a job done. Technical skills generally can be learned through high school course work and/or on-the-job training. However, you will need to have acquired the other skills—personal management, teamwork, and academic skills—before you are hired.

PERSONAL MANAGEMENT SKILLS

Personal management skills help a student become a responsible adult. They guide you in developing dependability and responsibility, setting and accomplishing goals, doing your best, making good decisions, acting honestly, and exercising self-control. The following skills are all positive attributes that will make you a more desirable potential employee.

Attend Work Daily and on Time

Those in the business world, whether employers, coworkers, or clients, assume that you will show up for work each day when you are supposed to. A mature, reliable person can be depended on to be on time. It is disrespectful to others to be inconsiderate of their time and to break commitments regarding time. Make plans, and have the discipline and the self-control to be on time. Make punctuality a goal. Successful people are on time!

BEING SUCCESSFUL IN A JOB IS EASY

You need to work hard, listen carefully, follow directions, think clearly, and have a good attitude. DeAnna used these basics to be successful on the job and in her life. DeAnna took seriously what she learned about employability skills and work. When DeAnna graduated from high school, she went straight to work at one of the banks in town. She wanted to go to college, but she had to earn money to pay for it first. DeAnna was hired as a teller, and she achieved success right away because of her interpersonal skills with the customers.

What set DeAnna apart from many of her coworkers were her teamwork skills. When her supervisor or other employees talked or gave directions, she listened intently to learn something new, to please them by performing her job well, and (most importantly) to show people that she respected them and what they had to say.

DeAnna wrote down specific directions so that she could perform her job just like her supervisors wanted it done. DeAnna was also good at asking questions. This helped her get things right the first time, but it also helped her to find out more about her job. DeAnna always thought things through before she did them, and she never took shortcuts.

Her dedication to doing things right not only helped her keep a good job; it got her a promotion and a raise. About six months after being hired right out of school, DeAnna became an assistant loan officer. Now she is using the same employability skills that she demonstrated as a teller to help people get loans for homes, small farms, and other businesses. She found this job so rewarding that she decided to major in business. The bank that employs her is even helping her pay for college!

DeAnna did not have some magic secret. She just worked hard, listened carefully, followed directions, thought things through, and maintained a good attitude. The skills outlined in this chapter should help you not only keep a job but also excel, just like DeAnna did.

Meet Work Deadlines

Meeting deadlines shows that you are responsible and organized enough to get work done on time. You must be able to devise a plan and organize your work so that you complete tasks on time. Meeting a deadline often requires that you work with others or coordinate your work with others' efforts. Failure to make a deadline may interfere with someone else's work and cause a breakdown in the flow of work. Failure to meet a deadline may lose the employer a customer and hence result in lost profits. Failure to meet deadlines could also cost you your job.

Develop Career Plans

Without goals, we have no focus or direction. In reality, happiness is determined by whether we achieve our goals. If we have no goals, how can we achieve happiness? Career plans include setting and accomplishing goals. Those who are successful in the workplace know where they are going and are executing their plans for how to get there.

Know Personal Strengths and Weaknesses

Do not accept a job if you do not have the requisite skills, unless you know that you will be trained on the job. It is best to look for a job that fits your talents and abilities, and in which you can use what you already know, so that you can be successful, productive, useful, and valuable to your employer. Aptitude testing can help you identify your strengths and weaknesses; identifying weaknesses shows where you need improvement. One way to find out is through your school's testing program. Your career and technical education counselor or guidance counselor can assist you with interpreting test results.

Demonstrate Self-Control

Situations may arise on the job that displease or upset you, but as an employee you will need to keep your composure and think before you react. You must know when to listen and when to speak, when to act and when to pause, when to work and when to play. Delaying a response to an uncomfortable or contentious situation will help you cool down and give you time to think through an appropriate response. Good timing and common sense are everything when it comes to self-control. Unresolved issues must often be taken to a supervisor for a solution.

Losing self-control with a supervisor is **insubordination** (disobedience or lack of submission to authority) and may cost you your job. Although company policies and procedures differ, most employers will give a reprimand for insubordination. The first time it happens, you may receive a verbal warning, which is noted in your employee file. The second time, you may be issued a warning **memorandum**, a formal written communication used in a business setting that becomes part of your employee file. The third time, you may be suspended from work without pay for several days. The next time, you may be terminated. In some companies, insubordination is grounds for immediate dismissal.

Pay Attention to Details

Paying attention to details helps you avoid many mistakes. Correcting errors wastes company time and money. It is your responsibility to do a task right the first time. It is also your responsibility to stay alert and ask questions if instructions or details are not clear. If you do not pay attention to details, you cannot produce the level of quality needed for the firm to compete in the global economy. Being observant of your surroundings and recognizing how your job fits into the total organization help you understand the importance of your job and your role in the organization. The details of your job will then become more meaningful to you.

Follow Written Instructions and Directions

Instructions are given for a reason: they transmit procedures that have been developed to promote production and to get quality work done efficiently and on time. Not following instructions causes delays and slows progress from one stage to the next, eventually causing confusion and disrupting the process. Not following instructions may cost you your job.

Whether you are a construction worker, a technician, or a government worker, you must be capable of reading, comprehending, and following written instructions and directions. If you do not understand something, ask questions. That is how smart people became smart. There is no such thing as a stupid question.

Follow Verbal Instructions and Directions

One of the most important personal management skills is the ability to communicate effectively. Most jobs require that you be able to communicate by speaking, listening, and writing. Your skill in communication (or lack thereof) may influence whether you get a promotion. It also makes a big difference in how you get along with your coworkers.

For many jobs, good communication ability is a vital part of the job. For instance, salespeople need to be able to speak well to describe merchandise to customers and influence clients to make purchases. Wait staff need to be able to write down what their customers are ordering as well as talk and listen to the customers.

Listening skills are another very important part of communication. A good listener looks at the person who is talking, asks questions as needed, avoids interrupting the speaker, and evaluates the message instead of the speaker. Do not let emotional words distract you, and try to rein in any mental wandering. Time can be wasted when you listen poorly. Employers want people who listen, understand, remember, and follow instructions.

Oral instructions require that you be able to listen and then perform as requested. If an employer tells you to do something, you should follow through to the best of your ability until the task is completed. Following instructions, listening, and getting all the steps right the first time is important. If you are not sure, ask questions.

In some jobs, you must follow detailed directions that require close attention. For other jobs, you will need few directions. Whether your job involves simple directions or complex directions, there are steps that can help you follow directions more easily:

Listen Carefully When your employer or supervisor is giving you directions, pay close attention to what he or she is saying. Do not let your mind wander. Concentrate on listening to all the directions.

Write Down the Directions, if Necessary You may be given a job that requires many steps to complete. When receiving lengthy directions, write them down. Do not depend on your memory. You may forget or misremember an important part of the directions.

Make Sure You Understand the Directions If you are not sure exactly what your supervisor means, ask questions. It is better to ask questions than to do the job wrong.

FIGURE 23–3 When a person is totally involved in her work, there is little need for supervision. *(Courtesy of University of Georgia—College of Agricultural & Environmental Sciences.)*

Think before Doing Before you begin a task, review the directions you have been given. Know the order in which things must be done. For maximum efficiency, gather any tools or materials you will need before starting. Break the directions down into steps and then decide in what order the steps should be taken. Thinking through these steps may save you time and mistakes.

Avoid Shortcuts If your supervisor tells you to do a job in a certain way, follow those directions exactly, as long as what you've been asked to do is ethical. If you have discovered a quicker or easier way to do the job that is different from the way you were directed, seek a meeting with your supervisor to discuss your ideas, but do not take shortcuts. Your supervisor probably has reasons for the directions she has given you. Later on, you might suggest ideas and changes to improve efficiency, but you must follow instructions until changes are officially approved and implemented.

Work without Supervision

Working without supervision means being able to perform your job without the supervisor standing over you. Your objective should be to please your boss by working to please yourself, like the person in Figure 23–3. If quality and achievement are your goals, work that satisfies you will almost certainly satisfy your employer. You earn the employer's trust when you show initiative and know your job well enough to do it on your own. Keep your mind on your work and do not waste company time. Stay with a task until it is finished. If you finish early, look around for other things that need to be done or assist others who need help.

Learn New Skills

It is important to know the skills needed to do your job well. However, employees also need to learn new skills to stay up to date with new technologies and processes. Being willing to learn new skills and applying them to your work may help you keep your job or get promotions.

Many employers that routinely train new employees look for people who are willing to learn. Every company has its own way of doing things. As a new employee, you will need to learn the company's procedures and policies as well as your job tasks. An employee who has been on the job for a while should be willing to learn more to do the job well, improve performance and productivity, and stay up to date.

Your employer may also ask you to change tasks or give you something to do in addition to your existing tasks. Changes in the company structure or focus may require you to learn new skills and take on new responsibilities. Having a desire to learn and understanding how to do so are traits valued by employers because of the dynamic nature of work in our technological age. If you are willing to learn new skills and tasks, you will save your employer the financial and time costs of hiring new employees or implementing expensive training programs.

Identify and Suggest New Ways to Do a Job

Some jobs will grow; others will not. One thing is for certain: the skills required of workers will change over their working lives. Do not be afraid to learn, identify, or suggest new methods to help get a job done more effectively or efficiently. Some companies offer bonuses or incentives for useful suggestions from workers.

Be Organized

You will need organizational skills for every type of job. You must be able to organize your time, tasks, and belongings. To give yourself enough time to get all your work done, you will need to plan ahead. For example, keeping your work supplies in order means that you need not waste time looking for the necessary tools and materials. Breaking large or complex tasks down into a series of smaller tasks makes these jobs more approachable and easier to handle.

Demonstrate Personal Values at Work

True leaders are not afraid to demonstrate their personal values by example in their everyday lives. You set a values-based example whenever you display honesty, trustworthiness, or respect for others on the job. Apply what you learned about values from Chapter 20 to determine what your values truly are. When you enter the workforce, you may revisit some values questions in relation to the job, asking yourself: What do I really believe in or stand for? What gives my life inner

quality? For some people, financial success gives meaning to their work. For others, value inheres mostly in efforts that give help or assistance to less fortunate others. Employers are likely to hire people whose values align with those of the company or the organization's founders; in turn, you will probably be most comfortable and effective in a job in which your personal values are not compromised.

TEAMWORK SKILLS

Teamwork skills help you function effectively and efficiently within a group. They include organizing, planning, listening, sharing, flexibility, and leadership.

A **team** is "a group of people, each with different skills and often with different tasks, who work together toward a common project, service, or goal with a meshing of their functions and with mutual support."[6] As members of a team, you must rely on each other and work together if all of you are to do your best and meet all your goals. "Teamwork involves working together to achieve something beyond the capabilities of individuals working alone."[7] Teamwork is managed because someone exercises control, whether officially or informally. Teamwork is planned because it results from preparation and organization. In the workplace, teams are organized so that individuals' talents and skills can be directed through group effort to the accomplishment of vital tasks and goals. Here are eight aspects of teamwork:

Active Participation in a Group

To gain the experience necessary to be effective in the workplace, you should start out by becoming active in school or community groups such as the FFA, VICA, band, choir, athletic team, yearbook staff, youth groups, and others. By participating in these kinds of groups, you develop and practice the teamwork skills necessary for success in the workplace. You also learn how to organize and plan group activities.

Following the Group's Rules and Values

When you get involved in a group, you must become familiar with the group's rules and values and operate in a way that supports them. You need to understand the team's norms, or expectations about how everyone acts on a team. For example, in basketball, you must not touch a person's arm while she is shooting the ball. If you do not follow this basic rule, your team will be heavily penalized and may lose the game. Likewise, in business, you must obey the rules of good business ethics and values. The penalties can include lawsuits, fines, loss of business, or loss of your job.[8]

Listening to Other Group Members

We have already stressed the importance of listening. Make it a priority to develop and cultivate this skill; it is tremendously valuable in the workplace, especially with groups. Good listening may benefit you in other ways, too, as a story shows: during a group meeting, a young teacher observed a brilliant professor and noticed that he did not take part in the group discussion. Afterward, the teacher asked the professor, "Why didn't you join the discussion?" The professor answered (only partly in jest), "I know what I know, and if I listen to others to find out what they know, I will become even smarter."

Expressing Ideas to Other Group Members

Although listening is a virtue, there is a time to speak. Your opinions are just as valid as the next person's. Be sure you have thought through your ideas before you speak; do not make contributions just to hear yourself talk or to allay your insecurities. The exchange of thoughts, ideas, and concepts in a group is a tremendous way of pulling together to solve a problem (Figure 23–4).

Sensitivity to Group Members' Ideas and Viewpoints

Everyone has feelings. When people share their feelings, ideas, and beliefs, they are allowing you to see their emotions and thought processes about an issue. This display of trust should be taken seriously. Earn and maintain trust by being kind, being sensitive to others' feelings, and working hard to find solutions to problems faced by the team while the walls are down.

Ability to Compromise (to Follow Democratic Majority Rule)

Do not be stubborn. When any group works together, there will be times when members must

FIGURE 23–4 Although listening is a virtue, there is a time to speak. The young woman in this photo is expressing her ideas to the group. *(© Monkey Business Images/Shutterstock.com)*

bend to an opposing opinion to enable group success in achieving the common goal. If the majority of a group holds a viewpoint contrary to your own, clearly explain your stance, explain your objections, and then allow the group to analyze and critique your position. Majority rule for teams is almost always the best policy.

Leading or Following to Accomplish the Goal

There are times to lead and times to follow. Knowing the right time to do each one is the mark of a good leader. If you know more about an issue than other members of the group, you should take the initiative and exercise leadership. If not, allow another person to take the lead. Do your best, both as a leader and as a follower, to accomplish the team's goal.

Flexibility and Acceptance, to Allow Work in Changing Settings and with Different People

In the 21st century, the U.S. workplace is becoming more diverse, mirroring changes in society at large. This means that you will end up working in different environments with a wide variety of people. You must be willing to embrace change and be open to experiencing new things. As you meet new people, you gain knowledge. When your knowledge increases, your self-concept improves, and you become worth more to your employer.

ACADEMIC AND TECHNICAL SKILLS

To meet the challenge of foreign competitors, American businesses currently impose high standards for reading, writing, computation, and technical skills and abilities. The fact that many new entrants into the workforce cannot meet these standards has become a major economic and competitive issue for U.S. companies.[9]

Reading has historically been considered *the* fundamental skill needed for the world of work. A person must be able to read to find out about available jobs, to get a job, to keep a job, to get ahead in a job, and to change jobs.

Writing skills today are important in almost every occupation and field. An employer's first opportunity to judge a potential employee's writing ability is by the quality of an application, cover letter, or résumé; that is why it is so important that these be error-free. Many employers regard the quality of letters, memorandums, progress reports, work orders, requisitions, recommendations, and instructions as an indicator of the overall quality of an employee's work.[10]

Computation, which is the third academic/technical competence skill, is no less important than the first two. Employers find great value in leaders with computation skills because they not only perform quantitative calculations, but also solve mathematical problems, analyze and interpret data, and apply sound decision making.[11] Technology and the dynamic nature of work these days make computation a highly sought-after skill.

Of course, an important requirement for job success is competence in occupation-related skills. **Competence** is the ability to do something well. **Occupation-related skills** are the skills that have direct bearing on and are needed in one's job or occupation. For example, an administrative assistant might need skills in typing, data entry, and shorthand. Successful workers are the ones who do the best they can to become competent at the job. They work to improve existing skills and to learn new things to increase their competence in skills needed for the job.[12] Here we briefly discuss important academic and technical skills.

Read and Comprehend Written Materials

One of the first and most important skills you must master is the ability to read and understand what you have read. Before you can move very far down the road to success, you must master reading, comprehension, and remembering. Successful people spend a great deal of time reading. Because they understand and remember what they have learned, they can put this new knowledge to work for them. The more you read, the smarter you get.

Interpret Charts and Graphs

Some basic information tools that have been around for a long time are still being used effectively today, both in school and in the workplace. These tools, known as charts and graphs, present information in a graphic form to make it more readily understandable. The data in a chart or graph are usually numeric or statistical, which are more easily comprehended and compared when displayed in this fashion. In fact, the high-tech tools now available make production of charts and graphs very easy; however, a worker must be able to interpret and understand the information and messages conveyed in these formats.

Perform Basic Mathematical Calculations

The ability to do arithmetic and simple math is a basic part of life and work. For example, cashiers and bank tellers need to be able to make correct change. Basic math consists of addition, subtraction, multiplication, and division. More advanced math may be used in certain jobs, especially for problem solving, design and construction, and engineering. You should be able to add a column of numbers in your head, work simple division without using pencil and paper, and have the multiplication tables through 10 memorized.

Solve Problems with Computation

Many of today's jobs require that you be able to compute cost, time, volume, percentages, and fractions. A worker should be able to add, subtract, multiply, divide whole numbers, use fractions and decimals, and calculate simple interest. You should also be able to calculate time, volume, area, weight, and distance (Figure 23–5). For example, truck drivers must be able to calculate their mileage and time spent on the road. Machinists and carpenters must be able to make measurements and calculate from them.

Use Research and Library Skills

When you are confronted with a problem, do not reinvent the wheel; much of what you must learn, know, or do has already been done. Save yourself time by developing good research and library skills. In some cases, you can download the information you have located from the Internet. If your library does not have an item that you need, it may be able to secure it from another library through an interlibrary loan. Reference materials can show you solutions, teach you methods, suggest new and productive approaches, and much more. If you are having trouble locating information, ask a reference librarian for help.

FIGURE 23–5 Many jobs require math and problem-solving skills. Numbers, figures, and calculations are all part of being a pilot. *(© Jacob Lund/Shutterstock.com)*

Use Tools and Equipment

The tools and equipment of the modern office include computers, calculators, wireless networks, fax machines, telephones, copiers, scanners, and printers, among many others. Information and communication technology alone are integral parts of almost every workplace. In addition, almost every field uses specialized high-tech tools and software for its particular applications, and some individual businesses have their own customized, proprietary versions of these tools. No matter what field you choose, you will have to learn how to use special tools and equipment. The sooner you can familiarize yourself with the tools of your chosen trade, the further ahead you will be when you start your job (Figure 23–6).

Speak the Language in Which Business Is Conducted

Every business has its own vocabulary. To be successful, you must learn the language of your occupation and be able to communicate well using it. For instance, many companies use acronyms as part of their normal terminology. Some examples of acronyms are USDA (United States Department of Agriculture), ERIC (Education Resource Information Consortium), DACUM (Developing a Curriculum), NASA (National Aeronautics and Space Administration), WIP (work in process), and NHL (National Hockey League).

Write the Language in Which Business Is Conducted

In the workplace, formal communications are written. For example, employees receive important information through e-mails, memorandums, handbooks, and manuals. Employees are expected to read and follow these instructions. Workers are also asked to fill out forms; write notes, letters, and reports; and keep records. Skillful written communications enable your coworkers and supervisor to understand you.

Solve Problems Using the Scientific Method

Problem solving and decision making are discussed at length in Chapter 13. We mention it here only briefly to show how problem-solving skills fit into and are used in the workplace. Problems will inevitably arise on the job, and as an employee

FIGURE 23–6 Certain jobs require the use of specialized equipment. This high school student is gaining experience with genetic identification equipment. *(© Nikita G. Sidorov/Shutterstock.com)*

you should be able to solve many of the problems you encounter. Being able to figure out the solution to a problem will make your work easier and make you more productive. In the workplace, creative thinking is generally expressed through the process of creative problem solving. Employers want employees to solve problems, to find workable solutions. Employees who are involved in problem solving generally feel better about their work and have greater job satisfaction because they were part of the solution. The scientific method (discussed in Chapter 13) is a very effective procedure for solving problems.

Use Specialized Knowledge to Get a Job Done

Almost every job requires a certain degree of specialized knowledge. **Technical knowledge** is the knowledge that is needed to perform a specific job. In our rapidly changing society, successful workers are the ones who engage in continuous learning activities to keep up to date on new technologies, new ways of doing things, and directions in which a particular industry may be heading. Successful workers increase and improve their technical and specialized knowledge through reading and research, professional development conferences, online coursework, social media, or even simulations and e-gaming.[13]

EMPLOYABILITY CHARACTERISTICS OF SUCCESSFUL WORKERS

As noted earlier, employers look for certain qualities in potential new employees. Understanding these qualities and their importance to employers is a step toward job success. *Building Life Skills*, a book by Liddell and Gentzler, addresses several qualities that contribute to making a successful employee and enhancing relationships: a positive attitude, cooperation, dependability, trustworthiness, working hard, respecting others, handling criticism, appropriate dress and grooming, showing initiative, diligence,[14] being capable, and showing commitment.

Positive Attitude

Your attitude affects your job success (along with everything else in your life). Employers hire and promote workers who display a positive attitude.

VIGNETTE 23-2 | **THE RIGHT STUFF**

BEING SUCCESSFUL AT WORK REQUIRES MORE THAN ONE TRAIT OR EVEN ONE SET OF TRAITS

You have to be good at many different things, especially if you want to be a leader at work. It is not enough to be good only at personal management skills in a world in which teams and groups are an everyday part of most jobs. It is not enough to have only teamwork skills and no technical ability. A good employee and especially a leader have to strive for a balance of ability among personal management skills, teamwork skills, and academic and technical skills. Maria, the most recent hire at CompuLeadership, is a perfect example of someone with the "right stuff."

Maria was fortunate enough to learn early how important it is to be on time, pay attention to details, set personal goals, and reflect on one's actions. These personal management skills helped her to get through school and find a job, and now they are helping her to become a valuable asset at the CompuLeadership Company.

The people skills she learned back in high school and by participating in activities in college are also helping her adjust to her new job. She listens to other employees and her supervisors as well as to the customers. She has been able to form relationships with the group members and learn their rules and values, as well as to share her own ideas while earning the respect of coworkers. Her knowledge of personality types, leadership styles, and communication styles has really given her an edge.

Maria would not have gotten her new job if she had not had the necessary technical skills. She had to be competent with computers just to get an interview; to keep the job and be successful in it, she has to do more. Maria reads constantly to stay up to date on the latest technology and developments in the field. She also works on developing her critical thinking and problem-solving ability so that she can efficiently and effectively troubleshoot technical issues.

Maria definitely has the right stuff. She understands that you have to strive for competence in personal management, human relations, and academic and technical skills. Do you have the right stuff yet? What is missing from your arsenal? What else do you believe is important to be successful on the job or in leading others at work?

They want workers who can get along with others, so employers tend to hire people with friendly, outgoing personalities, especially for supervisory or customer contact jobs. Your work attitude shows how you feel about your job and your coworkers. If you work hard and get along with others, you are showing a positive work attitude. If you show up late, complain, or do sloppy work, you are displaying a negative work attitude.

Your success in a job depends on your positive work attitudes.[15] Displaying the following attitudes will result in positive responses from your employer and fellow employees:

Show Enthusiasm and Pride Be happy and positive about the work you do and the place where you work. Enthusiasm helps you tackle big problems without getting discouraged. Like any attitude, enthusiasm is contagious: you can spread your enthusiasm to coworkers and thereby make the whole workplace better.

Be Cheerful Look on the bright side. Show your coworkers and employer that you are happy to be at work and glad to be doing the job you are doing. Smile!

Be Dependable Always arrive at work on time. Do what you are paid to do, and more if possible. Pull your own weight. Be willing to help coworkers when they need assistance.

Be Willing to Learn Nobody knows everything. If you are asked to take on a new task, look at it as an opportunity to learn a new skill and increase your chances for success.

Show Initiative In addition to being ready and willing to learn, actively seek out opportunities to learn new skills and to help others. Do not wait for someone to tell you what to do; find work that you can do and do it without being told.

Use Self-Control Show patience and tolerance for your coworkers. Do not get upset or angry when you are asked to do your share of an unpleasant task. Do not insist on always having things your way.

Be Cooperative Learn to work with others as a team to achieve a common goal; do not try to compete with coworkers. Remember, you succeed when you help others to succeed!

Accept Criticism Be open to suggestions from others. Do not think of criticism as negative; rather, accept it and use it to help you do a better job.

If you have a negative attitude, you can change it. It is not easy, but it is possible—and you are the only person who can truly change your attitude. To develop and maintain a positive general attitude, you must exert some effort. If you need help in changing your attitude, follow these suggestions:

Think Positive If you have a negative attitude about something, consciously look for something about it to like. Concentrate on the good points instead of the bad points. For example, if you have a bad attitude about school, think of the good things you get and benefits you derive from school. It may be hard at first to find good aspects, but say something positive about school to your friends every day. Soon you may find yourself liking school!

Avoid Negative Influences Do not let people with negative attitudes infect you with their bad attitudes. You control your own attitude, so protect yourself against negative influences from others. This may mean avoiding people who complain, whine, or make you feel bad about things.

Find Examples and Role Models Think of people you know who have positive attitudes. Watch what they do and say that exhibits this attitude. Try doing and saying some of these things; if people respond to you in a friendly, upbeat way, you have succeeded in showing a positive attitude.

Find Positive Influences Make friends and spend time with people who have positive attitudes. They will help you feel good about yourself and strengthen your optimistic outlook.

Respect Yourself Think of yourself in positive ways and concentrate on your good points. Believe that you have a lot going for you and that you are fortunate to be you. Do not get mad at yourself when failures occur; everyone encounters setbacks. If you blame yourself too much, you will lack the self-confidence to make an attempt the next time. Learn from your mistakes and think positively about the future. Respecting yourself by focusing on positive thinking also helps you manage and overcome stress, an ability that will also contribute to career success.[16]

Cooperation

Most jobs require you to work with other people. To get the work done, you must cooperate with these people. Developing good relations with coworkers makes your work more pleasant and enjoyable, and you get more work done. No matter how good you are at your job, you must get along with your coworkers to be successful. Some people are easy to get along with; others seem impossible to please. When interacting with people, you must figure out ways to get along with them.

Cooperative skills develop the ability to work or act together with people for a common purpose. These skills help you get the most out of your interactions with others. The skills you have already developed pertaining to leadership of groups and working with a team will be tremendously valuable to you in the workplace, even when you are not officially a leader. Remember the basics about developing good cooperative skills:[17]

Recognize and Accept Differences in Others People have different lifestyles and different values. Respect the way others choose to live and behave, within the bounds of rationality and law. Remember that your lifestyle and values may seem strange to others. Each individual is different; do not judge others by your standards. If you accept others as they are, they will tend to accept you.

Avoid Assumptions Do not make assumptions or judgments about others unless you know all the facts.

Do Your Share of the Work As a beginner, you will be taught how to do the job. Coworkers are usually happy to help you get started. Develop a good working relationship by listening carefully to instructions and asking questions related directly to the work you are doing. Once you have learned the job, do not expect others to do it for you; pull your own weight and do the job you were hired to do. Coworkers will respect you for doing your share of the work.

Keep Up Appearances The first thing others notice about you is your physical appearance. Create a good first impression by dressing appropriately and doing good personal grooming. Maintain a pleasing, appropriate appearance whenever you are at work, especially if you have contact with customers or clients.

Exhibit a Good Attitude People with positive attitudes are friendly and pleasant. Those with negative attitudes complain, whine, and become angry or aggressive. If you have a positive attitude, people will like being around you and working with you. This attitude shows that you care about your job and your coworkers—and it is contagious!

Avoid Gossip and Disputes Gossip is talking about the personal or private affairs of other people, usually in a negative way (spreading rumors, criticizing appearance or actions, making unfounded guesses about occurrences in their lives, etc.). Gossiping can deeply hurt your coworkers' feelings, cause them to get angry, and prompt them to gossip about you in return. To maintain good relations with coworkers, avoid gossip. If two or more coworkers have a dispute, stay out of it. Taking sides in a dispute may damage your working relationship with others in your group, whether or not they are directly involved in the argument.

Control Your Emotions At some point, you will feel hurt by or get angry with coworkers. You must learn not to react to these emotions on the job; instead, work out problems in a reasonable, rational manner. If you cannot control yourself, coworkers will avoid you and your work performance will suffer.

Learn to Compromise You cannot have things your way all the time. When you disagree with a coworker, learn to **compromise**: that is, both parties give a little or change their positions somewhat to come to a mutually satisfactory agreement. When there is disagreement, listen to what your coworkers have to say, think about their reasoning, and decide how much you are willing to change. Very little will get done unless you and your coworkers can agree to compromise.

Be Considerate and Sensitive Treat other people as you like to be treated. Consider their feelings and their points of view.

Choose Your Words Carefully There are many ways of saying the same thing. Be clear about what you mean, but use tact to get your message across and stay positive. Think about how the other person will react to what you say—before you say it.

Check Your Own Behavior Is it possible that poor human relations with another person may be your fault? Have you created a problem with this person? Sometimes you can unintentionally say or do things that hurt or offend others. Look at your behavior from the viewpoint of others. Is there room for improvement? You can do little to change another person, but you are always capable of and responsible for changing yourself.

The two workers in Figure 23–7 are exhibiting cooperative skills in their professional area.

FIGURE 23–7 To succeed at your job, you must be able to cooperate with employers, coworkers, and people outside your job. *(Courtesy of University of Georgia—College of Agricultural & Environmental Sciences.)*

Dependability

As an employee, you are most likely to succeed if you show **dependability**, that you can be trusted to accomplish a task. Your employer and coworkers depend on you to get your work done. You should not take too many breaks or expect others to do your work for you. Perform your job to the best of your ability. This quality relates closely to honesty and trustworthiness: you do an honest day's work for an honest day's pay, and you can be depended on (trusted) to do a thorough, complete job of good quality.

Part of being dependable is being punctual, which means you arrive on time at work and for appointments. A good employee is on time every time. Employers need to know that you will be at work when you are supposed to be, because they are depending on you to get a job done. If you are not at work, you cannot complete the tasks you are assigned. If you are punctual, your boss can depend on you to be there.[18]

Reliability is also a part of being dependable. Employers will not accept many excuses for tardiness to work. Rare exceptions such as a flat tire or a late bus are understandable and inevitable. Forgetting to set an alarm clock or oversleeping is not. Some employees mistakenly believe that their jobs are insignificant, and that their absence will therefore not make much difference. This is indeed a big mistake. When you are not on the job, others must take up the slack and do your work; this adds unnecessary stress to their lives and work quality may suffer. A good employee recognizes the importance of individual jobs to the smooth functioning of a team.

Before you are hired, most employers check into your background. What you have done in the past is a good indication of what you will do in the future. If you are a punctual and reliable student, you are likely to be a punctual and reliable employee. The same is true when you are considered for other jobs or promotions: your reputation precedes you. Being punctual, reliable, and dependable will help you achieve success on the job.

Trustworthiness

Another quality you need to be a good employee is **trustworthiness**, deserving of trust or confidence. Trustworthy people are honest, and honesty is important to every employer. Employers want to be sure that you will tell the truth and do the right thing.[19] For instance, the employer must be able to trust a cashier with the money in the cash register (Figure 23–8). Your boss will feel comfortable having you handle money if you are an honest person.

FIGURE 23–8 These workers are treating each other with mutual respect. They are treating each other like they would want to be treated. *(© baranq/Shutterstock.com)*

Trustworthiness also means doing your job and doing it right. Once you have been taught how to perform your job, in the way the employer wants it done, follow the instructions and work consistently. Wasting time and materials costs your employer money. The employer should be able to trust you to do quality work without having to go back and check to see whether you did it right. An honest, trustworthy employee is a great asset in any workplace.

Working Hard

The way to develop the habit of working hard now is always to do your best on school assignments, putting the necessary time and effort into each and every assignment. This includes both the classes you like and the classes that are not your favorites.[20] Be neat, accurate, and thorough. After all, right now school *is* your job!

Fulfilling your responsibilities as a family member can also help develop a willingness to work. Volunteering to do chores and being willing to help with duties wherever you are show that you are **ambitious**. In other words, being alert, energetic, willing to work, and caring about what you do help to develop a mindset that will carry through to your job—and will carry you through to success in the workplace.

Respecting Others

Just like being a successful group leader, to be a successful employee you must recognize and remember that all people want to be treated well. Always show respect for others. Respect in the workplace involves simple, common courtesy; staying within certain boundaries; and following the directions you are given. It also includes being aware of the rights of your coworkers[21] and not interfering while they do their jobs. Essentially, respecting others means treating others as you want to be treated at work.

Supervisors and bosses in particular should be treated with respect. The respect you owe your boss is much the same as the respect you owe your teachers. Your teachers and boss give you direction and help you learn. You respect them for their power, experience, knowledge, and the help they can give you in learning.[22] Knowing who has authority and respecting their positions will be a valuable asset in your quest for job success.

Handling Criticism

Handling criticism is difficult; no one likes to be wrong or make mistakes. However, we all make occasional mistakes. Take criticism constructively instead of personally, try to keep an open mind, and always be willing to make improvements and learn from mistakes. Employers want to help you be successful because your success benefits the company.

Knowing how to give criticism can also be helpful in the workplace. For instance, you may be asked to help a new employee learn how to do a job. When you criticize this person, you want to be tactful and constructive. Your mission is to help your fellow employee do a better job, not to hurt the person's feelings or attack him or her as a person. Feeling empathy for the coworker may help, so put yourself in his or her shoes. Try to say something positive before you say what or how the employee needs to improve.[23] There are many supportive, positive ways to give corrections and direction; use the skills you developed as a leader in school or club group work when you must criticize coworkers.

Appropriate Dress and Grooming

Well-groomed workers are likely to have a positive effect on other people at work (Figure 23–9). Good grooming includes being clean, maintaining good personal hygiene, and wearing clean, well-maintained clothes that are suitable for the

FIGURE 23–9 These coworkers are appropriately dressed, well groomed, and wearing smiles. *(© iStock/Squaredpixels)*

job. Presenting a good appearance makes you feel better about yourself and causes others to have a more positive attitude toward you.

Showing Initiative

When you show initiative, you are the first to start or introduce something. A good employee does not wait around to be told what to do or when to do it. Rather, the employee takes responsibility for knowing the job well and performs tasks without needing supervision. The employee with initiative looks for ways to improve performance and cares about the safe, efficient, productive operation of the workplace. Showing initiative requires being alert, working with urgency to get a job done, and looking for ways to assist others if needed. If good employees have finished an assignment, they find something else that needs to be done and then do it.

Good employees also learn more than they are required to learn. Many companies encourage employees to come up with ideas that will save time, save money, and improve productivity. Employees who are successful at their jobs may be rewarded with bonuses or promoted to jobs with greater responsibility.

Diligence

Most employers want employees who will stay on a job until it is finished. Sometimes you may have to stay late to get a job completed; this is called going beyond the call of duty. **Diligence** is constant, earnest, and persistent effort; it means sticking with an assignment no matter how difficult or frustrating it becomes, especially if it requires your continuous attention. A good employee does not quit in the middle of a project, but sees it through to completion. Diligence also involves dedicating yourself to doing actual work during work hours, getting tasks done in a timely manner, and avoiding interruptions or distractions that divert your attention from your job.

Be Capable

One of your major responsibilities as an employee is to do your job to the best of your ability. Employers hire you because they believe you are capable of doing certain jobs. Your **capability**, or your potential for being able to do the job, is important to the employer. The way you do your job affects the work of others in your workplace and ultimately affects the success of the business as a whole.[24]

When you begin your job, follow directions and do your work exactly as directed. You may not understand the reason for doing each task in this certain way, but your employer does have a reason for giving you those directions. After you have shown your ability, you may be able to suggest other ways of doing certain tasks, which indicates your capability as well as your initiative and creativity.[25]

Show Commitment

Similar to dependability and diligence is **commitment**, applying oneself to a task until it is completed. You can be counted on to attend meetings after hours or on weekends. If there is a special project, you will be there to do your best. Commitment is especially appreciated in community organizations. Nothing is more frustrating than planning an event and having people not follow through on what they have said they will do. Commitment means really staying with a task when the going gets tough; you do not support the team only when the team is winning!

EMPLOYER AND EMPLOYEE RESPONSIBILITIES

An employer should respect the needs of the employee. An employee should try to meet the goals and objectives of the employer. An employee should give it their best every day and never forget who signs the paycheck.

Employers Must Meet Employee Needs

The employer must attempt to fulfill the needs of every employee. In most cases, each employee needs more supervision in one area than others. The employer must lead employees, give them responsibility, and create a feeling of ownership and pride in the work. Employees must have a good understanding of working conditions.

Motivated employees who have faith and trust in their employer can make the difference between success and failure for a business. An employer conveys trust by delegating authority and giving employees opportunities to advance their careers. Employers should take an interest in the well-being of their employees and recognize achievements, if for no other reason than to maintain employee morale. Some other important aspects of a good employer–employee relationship are

- Giving corrections and criticism without embarrassing the employee or attacking the employee as a person
- Distributing unpleasant tasks among all employees, rather than always "dumping" them on a single person
- Anticipating needs and problems and being flexible when employees have to accommodate nonwork matters
- Asking for—and really listening to—feedback

Employees Should Perform to the Limits of Their Abilities

Employees should take an active role in maintaining a good relationship with the employer, recognizing that both employee and employer are part of the same team. Employers expect employees to be dependable and loyal to the organization. Employees also must respect themselves and the employer. Cooperation with the employer and with coworkers is also very important. Employees cannot always expect the employer to know there is a problem. Employees should be able to discuss conflicts and should expect feedback. A positive attitude can improve every situation, and it makes for a productive work environment. A good employee possesses and uses a number of desirable traits and qualities.

Honesty Honesty means more than just not taking things that do not belong to you. It means giving a fair day's work for a fair day's pay. It means holding up your end of the bargain. An employee's thoughts and actions should be honest. Nothing brings quicker dismissal or surer disgrace for an employee than dishonesty.

Loyalty Loyalty means supporting a person or an organization—and it is a two-way proposition. Employees expect employers to protect their interests, to provide steady work, and to promote them to better positions as openings occur. In addition to an honest day's work by each employee, the employer has a right to expect employees to support the company, to protect its interests, and to speak well of the firm to others. The employer also expects employees to confine minor problems to the workplace and to keep business matters confidential.

Willingness to Learn Production procedures and services differ from business to business. Employees are expected to learn their company's specific methods and procedures. When new machinery and pieces of equipment are installed, it is often necessary for even experienced employees to learn new skills and operating procedures. Employees who are promoted or given greater responsibilities are usually the ones who learned new skills and became more valuable to the firm.

Willingness to Assume Responsibility Most employers expect employees to recognize the task at hand and then do it. It is tiring and inefficient for supervisors to have to ask repeatedly about the progress of an assigned task. When an employee receives instructions about the performance of a task, he or she is expected to follow those directions. After the employer delegates or assigns a task, the employee should assume the responsibility and perform the duties without the need for further instruction or supervision.

Respect for Authority Authority is the responsibility and power to give commands, act, and make final decisions. Authority in agribusiness is usually entrusted to managerial personnel who are responsible for production, sales, or service. Even though many agribusinesses are relatively small and uncomplicated, it is important for employees to show respect for those in management positions.

Ability to Get Along with People Being able to get along with others is the most highly prized of all employee traits. This ability seems natural for some people, whereas others gain the ability only through thought, effort, training, and practice. One must learn to think of people as individuals, to view a situation from another person's viewpoint, and sincerely to desire to get along with others. The ability to create and maintain good interpersonal relationships pays big dividends in the form of profits. Both employers and coworkers appreciate good employee attitudes and temperaments.

Willingness to Cooperate One way to exhibit a cooperative attitude is to offer to assist other employees after you have completed your assigned tasks. In most workplaces, there are certain tasks no one likes to do. Employees who are willing to take on these tasks make a good impression on the supervisor and gain the cooperation of other workers. Employees should work as team members alongside supervisors and coworkers.

Observe Rules and Regulations People cannot work well together unless they understand the work. This includes knowing which tasks to do, the proper procedure for completing the tasks, the time required for completion, and who does what tasks. In any work situation, rules and regulations are a necessity, and all employees are expected to know and follow them.

Tardiness and Absenteeism Human beings are creatures of habit and tend to repeat a behavior (continue doing the same thing) unless the consequences are unpleasant. Some people develop the habit of being late early in their school days and just never grow out of it. This unfortunate habit will get a person into trouble on the job very quickly. The clock governs both business and industry. It is the employee's responsibility to use time so that he or she is ready for the next day's work and to be at work on time. When you are not at work you, you are costing your employer money and you are disrupting the organization.[26]

Of course, everyone occasionally has to take time off from work. Very few employers expect people to work when they are ill or having a serious family problem. However, employers have little patience with employees who are constantly absent from or duck out of work for relatively unimportant matters. This results in lost productivity, added burden on those who must try to get the job done with fewer employees, and often hard feelings from the coworkers who have had to pick up the slack.

Always notify your employer well in advance of a planned absence. In case of illness or an emergency that keeps him or her from work, the employee is negligent if he or she does not notify the employer as soon as possible. Simply failing to contact the employer is the absolute worst way to handle an absence: it leaves the employer without any information upon which to make decisions about reassigning jobs or rearranging the work flow.

RESPOND TO AUTHORITY

No matter how much you enjoy your job and your workplace, there will be times when you are asked to undertake tasks or duties you do not like. Your employers will expect you to do such tasks even if they know you do not want to. They expect you to accept your share of any unpleasant or difficult work and to do that work to the best of your ability. Your future is strongly affected by how you react to employers who assign tasks to you; that is, how you respond to authority.

Appropriate Response to Authority

Your supervisor, or boss, is the person of authority who has the responsibility of seeing that the work gets done. To do this, your boss must decide which workers will do the various tasks needed to get the work done. Do not complain or try to have unwanted jobs assigned to someone else. Be willing to cooperate with your supervisor and coworkers.

Not all bosses are kind, fair, and understanding. You may enjoy the job but dislike the way you are told to do your job. Your boss may shout at you, give impersonal orders, or make demands. Developing a cooperative working relationship with this authority figure may be difficult, but you must try to find ways to get along with this type of person: adapt your style to your supervisor's style. Follow the orders you are given exactly; smile and show the boss that you are willing to do what it takes to get the job done. Remember, the relationship you develop with your immediate supervisor is important to your job satisfaction and advancement.

When You Disagree with Your Supervisor

There may be times when you disagree with an authority figure. If you disagree with your supervisor, submit your suggestions for improvement or change through the proper channels using a professional attitude, and prepare yourself for a meeting with her.

FIGURE 23–10 When you disagree with your employer or supervisor, wait until work hours are over and ask for a private meeting. Talk tactfully and calmly, as these two people are doing. *(© iStock/annebaek)*

Supervisors are human, and good ones might be open to suggestions for a better way to do something. When you meet with your supervisor about your disagreement make sure you are prepared with justifications for your disagreement—justifications that address how your ideas will help the company or organization are most effective. It's also best to request a meeting after work hours are over, and ask to speak privately with your supervisor (Figure 23–10). Do not get emotional, angry, or abusive. Explain your feelings about the situation in a matter-of-fact manner and try to work out a solution.

Always remember that the company pays your salary and expects you to do the tasks you are assigned. Every job has situations with which you will disagree. The trick is to be thoughtful and respectful. Whatever the outcome of a discussion with your supervisor, complete your assignments to the best of your ability. How you handle a tough situation like this can go a long way towards building a good relationship with your boss and developing his trust.[27]

ETHICS IN THE WORKPLACE

In Chapter 20, we discussed ethics in the workplace. Remember, **ethics** refers to your standards and values—the basic principles you live by. Your concept of what is right or moral, and what is not, determines how you behave in various situations; it affects your conduct at work, at home, and in the community. People who live and behave according to their personal standards have integrity. To act ethically is to choose what is right. If you do what you know is wrong, you are behaving unethically. You must be willing to accept responsibility for all your choices. This is as true of workplace decisions as of choices in any other aspect of your life.

The Importance of Good Business Ethics

Your personal ethics and your employer's ethics are demonstrated in your products or services and create your reputation in the community. A good worker provides honest, capable service for value received. For example, an auto mechanic with good ethical standards competently determines what is wrong with a car, makes the necessary repairs, and charges a fair price. Customers know that they can trust and depend on this mechanic and will return to this shop when other repairs are needed, knowing that the mechanic treats all customers fairly. An unethical mechanic may make unnecessary repairs, do sloppy or inadequate work, and/or overcharge the customer. The auto mechanic who continues to treat customers unfairly and dishonestly will gain a reputation for unethical business practices and soon will have no customers. Good ethics are good business.

If your work reflects a high ethical code, customers will keep coming back, and they will recommend you to others as well. Following are ways you can demonstrate your high ethical standards through your job:

Be Honest If you sell a product, make sure it does what you say it will do. Do not claim that it is better than it really is just to make a sale. Put yourself in the customer's place: would you be satisfied with the product?

Provide Excellent Service When providing a service, do the best job you possibly can. If you install air conditioners, make sure everything is in working order and clean up before leaving the job. Do not overcharge for your time or services.

Have a Caring Attitude More than anything else, people respond to leaders and workers who are concerned with others and who have compassion. Although some customers and coworkers are hard to like, treat them as you would want to be treated, taking care to consider the financial, emotional, and long-term impacts of your decisions.[28]

JOB PROMOTIONS

Getting a job is the beginning of your career, not the end. One measure of success is your ability to improve your skills. A more obvious measure of success is job promotion. A promotion is an employer's way of indicating that a worker can handle increased responsibility. There are several things an employer looks for when deciding which workers to promote:

Job Mastery Job mastery, or knowing how to do your job well, is an important consideration in getting a promotion. Your employer will notice your skill at doing your job and your willingness to learn new skills. Very often, the best way to get a promotion is to show that you can already do the job to which you want to be promoted.

Mastering your job means your work reflects both quality and quantity. To be promoted, you may need additional training and education in your field to do the higher-level job. You may be able to get the training you need from the company in which you work, but you may also have to attend outside classes for this training. The successful worker never stops learning. Learning new skills leads to promotions.[29]

Initiative Initiative means doing tasks without being told. Initiative indicates a willingness to accept responsibility. If you show initiative on the job, your employer knows you can get the work done without much supervision.[30]

Good General Employability Skills If we took all the positive traits discussed in this chapter and gave each a ranking of one to five (with five being the best), you would need to rank at least a four on most of them to get promoted. You need to be a good follower before you can become a good leader.

Understanding Employer or Supervisor Responsibilities If your promotion moves you into a management or supervisory position, you will need to call on all the leadership abilities and skills you have learned and developed so far in life. You should also be aware that management has certain responsibilities that lower-level jobs do not. Sometimes these obligations are quite complex and carefully written out; other expectations and requirements may not be so formal or concrete. The sooner you understand them, the sooner you will be in a

Always be enthusiastic!

Never quit.

Use common sense.

Believe in yourself even when no one else does!

Work smart.

Smile!

Maintain a positive attitude at all times.

Think big!

Work hard.

Enjoy your life!

Plan your work and work your plan.

Stop making excuses!

Remember: When the going gets tough, the tough get going.

FIGURE 23–11 These techniques should help you get more out of your job and may even help you get a promotion.

position to get promoted. Some of the things you should be aware of are general company rules, as well as policies and procedures regarding the following:

- Safe working conditions
- Payroll and leaves
- Training and evaluation
- Hiring and promotion
- Dismissal and discipline
- Complaints, grievances, and appeals

A Word of Caution: A promotion may not always be the right thing for you. Just as bigger is not necessarily better, job promotions are not always an improvement in your work life or aligned with your career goals. Consider your long-term goals before you make any short-term decisions. The extra responsibility, time, and commitment may not complement your life, especially if you are still in school. Refer to Figure 23–11 for techniques that lead to more self-fulfilling jobs.

COMPLAINT AND APPEAL PROCESS

Most companies have employee handbooks. A good number of these handbooks contain a section detailing the organization's complaint and appeal process. Each company has its own rules

in this area, so make sure you read and understand your employer's policies. Despite the many variations that exist, all employee handbooks should clearly define the following five major areas of workers' rights:[31]

- Complaint and grievance rights
- Rights involving the procedures and nature of discipline
- Job security, including procedures and valid reasons for dismissal
- Rights concerning promotion, career advancement, training, and growth
- Rights concerning eligibility for insurance, pensions, time off, and other benefits

The following section presents one way to handle the complaint and appeal process.

Corporate Due Process

In general, a worker's complaint is initially handled through the employer's in-house complaint procedure. If that procedure does not produce a mutually satisfactory resolution, the worker can file a claim or grievance with an arbitration board. "At filing, each side [is] required to provide a deposit toward payment of the costs of arbitration, which [are] shared equally. The first and required effort at resolution [is] informal mediation, in which a professional mediator seeks to resolve the dispute directly and to reconcile the disputants."[32]

If mediation fails, the grievance goes to arbitration. Each party may represent itself, or it may be represented by the agent of its choice, such as "a labor relations expert, coworker, personnel official, union representative (whether or not the grievant was a union member), or lawyer." The parties select an arbitrator from lists of certified arbitrators or from rosters maintained by professional associations such as the American Arbitration Association. Each party usually has the right to strike potential arbitrators from the list; this is commonly done in private arbitrations. If the parties do not consent to use any of the arbitrators on the list, an arbitrator may be assigned by a labor board.

"The arbitrator would hear the grievance, the employer's response, and any other information either party thought relevant. Hearings would proceed under informal rules similar to those currently used in the arbitration of collective-bargaining agreements. The burden of proof would be on the employee to establish the validity of the claim." After consideration, the arbitrator issues a binding judgment.[33]

PERSONAL AND OCCUPATIONAL SAFETY

Personal and occupational safety in the workplace center around three major areas: awareness, being prepared and having a plan, and knowing your boundaries.[34] Employee handbooks usually include sections on personal and occupational safety standards and policies. Be sure to study this information when you begin employment. The Occupational Safety and Health Administration (OSHA) also has strict safety rules to which employers and employees must adhere.

Awareness

Simply being aware is the most important safety skill in any situation. You can develop the awareness skills that are important in your work setting by paying attention to the details of your environment.

When you begin work at a new job, your work setting and environment will most likely differ significantly from your home or school. You may work in a community other than where you live, where the customers, coworkers, and bosses are primarily strangers, at least initially.

On your first day of work, take a mental inventory of the physical surroundings. For example, if you work in a fast-food restaurant, find the locations of all entrances and exits. Note if there are cameras, alarms, or security guards (and where they are). Is your workplace small and confined, such as an office cubicle or a cashier's position at a grocery checkout? Is it spacious and open, like a sales clerk's in a department store?

Pay attention to the environment outside your workplace as you enter or leave. Are lots of people around, or is it generally quiet? Is it light or dark? What do the entrances and exits look like? Are you outside, near parking lots, alleys, sidewalks, trees, and bushes; or indoors, as in a mall? Is there heavy traffic or only an occasional car passing by? Make a mental checklist of these things on your first day of work. Once you have been there awhile, you may become familiar and comfortable with even a risky environment. If this familiarity dulls your awareness, your safety may be compromised.

Being Prepared and Having a Plan

No matter how secure your work environment appears to be, you are never guaranteed complete safety. Accidents and crimes occur in many settings, not just on the streets or in dark alleys. Therefore, your next step is to create a safety plan for your individual circumstances. If your workplace is a public business, robberies are possible; many fast-food restaurants and convenience stores are common targets for robbers. Most such establishments have some sort of security system, including surveillance cameras or silent alarm buttons by the registers that summon the police. If your workplace does not have any obvious safeguards and does not have a formal procedure, ask about getting one. If the employer does not provide one for you, mentally work out what to do if you are confronted or attacked. If your workplace is a warehouse or factory setting, know and follow the company safety policies—but also make your own plan to avoid accidents and know what to do if an accident does occur.

Know Your Boundaries

In almost any workplace, you will have contact with supervisors who have some authority over you and coworkers who are your peers. To learn how to avoid becoming the victim of a personal crime by someone you know, you must first understand the concept of **boundaries**. The lot on which your home sits may be marked off by a fence: that is a boundary. The basketball court in a gym is marked off by painted lines: these are also boundaries. People have boundaries, too.

Learning the extent or limits of your own boundary system is important to personal safety. Some people tolerate or accept some behaviors that you may find uncomfortable or threatening; conversely, you may not give a second thought to behavior that would be very upsetting to someone else. This is particularly true of personal contact. For example, if you grew up in a home where people often hugged you, swatted your behind, or pinched your cheek, you may believe that it is okay for people to touch you without asking permission. It is important to note that your boundary system is not right or wrong: it is what it is. What is important is that your beliefs and assumptions about your "comfort zone" may compromise your safety. In this area too, alertness and awareness of your surroundings, of other people, and of your own reactions will go a long way toward keeping you safe.

LEAVING A JOB

A letter of **resignation** is your official notice informing your current employer of your intent to leave your job. The letter should state a reason for leaving (e.g., taking a new job, moving away, etc.). This is not the place or time to complain or get even. Never burn any bridges; you never know when you may need to go back to a job or an employer. You may also need a reference from this employer in the future. Everything you have done in this job, including the way in which you handle your resignation, becomes part of the impression you leave behind.

Always give at least two weeks' notice, and more if you can. Write the resignation letter to your immediate supervisor. If you feel that other members of the organization need to be personally informed, send copies of the letter to them. If possible, say something personal and positive in your letter and express your regret about leaving.

CONCLUSION

How you feel about your job is reflected in the way you act and the willingness with which you do your duties. Successful workers show **pride** (a feeling of self-esteem arising from one's accomplishments) in and enthusiasm for their work. If you choose a job that matches your interests, aptitudes, and abilities, you will enjoy your work and take pride in what you are doing. This gives you a positive attitude about yourself and your work. If you give it your best, you should be a successful employee. As a last note, consider the following illustration of the importance of putting just a little more effort into whatever you pursue.

At 211 degrees, water is just hot water, powerless. At 212 degrees, water becomes steam, with more inherent power than it has ever been possible to harvest. At 211 degrees, the water in a locomotive boiler exerts not one ounce of pressure. At 212 degrees, the water in the boiler has the power to pull a long train of cars across a mountain pass.

So it is with human beings. Often the difference between success and failure is just turning up the heat one degree. You are the only person who determines your destiny. If you honestly want to succeed, you will find ways to do it. Remember, if your attitude is positive and your efforts match your goals, your progress will be limitless.

SUMMARY

Although rapid changes in technology have brought great change to the way many businesses operate, one thing that has not changed is employers' need for people with leadership skills, a good work ethic, human relations abilities, and other employability skills.

When employers list what they look for on an applicant's résumé and in interviews, among the more common characteristics are communication skills, positive attitude, dependability, punctuality, and responsibility. Maturity, stability, loyalty, honesty, sincerity, experience and good appearance, all rank higher than educational achievement.

Some of the job skills and personal attributes that employers are looking for when they hire people are achiever, activated, anticipative, attitude, commanding, committed, competitive, courageous, credible, dedicated, dependable, developer, disciplined, driven, empathetic, ethical, focused, ideational, knowledgeable, loyal, organized, responsible, self-confident, team player, and values-based.

The Michigan Employability Skills Task Force determined the general skills that every student should have for jobs at all levels. These include personal management skills, teamwork skills, and academic and technical skills.

Personal management skills include coming to work daily and on time, meeting work deadlines, developing career plans, knowing personal strengths and weaknesses, demonstrating self-control, paying attention to details, following written and oral instructions and directions, working without supervision, learning new skills, identifying and suggesting new ways to do a job, being organized, and demonstrating personal values at work.

Teamwork skills include actively participating in a group, following the group's rules and values, listening to other group members, expressing ideas, being sensitive to members' ideas and views, being willing to compromise if necessary to accomplish goals, being a leader or follower to accomplish goals, and being able to work in changing settings and with different people.

Academic and technical skills include reading and comprehending written materials, interpreting charts and graphs, doing basic math, computing to solve problems, using research and library skills, using tools and equipment, speaking and writing in the language in which business is conducted, using the scientific method to solve problems, and using specialized knowledge to get a job done.

You must use good human relationship skills with your fellow employees. Relationships are enhanced by a positive attitude, cooperation, dependability, trustworthiness, working hard, respecting others, handling criticism, exhibiting appropriate dress and grooming, showing initiative, diligence, being capable, and showing commitment.

Workers must also learn to respond to authority and practice ethics in the workplace. You can maintain and demonstrate a high ethical standard on the job by being honest, providing worthy service, and having a positive attitude.

Getting a job is the beginning of your career, not the end. An obvious measure of success is job promotion. There are several things an employer looks for when deciding which workers

to promote: job mastery, initiative, good general employability skills, and understanding a supervisor's responsibilities.

In return for their best possible efforts, an employer must attempt to fulfill the needs of each of its employees. A formal complaint and appeal process should be accessible by any employee who has a grievance, and discipline procedures should be fair and universally applied. Although the employer is legally responsible for many aspects of work safety, you should also take measures to maintain your personal safety and well-being in the workplace. Stay aware of what is going on around you, plan and be prepared for accidents and emergencies, and acknowledge your personal boundaries.

A letter of resignation is your official notice informing your employer of your intent to leave your job. Never burn any bridges; you never know when you may need to go back to a job.

You have the most control over your success on the job. If you honestly want to succeed, you will find ways to do it. Remember, if your attitude is positive and your efforts match your goals, your progress will be limitless.

REVIEW QUESTIONS

1. Define the Terms to Know.

2. List 25 job skills and personal attributes that employers are looking for in potential new hires.

3. Name 13 personal management skills needed by employees.

4. What are five steps that can help you follow directions?

5. Name eight teamwork skills needed by employees.

6. What are 10 academic and technical skills that help make you a good employee?

7. What are 12 qualities that a successful employee shows with fellow employees?

 Take It to the Net

Explore employability skills on the Internet. Use your favorite search engine and search until you find a site that you feel is both useful and authoritative. Summarize that site and tell why you feel it is useful. If you are having problems with the listed sites or just want more information, use the following search terms:

Search Terms

employability skills

work skills

work ethics

job skills

Chapter Exercises

8. List eight ways to exhibit or display positive work attitudes.

9. List five things you can do to change from a negative to a positive attitude.

10. What are 11 things you can do to help develop good cooperative skills with coworkers?

11. List four appropriate responses to authority.

12. How can you take care of the problem if you feel you have been treated unfairly by your employer?

13. What are three ways to show high ethical standards on the job?

14. What are four things an employer looks for when deciding which worker to promote?

15. When employers were asked to list the most important skills and characteristics they looked for when hiring new employees, they listed 10 things. What are they?

16. What 11 qualities do employers look for in résumés and during interviews?

17. What are four important aspects of a good employer–employee relationship.

18. What are 12 techniques for a more self-fulfilling job?

COMPLETION

1. Your attitude will affect your job _____.

2. Part of being dependable is being _____ and _____.

3. Respect involves simple, common _____.

4. Take criticism _____ instead of personally.

5. Breaking large tasks down into a _____ of _____ tasks will make jobs seem easier.

6. True leaders are not afraid to demonstrate personal_____ by example in their every-day lives.

7. _____ refer to your standards and values.

8. Employers first judge a prospective employee's writing ability by the quality of an _____., _____, or _____.

9. A _____ is an employer's way of indicating that a worker can assume increased responsibility.

10. Never burn any _____. You never know when you may need to go back to a job.

11. Technology is changing rapidly, but one thing that has not changed is the fact that employers are looking for employees with _____ ability, a good _____ _____, and _____ _____.

12. Employers primarily want good, _____, _____ people with _____ attitudes.

13. Make punctuality a goal. Successful people are _____ _____.

14. Failure to meet a deadline may cost the employer a customer and hence result in _____ _____.

15. Failure to meet deadlines could cost you _____ _____.

16. Those who are successful in the workplace know where they are going and are executing their plans to _____ _____.

17. Good timing and _____ _____ are everything when it comes to self control.

18. Your skills in communication (or lack thereof) may influence whether you get a _____.

19. Paying attention to _____ helps you avoid many mistakes.

20. If you do not understand something, _____ questions.

21. If your supervisor tells you to do a job in a certain way, follow those directions _____.

22. Your objective should be to please your boss by working to please _____.

23. When workers are willing to learn new tasks they save their employers both _____ and _____.

24. In the workplace, _____ are organized so that individuals' talents and skills can be directed through group effort to the accomplishment of vital tasks and goals.

25. In our rapidly changing society, successful workers are the ones who keep abreast of new _____ and the latest _____.

26. Like any attitude, _____ is contagious: you can _____ your _____ to coworkers and thereby make the whole workplace better.

27. You succeed when you help others to _____.

28. Learn from your _____, and think _____ about the future.

29. Do not make assumptions or judgments about others unless you know all the _____.

30. If two or more coworkers have a _____, stay out of it.

31. Treat other people as you like to be _____.

32. When you show _____, you are the first to start or introduce something.

33. _____ means really staying with a task when the going gets tough: you do not support the team only when they team is winning.

34. You need to be a good follower before you can become a good _____.

MATCHING

_____ 1. Juan always starts the workday in a pleasant mood.

_____ 2. After finishing the typing she was asked to do, Linda started straightening up the office.

_____ 3. Although his boss is rather unfriendly, Arnold accepts this and tries to be friendly to others.

_____ 4. Performing your job without the boss standing over you.

_____ 5. Pedro always gets to work on time.

_____ 6. Beyond being dependable and showing initiative, you can also be counted on because of your sincere desire for the organization to succeed.

A. honesty

B. initiative

C. willingness to accept change

D. dependability

E. willingness to accept criticism

F. cooperation

G. positive attitude

_____ 7. After Maria's employer showed her what she was doing wrong, she worked hard to improve that skill.

_____ 8. Miguel does not agree with his boss on how to get the job done, but he does know who the boss is.

_____ 9. Joe, who has been with the store for 20 years, worked hard to learn the new (and more difficult) inventory system.

_____ 10. Elena makes mistakes, but she knows what is right and wrong and is not swayed by negative influences at work.

_____ 11. Carmen realizes that she shortchanged a customer and will be 50 cents over in the cash drawer. Instead of putting the 50 cents in her pocket, she leaves the money in the drawer.

H. working without supervision

I. response to authority

J. commitment

K. ethics

ACTIVITIES

1. From the list of job skills and personal attributes that employers seek in employees, select five that will present the biggest challenge to you. Write a short paragraph on each, explaining why you need to improve and how you intend to do this.

2. This chapter discussed 13 qualities that can enhance relationships between an employer and employees. Interview an employer in your community, and ask what qualities he or she looks for in employees. Compare these answers with the qualities mentioned in this chapter.

3. You are interviewing for a job and the interviewer says, "Convince me that you can be a self-learner." Write a 100-word essay on what you will tell the interviewer.

4. From the list of personal management skills, select three that could be a problem for you. Write a plan of action for each of the three, explaining how to improve so that you can become a better employee.

5. Review the list of academic and technical skills in this chapter. Identify two that could prevent you from getting the job you want. Write a plan of action on how you can improve in these two areas.

6. To determine what work skills are needed in your area, ask your boss or some other authority where you work the 10 most important things he or she looks for in an employee. If you do not work, ask your instructor, someone in your neighborhood, or a friend of the family. Report these to your class.

7. Write a letter of resignation. Remember, the purpose of this letter is not to complain or get even. Say something personal and positive in your letter and express your regret about leaving.

8. Write a letter of recommendation for a friend. In this case, the friend is you. State in the letter why you would hire your "friend." Do not hold back. List as many positive attributes as you can.

9. List three positive potential employability skills for each member of your class.

NOTES

1. J. P. Robinson, "What Are Employability Skills?" _The Workplace_ (Alabama Cooperative Extension System Fact Sheet), 1(3) (September 15, 2000), retrieved October 3, 2016 from http://www.fremont.k12.ca.us/cms/lib04/CA01000848/Centricity/Domain/189/employability-skills.pdf.
2. Ibid.
3. S. Adams, "The 10 Skills Employers Most Want in 2015 Graduates," _Forbes_, 2014, retrieved October 3, 2016 from http://www.forbes.com/sites/susanadams/2014/11/12/the-10-skills-employers-most-want-in-2015-graduates/#6053465719f6.

4. D. Lee, *10 Qualities Interviewers Look For,* retrieved November 30, 2009, from http://www.goldsea.com /Career/Qualities/qualities.html.

5. P. Stemmer, B. Brown, and C. Smith, "The Employability Skills Portfolio," *Educational Leadership* 49(6): 32–35, at pp. 32, 33 (March 1992).

6. M. Hensey, *Collective Excellence: Building Effective Teams,* 2nd ed. (Reston, VA: ASCE Press, 2001), p. 3.

7. R. Lussier, *Human Relations In Organizations: Applications and Skill Building,* 10th ed., (New York: McGraw-Hill Education, 2017), p. 331.

8. Ibid, p. 334.

9. Carnevale, Gainer, and Meltzer, *Workplace Basics,* p. 18.

10. J. M. Lannon, *Technical Communication,* 10th ed. (Reading, MA: Addison-Wesley, 2005).

11. National Business Education Association: Educating for success in business and life, *Computation,* retrieved October 8, 2016 from https://www.nbea.org/newsite/curriculum/standards/computation.html.

12. S. Herring, "Transforming the Workplace: Critical Skills and Learning Methods for the Successful 21st Century Worker," *Big Think,* n. d., retrieved October 9, 2013 from http://bigthink.com/experts-corner /transforming-the-workplace-critical-skills-and-learning-methods-for-the-successful-21st-century-worker.

13. R. A. Noe, A. D. M. Clarke, and H. J. Klein, "Learning in the 21st Century Workplace," *The Annual Review of Organizational Psychology and Organizational Behavior,* 1, 2014, pp. 245–275, retrieved October 8, 2016 from http://www.annualreviews.org/doi/pdf/10.1146/annurev-orgpsych-031413-091321.

14. L. A. Liddell and Y. S. Gentzler, *Building Life Skills, 6th ed.* (Tinley Park, IL: Goodheart-Willcox Co., 2008).

15. B. L. Reece, *Effective Human Relations: Interpersonal and Organizational Application, 12th ed.,* 2014 (Mason, OH: South-Western), p. 131.

16. "Stress Management," *The Mayo Clinic,* 2014, retrieved October 9, 2016 from http://www.mayoclinic.org /healthy-lifestyle/stress-management/in-depth/positive-thinking/art-20043950?pg=2.

17. "Working with Others," *Skills for Success* series no. 12 (Nashville, TN: Tennessee Department of Education, 1993), pp. 209–211.

18. Liddell and Gentzler, *Building Life Skills,* pp. 516–517.

19. Liddell and Gentzler, *Building Life Skills,* p. 517.

20. Ibid., p. 516.

21. Ibid., p. 517.

22. Ibid.

23. Ibid.

24. N. Wehlage and M. Larson-Kennedy, *Goals for Living: Managing Your Resources* (Tinley Park, IL: Goodheart-Willcox, 2006).

25. Ibid.

26. M. C. Kocakulah, A. G. Kelley, K. M. Mitchell, and M. P. Ruggieri, "Absenteeism Problems and Costs: Causes, Effects and Cures," *International Business & Economics Research Journal,* 15(3), (2016), p. 89.

27. S. M. Heathfield, "How to Disagree with Your Boss—And Thrive: 10 Steps to Effective Disagreement in a Boss Relationship," *the balance,* 2015, retrieved October 10, 2016 from https://www.thebalance.com/how-to -disagree-with-your-boss-and-thrive-1917873.

28. T. Bennett, "12 Business Ethics Examples," *udemy blog,* 2014, retrieved October 10, 2016 from https://blog .udemy.com/business-ethics-examples/.

29. C. T. Nyakundi and J. Davidson, *How to Get a Promotion at Work* (J. D. Biz Publishing, 2015).

30. Bender Consulting Services, *Initiative—The Key to Becoming a Star Employee,* n. d., *retrieved* October 10, 2016 from http://www.benderconsult.com/articles/initiative-key-becoming-star-employee.

31. R. Edwards, *Rights at Work: Employment Relations in the Post-Union Era* (Washington, DC: The Brookings Institution, 1993), p. 226.

32. Ibid., p. 218.

33. Ibid.

34. D. Chaiet, *Staying Safe at Work* (New York: Rosen Publishing Group, 1995).

Appendix A

The "Shotgun Approach": Implementation of Key Principles of Learning

At the beginning of every unit, a list of terms is presented. The following explains how this material can best be taught. The purpose of presenting this information is simply to illustrate one method of teaching that is exciting, engaging, and student centered. It is not the only tool to use for teaching the dynamic information in this text, but it does integrate several different principles of learning.

One of the older methods of teaching is rote learning, or memorization through repetition. Many teachers used this in teaching spelling, states, and the state capitals. Agricultural education teachers have used this method in teaching the identification of breeds, shop tools, and other subject matter used in basic agricultural production courses. Was it effective? Generally, I'm sure that it was, depending on the teacher. The Shotgun Approach as a method of teaching is similar, but there is much more to this method than just rote drill. The latter part of this section will attempt to explain why this method should be used more often in education today. Now, an overview of the Shotgun Approach will be given.

The reason for the name *Shotgun Approach* is primarily the fact that it covers everything in one class period. In other words, the lesson is introduced, the subject matter is presented, and testing occurs within the same class period.

Another reason for the name is that it is hard to miss with this method if you are "aiming in the right direction."

The Shotgun Approach can be used in presenting almost any subject matter, but the author will use as an example teaching the definition of terms. The teaching objective will be for students to learn the definition of twenty new terms in a one-hour class period, with 80 percent of the class scoring 100 on a multiple-choice test at the end of the class period. For the sake of clarity, the following steps will be written in the form of a "job operations sheet." The steps are as follows.

1. Write all the terms on the board before the class starts. (Tell the class that you expect that at least 80 percent of them will score 100 at the end of the class and that they will possibly do even better.)
2. Ask a student, "Do you know the definition of the first term?" If the student does not know the answer, ask another student. Draw on information from the students.
3. Thoroughly build up students' interest and seek answers to questions by using problem solving–type questions. (The purpose here is to make the students think.)
4. After drawing information from the students, present the definition as it should be or restate what one of the students has said.

5. Present some extra information about the term, to make it relevant. (Present examples so that students can relate the information to something that is recorded in their perceptive field. Jesus made information relevant by using parables.)

6. Ask the question repeatedly to several students until you are sure that the information (term) has been thoroughly understood and perceived. (Ask the question in several different ways if necessary.)

7. Praise the students when they answer questions correctly. (When questions are not answered correctly, do not react with a negative comment. Thank the student for the response and make him/her feel that a positive contribution has been made.)

8. Proceed to the second term and repeat steps 2 through 7.

9. After thoroughly covering the second term, go back and ask students to define the first term again. (Do not proceed to the third term until you are sure everyone understands and perceives what is presented.)

10. Proceed to the third term and repeat steps 2 through 7. Do not continue with the next term until the third term is thoroughly understood and you have asked for and received the answers to questions about the first and second terms.

11. Proceed to the fourth term and repeat the same procedures already mentioned.

12. Continue through term 20, remembering always to ask the preceding questions "all the way back" to term number 1 before going on to the next question. (This goes faster than one would expect, once the teacher gets acclimated to the approach.)

13. After all twenty terms have been covered, ask for two volunteers. Try to select an overachiever and an underachiever without the students realizing the intended difference of selection.

14. Before accepting the volunteers, make sure that they know the conditions. (They are as follows: each student will be asked to stand in front of the room and each will be asked ten questions. If they get eight out of ten correct, the student will earn a score of 100. However, if only seven questions are

answered correctly, the student makes a score of zero.) Nevertheless, if the student earns a score of zero, he/she has the option of taking the test again when the other students take the test.

Note: The point in having the two volunteers in front of the classroom is to literally get the class on the edge of their chairs.

This can be done if the teacher manipulates the questions by asking the most difficult ones first so that the volunteer misses one or two questions out of the first six or seven. Total involvement and enthusiasm on the part of the teacher are mandatory so that the students will get excited. Also, the students will be thinking through the answers as the students in front of the class are participating. An optimum situation occurs when each student answers eight of ten questions correctly. Why the overachievers and underachievers? The overachievers can be asked the most difficult questions and the underachievers can be asked the less difficult questions. (Again, do not make this obvious to the students.)

Furthermore, since the purpose of the students being in front of the room and answering questions is for the remaining students to have their learning reinforced, the overachiever can add clarity to difficult questions while the accomplishment of the weaker student can add self-confidence to the remaining class members.

Note: This step is not mandatory for the Shotgun Approach to be effective as a teaching method. If you choose to use this step, it is not imperative that an overachiever and underachiever be selected. This step does add variety, however, and makes the class more interesting. Moreover, it increases the grades substantially.

15. After step 14, give the remaining students the test by giving a definition and having the students select the answer from the board. (It is unfair to assume that all students can learn to spell all twenty terms correctly in this amount of time.)

What are the results? Eighty percent of the students will hopefully score 100 if the approach is done correctly. However, it should be made clear that covering twenty questions is not the real goal. Fifteen questions may be all that can be covered. Allow approximately ten minutes for the testing. Stop at whatever term you are on at that point. The important thing is that the students thoroughly learn and understand the material that is presented.

WHY THE SHOTGUN APPROACH IS SUCCESSFUL

1. **Respect for the Student as a Self-Actualizing Human Being.** Praise, confidence, and the reduction of fear of failure are essential in order for maximum learning to take place. This helps students realize success, and success is a highly motivating force for learning. The use of the Shotgun Approach works under the philosophy that the teacher should be so effective that a) learning is reinforced, and b) students will score high on tests, even days after the teacher has presented the subject matter. Retention of material learned is high.

2. **Thinking, Understanding, and Perception Are Maximized.** Hammonds (1968) stated that "[u]nderstanding a thing consists largely in seeing why it is true. One cannot see why a thing is true unless he sees it in relation to other knowledge which he already has and understand, unless its relevant."[1] Ultimate understanding is full perception.

   The maximum observation of perception is one of the major factors in utilizing the Shotgun Approach. Perception goes further than understanding. For example, nobody has been able to teach or explain to this writer how the earth can be round without someone falling off. I know the earth is round. I have no doubts. I understand it. However, it still has not been perceived. No one has been able to relate this information to me so that "I can see it."

3. **Awareness of Student's Perceptive Field.** The Shotgun Approach emphasizes awareness and perception: a story or incident is told about each term so that the student can relate this to something in his or her perceptive field.

4. **Reinforcement and Repetition.** Reinforcement and repetition can be effective tools for retention of knowledge. In the Shotgun Approach, the more use the teacher makes of reinforcement and repetition, the more learning takes place. It is most effective here, when it fails in other circumstances, because it is combined with all of the other concepts being described herein.

5. **Problem-Solving Approach.** The problem-solving approach is utilized in the Shotgun Approach in that students are asked questions that make them think. The situations are real.

6. **Motivation.** Another crucial reason for the success of the Shotgun Approach is that it is motivating. The students get excited. Learning is a challenge to them, but it is fun and enjoyable.

7. **Attention to Talents Beyond Intelligence.** Our education system is primarily based upon the academic (cognitive) skills or intelligence of students. Beyond intelligence is the ability to think, produce, make decisions, communicate, make predictions, create, and plan. The Shotgun Approach brings all of a student's talents to the table.

Many other reasons could be given for the success or effectiveness of the Shotgun Approach as a method of teaching. However, the purpose of these comments is to introduce a method of teaching that applies to the best principles of teaching and learning as identified by some of the foremost educators in this century. Rosenshine and Furst (1971) list the following five criteria as the most important variables between teacher behavior and student achievement: clarity, variability, enthusiasm, task-orientation and/or businesslike behavior, and student opportunity to learn criterion material. You can be the judge as to whether the Shotgun Approach meets these criteria.[2]

In conclusion, the Shotgun Approach is used with maximum student learning as its goal. Schools exist for students to learn. Teaching is not a game. Teachers should never trick students or make learning a guessing game. The teacher should tell the students what he or she expects them to know or learn from the unit (objectives), and present the materials in such a fashion that optimum learning can take place. If the students score high on an evaluation, good. If they score low, the teacher in all likelihood is not teaching effectively. He or she is only presenting information or serving as a guide. This standard may seem high, but this is the first part of the twenty-first century: it is time for teachers to improve. Teachers must strive toward the goal that their students fully perceive any knowledge presented. Needless to say, the teacher cannot cover as much subject matter using the Shotgun Approach as s/he can with most other methods. However, if motivation, instilling confidence in students, making education enjoyable and, most of all, learning are the goals of the teacher, the Shotgun Approach should be used as an alternative teaching method.

NOTES

1. Hammonds, C., and Lamar, C. F. (1968). *Teaching vocations*. Danville, IL: The Interstate Printers and Publishers.
2. Rosenshine, B. and Furst, N. (1971). "Research on teacher performance criteria." In B. O. Smith (Ed.), *Research in Teacher Education* (pp. 37–72). Englewood Cliffs, NJ: Prentice Hall.

Appendix B

Parliamentary Procedure Career Development Event

IMPORTANT NOTE: *Please thoroughly read the Introduction Section at the beginning of this handbook for complete rules and procedures that are relevant to all National FFA Career Development Events.*

PURPOSE

To encourage students to learn to effectively participate in a business meeting and to assist in the development of their leadership, research, problem solving, and critical thinking skills.

OBJECTIVES

After completing this summary, the student should be able to:

1. Use parliamentary procedure to conduct an orderly and efficient meeting.
2. Demonstrate knowledge of parliamentary law.
3. Present a logical, realistic, and convincing debate on motions.
4. Evaluate minutes and organizational documents.
5. Use parliamentary resources to solve problems of organizational management and operations.

EVENT RULES

1. A team ... will consist of six members from the same chapter. All practicums will involve all six members.
2. It is highly recommended that participants wear FFA Official Dress.
3. The advisor will not consult with the team after entering the holding room prior to each round of the event.
4. Any participant in possession of an electronic device in the event area is subject to disqualification.

EVENT FORMAT

Equipment
Materials student must provide.

1. A minimum of two sharpened No. 2 pencils.
2. A copy of the current edition of *Robert's Rules of Order, Newly Revised*

Materials provided by the event committee.
1. A gavel will be supplied for the chair.
2. Teams may choose to use their own gavel if they so desire.
3. Paper and pencils will be provided to chair and secretary stations.

4. A searchable current edition of the *Robert's Rules of Order, Newly Revised* may be provided.

THE EVENT WILL HAVE FIVE PHASES

1. Written examination
2. An 11-minute team presentation of parliamentary procedure
3. Oral questions following the presentation
4. Team problem-solving practicum
5. Individual practicum focused on minutes and other records

WRITTEN TEST (200 POINTS)

Part 1

Five open-book parliamentary procedure research questions using the current edition of *Robert's Rules of Order, Newly Revised*. Participants will be allowed 30 minutes to complete Part 1 of the exam. All team members are required to provide their own copy of the most current edition of *Robert's Rules of Order Newly Revised*.

An example of one research question is outlined below:

- "The term *rules of order* refers to written rules of parliamentary procedure formally adopted by an assembly or an organization."

  Answer: *Robert's Rules of Order, Newly Revised*, beginning of page 15

Part 2

Forty-five multiple-choice questions taken *from Robert's Rules of Order, Newly Revised*. Participants will have one hour to complete Part 2 of the exam.

Note: References and materials cannot be used for this part.

Exam content will be guided by National Association of Parliamentarians Members and Leaders Body of Knowledge and the Society for Agricultural Education Parliamentarians (SAEP) accreditation processes.

Participants receiving a cumulative score of 80 percent or greater on the exam will be recognized

as Accredited Parliamentarians (APs) by the Society of Agricultural Education Parliamentarians (SAEP) and will be eligible for membership in the National Association of Parliamentarians (NAP) and American Institute of Parliamentarians (AIP). The average score of the six team members will be used to compute the total team score that will be used for each round.

PRESENTATION (500 POINTS)

1. The national event will have three rounds: a preliminary round, a semi-final round and a final round. The preliminary round will have six sections. A section shall be made up of six to nine teams. Two teams from each of the sections, for a total of twelve teams will advance to the semi-final round. Two teams in each semi-final section will advance to the final round of four teams.

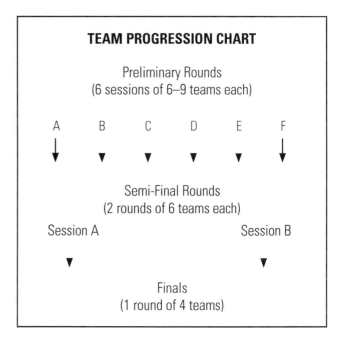

TEAM PROGRESSION CHART

Preliminary Rounds
(6 sessions of 6–9 teams each)

A   B   C   D   E   F

Semi-Final Rounds
(2 rounds of 6 teams each)

Session A   Session B

Finals
(1 round of 4 teams)

2. Seeding Process–Teams will be placed into preliminary and semi-final rounds based on the teams' exam scores, which is the average score of the six team members.
3. Item of Business–Each team will address a local chapter item of business that would normally be a part of a chapter's Program of Activities (e.g. Food for America, Project PALS, WEA, fundraisers, recreation, etc.).

Consult the Official FFA Manual and Student Handbook for specific activities and current programs. The motion will be specific and must be moved as an original main motion as it is written on the event card. Motions not on the chart of permissible motions, or secondary motions and debate applied to them, will not be scored.

4. Event Card–The event officials will select two subsidiary, two incidental, and one privileged or a motion that brings a question again before the assembly from the list of permissible motions. These motions will be on an index card and one will be randomly assigned to each team member. All teams in each session will be assigned the same motions. There are 25 permissible motions in the national FFA event. Team members will have one minute to review the main motion, the motions to be demonstrated and to identify his/her motion (which may be noted by bolding, underlining, or highlighting). Members may not confer, use nonverbal communications during the one-minute time period or during the demonstration.

SAMPLE CARD

Main Motion:

I move to sell citrus as a fundraiser.

Required Motions:

Lay on the Table
Amend
Suspend the Rules
Appeal
Reconsider

5. Opening and Closing the Demonstration– The team demonstrating shall assume that a regular chapter meeting is in progress and new business is being handled on the agenda. The chair shall start the presentation by saying, "Is there any new business?" time will stop when the chair declares the meeting adjourned.

  i. Original Main Motion: The event official will assign the main motion on an index card, no other original main motions that are not on the event card will result in a 50 point deduction from overall team presentation score.

  ii. The assigned original main motion is to be the first item of business presented, unless, take from the table, reconsider or rescind will be provided.

  iii. The person who makes the assigned main motion will be given credit for an additional motion.

6. Secondary Motions–There shall be no limitation to the number of subsidiary, incidental and privileged motions that the team must demonstrate. A member's required motion will not be counted as an additional motion for another member. No motion may count for an additional motion for more than one member. Incidental and privileged motions cannot be demonstrated as incidental motions.

7. Individual Member Recognition– A member may speak in debate on the main motion and conclude by offering a secondary motion. Judges will award points accordingly for both the debate and the secondary motion. Omission of the assigned motion by the assigned member on the event card will result in a 50 point deduction from overall team presentation score.

8. Motions that Bring a Question Again Before the Assembly–If the officials in charge designate rescind, reconsider, or take from the table, as a motion to be demonstrated, the scenario will be included on the event card. These motions shall not be used unless listed on the event card as a required motion.

9. Call for Orders of the Day–If the event officials designate call for orders of the day as a motion to be demonstrated, a scenario will be provided on the event card. Participants are to assume that a motion was postponed at the last meeting and made a special order for a time during the current demonstration.

10. Debate–The top four debates per member will be tabulated in the presentation score. No more than two debates per member per motion will be tabulated, even if the subsidiary motion to extend the limits of debate has been passed.

11. Time Limit and Deductions–A team, shall be allowed 11 minutes in which to demonstrate knowledge of parliamentary law. A deduction of two points/second for every second over 11 minutes will be assessed. Example: 11:05 = 10 point deduction. A timekeeper will furnish the time used by each team at the close of the event.

ORAL QUESTIONS (100 POINTS)

1. Individual Questions–The team members (not including the chair) will be asked a planned question, which may include one to three parts, relating to their assigned motion. No one may step forward to help another member answer their individual question. The chair will be asked a question relating to presiding, debate, assigning the floor or other general parliamentary procedures. Each member will be scored a maximum of 16 points for responses to questions. Chair will be scored at a maximum of 20 points.
2. Clarifying Questions–The judges will have three minutes to ask clarifying questions related to the team's demonstration that may impact other aspects of team demonstration scores. Questions may be directed to the team or an individual member. Team members may volunteer to answer the question for the team or to help another member. This round of questions are not scored separately.

TEAM PROBLEM-SOLVING PRACTICUM (150 POINTS)

In the preliminary and semi-final rounds, all teams will complete a team problem solving practicum. Teams advancing to the finals will carry with them an average of their scores in the first two rounds. Teams will be provided with a short parliamentary procedure scenario outlining a practical problem. Working as a team, they will have 30 minutes to research the problem and write a short solution with reference to specific page and line numbers in *Robert's Rules of Order, Newly Revised*. All team members are required to provide their own copy of the most current edition of *Robert's Rules of Order, Newly Revised*. Members of the National Association of Parliamentarians and the American Institutions of Parliamentarians will be invited to review and participate in this portion of the event. The most recent National FFA Career and Leadership Development Events Handbook has an excellent Sample Team Problem Solving Activity with Answers.

Teams may be provided with access to a computer to type their responses and access to a searchable database of the most current edition of *Robert's Rules of Order, Newly Revised*. The searchable database will be secured from the National Association of Parliamentarians online store: https://www.parliamentarians.org.

INDIVIDUAL PRACTICUM, MINUTES, AND OTHER RECORDS (50 POINTS)

Each team member will participate in a 30-minute practicum that addresses organizational minutes and other records. Participants will be provided materials and responses will be captured using a Scantron form. Reference materials will not be allowed during this practicum. The practicum will assess NAP Body of Knowledge for Leaders of Organizations Domain 2: Minutes and Other Records; and *Robert's Rules of Order, Newly Revised*, pages 354–355 and 468–480.

SCORING

Guidelines for Scoring Discussion (60 Points Per Member)

1. It is essential that each judge observe and maintain consistent criteria in scoring discussions for the duration of the event.
2. Judges must overlook personal opinions and beliefs and score discussion in an unbiased manner. All debate should be

NATIONAL FFA
CAREER AND LEADERSHIP
DEVELOPMENT EVENTS

Team Problem Solving Activity Scorecard
(Semi-Final and Final Rounds)

CHAPTER   STATE   TEAM NUMBER

SCORING CRITERIA	Possible Points	Points Earned
Reference • Team accurately identified the correct page(s) and line number(s) in *Robert's Rules of Order Newly*, Revised (11th ed.)	60	
Solution to the Problem • Team provided logical justification and reasoning to develop, using citations listed from above to solve the parliamentary procedure problem/issue.	75	
Grammar, Style and Legibility • Complete sentences • Correct spelling (deduction of 1 point/error) • Correct punctuation (deduction of 1 point/error) • Legibility and clarity	15	
TOTAL POINTS	150	

Comments:

_____

_____

_____

_____

_____

scored at the time it is delivered. Each time a participant in the presentation discusses any motion, they may earn a score. However, an individual may never earn more than 60 points in a given presentation. The top four debates per member will be tabulated in the presentation score. No more than two debates per member per motion will be tabulated even if an extension of debate is passed.

Characteristics of an Effective Debate

Characteristics of effective debate include the member's ability to state their position, to provide reason(s) supporting their position, and to tell the delegation how to vote. The delivery of the debate will include a) completeness of thought, b) logical reasoning, c) clear statement of speaker's position, d) conviction of delivery, and e) concise and effective statement of debate.

1. Good Debate – A good debate would be characterized by a presentation that includes the components of a good debate as well as the quality of delivery in which the debate is delivered. Those components are
   - States position
   - Provides more than one reason supporting their position
   - Tells delegation how to vote
2. Average Debate – An average debate would be characterized by a presentation that includes only one supporting reason or lacks in quality of delivery.
   - States position
   - Provides one reason supporting their position
   - Tells delegation how to vote
3. Poor Debate – A poor debate would be characterized by a lack of effective delivery, poor grammar, and faulty reasoning or substance, as well as the omission of one or more components of an effective debate.
4. Suggested grading scale for debates:
   - Good: 15–20 points
   - Average: 8–14 points
   - Poor: 0–7 points

GUIDELINES FOR SCORING THE CHAIR (80 POINTS)

The chair is evaluated by his/her ability to preside and his/her leadership.

1. Ability to preside—handling and stating of motions correctly, following rules of debate, keeping members informed, putting motions to a vote, announcing results of a vote, use of the gavel, and awareness of business on the floor. (65 points)

2. Leadership—stage presence, poise, self-confidence, politeness, and voice. (15 points)
3. Suggested grading scale for chair:
   - Excellent: 1–15 points
   - Good: 6–10 points
   - Poor: 0–5 points

The judges will use Form 2 to score the event. The top two teams will be ranked first and second based on the judges' lowest combined rank. The remaining teams will be designated silver or bronze awards.

GUIDELINES FOR SCORING TEAM EFFECT (20 POINTS)

1. Conclusions reached by the team: Main motion was well analyzed; this may include who, what, when, where, why, and how.
2. Team use of debate: degree to which debate was convincing, logical, realistic, orderly and efficient, germane, and free from repetition.
3. Team presence: voice, poise, expression, grammar, gestures, and professionalism.

TIEBREAKERS

Tiebreakers for teams will be as follows:

- Total final presentation score out of 500 possible points
- Team average score on the written exam
- Total team practicum, problem-solving score

AWARDS

Awards will be presented to teams at the awards ceremony based upon their rankings. Awards are sponsored by a cooperating industry sponsor(s) as a special project, and/or by the general fund of the National FFA Foundation. The first-place national team will be presented a trophy plaque. Each member of the first-place team will be presented with an individual team member plaque. A national gold plaque and individual medals will be presented to the top 12 teams competing in the event; silver plaques and individual medals to the middle; and remaining teams and individuals competing will receive bronze. The top four teams will each receive a designated gold plaque.

States vary in how they award teams and individuals.

Specialty Awards

Specialty awards may be given for the following:

- Outstanding chair
- Outstanding member
- Outstanding critical thinking team

- High average team exam score
- Perfect exam score

Reference: *Parliamentary Procedure 2017–2021, National FFA Career and Leadership Development Events Handbook.*

Appendix C

JOBS IN AGRICULTURE

Agricultural Advertiser Account Executive
Agricultural Agency Employee
Agricultural Attaché
Agricultural Chemical Salesperson
Agricultural Chemist
Agricultural Commodity Grader
Agricultural Commodity Inspector
Agricultural Commodity Warehousing Examiner
Agricultural Consultant
Agricultural Broker
Agricultural Business Administrator
Agricultural Economist
Agricultural Writer, Editor
Agricultural Educator
Agricultural Engineer
Agricultural Geographer
Agricultural Instructor
Agricultural Journalist
Agricultural Machinery Serviceperson
Agricultural Management Specialist
Agricultural Marketing Specialist
Agricultural Missionary
Agricultural Program Specialist
Agricultural Researcher
Agricultural Market Reporter
Agricultural Statistician
Agricultural Trade Magazine Editor

Agriculture Advertising Writer
Agriculturist (many countries)
Aerial Plant Nutrient Applicator
Agronomist
Animal Behaviorist
Animal Breeder
Animal Husbandperson
Animal Specialist
Aquatic Weed Specialist
Arachnologist (zoology of spiders)
Area Economist
Associate Buyer
Avian Specialist
Bacterial Pesticide Specialist
Bacteriologist
Baker Scientist Technologist, Manager
Bank Agricultural Representative
Banking Official
Beekeeper
Biochemist
Biological Pesticide Specialist
Biologist
Biophysicist
Biostatistician
Botanist
Breeding Technician
Cattle Manager
Cereal Chemist
Clay Mineralogist
Climatologist
College Agricultural Researcher
College Faculty Member

Commercial Cattle Feeder
Commodities Broker (Ag)
Conservationist—forest, range, soil, water, wildlife
Consumer Marketing Specialist
Cooperative Manager
County Extension Agent
County Extension 4-H Agent
Crop Insurance Specialist
Crop Physiologist
Crop Protection Specialist
Crop Researcher
Crop Specialist
Cytogeneticist (heredity through cells and genetics)
Dairyperson
Dairy Plant Manager
Dairy Technologist
Dendrologist (dates past events from tree rings)
Director of Ag Research
Earth Scientist
Ecologist
Electric Co-op Manager
Elevator Manager
Entomologist
Environmental Scientist
Extension Specialist
Exterminators—Insect
Farm Appraiser for a bank or agricultural lending institution
Farm Building Designer
Farm Credit Manager
Farmer
Farm Equipment Mechanic

Farm Machinery Dealer
Farm Machinery Designer
Farm Manager
Farm Planner
Farm Realtor
Farm Store Manager
Farm Superintendent
Farm Supply Cooperative Salesman
Feed Dealer
Feedlot Manager
Feed Mill Manager
Feed Salesperson
Feed Technologist
Fertilizer Industry Employee
Field Crop Grower
Field Representative of: Beef Breed Association
Beef/Pork Packing or Processing Company
Dairy Breed Association
Feed Company
Fruit Packing or Processing Company
Fuel Company
Insurance Company
Lumber Processing Company
Machinery Company
Plant Nutrient Company
Poultry Packing or Processing Company
Veterinary Supply Company
Fish Culturist
Fish and Wildlife Specialist
Fishery Biologist
Fishery Manager
Floral Designer

(continued)

Floriculturist
Florist
Food Inspector
Food Processor
Food Retailer
Food Scientist
Foreign Agricultural Affairs
    Officer
Forester
Forest Products Technologist
Forest Ranger
Forestry Aid and Technician
Fruit Grower
Game Warden
Garden Center Retailer
Geneticist—plant, animal
Geochemist
Geomorphologist (genesis of
    earth forms)
Golf Course Superintendent
Grain Buyer
Grain Miller
Grain Processor
Greenhouse Grower
Greenhouse Owner
Groundwater Geologist
Horticultural Supplies
    Worker
Horticultural Therapist
Horticulturist
Hydrologist
Ichthyologist (zoology of fish)
Industrial Agriculturist
Information Specialist
Inspector—food, feed
Insurance Broker
International Agriculture
    Specialist
Irrigation Engineer
Irrigation Manager
IVS Volunteer
Laboratory Technician

Land Appraiser
Land Economist
Landscape Architect
Land Surveyor
Land Utilization Specialist
Life Scientist
Livestock Breeder
Livestock Buyer
Livestock Feeder
Loan Specialist
Market Analyst
Market Research Workers
Meat Cutters
Meat Department Manager
Meat Grader
Meat Inspector
Microbiologist
Microscopist
Molecular Biologist
Mycologist (botany of fungi)
Naturalist
Nematologist (zoology of
    nematode)
Nursery Owner or Operator
Nurseryperson
Nutritionist—plant, animal
Organizational Fieldperson
Ornamental Plant Specialist
Ornithologist (zoology of
    birds)
Outdoor Recreation Specialist
Packinghouse Manager
Parasitologist
Park Manager
Park Naturalist
Park Ranger
Park Superintendent
Pathologist—plant, animal
Peace Corps Administrator
Peace Corps Volunteer,
    especially in Africa, Asia, or
    Latin America

Pest Control Manager
Pesticide Residue Analyst
Pet Food Processor
Plant Breeder
Plant Pathologist
Plant Physiologist
Plant Propagator
Plant Quarantine Inspector
Pest Control Inspector
Poultry Inspector
Poultryperson
Poultry Scientist
Produce Department Manager
Produce Development
    Specialist
Production Credit Fieldperson
Public Relations Manager
Purchasing Agent
Quality Control Specialist
Radio Farm Director
Rancher
Ranch Manager
Ranger—forest, park
Range Conservationist
Range Manager
Range Scientist
Range Specialist
Recreation Development
    Planner
Researcher, Specialist
Resource Economist
Rural Sociologist
Salesperson (of any one of
    many agricultural outputs
    or inputs)
Sales Representative
Sanitarian
Securities Salesman
Science Editor
Seed Broker
Seed Grower
Seed Technologist

Soil Analyst
Soil Chemist
Soil Conservationist
Soil Fertility Scientist
Soil Physicist
Soil Scientist
Soil Surveyor
Statistician
Taxocologist (classifies
    plants, animals)
Taxonomist
Technical Editor
Technical Writer
Textile Researcher
Textile Technologist
Timber Manager, Specialist
Transportation Manager
Turf Producer
Turf Specialist
TV Farm Director
Vegetable Grower
Virologist (viruses)
VISTA Volunteer
Vocational Agriculture
    Teacher
Waterfowl Specialist
Water-Life Management
    Specialist
Water Economist
Water Resources Administrator
    Engineer
Water Resources Development
    Official
Weed Science Specialist
Wildlife Administrator
Wildlife Biologist
Wildlife Specialist
Wildlife Writer, Editor
Youth Corps Conservation
    Director
Zoologist

Appendix D

National FFA Employment Skills Career Development Event

IMPORTANT NOTE: *Please thoroughly read the Introduction Section at the beginning of this handbook for complete rules and procedures that are relevant to all National FFA Career Development Events.*

4. Any participant in possession of an unapproved electronic device in the event area is subject to disqualification.
5. Job description, cover letter, and résumé must be uploaded by the designated deadline found at FFA.org.

PURPOSE

The National FFA Employment Skills Leadership Development Event is designed for FFA members to develop, practice, and demonstrate skills needed for seeking employment in the industry of agriculture. Each part of the event simulates, as closely as possible, real-world activities that will be used by real-world employers.

EVENT RULES

1. The National FFA Employment Skills Leadership Development Event will be limited to one participant per state.
2. Participants are strongly encouraged to wear Official FFA dress for this event.
3. All written materials, including cover letter, résumé, etc., will be the result of his or her own efforts.

EVENT FORMAT

The event is developed to help participants in their current job search (for SAE projects, internships, and part-time and full-time employment). Therefore, materials submitted by the participant must reflect their current skills and abilities and must be targeted to a job for which they would like to apply. In other words, participants cannot develop a fictitious résumé; they must utilize their actual experience. They are expected to target the résumé toward a real job for which they presently qualify.

EQUIPMENT

Participants are required to bring the following items to the event:
- Laptop or tablet capable of a WiFi connection
- Writing utensils

Participants may bring the following items:

- Blank paper
- Résumé
- Cover letter
- List of references
- Business cards
- Padfolio

The following items are not permitted:

- Letters of reference
- Samples of work
- Pictures
- Personal pages

ITEMS TO BE ELECTRONICALLY SUBMITTED BEFORE THE EVENT

For students competing in this event at the national level, all documents should be submitted in PDF form by September 1 at 5:00 pm EST.

- Job description
- Cover letter
- Résumé

States qualifying after the September 1 deadline will have 10 days from the state-qualifying event to submit all materials.

A penalty of 10 percent will be assessed for documents received after the September 1 deadline. If document is not received by seven days after deadline, the participant may be subject to disqualification.

Instructions for submitting electronic documents are always posted on the National FFA website.

ACTIVITIES

1. Job Description—The job description is required in order for the judges to score sections of the event. The job description will not be scored but is required for submission. Participants who fail to submit this component will be subject to disqualification. The job description should include a description of the position the student is applying for, desired qualifications and work experience. Sources for job descriptions can be found by looking in the newspaper or online through job search websites and company websites.

2. Cover Letter (Points—100)—The cover letter is to be typed, one page, single spaced, and left justified using Times, Times New Roman, or Arial 10- or 12-point minimum font. The letter is to be dated for the first day of the national event and addressed to Mark Kline, 6060 FFA Dr., P. O. Box 68960, Indianapolis, IN 46268-0960. Note: *Check* www.ffa.org *to make sure you are addressing the cover letter to the right person, as personnel changes are common in organizations.*

3. Résumé (Points—200)—The résumé should not exceed two pages total. The résumé must be non-fictitious and based upon actual work history. The résumé should be generated from the résumé generators at FFA.org.

ITEMS TO BE COMPLETED BEFORE CONVENTION

4. Electronic Employment Application (Points—100)—Participants will complete a standard electronic job application per instructions at the CDE/LDE website. The application will be open online from September 1–15.

5. Initial Phone Interview (Points—50)—The initial telephone contact will last three to five minutes. Students will sign up for a phone call time when they complete their job application online. The potential employer will contact the participant to arrange an interview time. The potential employer may ask questions regarding aspects of the participant's résumé.

ITEMS TO BE COMPLETED AT THE NATIONAL EVENT

Preliminary Round

6. Personal Interview (Points—500)—The preliminary round interview will be with a panel of judges. Each interview will last twenty minutes.

7. Follow-up Correspondence (Points—50)— Participants will submit follow-up correspondence after the interview. Participants will be provided with necessary information to compose a follow-up correspondence. Correspondence may include, but is not limited to, one of the following: email, handwritten note, or typed letter. Participants will have 30 minutes to complete the follow-up correspondence.

Semifinal Round

8. Personal Interview (Points—500)—The semifinal round will consist of a series of three one-on-one interviews with different judges than the preliminary round judges. Each interview will last a maximum of fifteen minutes. Scores will carry over to the final round.

Final Round

9. Networking Activity (Points—100)—Final participants will be given a networking scenario in which they will be expected to formulate a two- to three-minute extemporaneous response to one or more judges. Scenarios may include, but are not limited to meal function, a mixer, a career show, an elevator pitch, etc.

10. Telephone Job Offer (Points—100)— Participants will participate in a follow-up phone call where they will receive a job offer. They will be scored on their ability to collect information and negotiate. They will also be scored on their response to the offer and overall impression.

TIEBREAKERS

In the event of a tie in the preliminary round, the participant with the highest personal interview score shall receive the higher rank. If a tie still exists, the highest résumé score will receive the higher rank. In the event of a tie in the semifinal or final round the participant with the highest personal interview score shall receive the higher ranking. If a tie still exists, the highest résumé score will receive the highest ranking.

RUBRICS

This event is similar to the former Job Interview CDE. The intention was for the event to give you a more realistic experience in applying for and interviewing for a job. The following rubrics for each section of the new CDE/LDE not only show you how to be successful in the event, but they also give you insight into how to be successful in your real job search.

NATIONAL FFA
CAREER AND LEADERSHIP
DEVELOPMENT EVENTS

Cover Letter Rubric

100 points

NAME _____ MEMBER NUMBER _____

CHAPTER _____ STATE _____

INDICATOR	Very strong evidence of skill is present 5–4 points	Moderate evidence of skill is present 3–2 points	Weak evidence of skill is present 1–0 points	Points Earned	Weight	Total Points
Format and General Appearance	Does not exceed one page without overcrowding; margins are acceptable; font size and style is readable (10-12 pt); uses appropriate business format, date and address at top; addressed to appropriate person; appropriate signature block.	Does not exceed one page without overcrowding; margins are acceptable; font size and style is readable (10-12 pt); uses appropriate business format, date and address at top; not addressed to appropriate person; inappropriate signature block.	Exceeds one page; margins are inappropriate; font style is unreadable; font size is too small or too large; no signature; no date or address; no inside address; not in appropriate business format.		X 4	
Introductory Paragraph	Identifies position they are applying for; states how they heard about the position; states why they are interested in the position; uses wording to attract reader's attention.	Identifies position that are applying for; does not state how they found the job; vaguely describes why they are interested in the job; introduction is bland and not attention catching.	Does not clearly identify position they are seeking; no description of how you heard about the position; does not grab the reader's attention.		X4	
Skills and Experiences	Identifies two to three strongest qualifications for the job; indicates how education has prepared them for this job; states why you are interested in the position; skills and experiences are consistent with resume; makes reference to resume.	Identifies one to two qualifications for the job; indicates how education has prepared them for this job; provides a vague explanation of why interested in the job; skills and experiences are somewhat consistent with resume; makes reference to resume.	Does not identify relevant qualifications for the job; does not indicate how education has prepared them for this job; does not state why they are interested in the job; skills and experiences are not consistent with resume; does not mention resume.		X4	

FIGURE D-1A

Cover Letter Rubriccontinued

INDICATOR	Very strong evidence of skill is present 5–4 points	Moderate evidence of skill is present 3–2 points	Weak evidence of skill is present 1–0 points	Points Earned	Weight	Total Points
Closing Paragraph	Thanks reader for taking time to read; provides appropriate contact information, makes appropriate provisions for follow up.	Thanks reader for taking time to read; provides contact information, but makes reader to assume a follow up.	Does not thank reader; does not mention a plan for follow up; does not provide any contact information.		X3	
Spelling/ Grammar/ Punctuation	Spelling, grammar, and punctuation are extremely high quality with two or less errors in the document.	Spelling, grammar, and punctuation are adequate with three to five errors in the document.	Spelling, grammar, and punctuation are less than adequate with six or more errors in the document.		X5	
					TOTAL POINTS	

FIGURE D-1B

NATIONAL FFA
CAREER AND LEADERSHIP
DEVELOPMENT EVENTS

Resume Rubric
200 points

NAME

MEMBER NUMBER

CHAPTER

STATE

INDICATOR	Very strong evidence of skill is present 5–4 points	Moderate evidence of skill is present 3–2 points	Weak evidence of skill is present 1–0 points	Points Earned	Weight	Total Points
Contact Information	Includes name, address, email, and phone number; name stands out on resume; provides professional e-mail.	Name does not stand out; email is too casual.	Missing name, address, email, or phone number; email used is inappropriate or unprofessional.		X ☐	
Employment Objective	Focused objective that states how employee will help company achieve its goals.	Focused objective that states what you want from the company.	No objective identified.		X☐	
Education or Relevant Coursework	Contains complete information (listed in reverse chronological order) with relevant courses listed, dates formatted correctly, GPA listed in correct format (if appropriate), includes appropriate honors and awards.	Contains information (listed in reverse chronological order) with relevant courses listed, dates formatted correctly, may show gaps in work history; inappropriate GPA listed, includes appropriate honors and awards.	Information not listed in reverse chronological order, important information missing, information not listed in correct format.		X7	

FIGURE D-2A

Resume Rubriccontinued

INDICATOR	Very strong evidence of skill is present 5–4 points	Moderate evidence of skill is present 3–2 points	Weak evidence of skill is present 1–0 points	Points Earned	Weight	Total Points
Relevant Experience and Skills	Entries are listed in reverse chronological order; company name, title, location, and dates are included; strong action verbs used with correct verb tense; personal pronouns and extraneous words are omitted; bullets are concise, direct and indicate one's impact/ accomplishments; results are quantified; bullets are listed in order of importance.	Entries are listed in reverse chronological order; entries have a pattern of one type of error; action verbs are weak; verb tenses are inconsistent; bullets are not concise or direct and do not indicate impact; bullets are written in complete sentences.	Entries are not in reserve chronological order; most entries do not include company name, dates, location, or position title; bullets are written in complete sentences; verb tenses are inconsistent; bullets are wordy, vague, or do not indicate one's impact; bullets are not listed in order or importance to the reader; results are not quantified when appropriate; irrelevant or outdated information is listed.		X9	
Achievements and Honors	Appropriate and relevant achievements and honors listed; achievements and honors related to career goal; provides specific details for related to achievements and honors; listed in reserve chronological order.	Appropriate and relevant achievements and honors listed; achievements and honors related to career goal; lacks specific details for related to achievements and honors; listed in reserve chronological order.	Achievements and honors not listed in reverse chronological order; inappropriate or irrelevant achievements listed; no achievement or honors are listed.		X5	
References	Listed appropriate references and provide complete contact information for references.	References are listed but not all may be appropriate or not all contact information for references is included.	Inappropriate references are listed; no references listed; no contact information listed.		X2	
Spelling/ Grammar/ Punctuation	Spelling, grammar, and punctuation are extremely high quality with two or less errors in the document.	Spelling, grammar, and punctuation are adequate with three to five errors in the document.	Spelling, grammar, and punctuation are less than adequate with six or more errors in the document.		X5	
Format and General Appearance	Does not exceed two pages without overcrowding; margins are acceptable; font size and style is readable (10-12 point); headings reflect content and content substantiates headings; resume is targeted to job.	Does not exceed two pages; appears overcrowded; margins are acceptable; font size and style is readable (10-12 point); headings don't necessarily reflect content and content substantiates headings; resume is targeted to job.	Exceeds two pages; margins are inappropriate; font style is unreadable; font size is too small or too large.		X8	
			TOTAL POINTS			

FIGURE D-2B

NATIONAL FFA
CAREER AND LEADERSHIP
DEVELOPMENT EVENTS

Electronic Employment Application Rubric
100 points

NAME

MEMBER NUMBER

CHAPTER

STATE

INDICATOR	Very strong evidence of skill is present 5–4 points	Moderate evidence of skill is present 3–2 points	Weak evidence of skill is present 1–0 points	Points Earned	Weight	Total Points
Consistent with Resume	Name, education, experience and other personal information matches information provided on resume.	Name, education, experience and other personal information generally matches information provided on resume.	Name, education, experience and other personal information do not match information provided on resume.		X4	
Grammar/ Punctuation/ Spelling	Spelling, grammar and punctuation are extremely high quality with two or less errors in the document.	Spelling, grammar and punctuation are adequate with three to five errors in the document.	Spelling, grammar and punctuation are less than adequate with six or more errors in the document.		X6	
Form Completed	Entire application was completed with "N/A" indicated where appropriate.	Majority of the application was completed with few blank fields.	Several blank spaces and missing information.		X4	
Overall Impression	Application was consistent and appropriately highlighted candidates qualifications for the position.	Application was consistent and generally highlighted candidates qualifications for the position.	The application was not consistent and did not highlight candidates qualifications for the position.		X6	

TOTAL POINTS

FIGURE D-3

NATIONAL FFA
CAREER AND LEADERSHIP
DEVELOPMENT EVENTS

Initial Phone Interview Rubric
50 points

NAME _____   MEMBER NUMBER _____

CHAPTER _____   STATE _____

INDICATOR	Very strong evidence of skill is present 5–4 points	Moderate evidence of skill is present 3–2 points	Weak evidence of skill is present 1–0 points	Points Earned	Weight	Total Points
First Impression	Introduced self when answering the phone. Spoke articulately with no hesitation. Appropriate tone, speaks at right pace to be clear, pronunciation of words very clear and intent is apparent. Confident tone, no nervousness.	Incomplete introduction. Speaks articulately, but with some hesitation. Appropriate tone is usually consistent, speaks at right pace, but shows some nervousness. Pronunciation of words is usually clear, sometimes vague.	Did not introduce self upon answering the phone, Appropriate tone, but frequently hesitates, Has difficulty using appropriate tone, pace is too fast, nervous. Pronunciation of words is difficult to understand or unclear.		X 3	
Response to Questions	Confirmed date, time and location along with contact person/information. Provided complete, accurate and concise answers. Sold themselves without being pushy. Used correct terminology. Communicated knowledge of the related industry. Used time efficiently.	Did not confirm all needed information for interview. Provided some answers, some incomplete, rambled occasionally. Seemed off-putting at times in an attempt to sell themselves. Some question as to correct terminology. Seemed to have holes in knowledge of related industry.	Caller had to offer interview and provide information. Unable to answer questioned asked. Off-putting presentation (tried to sell self too hard). Used incorrect terminology for event. Did not have a firm knowledge of the related industry.		X5	
Overall Impression	Exhibited poise (cool under pressure). Was pleasant, professional and courteous. Ended call appropriately and smoothly (thanked caller, said good-bye). Did not have distracting mannerisms that affected their effectiveness.	Seemed nervous under pressure which impacted poise, pleasantness. Used incorrect grammar which distracted from interview. Mannerisms distracted from interview (use of "ums" and you know"). Ended call without thanking caller or somewhat appropriately (not sure what to do).	Very nervous, not poised (cracks under pressure). Ended call awkwardly and abruptly, did not thank caller or say good-bye, just hung up. Distracted from interview by mannerisms (excessive "ums" or "you know").		X2	

TOTAL POINTS [ ]

FIGURE D-4

NATIONAL FFA
CAREER AND LEADERSHIP
DEVELOPMENT EVENTS

Personal Interview Rubric
500 points

NAME _____ MEMBER NUMBER _____

CHAPTER _____ STATE _____

INDICATOR	Very strong evidence of skill is present 5–4 points	Moderate evidence of skill is present 3–2 points	Weak evidence of skill is present 1–0 points	Points Earned	Weight	Total Points
Appearance	**Professional dress/ groomed:** Follows standard dress code, polished shoes, clothes pressed, conservative accessories.	**Dress appropriate:** Just not as professional and "put together", shoes clean, but not polished.	**Very disheveled:** Dirty shoes, not wearing black shoes.		x 10	
First Impression	**Greeting:** Appropriate salutation and firm handshake. **Introduction:** States name and state association. **Body language:** Smiling and pleasant, does not sit until invited, confident in manner.	**Greeting:** Confident but uneasy, soft handshake. **Introduction:** States name only when asked. **Body language:** Rarely smiles, cologne or perfume is distracting.	**Greeting:** Does not use salutation, very informal. **Introduction:** Fails to introduce self, fails to shake hands with interviewer. **Body language:** Obnoxious cologne or perfume, chewing gum.		x 15	
Response to Questions	**Used appropriate language for career:** Cited relevant examples, evidence knowledge of career field (talk the talk), knows education and experience required for position, discussed skills gained through school or past jobs and how they are relevant to position applied, abilities described match the resume, responses concise and logically communicated, responses do not sound "canned" provided in-depth description of skills, not just a list, provides in-depth response to questions, not yes/no responses do responses provided establish a "theme" that overall describes their abilities.	**Seemed to know terms associated with career:** Some holes, cited several relevant examples, but list incomplete, knew about career, but conveyed incomplete picture unsure of education or experience required for position , incomplete list of skills gained through school and past jobs and relevance to position applied, abilities mostly match resume, Responses seemed rehearsed and somewhat disorganized, provided some depth to description of job skills, some listing, provided some depth to responses to question, some yes/no, was able to tie some abilities together to form a picture of qualifications.	**Knew some of the language of position, but used incorrectly or did not show understanding of terms:** Unable to cite or few relevant examples, position education and requirements not known or do not match applicants skill set, unable to relate skills learned in school or past jobs and relevance to position applied, abilities hardly match resume, responses seemed "canned" with little logical progression, mainly provided list of skills with little explanation, provided yes/no responses, unable to see an overall theme of persons abilities.		x 30	

FIGURE D-5A

Personal Interview Rubric continued

INDICATOR	Very strong evidence of skill is present 5–4 points	Moderate evidence of skill is present 3–2 points	Weak evidence of skill is present 1–0 points	Points Earned	Weight	Total Points
Communication Skills	**Persuasive:** Led the interview in a direction that enabled them to expand so their skills were expressed, took initiative to add information beyond question asked. **Confident:** Exhibited confidence in self with body language and verbally. **Appropriate volume:** Spoke with proper volume for room to be heard clearly; not too loud, not too soft. **Enunciation/grammar:** Avoided words like "git" versus "get and "agin" versus "again", used proper words when speaking (didn't use 10 dollar words when a five dollar word will do). **Concise:** Avoided run on sentences and answered with logical and organized thoughts. **Sincere:** Expressed true interest in the position they are seeking. **Poise:** avoids distracting mannerisms such as drumming fingers or overuse of "uhm" and "you know". **Discretion/Tact:** Shared appropriate information and did not create an awkward situation through responses.	**Persuasive:** Was able to expand somewhat on skills that are a fit for the position, volunteered some additional information to questions asked. **Confident:** Exhibited some nervousness, but covered well, voice and body language showed some uncertainty. **Appropriate volume:** Did not modulate volume to express answers, could hear sometimes, but quiet when unsure of response and hard to hear, **Enunciation/grammar:** Some language not appropriate for position applied, used some slang and exhibited some "dialect". **Concise:** Some questions answered in a rambling fashion, but point was able to be made. Thoughts were logical, but somewhat disorganized. **Poise:** Seemed comfortable with some nervousness, caught self before exhibiting distracting mannerisms, rarely used "uhm" or "you know". **Discretion/Tact:** Most professional in tone and shared information that created little if any awkwardness.	**Persuasive:** Answered yes orno to most questions, did not expand on skill set. **Confident:** Did not appear comfortable, nervous, slouched in chair. **Appropriate volume:** Hard to hear answers or volume too loud for room. **Enunciation/grammar:** Used overly complex or simplistic language, sprinkled in words like "git" versus "get" and "agin" versus "again". **Concise:** Rambled and used run on sentences. Answers were poorly organized and thought not clearly expressed. **Sincere:** Seemed uninterested in the position and distracted, **Poise:** demonstrated distracted mannerisms such as tapping foot, drumming fingers, cracking knuckles, etc., Excessive use of "uhm" and "you know". **Discretion/Tact:** Shared information that may be seen as personal about someone else creating awkwardness, appeared unprofessional.		x 30	
Conclusion	**Posed appropriate questions of interviewer:** e.g., when notification of selection will occur and how. Clarified next steps, inquired as to next step in interview process e.g., if there will be additional interviews, etc. **Appropriate thanks and exit:** Asked for business card, thanked interviewer, stands and shakes hands prior to exiting room.	**Questions posed were somewhat appropriate:** Some had no relevance to interview, Incomplete inquiry of the next steps in the interview process, Asked for business card, thanks interviewer and shook hand but seemed uncertain how to end the interview and exit.	**Asks no questions:** Questions asked (if asked, have no relevance to next steps in the interview process, Ends interview abruptly or awkwardly, exits without thanks or shaking hands.		x 15	

TOTAL POINTS [ ]

FIGURE D-5B

NATIONAL FFA
CAREER AND LEADERSHIP
DEVELOPMENT EVENTS

Follow Up Correspondence Rubric
50 points

NAME _____   MEMBER NUMBER _____

CHAPTER _____   STATE _____

INDICATOR	Very strong evidence of skill is present 5–4 points	Moderate evidence of skill is present 3–2 points	Weak evidence of skill is present 1–0 points	Points Earned	Weight	Total Score
Format	The document was directed to the appropriate person with an appropriate address and salutation. The level of formality was appropriate for the type of correspondence.	The document was directed to the appropriate person with an appropriate address and salutation with minor errors. The level of formality was generally appropriate for the type of correspondence.	The document was not directed to the appropriate person. No address or salutation was included. The level of formality was not appropriate.		X 2	
Content	Effectively expressed appreciation and appropriately reiterated their qualities. Expressed interest and appropriately stated provisions for follow-up.	Attempted to express appreciation and generally reiterated their qualities. Generally expressed interest and attempted to state provisions for follow-up.	Did not attempt to express appreciation. Did not attempt to reiterate their qualities. Did not attempt to express interest or state provisions for follow-up.		X3	
Grammar/ Punctuation/ Spelling	Spelling, grammar and punctuation are extremely high quality with two or less errors in the document.	Spelling, grammar and punctuation are adequate with three to five errors in the document.	Spelling, grammar and punctuation are less than adequate with six or more errors in the document.		X2	
Overall Impression	Writing (when appropriate) was legible and length was appropriate.	Writing (when appropriate) was difficult to read and length was generally appropriate.	Writing (when appropriate) was illegible. Length was inappropriate.		X3	
				TOTAL POINTS		

FIGURE D-6

NATIONAL FFA
CAREER AND LEADERSHIP
DEVELOPMENT EVENTS

Networking Activity Rubric

100 points

NAME   MEMBER NUMBER

CHAPTER   STATE

INDICATOR	Very strong evidence of skill is present 5–4 points	Moderate evidence of skill is present 3–2 points	Weak evidence of skill is present 1–0 points	Points Earned	Weight	Total Points
First Impression	Exhibited a clear, polite introduction, used correct posture and body language, initiated conversation clearly and professionally.	Had an introduction, somewhat exhibited correct posture and body language, attempted to maintain clear conversation.	Did not use proper posture and body language, struggled to maintain conversation, was not clear.		X 3	
Communication Skills	Clearly, confident, sincere and concise. Avoided rambling, is very engaging in the conversation and stays very detail oriented.	Rambled at times, attempted to engage in conversation, fairly detail oriented, fairly confident, sincere, and concise.	Unconfident, insincere, rambled, struggled to engage in conversation and vague.		X 7	
Making the Connection	Clearly connected interest to company/person, found commonalities with company/person, posed appropriate questions, made positive comments about company/person	Attempted to connect interest to company/person, find commonalities with company/person, posed questions, made positive comments about company/person	Struggled to connect interest to company/person, find commonalities with company/person, posed questions, made positive comments about company/person		X 7	
Conclusion	Proficiently used appropriate thanks, exchanged contact information, inquired about follow-up options (website, e-mail, company events), left positive impression upon exit.	Attempted to use appropriate thanks, exchange contact information, inquired about follow-up options (website, e-mail, company events), left neutral impression upon exit.	Struggled to use appropriate thanks, exchange contact information, inquired about follow-up options (website, e-mail, company events), left negative impression upon exit.		X 3	
				TOTAL POINTS		

NATIONAL FFA
CAREER AND LEADERSHIP
DEVELOPMENT EVENTS

Telephone Job Offer Rubric
100 points

NAME

MEMBER NUMBER

CHAPTER

STATE

INDICATOR	Very strong evidence of skill is present 5–4 points	Moderate evidence of skill is present 3–2 points	Weak evidence of skill is present 1–0 points	Points Earned	Weight	Total Points
Response to Offer	Expressed appreciation, upbeat, sincere, shows excitement for the offer.	Seemed caught off guard, attempted to be sincere and show excitement for offer.	Unengaged, insincere, shows little excitement for offer.		X 4	
Gathered appropriate information	Provisions for follow up expressed, Posed appropriate questions (start time, date, who to report to), got contact information.	Somewhat expressed provisions for follow up, attempted to pose appropriate questions (start time, date, who to report to), asked for contact information.	Poorly expressed provisions for follow up, did not pose appropriate questions (start time, date, who to report to), did not ask for contact information.		X 5	
Negotiating Points	Negotiating points appropriate. Exhibited appropriate poise and professionalism while negotiating points. Accepted results with an appropriate response and maturity.	Negotiating points were posed but werea little inappropriate. Exhibited some poise and professionalism while negotiating points. Accepted results with a response.	Negotiating points were inappropriate/ none were stated. Did not exhibit appropriate poise and professionalism. Was disgruntled with results.		X 8	
Overall Impression	Exhibited poise, was pleasant, professional, courteous, ended call appropriately.	Exhibited poise with some nervousness and attempted to be pleasant and courteous. Ended call with a thank you or just said bye.	Seemed nervous, forced conversation. Just hung up.		X 3	
			TOTAL POINTS			

FIGURE D-8

Appendix E

Supervised Agricultural Experience

Supervised Agricultural Experience (SAE) is an integral part of a total program of agricultural education. It is where you gain value entrepreneurial leadership skills in planning, conducting, documenting, and evaluating your experiential learning activities. It is where you can apply and test what you have learned about leadership. An SAE will help you become well-versed in record keeping for your own portfolio. It will also aid in your personal leadership development and in the strategic planning process for teams, groups, organizations, and programs.

TERMS TO KNOW

- Degree program
- Supervised Agricultural Experience (SAE)
- Ownership/Entrepreneurship SAE
- Placement/Internship
- Research SAE
- Exploratory SAE
- School-based Enterprise SAE
- Service Learning SAE
- Improvement Project
- Proficiency awards

OBJECTIVES

After completing this summary, the student should be able to:

- Define SAE
- List and explain the types of SAE
- Plan an SAE program
- Propose an SAE program
- Conduct an SAE program
- Document an SAE program
- Evaluate an SAE program
- Apply proper record-keeping skills
- Participate in youth leadership opportunities to create a well-rounded SAE
- Produce in a local program of activities using a strategic planning process
- Participate in a local program of activities using a strategic planning process

Supervised Agricultural Experience (SAE) is a program of experiential learning activities conducted outside of the regular agricultural education class time. It is designed to help you develop and apply the knowledge and skills you learn in your agricultural education classes and/or laboratories. This connection between content and experience especially happens with the leadership development you learned in this text. The SAE used to be called the Supervised Occupational Experience Program (SOEP), and the focus was on the occupation. These days the focus is on entrepreneurship and experiences, which will prepare you for a career. Think of SAE as the ultimate on-the-job training tool.

The SAE takes place under the direction of your agricultural education teacher. Often individual projects (e.g., showing pigs, raising a plant in the greenhouse, etc.) are considered to be SAEs, but the intent of a quality SAE is for it to be conducted as a series of experiences completed during your enrollment in agricultural education.

TYPES OF SAEs

SAE projects can be completed in a variety of categories, including ownership/entrepreneurship, placement/internship, research, exploratory, school-based enterprise, and service learning. Each chapter of this text has introduced you to concepts that will help you have the strongest SAE possible.

Ownership/Entrepreneurship—Work for Self

In the **Ownership/Entrepreneurship SAE**, you plan, implement, operate, and assume all or some of the financial risk in a productive or service activity or agriculture, food or natural resources-related business.[1] In this type of SAE, you are boss. Remember, part of being a good leader is understanding your leadership style, having the motivation to press on, and possessing the communication skills to market or sell your products. If you are the owner, failure is not an option, so implement everything you've learned in this text to succeed.

If you have this type of SAE, you own the materials and other inputs, and you keep financial records to determine return on investment. Some examples of SAE experiential learning activities that can make up an SAE program are listed below. At the end of this appendix is a much more comprehensive list.

Examples:

1. Growing grapes
2. Raising bees
3. Growing an acre of corn

Placement/Internship—Work for Someone Else

If you work for someone else on a farm or ranch, in an agricultural business, or in verified non-profit organization providing a "learning by doing" environment, your SAE qualifies for the **Placement/Internship SAE** type. This type of experience may be paid or non-paid.[2] Not everyone reading this can always have an ownership/entrepreneurship SAE, but almost everyone has the opportunity to seek out a paid or unpaid "working for someone else" experience for the purpose of learning and applying your agricultural knowledge and skills. Some examples are below, but many more examples are provided at the end of this appendix.

Examples:

1. Working in a flower shop
2. Working on Saturdays at a local stable
3. Working at a grocery store

When working in this type of SAE utilize your understanding of personality types for successful interactions with fellow employees, your boss, or even customers. Utilize what this text has taught you about communication skills to perform at your highest ability.

Research—Conduct an Experiment

The agriculture industry is becoming a highly scientific field full of problems to be solved by you and your classmates. Problems range from world hunger to climate change and even social ideals and political policies related to the agriculture industry. The **Research SAE** type helps you prepare for a long and productive career solving these important issues for our society. The research SAE involves a program of extensive activities where you plan and conduct experiments or other forms of scientific evaluation using the scientific process.

Exploratory SAE—Get It Started

This type of SAE works for everyone taking an agriculture course. It is designed to assist students in becoming knowledgeable in agriculture and/or becoming aware of potential careers in agriculture and agriscience. The Exploratory type of SAE results in the development of a plan to begin an SAE.[3] An SAE is a program of many different activities, and the Exploratory SAE is often the beginning. The chapter on Selecting a Career and Finding a Job will support your Exploratory SAE.

School-based Enterprise SAE—Manage It at School

The School-based Enterprise SAE is an entrepreneurial operation managed by you. Your operation, however, takes place in a school setting that provides not only facilities, but also goods and services that meet the needs of an identified market.[4] To give you the most educational value, this type of SAE should replicate the real world of work as much as possible. This type of SAE is usually cooperative in nature, with management decisions made by you in cooperation with your teacher.

Service Learning SAE—Plan a Service Project

Service learning is one of the highest forms of leadership. Our goal should always be to leave things better than we found them. The goal of the service learning type of SAE is to improve the community in which you live. This SAE type combines community service activities with structured reflection.

As part of this type of SAE, you can become involved in developing a needs assessment, planning goals, creating objectives and budgets, implementing the activity, and promoting and evaluating a chosen project. The project could be in support of a school, community organization, or nonprofit organization. You would be responsible for raising necessary funds for the project (if funds are needed). The project must stand by itself and not be part of an ongoing chapter project or community fundraiser. The project must be somewhat challenging and require the outstanding leadership you possess.[5]

Examples:

1. Test water wells in community for contamination
2. Design a web page for your FFA chapter or for a local organization
3. Design and install a landscape plan for a church

PLAN AN SAE PROGRAM

Planning is an important part of any team, business, project, or program. Your SAE program isn't any different. To properly plan, you'll need to understand the steps and guidelines involved in planning the SAE.

Steps in Planning the SAE

Step One—Identify your career interests in agriculture. Your SAE program and experiential learning activities and the career you choose someday need to be something about which you can get excited. If you like the SAE or your future career choice, you are more likely to stick with it and be successful.

Step Two—Review the job responsibilities of career interest areas you may have. You might like the idea of a particular project or career, but if you don't like specific tasks involved, you might need to choose another project or program. For example, if you like the idea of a Placement SAE as a Veterinarian Technical Assistant, but you don't want to clean out cages, you may have made a poor choice.

Step Three—Identify any SAE programs of interest by interviewing friends who have an SAE or by viewing suggestions found in various resources. Suggestions for certain experiential learning activities or projects are at the end of this appendix.

Step Four—Develop a timeline for your SAE program. In other words, which projects will happen first, second, third, etc.? What is the completion date for each activity? Electronic portfolios, record keeping systems or simple calendar systems can be used for this step.

Step Five—Building on step four, build a long-range plan for the SAE program. Remember, projects or activities happen in a shorter time span, but an SAE program, which is necessary to be competitive for FFA proficiency and Star awards, happens over a longer span of time and grows in scope and diversity.

Step Six—Develop the first-year (annual) plan. You have to actually start now. After you know the long-range plan and the timelines for the different experiential learning activities/projects, you are ready to put the first-year plan together. Again, think completion dates and specific strategies for reaching those goals.

Step Seven—Re-plan on a regular basis. Part of good planning is reviewing your activities in light of your plan and then adjusting your plan.

This should happen fairly often, but not so often that you are always planning instead of completing tasks.[6]

Parts of an Annual SAE Plan

A yearlong SAE program plan will keep you on track, and it will help you make decisions about your SAE. The annual plan consists of a calendar, description of projects, budget, improvement projects, and supplementary skills. For entrepreneurship/ownership projects, prepare a description of the size/scope of your enterprise, the location of your projects, the nature of the business or enterprise, partners involved, methods of marketing, facilities needed, and months involved in the project. For placement projects, detail the location, beginning and ending dates, and your pay schedule. It is also important to include a budget, and to keep up with income and receipts as the year progresses. You will also want to include a description of improvement projects you wish to complete and supplemental skills you would like to develop. An **improvement project** is an activity that improves the appearance, convenience, efficiency, safety, or value of a home, farm, ranch, agribusiness, or other agriculture facility. Your annual plan should include specific activities of the improvement project as well as hours of labor committed and an estimate of costs.[7]

Propose an SAE Program

Following your SAE plan, it is good practice to write a statement justifying your plans with a proposal. This proposal should talk about why you chose the SAE you did and why your plan is laid out the way it is. Your proposal should reiterate the importance of completion dates on your timeline. The SAE proposal should also detail specific agricultural or leadership concepts that you hope to achieve as a result of your SAE program.

CONDUCT AN SAE PROGRAM

The first step to conducting a quality SAE program is getting started. All of the planning and decision making can take time, but it is worth it. Once you've started, you are going to have a great time, but you have some work ahead of you as well. Specifically, you will need to become skilled at documenting your SAE program as well as evaluating it. As an entrepreneurial leader or manager of a small or large firm, you will be responsible for getting to work at some point. There is a lot of satisfaction in watching your goals come to life when those goals represent the ultimate success of your project or business.

Document an SAE Program

Documenting your SAE is important because it gives you an opportunity to see the planning that you've worked so hard on come to life. Documenting your SAE makes it a permanent record from which you can learn and improve your project. It shows the problems encountered and solved, which informs future activities and provides you with confidence to continue in your SAE.

Calendar—Using a calendar can help with more than planning. You can use a calendar, such as the one on your smartphone, to document hours invested, skills learned, and even tasks performed.

Journal or Portfolio—Another option rather than using a simple calendar system is to document your SAE with a journal or portfolio, either paper or electronic. A paper system could be kept using a notebook or ledger, but these days an electronic system is much easier to use. Many states have an online journal or portfolio that allows you to keep up with important documentation.

APPLY PROPER RECORD-KEEPING SKILLS

Record keeping is the process of keeping a journal or portfolio of everything you have done. As stated above, in your SAE you will need to make notes whenever you do or learn something new, and you need to document time and money spent on your activities, projects, and program. The skill of record keeping will serve you well in your career.[8] Entrepreneurial leaders keep detailed notes, especially when it comes to strategies that help keep more and lose less money.

At its core, record keeping is quite simple. For each project, you will need to capture the following for each activity or item:

- the date
- the name or description
- the hours you spent
- any expenses or income that went out or came in as a result

Be sure to add up the hours you've invested in all activities, because time is money, and each one of your experiences represents new knowledge and skills you've developed. You will also add up expenses and income to determine your profit and loss (P&L). It's really that simple. You can use state-endorsed, pre-programmed spreadsheets or websites to determine your P&L and track your hours invested or you can keep records on paper. It's up to your personal preference, available resources, and the wishes of your advisor or business partner.

Learning to keep records for and through your SAE has many benefits. Keeping accurate records can help you determine whether you made or lost money. It can help you keep others from cheating you out of what you have earned. Your knowledge and experience in record keeping will help you determine which parts of the business are doing well and which parts are not. Becoming proficient at record keeping will also help you make management decisions, document your net worth for loans, prepare tax returns, and plan for future events. Being a good record keeper for your SAE will also help document your activities for FFA recognitions and degree purposes. Good records can also protect you legally and help you plan a budget for the following year.[9]

Evaluate an SAE Program

When your teacher/advisor comes to visit you and observe your SAE, he/she is there to answer any questions you may have and to help you evaluate your SAE. The evaluation looks different for different types of SAEs. An entrepreneurial leader will evaluate their own business models

using many of the same data points as are used to evaluate an SAE program. Below are some of the metrics used to evaluate an Entrepreneurship/Ownership SAE.

1. Accuracy of Records
2. Neatness of Records
3. Dates of Records
4. Net Income (Total Income minus Total Expenses)
5. Are good management practices being used?
6. Efficiency factors (yield per acre, number of offspring raised, etc.)
7. Improvements made since the last observation
8. Cleanliness of facilities
9. Customer satisfaction
10. What skills were learned[10]

ENHANCING SAE WITH OTHER OPPORTUNITIES

SAE in not only financial record-keeping, but also a record of skills, knowledge, credentials, certifications, experiences, career planning, reflection, and leadership development.[11] A solid SAE will provide you the opportunity to receive FFA degrees, and compete for different proficiency awards.

The FFA **degree program** is FFA's primary recognition program. SAE helps students apply for each degree, and every degree has certain SAE requirements. There are five degrees of FFA membership, as follows: Discovery, Greenhand, Chapter, State, and American. These five degree areas recognize you for your overall participation in FFA. Everything you do in FFA, when combined, helps you move toward a degree.[12]

Proficiency awards are awards that recognize students' excellence in SAE.[13] Developing your SAE into a proficiency award is a time-consuming but rewarding task. You should apply for the proficiency award in the specific area and career pathway in which you are strongest and have the most experience. For instance, if you have worked for a livestock producer for three years and have raised goats for only one year, it would be best to apply in the placement area for Animal Systems rather than Entrepreneurship for the same system. The career pathways for your SAE and proficiency awards follow.

1. Agribusiness Systems—the study of business principles, including management, marketing, and finance, and their application to enterprises in agriculture, food, and natural resources.
2. Animal Systems—the study of animal systems, including life processes, health, nutrition, genetics, management, and processing, through the study of small animals, aquaculture, livestock, dairy, horses, and/or poultry.
3. Biotechnology Systems—the study of data and techniques of applied science for the solution of problems concerning living organisms.
4. Cluster Skills—the student will demonstrate competence in the application of leadership, personal growth, and career success skills necessary for a chosen profession while effectively contributing to society.
5. Environmental Service Systems—the study of systems, instruments, and technology used in waste management and their influence on the environment.
6. Food Products and Processing Systems— the study of product development, quality assurance, food safety, production, sales and service, regulation and compliance, and food service within the food science industry.
7. Natural Resources Systems—the study of the management of soil, water, wildlife, forests, and air as natural resources.
8. Plant Systems—the study of plant life cycles, classifications, functions, practices, through the study of crops, turf grass, trees and shrubs, and/or ornamental plants.
9. Power, Structure, and Technical Systems— the study of agricultural equipment, power systems, alternative fuel sources, and precision technology, as well as woodworking, metalworking, welding, and project planning for agricultural structures.

SUMMARY

Supervised Agricultural Experience (SAE) is a program of experiential learning activities conducted outside of the regular agricultural education class time. You can apply nearly everything you've learned in this leadership textbook through an SAE. There are different types of SAEs and related opportunities and rewards programs, so every student in agricultural education and FFA can learn and develop from the authentic experience. Record-keeping and documentation of the valuable experiences are very important, and will serve you and your future career well.

EXAMPLES OF PROJECTS FOR YOUR SAE

Agricultural Mechanics Project Examples

- build a patio for the home
- build frames for raised beds for gardeners
- build handicap ramps in local community
- build picnic tables/sell to schools and local community
- construct a utility building
- construct a hydro ram pump and calculate the efficiency and water delivery rate
- construct a wind powered generator and show its applications to agriculture
- construct and sell birdhouses and feeders
- construct and sell lawn furniture made of pvc
- construct compost bins to sell
- construct concrete projects for the home or farm
- construct or recondition a welding project (such as a trailer, cooker, etc.) at home or in school-provided facilities
- construct pre-fabricated wooden fence panels to sell to local hardware and building supply stores
- construct spray rigs for four wheelers
- construct and market woodworking projects (birdhouses, dog houses, etc.)
- construct metal projects
- contract with local EMCs or power companies to remove bolts, wire, etc. from old power poles (sell copper for recycling)

- contract with school system to maintain and service lawn care equipment
- cut out and paint lawn figures to sell
- electrical repair service
- install plumbing fixtures or plumbing system in your own building
- lawn mower maintenance service
- make craft items from wood, metal or concrete to sell at arts and crafts shows
- make personalized signs to sell
- paint the home, supervised by agricultural education teacher
- placement in a parts store
- provide a poultry house maintenance preparation business
- provide custom painted mailboxes and stands
- repair and rebuild damaged pallets for businesses
- start a chainsaw basic maintenance and service business
- start a custom vehicle refurbishing or painting business
- start a detailing business for cleaning farm equipment on the farm (wash, wax, clean, maintain)
- start an equipment locating business and match folks with something for sale with folks who want to buy something
- start a farm equipment tire disposal business (turn old tires into livestock feeders)
- start a farm fence maintenance business (cleaning fencerows, repairing)
- start a farm fencing company for custom work
- start a pallet manufacturing business
- start a small engine repair service
- wire a home shop, utility room, barn or tree house
- work as an agricultural mechanics aide
- work at a welding operation
- work at a building supply business
- work with a farm equipment dealer
- start an equipment trailer fabrication business

Agribusiness Sales and Service Project Examples

- become an agricultural news consultant for local radio or newspapers
- conduct a study of commodity trading over a period of time
- conduct general home maintenance
- contract with local Chamber of Commerce to conduct county tours for prospective businesses
- create a custom labor venture: mow pastures, remove undesirable weeds from crops, paint outbuildings, etc.
- design a computer application plan for some agricultural facility or program
- develop a marketing plan for an agricultural commodity
- fry pork rinds for local stores
- install electrical circuits or wiring system at home
- job placement in food distribution, restaurant, etc.
- job placement with local florist
- job-shadow agribusiness professionals, visits to agribusinesses to interview personnel, educational tours, etc.
- offer a custom parts and supplies delivery business to farms in your county
- pre-sell fresh meat to clients on a weekly basis
- pre-sell fresh seafood to clients on a weekly basis
- pre-sell fresh vegetables in family portions delivered weekly
- preserve food for home use
- process creamed corn in a food processing facility
- provide a co-op program for an agribusiness
- provide a custom barbecue service for the community
- provide custom feed for livestock (tap the organic, all-natural, no-chemical market)
- provide a hand weeding crew for local peanut/vegetable farmers
- provide a sausage making business at home; can be sold if regulations are met
- provide custom hay baling and/or hauling
- provide a farm sign business (manufacture, sell, install and maintain)
- provide livestock hauling
- provide small engine maintenance and repair service
- provide systematic maintenance and service on outdoor power equipment at home or at school-provided facilities

- purchase and resell aerial photographs from tax office to local landowners
- package fresh fruit or vegetable gift packs
- remove pesticide jugs monthly from farms and transport to landfill
- sell ready-to-freeze processed vegetables
- start a composting business by buying cow manure from local farmers, bagging for resale
- start a farm-sitting business for vacationing farmers
- start a kerosene route for homeowners (probably little demand in the summer time)
- start an MSDS compliance business by compiling and maintaining current sheets for farms and businesses in your county
- start a recycling business (collecting and selling newspapers and plastics to recycling plants)
- start an agricultural business promotion business (sell custom caps, t-shirts with farm or ag business names or logos to clients)
- start an agricultural photography service (animals, equipment, barns, families, children with animals, show animals)
- start a local farm produce sale paper and sell ads to farmers
- form a cooperative with other students and share in profits of a greenhouse crop
- write "how to" pamphlets to sell at local garden supply stores (example—How to Grow Tomatoes, etc.)
- Write news articles on agriculture or FFA for local newspapers

Agriscience Project Examples

- compare weight gain of chicks fed different feed rations
- conduct a plant growth and physiology experiment in school agriscience lab
- conduct a research project for agriscience fair (local and national)
- conduct a research project on a specific career; set up a business plan, including expenses, possible income, etc.
- conduct a supervised control burn and assess plant growth in the area
- conduct food science experiments

- grow crops with different mechanical/chemical applications, fertilizer, growth regulator, etc. and observe/report results
- monitor local air quality; record and report
- plant and maintain a research plot on different types of turf grasses
- plants raised beds and monitor the growth of the plants
- research project on how light intensity affects plant growth
- research project on how light quality affects plant growth
- research project on plant reproduction
- soil conservation project on private or public land
- study effect of fertilizer run-off into a stream or pond
- study effect of manure run-off into a stream or pond
- study effects of herbicide type and varying concentrations
- temperature effects on worms' food consumption
- work with agencies involved in research (USDA, etc.)
- conduct a plant growth and mineral deficiency experiment

Alternative Animals Project Examples

- provide a beehive rental service for farms and gardens
- raise a dog for show
- raise dairy goats
- raise dogs to sell
- raise fish in tanks or floating cages–research the rate of growth based on factors such as temperature and amount of feed given
- raise llamas
- raise market goats for show
- raise meat birds (chickens, turkeys, ducks) to the desired weight and sell to customers
- raise meat goats
- raise mice, hamsters or gerbils
- raise miniature cattle
- raise miniature horses
- raise quail or other game birds for flight and meat
- raise rabbits for pets or meat animals
- raise special breeds of dogs

- raise tropical fish in aquariums
- raise worms; collect and sell to bait stores
- start a crawfish farm
- start a cricket ranch
- start a dog exercising business for elderly folks or sick people
- start a dog obedience school
- start a fish bait farm (mealworms, golden grubs, etc.)
- start a gopher tortoise relocation service for landowners
- start a honey production business (would work well with hive rental)
- start a pet grooming business
- start a turtle farm (sell to pet stores and pond owners)
- train sporting dogs (quail, rabbit and retriever dogs)
- work at a dog kennel
- work at a pet store
- work at a veterinary hospital

Animal Science Project Examples

- board horses
- build a backyard poultry research project
- contract finish swine
- develop a cow-calf operation
- develop a small swine operation
- develop a stocker cattle operation
- raise replacement heifers
- raise dairy replacement heifers
- produce feeder pigs
- provide a deer processing service
- provide a home animal care service
- provide a horse training service
- provide a horseshoeing service
- provide a meat processing service
- provide a poultry processing service
- raise a beef heifer for show
- raise a horse for show
- raise a market hog for show
- raise a market steer for show
- raise breeding sheep for show
- raise breeding swine for show or breeding
- raise dairy heifers for show
- raise market lambs for show
- raise poultry for show
- start a small animal care business
- start an Easter egg business
- work at a horse operation or stables

- work at a poultry processing operation
- work in the egg industry–packaging and distribution
- work on a beef cattle operation
- work on a dairy operation
- work on a poultry operation
- work on a sheep operation
- work on a swine operation
- operate a pay-to-fish business
- provide fish pond management
- raise catfish in cages
- raise fish in an aquaculture system
- raise fish in cages in a pond or other body of water
- care and incubation of hatching eggs

Crops Project Examples

- organic vegetable production
- produce vegetables for decoration–Indian corn, mini pumpkins, gourds, etc.
- produce farm crops at home or at school-provided facilities
- product forage crops at home or at school-provided facilities
- produce watermelons
- scout cotton or peanuts for producers

Forestry Project Examples

- bale and market pine straw
- buy unusable lumber from builders' supply and building sites; grind up and chip for mulch to sell
- collect green pine cones (for seed in the fall)
- collect/market natural supplies (i.e., pine cones, acorns, nuts, corn shucks, etc.) to sell to craft stores
- container pine seedling production
- contract with a tree removal service to cut firewood and remove fallen trees
- contract with local timber companies and landowners to maintain boundary lines by painting and chopping
- cut and sell firewood provided free by national forests and state and local parks
- cutting and/or marketing firewood
- grow longleaf pine seedlings
- measure timber on school forestry plot; determine volume and establish a management plan

- provide a soil sampling service for farms and lawns
- purchase bulk pine bark from sawmill; bag and resell
- purchase seedlings from Georgia Forestry Commission and pot and grow out to sell
- remove lightning strike trees (insect damaged, mechanical injuries) for landowners
- start a custom forest herbicide application crew; (must have forest commercial pesticide license)
- start a forest tree planting business
- start an ornamental tree care service

Horticulture Project Examples

- adopt a community building for beautification
- adopt an area of the school campus for beautification
- collect and sell dry/preserved native plant materials (acorns, leaves, wiregrass), especially for floral design retail/wholesale
- collect, press, mount and identify plants that are growing on campus
- construct a garden arbor
- construct backyard water gardens
- container gardening ornamental plants
- container gardening vegetables
- create and market custom floral designs
- develop a business making dried arrangements to sell
- grow herbs
- produce daylilies
- develop a park on public property
- entrepreneurship in floral design
- establish a community roadside wildflower planting
- establish a garden plot at home or at school; produce crops to market
- grow and sell mushrooms
- grown and sell produce crops
- grow greenhouse plants on rented school greenhouse/cold frame space
- grow, harvest and can or preserve fruits and vegetables
- grow organic cut flowers for farmer's markets
- horticulture therapy
- indoor plant rentals and care service for businesses and offices

- landscape maintenance
- landscape pruning enterprise
- native plant materials
- offer a shrub care service (pruning, trimming and cutting back shrubs, fertilization)
- produce fruit crops (at home or school-provided facility) i.e., watermelons
- produce greenhouse crop (at home or school-provided facility) i.e., ferns
- produce perennials from seed
- produce turf grass (at home or school-provided facility)
- propagate and market shrubs
- provide a fruit tree pruning service
- provide a mulching service for urban gardeners
- provide landscaping materials for local businesses (pine straw, rocks, etc.)
- raise a trial garden plot on school grounds (similar to UGA); seed companies may donate seed/plugs
- raise tomato seedlings and replant into one-gallon pots to sell
- rent indoor plants to teachers in your school
- rent houseplants to homeowners (care for plants and change plants weekly)
- Rent-A-Plant; rent plants for weddings, banquets, parties–i.e., ferns and tropicals
- start a commercial flower up-keep business; change hanging baskets, potted plants and window boxes for businesses
- start a floral design business by creating table centerpieces to sell at farmers markets, grocery stores and vegetable stands
- start a garden photography business
- start a hydroponics vegetable business
- start a lawn irrigation installation business
- start a houseplant renovation business
- start a turfgrass establishment business (seedlings, sodding, hydroseeding, etc.)
- start a vegetable transplant seedling business
- work at a florist
- work at a garden center
- work in a nursery business

Natural Resources Project Examples

- adopt a local stream to monitor water quality
- collect water run-off from school parking lot and analyze for various pollution indicators

- collect, mount and identify insects found on school campus
- conduct a research project on how to prevent deer damage to a home garden
- conduct a water quality study on area lakes or streams
- conduct endangered plant surveys for landowners
- construct deer stands to sell (portable and stationary)
- construct duck nesting boxes to sell to landowners
- construct turtle traps for pond owners (use this in conjunction with turtle farm as a source of breeding stock)
- develop a backyard bird habitat
- develop a backyard wildlife habitat
- develop a schoolyard wildlife habitat
- develop and/or maintain a wildlife food plot on private or public land
- develop and/or maintain wetland area on private or public land
- measure land for the local FSA office
- monitor success rate of bluebird houses
- plan and develop a school nature trail
- plan and develop an outdoor classroom
- plant a butterfly garden at school
- provide a debris removal service along rivers and streams; sell driftwood and other items to consumers
- provide pond fertilization and testing service
- provide custom dove shoots or quail hunts
- raise mallard or wood ducks to sell to pond owners
- raise popular games birds; sell them for meat and as taxidermy products
- start a bullfrog farm; sell fresh frog legs to local restaurants
- start a fish fingerling nursery (catfish, trout, bream)
- start a red cockaded woodpecker relocation service
- start a rock store; sell for landscaping purposes (gravel, pebbles, stones)
- start a wildlife food plot and native plant enhancement business for local landowners and hunting clubs
- trap nuisance animals
- provide non-game wildlife management[14]

Search Terms

portfolio

NOTES

1. National FFA Organization, *Supervised Agricultural Experiences*, accessed February 18, 2016. https://www.ffa.org/about/supervised-agricultural-experiences.
2. Ibid.
3. National FFA Organization.
4. "Philosophy and Guiding Principles for Execution of the Supervised Agricultural Experience Component of the Total School-Based Agricultural Education Program," *The Council: A National Partner for Excellence in Agriculture and Education*, National Council for Agricultural Education. Web. 18 Feb. 2016. https://www.ffa.org/thecouncil/sae, pp. 2–3.
5. Ibid., p. 3.
6. California Core Agriscience Lesson Library, *Lesson 612a: Planning your SAE program*, retrieved February 20, 2016 from http://calaged.csuchico.edu/ResourceFiles/Curriculum/CoreAgriscience/CD_old/Lessons/612a.pdf
7. Ibid., p. 4.
8. National FFA Organization, ed. *LifeKnowledge, Lesson MS.69: Record Keeping*, retrieved February 22, 2016 from http://harvest.cals.ncsu.edu/site/WebFile/MS69.PDF, p. 3.
9. National FFA Organization, ed. *SAE Handbook, Lesson 7: Why Do We Keep Records?* retrieved February 23, 2016 from http://harvest.cals.ncsu.edu/site/WebFile/IIB7.pdf, p. 4.
10. "Philosophy and Guiding Principles. . ." p. 4.
11. National FFA Organization, ed. *SAE Handbook, Lesson MS. 70: Proficiency Awards and SAE*, retrieved February 23, 2016 from http://harvest.cals.ncsu.edu/site/WebFile/IIB8.pdf, p. 3.
12. Ibid., p. 3.
13. National FFA Organization, ed. *National FFA Agricultural Proficiency Awards: A Special Project of the National FFA Foundation*, retrieved February 23, 2016 from https://www.ffa.org/sitecollectiondocuments/prof_handbook.pdf, p. 11.
14. J. Ricketts. *Project Workshop*. SAE Handbook. Unpublished document for teacher training in the Republic of Georgia.

Glossary/Glosario

A

abilities—competencies in an activity or occupation

habilidades—las competencias en una actividad o una profesión

academic skills—skills and subject-matter material learned in a formal school situation, such as science, English, and math

habilidades académicas—las habilidades y una materia de estudio aprendido en una situación escolar formal, tales como las ciencias, el inglés, y las matemáticas

accommodation—learning style characterized by learning by doing, or trial and error; learning only what is necessary to accomplish a particular task

alojamiento—el estilo de aprendizaje caracterizado por aprender haciendo, o ensayo y error; aprender de sólo lo necesario para realizar una tarea determinada.

accommodator—person who learns by doing and feeling; also, a conflict-resolution style in which the person sacrifices his or her interests and concerns while helping others to achieve their interests

acomodador—quien que aprende haciendo y sintiendo emocionalmente; además, un estilo de resolución de conflictos en los que la persona sacrifica sus intereses e inquietudes, al mismo tiempo, ayudando a otros a lograr sus intereses

accountability—being answerable or capable of being explained

responsibilidad personal—ser responsible por sus acciones o capaz de ser explicado

acronym—a word formed from the initial letters of a phrase or title, such as FFA, NASA, USA

acrónimo—una palabra formada por la primera letra de una frase o título, tales como FFA, NASA, EE.UU.

action—a state of motion, either physical or mental

Acción—un estado de movimiento, bien sea físico o mental

ad hoc committees—temporary task groups formed for a specific purpose

comités ad hoc—grupos de tarea temporal formados para un propósito específico

adjourned meeting—a continued meeting set to meet again at a certain time; similar to recess

reunión aplazada—una reunión continuada planeada para reunirse de nuevo en un momento determinado; similar a una suspensión de actividades

advancement—in the context of this book, the act of being raised to a higher rank or position; promotion

avance—en el contexto de este libro, el acto de ser elevado a un rango más alto o una posición más alta; una promoción

affective attitudes—the emotions attached to a person, object, or event

actitudes afectivas—las emociones adjuntas a una persona, a un objeto o a un evento

affective learning—learning that has to do with intelligence in the personality or human relations arena

aprendizaje afectivo—el aprendizaje que tiene que ver con la inteligencia en la personalidad o la arena de relaciones humanas

affective skills—skills that help you behave appropriately in the social situations of life

habilidades afectivas—las habilidades que le ayudan a comportarse adecuadamente en las situaciones sociales de la vida

affinity—attraction among or similarity of group members

afinidad—la atracción o similaridad entre los miembros del grupo

affirmation statement—a positive declaration about who we are and what we can become

afirmación declaración—una declaración positiva sobre quiénes somos y lo que podemos ser

agenda—a list, plan, or the things to be done; matters to be acted or voted on during a meeting

agenda—una lista, un plan, o las cosas que hay que hacer; los asuntos a ser actuado o que debe ser votado durante una reunión

agricultural education—program of instruction in agriculture in high schools

educación agrícola—un programa de instrucción en la agricultura en las escuelas secundarias

alternatives—different courses of action that one might take; solutions

alternativas—los cursos distintos de acción que alguien puede tomar; las soluciones posibles

ambiguous—having more than one possible meaning

ambiguo—tener más de un significado posible

ambitious—showing a desire to achieve or obtain power, superiority, or distinction; being alert, energetic, willing to work, and caring about what one does

ambicioso—mostrando un deseo de lograr o conseguir poder, la superioridad o una distinción; estar alerta, enérgico, dispuestos a trabajar, y preocuparse por lo que alguien hace

amendable—modifiable or changeable; in parliamentary procedure, when it is possible to modify or change the wording, and in some cases the meaning, of the motion to which an amendment is applied

enmendable—modificables o cambiantes; en el procedimiento parlamentario, cuando es posible modificar o cambiar la redacción y, en algunos casos, el sentido de la marcha a la que se aplica una enmienda

amendment—a term used for a *motion to amend*, or to modify or change the wording, and in some cases the meaning, of the motion to which it is applied

enmienda—un término utilizado para una *moción para enmendar*, o modificar o cambiar la redacción, y en algunos casos, el sentido de la marcha a la que se aplica

animate—possessing or characterized by life

animar—poseer o caracterizada por la vida

anxiety—an uncomfortable feeling or uneasiness about a situation or event

ansiedad—una sensación incómoda o inquietud acerca de una situación o evento

apathy—lack of interest or concern

apatía—Falta de interés o preocupación

applicants—those who apply for a job

solicitantes—quienes aplican para un trabajo

application form—a general form that applicants fill out when seeking employment

formulario de aplicación—una forma general que los solicitantes completen cuando buscan empleo

aptitude—quickness in learning and understanding; a natural or acquired talent, ability, inclination, or intelligence

aptitud—rapidez en el aprendizaje y la comprensión; un talento natural o adquirido, una habilidad, una inclinación, o una inteligencia

ardently—with warm or intense feeling

ardientemente—con una sensación tibia o intense

articulation—refers to the way the tongue, teeth, palate, and lips are moved and used to produce the crisp, clear sounds of good speech; similar to diction

articulación—se refiere al modo en que la lengua, los dientes, el paladar y los labios se mueven y se utilizan para producir los sonidos nítidos y claros de buen discurso; similar a la dicción

assimilator—person who learns by observing and thinking

asimilador—quien aprende observando y pensando

atmosphere—the mood, tone, or feeling that permeates the group

atmósfera—el humor, tono, o sensación de que impregna el grupo

attitude—disposition toward others and ourselves; strong belief or feeling about people, things, and situations

actitud—disposición hacia los demás y a nosotros mismos; fuerte creencia o sentimiento sobre las personas, cosas y situaciones

attributes—qualities or characteristics

atributos—las cualidades o características

authentic leaders—charismatic leaders who use their power ethically to ensure that they and their followers pursue and achieve positive and moral goals that advance both the organization and society as a whole.

auténticos líderes—líderes carismáticos que use su poder éticamente para garantizar que ellos y sus seguidores perseguir y alcanzar objetivos positivos y morales que fomenten tanto la organización y la sociedad en su totalidad.

authenticity—genuineness; realness

autenticidad—la autenticidad; la genuinidad

authoritarian—behavioral leadership style in which the leader leads and makes decisions regardless of the wishes of the group

autoritarios—estilo de liderazgo de comportamiento en el que el líder conduce y toma decisiones independientemente de los deseos del grupo

autocratic leadership style—a leadership style in which the leader makes decisions independent of the group

estilo de liderazgo autocrático—un estilo de liderazgo en el que el líder toma decisiones independientes del grupo

avocational (part-time) job placement—a job or occupation that supplements a full-time job

avocational (a tienpo parcial) de colocación laboral—un empleo u ocupación, que complementa un trabajo de tiempo completo

avoidance reinforcement—negative reinforcement; also known as *avoidance learning*

refuerzo de evitación—un refuerzo negativo; también se conoce como *aprendizaje por evitación*

avoider—a conflict-resolution style that denies the very existence of a conflict, changes the subject, ducks discussion, and is noncommittal

evasor—un estilo de resolución de conflictos que niega la existencia de un conflicto, se cambia el asunto, evita la discusión y son evasivos

B

barriers—obstructions or hindrances; for example, mental obstructions that keep us from communicating clearly

barreras—obstáculos o impedimentos; por ejemplo, obstrucciones mentales que nos impiden comunicar con claridad

behavior—the actions or reactions of persons or things under specified circumstances

Comportamiento—las acciones o reacciones de personas o cosas en circunstancias especificadas

behavioral attitudes—the tendency to act in a particular way toward a person, object, or event

actitudes de comportamiento—la tendencia a actuar de una manera particular hacia una persona, objeto o evento

behavioral leadership—leadership according to one's personality style; in the context of this book, refers to authoritarian or democratic leadership

liderazgo conductual—liderazgo hecho según el estilo de la personalidad; en el contexto de este libro, se refiere al liderazgo democrático o autoritario

belief—the state of believing that certain things are true or real; an individual's conviction or acceptance that something is true or real

creencia—el estado de creer que ciertas cosas son verdaderas o reales; un individuo condena o acepta que algo es verdadero o real

boards—a type of standing committee of directors or trustees of corporations, whose task is to oversee management of the organization's assets, set or approve policies and objectives, and review progress toward the objectives

juntas—un tipo de comité permanente de directores o administradores de empresas, cuya tarea es supervisar la gestión de los activos de la organización, fijar o aprobar políticas y objetivos, y examinar los progresos realizados en la consecución de los objetivos

bodily-kinesthetic intelligence—involves the skillful control of one's body, especially when handling objects, and includes the timing and reflexes related to manipulating those objects

corporal inteligencia kinestésica—involucra la habilidad de tener control de su cuerpo, especialmente cuando manipulando los objetos a mano, e incluye la distribución y los reflejos relacionados con la manipulación de los objetos

body—(1) the bulk of the speech, which carries the central theme and main ideas; (2) the group members of an assembly

cuerpo—(1) la mayoria del discurso, que lleva el tema central y las ideas principales; (2) los miembros del grupo de una asemblea

body language—the nonverbal way one communicates

lenguaje corporal—una forma de comunicación que no es verbal

brainstorming—the unrestrained offering of ideas or suggestions

la tormenta de ideas—la oferta ilimitada de ideas o sugerencias

brainstorming method—the process of suggesting many alternatives, without evaluation, to solve a problem; most effective for discovering numerous possible solutions to a problem

Método de la tormenta de ideas—el proceso de sugerir muchas alternativas, sin evaluación, para resolver un problema; más eficaces para descubrir numerosas posibles soluciones a un problema

breadth—when talking about participation in a group, refers to the number of members who participate

amplitud—cuando se habla de su participación en un grupo, se refiere al número de miembros que participan

buzz groups—groups of six to eight members, each with a leader and a recorder, with the purpose of generating ideas, solutions, and possibly common ground in a given amount of time

grupos buzz—grupos de seis a ocho miembros, cada uno con un líder y un registrador, con el propósito de generar ideas, soluciones y, posiblemente, un terreno común en una cantidad determinada de tiempo

bylaws—standing rules governing the regulation of an organization's internal affairs

estatutos—las reglas permanentes sobre la regulación de los asuntos internos de una organización.

C

capability—one's potential for being able to do the job

capacidad—su potencial para poder hacer el trabajo

captivate—hold an audience's attention; fascinate or charm

cautivar—retener la atención de una audiencia; fascinar o encantarse a alguien

career—a series of jobs that is pursued more or less in sequence to achieve the ultimate occupation desired by the individual; something you really want to do—and be—for the rest of your life

carrera—una serie de trabajos que se persiguen más o menos en orden para lograr la máxima ocupación deseada por el individuo; algo que realmente quiere hacer y ser para el resto de su vida.

career planning—mapping out your life mission, or what you see as the meaning of your life

la planificación de la carrera—mapear su misión de vida, o lo que ve como el significado de tu vida

causal order—a method of organizing a speech in which the speaker identifies certain causes and then discusses the results or consequences that follow from them

orden causal—un método de organizar un discurso en que el orador identifica algunas causas y luego analiza los resultados o consecuencias que se derivan de ellos

chairperson—the presiding officer of a meeting; also called the chair

presidente(a)—quien presida una reunión; también llamada la silla

channel—the means by which the sender communicates the message

canal—el medio por el cual el emisor comunica el mensaje

chapter—a local branch of an organization or club

capítulo (delegación)—una sucursal local de una organización o club

charismatic—a leader who inspires others by having great personal charm and appeal

carismático—un líder que inspira a otros con gran encanto personal y apelación

choleric—personality type with a bossy, quick, active, and strong-willed temperament

colérico—el tipo de personalidad con un autoritario, rápido, activo y temperamento de carácter fuerte.

classified ad—a newspaper job advertisement; the section of the newspaper in which job advertisements are collected

anuncio clasificado—un anuncio de trabajo en el periódico; la sección del periódico en el que se recogen los anuncios de trabajo

cliques—informal groups of people who come together voluntarily because of similar interests

camarillas—grupos informales de personas que se unen voluntariamente a causa de intereses similares.

clone—an exact copy

clon—una copia exacta

coercive power—influence based on an individual's ability to prevent someone from obtaining deserved rewards, or to take away something the person already has.

poder coercitivo—influencia basada en la capacidad de una persona para impedir que alguien de obtener recompensa merecida, o para quitarle algo que la persona ya tiene.

cognitive attitudes—set of values and beliefs that an individual has toward a person, object, or event

actitudes cognitivas—un conjunto de valores y creencias que una persona tiene hacia una persona, objeto, o evento

cognitive learning—learning based on theoretical symbols (numbers, words) as well as logical reasoning; academic learning

aprendizaje cognitivo—aprendizaje teórico basado en símbolos (números, palabras), así como el razonamiento lógico; el aprendizaje académico

coherence—sticking together; an orderly or logical relation of parts that affords comprehension or recognition

coherencia—mantenerse unida ; una relación ordenada o lógica de piezas que ofrece la comprensión o el reconocimiento

cohesiveness—the bonds among members as well as members' bonds with and closeness to the group as an entity

cohesión—los lazos entre los miembros, así como los lazos con los miembros y cercanía al grupo como una entidad

collaborate—work together to reach a decision that brings mutual gain

colaborar—trabajando juntos para llegar a una decisión que trae el beneficio mutuo

collaborative individualism—a concept of working in teams in which individuals are not bound by the group's limits, but cooperate with other team members while maintaining their personal, internal motivation and conflict management skills

individualismo cooperativo—Un concepto de trabajo en equipos en los que los individuos no están obligados por los límites del grupo, sino cooperar con otros miembros del equipo y manteniendo su motivación personal interno y aptitudes para el manejo de conflictos

collaborator—the conflict-resolution style of an active listener who focuses on the issues

colaborador—el estilo de resolución de conflictos de un oyente activo que se centra en los temas

combination leadership—leading by combining all styles of leadership

liderazgo de combinación—principales combinando todos los estilos de liderazgo

commitment—applying oneself to a task until it is completed; dedication

compromiso—aplicarse a una tarea hasta que se haya completado; la dedicación

committee—a chosen group of members with specified responsibilities

comité—un grupo escogido de miembros con responsabilidades específicas

committee of the whole—a device in which all of the members of an organization or assembly sit as a single committee

comité plenario—un dispositivo en el que todos los miembros de una organización o conjunto sentarse como un único comité

committees—task groups established by an organization to perform specific functions

comités—Grupos de tareas establecidos por una organización para llevar a cabo funciones específicas

communication—the process of sending and receiving messages through which two or more people reach understanding

comunicación—el proceso de envío y la recepción de mensajes mediante el cual dos o más personas llegan a una comprensión

competence—the ability to do something well

competencia—la capacidad de hacer algo bien

competitor—a conflict-resolution style characterized by aggressive, self-focused, forceful, verbally assertive, and uncooperative behavior that is intended to achieve one's own interests at the expense of others

competidor—un estilo de resolución de conflictos caracterizado por un comportamiento agresivo, auto-centrado, enérgico, asertivo verbalmente, y no cooperativo que se destina para lograr sus propios intereses a expensas de los demás

compromise—in a disagreement, both parties give a little or change their positions somewhat to come to a mutually satisfactory agreement

avenencia—en un desacuerdo, ambas partes dan un poco o cambiar sus posiciones algo para llegar a un acuerdo mutuamente satisfactorio

compromiser—a conflict-resolution style that tries to get all parties to work together to settle for partial satisfaction of their interests

conciliador—un estilo de resolución de conflictos que intenta llegar a todas las partes a que trabajen juntos para arreglar para la satisfacción parcial de sus intereses

conceit—excessive regard for or estimation of one's own worth

vanidad—respecto excesivo o estimación de su propio valor

conceptualize—to visualize an abstract concept

conceptualizar—visualizar un concepto abstracto

conceptual leadership skills—thinking skills that can be taught, such as problem solving, decision making, and delegation

habilidades de liderazgo conceptual—habilidades de pensamiento que puede ser enseñado, tales como la resolución de problemas, la toma de decisiones y la delegación.

condescend—to do something one regards as beneath one's dignity

dignarse—hacer algo se considera como bajo su dignidad

confirmation behaviors—positive behaviors that enhance the recipient's feelings of self-worth

comportamientos de confirmación—los comportamientos positivos que mejoran los sentimientos del destinatario de la autoestima

conflict—what occurs when two or more groups or individuals have or think they have different goals, values, beliefs, or ideas about the way things should be

conflicto—lo que ocurre cuando dos o más grupos o personas tienen o piensan que tienen objetivos distintos, valores, creencias, o ideas acerca de la manera en que las cosas deben ser

conflict resolution—tools and processes used by a leader to assess conflict and solve disputes and problems that exist between individuals and groups

resolución de conflictos—las herramientas y procesos utilizados por un líder para evaluar el conflicto y resolver las controversias y problemas que existen entre los individuos y los grupos

confront—to approach a problem head-on or face to face

enfrentar—para abordar un problema de frente o cara a cara.

consensus method—problem-solving technique when the group comes to substantial agreement on a solution

método de consenso—una técnica para la solución de problemas cuando el grupo llega a un acuerdo sustancial sobre una solución

consistent style—tendency to know the appropriate amount of information to consider and evaluate before making a decision

estilo coherente—la tendencia a saber la cantidad apropiada de información para considerar y evaluar antes de hacer una decisión

consultative leadership style—a leadership style in which the leader goes to individual group members seeking additional information that will help him or her solve the problem or make the decision

estilo de liderazgo consultivo—un estilo de liderazgo en la que el líder va a los miembros individuales del grupo solicitando información adicional que le ayude a resolver el problema o tomar la decisión

content theories—psychological suppositions that focus attention on the factors within an individual that cause the individual to behave in a particular manner

contenido teorías—los supuestos psicológicos que centren su atención en los factores dentro de un individuo que provocan que el individuo se comporte de una manera determinada.

contingency leadership—leadership style whose emergence or effectiveness is dependent or contingent on the situation or immediate circumstances in which the leader is operating. See also *situational leadership*

liderazgo de contingencia—el estilo de liderazgo cuya aparición o la eficacia es dependiente o contingente sobre la situación o circunstancias inmediatas, en el que el líder está en funcionamiento. Véase también *el liderazgo situacional*

continuum—a series connecting two extremes with an infinite number of variations in between (e.g., the decision-making continuum)

continuo—una serie conectando dos extremos con un número infinito de variaciones intermedias (por ejemplo, el continuo de la toma de decisiones)

conventional method—in problem solving, there is group discussion, but the discussion is typically dominated by one or a very few individuals; a vote is taken on a single solution, which is put in place if majority votes for it

método convencional—en la solución de los problemas, hay discusión de grupos, pero la discusión es típicamente dominadas por uno o muy pocos individuos; una votación en una sola solución, que se pone en marcha si la mayoría de votos para eso

converger—a person who learns best by both doing and thinking

converger—una persona que aprende mejor por hacer y también pensar

converging—learning style characterized by the integration of theory and practice, or thinking and doing

convergente—el estilo de aprendizaje caracterizado por la integración de la teoría y la práctica, o el hacer y el pensar

convey—to communicate or make known

transmitir—para comunicar o dar a conocer

cooperative genius approach—leadership technique that involves making use of the combined intelligence of a group to solve problems. See also *synergy*

estrategia genia cooperativa—dirección técnica que implica hacer uso de la combinación de un grupo de inteligencia para resolver problemas. Véase también la *sinergia*

cooperative skills—the ability to work or act together with people for a common purpose

habilidades cooperativas—la capacidad de trabajar o actuar conjuntamente con otras personas para un propósito común

creative problems or decisions—problems or decisions that require a plan or design to help decision makers come to a solution

problemas o decisiones creativas—los problemas o decisiones que requieren un plan o diseño para ayudar a los tomadores de decisiones a llegar a una solución

credence—belief

credibilidad—una creencia

criteria—standards, rules, measures, or tests by which something can be judged

criterios—las normas, las reglas, las medidas o las pruebas por las cuales algo puede juzgarse

critical thinking—using objective analysis and evaluation of a situation or issue to form a judgment

pensamiento crítico—mediante el análisis y evaluación objetiva de una situación o problema para formar un juicio

culture—socially transmitted behavior patterns, beliefs, and all other products of human work and thought patterns of a society, community, or population

cultura—patrones de comportamiento transmitida socialmente, las creencias, y todos los demás productos del trabajo humano y los patrones de pensamiento de una sociedad, la comunidad o la población

curriculum—course of study; in this text, agricultural education

plan de estudios—un curso de estudio; en este libro de texto, la educación agrícola

D

daily log—diary, or record and schedule of activities and appointments throughout the day

registro diario—un diario o registro y calendario de actividades y citas de todo el día

debatable—open to discussion, as with an item being considered by a group

discutible—abierto a la discusión, como con un asunto que está siendo examinado por un grupo

decision making—the process by which a new or different course of action is selected

La toma de decisiones—el proceso por el cual un curso de acción seleccionado nuevo o diferente

decoding—receiving the message and converting it into an understandable form (in communication)

descodificar—recibir el mensaje y convertirlo en una forma comprensible (en la comunicación)

decorum—propriety of debate

decoro—la corrección de debate

defamation of character—making cruel statements that may harm another's reputation

la difamación de carácter—haciendo declaraciones antipáticas que pueda dañar la reputación de la otra persona

deliberate—careful and intentional, as in the delivery of a speech

deliberada—cuidado e intencional, como en la entrega de un discurso

delivery—conveying, or the method of making a speech

Entrega—transmitir, o el método de hacer un discurso

Delphi method—a decision-making technique that polls a group through a series of anonymous questionnaires repeatedly given to a group until it reaches a position that is acceptable to all or nearly all members

Método Delphi—la técnica para la adopción de una decisión que sondea un grupo a través de una serie de cuestionarios anónimos dada repetidamente a un grupo hasta que alcanza una posición que sea aceptable para todos o casi todos los miembros

democratic—behavioral leadership style in which the leader leads and makes decisions based on the input of the group

democrático—estilo de liderazgo de comportamiento en el que el líder dirige y toma decisiones basadas en los comentarios del grupo

dependability—can be trusted to accomplish a task

fiabilidad—puede ser de confianza para llevar a cabo una tarea

desire—a state of mind, a longing or hope for something

deseo—un estado de la mente, un anhelo o una esperanza de algo

devil's advocate method—discussion or decision-making technique in which a person may propose a solution he or she does not really support, just to make others think and react

método de abogado del diablo—la discusión o una decisión técnica en la cual una persona puede proponer una solución que él o ella realmente no apoyo, sólo para hacer otros piensan y reaccionan

dialogue—when two or more people engage in a conversation from which everyone learns something

diálogo—cuando dos o más personas participen en una conversación en la cual cada uno aprende algo

diligence—constant, earnest, and persistent effort

diligencia—un esfuerzo constante, serio, y persistente

dimensions—types of conflict

dimensiones—los tipos de conflicto

directors—task-oriented people who enjoy telling others what to do (choleric personality type)

directores—la gente orientada hacia las tareas quien disfruta de decirle a los demás qué hacer (la personalidad de tipo colérico)

discipline—the ongoing process of bringing oneself under control; orderly or prescribed pattern of behavior

disciplina—el proceso de traer a sí mismo bajo control; ordenado o un patrón de conducta prescrita

discontent—dissatisfied

descontento—el insatisfecho

discreet—modest, reserved

discreto—modesto, reservado

discuss—to talk about; in parliamentary procedure, to talk about a motion after being properly recognized

discutir—hablar sobre algo; en el procedimiento parlamentario, para hablar sobre una moción después de ser reconocida adecuadamente

distributive negotiation—a bargaining form of negotiation in which the negotiators elevate their own self-interests over any collaborative answers to conflict

la negociación distributiva—una negociación en forma de negociación en la cual los negociadores elevan sus propios intereses por encima de cualquier respuestas colaborativas al conflicto

diverger—person who learns by observing or feeling

divergente—quien aprende observando o por sensación

diverging—learning style characterized by observing, feeling, and the generation of new ideas

divergentes—estilo de aprendizaje caracterizado por la observación, el sentimiento, y la generación de ideas nuevas

doers—people who learn by acting out new information immediately rather than thinking about it

hacedores—las personas que aprenden por actuar con información nueva de forma inmediata, en lugar de pensar en ella

domains—in the context of this book, major areas of learning (cognitive, affective, and psychomotor)

dominios—en el contexto de este libro, los principales ámbitos de aprendizaje (cognitivo, afectivo y psicomotor)

doodling—drawing or scribbling idly

doodles—el dibujo o garabatos indiferentes

doubt—state of questioning one's ability to learn, think creatively, make decisions, accomplish, and succeed; uncertainty

duda—el estado de cuestionar la capacidad de aprender, pensar creativamente, tomar decisiones, lograr, y triunfar; incertidumbre

E

earnestness—sincerity or seriousness

seriedad—la sinceridad o la gravedad

efficiency—producing the desired effect without waste of time or materials

eficacia—produciendo el efecto deseado sin la pérdida de tiempo o de las materiales

emotions—personal feelings that can fuel conflict

emociones—los sentimientos personales que pueden alimentar los conflictos

empathize—to feel or understand as someone else does, as if you were that person

empatizar—para sentir o entender como alguien se siente, como si fuera esa persona

empathy—ability to involve oneself, imaginatively and sym- pathetically, with the feelings, thoughts, or attitudes of other people

empatía—la capacidad de involucrarse, de manera imaginativa y simpatéticamente, con los sentimientos, pensamientos, o actitudes de otras personas

employability skills—those basic skills necessary for getting, keeping, and doing well on a job

las competencias para la empleabilidad—los conocimientos básicos necesarios para obtener, mantener, y hacer las cosas bien en un trabajo

empowerment—giving those around you a sense of power to do their best by trusting that they will perform to the best of their abilities

la potenciación—dando a aquellos a su alrededor una sensación de poder para hacer sus mejores y confiando en que se llevará a cabo a lo mejor de sus capacidades

emulate—to copy or simulate others

Emular—Copiar o simular otros

encoded—put into a form that the sender believes the receiver will understand

codificado—poner en una forma que el remitente cree que el receptor va a entender

energy cycle—in the context of this book, scheduling things at certain times in a meeting in order to be the most productive

ciclo energético—en el contexto de este libro, la programación de ciertas cosas en determinados momentos en una reunión a fin de ser más productivos

enlists others—a leader who is able to get others to agree with his or her vision

alista a otros—un líder que es capaz de atraer a otros para estar de acuerdo con la vision de él o ella

enthusiastic—leadership quality characterized by feeling or showing strong excitement about something

entusiasta—una calidad de liderazgo caracterizado por el sentimiento o por demonstrar un entusiasmo fuerte por algo

equity—fairness; an individual's belief that he or she is being treated fairly in relation to relevant others

equidad—la justicia; un individuo cree que él o ella está siendo tratada justamente en relación a otras entidades pertinentes

E.R.G.—existence, relatedness, and growth; a theory of needs that says people have three types of needs: existence (material needs such as food, water, pay, benefits, and so on); relatedness (relationship needs; relationships with "significant others" such as coworkers, family, friends, and so on); and growth (the desire for unique personal development)

E.R.G.—la existencia, la relación y el crecimiento; una teoría de las necesidades que dice que la gente tiene tres tipos de necesidades: las necesidades materiales de existencia (tales como alimentos, el agua, los salarios, las prestaciones, etc.); la relación (las necesidades de relación; las relaciones con los "otros significativos", tales como compañeros de trabajo, familiares, amigos, etc.); y el crecimiento económico (el deseo de desarrollo personal único)

esprit de corps—term for group solidarity or morale; feelings of loyalty, enthusiasm, and devotion to a group among its members

el espiritu del grupo—un término para la solidaridad de grupo o moral; los sentimientos de lealtad, el entusiasmo y la devoción a un grupo entre sus miembros

esteem needs—needs that must be met to help people feel good about themselves; include self-respect, as well as respect, recognition, and affirmation from others

las necesidades de estima—necesidades que deben satisfacerse para ayudar a la gente a sentirse bien acerca de sí mismos; incluyen la autoestima, así como el respeto, el reconocimiento y la afirmación de los demás

ethical—conforming to the standards of conduct of a given profession or group

ético—conforme a las normas de conducta de una profesión determinada o un grupo

ethics—rules of conduct that reflect the character and sentiment of a community or group

las éticas—las reglas de conducta que reflejan el carácter y el sentimiento de una comunidad o un grupo

etiquette—manners, practices, and customs prescribed by a culture or society

etiqueta—las costumbres, las prácticas y costumbres prescritas por una cultura o la sociedad

exact-reasoning problems or decisions—problems or decisions that usually have one definite (usually factual) answer

los problemas de razonamiento exacto o decisiones—los problemas o las decisiones que suelen tener una respuesta determinada (generalmente) fáctica

exaggeration—overstatement of the importance of a person, thing, or occurrence

exagerado—la exageración de la importancia de una persona, una cosa, o un acontecimiento

Executive Committee—group composed of chapter officers and those who chair the major committees

Comité Ejecutivo—un grupo compuesto de oficiales del capítulo y quienes presidir los comités principales

existential intelligence—the capacity to raise and reflect on philosophical questions about life, death, and ultimate realities

inteligencia existencial—la capacidad de iniciar y reflexionar sobre cuestiones filosóficas sobre la vida, la muerte, y las realidades últimas

expert power—a form of power based on an individual's expertise in situations where others depend on the individual for advice, knowledge, or assistance,

poder experto—una forma de poder basada en la experiencia de una persona en situaciones en las que otros dependen del individuo para el asesoramiento, información, o asistencia,

extemporaneous—a type of speech for which the speaker prepares ideas but does not memorize exact words

extemporánea—un tipo de discurso para que el orador prepara ideas pero no memorizar palabras exactas

external motivation—outer force or power that causes a person to want to achieve a goal or urges the person to action

La motivación externa—fuerza exterior o el poder que hace que una persona desea lograr un objetivo o alenta a la persona a la acción

extinction—elimination of an undesirable behavior by removal of positive reward, such as by withholding rewards when the behavior occurs

extinción—la eliminación de un comportamiento indeseable por remover una recompensa positiva, como mediante la retención de recompensas cuando el comportamiento ocurre

F

facilitator—someone who works with a group to make things run effectively and efficiently

facilitador—alguien que trabaja con un grupo para hacer que las cosas funcionen de manera eficaz y eficiente

fear—overwhelming anticipation or awareness of danger

miedo—gran anticipación o conciencia de peligro

feedback—verbal or nonverbal response to the message

realimentación—la respuesta verbal o no verbal del mensaje

feeling—making meaning from experiences or information by empathy, or the feelings evoked by the actual experience or information. See also *sensing*

sensación—hacer el significado a partir de las experiencias o la información por empatía o los sentimientos evocados por la experiencia real o información. Véase también *detectado*

FFA Creed—philosophy statement on agriculture written by E. M. Tiffany

FFA Credo—la declaración de la filosofía sobre agricultura escrita de E. M. Tiffany

fidelity—the quality of being loyal, dedicated, faithful, with honest and firm attachment

fidelidad—La calidad de ser leal, dedicado, fiel, honesto y firme apego

filtering—manipulating a message (information) to reflect the receiver's desire or objective

Filtrado—la manipulación de un mensaje (la información) para reflejar el deseo del receptor o del objetivo

fixed interval schedule—rewards behavior at regular, set times

horario de intervalo fijo—las recompensas regulares de comportamiento, hechas en tiempos establecidos

fixed ratio schedule—rewards behavior after a predetermined number of actual outputs

relación fija agenda—el comportamiento de recompensa después de un número determinado de salidas reales

floor—recognition from the chairperson of the right to speak

piso—el reconocimiento por parte de la presidenta de permiso de hablar

fluency—smooth and easy flow of speech

fluidez—el flujo suave y fácil de discurso

follow-up letter—a written communication sent to the potential employer or interviewer after an interview to thank the interviewer and express continued interest in the job

carta de seguimiento—una comunicación escrita enviada al empleador potencial o a un entrevistador después de una entrevista para agradecerle al entrevistador y expresarle interés continuado en el trabajo

force—in speaking, volume and variations in volume for emphasis

fuerza—cuando hablando, el volumen y las variaciones en el volumen de énfasis

frames—images of leadership (authoritarian, democratic, political, and traditional) that result from the blending of elements of all leadership models and styles that occurs with combination leadership

marcos—los imágenes de liderazgo (el autoritario, el democrático, el político y el tradicional) que resultan de la combinación de elementos de todos los modelos y estilos de liderazgo que ocurre con una combinación de liderazgo

functional groups—formal groups with an indeterminate life span

grupos funcionales—los grupos formales con una duración indeterminada

G

general (unanimous) consent—decision-making technique in which everyone agrees and no vote is necessary

el consentimiento general (por unanimidad)—la técnica de la toma de decisiones en el que todo el mundo está de acuerdo y ningún voto es necesario

germane—closely related to or having bearing on the subject of a motion

pertinentes—estrechamente relacionados o tener influencia en la presentación de una moción

goals—the ends toward which effort is directed

objetivos—los extremos hacia la que se dirige el esfuerzo

gofer delegation—delegation of specific simple tasks ("go for" this item, do this for the team, and report when you have finished); used by micromanagers

la delegación gofer—la delegación de tareas determinadas simples ("vaya por" este artículo, hágalo para el equipo, y reportar cuando haya terminado); usado por micro-gestor

gossip—talking about the personal or private affairs of other people, usually in a negative way

chismes—hablar sobre los asuntos privados o personales de otras personas, usualmente en una forma negativa

Greenhand degree—first of four degrees of membership in the FFA

Greenhand grado—primero de los cuatro grados de membresía en el FFA

gross national product (GNP)—the yardstick that measures the value of goods and services that a country produces in one year

El producto nacional bruto (PNB)—el método que mide el valor de los bienes y los servicios que un país produce por todo un año

group dynamics—pattern of interactions within a group

dinámicas de grupo—modelo de interacciones dentro de un grupo

groupthink—a tendency toward a herd mentality; members becoming content with easy solutions and failing to question or challenge ideas and recommendations; a numb and unthinking acceptance of what most of the group appears to want or believe

pensamiento grupal—una tendencia hacia una mentalidad de manada; miembros poniéndose contentos con soluciones fáciles y no cuestionar o desafiar las ideas y las recomendaciones; un insensibilidad y la aceptación irreflexiva de lo que la mayoría del grupo parece querer o creer

H

hearing—receiving sound

la audición—la recepción de sonido

hierarchy—ranked order that denotes the relative importance of each item

jerarquía—el orden de clasificación que indica la importancia relativa de cada elemento

holistic—emphasizing the importance of the whole and the interdependence of its parts

holístico—destacando la importancia del conjunto y la interdependencia de sus partes

huddle/discussion 66 method—a group discussion technique based on the concept of six people discussing a subject for six minutes

grupito/Método de discusión 66—una técnica de discusión de grupo basada en el concepto de seis personas discutiendo un tema durante seis minutos

human relations skills—skills needed to understand and work with others

las habilidades de relaciones humanas—los conocimientos necesarios para comprender y trabajar con otros

human resource frame—relates to the needs of members, without the strong emphasis on procedure and policy found in the structural frame

recursos humanos—bastidor-se relaciona con las necesidades de sus miembros, sin el fuerte énfasis sobre el procedimiento y la directiva que se encuentra en el marco estructural

hygienes—conditions and practices that promote or preserve health or psychological well-being; in the context of this book, hygienes include salary, job security, working conditions, relationships, status, company procedures, and quality of supervision

higienes—las condiciones y prácticas que promueven o conservar la salud o el bienestar psicológico; en el contexto de este libro, los higienes incluyen salarios, la seguridad en el empleo, las condiciones de trabajo, las relaciones con el estado, la empresa, los procedimientos, y la calidad de la supervisión

hyperbole—exaggeration for effect that is not meant to be taken literally

hipérbole—la exageración para efecto sin el intento de ser tomado literalmente

hypothetical—a made-up situation used to convey a point or illuminate an issue

hipotética—un situación inventada utilizada para transmitir un punto o iluminar un tema

I

immediate goal—goal that is set to happen within a day or week

meta inmediata—el objetivo que va a suceder dentro de un día o una semana

immediate job placement—job market entered immediately after high school graduation

la colocación inmediata—entrando en el mercado de trabajo inmediatamente después de su graduación de la escuela secundaria

immediately pending motion—a formal proposal that is under consideration on the motion being discussed

inmediatamente moción pendiente—una propuesta oficial que está bajo la consideración de que la moción siendo discutido

impetuousness—rushing into action with little forethought

impetuosidad—apurarse en acción con poca previsión

important—having great value, significance, or consequence and having to do with results

Importante—tener gran valor, significado o consecuencia y que tienen que ver con resultados

impromptu—a type of speech delivery in which the speaker talks "off the cuff," with no chance for preparation

impromptu—un tipo de entrega de discurso en que el orador habla "de improviso", sin la posibilidad de preparación

inauthentic leaders—charismatic leaders who use their power unethically to exploit their followers' loyalty by pursuing evil or immoral goals that primarily benefit the leader

líderes inauténticos—los líderes carismáticos que usan su poder contra la ética para explotar "la lealtad" de sus seguidores aplicando metas males o inmorales que benefician principalmente al líder

incidental motion—a motion that arises out of other motions and takes precedence over other motions when appropriate

petición incidental—una moción que surge de otras mociones y toma precedencia sobre otras mociones cuando corresponda

incontrovertible solution—a solution that is indisputable

solución incontrovertible—una solución que es indiscutible

inequity—injustice or unfairness; an individual's belief that he or she is being treated unfairly in relation to relevant others

inequidad—las injusticias e indebidas; las creencias de un individuo que él o ella está siendo tratada injustamente en relación a otras entidades pertinentes

inertia—tendency of matter to remain at rest if at rest or, if moving, to keep moving in the same direction

inercia—la tendencia de la materia a permanecer en reposo si en reposo, o, si se mueve, para seguir avanzando en la misma dirección

inflections—the tone or pitch of a voice

inflexiones—el tono o agudez de voz

influence leadership—leading by convincing others of an idea, so that they will follow of their own free will

influencia liderazgo—la gestión por convencer a los demás de una idea, de modo que sigan según su propia libre voluntad

initiative—ability to take action and be a self-starter

iniciativa—la capacidad para actuar y ser alguien con espíritu emprendedor

innate—inborn or inherent (such as characteristics or attributes)

innata—inherentes o innatos (tales como características o atributos)

innovativeness—the ability to see something that others do not; the ability to generate new ideas

innovación—la capacidad de ver algo que otros no pueden ver; la capacidad de generar nuevas ideas

input-based communication—when a person is receiving communication

entrada basada en la comunicación—cuando una persona está recibiendo comunicación

insecure—lacking self-confidence; uncertain

inseguridad—la falta autoconfianza; incerteza

inspiring—leadership quality that causes people to want to do or create something better

inspirador—la calidad de liderazgo que hace que la gente quiere hacer o crear algo mejor

instrumental values—values that reflect the way you prefer to behave, such as ambitious and hardworking

valores instrumentales—los valores que reflejan la forma en que prefiere comportarse como ambicioso/a y trabajador/a

insubordination—disobedience or lack of submission to authority (such as disobeying or disrespecting a supervisor)

insubordinación—la desobediencia o falta de sumisión a la autoridad (como la desobediencia o desacato a un supervisor)

intangible—cannot be appraised for value; having no material form or substance

intangible—no puede ser evaluado para el valor; no tiene forma material o sustancia

integrative negotiation—a constructive, problem-solving form of negotiation

la negociación integrativa—un constructivo, forma de negociación que centra en resolviendo problemas

integrity—striving to live up to a code of rules or moral values

integridad—queriendo vivir hasta un código de normas o valores morales

intelligence—the ability to respond successfully to new situations and the capacity to learn from past experiences

inteligencia—la capacidad de responder con éxito a las nuevas situaciones y la capacidad para aprender de las experiencias pasadas

intensity—when talking about group participation, refers to a measure of how often individual members take part and how emotionally involved they become

intensidad—cuando hablamos acerca de la participación del grupo, se refiere a una medida de cuán a menudo los miembros individuales toman parte y cómo se vuelven involucrados emocionalmente

interference—anything that hinders the sender's ability to make a message understood

interferencias—todo lo que obstaculiza la capacidad del remitente para que un mensaje sea entendido

intergroup—among or between different groups (such as in competitions in which group members act together to beat rival groups)

intergrupo—entre distintos grupos (como en competiciones en las que los miembros del grupo actúan juntos para derrotar a los grupos rivales)

internal motivation—an inner force or power that spurs a person to want to achieve a goal or stimulates the person to action

motivación interna—una potencia o fuerza interior que impulsa a una persona a querer lograr un objetivo o estimula a la persona a la acción

interpersonal communication barriers—differences and personal characteristics of senders and receivers that hinder communication

las barreras de la comunicación interpersonal—las diferencias y las características personales de los emisores y receptores que obstaculizan la comunicación

interpersonal intelligence—the ability to relate to, understand, appreciate, and get along with other people

inteligencia interpersonal—la capacidad para relacionar, comprender, apreciar y relacionarse con otras personas

interpersonal relations—term referring to skills needed to get along with others

relaciones interpersonales—término que se refiere a las habilidades necesarias para relacionarse con los demás

interview—a meeting between an employer and a job applicant, for the purpose of deciding whether the applicant will be hired

entrevista—un encuentro entre un empleador y un solicitante de empleo, con el fin de decidir si el solicitante será contratado

interviewee—the person being interviewed

entrevistado—la persona entrevistada

interviewer—the person conducting an interview

entrevistador—la persona que lleva a cabo una entrevista

intragroup—within a group (such as when FFA chapter members compete among themselves)

intragrupo—dentro de un grupo (como cuando capitulares de FFA compiten entre sí)

intrapersonal intelligence—the ability to understand ourselves, be sensitive to our own values, and know who we are and how we fit into the greater scheme of the universe

inteligencia intrapersonal—la capacidad para comprendernos a nosotros mismos, ser sensible a nuestros propios valores y saber quiénes somos y cómo encajamos en el gran esquema del universo

intrapersonal relations—term referring to thinking skills needed for self-reflection and self-evaluation

relaciones intrapersonales—un término que se refiere a las habilidades de pensamiento necesarias para la auto-reflexión y auto-evaluación

introduction—the beginning thought to get the audience interested in a speech

Introducción—el pensamiento principio para llegar al público interesado en un discurso

introspective—a person who examines a situation through the lens of his or her own thoughts and feelings

introspectivo—una persona que examina una situación a través de la lente de sus propios pensamientos y sentimientos

irony—humor based on use of words in a way that suggests the opposite of their literal meaning

ironía—el humor basado en el uso de palabras en una forma que sugiere lo contrario de su significado literal

J

job—a paid position at a specific place or in a specific setting

un trabajo—una posición pagada en un lugar específico o en una configuración específica

job lead—information about a job opening

las pistas de empleo—la información sobre la apertura de un trabajo

job-related skills—the capabilities, abilities, characteristics, and knowledge (skills) you need to perform a specific job

aptitudes relacionadas con el trabajo—las capacidades, las habilidades, las características y los conocimientos (destrezas) que necesita para realizar un trabajo específico

judgment problems or decisions—problems or decisions that requires one to consider many factors, list and compare alternatives, and evaluate the alternatives before reaching a decision

problemas de juicio o tomar decisiones—problemas o decisiones que requiere alguien de considerar muchos factores, hacer listas y comparar alternativas, y evaluar las alternativas antes de tomar una decisión

justice—fairness, equity

justicia—la imparcialidad, la equidad

K

Kaizen—Japanese management style in which teams of workers take on tasks previously done only by leaders or managers. Adapted to the American workplace as the *quality circle*

Kaizen—un estilo de gestión japonés en el que los equipos de trabajadores asumen tareas anteriormente realizadas solamente por dirigentes o gerentes. Adaptada a la fuerza laboral norteamericana como el *círculo de calidad*

kinesics—the study of communication through body motion

la kinésica—el estudio de la comunicación a través del movimiento del cuerpo

kinesthetic intelligence—learning by interacting with space, objects, and tangible materials

inteligencia kinestésica—el aprendizaje interactuando con el espacio, objetos, y materiales tangibles

L

laissez-faire leadership—leading so that the group can make its own decisions without the leader or, at least, with very little input from the leader

liderazgo laissez-faire—los principales para que el grupo pueda tomar sus propias decisiones sin el líder o, por lo menos, con muy poca participación del líder

laissez-faire leadership style—a leadership style in which the group, not the leader, solves the problem or makes the decision

estilo de liderazgo laissez-faire—un estilo de liderazgo en los que el grupo, el líder, no resuelve el problema o toma la decisión

leadership—the ability of a person—the leader—to move an organization or group toward the achievement or accomplishment of its goals and objectives, using whatever style is the most effective in each situation; the ability to influence others

liderazgo—la capacidad de una persona—el líder—mover una organización o grupo hacia el logro o cumplimiento de sus metas y objetivos, utilizando cualquier estilo es el más eficaz en cada situación; la capacidad de influir en los demás

learned needs theory—psychological theory that states that people can acquire needs from society itself; when such a need is strong enough in a person, the person is motivated to a behavior that will satisfy the need

teoría de las necesidades aprendidas—teoría psicológica que afirma que las personas pueden adquirir las necesidades de la propia sociedad; cuando esa necesidad es suficientemente fuerte en una persona, la persona está motivada para un comportamiento que satisfaga su necesidad

lectern—speaker's stand or podium

atril—el facistol del altavoz o podio

left-brain people—persons who have characteristics of a melancholy personality; tend to be logical, detailed, active, and objectives oriented; prefer jobs that require precision, detail, or repetition

las personas del hemisferio izquierdo del cerebro—las personas que tienen características de personalidad de la melancolía; tienden a ser lógico, detallado, activo y orientado a objetivos; prefieren trabajos que requieren la precisión, el detalle o la repetición

legitimate power—power that stems from formal authority

poder legítimo—el poder que procede de una autoridad formal

letter of application—a written communication used to apply for a job; when accompanied by a résumé, often called a *cover letter*

carta de solicitud—una comunicación escrita utilizada para solicitar un puesto de trabajo; cuando van acompañadas de un curriculum vitae, a menudo llamado una *carta de presentación*

linguistic intelligence—verbal abilities; sensitivity to language, meanings, and the relationships among words

inteligencia lingüística—las habilidades verbales; la sensibilidad al idioma, los significados y las relaciones entre palabras

listening—a conscious mental effort to understand what is heard

Escuchando—un esfuerzo mental consciente para comprender lo que se escucha

logical-mathematical intelligence—abilities with numbers, patterns, abstract thought, precision, counting, organization, and logical structure

inteligencia lógico-matemático—las habilidades con números, los patrones, el pensamiento abstracto, la precisión, el recuento, la organización, y la estructura lógica

long-term goals—goals that are set to be reached two or more years in the future

metas a largo plazo—las metas que se han fijado para ser alcanzado dos años o más en el futuro

M

main motion—introduces new business before the assembly

motion principal—introduce negocios nuevos ante la asamblea

maintenance roles—functions that help the group members work together smoothly as a unit

funciones de mantenimiento—las funciones que ayudan a los miembros del grupo trabajan juntos sin problemas como una unidad

majority vote—a decision by more than half of the votes cast by those legally entitled to vote

voto mayoritario—una decisión de más de la mitad de los votos emitidos por las personas legalmente facultados para votar

managing ourselves—the fourth generation of time management; time management that focuses on improving personal and professional results and relationships instead of focusing on things and deadlines

la gestión de nosotros mismos—la cuarta generación de la gestión del tiempo, la gestión del tiempo, que se centra en la mejora de resultados personales y profesionales y de las relaciones en lugar de centrarse en las cosas y plazos

mass media—any form of communication that reaches a very large audience, such as the Internet, television, newspapers, radio, podcasts, magazines, and outdoor advertisements

la mediática—cualquier forma de comunicación que llega a un público muy numeroso, tales como el Internet, la televisión, los periódicos, la radio, los podcasts, las revistas y la publicidad al aire libre

mastery—high accomplishment and skill in a given area

maestría—los logros altos y las habilidades en un área determinada

matrix—a rectangular array or network of intersections

la matriz—una matriz rectangular o red de intersecciones

mediation—conflict-resolution method by which a third party intervenes in a conflict or disagreement to help clarify the problem and resolution options

mediación—un método de resolución de conflictos mediante el cual un tercero interviene en un conflicto o desacuerdo para ayudar a aclarar el problema y las opciones de resolución

melancholy—personality type that is analytical, self- sacrificing, gifted, and perfectionist

melancolía—un tipo de personalidad que es analítico, auto-sacrificio, talentoso, y perfeccionista

member role—a term for the general expectation of the group member within the group. Clearly defined member roles aid greater goal achievement

Rol de miembro—un término para la expectativa general del miembro del grupo dentro del mismo grupo. Una definición clara de los roles de los miembros mayores de ayuda el logro de los objetivos

memorandum—a formal written communication used in a business setting

memorando—una comunicación escrita formal utilizada en un negocio

mentalist—term for a person who exhibits a high degree of interpersonal intelligence, or the ability to "read" the intentions and desires of others without linguistic communication

mentalista—plazo para una persona que presenta un alto grado de inteligencia interpersonal, o la capacidad para "leer" las intenciones y deseos de otros sin comunicación lingüística

message—whatever one person communicates to another

mensaje—cualquiera que una persona comunica a otra

metaphor—figure of speech containing an implied comparison that omits the word *like* or *as*, in which a word or phrase ordinarily and primarily used in relation to one thing is applied to another

metáfora—figura de expresión que contenga una comparación implícita que omite la palabra *como* o, en la que una palabra o frase normalmente y se utiliza principalmente en relación a una cosa es aplicado a otro

minimizing approach—problem-solving method in which a person opts for the first solution available to solve the problem or make the decision, even though this solution may not necessarily be the best

la estrategia de la reducción al mínimo—método de solución de problemas en los que una persona opta por la primera solución disponible

para solucionar el problema o tomar la decisión, aunque esta solución puede no ser necesariamente el mejor

minutes—the official written record of the proceedings of the meetings of an organization, reflecting all actions taken and meaningful discussion on important issues

Minutos—el registro oficial escrito de las actas de las reuniones de la organización, en el que se recogen todas las medidas adoptadas y el debate sobre cuestiones importantes

misconceptions—false or mistaken ideas or thoughts that arise from a lack of information

malentendidos—las ideas o los pensamientos falsos o erróneos que surgen de la falta de información

mnemonic—of or relating to memory; a device (such as a word, phrase, or song) that helps a person remember something

mnemónico—referente a la memoria; un dispositivo (como una palabra, una frase o una canción) que ayuda a la persona a recordar algo

momentum—the quantity of motion of a moving object, equal to the product of its mass and its velocity

impulso—la cantidad de movimiento de un objeto en movimiento, igual al producto de su masa y su velocidad

monotone—speaking without fluctuations or inflections in the voice

monotono—hablando sin fluctuaciones o inflexiones en la voz.

motion to commit—alternative term for *motion to refer*

moción para cometer—otro término para *una moción de devolución*

motion to refer—a motion to refer an item of business to a committee for further study and/or action

motion para referirse—una moción para consultar un tema de negocio a una comisión para su estudio y/o acción

motion to postpone to a certain time—a motion that puts off action on a pending motion and fixes a definite time for its future consideration. Also known as a *motion to postpone definitely*

moción de aplazamiento para un tiempo determinado—una moción en la que se posterga la acción sobre una moción pendiente y fija un plazo para su consideración en el futuro. También conocido como una *moción para aplazar definitivamente*

motivation—an individual's desire to demonstrate constructive behavior, and reflects a person's willingness to expend effort; also, the focus of the need or desire to act

motivación—el deseo de un individuo de demostrar una actitud constructiva, y refleja una persona está dispuesta a invertir su esfuerzo; asimismo, el foco de la necesidad o el deseo de actuar

motivator—term describing a leader who gives others a reason for doing what the leader requests

motivador—un término que describe a un líder que da a otros una razón para hacer lo que el líder pide

motivators—higher-level needs that include achievement, recognition, responsibility, advancement, work itself, and the possibility of growth

motivadores—las necesidades de alto nivel que incluyen el logro, el reconocimiento, la responsabilidad, la promoción, el propio trabajo, y la posibilidad de crecimiento

motives—needs that cause us to act in certain ways

motivos—las necesidades que nos llevan a actuar en ciertas formas

multidimensional—having several sides or parts

multidimensionales—tener varias partes o piezas

musical intelligence—sensitivity to pitch and rhythm, and to the emotional power and complex organization of music

inteligencia musical—la sensibilidad al tono y ritmo, y el poder emocional y la organización compleja de música

N

naturalist—a person with expertise in the recognition, investigation, and classification of plants and animals

naturalista—una persona con conocimientos técnicos en el reconocimiento, la investigación y la clasificación de plantas y animales.

naturalistic intelligence—intelligence associated with observing, understanding, and using patterns in the natural environment

inteligencia naturalista—la inteligencia asociada con la observación, la comprensión y el uso de patrones en el ambiente natural

need—a physiological or psychological requirement for well-being; motivation

necesidad—un requisito fisiológico o psicológico para el bienestar; la motivación

negative reinforcement—method of using unpleasant consequences or punishments to discourage undesirable behavior

refuerzo negativo—el método de usar las consecuencias desagradables o los castigos para desalentar el comportamiento indeseable negotiation—a tool of conflict resolution that simply helps people to work together

negociación—una herramienta de resolución de conflictos que simplemente ayuda a la gente a trabajar juntos

network—an informal group of people and contacts that can be used to help one get a job

red—un grupo informal de personas y contactos que pueden ser utilizados para ayudar a conseguir un trabajo

nominal group method—the process of generating and evaluating alternatives through a structured voting method

Método de grupo nominal—el proceso de la generación y la evaluación de alternativas a través de un método de votación estructurada

nonverbal communication—messages conveyed by physical behaviors, including gestures, body language, and the physical environment

comunicación no verbal—los mensajes transmitidos por comportamientos físicos, incluidos gestos, el lenguaje corporal, y el entorno físico

norms—the group's shared expectations regarding member behavior

normas—las expectativas que el grupo comparte sobre el comportamiento de los miembros

not applicable (N/A)—written in the blank on a job application when an item does not apply to you

No aplicable (N/A)—escrito en el blanco en una solicitud de trabajo cuando un elemento no se aplique a usted

O

objectives—measurable goals

objetivos—los objetivos medibles

observers—people who make meaning from experiences or information by reflection, or filtering them through the lens of their own experience

observadores—las personas que hacen el significado de las experiencias o la información por parte de la reflexión, o el filtrado a través de la lente de su propia experiencia

occupation—a group of similar or related tasks that a person performs for pay

ocupación—un grupo de tareas similares o afines que una persona realiza para un salario

occupation-related skills—skills that have direct bearing on and are needed in one's job or occupation

las habilidades relacionadas con la ocupación—las habilidades que tienen influencia directa sobre y son necesarias en el empleo o su ocupación

operant conditioning—a way to modify an individual's behavior through the appropriate use of immediate rewards or punishments

acondicionamiento operante—Una manera de modificar el comportamiento de una persona mediante el uso apropiado de las recompensas o castigos inmediatos

opportunists—persons who take advantage of unplanned situations, coincidences, and serendipity and turn them into positives

oportunistas—las personas que se aprovechan de situaciones imprevistas, las coincidencias y la serendipia y convertirlos en algo positivo

optimizing approach—problem-solving technique in which the person takes the time to review many different alternatives and solutions before making a decision, in order to choose the most effective, appropriate, or helpful solution

el enfoque de optimización—una técnica de la solución de problemas en la cual la persona se toma el tiempo para revisar muchas diferentes alternativas y soluciones antes de tomar una decisión, a fin de escoger el más eficaz, adecuada, o solución útil

order of business—an agenda (e.g., for a meeting)

Orden de negocios—un programa (por ejemplo, para una reunión)

order of the day—an item of business on the agenda

orden del día—un tema en la agenda

out of order—inappropriate action in parliamentary procedure

fuera del orden—una acción inapropiada en el procedimiento parlamentario

output-based communication—when a person produces a communication

la comunicación basada en el output—cuando una persona produce una comunicación

P

pangs—sudden, sharp, and brief pains; may be physical or emotional

las punzadas—dolores intensos, bruscos y breves; pueden ser físicos o emocionales.

paraphernalia—the gavel, owl, plow, flag, secretary's book, treasurer's book, and other items used in an FFA meeting

parafernalia—el martillo, el buho, el arado, la bandera, el libro del secretario, el linbro del tesorero, y otros elementos utilizados en una sesión de FFA

Pareto's principle—theory stating that 20 percent of tasks that are high priority (important rather than urgent) can generate 80 percent of the results needed to reach goals

Principio de Pareto—la teoría afirmando que el 20 por ciento de las tareas que son de alta prioridad (importante más que urgente) puede generar el 80 por ciento de los resultados necesarios para alcanzar los objetivos

parliamentarian—an expert on rules governing meetings who serves as an adviser to the chairperson

parlamentario—un experto sobre las normas que rigen las reuniones que sirve como asesor de la presidenta

parliamentary inquiry—an incidental motion that requests information regarding the correct usage of parliamentary procedure or the rules of the organization

investigación parlamentaria—una petición incidental que solicita información sobre el uso correcto del procedimiento parlamentario o las reglas de la organización

parliamentary procedure—a set of rules and procedures for keeping a meeting orderly and harmonious, and guaranteeing that all persons have equal opportunity to express themselves

procedimiento parlamentario—un juego de normas y procedimientos para mantener una reunión armoniosa y ordenada, y garantizar que todas las personas tengan oportunidades iguales para expresarse

participative leadership—leading by gathering and considering input from group members

liderazgo participativo—los principales mediante la recolección y teniendo en cuenta las aportaciones de los miembros del grupo

participative leadership style—a leadership approach in which the leader has a tentative solution or decision in mind but goes to the group seeking its ideas and opinions before making a final choice

estilo de liderazgo participativo—un enfoque de liderazgo en que el líder tiene una solución provisional o decisión en mente pero va al grupo que buscan sus ideas y opiniones antes de hacer una elección definitiva

participative management—involving employees or coworkers in decision making/problem solving

gestión participativa—involucrando a empleados o co-trabajadores en la toma de decisiones y la solución de problemas

peer groups—groups of people who share rank, class, or age

grupos de compañeros—los grupos de gente que comparte el rango de clase o edad

people-mover—a leader who inspires and motivates others to fulfill their potential

transportador de personas—un líder que inspira y motiva a otros a cumplir su potencial

perceive—to make meaning from a message or experience based on our own beliefs, knowledge, and ways of organizing information

perciben—para el significado de un mensaje o una experiencia basada en nuestras propias creencias, conocimientos y formas de organizar la información

perception—way of understanding a message, based on our own beliefs, knowledge, and ways of organizing information; also, a personal view of a conflict

percepción—el modo de entender un mensaje, basado en nuestras propias creencias, conocimientos y formas de organizar la información; asimismo, una visión personal de un conflicto

perceptions—ways of understanding a message, based on our own beliefs, knowledge, and ways of organizing information; how we view a problem

percepciones—las formas de entender un mensaje, basado en nuestras propias creencias, conocimientos y formas de organizar la información; cómo vemos un problema

perpetuating—continuing in existence; keeping alive or active

perpetuar—continuando en existencia; mantener vivo o activo

personal identification—the process by which followers connect with charismatic leaders by making the leader the vision of the follower's ideal self, which he or she is striving to become

identificación personal—el proceso mediante el cual los seguidores conectar con líderes carismáticos, haciendo que el líder en la visión del ser ideal del seguidor, que él o ella está tratando de ser

personal leadership—a person's ability to establish a specific direction for his or her life, to commit to moving in that direction, and to take considered, determined action to acquire, accomplish, or become whatever that goal demands

liderazgo personal—la capacidad de una persona para establecer una dirección específica para su vida, con el fin de comprometerse a avanzar en esa dirección, y tomar acción decidida y considerado para adquirir, lograr, o convertirse en cualquier cosa que ese exige el objetivo

personal management skills—intrapersonal skills required to be a good employee, such as personality, attitude, honesty, enthusiasm, and an ability to meet deadlines; same as self-management or adaptive skills

las habilidades de gestión personal—las habilidades intrapersonales necesarias para ser un buen empleado, tales como la personalidad, las actitudes, la honestidad, el entusiasmo y la capacidad para cumplir los plazos de entrega; el mismo como la autogestión o las habilidades adaptativas

personal time—free or leisure time

tiempo personal—el tiempo libre o el tiempo de ocio

personification—a figure of speech in which a thing, quality, or idea is represented as a person

personificación—una figura retórica por la cual una cosa, la calidad o la idea es representado como una persona

personnel office—the department in a company that handles all staffing issues

oficina de personal—El departamento de una empresa que se encarga de todas las cuestiones de personal

philosophy—a system of values by which one lives; also, inquiry into the nature of things based on logical reasoning rather than empirical investigation

filosofía—un sistema de valores por los que uno vive; además, la investigación sobre la naturaleza de las cosas sobre la base de un razonamiento lógico en lugar de investigación empírica

phlegmatic—personality type with a calm, cool, slow, easygoing, well-balanced temperament

flemático—el tipo de personalidad con un temperamento calmo, tranquilo, lento, campechano, equilibrado

physical strokes—recognition that occurs physically, such as a high five or a pat on the back

trazos físicos—reconocimiento que ocurre físicamente, como un alto cinco o una palmadita en la espalda

pitch—the height or depth of the tone of voice

tono—la altura o la profundidad del tono de voz

poise—composure under pressure or stress

el aplomo—la compostura bajo presión o estrés

policy—a rule, action, or norm of an organization, usually adopted for the sake of consistency, equity, expediency, and/or efficiency

política—una regla o norma, la acción de una organización, generalmente adoptado en aras de la coherencia, la equidad, la conveniencia, y/o la eficiencia

political frame—persuasive process within groups that involves efforts by group members to gain or consolidate power or to protect existing power sources

marco político—el proceso de la persuasión dentro de grupos que se involucran en los esfuerzos realizados por los miembros del grupo para ganar o consolidar el poder o para proteger las fuentes de poder

popularity (perceived) leadership—leadership position bestowed on someone because of celebrity or group perceptions rather than the leader's actual ability

el liderazgo de popularidad (percibido)—el liderazgo otorgado a alguien debido a su celebridad o las percepciones de grupo en lugar de su capacidad real de ser un líder

portability—ability to be carried or moved easily

portabilidad—la capacidad para ser transportado o movido fácilmente

portfolio—collection of tangible (and usually visual) evidence of a person's accomplishments and performance, such as samples of a journalist's articles or a photographer's pictures

portafolio—la colección de tangibles (y generalmente) la evidencia visual de una persona de logros y rendimiento, tales como muestras de los artículos de un periodista o las fotografías de un fotógrafo

position power—potential influence based on an individual's position in an organization, and the resulting ability to control the organization's resources.

posición power—la influencia potencial sobre la base de la posición de un individuo en una organización, y la consiguiente capacidad para controlar los recursos de la organización.

positive reinforcement—a method of using pleasant consequences or rewards to encourage performance and repetition of a desired behavior

refuerzo positivo—un método del uso de las consecuencias o las recompensas agradables para fomentar el rendimiento y la repetición de una conducta deseada

positive written communication—any written materials that express appreciation for a well-performed behavior

comunicación positiva escrita—cualquier material escrito que expresa agradecimiento por un comportamiento bien ejecutadas

postponed job placement—careers available after advanced education or training beyond high school, such as two-year vocational schools or college

una colocación aplazada—las carreras disponibles después de la educación o la capacitación avanzada más allá de la escuela secundaria, tales como escuelas de formación profesional de dos años o la universidad

power leadership—leading by force, with the group submitting to the leader (perhaps against its will)

liderazgo poderoso—la gestión por la fuerza, con el grupo sometiéndose al líder (quizás contra su voluntad)

praise—expressed approval of something (e.g., a behavior); to express approval of something

alabanza—la aprobación expresada de algo (por ejemplo, un comportamiento); para expresar la aprobación de algo

precedence—priority; the right to come before or act first

Precedencia—la prioridad; el derecho a venir antes o actuar primero

preparation—the study and practice required to achieve mastery of something

preparación—el estudio y la práctica necesaria para lograr al dominio de algo

previous question—a formal motion requesting a vote and requiring two-thirds vote for passage

pregunta anterior—una moción solicitando una votación formal y que requieren un voto de dos tercios para su aprobación

pride—a feeling of self-esteem arising from one's accomplishments

orgullo—un sentimiento de autoestima derivados de los logros

primary amendment—an amendment applied to any amendable motion, except the motion to amend

enmienda principal—una enmienda aplicada a cualquier movimiento enmendable, excepto la moción de enmienda

prime time—the hours of the day when you get the most done

la hora óptima—las horas del día cuando usted consigue más cosas acabadas

prioritization—arranging or dealing with in order of importance

Priorización—organizar o tratar en orden de la importancia

privileged motion—has nothing to do with the pending motion, but is of such urgency and importance that it is allowed to interrupt the consideration of other questions

moción privilegiada—no tiene nada que ver con la moción pendiente, pero es de tal importancia y urgencia que es permitido para interrumpir la consideración de otras cuestiones

problem—difference between what is actually happening and what an individual or group wants to have happen

problema—la diferencia entre lo que está sucediendo y lo que un individuo o un grupo quiere que suceda

problem solving—the method of arriving at a decision, answer, or solution

resolución de problemas—el método para llegar a una decisión, una respuesta o una solución

process—to integrate information or experience into the individual's overall makeup (i.e., to learn) either by observation and reflection or by doing

Proceso—para integrar la información o la experiencia en la composición global del individuo (es decir, para aprender), ya sea mediante la observación y la reflexión o por haciéndolo

process theories—psychological approaches that examine how behavior is motivated

procesar las teorías—los enfoques psicológicos que examinan cómo se motiva el comportamiento

procrastination—putting off doing something until a later time (sometime in the future)

La postergación—la dilación de algo hasta un momento más tarde (en algún momento en el futuro)

professional time—time spent at or relating to school or employment

tiempo profesional—el tiempo invertido en o relativo a la escuela o el trabajo

Program of Activities—a document within the FFA organization that records the chapter's goals and the ways and means to achieve them

Programa de Actividades—un documento dentro de la organización FFA que registra los objetivos del capítulo y los medios para alcanzarlos

pronunciation—the form and accent a speaker gives to the syllables of a word

pronunciación—la forma y el énfasis que un orador pone en las sílabas de una palabra

proxemics—the science of how spacing and placement send messages

proxémica—la ciencia del espacio y la colocación de cómo enviar mensajes

prudence—common sense; judiciousness or thinking about consequences and possibilities before you act

prudencia—el sentido común; la juridicidad o pensar en las consecuencias y las posibilidades antes de actuar

psychic income—praise or positive things that happen to us that give us a psychological boost or make us feel good

Ingreso psíquico—la alabanza o cosas positivas que nos sucede que nos dé un impulso psicológico o nos hace sentir bien

psychomotor learning—learning that has to do with intelligence in the area of manual dexterity and physical doing

Aprendizaje psicomotor—el aprendizaje que tiene que ver con la inteligencia en el ámbito de la física y la destreza manual

psychosomatic—bodily symptoms experienced as a result of mental conflict or upset

psicosomáticos—los síntomas corporales experimentadas como resultado del conflicto mental o de malestar

punishment—a negative consequence for an undesirable behavior

castigo—una consecuencia negativa para un comportamiento indeseable

Pygmalion—in Greek mythology, the sculptor whose love for a statue of the goddess Aphrodite he had created was so strong that the beloved statue came to life; the idea that one's expectations shape the behavior of others

Pygmalion—en la mitología griega, el escultor cuyo amor por una estatua de la diosa Afrodita que había creado era tan fuerte que la estatua amada llegó a la vida; la idea de que las expectativas de uno forma el comportamiento de los demás

Q

qualified motion to recess—a privileged motion to recess that is made while another question is pending

la moción calificada de hacer un receso—una moción privilegiada para el receso que se realiza mientras otra cuestión es pendiente

qualifier—a word or phrase that qualifies, limits, or modifies something

cualificador—una palabra o frase que califica, limita o modifica algo

qualities—characteristics, innate or acquired, that determine the nature and behavior of a person or thing

cualidades—las características innatas o adquiridas, que determinan la naturaleza y el comportamiento de una persona o una cosa

quality—term referring to tone of voice

calidad—un término refiriéndose al tono de voz

quality circle—the most widely-known type of problem-solving group, first used by the Japanese, in which rank-and-file employees with hands-on jobs in a production process meet to discuss ways of improving product quality, operational efficiency, and the work environment. These recommendations are then proposed to management for approval

círculo de calidad—el tipo de problema-solución de grupo más ampliamente conocido, utilizado por primera vez por el japonéses, en el que los empleados en puestos de trabajo en un proceso de producción se reúnen para discutir maneras de mejorar la calidad del producto, la eficacia operativa y el entorno de trabajo. A continuación se proponen estas recomendaciones a la administración para su aprobación

question of privilege—a privileged motion that requests immediate action on some urgent matter related to the comfort, convenience, rights, or privileges of the group or assembly as a whole, or of a single group member

cuestión de privilegio—una moción privilegiada que pide acción inmediata sobre algún asunto urgente relacionado con el confort, l acomodidad, los derechos o los privilegios del grupo o conjunto completo o de un solo miembro del grupo.

quotations—use of words or phrases written or spoken by another person

citas textuales—el uso de las palabras o frases escritas o habladas por otra persona

R

rapport—a feeling of warmth, harmony, and alignment between speaker and listener

compenetración—una sensación de calidez, armonía y la alineación entre los altavoces y el oyente

reasoning—using logic to come up with the best possible solution to a problem

razonamiento—usando la lógica para llegar a la mejor solución posible a un problema

receiver—one for whom the message is intended

receptor—la persona para quien se dirige el mensaje

recess—a short intermission in a meeting, during which the meeting remains open; after the recess, business starts at exactly the point at which it was interrupted

receso—un corto descanso en una reunión, durante la cual la reunión permanece abierto; después del receso, negocio comienza exactamente en el punto en que fue interrumpido

recognition—special notice or attention; acknowledging individuals and their accomplishments

reconocimiento—un aviso especial o atención; reconociendo a los individuos y sus logros

recognition from the chair—permission for a member to speak, given when the chair is properly addressed

reconocimiento de la Presidencia—el permiso para un miembro de hablar, dada cuando la silla está correctamente dirigida

reference—personal respect and loyalty earned by an individual from others. The basis for *referent power*.

referencia—respeto personal y lealtad ganado por un individuo de los demás. La base para *poder referente*.

reference groups—groups, such as college fraternities or sororities, that are sources of attitude formation and influence for young adults; may act as a point of comparison and a source of information for individuals

grupos de referencia—grupos, tales como las fraternidades o hermandades universitarias, que son fuentes de la formación de actitud y tienen influencia para los adultos jóvenes; puede actuar como un punto de comparación y una fuente de información para los individuos

references—persons listed on a résumé who may be contacted by a prospective employer to inquire about an applicant's qualifications

Referencias—las personas enumeradas en un currículum vita que puede ser contactado por un empleador para preguntar acerca de una calificaciones del solicitante

referent power—power based on the desire of others to please an individual for whom they feel strong affection or to whom they are loyal

el poder referente—la alimentación basada en la voluntad de los demás para complacer a un individuo para quien se siente cariño fuerte o a quienes son leales

reflective style—problem-solving approach in which one identifies, analyzes, and evaluates as many alternatives as possible for solving a problem or making a decision

estilo reflexivo—el enfoque de resolución de problemas en la cual se identifica, analiza y evalúa el mayor número posible de alternativas para solucionar un problema o tomar una decisión

reflexive style—reactive problem-solving approach in which one makes quick and sometimes unthinking decisions or choices of solutions

estilo reflexivo—el enfoque reactivo de la resolución de problemas en los que uno se hace rápido y a veces decisiones precipitadas u opciones de soluciones

reinforcement—the motivation and control of behavior by the use of rewards

refuerzo—la motivación y el control de comportamiento mediante el uso de recompensas

reinforcement scheduling—timing of rewards or punishments

Programación de refuerzo—la sincronización de recompensas o castigos

relaters—people who avoid risk and seek tranquility, calmness, and peace (phlegmatic personality type)

relator—los que evite los riesgos y las personas que buscan la tranquilidad, la serenidad y la paz (tipo flemático de personalidad)

relevant others—people (such as peers, coworkers, and so on) with whom an individual can compare him-or herself

otros pertinentes—personas (tales como amigos, compañeros de trabajo, etc.) con las cuales un individuo puede comparar con sí mismo

reprimanding—giving a severe, usually formal, reproof, rebuke, or correction

reprender—dando una severa, generalmente formal, represión, regaño, o corrección

resignation—an official notice informing the current employer of the intent to leave a job by the job holder

renuncia—un anuncio oficial informando el patrón actual de su intención de dejar un trabajo

resilient—able to bounce back after a setback or failure

resistentes—capaz de recuperarse después de un revés o fracaso

resources—something that can be drawn on for aid

Recursos—algo que puede ser utilizado para ayudar

responsibility—the state of being responsible; being in control of something; accountability

responsabilidad—el estado de ser responsable; estar en control de algo; la rendición de cuentas

résumé—a one- or two-page description of a job applicant that lists his or her educational background, experiences, skills, and qualifications

curriculum vita—una descripción de uno o dos páginas de un solicitante de empleo que enumera sus antecedentes educativos, sus experiencias, sus habilidades y sus cualificaciones

reward power—influence based on an individual's ability to control an organization's resources. See also *position power*

poder de recompensa—la influencia basada en la capacidad de una persona para controlar los recursos de una organización. Vea también el *poder de posición*

right-brain people—have characteristics of sanguine or choleric personality type; tend to be spontaneous, emotional, holistic, physical (nonverbal), and visual, and like jobs without repetition or routines or that require them to generate ideas

las personas del hemísfero derecho de cerebro—tienen las características de sanguina o tipo de personalidad colérica; tienden a ser espontánea, emocional, holística, físicas (no verbal) y visual, y prefieren trabajos sin repetición o rutinas o que les obligan a generar ideas

role model—someone you admire who has an influence on you and whom you seek to emulate

modelo de rol—alguien que admira que tiene una influencia sobre usted y a quien usted busca a emular

role-playing—acting out a situation relating to a real-life problem

juego de rol—interpretenado una situación relativa a un problema de la vida real

rudiments—basic principles

rudimentos—los principios básicos

S

salutation—the beginning of a speech with a greeting, either addressing or welcoming

saludo—el comienzo de un discurso con un saludo, ocupándose con el problema o de bienvenida

sanguine—personality type with a warm, lively, cheerful temperament

sanguina—un tipo de personalidad con un temperamento cálido, vivo, y alegre

scantily—scarcely; barely sufficient

escasamente—por poco; apenas suficiente

schoolhouse giftedness—the type of intelligence possessed by an individual who does well on taking tests and learning lessons

superdotado escolarmente—el tipo de inteligencia que posee un individuo que hace bien en tomar pruebas y aprender lecciones

secondary amendment—an amendment applied to a motion to amend

enmienda secundaria—una enmienda aplicada a una moción para enmendar

self-communication—the ability to answer questions that you ask yourself honestly and deal with controversial or difficult issues

auto-comunicación—la capacidad para responder a las preguntas que usted se pregunta honestamente y abordar cuestiones difíciles o controvertidos

self-concept—the way one perceives and feels about oneself

auto-concepto—la manera en que se percibe y se siente sobre sí mismo

self-confidence—belief in oneself or one's abilities

auto-confianza—la creencia en uno mismo o sus habilidades

self-deprecating—devaluing or downplaying one's own abilities or accomplishments

auto-denunciamos—devaluar o menospreciando las propias capacidades o logros

self-determination—inner motivation to achieve goals

auto-determinación—la motivación interior para lograr objetivos

self-disclosure—group members' willingness to engage in open communication with other group members

auto-revelación—los miembros del grupo estaban dispuestos a entablar una comunicación abierta con otros miembros del grupo

self-esteem—belief in one's abilities and respect for oneself

autoestima—la creencia en las propias capacidades y el respeto por sí mismo

self-fulfilling prophecy—concept that people tend to behave in a way that both reflects and supports what they believe about themselves; what you think is what you will make come true

auto-profecía—el concepto de que las personas tienden a comportarse de una manera que refleja a lo que creen acerca de sí mismos; lo que crees es lo que va a hacer la realidad

self-image—the idea, concept, or mental picture one has of oneself

auto-imagen—la idea, el concepto o el imagen mental que uno tiene de sí mismo

self-management (adaptive) skills—skills that are required to be a good worker, such as diligence, honesty, enthusiasm, and ability to learn

autogestión (adaptive) habilidades—las habilidades que se requieren para ser un buen trabajador, tales como la diligencia, la honestidad, el entusiasmo y la capacidad para aprender

self-motivated—having strong internal motivation

auto-motivado—contando con una fuerte motivación interna

self-responsibility—accepting the consequences of any effort, good or bad

auto-responsabilidad—aceptando las consecuencias de cualquier esfuerzo, bueno o malo

semantics—the study of the shifting meanings of language, by which a particular word or phrase may not mean the same thing to one person as it does to someone else

semántica—el estudio de cambios de significados de la lengua, por lo que una palabra o frase determinada no puede significar lo mismo a una persona como a alguien

sender—person who wishes to communicate with (sends a message to) someone else

remitente—la persona quien desea comunicarse con (envía un mensaje) alguien más

sensing—making meaning from experience or information by empathy, or the feelings evoked by the actual experience or information. See also *feeling*

detección—haciendo el significado de la experiencia o la información por empatía o los sentimientos evocados por la experiencia real o la información. Véase también el *sentimiento*

sensitive—keenly aware of the needs of others

sensible—plenamente consciente de las necesidades de los demás

sensitivity—the ability to understand the needs of others

sensibilidad—la habilidad para entender las necesidades de los demás

short-term goals—goals that are set to happen immediately or very soon

objetivos de corto plazo—las metas que se han fijado para suceder inmediatamente o muy pronto

simile—figure of speech that presents a brief comparison of two basically unlike items using the word *like* or *as*

símil—la figura de discurso que presenta una breve comparación de dos elementos básicamente diferentes utilizando la palabra *como*

simultaneous—happening at the same time

simultánea—sucediendo al mismo tiempo

situation—when and where a communication takes place

situación—cuando y donde tiene lugar una comunicación

situational engineering—a form of power based on ability to influence the situations of others through control over the physical environment, technology, and any organization of the workplace.

ingeniería situacional—una forma de poder basada en la capacidad de tener influencia sobre las situaciones de los demás mediante el control sobre el entorno físico, la tecnología, y cualquier organización del lugar de trabajo.

situational leadership—leadership style that depends on or is contingent on the situation or immediate circumstances in which the leader is operating

liderazgo situacional—el estilo de liderazgo que depende o es contingente sobre la situación o las circunstancias inmediatas, en el que el líder está en funcionamiento

situational-timing barriers—hindrances to understanding raised by the time and place communication occurs

barreras situacionales de sincronización—los obstaculos a la comprensión generadas por el tiempo y el lugar se produce la comunicación

skills—practiced abilities

habilidades—las habilidades practicadas

slovenliness—carelessness, sloppiness, or messiness

descuido—la flojera, la negligencia, o el desaseo

SMART—acronym for specific, measurable, attainable, relevant, timebound (used to describe goals)

SMART—acrónimo de específicos, Medible, alcanzable, relevante, *delimitadas en el tiempo* (utilizado para describir metas)

smooth flow—organization of material that preserves unity of thought through the introduction, body, and conclusion

fluidez—la organización de material que preserva la unidad de pensamiento mediante la introducción, el cuerpo y la conclusión

social control—the process of ensuring conformity to the expectations of group members; rewards group members for meeting group standards

control social—el proceso de garantizar la conformidad con las expectativas de los miembros del grupo; recompensa a los miembros del grupo para cumplir con las normas del grupo

social media—the various electronic tools, technologies, and applications that facilitate interactive communication and content exchange

Medios sociales—las distintas herramientas electrónicas, las tecnologías, y las aplicaciones que faciliten la comunicación interactiva y el intercambio de contenido

socialization—learning to get along with others

socialización—aprendiendo a relacionarse con los demás

socializers—relationship-oriented people who appear to need the approval of those around them (sanguine personality type)

socializadores—la relación que es orientadas a personas que parecen contar con la aprobación de quienes les rodean (tipo de personalidad optimista)

socializing—taking part in fun, entertainment, or social activities

socializando—tomar parte en la diversión, el entretenimiento, o las actividades sociales

space order—a method of organizing a speech in which events are organized in terms of direction (east to west, top to bottom, front to rear, and so on)

el orden según espacio—un método de organizar un discurso en el que los eventos se organizan en términos de dirección (de este a oeste, de arriba a abajo, de delante a atrás, y así sucesivamente)

spatial intelligence—the ability to think in vivid mental pictures, make keen observations, and mentally revise or re-create a given image or situation

la inteligencia espacial—la capacidad de pensar en imágenes mentales vívidas, hacer observaciones, entusiasta y mentalmente revisar o volver a crear una imagen determinada o situación

special committee—a committee created for a specific, special purpose that goes out of existence as soon as it has completed its task

comité especial—un comité creado para un propósito especial, que sale de la existencia tan pronto como haya terminado su tarea

spin—manipulating a message (information) to reflect the receiver's desire or objective. Also called *filtering*

la torción—la manipulación de un mensaje (la información) para reflejar el deseo del receptor u objetivo. También se denomina *filtrando*

spontaneity—action occurring or arising without apparent premeditation

espontaneidad—la acción ocurriendo o derivados sin premeditación aparente

standing committee—a committee that has a continuing existence

comité permanente—un comité que tiene una existencia continuada

standing committees—permanent groups that exist to deal with year-to-year issues

comités permanentes—los grupos permanentes que existen para manejar los asuntos de cada año

statistics—facts or data used in a speech to present information of a numerical or empirical nature

Estadísticas—los hechos o los datos utilizados en un discurso para presentar información numérica o de una naturaleza empírica

status—perceived ranking of one member relative to other members of the group

estatus—el rango percibido de un miembro en relación con otros miembros del grupo.

status quo—the way it already is or has been so far

status quo—la manera en que existe todo o como ha sido hasta ahora

stewardship delegation—delegation in which the leader and assignee agree up front on the desired result, general guidelines, available resources, and responsibility for results, after which the assignee is allowed to decide how he or she wants to proceed

stewardship delegación—la delegación en la que el líder y el cesionario convienen por adelantado sobre el resultado deseado, las directrices generales, los recursos disponibles, y la responsabilidad por los resultados, después de que el cesionario pueda decidir cómo él o ella quiere continuar

stroke deficit—lack of recognition from others

el deficit de caricias—la falta de reconocimiento de los demás

stroking—term used to describe various ways (physical or verbal) in which a person may recognize another person's behavior

acariciando—un término utilizado para describir diferentes formas (física o verbal) en la cual una persona puede reconocer el comportamiento de otra persona

structural frame—relates to relationships and formal roles in the organization, such as organizational charts, policies, and procedures

marco estructural—se refiere a las relaciones y las funciones formales en la organización, tales como los gráficos de la organización, las políticas y los procedimientos

structure—external framework of a problem

estructura—el marco externo de un problema

subsidiary motions—motions that relate to some other motion on the floor and can change or alter the main motion

movimientos auxiliares—las mociones que se refieren a algún otro movimiento en el piso y puede cambiar o alterar el movimiento principal

subtlety—having or showing keenness about small differences, particularly in meaning

sutileza—teniendo o mostrando entusiasmo sobre las pequeñas diferencias, especialmente en sentido

superleadership—leading people to lead themselves; creating and fostering other leaders

superliderazgo—guiando a la gente a dirigirse por sí mismos; crear y fomentar otros líderes

supplementary (tributary) job placement—nonagricultural jobs attained through leadership and personal development skills or other skills developed through a quality agricultural education program (e.g., attorneys, legislators, supervisors, salespersons)

la colocación del empleo suplementarios (afluente)—los empleos no agrícolas alcanzados a través del liderazgo y las habilidades de desarrollo personal u otras habilidades desarrolladas a través de un programa de educación agrícola de calidad (por ejemplo, los abogados, los legisladores, los supervisores, los vendedores)

surface analysis—judging the actual appearance rather than the inner nature

análisis de superficie—juzgando la apariencia real en lugar de la naturaleza interior

symbiotic—the relationship of two or more different organisms in a close association that benefits each

simbiótica—la relación de dos o más organismos diferentes en una asociación estrecha que beneficia cada uno

symposium—a group of talks, lectures, or speeches presented by several individuals on the various aspects of a single subject

simposio—un grupo de charlas, conferencias o ponencias presentadas por varios individuos sobre los diversos aspectos de un solo tema.

synectics—the group problem-solving process of generating creative alternatives through role-playing and fantasizing; use of analogies to stimulate mental images

la sinéctica—el proceso grupero de resolver problemas por generar alternativas creativas a través de juegos de rol y fantasiosa; uso de analogías para estimular imágenes mentales

synergy—the power of a group of people working together; combined actions of individuals that, when taken together, increase one another's effectiveness

sinergia—el poder de un grupo de personas que trabajan juntas; combinando acciones de individuos que, tomados en conjunto, aumentar la eficacia del otro

T

talent area—natural ability or predilection to learn or do something

área de talento—la habilidad natural o una predilección para aprender o hacer algo

tangible—having actual form and substance; can be appraised for value

tangible—teniendo forma y sustancia verdadera; puede apreciarse por el valor

task groups—committees; formal groups established by an organization to perform specific functions

grupos de tareas—comités; los grupos formales establecidos por una organización para llevar a cabo funciones específicas

task roles—parts that members in a group play in an effort to get things done

Funciones de tareas—las piezas que desempeñan los miembros de un grupo en un esfuerzo por hacer las cosas bien

tasks—things to be done

Tareas—las cosas que hacer

teamwork skills—the ability to work well with others in a group situation

habilidades para trabajar en equipo—la capacidad de trabajar bien con los demás en una situación de grupo

technical knowledge—knowledge that is needed to perform a specific job, including the "why" and "how"

conocimiento técnico—los conocimientos que se necesitan para realizar un trabajo específico, incluido el "por qué" y "cómo"

technical leadership skills—leadership abilities that involve doing, such as public speaking or time management

habilidades de liderazgo técnico—las habilidades de liderazgo que se refieren a hacer, como hablar en público o la gestión del tiempo

technical skills—psychomotor skills, such as typing, welding, or drafting; generally learned through high school course work and/or on-the-job training

habilidades técnicas—las habilidades psicomotoras, tales como escribir a máquina, soldar o la elaboración; generalmente aprenden a través de la escuela secundaria el trabajo del curso y/o la capacitación en el trabajo

temperance—allowing just the right amount, at the right time, of any pleasure

templanza—permitiendo sólo la cantidad correcta, en el momento oportuno, de cualquier placer

terminal values—goals you strive to accomplish before you die, such as security and self-respect

valores terminal—los objetivos que pretenden lograr antes de morir, como la seguridad y la autoestima

think—to make meaning of experiences or information by using reason and logic to analyze them

pensar—para hacer sentido de las experiencias o la información mediante el uso de la razón y la lógica para analizarlos

thinkers—people who enjoy solitary, intellectual, and philosophical challenges (melancholy personality type)

pensadores—quienes disfrutan estar solitario, son intelectuales y disfrutan los desafíos filosóficos (tipo melancolía de personalidad)

tie vote—the same number for and against a motion

el voto empatado—el mismo número a favor y en contra de una moción

time management—planning how to control your time so that you can do the things you need and want to do

gestión del tiempo—planificando cómo controlar su tiempo para que usted pueda hacer las cosas que necesitan y desean hacer

time sequence—a method of organizing a speech in which events are discussed in terms of the hour, day, month, or year, moving forward or backward from a certain time

secuencia de tiempo—un método de organizar un discurso en el que los acontecimientos se examinan en términos de la hora, día, mes o año, moviendo hacia delante o hacia atrás a partir de una hora determinada

time-value ratio—alternative term for *Pareto's principle*; performing 20 percent of the highest priority tasks can generate 80 percent of the results needed to achieve a particular goal

tiempo- elcoeficiente de valor—un término alternativo para *el principio de Pareto*; realizar el 20% de las tareas de mayor prioridad puede generar el 80 por ciento de los resultados necesarios para alcanzar un objetivo concreto

tone—the particular or relative pitch of a word, phrase, or sentence and the modulations of the voice

tono—la particular o pariente tono de una palabra, frase u oración y las modulaciones de la voz

traditional leadership—leading in a particular way because it has always been done that way; sometimes referred to as *cultural* or *symbolic leadership*

liderazgo tradicional—dirigiendo de una manera particular, porque siempre se ha hecho así; a veces se denomina *liderazgo cultural o simbólico*

trait leadership—leadership based on a person's distinctive qualities or characteristics (such as intelligence) that predispose the person to be a leader

liderazgo de atributo—el liderazgo basado en una persona de cualidades o características (como la inteligencia) que predisponen a la persona para ser un líder

traits—distinctive qualities or characteristics

rasgos—las cualidades o características distintivas

transferable skills—skills that you can use in many jobs (e.g., writing clearly), though the importance of each skill may vary according to the job application

habilidades transferibles—las habilidades que puede utilizar en muchos trabajos (por ejemplo, escribiendo claramente), aunque la importancia de cada habilidad puede variar de acuerdo a la solicitud de trabajo

transition—the way in which the speaker moves from one part of a speech to the next, preserving unity of thought and presenting points in a logical order

Transición—la forma en la que el altavoz se mueve de una parte de un discurso a la siguiente, preservando la unidad de pensamiento y presentando los puntos en un orden lógico

trustworthiness—deserving of trust or confidence; being honest, faithful, reliable, and dependable

confiabilidad—mereciendo la confianza o la dependencia; siendo honesto, fiel, confiable y fiable

truth—the quality or state of being correct, accurate, or true; agreement with a standard, a rule, or an established or verified fact or principle

verdad—La calidad o el estado de ser correctos, exactos, o verdadero; de acuerdo con una norma, una regla, un hecho establecido o verificado o un principio

tunnel vision—a narrow perspective

Visión de túnel—una perspectiva estrecha

two-thirds vote—decision by at least two-thirds of those who are legally entitled to vote

el voto de dos tercios—la decisión por al menos dos tercios de aquellos que están legalmente facultados para votar

U

unclassified motions—motions (such as to bring back, to take from the table, to rescind, to ratify, to reconsider, and so on) that cannot conveniently be classified as either main, subsidiary, incidental, or privileged)

mociones sin clasificación—las mociones (como para traer de vuelta, para tomar de la tabla, para revocar, a ratificar, a reconsiderar, etc.) que no pueden ser convenientemente clasificados como principales y subsidiarios, incidentales, o privilegiados)

unqualified motion to recess—a motion to recess that is made while no other question is pending. It is treated as a main motion

moción hasta el receso—una moción a tener un receso que se hace cuando no hay otra cuestión pendiente. Es tratada como un movimiento principal

unstated subsidiary motion—a subsidiary motion that may not have been seconded and has not been stated by the chair

moción subsidiaria tácita—una moción subsidiaria que pueden no haber sido secundada y no ha sido declarado por el presidente

urgent—requiring immediate attention

urgente—requieren la atención inmediata

V

value-added—processing and packaging of food and fiber purchased from farmers and ranchers that increases monetary worth

valor añadido—el embalaje y procesamiento de alimentos y fibras de comprar a los agricultores y rancheros que aumenta el valor económico

value conflict—problem that arises when one is torn between two competing priorities or value systems

conflicto de valor—el problema que surge cuando uno está desgarrado entre dos prioridades que compiten entre sí o los sistemas de valores

value system—an individual's set of beliefs

Sistema de valores—un conjunto de creencias del individuo.

values—social principles, goals, or standards held or accepted by an individual, class, or society

valores—los principios sociales, metas o normas celebró o aceptados por un individuo, una clase o la sociedad

variable interval schedule—rewards a desired behavior after a varying period of time

programa de intervalo variable—las recompensas una conducta deseada después de un período variable de tiempo

variable ratio schedule—rewards a desired behavior after a varying number of outputs

Programación de tasa variable—las recompensas una conducta deseada después de un número variable de salidas

verbal strokes—recognition that occurs verbally, such as words of thanks and appreciation

caricias verbales—el reconocimiento que ocurre de manera verbal, como las palabras de agradecimiento y reconocimiento

verbose—using too many words

ampuloso—utilizando demasiadas palabras

violation—misuse of proper parliamentary procedure

violación—abuso de procedimiento parlamentario adecuado

vision—an individual's clear mental picture of what he or she wants to attain

visión—una clara imagen mental del individuo de lo que él o ella quiere alcanzar

vitality—physical or mental strength or energy

vitalidad—la fuerza física o mental o energía

vivacity—being spirited, full of life and energy; liveliness

vivacidad—siendo vibrante, llena de vida y energía; la vivacidad

vote—method by which members express approval of or opposition to a particular action or motion

votar—un método mediante el cual los miembros expresan su aprobación o la oposición a una determinada acción o movimiento

vote by voice (viva voce)—a vote in which the chairperson asks for verbal responses ("aye" from those supporting a motion and "nay" from those opposing a motion)

votar por voz (viva voce)—una votación en la cual la Presidenta pide respuestas verbales ("aye" de aquellos que apoyan un movimiento y "nay" de quienes se oponen a un movimiento)

W

ways and means—a step-by-step procedure detailing how goals will be accomplished

medios y arbitrios—un procedimiento paso a paso detallando cómo metas serán cumplidas

winner—someone who achieves victory over others in a competition or success in an endeavor

ganador—alguien que alcanza la victoria sobre los demás en una competición o el éxito en una empresa

withdraw—to remove a motion from consideration before a vote is taken

Retirar—para quitar una moción del examen antes de la votación

work—activity, directed toward a purpose or goal, that produces something of value to oneself and/or society; may or may not be compensated monetarily

trabajo—la actividad, dirigida hacia un objetivo o meta, que produce algo de valor a sí mismo y/o de la sociedad, puede o no ser compensado monetariamente

work team—a group (two or more) of individuals with a common purpose

equipo de trabajo—un grupo (dos o más) de los individuos con un propósito común

worst-case scenario—the least desirable thing that could possibly happen in a given situation

peor escenario posible—la menos deseable de lo que podría suceder en una determinada situación

Index

Note: Page numbers followed by an "f" indicate figures.